Lecture Notes in Electrical Engineering

Volume 359

Board of Series editors

Leopoldo Angrisani, Napoli, Italy
Marco Arteaga, Coyoacán, México
Samarjit Chakraborty, München, Germany
Jiming Chen, Hangzhou, P.R. China
Tan Kay Chen, Singapore, Singapore
Rüdiger Dillmann, Karlsruhe, Germany
Haibin Duan, Beijing, China
Gianluigi Ferrari, Parma, Italy
Manuel Ferre, Madrid, Spain
Sandra Hirche, München, Germany
Faryar Jabbari, Irvine, USA
Janusz Kacprzyk, Warsaw, Poland
Alaa Khamis, New Cairo City, Egypt
Torsten Kroeger, Stanford, USA
Tan Cher Ming, Singapore, Singapore
Wolfgang Minker, Ulm, Germany
Pradeep Misra, Dayton, USA
Sebastian Möller, Berlin, Germany
Subhas Mukhopadyay, Palmerston, New Zealand
Cun-Zheng Ning, Tempe, USA
Toyoaki Nishida, Sakyo-ku, Japan
Bijaya Ketan Panigrahi, New Delhi, India
Federica Pascucci, Roma, Italy
Tariq Samad, Minneapolis, USA
Gan Woon Seng, Nanyang Avenue, Singapore
Germano Veiga, Porto, Portugal
Haitao Wu, Beijing, China
Junjie James Zhang, Charlotte, USA

About this Series

"Lecture Notes in Electrical Engineering (LNEE)" is a book series which reports the latest research and developments in Electrical Engineering, namely:

- Communication, Networks, and Information Theory
- Computer Engineering
- Signal, Image, Speech and Information Processing
- Circuits and Systems
- Bioengineering

LNEE publishes authored monographs and contributed volumes which present cutting edge research information as well as new perspectives on classical fields, while maintaining Springer's high standards of academic excellence. Also considered for publication are lecture materials, proceedings, and other related materials of exceptionally high quality and interest. The subject matter should be original and timely, reporting the latest research and developments in all areas of electrical engineering.

The audience for the books in LNEE consists of advanced level students, researchers, and industry professionals working at the forefront of their fields. Much like Springer's other Lecture Notes series, LNEE will be distributed through Springer's print and electronic publishing channels.

More information about this series at http://www.springer.com/series/7818

Yingmin Jia · Junping Du · Hongbo Li
Weicun Zhang
Editors

Proceedings of the 2015 Chinese Intelligent Systems Conference

Volume 1

Springer

Editors
Yingmin Jia
Beihang University
Beijing
China

Hongbo Li
Tsinghua University
Beijing
China

Junping Du
Beijing University of Posts and Telecommunications
Beijing
China

Weicun Zhang
University of Science and Technology Beijing
Beijing
China

ISSN 1876-1100 ISSN 1876-1119 (electronic)
Lecture Notes in Electrical Engineering
ISBN 978-3-662-51705-5 ISBN 978-3-662-48386-2 (eBook)
DOI 10.1007/978-3-662-48386-2

Springer Heidelberg New York Dordrecht London
© Springer-Verlag Berlin Heidelberg 2016
Softcover re-print of the Hardcover 1st edition 2016
This work is subject to copyright. All rights are reserved by the Publisher, whether the whole or part of the material is concerned, specifically the rights of translation, reprinting, reuse of illustrations, recitation, broadcasting, reproduction on microfilms or in any other physical way, and transmission or information storage and retrieval, electronic adaptation, computer software, or by similar or dissimilar methodology now known or hereafter developed.
The use of general descriptive names, registered names, trademarks, service marks, etc. in this publication does not imply, even in the absence of a specific statement, that such names are exempt from the relevant protective laws and regulations and therefore free for general use.
The publisher, the authors and the editors are safe to assume that the advice and information in this book are believed to be true and accurate at the date of publication. Neither the publisher nor the authors or the editors give a warranty, express or implied, with respect to the material contained herein or for any errors or omissions that may have been made.

Printed on acid-free paper

Springer-Verlag GmbH Berlin Heidelberg is part of Springer Science+Business Media (www.springer.com)

Contents

Chapter 1
Observer Design for Discrete-Time Switched Lipschitz Nonlinear Singular Systems with Time Delays and Unknown Inputs

Jinxing Lin

Abstract In this paper, the state estimation problem for a class of discrete-time switched nonlinear singular systems simultaneously subject to Lipschitz constraints, state delays, unknown inputs (UIs), and arbitrary switching sequences is considered. A mode-dependent observer is constructed and, based on the idea of exact state and UI decoupling, sufficient conditions for the existence of the proposed observer are given in terms of linear matrix inequalities. By defining a decay-rate-dependent switched Lyapunov function, the convergence rate of the state estimation error is proved to be exponential.

Keywords Switched singular systems · Discrete-time · Unknown inputs · Time-delay · State estimation · Decoupling

1.1 Introduction

During the last several decades, tremendous research activities have been developed to deal with observer design for dynamic systems with unknown inputs (UIs). The UI observers have many practical applications such as fault detection and isolation, secure communications, data reconciliation, etc. [1]. A variety of efficient approaches have been developed to design both full- and reduced-order UI observers for linear time-invariant systems; for instance, generalized inverse method [2–4], algebraic method [5] and sliding-mode algorithm [6].

Switched systems are an important class of hybrid dynamic systems, which can be used to describe a wide range of modern engineering systems [7]. They comprise a collection of subsystems and a switching rule that specifies the switching between the subsystems. The interaction of continuous and discrete dynamics in a switched system leads to rich dynamical behavior not encountered in purely continuous- or

J. Lin (✉)
College of Automation, Nanjing University of Posts and Telecommunications, Nanjing 210023, China
e-mail: jxlin2004@126.com

© Springer-Verlag Berlin Heidelberg 2016

Y. Jia et al. (eds.), *Proceedings of the 2015 Chinese Intelligent Systems Conference*, Lecture Notes in Electrical Engineering 359,
DOI 10.1007/978-3-662-48386-2_1

discrete-time systems. On the other hand, time delays are the inherent features of many physical process and often a source of instability and poor performances. Since 1990s, switched systems with or without time delays have been extensively studied, and many useful results on observability and observer design have been obtained; see [8, 9] and the references therein. However, few efforts are devoted to design observers for switched systems with UIs. In [10], a switched version of the general structured unknown input observer [4] was proposed for discrete-time switched linear systems with UIs, and sufficient conditions for the existence of the observer were derived by using the generalized-inverse method and the switched Lyapunov function method. But the considered UIs do not directly influence the system output. For a general discrete-time switched linear systems with UIs affecting both system states and outputs, a switched unknown input observer with delayed measurements was constructed by using the system invertibility [11]. In [12], a switched Luenberger-like observer was presented for a class of discrete-time switched, superdetectable linear parameter varying systems with uncertainty and UIs, and both deterministic and randomized algorithms were given to design the observer. For continuous-time context, under the assumption of strong detectability for all subsystems, a reduced-order unknown input switched observer was designed in [13]. In [14], by performing a state and output coordinates transformation and introducing a piecewise time-varying Lyapunov function, a switched unknown input switched observer was designed. It should be pointed that the systems studied in [10–14] do not contain time delays and are all confined to regular state-space description.

More recently, a new type of switched time-delay systems with singular (also known as descriptor and implicit) description, called switched singular time-delay systems (SSTDSs), has been modeled in [15]. Such systems are more convenient and natural than the switched regular state-space ones, because the singular description can describe both the dynamic relationships and static constrains of the system. Since the solution of a singular system model may include the distributed terms, the problem of observing the state for singular systems with UIs is more important and challenging than that for regular ones [16]. It is worth noting that in *Koenig et al.'s* recent work [17], the results in [10] were successfully extended to switched Lipschitz nonlinear singular systems. However, to the authors' knowledge, there are no research result on unknown input observer design for discrete-time SSTDSs up to now.

In this paper, we discuss the state observation problem for a class of discrete-time nonlinear SSTDSs with Lipschitz constraints, UIs and arbitrary switching sequences. A mode-dependent observer is constructed and, based on the generalized-inverse method, necessary conditions for the existence of the proposed observer are given in terms of linear matrix inequalities. Different from the asymptotical observers presented in [10, 17], by constructing decay-rate-dependent switched Lyapunov function, the observers designed gives exponential state estimation.

Notations: $\mathbb{R}^n$ is the n-dimensional real Euclidean space. M^T is the transpose of the matrix M. $M > 0$ ($M \geq 0$) means M is positive definite (semi-positive definite). $\lambda_{\min}(M)$ ($\lambda_{\max}(M)$) denotes the minimum (maximum) eigenvalue of symmetric matrix M. The notation '$*$' in a matrix always denotes the symmetric block in the matrix. diag$\{\cdots\}$ stands for a block-diagonal matrix. $(\cdot)^+$ denotes any generalized inverse of the matrix $(\cdot)$, where $(\cdot)^+(\cdot)(\cdot)^+ = (\cdot)^+$ and $(\cdot)(\cdot)^+(\cdot) = (\cdot)$. $(\cdot)_{k_-}$ and $(\cdot)_{k_+}$ stand for $(\cdot)_{\sigma(k-1),\sigma(k)}$ and $(\cdot)_{\sigma(k),\sigma(k+1)}$, respectively, for example, $T_{k_+} = T_{\sigma(k),\sigma(k+1)}$ and $T_{\sigma(k-\tau),k_+} = T_{\sigma(k-\tau),\sigma(k),\sigma(k+1)}$.

1.2 Problem Statement

Consider the following discrete-time nonlinear switched singular system with state delay

$$\begin{cases} E_{\sigma(k+1)}x_{k+1} = A_{\sigma(k)}x_k + A_{d\sigma(k)}x_{k-\tau} \\ \qquad\qquad + B_{\sigma(k)}u_k + F_{\sigma(k)}d_k + J_{\sigma(k)}f_k \\ y_k = C_{\sigma(k)}x_k + G_{\sigma(k)}d_k \\ x_k = \phi_k, \quad k = -\tau, \ldots, -1, 0 \end{cases} \tag{1.1}$$

where $x_k \in \mathbb{R}^n$ is the state vector, $u_k \in \mathbb{R}^m$ is the control input vector, $d_k \in \mathbb{R}^q$ is unknown input vector, $f_k = f(x_k, u_k, k) : \mathbb{R}^n \times \mathbb{R}^m \times \mathbb{N} \to \mathbb{R}^{n_f}$ is the nonlinearity vector, and $y_k \in \mathbb{R}^p$ is the output vector. $\sigma(k) : \mathcal{Z}^+ \to \mathcal{I}$ is a piecewise constant switching signal, where $\mathcal{Z}^+ = \{0, 1, \ldots\}$ and $\mathcal{I} = \{1, 2, \ldots, N\}$ is an index set. At a switching time k, we have $\sigma(k-1) \neq \sigma(k)$. As often assumed in the literature, we exclude Zeno behavior for the switching signal here, i.e. the switching is 'finite time finite switching'. $E_i \in \mathbb{R}^{n' \times n}$, $A_i \in \mathbb{R}^{n' \times n}$, $A_{di} \in \mathbb{R}^{n' \times n}$, $B_i \in \mathbb{R}^{n' \times m}$, $F_i \in \mathbb{R}^{n' \times q}$, $J_i \in \mathbb{R}^{n' \times n_f}$, $C_i \in \mathbb{R}^{p \times n}$, and $G_i \in \mathbb{R}^{p \times q}$ are known constant matrices for $i \in \mathcal{I}$ and $i = \sigma(k)$, $n' \leq n$. E_i may be rectangular and rank$E_i = r < n$. Moreover, when $\sigma(k) = i$ and $\sigma(k+1) = j$, the matrices $(E_j, A_i, A_{di}, B_i, F_i, C_i, G_i)$ are activated. τ denotes the known state delay, and ϕ_k, $k = -\tau, \ldots, -1, 0$ is a given initial condition sequence.

In the sequel, the following assumptions are made:

Assumption 1 The nonlinearity $f_k = f(x_k, u_k, k)$ is globally Lipschitz in x with Lipschitz constant γ, i.e.,

$$\| f(x_k, u_k, k) - f(\hat{x}_k, u_k, k) \| \leq \gamma \| x_k - \hat{x}_k \|, \forall u_k \in \mathbb{R}^m.$$

Assumption 2 The switching time sequence, i.e., the ordered sequence of the switching signal, is real-time accessible.

Assumption 3 (1) rank $\begin{bmatrix} F_{\sigma(k)} \\ G_{\sigma(k)} \end{bmatrix} = q$, rank$\begin{bmatrix} C_{\sigma(k)} & G_{\sigma(k)} \end{bmatrix} = p$;

(2) rank$\begin{bmatrix} E_{\sigma(k+1)} & F_{\sigma(k)} & 0 \\ C_{\sigma(k+1)} & 0 & G_{\sigma(k+1)} \\ 0 & G_{\sigma(k)} & 0 \end{bmatrix} = n + \text{rank} \begin{bmatrix} F_{\sigma(k)} \\ G_{\sigma(k)} \end{bmatrix} + \text{rank}G_{\sigma(k+1)}$.

Definition 1 ([18]) System (1.1) with $E_{\sigma(k)} = I$ and $u_k = d_k = 0$ is said to be exponentially stable under switching signal $\sigma(k)$ if its solution x_k satisfies: $\|x_k\| \leq c\lambda^{-(k-k_0)}\|\phi\|_L$, $k \geq k_0$, for any initial conditions ϕ_k, $k = k_0 - \tau, \ldots, k_0$, where $\|\phi\|_L = \sup_{k_0-\tau\leq l\leq k_0}\|\phi_l\|$, k_0 is the initial time step, $c > 0$ is the decay coefficient, and $\lambda > 1$ is the decay rate.

Here, our aim is to design a full order switched observer as the following form

$$\begin{cases} z_{k+1} = L_{\sigma(k-\tau),k_+}z_k + L_{d\sigma(k-\tau),k_+}z_{k-\tau} + T_{\sigma(k-\tau),k_+}B_{\sigma(k)}u_k \\ \qquad + H_{\sigma(k-\tau),k_+}y_k + K_{\sigma(k-\tau),k_+}y_{k-\tau} + T_{\sigma(k-\tau),k_+}J_{\sigma(k)}f(\hat{x}_k, u_k, k) \\ \hat{x}_k = z_k + N_{\sigma(k-\tau-1),k_-}y_k \end{cases} \tag{1.2}$$

with the initial state $z_k = \phi'_k$, $k = -\tau, \ldots, -1, 0$, where $z_k \in \mathbb{R}^n$ and $\hat{x}_k \in \mathbb{R}^n$ are respectively the observer state and the estimation of x_k,

$$H_{\sigma(k-\tau),k_+} = H_{1\sigma(k-\tau),k_+} + L_{\sigma(k-\tau),k_+}N_{\sigma(k-\tau-1),k_-} \tag{1.3}$$

$$K_{\sigma(k-\tau),k_+} = K_{1\sigma(k-\tau),k_+} + L_{d\sigma(k-\tau),k_+}N_{\sigma(k-2\tau-1),\sigma(k-\tau-1),\sigma(k-\tau)} \tag{1.4}$$

and $L_{\sigma(k-\tau),k_+}, L_{d\sigma(k-\tau),k_+}, T_{\sigma(k-\tau),k_+}, H_{1\sigma(k-\tau),k_+}, K_{1\sigma(k-\tau),k_+}, N_{\sigma(k-\tau-1),k_-}$ are matrices to be determined such that $\hat{x}_k \in \mathbb{R}^n$ exponentially converges to x_k.

Remark 1 The observer presented in (1.2) is of delay type. The introduction of delayed terms $z_{k-\tau}$ and $y_{k-\tau}$ provides more freedom for the design of observer.

1.3 Main Results

The following proposition gives the structure of the observer.

Proposition 1 *Under* Assumption 3, *there exist matrices* $L_{\sigma(k-\tau),k_+}, L_{d\sigma(k-\tau),k_+}, T_{\sigma(k-\tau),k_+}, H_{1\sigma(k-\tau),k_+}, K_{1\sigma(k-\tau),k_+}$, *and* $N_{\sigma(k-\tau),k_+}$ *such that*

$$T_{\sigma(k-\tau),k_+}E_{\sigma(k+1)} + N_{\sigma(k-\tau),k_+}C_{\sigma(k+1)} = I_n \tag{1.5}$$

$$L_{\sigma(k-\tau),k_+} = T_{\sigma(k-\tau),k_+}A_{\sigma(k)} - H_{1\sigma(k-\tau),k_+}C_{\sigma(k)} \tag{1.6}$$

$$L_{d\sigma(k-\tau),k_+} = T_{\sigma(k-\tau),k_+}A_{d\sigma(k)} - K_{1\sigma(k-\tau),k_+}C_{\sigma(k-\tau)} \tag{1.7}$$

$$T_{\sigma(k-\tau),k_+}F_{\sigma(k)} - H_{1\sigma(k-\tau),k_+}G_{\sigma(k)} = 0 \tag{1.8}$$

$$K_{1\sigma(k-\tau),k_+}G_{\sigma(k-\tau)} = 0 \tag{1.9}$$

$$N_{\sigma(k-\tau),k_+}G_{\sigma(k+1)} = 0 \tag{1.10}$$

and the difference of the state estimation error $e_k = x_k - \hat{x}_k$ satisfies

$$e_{k+1} = L_{\sigma(k-\tau),k_+}e_k + L_{d\sigma(k-\tau),k_+}e_{k-\tau} + T_{\sigma(k-\tau),k_+}J_{\sigma(k)}\hat{f}_k \tag{1.11}$$

where $L_{\sigma(k-\tau),k_+} = \Psi\Theta^+_{\sigma(k-\tau),k_+}\varphi_{\sigma(k)} - Z_{\sigma(k)}\Theta^\perp_{\sigma(k-\tau),k_+}\varphi_{\sigma(k)}$, $L_{d\sigma(k-\tau),k_+} = \Psi\Theta^+_{\sigma(k-\tau),k_+}\psi_{\sigma(k-\tau),\sigma(k)} - Z_{\sigma(k)}\Theta^\perp_{\sigma(k-\tau),k_+}\psi_{\sigma(k-\tau),\sigma(k)}$,

$$\hat{f}_k = f(x_k, u_k, k) - f(\hat{x}_k, u_k, k)$$
$$T_{\sigma(k-\tau),k_+}J_{\sigma(k)} = \Psi\Theta^+_{\sigma(k-\tau),k_+}\varsigma_{\sigma(k)} - Z_{\sigma(k)}\Theta^\perp_{\sigma(k-\tau),k_+}\varsigma_{\sigma(k)}$$

with $Z_{\sigma(k)}$ is an arbitrary matrix of appropriate dimension, and

$$\Theta^\perp_{\sigma(k-\tau),k_+} = I_{n'+3p+2n} - \Theta_{\sigma(k-\tau),k_+}\Theta^+_{\sigma(k-\tau),k_+}$$
$$\Psi = \begin{bmatrix} I_n \; 0\; 0\; 0\; 0\; 0 \end{bmatrix}, \; \varphi_{\sigma(k)} = \begin{bmatrix} A^T_{\sigma(k)} \; 0 \; -C^T_{\sigma(k)} \; 0\; 0\; 0 \end{bmatrix}^T$$

$$\psi_{\sigma(k-\tau),\sigma(k)} = \begin{bmatrix} A^T_{d\sigma(k)} \; 0\; 0 \; -C^T_{\sigma(k-\tau)} \; 0\; 0 \end{bmatrix}^T, \; \varsigma_{\sigma(k)} = \begin{bmatrix} J^T_{\sigma(k)} \; 0\; 0\; 0\; 0\; 0 \end{bmatrix}^T$$
$$\Theta_{\sigma(k-\tau),k_+} = \begin{bmatrix} E_{\sigma(k+1)} & A_{\sigma(k)} & A_{d\sigma(k)} & F_{\sigma(k)} & 0 & 0 \\ C_{\sigma(k+1)} & 0 & 0 & 0 & 0 & G_{\sigma(k+1)} \\ 0 & -C_{\sigma(k)} & 0 & -G_{\sigma(k)} & 0 & 0 \\ 0 & 0 & -C_{\sigma(k-\tau)} & 0 & G_{\sigma(k-\tau)} & 0 \\ 0 & -I_n & 0 & 0 & 0 & 0 \\ 0 & 0 & -I_n & 0 & 0 & 0 \end{bmatrix}.$$

Proof Suppose that (1.5) holds, then $e_{k+1} = x_{k+1} - \hat{x}_{k+1}$ becomes

$$e_{k+1} = T_{\sigma(k-\tau),k_+}E_{\sigma(k+1)}x_{k+1} - z_{k+1} - N_{\sigma(k-\tau),k_+}G_{\sigma(k+1)}d_{k+1}.$$

By (1.1) and (1.2), e_{k+1} can be further rewritten as

$$\begin{aligned} e_{k+1} &\\ = \Big(&T_{\sigma(k-\tau),k_+}A_{\sigma(k)} + L_{\sigma(k-\tau),k_+}N_{\sigma(k-\tau-1),k_-}C_{\sigma(k)} - H_{\sigma(k-\tau),k_+}C_{\sigma(k)} \\ &- L_{\sigma(k-\tau),k_+}\Big)x_k + \Big(T_{\sigma(k-\tau),k_+}A_{d\sigma(k)} + L_{d\sigma(k-\tau),k_+} \\ &\times N_{\sigma(k-2\tau-1),\sigma(k-\tau-1),\sigma(k-\tau)}C_{\sigma(k-\tau)} - K_{\sigma(k-\tau),k_+}C_{\sigma(k-\tau)} \\ &- L_{d\sigma(k-\tau),k_+}\Big)x_{k-\tau} + \Big(T_{\sigma(k-\tau),k_+}F_{\sigma(k)} + L_{\sigma(k-\tau),k_+}N_{\sigma(k-\tau-1),k_-} \\ &\times G_{\sigma(k)} - H_{\sigma(k-\tau),k_+}G_{\sigma(k)}\Big)d_k + \Big(L_{d\sigma(k-\tau),k_+}N_{\sigma(k-2\tau-1),\sigma(k-\tau-1),\sigma(k-\tau)} \\ &\times G_{\sigma(k-\tau)} - K_{\sigma(k-\tau),k_+}G_{\sigma(k-\tau)}\Big)d_{k-\tau} - N_{\sigma(k-\tau),k_+}G_{\sigma(k+1)}d_{k+1} \\ &+ T_{\sigma(k-\tau),k_+}J_{\sigma(k)}\hat{f}_k + L_{\sigma(k-\tau),k_+}e_k + L_{d\sigma(k-\tau),k_+}e_{k-\tau}. \end{aligned} \tag{1.12}$$

Substituting (1.3) and (1.4) into (1.12) and using the relations in (1.5)–(1.10) results in $e_{k+1} = L_{\sigma(k-\tau),k_+} e_k + L_{d\sigma(k-\tau),k_+} e_{k-\tau} + T_{\sigma(k-\tau),k_+} J_{\sigma(k)} \hat{f}_k$.

Define

$$\theta_{\sigma(k-\tau),k_+} = \left[T_{\sigma(k-\tau),k_+} \; N_{\sigma(k-\tau),k_+} \; H_{1\sigma(k-\tau),k_+} \; K_{1\sigma(k-\tau),k_+} \; L_{\sigma(k-\tau),k_+} \; L_{d\sigma(k-\tau),k_+} \right].$$

Rewriting (1.11) and (1.5)–(1.10) by known system matrices and $\theta_{\sigma(k-\tau),k_+}$, respectively, gives

$$e_{k+1} = \theta_{\sigma(k-\tau),k_+} \left(\varphi_{\sigma(k)} e_k + \psi_{\sigma(k-\tau),\sigma(k)} e_{k-\tau} + \varsigma_{\sigma(k)} \hat{f}_k \right) \tag{1.13}$$

$$\Psi = \theta_{\sigma(k-\tau),k_+} \Theta_{\sigma(k-\tau),k_+}. \tag{1.14}$$

Under *Assumption* 2, we have rank $\begin{bmatrix} \Theta_{\sigma(k-\tau),k_+} \\ \Psi \end{bmatrix} = \text{rank}\Theta_{\sigma(k-\tau),k_+}$. Then, the solution $\theta_{\sigma(k-\tau),k_+}$ of (1.14) exists and is given by [19]

$$\theta_{\sigma(k-\tau),k_+} = \Psi \Theta^{+}_{\sigma(k-\tau),k_+} - Z_{\sigma(k)} \Theta^{\perp}_{\sigma(k-\tau),k_+} \tag{1.15}$$

where $\Theta^{\perp}_{\sigma(k-\tau),k_+} = I_{n'+3p+2n} - \Theta_{\sigma(k-\tau),k_+} \Theta^{+}_{\sigma(k-\tau),k_+}$ and $Z_{\sigma(k)}$ is an arbitrary matrix of appropriate dimension. Substituting (1.15) into (1.13) yields (1.11), where $L_{\sigma(k-\tau),k_+}$, $L_{d\sigma(k-\tau),k_+}$ and $T_{\sigma(k-\tau),k_+} J_{\sigma(k)}$ are determined by known system matrices and $Z_{\sigma(k)}$.

Remark 2 The matrices $Z_{\sigma(k)}$ in (1.15) play the role in parametrization. In some special cases, an arbitrary choice of Z_i, $i \in \mathcal{I}$, may involve a loss of system performance such as observability; see [2] for the linear case.

Now, the observer design is reduced to find matrices Z_i, $i \in \mathcal{I}$, ensuring the stability of system (1.11) under arbitrary switching signals. The condition of global exponential stability of (1.11) is stated in the following theorem.

Theorem 1 *For given scalar* $0 < \alpha < 1$, *if there exist matrices* $P_i > 0$, $Q_i > 0$, U_i *and* R_i, $i \in \mathcal{I}$, *and scalar* $\epsilon > 0$ *such that*

$$\begin{bmatrix} P_j - R_i - R_i^T & (1,2) & (1,3) & (1,4) \\ * & (2,2) & 0 & 0 \\ * & * & -(1-\alpha)^{\tau} Q_l & 0 \\ * & * & * & -\epsilon I \end{bmatrix} < 0, \forall (l,i,j) \in \mathcal{I} \times \mathcal{I} \times \mathcal{I} \tag{1.16}$$

where $(1,2) = R_i \Psi \Theta^{+}_{l,i,j} \varphi_i - U_i \Theta^{\perp}_{l,i,j} \varphi_i$, $(1,3) = R_i \Psi \Theta^{+}_{l,i,j} \psi_{l,i} - U_i \Theta^{\perp}_{l,i,j} \psi_{l,i}$, $(1,4) = R_i \Psi \Theta^{+}_{l,i,j} \varsigma_i - U_i \Theta^{\perp}_{l,i,j} \varsigma_i$, *and* $(2,2) = -(1-\alpha) P_i + Q_i$, *then the state estimation error system (1.11) is exponentially stable and ensures a decay rate* $\frac{1}{1-\alpha}$. *Moreover, the resulting observer gains are given by (1.15) with* $Z_i = R_i^{-1} U_i$, $\forall \sigma(k) = i \in \mathcal{I}$.

Proof Consider the following switched Lyapunov function for system (1.11)

$$V_k = e_k^T P_{\sigma(k)} e_k + \sum_{l=k-\tau}^{k-1} e_l^T (1-\alpha)^{k-1-l} Q_{\sigma(l)} e_l. \tag{1.17}$$

Defining

$$\triangle V_k = V_{k+1} - (1-\alpha) V_k \tag{1.18}$$

and taking the forward difference $\triangle V_k$ along the solution of (1.11) yields

$$\begin{aligned}
&\triangle V_k \mid_{(11)} \\
&= e_{k+1}^T P_{\sigma(k+1)} e_{k+1} - (1-\alpha) e_k^T P_{\sigma(k)} e_k + e_k^T Q_{\sigma(k)} e_k \\
&\quad - (1-\alpha)^\tau e_{k-\tau}^T Q_{\sigma(k-\tau)} e_{k-\tau} \\
&= \eta_k^T \left(\begin{bmatrix} L_{\sigma(k-\tau),k_+}^T \\ L_{d\sigma(k-\tau),k_+}^T \\ J_{\sigma(k)}^T T_{\sigma(k-\tau),k_+}^T \end{bmatrix} P_{\sigma(k+1)} \begin{bmatrix} L_{\sigma(k-\tau),k_+} & L_{d\sigma(k-\tau),k_+} & T_{\sigma(k-\tau),k_+} J_{\sigma(k)} \end{bmatrix} \right. \\
&\quad \left. + \begin{bmatrix} -(1-\alpha)P_{\sigma(k)} + Q_{\sigma(k)} & 0 & 0 \\ * & -(1-\alpha)^\tau Q_{\sigma(k-\tau)} & 0 \\ * & * & 0 \end{bmatrix} \right) \eta_k
\end{aligned}$$

under the constraint in *Assumption* 1, where $\eta_k = \begin{bmatrix} e_k^T & e_{k-\tau}^T & \hat{f}_k^T \end{bmatrix}^T$. On the other hand, from *Assumption* 1 and $e_k = x_k - \hat{x}_k$, we have

$$\Gamma_k := \hat{f}_k^T \hat{f}_k - \gamma^2 e_k^T e_k \leq 0. \tag{1.19}$$

Therefore, we can deduce from the well known S-procedure lemma in [20] that the Eq. (1.11) is stable if there exists a scalar $\epsilon > 0$ such that $\triangle V_k \mid_{(11)} - \Gamma_k < 0$, that is,

$$\begin{aligned}
&Y_{\sigma(k-\tau),k_+} \\
&= \left[\begin{matrix} \begin{pmatrix} L_{\sigma(k-\tau),k_+}^T P_{\sigma(k+1)} L_{\sigma(k-\tau),k_+} \\ -(1-\alpha)P_{\sigma(k)} + Q_{\sigma(k)} + \epsilon\gamma^2 I \end{pmatrix} & L_{\sigma(k-\tau),k_+}^T P_{\sigma(k+1)} L_{d\sigma(k-\tau),k_+} \\ * & \begin{pmatrix} L_{d\sigma(k-\tau),k_+}^T P_{\sigma(k+1)} L_{d\sigma(k-\tau),k_+} \\ -(1-\alpha)^\tau Q_{\sigma(k-\tau)} \end{pmatrix} \\ * & * \end{matrix} \right. \\
&\qquad \left. \begin{matrix} L_{\sigma(k-\tau),k_+}^T P_{\sigma(k+1)} T_{\sigma(k-\tau),k_+} J_{\sigma(k)} \\ L_{\sigma(k-\tau),k_+}^T P_{\sigma(k+1)} T_{\sigma(k-\tau),k_+} J_{\sigma(k)} \\ \begin{pmatrix} J_{\sigma(k)}^T T_{\sigma(k-\tau),k_+}^T P_{\sigma(k+1)} \\ \times T_{\sigma(k-\tau),k_+} J_{\sigma(k)} - \epsilon I \end{pmatrix} \end{matrix} \right] < 0, \forall (l,i,j) \in \mathcal{I} \times \mathcal{I} \times \mathcal{I}.
\end{aligned}$$

Since $Y_{\sigma(k-\tau),k_+} < 0$ has to be satisfied under arbitrary switching laws, it follows that this has to hold for special configuration $\sigma(k+1)=j$, $\sigma(k)=i$ and $\sigma(k-\tau)=l$, for all η_k, we get

$$\begin{bmatrix} \begin{pmatrix} L_{l,i,j}^T P_j L_{l,i,j} + Q_i \\ -(1-\alpha)P_i + \epsilon\gamma^2 I \end{pmatrix} & L_{l,i,j}^T P_j L_{dl,i,j} & L_{l,i,j}^T P_j T_{l,i,j} J_i \\ * & \begin{pmatrix} -(1-\alpha)^\tau Q_l \\ +L_{dl,i,j}^T P_j L_{dl,i,j} \end{pmatrix} & L_{dl,i,j}^T P_j T_{l,i,j} J_i \\ * & * & J_i^T T_{l,i,j}^T P_j T_{l,i,j} J_i - \epsilon I \end{bmatrix} < 0$$

$$\forall (l,i,j) \in \mathcal{I}\times\mathcal{I}\times\mathcal{I}. \tag{1.20}$$

which, by Schur complement formula, is equivalent to:

$$\begin{bmatrix} -P_j^{-1} & L_{l,i,j} & L_{dl,i,j} & T_{l,i,j} J_i \\ * & \begin{pmatrix} -(1-\alpha)P_i \\ +Q_i + \epsilon\gamma^2 I \end{pmatrix} & 0 & 0 \\ * & * & -(1-\alpha)^\tau Q_l & 0 \\ * & * & * & -\epsilon I \end{bmatrix} < 0, \forall (l,i,j) \in \mathcal{I}\times\mathcal{I}\times\mathcal{I} \tag{1.21}$$

Now, suppose (1.16) is feasible, then it is easy to see that $P_j - R_i - R_i^T < 0$, $\forall i,j \in \mathcal{I}\times\mathcal{I}$. This means that R_i, $\forall i \in \mathcal{I}$, is of full rank. Since P_j, $\forall j \in \mathcal{I}$, is strictly positive definite, we have $(P_j - R_i)P_j^{-1}(P_j - R_i)^T \geq 0$, $\forall i,j \in \mathcal{I}\times\mathcal{I}$, which implies $-R_i P_j^{-1} R_i^T \leq P_j - (R_i^T + R_i)$, $\forall (i,j) \in \mathcal{I}\times\mathcal{I}$. Then, it follows from (1.16) that $\begin{bmatrix} -R_i P_j^{-1} R_i^T & (1,2) & (1,3) & (1,4) \\ * & (2,2) & 0 & 0 \\ * & * & -(1-\alpha)^\tau Q_l & 0 \\ * & * & * & -\epsilon I \end{bmatrix} < 0$, $\forall (l,i,j) \in \mathcal{I}\times\mathcal{I}\times\mathcal{I}$. Premultiplying $\mathrm{diag}\{R_i^{-1}, I, I, I\}$, postmultiplying $\mathrm{diag}\{R_i^{-T}, I, I, I\}$ and substituting $U_i = R_i Z_i$ to the above inequality result in (1.21). Therefore, we have that $\triangle V_k \mid_{(11)} < 0$ for any $\eta_k \neq 0$, which by (1.18) implies

$$V_k \mid_{(11)} < (1-\alpha)^k V_0. \tag{1.22}$$

From the piecewise Lyapunov-like function (1.17), it follows that

$$a\|e_k\|^2 \leq V_k \mid_{(11)},\ V_0 \mid_{(11)} \leq b\|e_0\|_\tau^2 \tag{1.23}$$

where $a = \min_{i\in\mathcal{I}} \lambda_{\min}(P_i)$ and $b = \max_{i\in\mathcal{I}}(\lambda_{\max}(P_i) + \tau\lambda_{\max}(Q_i))$. Combining (1.22) and (1.23) gives $\|e_k\| \leq \sqrt{\frac{b}{a}}\left(\frac{1}{1-\alpha}\right)^{-k} \|e_0\|_\tau$ for all $k \geq 0$, Therefore, it is concluded from Definition 1 with $k_0 = 0$ that system (1.11) is exponentially stable and ensures the decay rate $\frac{1}{1-\alpha}$.

1.4 Conclusions

The state observation problem for a class of discrete-time switched Lipschitz nonlinear singular time-delay systems with UIs and arbitrary switching sequences has been addressed. Sufficient conditions for the existence of the observer have been given in terms of LMIs, and the exponential convergence of the state estimation error system has been established by appropriate Lyapunov analysis.

Acknowledgments This work was supported by the National Natural Science Foundation of China (61473158) and the Natural Science Foundation of Jiangsu Province (BK20141430).

References

1. Teh PS, Trinh H (2013) Design of unknown input functional observers for nonlinear systems with application to fault diagnosis. J Process Control 23(8):1169–1184
2. Kudva P, Viswanadham N, Ramakrishna A (1980) Observers for linear systems with unknown inputs. IEEE Trans Autom Control 25(1):113–115
3. Darouach M, Zasadzinski M, Xu S (1994) Full-order observers for linear systems with unknown inputs. IEEE Trans Autom Control 39(3):606–609
4. Chang S-K, You W-T, Hsu P-L (1997) Design of general structured observers for linear systems with unknown inputs. J Frankl Inst 334(2):213–232
5. Valcher ME (1999) State observers for discrete-time linear systems with unknown inputs. IEEE Trans Autom Control 44(2):397–401
6. Walcott B, Żak SH (1987) State observation of nonlinear uncertain dynamical systems. IEEE Trans Autom Control 32(2):166–170
7. Liberzon D (2003) Switching in systems and control. Birkhauser, Boston
8. Djemai M, Defoort M (eds) (2015) Hybrid dynamical systems observation and control. Springer International Publishing, Switzerland
9. Li S, Xiang Z, Karimi HR (2014) Mixed l_-/l_1 fault detection observer design for positive switched systems with time-varying delay via delta operator approach. Int J Control Autom Syst 12(4):709–721
10. Millerioux G, Daafouz J (2004) Unknown input observers for switched linear discrete time systems. Proceedings of the american control conference, June 2004, pp. 5802–5805
11. Sundaram S, Hadjicostis CN (2006) Designing stable inverters and state observers for switched linear systems with unknown inputs. Proceedings of the 45th IEEE conference on decision and control, December 2006, pp. 4105–4110
12. Chen J, Lagoa CM (2006) Robust observer design for a class of switched systems. Proceedings of the 45th IEEE conference on decision and control, December 2006, pp. 1659–1664
13. Bejarano FJ, Pisano A (2011) Switched observers for switched linear systems with unknown inputs. IEEE Trans Autom Control 56(3):681–686
14. Huang G-J, Chen W-H (2014) A revisit to the design of switched observers for switched linear systems with unknown inputs. Int J Control Autom Syst 12(3):954–962
15. Ma S, Zhang C, Wu Z (2008) Delay-dependent stability and H_∞ control for uncertain discrete switched singular systems with time-delay. Appl Math Comput 206(1):413–424
16. Ezzine M, Darouach M, Ali HS, Messaoud H (2013) Unknown inputs functional observers designs for delay descriptor systems. Int J Control 86(10):1850–1858
17. Koenig D, Marx B, Jacquet D (2008) Unknown input observers for switched nonlinear discrete time descriptor systems. IEEE Trans Autom Control 53(1):373–379

18. Lin J, Fei S, Gao Z, Ding J (2013) Fault detection for discrete-time switched singular time-delay systems: an average dwell time approach. Int J Adapt Control Signal Process 27(7):582–609
19. Rao CR, Mitra SK (1971) Generalized Inverse of Matrices and its Applications. Wiley, New York
20. Boyd S, El Ghaoui L, Feron E, Balakrishnan V (1994) Matrix Inequalities in System and Control Theory. SIAM, Philadelphia

Chapter 2
Anti-disturbance Control for Nonlinear Systems with Mismatched Disturbances Based on Disturbance Observer

Lingyan Zhang and Xinjiang Wei

Abstract An anti-disturbance control scheme based on disturbance observer is proposed for a class of nonlinear system with mismatched disturbances. With such control approach, the disturbances can be rejected and the semi-global uniformly ultimate bounded (SGUUB) stability of the closed-loop system can be achieved. Finally, simulations for a numerical example are given to demonstrate the feasibility and effectiveness of the proposed scheme.

Keywords Anti-disturbance control · Disturbance observer · Mismatched disturbances

2.1 Introduction

Disturbance-observer-based control (DOBC) strategy appeared in the late 1980s [1] and has been applied in many control fields [2]. In recent years, a composite DOBC and other control approaches has been proposed, such as H_∞ control [3], sliding mode control [4],adaptive control [5], fuzzy control [6]. Back-stepping method is one of the most important design techniques in the nonlinear control area [7]. Recently, back-stepping method has been integrated with other control approaches, such as adaptive control [8], sliding mode control [9], H_∞ control [10], fuzzy control [11].

However, there are few reports about composite DOBC scheme and back-stepping method. In [12], a control scheme combining disturbance observer technique and back-stepping method was proposed for a class of nonlinear system with disturbances and nonlinear functions. Considering that it is linear for the main part of the system in [12], we will extend to a class of more general nonlinear systems with mismatched disturbances.

L. Zhang · X. Wei (✉)
School of Mathematics and Statistics Science, Ludong University, Yantai, China
e-mail: weixinjiang@163.com

© Springer-Verlag Berlin Heidelberg 2016

Y. Jia et al. (eds.), *Proceedings of the 2015 Chinese Intelligent Systems Conference*, Lecture Notes in Electrical Engineering 359,
DOI 10.1007/978-3-662-48386-2_2

The main contributions of this paper are summarized as follows:

(1) The disturbances with partially-known information represented by exogenous systems are mismatched disturbances, which appear in different channels as the control inputs.
(2) A composite anti-disturbance control strategy with nonlinear disturbance observer and back-stepping method is proposed to apply in a class of more general nonlinear systems with mismatched disturbances.

2.2 Formulation of the Problem

The following nonlinear system with mismatched disturbances is described as

$$\begin{aligned} \dot{x}_1 &= f_1(x_1) + g_1(x_1)x_2 + h_1(x_1)d_1(t), \\ \dot{x}_2 &= f_2(x_1, x_2) + g_2(x_1, x_2)x_3 + h_2(x_1, x_2)d_2(t), \\ &\vdots \\ \dot{x}_n &= f_n(x_1, \ldots, x_n) + g_n(x_1, \ldots, x_n)u + h_n(x_1, \ldots, x_n)d_n(t), \\ y &= s(x), \end{aligned} \tag{2.1}$$

where $x = [x_1, x_2, \ldots, x_n]^T \in R^n$, $u \in R$, y are the system states, the control input and the system output, respectively. f_i, g_i, h_i, s are smooth functions and are differentiable with $f_i(0) = 0$ and $g_i(\cdot) \neq 0$ for $i = 1, 2, \ldots, n$. $d_i(t) \in R$ represents the external disturbance.

Assumption 1 The disturbance $d_i(t)$ can be described by the following exogenous system:

$$\begin{aligned} \dot{w}_i(t) &= A_i w_i(t), \\ d_i(t) &= C_i w_i(t), \end{aligned} \tag{2.2}$$

where $A_i \in R^{m\times m}$ and $C_i \in R^{1\times m}$ are proper matrices. Generally, the exogenous system (2.2) is deemed to be neutral stable.

2.3 Nonlinear Disturbance Observer

A new nonlinear disturbance observer is introduced as follows:

$$\begin{aligned} \dot{q}_1 &= [A_1 - l_1(x_1)h_1(x_1)C_1]q_1 + A_1 p_1(x_1) - l_1(x_1)[h_1(x_1)C_1 p_1(x_1) + f_1(x_1) \\ &\quad + g_1(x_1)x_2], \\ \dot{q}_2 &= [A_2 - l_2(x_2)h_2(x_1, x_2)C_2]q_2 + A_2 p_2(x_2) - l_2(x_2)[h_2(x_1, x_2)C_2 p_2(x_2) \end{aligned}$$

$$
\begin{aligned}
&+f_2(x_1,x_2)+g_2(x_1,x_2)x_3],\\
&\vdots\\
\dot{q}_n &= [A_n - l_n(x_n)h_n(x_1,\dots,x_n)C_n]q_n + A_np_n(x_n)\\
&\quad - l_n(x_n)[h_n(x_1,\dots,x_n)C_np_n(x_n)+f_n(x_1,\dots,x_n)+g_n(x_1,\dots,x_n)u],\\
\hat{w}_i &= q_i + p_i(x_i),\\
\hat{d}_i &= C_i\hat{w}_i,
\end{aligned}
\tag{2.3}
$$

where $\hat{w}_i$ is the estimation of w_i, $q_i \in R^{m\times 1}$ is the auxiliary vector and $p_i(x_i) \in R^{m\times 1}$ is an nonlinear function to be designed. The nonlinear observer gain $l_i(x_i)$ is determined by $l_i(x_i) = \frac{\partial p_i(x_i)}{\partial x_i}$. The estimation error is denoted as $e_i = w_i - \hat{w}_i$. Then the estimation errors dynamics are presented as

$$
\dot{e} = [A - l(x)h(x)C]e, \tag{2.4}
$$

where $e = [e_1, e_2, \dots, e_n]^T$, $A = diag\{A_1, A_2, \dots, A_n\}$, $l(x) = diag\{l_1(x_1), l_2(x_2), \dots, l_n(x_n)\}$, $h(x) = diag\{h_1(x_1), h_2(x_1,x_2), \dots, h_n(x_1,\dots,x_n)\}, C = diag\{C_1, C_2, \dots, C_n\}$.

Assumption 2 Suppose that the relative degree from the disturbance d_i to the output $r_i \geq 1$ such that $m_i(x) = L_{H_i}L_F^{r_i-1}s(x) \neq 0$, where L denotes Lie derivatives, $m_i(x)$ is bounded with respect to x in its operation region.

Select $l_i(x_i) = K_i\frac{\partial L_F^{r_i-1}s(x)}{\partial x_i}$, then $l_i(x_i)h_i(x_1,\dots,x_i) = K_im_i(x)$. If there exists a constant m_{0i} and a bounded nonlinear function $m_{ii}(x)$ satisfying $m_i(x) = m_{0i} + m_{ii}(x)$, where $m_{ii}^2(x) \leq \bar{m}_{ii}^2$, $\bar{m}_{ii}$ is known constant, then the disturbance estimation errors system (2.4) can be rewritten as

$$
\dot{e} = [\bar{A} - KM(x)C]e, \tag{2.5}
$$

where $\bar{A} = A - KMC$, $K = diag\{K_1, K_2, \dots, K_n\}$, $M = diag\{m_{01}, m_{02}, \dots, m_{0n}\}$, $M(x) = diag\{m_{11}(x), m_{22}(x), \dots, m_{nn}(x)\}$. Defining matrix $\bar{M} = diag\{\bar{m}_{11}^{-1}, \bar{m}_{22}^{-1}, \dots, \bar{m}_{nn}^{-1}\}$, based on Assumptions 1 and 2, the following result can be obtained.

Theorem 1 *For given matrix* M, $\bar{M}$, *if there exist* $P > 0$ *and* Q *satisfying*

$$
\begin{bmatrix} A^TP + PA - C^TM^TQ^T - QMC + C^TC & Q \\ Q^T & -\bar{M}^2 \end{bmatrix} < 0, \tag{2.6}
$$

where $K = P^{-1}Q$, *then based on the nonlinear disturbance observer (2.3), the disturbance estimation errors system (2.5) is asymptotically stable.*

2.4 Back-Stepping Controller Design and Stability Analysis

Based on the back-stepping method, the anti-disturbance controller u is constructed as

$$
\begin{aligned}
u &= \alpha_n(x_1, \ldots, x_n, \hat{d}_1, \ldots, \hat{d}_n) \\
&= \frac{1}{g_n}\{-z_{n-1}g_{n-1} - f_n - h_n\hat{d}_n + \sum_{j=1}^{n-1}[\frac{\partial \alpha_{n-1}}{\partial x_j}(f_j + g_j x_{j+1} + h_j\hat{d}_j) \\
&\quad + \frac{\partial \alpha_{n-1}}{\partial \hat{d}_j} C_j A_j \hat{w}_j] - k_n z_n + r_n\},
\end{aligned} \tag{2.7}
$$

where $k_n > 0$ is an adjustable controller parameter, and $r_n = -\sum_{j=1}^{2n-1} \delta_{nj} z_n$, $\delta_{nj} > 0$, $j = 1, 2, \ldots, 2n-1$ are adjustable parameters. Thus,

$$
\begin{aligned}
\dot{V}_n &= \sum_{j=1}^{n}(-k_j z_j^2 + z_j h_j C_j e_j + z_j r_j) - \sum_{t=2}^{n}[z_t \sum_{j=1}^{t-1}(\frac{\partial \alpha_{t-1}}{\partial x_j} h_j C_j e_j \\
&\quad + \frac{\partial \alpha_{t-1}}{\partial \hat{d}_j} C_j l_j h_j C_j e_j)] \\
&\le -\sum_{j=1}^{n} k_j z_j^2 + \varepsilon_n,
\end{aligned} \tag{2.8}
$$

where $\varepsilon_n = \frac{1}{\delta_{n1}} h_n^2 C_n e_n e_n^T C_n^T + \frac{1}{\delta_{n2}}(\frac{\partial \alpha_{n-1}}{\partial x_1} h_1)^2 C_1 e_1 e_1^T C_1^T + \cdots + \frac{1}{\delta_{n(n+1)}}(\frac{\partial \alpha_{n-1}}{\partial \hat{d}_1} h_1)^2 C_1 l_1 C_1 e_1 e_1^T C_1^T l_1^T C_1^T + \cdots + \varepsilon_{n-1}$.

The inequality (2.8) can be further rewritten as

$$
\dot{V}_n \le -\lambda V_n + \varepsilon_n, \tag{2.9}
$$

where $\lambda = \min\{k_1, k_2, k_3, \ldots, k_n\}$. The inequality (2.9) is equivalent to

$$
0 < V_n \le \frac{\varepsilon_n}{\lambda} + [V_n(0) - \frac{\varepsilon_n}{\lambda}]e^{-\lambda t}. \tag{2.10}
$$

From (2.10) and in the same proof as [9, 11], it can be shown that the closed-loop system is semi-global uniformly ultimate bounded (SGUUB). According to the above design and analysis, the stability of the closed-loop system is summarized in the following theorem:

Theorem 2 *Consider nonlinear system (2.1) with mismatched disturbances under Assumptions 1 and 2. By designing the nonlinear disturbance observer (2.3) and the anti-disturbance controller (2.7), the closed-loop system is semi-global uniformly ultimate bounded (SGUUB).*

2.5 Simulation Example

The mathematical model of a second-order nonlinear system with mismatched disturbances is described as follows:

$$\begin{aligned}
\dot{x}_1 &= x_1^3 + 2x_1 + x_2 + (1 + x_1^2)d_1(t), \\
\dot{x}_2 &= -x_1 + x_2 + (2 + 3x_1^2 + x_2^2)u + (1 + x_1^2 + x_2^2)d_2(t), \\
y &= x_1 + x_2,
\end{aligned} \tag{2.11}$$

where $x = [x_1, x_2]^T, u \in R$ are the system states and input, respectively. $d_i(t), i = 1, 2$ is assumed to be an unknown harmonic disturbance described by (2.2) with

$$A_1 = \begin{bmatrix} 0 & -5 \\ 5 & 0 \end{bmatrix}, \; A_2 = \begin{bmatrix} 0 & -2 \\ 2 & 0 \end{bmatrix}, \; C_1 = \begin{bmatrix} 2 & 0 \end{bmatrix}, \; C_2 = \begin{bmatrix} 3 & 0 \end{bmatrix}.$$

The relative degree from the disturbance $d_i, i = 1, 2$ to the output is calculated as $(r_1, r_2) = (2, 2)$. And $m_1(x) = 3x_1^4 + 4x_1^2 + 1$, $m_2(x) = x_1^2 + x_2^2 + 1$. We can choose $m_{01} = 1$, $m_{02} = 1$, $\bar{m}_{11} = 0.5$, $\bar{m}_{22} = 0.5$. According to Theorem 1, it can be obtained that

$$K = \begin{bmatrix} 0.5324 & 0 \\ -0.0384 & 0 \\ 0 & 0.5277 \\ 0 & -0.0973 \end{bmatrix}.$$

The controller is obtained by the proposed method in Sect. 2.4,

$$\begin{aligned}
u = & \frac{1}{2 + 3x_1^2 + x_2^2}\{-x_2 - (1 + x_1^2 + x_2^2)\hat{d}_2 + \frac{\partial \alpha_1(x_1, \hat{d}_1)}{\partial x_1}[x_1^3 + 2x_1 + x_2 \\
& + (1 + x_1^2)\hat{d}_1] + \frac{\partial \alpha_1(x_1, \hat{d}_1)}{\partial \hat{d}_1} C_1 A_1 \hat{w}_1 - k_2 z_2 + r_2\},
\end{aligned} \tag{2.12}$$

with $\frac{\partial \alpha_1(x_1, \hat{d}_1)}{\partial x_1} = -3x_1^2 - 2x_1\hat{d}_1 - (k_1 + \delta_{11} + 2)$, $\frac{\partial \alpha_1(x_1, \hat{d}_1)}{\partial \hat{d}_1} = -x_1^2 - 1$, $z_2 = x_2 + x_1^3 + (1 + x_1^2)\hat{d}_1 + (k_1 + \delta_{11} + 2)x_1$, $r_2 = -(\delta_{21} + \delta_{22} + \delta_{23})z_2$, where k_1, k_2 are positive numbers, $\delta_{11} > 0$, $\delta_{2j} > 0, j = 1, 2, 3$ are adjustable parameters. In simulation, the initial value of the states are set to be $x_1(0) = -1$, $x_2(0) = 4$ and the design parameters are chosen as $k_1 = 0.1$, $k_2 = 0.1$, $\delta_{11} = 1$, $\delta_{21} = 1$, $\delta_{22} = 1$, $\delta_{23} = 1$.

The simulation results are shown in Figs. 2.1, 2.2 and 2.3. Figure 2.1 demonstrates the system performance using the proposed control scheme. Figure 2.2 illustrates the estimation error for system disturbances with the anti-disturbance approach. Figure 2.3 shows the responses of the control input signal using the proposed method.

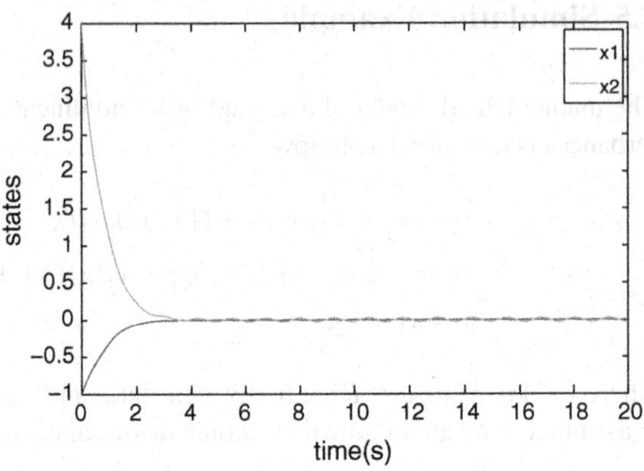

Fig. 2.1 System performance for disturbances

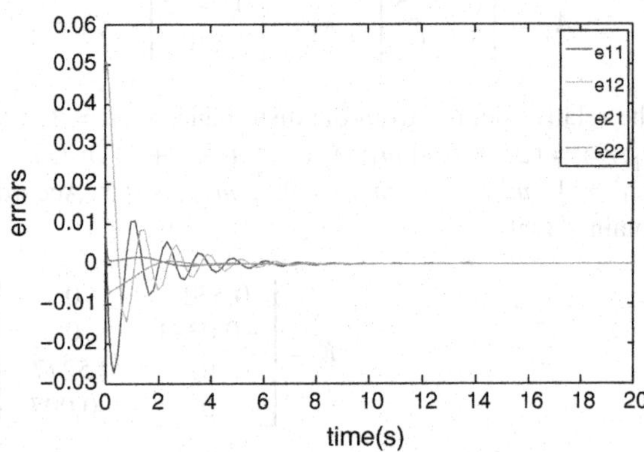

Fig. 2.2 Estimation errors for disturbances

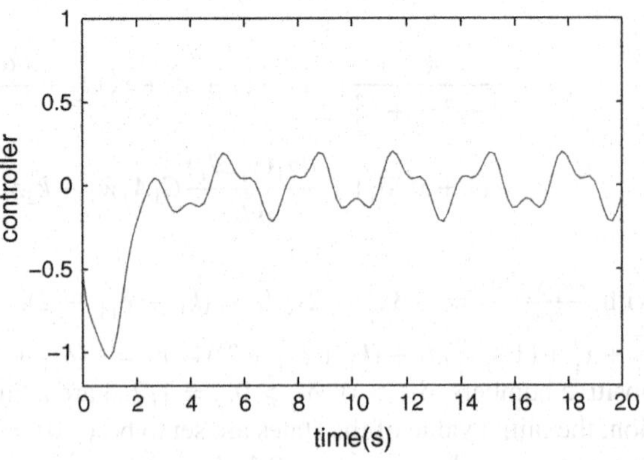

Fig. 2.3 The trajectory of control input

2.6 Conclusions

It is shown that the proposed anti-disturbance control scheme based on nonlinear disturbance observer is an effective control approach for a class of general nonlinear system subject to mismatched disturbances. However, in practical engineering, many complex systems can be described as mathematical models with multiple disturbances. If the disturbances in the nonlinear system (2.1) are multiple disturbances with unknown parameters, the situations will be more sophisticated. Consequently, new anti-disturbance control approaches are required to be considered and further research is needed in the future.

Acknowledgments The work is supported by National Science Foundation of China 61374108; Shandong Provincial Natural Science Foundation, China ZR2011FM016.

References

1. Nakao M, Ohnishi K, Miyachi K (1987) A robust decentralized joint control based on interference estimation. In Proceedings of the IEEE international conference robotics and automation, pp. 326–331
2. Iwasaki M, Shibata T, Matsui N (1999) Disturbance-observer-based nonlinear friction compensation in the table drive systems. IEEE/ASME Trans Mechatron 4(1):3–8
3. Wei X, Guo L (2010) Composite disturbance-observer-based control and H_∞ control for complex continuous models. Int J Robust Nonlinear Control 20(1):106–118
4. Wei X, Guo L (2009) Composite disturbance-observer-based control and terminal sliding mode control for nonlinear systems with disturbances. Int J Control 82(6):1082–1098
5. Guo L, Wen X (2011) Hierarchical anti-disturbance adaptive control for nonlinear systems with composite disturbances and applications to missile systems. Trans Inst Meas Control 33(8):942–956
6. Wei X, Chen N (2014) Composite hierarchical anti-disturbance control for nonlinear systems with DOBC and fuzzy control. Int J Robust Nonlinear control 24(2):362–373
7. Liu X, Lin Z (2012) On the back-stepping design procedure for multiple input nonlinear systems. Int J Robust Nonlinear Control 22(8):918–932
8. Zhang T, Ge SS, Hang CC (2000) Adaptive neural network control for strict-feedback nonlinear systems using back-stepping design. Automatica 36(12):1835–1846
9. Zhang C, Chen Z, Wei C (2014) Sliding mode disturbance observer-based back-stepping control for a transport aircraft. Sci China Inf Sci 57(5):1–16
10. Ma R, Dimirovski GM, Zhao J (2013) Back-stepping robust H_∞ Control for a class of uncertain switched nonlinear systems under arbitrary switchings. Asian J Control 15(1):41–50
11. Tong S, Li Y (2014) Observer-based adaptive fuzzy back-stepping control of uncertain nonlinear pure-feedback systems. Sci China Inf Sci 57(1):1–14
12. Sun H, Guo L (2014) Composite adaptive disturbance observer based control and back-stepping method for nonlinear system with multiple mismatched disturbances. J Frankl Inst 351(2):1027–1041

Chapter 3
Distributed Optimization for Continuous-Time Multi-Agent Systems with External Disturbance and Discrete-Time Communication

Zhenhua Deng and Yiguang Hong

Abstract In this paper, distributed optimization problem for continuous-time multi-agent systems with external disturbance and discrete-time communication is considered. A distributed algorithm is developed to achieve the exact optimal solution by completely rejecting the disturbance. An upper bound for the discrete-time communication period is obtained to ensure the exponential convergence for this optimization problem. Finally, a numerical example is given to illustrate the effectiveness of the proposed algorithm.

Keywords Distributed optimization · Disturbance rejection · Internal model · Multi-agent system

3.1 Introduction

Distributed optimization is an important topic of multi-agent systems [1, 2], which is applicable to various fields, such as distributed parameter estimation [3], distributed convex computation [4], and distributed optimal resource allocation [5]. In a standard distributed optimization setup, each agent only knows itself local cost function and communicates with its neighbors, but their aim is optimizing the global cost function which is the sum of all the local cost functions.

Since the controlled system usually is continuous-time in practical applications, such as robots and unmanned aircrafts, distributed continuous-time optimization algorithms have been paid more and more attention recently. For example, a distributed continuous-time coordination algorithm to solve network optimization problems has been introduced in [6]. Moreover, a distributed optimization problem has

Z. Deng (✉) · Y. Hong
Key Laboratory of System and Control, Academy of Mathematics
and System Science, Chinese Academy of Sciences, Beijing 100190, China
e-mail: zhdeng@amss.ac.cn

© Springer-Verlag Berlin Heidelberg 2016
Y. Jia et al. (eds.), *Proceedings of the 2015 Chinese Intelligent Systems Conference*, Lecture Notes in Electrical Engineering 359,
DOI 10.1007/978-3-662-48386-2_3

been studied for multi-agent system to deal with a class of deterministic external disturbances in [7].

In reality, many concerns have to be taken into consideration for practical implementation of the optimization algorithms. For example, the communication among the agents is usually carried out in discrete time, and external disturbance may influence the agents when these agents move in a complicated environment. For discrete-time communication constraints, sampled-data methods can be employed, and in fact, sampled-data multi-agent consensus has been studied for various connectivities in recent years [8, 9]. On the other hand, the internal model design is one of the most effective strategies to deal with deterministic external disturbance and has been also widely used in the coordination control of multi-agent systems [10, 11].

The objective of this paper is to discuss a distributed optimization problem for continuous-time multi-agent system with external disturbance and discrete-time communication. Here we propose a distributed algorithm to achieve the exact optimization with communication sampling constraints, and then give an estimation for the communication period of discrete-time communication and control design based on Lyapunov function and internal model. With quite mild assumptions, we prove that our algorithm can optimize the global cost function with disturbance rejection.

This paper is organized as follows. In Sect. 3.2, basic concepts and problem formulation are presented. The main result is proved in Sect. 3.3, and a simulation example is shown in Sect. 3.4 to demonstrate the effectiveness of the proposed algorithm. In Sect. 3.5, concluding remarks are given.

Notations: $\mathbb{R}$ and $\mathbb{N}$ stand for the set of real and natural numbers, respectively. $\mathbb{R}^n$ is n-dimension Euclidean space. $\otimes$ and $\|\cdot\|$ denote the Kronecker product and the standard Euclidean norm, respectively. A^T and $\|A\|$ are the transpose and the spectral norm of matrix A, respectively. I_n is $n \times n$ identity matrix. 1_n and 0_n are the column vectors of n ones and zeros, respectively.

3.2 Preliminaries

In this section, we first introduce basic concepts about convex analysis and graph theory, and then formulate the problem.

A function $f : \mathbb{R}^n \to \mathbb{R}$ is said to be m-strong convex ($m > 0$) over a convex set $C \in \mathbb{R}^n$ if

$$(X - Y)^T(\nabla f(X) - \nabla f(Y)) \geq m\|X - Y\|^2, \forall X, Y \in C.$$

A function $f : \mathbb{R}^n \to \mathbb{R}^n$ is said to be M-Lipschitz ($M > 0$) over a set $C \in \mathbb{R}^n$ if

$$\|f(X) - f(Y)\| \leq M\|X - Y\|, \forall X, Y \in C.$$

More details can be found in [12].

Next, let us recall some basic knowledge about graph theory from [13]. $\mathcal{G} := \{\mathcal{V}, \mathcal{E}, \mathcal{A}\}$ denotes an undirected graph, where $\mathcal{V} = \{1, 2, \ldots, N\}$, $\mathcal{E} \in \mathcal{V} \times \mathcal{V}$ and $\mathcal{A} = [a_{ij}]_{i,j=1,\ldots,N}$ are the node set, the edge set, and the adjacency matrix, respectively. An edge of $\mathcal{G}$ is denoted by a pair of nodes $(i, j) \subset \mathcal{E}$ if j is a neighbor of i. A graph is undirected if and only if $a_{ij} = a_{ji}$ for all $i, j \in \mathcal{V}$. The undirected graph is connected if, for every pair of nodes, there is a path that has them as its end nodes. $\mathcal{A}$ is a nonnegative matrix with the property that $a_{ij} > 0$ if $(i, j) \subset \mathcal{E}$ and $a_{ij} = 0$, otherwise. The degree of node i are defined by $\bar{d}_i = \sum_{j=1}^{N} a_{ij}$. The Laplacian matrix of graph $\mathcal{G}$ is $\mathcal{L} = \mathcal{D} - \mathcal{A}$ with $\mathcal{D} = diag\{\bar{d}_1, \ldots, \bar{d}_N\}$. It is obvious that $\mathcal{L}1_N = 0$. The eigenvalues of $\mathcal{L}$ are denoted by $\lambda_1, \ldots, \lambda_N$, and $\lambda_i \leq \lambda_j$ if $i \leq j$. It is known that the undirected graph is connected if and only if $\lambda_2 > 0$.

Consider a network of N agents with interaction topology described by a graph $\mathcal{G}$. Agent i is endowed with a local cost function $f_i : \mathbb{R}^n \to \mathbb{R}$ and a dynamics

$$\dot{x}_i = u_i + d_i, \quad i = 1, ..., N, \tag{3.1}$$

where $x_i \in \mathbb{R}^n$ and u_i are state variable and control input, respectively, and d_i is the local disturbance generated by the following system:

$$d_i = B\omega_i(t), \dot{\omega}_i = S\omega_i, \tag{3.2}$$

where $\omega_i(t) \in \mathbb{R}^p$. Assume all eigenvalues of $S \in \mathbb{R}^{p\times p}$ are distinct lying on the imaginary axis, which means the boundedness of the disturbance.

The global cost function $f : \mathbb{R}^n \to \mathbb{R}$ is the sum of the all local cost functions, i.e.,

$$f(x) = \sum_{i=1}^{N} f_i(x).$$

Our problem in this paper is to design a control protocol u_i with discrete-time communication such that the multi-agent system solves the optimization problem $x^* = \arg\min_{x\in\mathbb{R}^n} f(x)$ by driving x_i to x^*. If we can admit a continuous-time communication, then our problem becomes that discussed in [7].

Assumption 3.1

(a) The undirected graph $\mathcal{G}$ is connected.
(b) The local cost function f_i is m_i-strongly convex and differentiable, and its gradient is M_i-Lipschitz on $\mathbb{R}^n$.

3.3 Main Result

In this section, we design a distributed optimization algorithm for multi-agent system with external disturbance and discrete-time communication.

Similar to [7], we first let $p(\lambda) = \lambda^s + p_1\lambda^{s-1} + \cdots + p_s$ be the minimal polynomial of S. Let $\mu_i = \left[\mu_{i1}^T \cdots \mu_{in}^T\right]^T$, $\psi = \left[1|0_{1\times(s-1)}\right]$ and $\phi = \left[\begin{array}{c|c} 0 & I_{s-1} \\ \hline -p_s & -p_{s-1} \cdots -p_1 \end{array}\right]$, where $\mu_{ij} = \left[d_{ij}(t)\ \frac{dd_{ij}(t)}{dt}\ \cdots\ \frac{d^{s-1}d_{ij}(t)}{dt^{s-1}}\right]^T, j = 1, \ldots, n$. It is clear that

$$\dot{\mu}_i = (I_n \otimes \phi)\mu_i, \quad d_i(t) = (I_n \otimes \psi)\mu_i. \tag{3.3}$$

There exists a matrix G such that $F = \phi + G\psi$ is Hurwitz because the pair (ψ, ϕ) is observable. Furthermore, there exists a positive definite symmetric matrix P such that $F^T P + PF = -2I_s$.

Due to the discrete-time communication, the control protocol is proposed as follows:

$$\begin{cases} \dot{v}_i = \alpha\beta \sum\limits_{j=1}^{N} a_{ij}(\widehat{x}_i - \widehat{x}_j), \\ \dot{\eta}_i = (I_n \otimes F)\eta_i + (I_n \otimes G)u_i, \\ u_i = -\alpha\nabla f_i(x_i) - v_i - (I_n \otimes \psi)\eta_i - \beta \sum\limits_{j=1}^{N} a_{ij}(\widehat{x}_i - \widehat{x}_j), \end{cases} \tag{3.4}$$

where $\widehat{x}_i$ is the last known state of agent $i \in \{1, \ldots, N\}$, α and β are the designed parameters of controller. In this paper, assume control protocol (3.4) communicates over $\mathcal{G}$ synchronously every Δ (called communication period) starting at $t_1 = 0$, i.e., $t_k = \Delta k$ for all $i \in \{1, \ldots, N\}$ and $k \in \mathbb{N}$.

Let $m = \min\{m_1, \cdots, m_N\}$, $M = \max\{M_1, \cdots, M_N\}$ and $\delta = \alpha^2(r+1)m - \alpha\sqrt{G^T F^T PPFG}(r+1) + \alpha(r+1)\psi G - \alpha^2 M^2 - (\psi G)^2 - (1+\gamma)G^T F^T PPFG$, where α, γ and r are positive numbers.

Remark 3.1 Once G is determined, $G^T F^T PPFG$ and ψG are constants in δ. Besides, m and M are also constants for a given system. For any $r > \frac{M^2}{m} - 1$, δ is a convex function about α. Therefore, there are appropriate positive numbers α, γ and r such that $\delta > 0$.

It is known from [7] that the whole system becomes continuous time and the convergence can be achieved if the communication period $\Delta = 0$. Of course, large Δ may fail the convergence. Therefore, our concern is to give an estimation to make sure how large Δ can still guarantee the convergence. Let $0 < \varepsilon < \frac{2}{\lambda_N}, 0 < \lambda < 0.5$. The following theorem provides a sufficient condition for system (3.1) with protocol (3.4).

Theorem 3.1 *Under Assumption 3.1, given $\alpha, \beta, \gamma, r > 0$ such that $\delta > 0$, if $\Delta \in (0, \tau)$, then the protocol (3.4) makes the $x_i \to x^*$ exponentially for $i \in \{1, \ldots, N\}$ and all initial conditions $x_i(0), v_i(0) \in \mathbb{R}^n$ with $\sum_{i=1}^{N} v_i(0) = 0_n$, where*

$\tau = \frac{1}{b}\ln\left(1+\frac{b\xi}{a(\xi+1)+b}\right)$, $a = \beta\lambda_N\sqrt{1+\alpha^2}$, $b = 2+\alpha M+|\psi G|+\|F\|+\sqrt{G^T F^T F G}$, $\xi^2 = \frac{\varepsilon\lambda\theta}{\alpha\beta r\lambda+\varepsilon\alpha^2(r+1)^2}$, $\theta = \min\{2\delta, 2\gamma, 0.5-\lambda\}$.

Proof Let $v = \left[v_1^T \cdots v_n^T\right]^T$, $\bar{\eta} = \left[\bar{\eta}_1^T \cdots \bar{\eta}_n^T\right]^T$, $x = \left[x_1^T \cdots x_n^T\right]^T$, $\hat{x} = \left[\hat{x}_1^T \cdots \hat{x}_n^T\right]^T$ and $\tilde{f}(x) = \sum_{i=1}^N f_i(x_i)$ with $\bar{\eta}_i = \eta_i - \tau_i$. Thus, based on $\phi = F - G\psi$, (3.1) and (3.4), the following equations are established.

$$\begin{cases} \dot{x} = -\alpha\nabla\tilde{f}(x) - v - \beta(\mathcal{L}\otimes I_n)\hat{x} - (I_n\otimes\psi)\bar{\eta}, \\ \dot{v} = \alpha\beta(\mathcal{L}\otimes I_n)\hat{x}, \\ \dot{\bar{\eta}} = (I_n\otimes F)\bar{\eta} - (I_n\otimes G)(\alpha\nabla\tilde{f}(x) + v + (I_n\otimes\psi)\bar{\eta} + \beta(\mathcal{L}\otimes I_n)\hat{x}). \end{cases} \tag{3.5}$$

Let x^0, v^0 and $\bar{\eta}^0$ be the equilibrium of (3.5). Once the equilibrium point is achieved, $\hat{x} = x^0$. Therefore,

$$\begin{cases} \alpha\nabla\tilde{f}(x^0) + v^0 + \beta(\mathcal{L}\otimes I_n)x^0 + (I_n\otimes\psi)\bar{\eta}^0 = 0, \\ \alpha\beta(\mathcal{L}\otimes I_n)x^0 = 0, \\ (I_n\otimes F)\bar{\eta}^0 - (I_n\otimes G)(\alpha\nabla\tilde{f}(x^0) + v + (I_n\otimes\psi)\bar{\eta}^0 + \beta(\mathcal{L}\otimes I_n)x^0) = 0, \end{cases}$$

which indicates $(\mathcal{L}\otimes I_n)x^0 = 0, (I_n\otimes F)\bar{\eta}^0 = 0$ and $v^0 = -\alpha\nabla\tilde{f}(x^0)$. Because of Assumption 3.1a, $1_N^T\mathcal{L} = 0$ and $\mathcal{L}1_N = 0$, which implies that $(1_N^T\otimes I_n)\dot{v} = 0$.

Since $\mathcal{L}1_N = 0$, $\sum_{i=1}^N v_i(0) = 0$ and $(\mathcal{L}\otimes I_n)x^0 = 0$, we have $(1_N^T\otimes I_n)v = 0_n$ and $x^0 = 1_N\otimes x^*$, where $x^*\in\mathbb{R}^n$, Furthermore, $(1_N^T\otimes I_n)\nabla\tilde{f}(x^0) = 0_n$, i.e., $\sum_{i=1}^N \nabla f_i(x^*) = 0_n$. It follows that the optimal point of the global cost function is also reached when the state variables reach the equilibrium point.

Transfer the equilibrium point to the origin, and let $\tilde{x} = x - x^0$, $\tilde{v} = v - v^0$, $\tilde{\eta} = \bar{\eta} - (I_{Nn}\otimes G)\tilde{x}$ and $h = \nabla\tilde{f}(\tilde{x}+x^0) - \nabla\tilde{f}(x^0)$, which results in

$$\begin{cases} \dot{\tilde{x}} = -\alpha h - \tilde{v} - \beta(\mathcal{L}\otimes I_n)\hat{x} - (I_{Nn}\otimes\psi)(\tilde{\eta} + (I_{Nn}\otimes G)\tilde{x}), \\ \dot{\tilde{v}} = \alpha\beta(\mathcal{L}\otimes I_n)\hat{x}, \\ \dot{\tilde{\eta}} = (I_{Nn}\otimes F)\tilde{\eta} + (I_{Nn}\otimes FG)\tilde{x}. \end{cases} \tag{3.6}$$

There is a matrix $R\in\mathbb{R}^{N\times(N-1)}$ such that $1_N^T R = 0$, $R^T R = I_{N-1}$, and $RR^T = I_N - \frac{1}{N}1_N 1_N^T$ [13]. Let $\chi = \left[\chi_1^T\ \chi_{2:N}^T\right]^T = (T\otimes I_n)\tilde{x}$, $\hat{\chi} = \left[\hat{\chi}_1^T\ \hat{\chi}_{2:N}^T\right]^T = (T\otimes I_n)\hat{x}$ and $\vartheta = \left[\vartheta_1^T\ \vartheta_{2:N}^T\right]^T = (T\otimes I_n)\tilde{v}$, where $T^T = \left[\frac{1}{\sqrt{N}}1_N\ R\right]$. Since $(1_N^T\otimes I_n)v = 0_n$,

$\vartheta = \begin{bmatrix} 0 \\ (R^T \otimes I_n)(v - v^0) \end{bmatrix}$, which means $\vartheta_1 = 0$. Let $\bar{\chi}_{2:N}(t) = \widehat{\chi}_{2:N}(t_k) - \chi_{2:N}(t), t \in [t_k, t_{k+1})$. Because $\widehat{\chi}_{2:N}(t_k) = \chi_{2:N}(t_k)$, $\bar{\chi}_{2:N}(t_k) = 0$. Thus, we obtain

$$\begin{cases} \dot{\chi}_1 = -(\frac{1}{\sqrt{N}} 1_N^T \otimes I_n)(\alpha h + Z), \\ \dot{\chi}_{2:N} = -\alpha(R^T \otimes I_n)h - \vartheta_{2:N} - \beta(R^T \mathcal{L} R \otimes I_n)(\chi_{2:N} + \bar{\chi}_{2:N}) - (R^T \otimes I_n)Z, \\ \dot{\vartheta}_1 = 0, \\ \vartheta_{2:N} = \alpha\beta(R^T \mathcal{L} R \otimes I_n)(\chi_{2:N} + \bar{\chi}_{2:N}), \end{cases} \tag{3.7}$$

where $Z = (I_{Nn} \otimes \psi)(\tilde{\eta} + (I_{Nn} \otimes G)(T^{-1} \otimes I_n)\chi)$.

Take

$$\begin{aligned} V_1 &= \frac{1}{2}\alpha(r+1)\chi_1^T \chi_1 + \frac{1}{2}\alpha r \chi_{2:N}^T \chi_{2:N} + \frac{1}{2\alpha}(\alpha\chi_{2:N} + \vartheta_{2:N})^T(\alpha\chi_{2:N} + \vartheta_{2:N}) \\ &\quad + \frac{1}{2\beta}(r+1)\vartheta_{2:N}^T((R^T \mathcal{L} R)^{-1} \otimes I_n)\vartheta_{2:N}, \\ V_2 &= \tilde{\eta}^T(I_{Nn} \otimes P)\tilde{\eta}. \end{aligned}$$

Since $\chi = (T \otimes I_n)\tilde{x}$, $\vartheta = (T \otimes I_n)\tilde{v}$, $T^T = \begin{bmatrix} \frac{1}{\sqrt{N}} 1_N & R \end{bmatrix}$ and $T^{-1} = T^T$, based on (3.7), we have

$$\begin{aligned} \dot{V}_1 &= -\alpha^2(r+1)\tilde{x}^T h - \alpha(r+1)\tilde{x}^T Z \\ &\quad - \alpha\beta r \chi_{2:N}^T(R^T \mathcal{L} R \otimes I_n)\chi_{2:N} - \alpha\beta r \chi_{2:N}^T(R^T \mathcal{L} R \otimes I_n)\bar{\chi}_{2:N} \\ &\quad - \vartheta_{2:N}^T(R^T \otimes I_n)(\alpha h + Z) - \vartheta_{2:N}^T\vartheta_{2:N} + \alpha(r+1)\vartheta_{2:N}^T\bar{\chi}_{2:N}, \\ \dot{V}_2 &= 2\tilde{\eta}^T(I_{Nn} \otimes PF)\tilde{\eta} + 2\tilde{\eta}^T(I_{Nn} \otimes PFG)\tilde{x}. \end{aligned}$$

By the strong convexity of the local cost function and $\|\chi\| = \|\tilde{x}\|$, we have $\tilde{x}^T h \geq m\|\chi\|^2$. Owning to $T^T T = I_N$,

$$-\alpha\chi^T\left(T \otimes I_n\right)\left(I_{Nn} \otimes \psi\right)\tilde{\eta} \leq \frac{1}{2}\varepsilon_1\alpha^2\chi^T\chi + \frac{1}{2\varepsilon_1}\tilde{\eta}^T\tilde{\eta}.$$

It results from $(R^T \mathcal{L} R \otimes I_n)(\left(R^T \mathcal{L} R\right)^T \otimes I_n) = R^T \mathcal{L}^2 R \otimes I_n$ that

$$-\chi_{2:N}^T(R^T \mathcal{L} R \otimes I_n)\bar{\chi}_{2:N} \leq \frac{1}{2}\varepsilon_2\chi_{2:N}^T(R^T \mathcal{L}^2 R \otimes I_n)\chi_{2:N} + \frac{1}{2\varepsilon_2}\bar{\chi}_{2:N}^T\bar{\chi}_{2:N}.$$

Because of the gradient of f_i is M_i-Lipschitz, $\|R^T \otimes I_n\| = 1$ and $\|h\|^2 \leq M^2\|\tilde{x}\|^2$, $-\alpha\vartheta_{2:N}^T(R^T \otimes I_n)h \leq \frac{1}{2}\varepsilon_3\vartheta_{2:N}^T\vartheta_{2:N} + \frac{\alpha^2}{2\varepsilon_3}M^2\|\chi\|^2$.

Moreover,

$$-\vartheta_{2:N}^T\left(R^T\otimes I_n\right)\left(I_{Nn}\otimes\psi\right)\tilde{\eta}\le\frac{1}{2}\varepsilon_4\vartheta_{2:N}^T\vartheta_{2:N}+\frac{1}{2\varepsilon_4}\tilde{\eta}^T\tilde{\eta},$$

$$-\vartheta_{2:N}^T\left(R^T\otimes I_n\right)\left(T^{-1}\otimes\psi GI_n\right)\chi\le\frac{1}{2}\varepsilon_5\vartheta_{2:N}^T\vartheta_{2:N}+\frac{(\psi G)^2}{2\varepsilon_5}\chi^T\chi,$$

$$\alpha(r+1)\vartheta_{2:N}^T\bar{\chi}_{2:N}\le\frac{\varepsilon_6}{2}\vartheta_{2:N}^T\vartheta_{2:N}+\frac{1}{2\varepsilon_6}\alpha^2(r+1)^2\bar{\chi}_{2:N}^T\bar{\chi}_{2:N}.$$

Here $\varepsilon_1,\varepsilon_2,\varepsilon_3,\varepsilon_4,\varepsilon_5$ and ε_6 are positive real numbers.

In addition, $2\tilde{\eta}^T(I_{Nn}\otimes PF)\tilde{\eta}+2\tilde{\eta}^T(I_{Nn}\otimes PFG)\tilde{x}\le-\tilde{\eta}^T\tilde{\eta}+G^TF^TPPFG\tilde{x}^T\tilde{x}$.

Let $V=V_1+(\frac{r+1}{2\varepsilon_1}+\frac{1}{2\varepsilon_4}+\gamma)V_2$, $W=\left[\chi_1^T\ \chi_{2:N}^T\ \vartheta_{2:N}^T\ \tilde{\eta}^T\right]^T$, $\varepsilon_1=\frac{\sqrt{G^TF^TPPFG}}{\alpha}$, $\varepsilon_2=\varepsilon,\varepsilon_3=\frac{1}{2},\varepsilon_4=\frac{1}{2},\varepsilon_5=\frac{1}{2}$, and $\varepsilon_6=\lambda$, and hence,

$$\begin{aligned}\dot{V}\le&-\frac{1}{2}\theta\left(\|W\|^2-\xi^{-2}\|\bar{\chi}_{2:N}\|^2\right)\\&-\alpha\beta r\chi_{2:N}^T\left((R^T\mathcal{L}R\otimes I_n)-\frac{\varepsilon}{2}(R^T\mathcal{L}^2R\otimes I_n)\right)\chi_{2:N}.\end{aligned}$$

Let $y=\frac{\|\bar{\chi}_{2:N}\|}{\|W\|}$, and based on $\bar{\chi}_{2:N}(t_k)=0$, we have $y(t_k)=0$. Because $\dot{\bar{\chi}}_{2:N}=-\dot{\chi}_{2:N}$ and $\|\dot{\chi}\|\le\|\dot{W}\|$, $\dot{y}\le(1+y)\frac{\|\dot{W}\|}{\|W\|}$. Moreover,

$$\dot{W}=\underbrace{\begin{bmatrix}-(\frac{1}{\sqrt{N}}1_N^T\otimes I_n)(\alpha h+Z)\\-(R^T\otimes I_n)(\alpha h+Z)\\0\\0\end{bmatrix}}_{w_1}+\underbrace{\begin{bmatrix}0\\-\beta(R^T\mathcal{L}R\otimes I_n)(\chi_{2:N}+\bar{\chi}_{2:N})\\\alpha\beta(R^T\mathcal{L}R\otimes I_n)(\chi_{2:N}+\bar{\chi}_{2:N})\\0\end{bmatrix}}_{w_2}$$

$$+\underbrace{\begin{bmatrix}0\\-\vartheta_{2:N}\\0\\0\end{bmatrix}}_{w_3}+\underbrace{\begin{bmatrix}0\\0\\0\\(I_{Nn}\otimes F)\tilde{\eta}\end{bmatrix}}_{w_4}+\underbrace{\begin{bmatrix}0\\0\\0\\(I_{Nn}\otimes FG)\tilde{x}\end{bmatrix}}_{w_5}.$$

It is clear that $\|w_1\|=\alpha\|h\|+\|Z\|$. Due to $R^T\mathcal{L}^2R\le\lambda_N^2I_{N-1}$, we have $\left\|\begin{bmatrix}-\beta(R^T\mathcal{L}R\otimes I_n)\\\alpha\beta(R^T\mathcal{L}R\otimes I_n)\end{bmatrix}\right\|\le\sqrt{1+\alpha^2}\beta\lambda_N$, which results in

$$\|w_2\|\le\beta\lambda_N\sqrt{(1+\alpha^2)}\left(\|\chi_{2:N}\|+\|\bar{\chi}_{2:N}\|\right).$$

Besides, $\|w_3\| = \|\vartheta_{2:N}\|$, $\|w_4\| \leq \|F\| \, \|\tilde{\eta}\|$ and $\|w_5\| = \sqrt{G^T F^T F G} \, \|\chi\|$. From the expression of Z, we know that $\|Z\| \leq \|\tilde{\eta}\| + |\psi G| \, \|\chi\|$. Therefore,

$$\begin{aligned}\dot{W} &\leq (2 + \alpha M + |\psi G| + \|F\| + \sqrt{G^T F^T F G} + \beta \lambda_N \sqrt{(1 + \alpha^2)}) \, \|W\| \\ &\quad + \beta \lambda_N \sqrt{(1 + \alpha^2)} \, \|\bar{\chi}_{2:N}\| \, .\end{aligned}$$

Furthermore, $\dot{y} \leq a(1 + y)^2 + b(1 + y)$. From the Comparison Lemma [14, Lemma 3.4], $y \leq \varphi$, where φ is the solution of $\dot{\varphi} = a(1 + \varphi)^2 + b(1 + \varphi)$. Because $y(t_k) = 0$, let $\varphi(0) = 0$. Then, $y \leq \frac{(a+b)(e^{bt}-1)}{-ae^{bt}+a+b}$.

For a given system, by Remark 3.1, there are appropriate constants α and r such that $\delta > 0$. As a result, when $\varepsilon \in \left(0, \frac{2}{\lambda_N}\right)$ and $\lambda \in (0, 0.5)$, we have $\theta > 0$ and $\chi_{2:N}^T \left((R^T \mathcal{L} R \otimes I_n) - \frac{\varepsilon}{2} (R^T \mathcal{L}^2 R \otimes I_n) \right) \chi_{2:N} > 0$. If $t_{k+1} - t_k < \tau$ for all $k \in \mathbb{N}$, $\|W\|^2 > \xi^{-2} \|\bar{\chi}_{2:N}\|^2$. Therefore, $\dot{V} < 0$. Consequently, $\chi \to 0$ as $t \to \infty$, which is equivalent to $x_i \to x^*$ as $t \to \infty$. Because the Lyapunov function V and the upper bound of its derivative are quadratic, the convergence is exponential. This completes the proof.

Note that the conditions given in the theorem are only sufficient, and the estimated bounds for the control parameters may be too conservative. In fact, in practice, those parameters, when do not satisfy the conditions, may still make the optimization design work.

3.4 Simulation

In this section, a numerical simulation is done to verify the effectiveness of the proposed algorithm.

A network of 5 agents is considered, which is depicted by Fig. 3.1, with their respective local cost functions as follows:

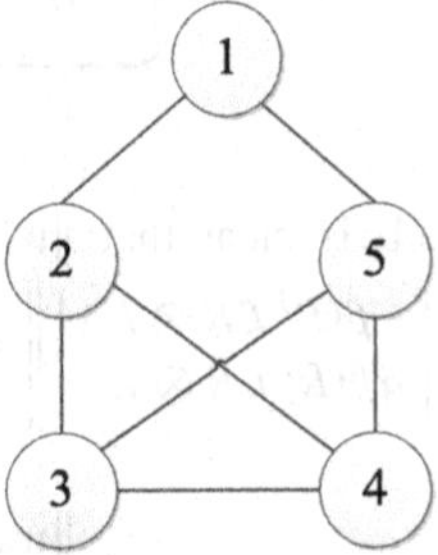

Fig. 3.1 The interaction topology of system

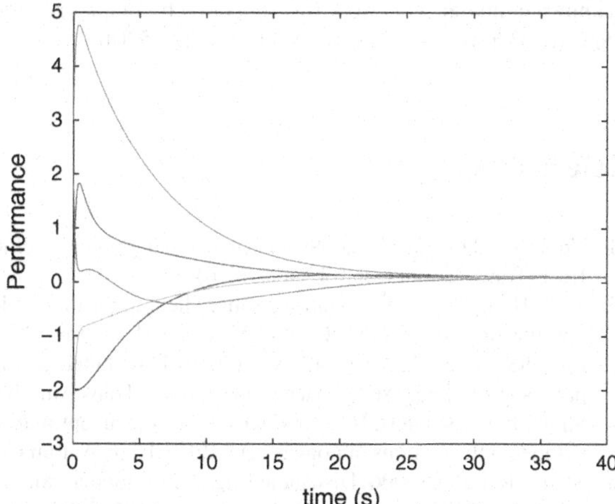

Fig. 3.2 The simulation result ($\alpha = 4, \beta = 1, \Delta = 0.001$ s)

$$f_1(x) = (x+2)^2, f_2(x) = (x-5)^2, f_3(x) = (x-2)^2 + 3x + 1,$$
$$f_4(x) = (x+3)^2 - x + 1, f_5(x) = x^2 + x + 6.$$

It is obvious that $m = 2$ and $M = 2$. All edges weights are set to be 0.1. Therefore, the requirements of Theorem 3.1 are satisfied. As given in [7], the disturbance of each agent is $d_i(t) = A_i \sin\left(\omega t + c_i\right)$, which is generated by (3.2) with

$$S = \begin{bmatrix} 0 & \omega \\ -\omega & 0 \end{bmatrix}, B = \begin{bmatrix} 1 & 0 \end{bmatrix}, \omega_i(0) = \begin{bmatrix} A_i \sin c_i \\ A_i \cos c_i \end{bmatrix},$$

where $\omega = 1$, $\left[A_1, A_2, A_3, A_4, A_5\right]^T = \left[\sqrt{2}, 3\sqrt{2}, 2\sqrt{2}, 2\sqrt{2}, \sqrt{2}\right]^T$ and $c_i = \frac{\pi}{4}$. Further, we have $\phi = \begin{bmatrix} 0 & 1 \\ -1 & 0 \end{bmatrix}$ and $\psi = \begin{bmatrix} 1 & 0 \end{bmatrix}$. Let $G = \begin{bmatrix} -1 & \frac{4}{5} \end{bmatrix}^T$ such that F is Hurwitz. According to Theorem 3.1, we can take $\alpha = 4, \beta = 1$ and $\Delta = 0.001$ s. Then Fig. 3.2 shows the simulation results, where the state of each agent converge to the global optimization point $x^* = 0.1$, which verifies the correctness of Theorem 3.1.

3.5 Conclusions

Distributed optimization algorithm has been presented in this paper for the optimization problem of a continuous-time multi-agent system with external disturbance and discrete-time communication. A sufficient condition on communication period has been given, under which the proposed algorithm can guarantee the system exponentially converges to the optimization point. Then the correctness of the proposed algorithm has been verified by simulation.

Acknowledgments This work was supported by Beijing Natural Science Foundation (4152057), NSFC (61333001), and Program 973 (2014CB845301/2/3).

References

1. Nedić A, Ozdaglar A (2009) Distributed subgradient methods for multi-agent optimization. IEEE Trans Autom Control 54(1):48–61
2. Yi P, Hong Y (2014) Quantized subgradient algorithm and data-rate analysis for distributed optimization. IEEE Trans Control Netw Syst 1(4):380–392
3. Ram SS, Nedić A, Veeravalli VV (2010) Distributed and recursive parameter estimation in parametrized linear state-space models. IEEE Trans Autom Control 55(2):488–492
4. Shi G, Johansson KH, Hong Y (2013) Reaching an optimal consensus: dynamical systems that compute intersections of convex sets. IEEE Trans Autom Control 58(3):610–622
5. Madan R, Lall S (2006) Distributed algorithms for maximum lifetime routing in wireless sensor networks. IEEE Trans Wirel Commun 5(8):2185–2193
6. Kia SS, Cortes J, Martinez S (2015) Distributed convex optimization via continuous-time coordination algorithms with discrete-time communication. Automatica 55:254–264
7. Wang X, Yi P, Hong Y (2014) Dynamical optimization for multi-agent systems with external disturbances. Control Theory Technol 12(2):132–138
8. Gao Y, Wang L, Xie G, Wu B (2009) Consensus of multi-agent systems based on sampled-data control. Int J Control 82(12):2193–2205
9. Chen X, Chen Z (2015) Sampled measurement output feedback control of multi-agent systems with jointly connected topologies, IEEE Trans Autom Control, to appear in 2015
10. Hong Y, Wang X, Jiang Z (2013) Distributed output regulation of leadercfollower multi-agent systems. Int J Robust Nonlinear Control 23(1):48–66
11. Su Y, Huang J (2013) Cooperative adaptive output regulation for a class of nonlinear uncertain multi-agent systems with unknown leader. Syst Control Lett 62(6):461–467
12. Nesterov Y (2004) Introductory Lectures on Convex Optimization: A Basic Course. Kluwer Academic Publishers, Boston
13. Godsil C, Royle G (2001) Algebraic Graph Theory. Springer, New York
14. Khalil HK (2002) Nonlinear Systems. Prentice-Hall, New Jersey

Chapter 4
Camera Calibration Implementation Based on Zhang Zhengyou Plane Method

Pingping Lu, Qing Liu and Jianming Guo

Abstract Camera calibration is a crucial step in computer vision, the main determinant of the visual measurement effect, laying the basis for three-dimensional reconstruction. In order to know calibration's precision exactly, pinhole model and relations of the four coordinates are used, camera's internal and external parameter matrices can be solved by Zhang Zhengyou plane calibration method, camera's distortion coefficients are then easily solved, further considered radial distortion and tangential distortion. The paper establishes a simple but clear error assessment system to evaluate the accuracy of the results and compares them with MATLAB toolbox. The experiment demonstrates that the method has high accuracy, establishing a foundation for seeking depth by binocular stereo vision.

Keywords Camera calibration · Zhang zhengyou calibration method · Image distortion · MATLAB toolbox

4.1 Introduction

In computer vision research which is rapidly developing, we could extract geometric information (such as the position, shape, location, etc.) of objects in 3D space in accordance with images the camera captured, then reconstruct and perceive objects according to this information [1]. A point on the surface of the three-dimensional object can be mapped to the corresponding point on the image by the geometric model (such as pinhole model) of the camera. The geometric parameters are named camera parameters [2]. Usually under most conditions they can be calculated by experiments, the process is camera calibration [3].

Those current camera calibration techniques can be roughly classified into two categories: traditional calibration and self-calibration [4].

P. Lu · Q. Liu (✉) · J. Guo
School of Automation, Wuhan University of Technology,
Wuhan 430070, Hubei, China
e-mail: qliu2000@163.com

© Springer-Verlag Berlin Heidelberg 2016

Y. Jia et al. (eds.), *Proceedings of the 2015 Chinese Intelligent Systems Conference*, Lecture Notes in Electrical Engineering 359,
DOI 10.1007/978-3-662-48386-2_4

Traditional calibration: based on camera imaging model (such as pinhole or fish-eye model), under these conditions that the shape and size of the calibration target is fixed and must have been known, through image processing and a series of mathematical transformations (linear calculation and nonlinear optimization), to calculate camera parameters [5]. Self-calibration: using the corresponding relationships between some specific quantities of the two images imaged before and after the camera rotation or translation, to complete camera calibration. Because the camera's main point and effective focal length have inherent constraints on the basis of certain camera imaging model, and these constraints usually have nothing to do with the surrounding environment and the movement of the camera. Therefore, self-calibration can take advantage of this [6].

Camera calibration has a long history, as early as 1986, R. Tsai [7] has created a classical Tsai's camera model, putting forward the two-step calibration strategy which belongs to traditional calibration. This camera model can compensate camera radial distortion. The two-step calibration strategy [8] establishes equations by using RAC (radial alignment constraint), through direct linear operation and nonlinear optimization, to seek for the internal and external parameters. But this method has complex calculation process and high equipment accuracy requirement so that not fit with simple experimental conditions. Moreover, it is hard to detect feature points and measure data [9].

Professor Zhang [10] improved the two-step calibration strategy and proposed method based on planar template. At first, a set of images are obtained by observing a planar template at a few (at least two) different orientation. Then, the procedure consists of a closed-form solution, followed by a nonlinear refinement based on the maximum likelihood criterion.

Compared the above-mentioned methods, we adopt the second. This paper is organized as follows: Sect. 4.2 describes the basic principle of camera calibration. Section 4.3 describes the calibration procedure. We make the planar template in front of a camera rotate or translate two times or above; or camera rotate or translate two times or above while planar template is fixed. Specific parameters of planar template moving need not be known. Section 4.4 provides the experimental results. Finally, Sect. 4.5 presents a brief summary.

4.2 The Basic Principle of Camera Calibration

4.2.1 Four Kinds of Coordinate Systems

There are many kinds of coordinate systems in computer vision. The following four kinds of coordinate systems are often used while calibrating.

(1) image pixel coordinate system
As shown in Fig. 4.1, there is a two-dimensional orthogonal coordinate system $\boldsymbol{O_0uv}$, whose origin $\boldsymbol{O_0}$ is on the top left corner of the digital

image [2]. (u, v) is the coordinate of each pixel, indicating the v-th row and the u-th column element in the array. $\boldsymbol{O_0uv}$ is called image pixel coordinate system [11].

(2) image physics coordinate system

As shown in Fig. 4.1, an orthogonal coordinate system $\boldsymbol{O_1xy}$ is built on the image plane, whose origin $\boldsymbol{O_1}$ is the main point of the image [1]. Its x and y axes parallel to u and v axes respectively. Assumed the image pixel coordinate of the origin $\boldsymbol{O_1}$ is (u_0, v_0), the distances between every two pixel along x and y axes direction are dx and dy respectively (The unit can assume to be *mm*), so the relation between image pixel coordinate (u, v) and image physics coordinate (x, y) of any point in image is:

$$u = \frac{x}{dx} + u_0 \quad v = \frac{y}{dy} + v_0 \tag{4.1}$$

We represent the relation with homogeneous coordinates and the matrix. The expression is:

$$\begin{bmatrix} u \\ v \\ 1 \end{bmatrix} = \begin{bmatrix} 1/dx & 0 & u_0 \\ 0 & 1/dy & v_0 \\ 0 & 0 & 1 \end{bmatrix} \begin{bmatrix} x \\ y \\ 1 \end{bmatrix} \tag{4.2}$$

(3) camera coordinate system

The geometric relationship about camera imaging is shown in Fig. 4.2. The camera coordinate system $\boldsymbol{O_CX_CY_CZ_C}$ has its origin $\boldsymbol{O_C}$ at the center

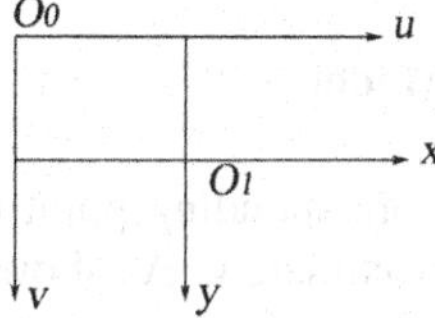

Fig. 4.1 Image pixel coordinate system and image physics coordinate system

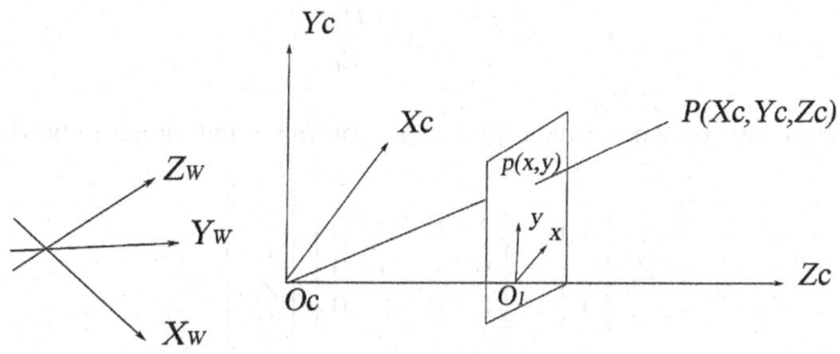

Fig. 4.2 Camera coordinate system and world coordinate system

of projection, its Z_C axis along the optical axis, and its X_C and Y_C axes parallel to the x and y axes of image physics coordinate system respectively [5]. The distance from image plane to the core of camera $\boldsymbol{O_C O_1}$ is the camera's focal length f [5].

(4) world coordinate system

Because the position of camera and target is not fixed in the space, we can only describe their relative position through the establishment of reference coordinate system. This is the world coordinate system $\boldsymbol{O_W X_W Y_W Z_W}$, as shown in Fig. 4.2 [2]. The relationship between the camera coordinate and the world coordinate can be represented by a rotation matrix $\boldsymbol{R}$ and a translation vector $\boldsymbol{t}$ [2]. Assumed that a specific point $\mathbf{P}$ in 3D space, whose camera coordinate and world coordinate are $(X_C, Y_C, Z_C)^T$, $(X_W, Y_W, Z_W)^T$, the relationship between them is:

$$\begin{bmatrix} X_C \\ Y_C \\ Z_C \\ 1 \end{bmatrix} = \begin{bmatrix} \boldsymbol{R} & \boldsymbol{t} \\ \boldsymbol{0}^T & 1 \end{bmatrix} \begin{bmatrix} X_W \\ Y_W \\ Z_W \\ 1 \end{bmatrix} = \boldsymbol{M}_2 \begin{bmatrix} X_W \\ Y_W \\ Z_W \\ 1 \end{bmatrix} \tag{4.3}$$

$\boldsymbol{R}$ is 3×3 unit orthogonal matrix and $\boldsymbol{t}$ is translation vector, $\boldsymbol{0} = (0, 0, 0)^T$, $\boldsymbol{M}_2$ is 4×4 matrix, standing for the relationship between the camera coordinate and the world coordinate.

4.2.2 *The Camera Model*

4.2.2.1 The Linear Camera Model

A 3D point $\mathbf{P}$ projected to the corresponding point $\mathbf{p}$ on the image plane can be expressed by pinhole model approximately. As shown in Fig. 4.2, an image point $\mathbf{p}$ is intersection between image plane and the attachment of camera's core $\boldsymbol{O_C}$ and $\mathbf{P}$ [2]. We call this model center projection or perspective projection [2]. According to the principle of similar triangles, we get:

$$x = \frac{fX_C}{Z_C} \quad y = \frac{fY_C}{Z_C} \tag{4.4}$$

The relationship between image physical coordinate and camera coordinate is:

$$Z_C \begin{bmatrix} x \\ y \\ 1 \end{bmatrix} = \begin{bmatrix} f & 0 & 0 & 0 \\ 0 & f & 0 & 0 \\ 0 & 0 & 1 & 0 \end{bmatrix} \begin{bmatrix} X_C \\ Y_C \\ Z_C \\ 1 \end{bmatrix} \tag{4.5}$$

We put Eqs. 4.2 and 4.3 into Eq. 4.5, then get the relationship between world coordinate and image pixel coordinate:

$$Z_C\begin{bmatrix}u\\v\\1\end{bmatrix}=\begin{bmatrix}f/dx & 0 & u_0 & 0\\0 & f/dy & v_0 & 0\\0 & 0 & 1 & 0\end{bmatrix}\begin{bmatrix}\boldsymbol{R} & \boldsymbol{t}\\\boldsymbol{0}^T & 1\end{bmatrix}\begin{bmatrix}X_W\\Y_W\\Z_W\\1\end{bmatrix}=\boldsymbol{M}_1\boldsymbol{M}_2\begin{bmatrix}X_W\\Y_W\\Z_W\\1\end{bmatrix}=\boldsymbol{M}\begin{bmatrix}X_W\\Y_W\\Z_W\\1\end{bmatrix} \tag{4.6}$$

$Fx = f/dx, fy = f/dy$ are expressed as the effective focal length of the camera in the x and y axes. $\boldsymbol{M}$ is 3×4 matrix, matrix $\boldsymbol{M}_1$ is only related to the camera's internal structure, which is called camera intrinsic parameters matrix defined by fx, fy, u_0, v_0. While matrix $\boldsymbol{M}_2$ is only related to the camera's external parameters, which is called camera extrinsic parameters matrix.

4.2.2.2 The Non-Linear Camera Model

When the camera lens are wide-angle lens or the production of camera is not standard, it occurs the distortion on the edge of image, what's more, the more close to the edge, the more serious distortion phenomenon. Therefore, if to use linear model to calibrate camera, image point **p** will deviate from the original position and produce very large error. So the non-linear camera model is used [2], as follows:

$$\begin{cases}\delta_x = k_1x(x^2+y^2)+k_2x(x^2+y^2)^2+k_3x(x^2+y^2)^3+p_2(3x^2+y^2)+2p_1xy\\\delta_y = k_1y(x^2+y^2)+k_2y(x^2+y^2)^2+k_3y(x^2+y^2)^3+p_1(3x^2+y^2)+2p_2xy\end{cases} \tag{4.7}$$

where k_1, k_2 are radial distortion coefficients, p_1, p_2 are tangential distortion coefficients, δ_x, δ_y are distortion errors along x and y axes.

4.3 Camera Calibration Method

4.3.1 Solving Internal and External Parameters

At first, to capture 3 or more images in the camera's view. Planar template can be translated or rotated at random, but the Z_W of the world coordinate system of each image is chosen to perpendicular to the plane, the first corner detected is perceived as the world coordinate system's origin, and $Z_W = 0$ in the plane.

Let's denote the i-th column of the rotation matrix $\boldsymbol{R}$ by r_i. So Eq. 4.5 can be written as:

$$\begin{bmatrix} u \\ v \\ 1 \end{bmatrix} = s\boldsymbol{M}_1 \begin{bmatrix} r_1 & r_2 & r_3 & \boldsymbol{t} \end{bmatrix} \begin{bmatrix} X_W \\ Y_W \\ 0 \\ 1 \end{bmatrix} = s\boldsymbol{M}_1 \begin{bmatrix} r_1 & r_2 & \boldsymbol{t} \end{bmatrix} \begin{bmatrix} X_W \\ Y_W \\ 1 \end{bmatrix} = \boldsymbol{H} \begin{bmatrix} X_W \\ Y_W \\ 1 \end{bmatrix} \quad (4.8)$$

where s is an arbitrary scale factor. Let

$$\boldsymbol{H} = \begin{bmatrix} h_{11} & h_{12} & h_{13} \\ h_{21} & h_{22} & h_{23} \\ h_{31} & h_{32} & 1 \end{bmatrix} \quad (4.9)$$

$$\boldsymbol{h} = \begin{bmatrix} h_{11} & h_{12} & h_{13} & h_{21} & h_{22} & h_{23} & h_{31} & h_{32} & 1 \end{bmatrix}^T \quad (4.10)$$

Then obtain:

$$\begin{pmatrix} X_W & Y_W & 1 & 0 & 0 & 0 & -uX_W & -uY_W & -u \\ 0 & 0 & 0 & X_W & Y_W & 1 & -vX_W & -vY_W & -v \end{pmatrix} \boldsymbol{h} = 0 \quad (4.11)$$

Each point can be shown above the two equations. One image has N points, and $2 \times N$ equations can be obtained, written as $\boldsymbol{Sh} = 0$. The solution of $\boldsymbol{h}$ is the corresponding eigenvector to the minimum eigenvalue of the equation $\boldsymbol{S}^T\boldsymbol{S} = 0$ [2], $\boldsymbol{H}$ can be solved after the vector $\boldsymbol{h}$ is normalized. Nonlinear least square method can be used to solve maximum likelihood estimation of $\boldsymbol{H}$, here is Levenberg-Marquardt algorithm [12].

Each image has homography matrix $\boldsymbol{H}$, written as column vectors, $\boldsymbol{H} = [h_1\ h_2\ h_3]$, where each h is 3×1 vector. $\boldsymbol{H} = [h_1\ h_2\ h_3] = s\boldsymbol{M}_1[r_1\ r_2\ \boldsymbol{t}]$ can be decomposed as:

$$h_i = s\boldsymbol{M}_1 r_i \text{ or } r_i = \lambda \boldsymbol{M}_1^{-1} h_i \quad (4.12)$$

where $\lambda = 1/s$, $i = 1, 2, 3$.

Rotation matrix $\boldsymbol{R}$ is unit orthogonal matrix, so r_1 and r_2 are orthogonal. We have two constraints:

$$h_1^T \boldsymbol{M}_1^{-T} \boldsymbol{M}_1^{-1} h_2 = 0 \quad (4.13)$$

$$h_1^T \boldsymbol{M}_1^{-T} \boldsymbol{M}_1^{-1} h_1 = h_2^T \boldsymbol{M}_1^{-T} \boldsymbol{M}_1^{-1} h_2 \quad (4.14)$$

Let $\boldsymbol{B} = \boldsymbol{M}_1^{-T}\boldsymbol{M}_1^{-1}$, we can get:

$$\boldsymbol{B} = \boldsymbol{M}_1^{-T} \boldsymbol{M}_1^{-1} = \begin{bmatrix} B_{11} & B_{12} & B_{13} \\ B_{21} & B_{22} & B_{23} \\ B_{31} & B_{32} & B_{33} \end{bmatrix} \quad (4.15)$$

In fact, there is general closed-form of matrix $\boldsymbol{B}$:

$$\boldsymbol{B} = \begin{bmatrix} \frac{1}{f_x^2} & 0 & \frac{-u_0}{f_x^2} \\ 0 & \frac{1}{f_y^2} & \frac{-v_0}{f_y^2} \\ \frac{-u_0}{f_x^2} & \frac{-v_0}{f_y^2} & \frac{u_0}{f_x^2} + \frac{v_0}{f_y^2} + 1 \end{bmatrix} \tag{4.16}$$

The two constraints have their general form $h_i^T \boldsymbol{B} h_j$ via matrix $\boldsymbol{B}$. We could obtain each element just making sure the six elements of $\boldsymbol{B}$ as matrix $\boldsymbol{B}$ is symmetric matrix. The six elements are written as a column vector:

$$h_i^T \boldsymbol{B} h_j = v_{ij}^T \boldsymbol{b} = \begin{bmatrix} h_{i1}h_{j1} \\ h_{i1}h_{j2} + h_{i2}h_{j1} \\ h_{i2}h_{j2} \\ h_{i3}h_{j1} + h_{i1}h_{j3} \\ h_{i3}h_{j2} + h_{i2}h_{j3} \\ h_{i3}h_{j3} \end{bmatrix} \begin{bmatrix} B_{11} \\ B_{12} \\ B_{22} \\ B_{13} \\ B_{23} \\ B_{33} \end{bmatrix} \tag{4.17}$$

The two constraints can be written as Eq. 4.18 via the definition of v_{ij}^T

$$\begin{bmatrix} v_{12}^T \\ (v_{11} - v_{22})^T \end{bmatrix} \boldsymbol{b} = 0 \tag{4.18}$$

If we get K images at the same time, we can get $\boldsymbol{V}\boldsymbol{b} = 0$ where $\boldsymbol{V}$ is a $2K \times 6$ matrix.

If $K \geq 2$, it has solution. At last, we compute the internal parameters.

$$f_x = \sqrt{\lambda / B_{11}} \tag{4.19}$$

$$f_y = \sqrt{\lambda B_{11} / (B_{11}B_{22} - B_{12}^2)} \tag{4.20}$$

$$u_0 = -B_{13} f_x^2 / \lambda \tag{4.21}$$

$$v_0 = (B_{12}B_{13} - B_{11}B_{23}) / (B_{11}B_{22} - B_{12}^2) \tag{4.22}$$

$$\lambda = B_{33} - \left[B_{13}^2 + v_0(B_{12}B_{13} - B_{11}B_{23})\right] / B_{11} \tag{4.23}$$

Then we compute the external parameters by $\boldsymbol{B}$:

$$r_1 = \lambda \boldsymbol{M}_1^{-1} h_1 \tag{4.24}$$

$$r_2 = \lambda M_1^{-1} h_2 \quad (4.25)$$

$$r_3 = r_1 \times r_2 \quad (4.26)$$

$$t = \lambda M_1^{-1} h_3 \quad (4.27)$$

$$\lambda = 1 / \| M_1^{-1} h_1 \| \quad (4.28)$$

But if let r_1, r_2, r_3 combine into rotation matrix $\boldsymbol{R}$, there may be a large error because $\boldsymbol{R}$ is not a positive definite matrix in the actual process, that is to say $\boldsymbol{R}^T\boldsymbol{R} = \boldsymbol{R}\boldsymbol{R}^T = \boldsymbol{I}$ is not founded.

To solve this problem, we can choose to use singular value decomposition (SVD), to make $\boldsymbol{R} = \boldsymbol{U}\boldsymbol{D}\boldsymbol{V}^T$ is founded. $\boldsymbol{U}$ and $\boldsymbol{V}$ are orthogonal matrices, $\boldsymbol{D}$ is diagonal matrix. Moreover, because r_1, r_2, r_3 are orthogonal to each other, matrix $\boldsymbol{D}$ must be identity matrix $\boldsymbol{I}$ so that $\boldsymbol{R} = \boldsymbol{U}\boldsymbol{I}\boldsymbol{V}^T$. So we first solve singular value decomposition of $\boldsymbol{R}$, then $\boldsymbol{D}$ to be set as identity matrix, finally multiply $\boldsymbol{U}$ and $\boldsymbol{V}$ to solve rotation matrix R′ which is matching the requirement.

4.3.2 *Maximum Likelihood Estimation*

We adopt Levenberg-Marquardt algorithm to optimize these parameters after they are solved. Evaluation function is expressed as:

$$C = \sum_{i=1}^{N} \sum_{j=1}^{K} \left\| m_{ij} - m(\boldsymbol{M}_1, \boldsymbol{R}_i, \boldsymbol{t}_i, \boldsymbol{M}_{ij}) \right\|^2 \quad (4.28)$$

where N is the total number of image, K is the total number of points in each image, m is image pixel coordinate, $\boldsymbol{M}$ is world coordinate, $m(\boldsymbol{M}_1, \boldsymbol{R}_i, \boldsymbol{t}_i, \boldsymbol{M}_{ij})$ is image pixel coordinate computed by these known parameters

4.3.3 *Solving Camera Distortion*

We have not dealt with camera distortion so far. We use camera's internal and external parameters and all distortion coefficients set to zero as initial values to compute them. Let (x_p, y_p) is the location of point, (x_d, y_d) is the distortion location of point, so

$$\begin{bmatrix} x_p \\ y_p \end{bmatrix} = \begin{bmatrix} f_x X_W / Z_W + u_0 \\ f_y Y_W / Z_W + v_0 \end{bmatrix} \quad (4.29)$$

Combine with Eq. 4.7, we get

$$\begin{bmatrix} x_p \\ y_p \end{bmatrix} = (1 + k_1 r^2 + k_2 r^4 + k_3 r^6) \begin{bmatrix} x_d \\ y_d \end{bmatrix} + \begin{bmatrix} 2p_1 x_d y_d + p_2(r^2 + 2x_d^2) \\ p_1(r^2 + 2y_d^2) + 2p_2 x_d y_d \end{bmatrix} \tag{4.30}$$

where $r^2 = x_d^2 + y_d^2$. We can get a large number of equations and solve them to obtain distortion coefficients.

4.4 Analysis of Experimental Results

4.4.1 Error Assessment Method

Camera calibration results are difficult to evaluate whether accurate or not, there is no objective criteria. We can use world coordinate (X_W, Y_W, Z_W) and the camera parameter matrix of a 3D point **P**, to get the back-projection value $(\tilde{u}, \tilde{v})$ after matrix multiplication, then compare it with origin value (u, v) detected actually, get average error E_{uv}, standard deviation e_{uv} and maximum error e in the image pixel coordinate system.

$$E_{uv} = \frac{\sum_{i=1}^{K} (|\widetilde{u}_i - u_i| + |\widetilde{v}_i - v_i|)}{K} \tag{4.31}$$

$$e_{uv} = \sqrt{\frac{\sum_{i=1}^{K} (|\widetilde{u}_i - u_i|^2 + |\widetilde{v}_i - v_i|^2)}{K}} \tag{4.32}$$

$$e = \min(\sum_{i=1}^{K} |\widetilde{u}_i - u_i|, |\widetilde{v}_i - v_i|) \tag{4.33}$$

4.4.2 The Introduction of Experiment System

The calibration template is checkerboard with nine corners along length and six corners along width. The size of each black square is 20 mm × 20 mm, there are 54 angular points, the first corner of each image will be origin of the world coordinate system, and world coordinates of other corners will be computed.

Adopting microscopical industrial camera, whose resolution is 640 × 480 pixels, and focal length is 12 mm, the physical size is 1/3″, the distance of every two pixel is 3.2 μm. The flow chart of camera calibration algorithm is shown in Fig. 4.3.

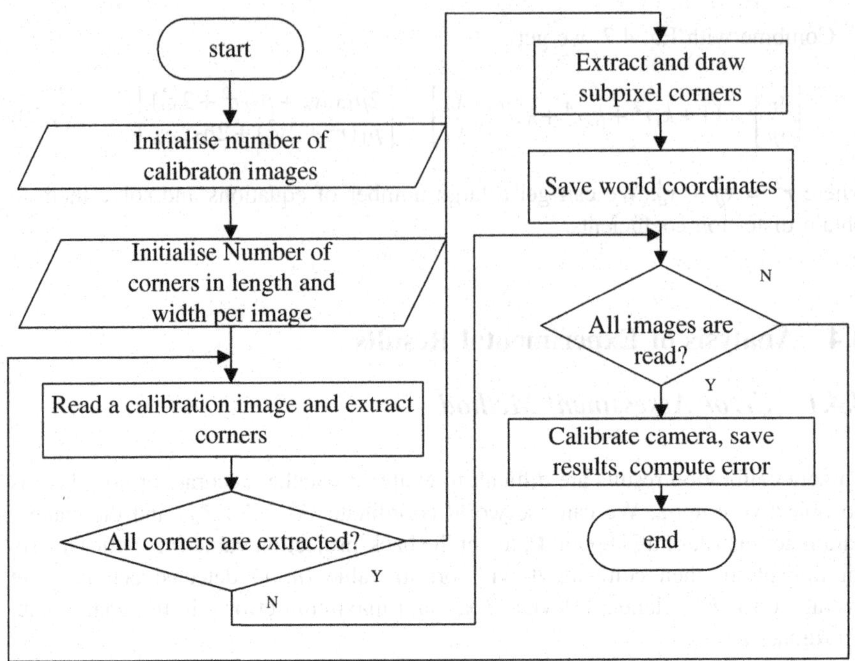

Fig. 4.3 The flow chart of camera calibration

4.4.3 Analysis of Experimental Results

To complete calibration through 5, 10, 15 and 20 images respectively, using Eqs. 4.31–4.33 to get average error, standard deviation and maximum error (See Fig. 4.4).

With the number of calibration image increasing, the average error and standard deviation gradually decrease and stabilize to a certain value. We should take 20–25 image at least. In order to make the errors seem more convenient, we describe them as a scatter plot (See Fig. 4.4).

We choose an image to calculate the camera parameters. The rotation matrix and translation vector of 13-rd image are:

$$\boldsymbol{R} = \begin{bmatrix} -0.614031 & 0.789249 & -0.007166 \\ 0.766653 & 0.598560 & 0.232309 \\ 0.187639 & 0.137151 & -0.972615 \end{bmatrix}$$

$$\boldsymbol{t} = [\,-99.62857 \quad -138.42137 \quad 1409.07778\,]$$

In order to verify the accuracy of the calibration results, we use MATLAB toolbox to calibrate the 20 images (See Tables 4.1 and 4.2).

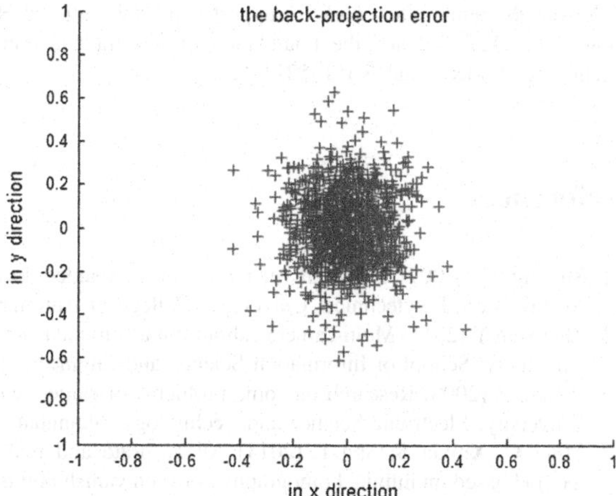

Fig. 4.4 The scatter plot of back-projection error

Table 4.1 The error comparison of calibration precision

Images number	Average error	Standard deviation	Maximum error
5	0.287452	0.2255661	2
10	0.237154	0.220006	2
15	0.2122855	0.200457	1
20	0.204852	0.194444	1

Table 4.2 The comparison of calibration results

Parameters	20 images	MATLAB toolbox
f_x	1741.430026	1746.504035
f_y	1743.25123	1748.425046
u_0	340.4265682	347.7840171
v_0	268.8052343	270.0848203
k_1	−0.56386008	−0.59154964
k_2	4.865922599	5.322022256
p_1	0.001976323	0.001936653
p_2	−0.011511625	−0.012229335

4.5 Conclusion

Experiments show that Zhang Zhengyou plane calibration method not only has low requirement to the experimental equipment just with a camera and a calibration template, but also has high precision, which is a transition method between the traditional calibration and the self-calibration. In simple experimental conditions, we can accurately obtain the camera parameters, and use the theory of binocular vision to compute depth of field. At the same time, the software implementation compared with MATLAB toolbox is simple, do not needs to extract angular point manually.

Acknowledgments This work is supported by the National Natural Science Foundation under Grant No. 51279152 and the Foundation of postgraduate innovation of Wuhan University of Technology under Grant No. 155211005.

References

1. Shujing W (2007) Research on passive vision aided location of multiple observers [D]. Northwestern Polytechnical University, College of Automation, Xi'an
2. Zhenxian Y (2011) Multi-camera calibration technique research and its application [D]. Hunan University: School of Information Science and Engineering, Changsha
3. Jianlan Z (2008) Research on some problems of camera calibration [D]. East China Normal University: Electronic Science and Technology, Shanghai
4. Ming C, Xiuxia S, Shu L (2014) An accurate and real-time focal-length self-calibration method based on infinite homography between vanish points [J]. Acta Optica Sinica 34(5):1–6
5. Qinglei Q (2010) Research and implementation on three-dimensional position technology based on binocular stereo vision [D]. Northeastern University: School of Information Science and Engineering, Shenyang
6. Qianqian Y (2010) The research on automatic finding the box-parts' position and processing based on the machining center [D]. Lanzhou University of Technology: Electromechanic Engineering College, Lanzhou
7. Tsia RY (1986) An efficient and accurate camera calibration technique for 3D machine vision [J]. Proc CV PR 364–374
8. Jianfei M, Xiyong Z, Jing Z (2004) Improved two-stage camera calibration from a plane [J]. J Image Graph 9(7):846–852
9. Bo L, Xiao-tong W, Xiao-gang X (2006) A linear three-step approach for camera calibration [J]. J Image Graph 11(7):928–932
10. Zhengyou Z (1999) A flexible camera calibration by viewing a plane from unknown orientation [J]. Comput Vis, the proceedings of the seventh IEEE international conference 1:666–673
11. Ruxin G, Junmeng W (2014) Target recognition and location based on binocular vision [J]. J Henan Polytech Univ (Nat Sci) 33(4):495–500
12. Shiqi Y, Ruizhen L (2009) Learning OpenCV [M]. Tsinghua University Press, Beijing, pp 406–440

Chapter 5
Online Identification and Robust Adaptive Control for Discrete Hysteresis Preisach Model

Xuehui Gao, Xuemei Ren, Chengyuan Zhang and Changsheng Zhu

Abstract By using the concept of online goal adaptation, we develop a new online identification and a robust adaptive control for hysteresis system which is described by discrete Preisach model. The proposed identification requires triangle matrices to simply the calculation for Preisach model. In addition, a robust adaptive control is adopted without inverse Preisach model based on the weighting factor recursive least square (WFRLS) method. A Lyapunov function candidate guarantees the stability of this hysteresis system. Finally, simulations perform on a typical system which clarify the validity of the proposed approach.

Keywords Preisach · Hysteresis · Weighting factor recursive least square

5.1 Introduction

Hysteresis exists in magnetic circuits, precise servo systems and smart materials, etc. Since the memory property and nonlinearity, hysteresis can lead to inaccuracies, oscillations, or even instabilities. Thus, modeling and controller to hysteresis has been widely investigated for many years.

Hysteresis models can be roughly classified into two categories: mathematical models and physical models. physical models concern specific physical parameters of the materials (sensors/actuators). Jiles-Atherton model [1] and bounc-Wen

This work is supported by National Natural Science Foundation of China (No. 61433003, No. 61273150 and No. 61321002), the Research Fund for the Doctoral Program of Higher Education of China (No. 20121101110029).

X. Gao · X. Ren (✉)
School of Automation, Beijing Institute of Technology, Beijing 10081, China
e-mail: xmren@bit.edu.cn

X. Gao · C. Zhang · C. Zhu
Department of Mechanical and Electrical Engineering, Shandong University of Science and Technology, Tai'an 271019, China
e-mail: xhgao@163.com

© Springer-Verlag Berlin Heidelberg 2016

Y. Jia et al. (eds.), *Proceedings of the 2015 Chinese Intelligent Systems Conference*, Lecture Notes in Electrical Engineering 359,
DOI 10.1007/978-3-662-48386-2_5

[2] model are common physical models in practice. Otherwise, mathematic models are universal models which only involve input, output and other common parameters but don't need specific physical parameters. Popular mathematic models include Preisach model [3–5], Prandtl-Ishlinskii (PI) model [6–8] and Krasnosel'skii-Pokrovskii (KP) model [9].

Many researchers focus on the Preisach model in recent years. Some identifications and control strategies were proposed for continuous or discrete Preisach models. Tan and Baras [10] addressed a recursive identification and adaptive inverse control in smart material actuators, where hysteresis was modeled by a Preisach operator with Piecewise uniform density function. Gao et al. [3] proposed a determinate identification for discrete Preisach model which depended on triangle matrices of Preisach density function. An inverse compensation approach for Preisach model using the inverse multiplicative structure was developed in Li et al. [4]. Xiao and Li [5] presented a novel modified inverse Preisach model to compensate the hysteresis for a piezoelectric actuator at varying frequency ranges whose the Preisach density function was identified by least square method. All these approaches need either complex computation or off-line identification. Thus, the identification for Preisach model should be investigated much more.

Different control strategies have been proposed to precisely control the hysteresis systems due to the existence of the nonlinearities. For Preisach model, the control strategy can be classified as feedback control strategy [11, 12], feedforward control strategy [10, 13] and Feedback-Feedforward strategy [14].

Rosenbaum et al. [13] considered hysteresis models in the scalar form and applied inverse feedforward control for electromagnetic actuators. An adaptive output feedback controller was presented for a class of single-input-single-output (SISO) nonlinear systems in Li and Tan [11], where the unknown hysteresis nonlinearity was represented by the Preisach model. Liu et al. [14] proposed a discrete composite control strategy with a feedforward-feedback structure where the feedforward controller was the inverse model and the feedback controller was secondary proportional-integral (PI) controller to control the hysteresis system accurately.

Above all, most controllers need inverse model to compensate the hysteresis in control strategies, which lead to complex calculation and weak performance. In this paper, we propose an online identification for discrete Preisach model and a robust adaptive control without inverse model to precisely control the hysteresis system. Firstly, we expand the results of [3] and propose an online discrete identification which also depend on the triangle matrices for matrix calculation simplified than conventional integral compute. Then, a new robust adaptive controller is proposed to suppress the nonlinearity of the hysteresis by the results of the online identification. In addition, the new proposed robust adaptive control strategy doesn't require inverse Preisach model which different from other control strategies as aforementioned, which can lead to a simply calculation.

This paper is organized as follows. The online identification is given in Sect. 5.2 and the robust adaptive controller is proposed in Sect. 5.3. Finally, Sect. 5.4 provides some numerical simulation and some conclusions are provided in Sect. 5.5.

5.2 Online Identification of Discrete Preisach Model

In this section we review the discrete Preisach model and the off-line identification [3]. Based on our prior work, we expand the results of [3] and propose a new online identification with triangle matrices for discrete hysteresis Preisach model.

5.2.1 Discrete Preisach Model

Considering a pair of thresholds (α, β), with $\alpha \geqslant \beta$, and $u \in C([0, T])$, $\zeta \in 1, 0$, then the Preisach operator $\hat{\gamma}_{\alpha,\beta}[u, \zeta]$ is defined as:

$$\hat{\gamma}_{\alpha,\beta}[u,\zeta] = \begin{cases} 1 & \text{if } u(t) < \beta, \\ 0 & \text{if } u(t) > \alpha, \\ \hat{\gamma}_{\alpha,\beta}[u,\zeta](t^-) & \text{if } \beta \leqslant u(t) \leqslant \alpha. \end{cases} \tag{5.1}$$

where $t \in [0, T]$, $\hat{\gamma}_{\alpha,\beta}[u, \zeta](0^-) = \zeta$ and $t^- = \lim\limits_{\varepsilon>0,\varepsilon\to 0} t \to \varepsilon$.

Thus, the discrete Preisach model can be expressed as [10]:

$$y(k) = \sum_{i=1}^{k} \sum_{j=1}^{i} \omega_{ij}(k)\mu_{ij}(k) \tag{5.2}$$

where $\omega_{ij}(k)$ is the Preisach operator at time k, i.e. $\omega_{ij}(k) = \hat{\gamma}_{\alpha,\beta}[u, \zeta(\alpha, \beta)](k)$, $\mu_{ij}(k)$ is the Preisach density function and $k \in [0, T]$.

5.2.2 Online Identification of Preisach Model

Literature [3] adopted triangle matrices to identify discrete Preisach model, but that was an off-line approach. In order to continue investigating this issue, A lemma [3] is introduced as follows:

Lemma 1 *Discrete Preisach model is expressed as* (5.2). *For each monotonous section of hysteresis, the form of Preisach model with triangle matrices can be described as follows:*

$$y = \omega\tilde{u}\mu \tag{5.3}$$

where

$$\omega = \begin{bmatrix} 1 & 0 & 0 & \dots & 0 \\ 1 & 1 & 0 & \dots & 0 \\ 1 & 1 & 1 & \dots & 0 \\ \dots & \dots & \dots & \dots & \dots \\ 1 & 1 & 1 & \dots & 1 \end{bmatrix}, \; \mu = \begin{bmatrix} \mu_{11} & 0 & 0 & \dots & 0 \\ \mu_{21} & \mu_{22} & 0 & \dots & 0 \\ \mu_{31} & \mu_{32} & \mu_{33} & \dots & 0 \\ \dots & \dots & \dots & \dots & \dots \\ \mu_{k1} & \mu_{k2} & \mu_{k3} & \dots & \mu_{kk} \end{bmatrix},$$

$$\tilde{u} = \begin{bmatrix} \tilde{u}_{11} & 0 & 0 & \dots & 0 \\ \tilde{u}_{21} & \tilde{u}_{22} & 0 & \dots & 0 \\ \tilde{u}_{31} & \tilde{u}_{32} & \tilde{u}_{33} & \dots & 0 \\ \dots & \dots & \dots & \dots & \dots \\ \tilde{u}_{k1} & \tilde{u}_{k2} & \tilde{u}_{k3} & \dots & \tilde{u}_{kk} \end{bmatrix}, \; \tilde{u}_{ij} = \frac{u_i - u_{i-1}}{i}, \; i = 1, 2, \dots k, \; j = 1, 2, \dots, i.$$

ω is lower triangle identity matrix. μ means Preisach density function and $\tilde{u}$ indicates input. k is the discrete level of one monotonous section, e.g. continuous input u of one monotonous section is discretized k levels: $u_1, u_2, \dots u_k$.

Rewrite μ as vector form:

$$\tilde{\mu} = \left[\sum_{j=1}^{k} \mu_{1j} \; \sum_{j=1}^{k} \mu_{2j} \; \sum_{j=1}^{k} \mu_{3j} \; \cdots \; \sum_{j=1}^{k} \mu_{kj} \right]^T, \tag{5.4}$$

and let $\theta = \tilde{\mu}$, then

$$\theta = \left[\theta_1 \; \theta_2 \; \theta_3 \; \dots \; \theta_k \right]^T. \tag{5.5}$$

Considering θ and input of one monotonous section, online identification is expressed as Theorem 1.

Theorem 1 *Discrete Preiach model is represented as* (5.2) *and θ is defined as* (5.5), *then*

$$y_i = y_{i-1} + \frac{u_i - u_{i-1}}{i} \sum_{j=1}^{i-1} \theta_j \tag{5.6}$$

where i=1,2,...k.

Proof From Eq. (5.3), the following hold.

$$\begin{bmatrix} y_1 \\ y_2 \\ y_3 \\ \dots \\ y_k \end{bmatrix} = \omega \tilde{u} \mu = \begin{bmatrix} 1 & 0 & 0 & \dots & 0 \\ 1 & 1 & 0 & \dots & 0 \\ 1 & 1 & 1 & \dots & 0 \\ \dots & \dots & \dots & \dots & \dots \\ 1 & 1 & 1 & \dots & 1 \end{bmatrix} \begin{bmatrix} \tilde{u}_{11} & 0 & 0 & \dots & 0 \\ \tilde{u}_{21} & \tilde{u}_{22} & 0 & \dots & 0 \\ \tilde{u}_{31} & \tilde{u}_{32} & \tilde{u}_{33} & \dots & 0 \\ \dots & \dots & \dots & \dots & \dots \\ \tilde{u}_{k1} & \tilde{u}_{k2} & \tilde{u}_{k3} & \dots & \tilde{u}_{kk} \end{bmatrix} \begin{bmatrix} \mu_{11} & 0 & 0 & \dots & 0 \\ \mu_{21} & \mu_{22} & 0 & \dots & 0 \\ \mu_{31} & \mu_{32} & \mu_{33} & \dots & 0 \\ \dots & \dots & \dots & \dots & \dots \\ \mu_{k1} & \mu_{k2} & \mu_{k3} & \dots & \mu_{kk} \end{bmatrix} \tag{5.7}$$

Substituting (5.5) into (5.7), we obtain

$$\begin{bmatrix} y_1 \\ y_2 \\ y_3 \\ \dots \\ y_k \end{bmatrix} = \begin{bmatrix} \sum_{i=1}^{1} \tilde{u}_{i1} & 0 & 0 & \dots & 0 \\ \sum_{i=1}^{2} \tilde{u}_{i1} & \sum_{i=2}^{2} \tilde{u}_{i2} & 0 & \dots & 0 \\ \sum_{i=1}^{3} \tilde{u}_{i1} & \sum_{i=2}^{3} \tilde{u}_{i2} & \sum_{i=3}^{3} \tilde{u}_{i3} & \dots & 0 \\ \dots & \dots & \dots & \dots & \dots \\ \sum_{i=1}^{k} \tilde{u}_{i1} & \sum_{i=2}^{k} \tilde{u}_{i2} & \sum_{i=3}^{k} \tilde{u}_{i3} & \dots & \sum_{i=k}^{k} \tilde{u}_{ik} \end{bmatrix} \begin{bmatrix} \theta_1 \\ \theta_2 \\ \theta_3 \\ \dots \\ \theta_k \end{bmatrix} = \begin{bmatrix} \sum_{i=1}^{1} \sum_{j=1}^{i} \tilde{u}_{ij}\theta_i \\ \sum_{i=1}^{2} \sum_{j=1}^{i} \tilde{u}_{ij}\theta_i \\ \sum_{i=1}^{3} \sum_{j=1}^{i} \tilde{u}_{ij}\theta_i \\ \dots \\ \sum_{i=1}^{k} \sum_{j=1}^{i} \tilde{u}_{ij}\theta_i \end{bmatrix}. \tag{5.8}$$

Considering Eq. (5.8) and $\tilde{u}_{ij} = \frac{u_i - u_{i-1}}{i}$ which is defined in Lemma 1, we have

$$y_i = y_{i-1} + \frac{u_i - u_{i-1}}{i} \sum_{j=1}^{i-1} \theta_j. \tag{5.9}$$

This completes the proof.

5.3 Design of Controller

5.3.1 Problem Statement

Consider a discrete hysteresis system as follows :

$$A(z^{-1})y(k) = B(z^{-1})u(k) + \varpi(k) \tag{5.10}$$

where $A(z^{-1}) = 1 + a_1 z^{-1} + \dots + a_n z^{-n}, B(z^{-1}) = b_1 z^{-1} + b_2 z^{-2} + \dots + b_n z^{-n}$, n is the system order. $\varpi(k)$ means hysteresis which is described by Eq. (5.6).

Rewrite Eq. (5.10) as least squares(LS) form:

$$y(k) = \psi^T(k)\vartheta \tag{5.11}$$

where $\psi(k) = [-y(k-1), \dots, -y(k-n), u(k), u(k-1), \dots, u(k-n)]^T$ and $\vartheta = [a_1, a_2, \dots, a_n, \frac{i}{k} \sum_{j=1}^{k-1} \theta_j, b_1 - \frac{1}{k} \sum_{j=1}^{k-1} \theta_j, \dots, b_n]^T$. For Eq. (5.11), we adopt weighting factor recursive least square (WFRLS) to evaluate the parameters. The WFRLS can be designed as follows [15]:

$$\begin{cases} \hat{\vartheta}(k) & = \hat{\vartheta}(k) + W(k)P(k)\psi(k-1)e(k) \\ P(k) & = P(k-1) - \frac{P(k-1)\psi(k-1)\psi^T(k-1)P(k-1)}{\frac{1}{W(k)} + \psi^T(k-1)P(k-1)\psi(k-1)} \end{cases} \tag{5.12}$$

where $W(k)$ means weighting factor and it is defined as follows:

$$W(k) = \begin{cases} 1, & |e(k)| \leqslant e \\ \frac{e}{|e(k)|}, & |e(k)| > e \end{cases} \tag{5.13}$$

The error $e(k)$ defines $e(k) = y(k) - \psi^T(k-1)\hat{\vartheta}(k-1)$ and e indicates the threshold of Huber function. $P(0) = \alpha I, \vartheta(0) = \varepsilon$, where I, ε denote identity vector and zero vector, respectively, and α is sufficiently large positive real number.

5.3.2 Design of Controller

In this paper, the controller is picked as follows:

$$u(k) = \frac{k}{\rho + \sum_{j=1}^{k-1}\theta_j}\left[\hat{y}(k) + a_1\hat{y}(k-1) + \cdots + a_n\hat{y}(k-n) + \left(\frac{1}{k}\sum_{j=1}^{k-1}\theta_j - b_1\right)u(k-1) + \cdots + b_n u(k-n)\right] \tag{5.14}$$

where ρ is a sufficiently small positive integer, which guarantees the equation non-singularity when $\sum_{j=1}^{k-1}\theta_j = 0$.

Considering the discrete hysteresis system (5.10), if the controller is chosen as (5.14), the following Theorem holds.

Theorem 2 *Considering a discrete system is described as Eq.* (5.10), *when the parameters are estimated by Eqs.* (5.12), (5.13) *and the controller is chosen as Eq.* (5.14), *the discrete system is stabilized.*

Proof Select a Lyapunov function candidate as:

$$V(k) = \tilde{\vartheta}^T(k)P^{-1}(k)\tilde{\vartheta}(k) \tag{5.15}$$

where $\tilde{\vartheta}(k) = \hat{\vartheta}(k) - \vartheta$.

Since $\tilde{\vartheta}(k) = \hat{\vartheta}(k) - \vartheta$, by using $\hat{\vartheta}(k) = \hat{\vartheta}(k-1) + W(k)P(k)\psi(k-1)e(k)$ in Eq. (5.12), we have

$$\tilde{\vartheta}(k) = \tilde{\vartheta}(k-1) + W(k)P(k)\psi(k-1)e(k). \tag{5.16}$$

Considering $e(k) = y(k) - \psi^T(k-1)\hat{\vartheta}(k-1)$ and $\tilde{\vartheta}(k) = \hat{\vartheta}(k) - \vartheta$, the error is expressed as:

$$e(k) = -\psi^T(k-1)\tilde{\vartheta}(k-1). \tag{5.17}$$

Substituting Eq. (5.17) into (5.16), we have

$$\tilde{\vartheta}(k) = P(k)(P^{-1}(k) - W(k)\psi(k-1)\psi^{-1}(k-1))\tilde{\vartheta}(k-1). \tag{5.18}$$

By using Eqs. (5.12), (5.18) can be rewritten as:

$$\tilde{\vartheta}(k) = P(k)P^{-1}(k-1)\tilde{\vartheta}(k-1). \tag{5.19}$$

Therefore, from (5.15), (5.16), (5.17) and (5.19), we have

$$\begin{aligned}
\Delta V(k) &= \tilde{\vartheta}^T(k)P^{-1}(k)\tilde{\vartheta}(k) - \tilde{\vartheta}^T(k-1)P^{-1}(k-1)\tilde{\vartheta}(k-1) \\
&= \tilde{\vartheta}^T(k)P^{-1}(k)P^{-1}(k-1)\tilde{\vartheta}(k-1) - \tilde{\vartheta}^T(k-1)P^{-1}(k-1)\tilde{\vartheta}(k-1) \\
&= e(k)W(k)\psi^T(k-1)P^T(k)P^{-1}(k-1)\tilde{\vartheta}(k-1) \\
&\leqslant e(k)W(k)\psi^T(k-1)P(k-1)P^{-1}(k-1)\tilde{\vartheta}(k-1) \\
&= -W(k)e^2(k).
\end{aligned} \tag{5.20}$$

Equations (5.15) and (5.20) imply that V is nonincreasing. Hence, the discrete system is stabilized.

This completes the proof.

5.4 Simulation Studies

In this section, we illustrate the above methods on a line motor system. For this system, we ignore the dynamic characteristics of current and only consider the effect of hysteresis. We have the discrete system model as:

$$y(2) + \gamma z^{-1}y(1) = \frac{1}{M}z^{-1}u(2) - \frac{\gamma}{M}z^{-2}u(1) + \varpi \tag{5.21}$$

where y indicates position of load and u is input voltage. M, γ and ϖ means total mass of line motor system, friction coefficient and hysteresis, respectively.

In this simulation, M, γ is chosen as $M = 1.4, \gamma = 0.23$, respectively. Let $\eta_1 = \frac{1}{k}\sum_{j=1}^{k-1}\theta_j - b_1, \eta_2 = \frac{1}{k}\sum_{j=1}^{k-1}\theta_j - b_2$ and the controller is designed as Eq. (5.14) where ρ is defined as $\rho = 0.001$. We choose the reference input as $r(i) = e^{10i}sin(\frac{\pi}{2-i})$. The simulation results are illustrated in Figs. 5.1 and 5.2a, b.

Figure 5.1 shows the identification results of the parameters in Eq. (5.21) with WFRLS and Fig. 5.2a illustrates the tracking curve of the proposed controller with the input $r(i)$. Figure 5.2b shows the tracking errors of the reference signal and controlled results. From Fig. 5.1, it is shown that the identification results can convergence after 20 iterations and the controller can good track the reference signal from Fig. 5.2a. We can see that the errors of reference signal and controlled results in

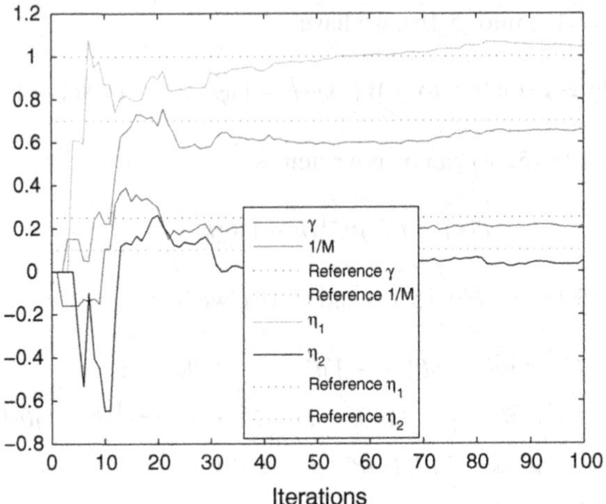

Fig. 5.1 The identification results of parameters

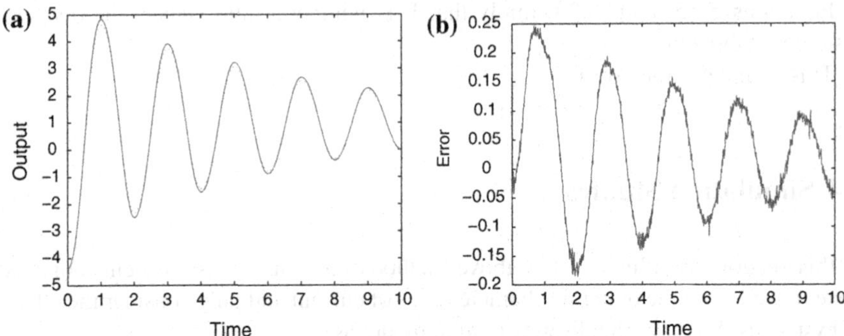

Fig. 5.2 The contrast of reference output and control results. **a** The control results. **b** The tracking errors

Fig. 5.2b, which the maximum error is 0.25, the mean error is 0.0316 and mean absolute error (MAE) is 0.0903. That indicated the proposed control scheme has good tracking performance and robustness.

5.5 Conclusion

An online identification approach was proposed for discrete Preisach model of hysteresis nonlinearity in this paper. The identification computation was simplified by matrix calculation. Then, a WFRLS was adopted on this discrete system to estimate the parameters and a robust adaptive control scheme was presented based on the identification without inverse Preisach model to compare with other control strategies. In addition, a Lyapunov function candidate was designed to guarantee the stability of the controlled system. Finally, simulations demonstrated the effectiveness of the proposed approaches.

References

1. Andrei P, Dimian M (2013) Clockwise jiles-atherton hysteresis model. IEEE Trans Magn 49(7):3183–3186
2. Laudani A, Fulginei F, Salvini A (2014) Bouc-wen hysteresis model identification by the metric-topological evolutionary optimization. IEEE Trans Magn 50(2):621–624
3. Gao X, Ren X, Gong X, Huang J (2013) The identification of preisach hysteresis model based on piecewise identification method. In: Control Conference (CCC), 2013 32nd Chinese, pp 1680–1685
4. Li Z, Su CY, Chai T (2014) Compensation of hysteresis nonlinearity in magnetostrictive actuators with inverse multiplicative structure for preisach model. IEEE Trans Autom Sci Eng 11(2):613–619
5. Xiao S, Li Y (2013) Modeling and high dynamic compensating the rate-dependent hysteresis of piezoelectric actuators via a novel modified inverse preisach model. IEEE Trans Control Syst Technol 21(5):1549–1557
6. Janaideh Al, Krejci MP (2013) Inverse rate-dependent prandtl-ishlinskii model for feedforward compensation of hysteresis in a piezomicropositioning actuator. IEEE/ASME Trans Mechatron 18(5):1498–1507
7. Gu GY, Zhu LM, Su CY (2014) Modeling and compensation of asymmetric hysteresis nonlinearity for piezoceramic actuators with a modified prandtl-ishlinskii model. IEEE Trans Ind Electron 61(3):1583–1595
8. Liu S, Su CY, Li Z (2014) Robust adaptive inverse control of a class of nonlinear systems with prandtl-ishlinskii hysteresis model. IEEE Trans Autom Control 59(8):2170–2175
9. Zhifu L, Peng Y, Yueming H, Tiemei C (2011) Adaptive control of a class of uncertain nonlinear systems with unknown input hysteresis. In: Information and Automation (ICIA), 2011 IEEE International Conference on, pp 141–146
10. Tan X, Baras J (2005) Adaptive identification and control of hysteresis in smart materials. IEEE Trans Autom Control 50(6):827–839
11. Li CT, Tan YH (2005) Adaptive output feedback control of systems preceded by the preisach-type hysteresis. IEEE Trans Syst Man Cybern Part B: Cybern 35(1):130–135
12. Ruderman M, Bertram T (2014) Control of magnetic shape memory actuators using observer-based inverse hysteresis approach. IEEE Trans Control Syst Technol 22(3):1181–1189
13. Rosenbaum S, Ruderman M, Strohla T, Bertram T (2010) Use of jiles-atherton and preisach hysteresis models for inverse feed-forward control. IEEE Trans Magn 46(12):3984–3989
14. Liu L, Tan KK, Chen S, Teo CS, Lee TH (2013) Discrete composite control of piezoelectric actuators for high-speed and precision scanning. IEEE Trans Ind Informatics 9(2):859–868
15. Zhao M (1992) Study on a novel robust discrete adaptive control method (in chinese). J Wuhan Iron and Steel Univ 15(2):178–184

Chapter 6
Switching Control for Multi-motor Driving Servo System with Uncertain Parameters

Wei Zhao and Xuemei Ren

Abstract This paper presents a novel switching control for multi-motor driving servo systems with uncertain parameters, such that it is sufficient to achieve speed synchronization among motors and load tracking. In order to solve the problem of complex coupling relationship between synchronization and tracking, a switching plane (SP) is introduced to implement switching between speed synchronization control and load tracking control. In design of synchronization control, the adaptive scheme based on speed negative feedback is proposed to achieve synchronization errors convergence in finite time. Due to the properties of robustness, the adaptive robust algorithm is utilized to attain load tracking accurately, where dual-observer and adaptive laws are presented to estimate unknown friction state and uncertain parameters. Comparative simulation results are included to verify the reliability and effectiveness.

Keywords Multi-motor driving servo systems · Synchronization control · Tracking control · Parameter estimation · Friction compensation

---

This work is supported by National Natural Science Foundation of China (No. 61433003, No. 61273150 and No. 61321002), the Research Fund for the Doctoral Program of Higher Education of China (No. 20121101110029). The corresponding author is Xuemei Ren (E-mail: xmren@bit.edu.cn).

W. Zhao · X. Ren (✉)
School of Automation, Beijing Institute of Technology, Beijing, People's Republic of China
e-mail: xmren@bit.edu.cn

W. Zhao
e-mail: zw198603@126.com

© Springer-Verlag Berlin Heidelberg 2016

Y. Jia et al. (eds.), *Proceedings of the 2015 Chinese Intelligent Systems Conference*, Lecture Notes in Electrical Engineering 359,
DOI 10.1007/978-3-662-48386-2_6

6.1 Introduction

In recent years, large inertia systems have been widely used in industry and military, such as steel rolling systems, radar control systems and so on. Due to the limitations, single motor can not drive large inertia systems with preferable performances. Considering large driving force, multi-motor is attracting wide attentions from different communities. Moreover, the speed synchronization among multi-motor is a fundamental issue to have influences on the performances of multi-motor systems, i.e., stability, precision and response speed. Simultaneously, system parameters are affected by environment and may have different values in various environments, which affect the system performances and need to be eliminated. Thus, it is necessary to investigate a novel controller based on adaptive parameter estimation, such that the high performances of speed synchronization and load tracking can be achieved.

Considering the issue of load position tracking, many different advanced schemes have been proposed to attain load position tracking precisely, such as H_∞ control [1, 2] and intelligent control [3]. But, most of above algorithms paid close attentions to single motor driving systems, which can not be directly applied on multi-motor driving systems. Speed asynchronization will result in motors collision and system performances weakening, thus speed synchronization among multi-motor becomes a fundamental issue and has been investigated by some schemes [4, 7, 10]. Whereas, most of the literatures only studied speed synchronization without load position tracking, which may cause uncontrollability of system output. To obtain highly precise results, this paper proposes a switching controller to achieve finite-time speed synchronization and load position tracking.

Due to influence of external environments (e.g. temperature, humidity, etc.), the values of system parameters are not invariable, which will be a handicap to accurate control. To compensate effects of uncertain parameters, two approaches, namely robust control (RC) and adaptive control (AC) [5, 6, 8], are mostly utilized for systems with uncertain parameters. RC has guaranteed transient and steady-state performances, but this scheme may result in switching or infinite gain feedback for asymptotic tracking [6]. In contrast to RC, AC is able to achieve asymptotic tracking without resorting to infinite gain feedback. Although above AC schemes can eliminate the nonlinear effects of uncertain parameters, the robustness are not considered.

The main contributions of this paper are listed as follows: Firstly, a novel SP is proposed based on synchronization error for control design to achieve speed synchronization and load tracking. Secondly, an adaptive method based on speed negative feedback is presented to eliminate the nonlinear effects of uncertain parameters and achieve speed synchronization in finite time. Finally, an adaptive robust control incorporated with dual-observer is studied to attain load tacking, where dual-super twisting (dual-ST) observer and adaptive laws are given to estimate the LuGre friction state and uncertain parameters with bounded errors.

6.2 Problem Formulation

The dynamics of multi-motor driving servo systems can be described as

$$\begin{cases} J_i\ddot{\theta}_i + b_i\dot{\theta}_i &= u_i - \tau_i - (-1)^i\bar{T} + w_i \; i = 1, \ldots, n \\ J_m\ddot{\theta}_m + f_m(\dot{\theta}_m) &= \sum_{i=1}^n \tau_i \end{cases} \tag{6.1}$$

where θ_i and θ_m describe angle positions of the motor i and the load, $\dot{\theta}_i$ and $\dot{\theta}_m$ are their velocities respectively, J_i and b_i are the moment of inertia and the viscous friction coefficient of motor i, J_m denotes the moment of inertia of load, u_i is the system input torque, w_i represents bounded disturbance and $|w_i| \le \bar{w}$, f_m is the friction torque, $\bar{T} > 0$ is the constant bias torque, the transmission torque τ_i is defined as

$$\tau_i = \kappa_i D(\theta_i - \theta_m) \begin{cases} \kappa_i(\theta_i - \theta_m + \alpha), \; \theta_i - \theta_m < -\alpha \\ 0, \qquad\qquad -\alpha \le \theta_i - \theta_m \le \alpha \\ \kappa_i(\theta_i - \theta_m - \alpha), \; \theta_i - \theta_m > \alpha \end{cases} \tag{6.2}$$

with κ_i and α being torsional coefficient and backlash width.

Assumption 1 Each moment of inertia of motors is uniform, i.e., $J_1 = J_2 = \cdots = J_n = J$. All of J, J_m and b_i are not accurately known. The transmission torque τ_i and friction f_m are immeasurable.

In this paper, the LuGre model is selected as the friction model of servo systems, and the expression is given as follows:

$$f_m = \sigma_0\mu + \sigma_1\dot{\mu} + \sigma_2\dot{\theta}_m, \quad \dot{\mu} = \dot{\theta}_m - \frac{|\dot{\theta}_m|}{h(\dot{\theta}_m)}\mu, \quad h(\dot{\theta}_m) = F_c + (F_s - F_c)e^{-(\dot{\theta}_m/\dot{\theta}_s)^2} \tag{6.3}$$

where μ represents the deformation of the bristle, F_c stands for the Coulomb friction level, F_s denotes the static force level and $F_s \ge F_c$, $\dot{\theta}_s$ is Stribeck velocity, σ_0 denotes the stiffness, σ_1 is the damping coefficient and σ_2 denotes viscous friction coefficient. In (6.3), σ_0, σ_1, σ_2 and μ are all unknown.

According to Assumption 1, the following states are denoted as

$$\begin{cases} x_1 = \theta_m \\ x_2 = \dot{\theta}_m \end{cases} \quad \begin{cases} x_{3i} = \theta_i \\ x_{4i} = \dot{\theta}_i, \end{cases} \quad i = 1, 2, \ldots, n \tag{6.4}$$

and the dynamics of servo systems (6.1) can be deduced as

$$\begin{cases} \dot{x}_1 = x_2 \\ \dot{x}_2 = a_1 \sum_{i=1}^n \tau_i - a_1 f_m \end{cases} \qquad \begin{cases} \dot{x}_{3i} = x_{4i} \\ \dot{x}_{4i} = a_2(u_i - \tau_i - (-1)^i\bar{T} + w_i) - a_{3i}x_{4i} \end{cases} \tag{6.5}$$

where $a_1 = 1/J_m$, $a_2 = 1/J$, $a_{3i} = b_i/J$, $i = 1, 2, \ldots, n$.

In following section, a novel switching control is dedicated to achieve finite-time speed synchronization control and load tracking control.

6.3 Design of Switching Control

In order to achieve speed synchronization and load tracking, the adaptive switching controller is given in this section. A SP is introduced to study the stability of servo systems

$$S_0 = \left\{(s_1^T, s_2)|s_1^T s_1 = 0\right\}. \tag{6.6}$$

where $s_1^T s_1 = 0$ holds whenever the systems stay on the S_0.

The synchronization error s_1 is defined as follows

$$s_1 = [x_{41} - x_{42}, \ldots, x_{4i} - x_{4i+1}, \ldots, x_{4n} - x_{41}]^T = [e_{s1}, \ldots, e_{si}, \ldots, e_{sn}]^T \tag{6.7}$$

and s_2 is denoted as the filtering value of load tracking error $e_t = x_1 - y_d$

$$s_2 = \dot{e}_t + \lambda_1 e_t + \lambda_2 |e_t|^{\lambda_3} sign(e_t) \quad (\lambda_1, \lambda_2 \geq 0) \tag{6.8}$$

where y_d is the continuous reference signal, $\lambda_3 = q/p$ and the positive odd constants $p > q > 0$ satisfy $\frac{1}{2} \leq q/p \leq 1$.

In following subsections, the switching controller u_i based on SP is presented to achieve speed synchronization and load tracking.

$$u_i = u_{si} + u_{ti}, \tag{6.9}$$

which consists of two parts: speed synchronization u_{si} and tracking control u_{ti}.

6.3.1 Design of Speed Synchronization Controller

In order to eliminate the influence on synchronization control, u_{ti} is chosen as $u_{t1} = u_{t2} = \cdots = u_{tn}$. Then synchronization error (6.7) can be deduced that

$$J\dot{e}_{si} = u_{si} - (-1)^i \bar{T} + w_i - \rho_i^T \phi_i - (u_{si+1} - (-1)^{i+1} \bar{T} + w_{i+1} - \rho_{i+1}^T \phi_{i+1})$$
$$i = 1, 2, \ldots n - 1 \tag{6.10}$$

$$J\dot{e}_{sn} = u_{sn} - (-1)^n \bar{T} + w_n - \rho_n^T \phi_n - (u_{s1} + \bar{T} + w_1 - \rho_1^T \phi_1). \tag{6.11}$$

where the vectors ρ_i and ϕ_i are defined as

$$\rho_i = [\kappa_i, b_i]^T, \quad \phi_i = [D(x_{4i} - x_2), x_{4i}]^T, \tag{6.12}$$

and J, b_i, κ_i, $i = 1, 2, \ldots, n$ are all uncertain parameters, the $D(x_{4i} - x_2)$ is given in (6.2).

Based on speed negative feedback, the adaptive controller u_{si} is proposed to keep speed synchronization

$$u_{si} = -(k_1 x_{4i} - \frac{k_1}{n} \sum_{j=1}^{n} x_{4j}) - \vartheta_i + \hat{\rho}_i^T \phi_i \tag{6.13}$$

where the control gain $k_1 > 0$ and $\hat{\rho}_i$ is the estimation of ρ_i and $\hat{\rho}_i = [\hat{\kappa}_i, \hat{b}_i]^T$.

Moreover, the function ϑ_i is selected to guarantee finite-time convergence of synchronization error

$$\begin{aligned} \vartheta_1 &= (k_2 |e_{s1} - e_{sn}|^r + k_3) sign(e_{s1} - e_{sn}) \\ \vartheta_i &= (k_2 |e_{si} - e_{si-1}|^r + k_3) sign(e_{si} - e_{si-1}) \quad i = 2, 3, \ldots, n \end{aligned} \tag{6.14}$$

where k_2 and k_3 are positive gains and $0 < r < 1$.

And the adaptive laws are defined as

$$\dot{\hat{\rho}}_1 = -\Gamma_1^{-1} \phi_1 (e_{s1} - e_{sn}), \quad \dot{\hat{\rho}}_i = -\Gamma_i^{-1} \phi_i (e_{si} - e_{si-1}) \quad i = 2, 3, \ldots, n \tag{6.15}$$

where $\Gamma_i \in R^{n \times n}$ $i = 1, 2, \ldots, n$ are positive definite matrices.

Theorem 6.1 *Consider multi-motor driving servo systems* (6.5). *The controller is design as* (6.13) *with adaptive laws* (6.15), *and then speed synchronization among motors is achieved in finite time. Moreover, the uncertain parameters κ_i and b_i are both estimated with arbitrarily little errors.*

Proof Select the Lyapunov function candidate V_s as

$$V_s = \frac{1}{2} J s_1^T s_1 + \frac{1}{2} \sum_{i=1}^{n} \tilde{\rho}_i^T \Gamma_i \tilde{\rho}_i = \frac{1}{2} \sum_{i=1}^{n} J e_{si}^2 + \frac{1}{2} \sum_{i=1}^{n} \tilde{\rho}_i^T \Gamma_i \tilde{\rho}_i \tag{6.16}$$

where $\tilde{\rho}_i = \hat{\rho}_i - \rho_i$.

Considering control (6.13)~(6.15) and choosing $k_3 > \bar{T} + \bar{w}$, the derivative of V_s is given by

$$\begin{aligned}\dot{V}_s &= \sum_{i=1}^n (Je_{si}\dot{e}_{si} + \tilde{\rho}_i^T \Gamma_i \dot{\hat{\rho}}_i)\\ &= (e_{s1} - e_{sn})(u_1 + \bar{T} + w_1 - \rho_1^T \phi_1) + \sum_{i=1}^n (\tilde{\rho}_i^T \Gamma_i \dot{\hat{\rho}}_i)\\ &\quad + \sum_{i=2}^n [(e_{si} - e_{si-1})(u_i - (-1)^i \bar{T} + w_i - \rho_i^T \phi_i)]\\ &\le (e_{s1} - e_{sn})(-k_1 x_{41} + \tfrac{k_1}{n} \sum_{j=1}^n x_{4j} - k_2 |e_{s1} - e_{sn}|^r sign(e_{s1} - e_{sn}) + \tilde{\rho}_1^T \phi_1)\\ &\quad + \sum_{i=2}^n [(e_{si} - e_{si-1})(-k_1 x_{4i} + \tfrac{k_1}{n} \sum_{j=1}^n x_{4j} - k_2 |e_{si} - e_{si-1}|^r sign(e_{si} - e_{si-1})\\ &\quad + \tilde{\rho}_i^T \phi_i)] - \tilde{\rho}_1^T \phi_1 (e_{s1} - e_{sn}) - \sum_{i=2}^n \tilde{\rho}_i^T \phi_i (e_{si} - e_{si-1})\\ &\le -k_1 (e_{s1} - e_{sn}) x_{41} - k_1 \sum_{i=2}^n (e_{si} - e_{si-1}) x_{4i}\\ &\quad - k_2 |e_{s1} - e_{sn}|^{r+1} - k_2 \sum_{i=2}^n |e_{si} - e_{si-1}|^{r+1}\\ &\le -k_1 \sum_{i=1}^n e_{si}^2 - k_2 \left(\sum_{i=1}^n e_{si}^2\right)^{\frac{r+1}{2}} \le 0.\end{aligned} \tag{6.17}$$

From (6.17), one obtains that the errors of speed synchronization and parameter estimations are decreased with increasing of time t. Moreover, both errors finally converge to zeros.

Thus, it is able to deduce that $\tilde{\rho}_i^T \phi_i$ are bounded and $max(\tilde{\rho}_i^T \phi_i) \le L$ with $L > 0$. As $k_3 > \bar{T} + \bar{w} + L$, the derivative of $V_{s1} = \frac{1}{2} J s_1^T s_1$ is given by

$$\dot{V}_{s1} \le -k_1 \sum_{i=1}^n e_{si}^2 - \left(k_2 \sum_{i=1}^n e_{si}^2\right)^{\frac{r+1}{2}} \le -k_1 V_{s1} - k_2 V_{s1}^{\frac{r+1}{2}}. \tag{6.18}$$

From the finite-time theorem [9], it is found that the speed synchronization among motors is attained in finite time. This ends the proof of Theorem 6.1.

6.3.2 Design of Load Tracking Controller

As speed synchronization is achieved, i.e., $x_{4i} = \sum_{j=1}^n x_{4j}/n$, the synchronization error e_{si} tends to 0. Then the synchronization controller u_{si} is equal to

$$u_{si} = \hat{\rho}_i^T \phi_i. \tag{6.19}$$

From (6.10), (6.11) and (6.19), it is found that $\dot{e}_{si} = 0$ holds whenever

$$u_{ti} = -\hat{\rho}_i^T \phi_i + \tau_i - (-1)^i \bar{T} - w_i + b_i x_{4i}, \tag{6.20}$$

which implies that

$$\sum_{i=1}^n \tau_i = n u_{ti} + \sum_{i=1}^n \hat{\rho}_i^T \phi_i - \sum_{i=1}^n (-(-1)^i \bar{T} - w_i + b_i x_{4i}). \tag{6.21}$$

In order to guarantee system on the S_0, i.e., $e_{si} = 0$ all the time, the tracking controller u_{ti} will be designed based on (6.21) in the following.

Due to Theorem 6.1, it is found that $\sum_{i=1}^{n} b_i x_{4i}$ can be approximated by $\sum_{i=1}^{n} \hat{b}_i x_{4i}$ with arbitrary error. Then, (6.21) is simplified as

$$\sum_{i=1}^{n} \tau_i = nu_{ti} + \sum_{i=1}^{n} (\hat{\tau}_i + (-1)^i \bar{T} + w_i). \tag{6.22}$$

Moreover, the dynamic of load is changed as

$$\dot{x}_1 = x_2, \quad J_m \dot{x}_2 = nu + \sum_{i=1}^{n} (w_i + (-1)^i \bar{T}) - \sigma_0 \mu + \sigma_1 \frac{|x_2|}{h(x_2)} \mu - \sigma x_2 \tag{6.23}$$

where $\sigma = \sigma_1 + \sigma_2$, $u = u_{ti} + \sum_{i=1}^{n} \hat{\tau}_i / n$ and $\hat{\tau}_i = \hat{\kappa}_i D(x_{4i} - x_2)$.

Due to the existence of uncertain friction state μ, the estimation of μ needs to be discussed. Thus, inspired by [5], the dual-ST observer is proposed as follows:

$$\dot{\hat{\mu}}_0 = -\frac{|x_2|}{h(x_2)} \hat{\mu}_0 + x_2 - g_1 |s_2|^{1/2} sign(s_2), \quad \dot{\hat{\mu}}_1 = -\frac{|x_2|}{h(x_2)} \hat{\mu}_1 + x_2 - g_2 sign(s_2) \tag{6.24}$$

where g_1 and g_2 are positive gains.

Afterwards, the u_{ti} will be designed to attain load tracking. Considering remarkable properties, the adaptive robust method is used to design tracking control u_{ti}, which is defined as

$$u_{ti} = u - \sum_{i=1}^{n} \frac{\hat{\tau}_i}{n} \tag{6.25}$$

and

$$u = -\frac{1}{n} \left(r_1 \varphi_1(s_2) + r_2 tanh(\xi s_2) - \hat{p}_m^T \phi_m \right) \tag{6.26}$$

where

$$\varphi_1(s_2) = |s_2|^{\frac{1}{2}} sign(s_2) \tag{6.27}$$

$$\hat{p}_m = [\hat{\sigma}_0, \hat{\sigma}_1, \hat{\sigma}, \hat{J}_m]^T, \quad \phi_m = [\hat{\mu}_0, -\frac{|x_2|}{h(x_2)} \hat{\mu}_1, x_2, x_{eq}]^T \tag{6.28}$$

$$\dot{\hat{p}}_m = -s_2 \Gamma_m^{-1} \phi_m, \quad x_{eq} = \ddot{y}_d - \lambda_1 \dot{e}_t - \lambda_2 \lambda_3 |e_t|^{\lambda_3 - 1} sign(e_t) \dot{e}_t, \tag{6.29}$$

and r_1, r_2, ξ are positive constants, $\Gamma_m \in R^{n \times n}$ is positive definite matrix, and $\tilde{p}_m = \hat{p}_m - p_m$.

Then, the stability of tracking control is analyzed in Theorem 6.2.

Theorem 6.2 *Consider multi-motor driving servo systems* (6.5). *The controller is selected as* (6.25) *with* (6.26). *As the speed synchronization driven by* (6.13) *is achieved, the load tracking and estimation of friction state are attained with bounded errors.*

Proof Choose the Lyapunov function candidate as

$$V_t = \frac{1}{2}J_m s_2^2 + \frac{1}{2}\tilde{\rho}_m^T \Gamma_m \tilde{\rho}_m + \frac{1}{2}\sigma_0 \tilde{\mu}_0^2 + \frac{1}{2}\sigma_1 \tilde{\mu}_1^2 \tag{6.30}$$

where $\tilde{\rho}_m = \hat{\rho}_m - \rho_m$, $\tilde{\mu}_0 = \hat{\mu}_0 - \mu$, $\tilde{\mu}_1 = \hat{\mu}_1 - \mu$.

Considering the equations $\hat{\sigma}_0\hat{\mu}_0 - \sigma_0\mu_0 = \tilde{\sigma}_0\hat{\mu}_0 + \sigma_0\tilde{\mu}_0$, $\hat{\sigma}_1\hat{\mu}_1 - \sigma_1\mu_1 = \tilde{\sigma}_1\hat{\mu}_1 + \sigma_1\tilde{\mu}_1$, and observer (6.24) with control (6.26)~(6.29), $\dot{V}_t$ is deduced as

$$\begin{aligned}
\dot{V}_t &= J_m s_2 \dot{s}_2 + \tilde{\rho}_m^T \Gamma_m \dot{\hat{\rho}}_m + \sigma_0 \tilde{\mu}_0 \dot{\hat{\mu}}_0 + \sigma_1 \tilde{\mu}_1 \dot{\hat{\mu}}_1 \\
&= s_2[nu - (\sigma_0\mu - \sigma_1 \tfrac{|x_2|}{h(x_2)}\mu + \sigma x_2 - J_m x_{eq})] + \tilde{\rho}_m^T \Gamma_m \dot{\hat{\rho}}_m + \sigma_0 \tilde{\mu}_0 \dot{\hat{\mu}}_0 + \sigma_1 \tilde{\mu}_1 \dot{\hat{\mu}}_1 \\
&= -s_2\left(r_1\varphi_1(s_2) + r_2 tanh(\xi s_2)\right) + s_2 \textstyle\sum_{i=1}^n (w_i + (-1)^i \bar{T}) \\
&\quad + s_2[(\tilde{\sigma}_0\hat{\mu}_0 + \sigma_0\tilde{\mu}_0) - \tfrac{|x_2|}{h(x_2)}(\tilde{\sigma}_1\hat{\mu}_1 + \sigma_1\tilde{\mu}_1) + \tilde{\sigma}x_2 + \tilde{J}_m x_{eq}] - s_2\tilde{\rho}_m^T\phi_m \\
&\quad - \sigma_0\tilde{\mu}_0(\tfrac{|x_2|}{h(x_2)}\tilde{\mu}_0 + g_1|s_2|^{1/2} sign(s_2)) - \sigma_1\tilde{\mu}_1(\tfrac{|x_2|}{h(x_2)}\tilde{\mu}_1 + g_2 sign(s_2)).
\end{aligned} \tag{6.31}$$

By choosing r_2 to eliminate the term $s_2 \sum_{i=1}^n (w_i + (-1)^i \bar{T})$ and $g_1 = |s_2|^{1/2}$, $g_2 = |x_2||s_2|/h(x_2)$, it is found that (6.31) is transformed as

$$\dot{V}_t \le -r_1 s_2 \varphi_1(s_2) - \sigma_0 \frac{|x_2|\tilde{\mu}_0^2}{h(x_2)} - \sigma_1 \frac{|x_2|\tilde{\mu}_1^2}{h(x_2)} \le 0. \tag{6.32}$$

where $|x_2|/h(x_2) \ge 0$.

From (6.32), it is deduced that s_2 tends to zero, which implies that tracking error can converge to origin with bounded error. This ends proof of Theorem 6.2.

6.4 Numerical Simulations

In this section, in order to further illustrate the performances of the proposed algorithm, we consider the following four-motor driving servo system:

$$\begin{cases} \dot{x}_1 = x_2 \\ \dot{x}_2 = \frac{1}{0.028}\sum_{i=1}^4 \tau_i - \frac{1}{0.028} f_m \end{cases} \quad \begin{cases} \dot{x}_{3i} = x_{4i} \\ \dot{x}_{4i} = \frac{1}{0.185}(u_i - (-1)^i \bar{T} - \tau_i + w) - \frac{1.2}{0.185} x_{4i} \end{cases} \tag{6.33}$$

where the driving motors $i = 1, 2, 3, 4$ are selected with same parameters (i.e., $\alpha = 0.2$, $\kappa_i = 560$, $\bar{T} = 0.8$) and the parameters of LuGre friction f_m are chosen as $F_s = 2.4$, $F_c = 0.8$, $\dot{\theta}_s = 1.05$, $\sigma_0 = 1.3$, $\sigma_1 = 1.5$, $\sigma_2 = 0.8$. The reference position of load is chosen as $y_d = 2sin(\frac{2}{5}\pi t) + 1.3cos(\frac{4}{5}\pi t)$.

To verify the effectiveness of the proposed switching control, the comparative simulations among the switching controller, PID and NPID are proposed. The parameters of proposed controller are listed as $k_1 = 10$, $k_2 = 5$, $k_3 = 15$, $r_1 = 20$, $r_2 = 5$,

$r = 1/2$, $\xi = 500$, $\lambda_1 = 3$, $\lambda_2 = 1$, $\lambda_3 = 7/11$. The PID is designed as $k_p = 50$, $k_i = 2$ and $k_d = 5$. And the NPID is selected as $k_p = 18$, $k_i = 2$.

The performance curves with proposed switching control are listed in Fig. 6.1. It can be found that the proposed method guarantees speed synchronization and load tracking with bounded errors. Moreover, both errors converge to bounded regions around origin in finite time.

Figure 6.2 gives the comparative results of tracking performance. From the curves, it is obvious that all of three algorithms can achieve load tracking with bounded errors. However, compared with PID and NPID, the proposed scheme gives faster convergence speed and smaller steady-state error, which implies that the proposed switching control has better performance.

Figure 6.3 depicts the estimations of friction state $\mu(t)$ and parameters. The curves show that the friction state $\mu(t)$ can be approximated by both state estimations μ_0 and μ_1 with bounded errors. Moreover, it is evident that the parameters of J_m, b_i, σ_0, σ_1, σ can all be accurately estimated by proposed adaptive laws.

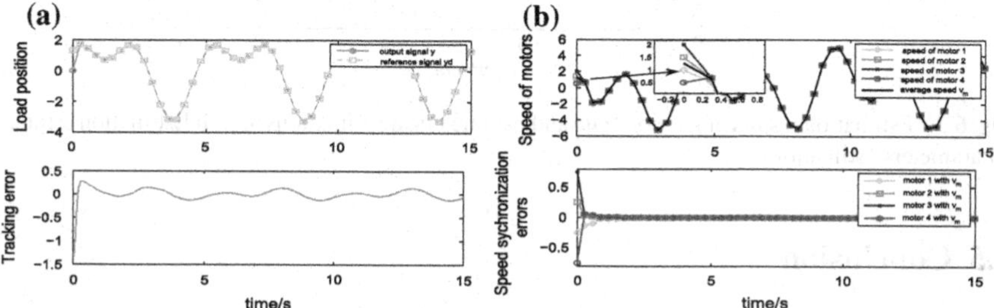

Fig. 6.1 The curves with proposed controller. **a** Load tracking curves. **b** Speed synchronization curves

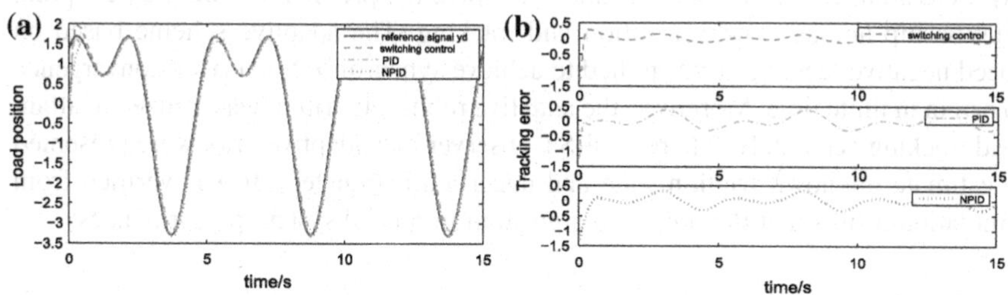

Fig. 6.2 Tracking performance comparisons with different controllers. **a** Tracking performances. **b** Tracking errors comparison

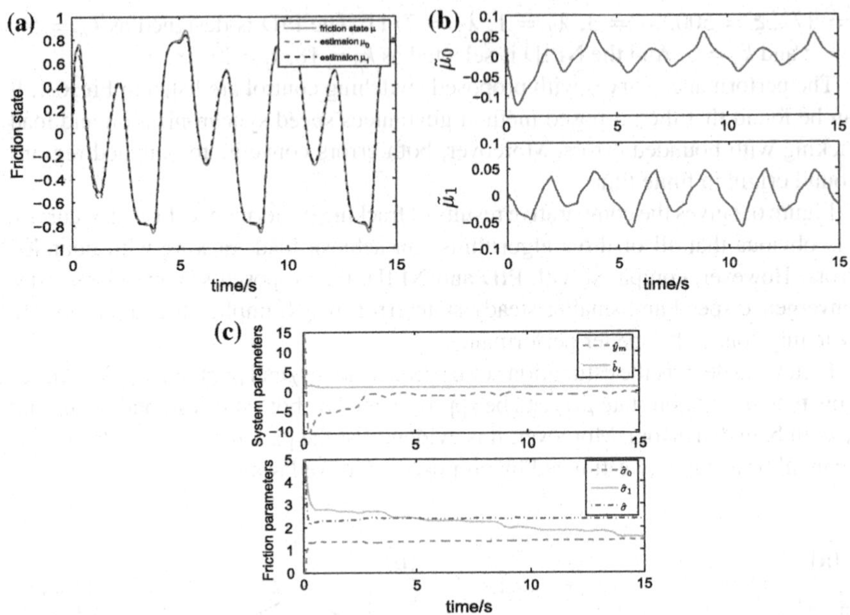

Fig. 6.3 Estimation results of friction state and parameters. **a** Estimations of μ. **b** Estimation errors. **c** Parameters estimations

6.5 Conclusion

A novel switching control was provided for multi-motor driving servo systems with uncertain parameters, such that speed synchronization among motors and load tracking were attained. A SP was introduced to solve the problem of complex coupling relationship among synchronization and tracking. The adaptive scheme based on speed negative feedback was studied to achieve synchronization errors convergence to origin in finite time. Moreover, the adaptive robust algorithm was utilized to attain load tracking accurately, where dual-ST observer and adaptive laws were presented to estimate unknown friction state and uncertain parameters. It was verified from simulation results that the proposed algorithm improved system performances.

References

1. El-Sousy FF (2010) Hybrid-based wavelet-neural-network tracking control for permanent-magnet synchronous motor servo drives. IEEE Trans Ind Electron 57(9):3157–3166
2. El-Sousy FF (2011) Hybrid recurrent cerebellar model articulation controller-based supervisory H_∞ motion control system for permanent-magnet synchronous motor servo drive. Electric Power Appl, IET 5(7):563–579

3. El-Sousy FF (2013) Intelligent optimal recurrent wavelet elman neural network control system for permanent-magnet synchronous motor servo drive. IEEE Trans Ind Inf 9(4):1986–2003
4. Liu FC, Zhang X, Liu L (2002) Synchronous control theory and practical study of multi-motor synchronous driving system. Basic Autom 4:027
5. Liu Z, Wang J, Zhao J (2009) Friction compensation using dual observer for 3-axis turntable servo system. In: IEEE international conference on automation and logistics, ICAL'09, 2009. IEEE, pp 658–663
6. Lu L, Yao B, Wang Q, Chen Z (2009) Adaptive robust control of linear motors with dynamic friction compensation using modified LuGre model. Automatica 45(12):2890–2896
7. Sun G, Su Y, Hong M, Long F (2013) Application of fuzzy control algorithm with variable scale factor to multi-motor synchronization control system. In: Proceedings of the 8th IEEE conference on industrial electronics and applications (ICIEA), 2013. IEEE, pp 307–310
8. Xu L, Yao B (2000) Adaptive robust control of mechanical systems with nonlinear dynamic friction compensation. In: American control conference, 2000. Proceedings of the 2000, vol 4. IEEE, pp 2595–2599
9. Yu S, Yu X, Shirinzadeh B, Man Z (2005) Continuous finite-time control for robotic manipulators with terminal sliding mode. Automatica 41(11):1957–1964
10. Zhang Z, Chau K, Wang Z (2012) Chaotic speed synchronization control of multiple induction motors using stator flux regulation. IEEE Trans Magn 48(11):4487–4490

Chapter 7
Method of Fault Diagnosis Based on SVDD-SVM Classifier

Feng Lv, Hua Li, Hao Sun, Xiang Li and Zeyu Zhang

Abstract Aiming at the problem of incomplete fault data samples, a fault diagnosis method based on Support vector data description and Support vector machine (SVDD-SVM) is presented. First, the data description model is build based on the normal data samples and known fault data samples, and SVM model is built based on known fault data samples. Then the test data samples are tackled by the data description model to reject or accept. The specific categories of accepted samples are diagnosed by the SVM model and the rejected samples are unknown fault types. Tests show that this method can efficiently solve the fault diagnosis problem of incomplete fault samples.

Keywords Support vector data description · Support vector machine · Fault diagnosis

7.1 Introduction

Fault diagnosis is essentially a classification process of fault samples. In the field of fault diagnosis, it is difficult to collect all types of fault data, which brings difficulties for the traditional fault diagnosis methods. Support Vector Machine (SVM) [1] is proposed in the 1990s. The traditional SVM algorithm is originally designed for binary classification, and two kinds of samples are required for training the model. SVM convert the original nonlinear problems in low dimensional space

F. Lv · X. Li · Z. Zhang
Electronic Department, Hebei Normal University, 050024 Shijiazhuang, China

H. Li (✉)
College of Electrical and Information Engineering, Hunan University, 410082 Changsha, Hunan, China
e-mail: lihuax173@163.com

H. Sun
Shijiazhuang Power Supply Company, 050010 Shijiazhuang, China

© Springer-Verlag Berlin Heidelberg 2016

Y. Jia et al. (eds.), *Proceedings of the 2015 Chinese Intelligent Systems Conference*, Lecture Notes in Electrical Engineering 359,
DOI 10.1007/978-3-662-48386-2_7

to linear classification problems in higher dimensional, and thus the calculation process is simplified. SVM has been widely used in fault diagnoses and shows excellent performance [2, 3]. Support vector data description (SVDD) [4] is a single classification algorithm proposed by Tax, which has strong robustness and high computation speed [5].

In this paper, a novel diagnosis method combining SVDD and SVM is proposed. The method can diagnose fault types with incomplete samples and has some practical significance.

7.2 Support Vector Data Description

The idea of SVDD is to create a compactly closed hypersphere according to the target samples. When using the target samples to construct the hypersphere, the radius of hypersphere are required to be as small as possible and the target samples should be enclosed in its interior as more as possible. Figure 7.1 shows the description of SVDD in two-dimension space.

In Fig. 7.1, $\boldsymbol{a}$ and R are the centre and the radius of hypersphere, respectively, $\boldsymbol{v_1}$ is the target sample space and $\boldsymbol{v_2}$ is the non-target sample space.

Assuming the training samples contain N target samples, $\boldsymbol{x_i}$, $\boldsymbol{i} = 1, 2, \ldots, N$. The target samples are often not concentrated, and the distance between individual sample and $\boldsymbol{a}$ is much larger. With all samples included, it will include some unusual data (non-target sample) as hypersphere is constructed. In order to enhance the classification robustness, slack variables ξ_i is introduced. The corresponding mathematical description can be described by Eq. (7.1).

$$\begin{cases} \min\colon R^2 + C\sum_{i}^{N} \xi_i \\ s.t.\, R^2 + \xi_i - \|x_i - a\|^2 \geq 0, i = 1, 2, \ldots, N \end{cases} \tag{7.1}$$

where, C is the penalty factor.

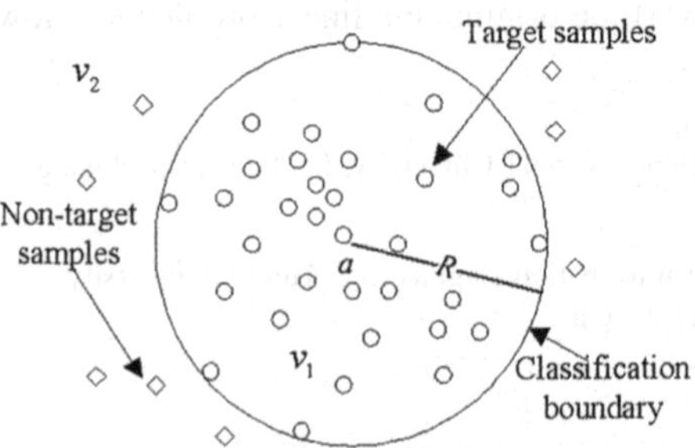

Fig. 7.1 Schematic of Support vector data description in two-dimension space

Equation (7.1) is a typical quadratic programming problem, which can be solved by introducing Lagrange multipliers. Equation (7.1) is rewritten in the form of Lagrange function as:

$$L_p(a,R,\alpha,\beta)=R^2+C\sum_{i=1}^{N}\xi_i-\sum_{i=1}^{N}\alpha_i(R^2+\xi_i-\|x_i-a\|^2)-\sum_{i=1}^{N}\beta_i\xi_i \tag{7.2}$$

where, α_i and β_i are the Lagrange coefficients. The dual form of Eq. (7.1) is:

$$\begin{cases} \max\colon L=\sum_i \alpha_i(x_i\cdot x_j)-\sum_{i,j}\alpha_i\alpha_j(x_i\cdot x_j) \\ s.t.\ \sum_i \alpha_i=1,\ 0\le\alpha_i\le C,\ i=1,2,\ \ldots,N \end{cases} \tag{7.3}$$

The samples corresponding to nonzero $\boldsymbol{\alpha_i}$ are the support vectors. $\boldsymbol{a}$ and R are only related to support vectors and they can be solved by Eq. (7.4).

$$\begin{cases} a=\sum_{i=1}^{N}\alpha_i x_i \\ R^2=\|x_u-a\|^2=\left\|x_u-\sum_{i=1}^{N}\alpha_i x_i\right\|^2=(x_u,x_u)-2\sum_{i=1}^{N}\alpha_i(x_u,x_i)+\sum_{i=1}^{N}\sum_{j=1}^{N}\alpha_i\alpha_j(x_i,x_j) \end{cases} \tag{7.4}$$

If test samples meet Eq. (7.5),

$$\|z-a\|^2=(z\cdot z)-2\sum_i\alpha_i(z\cdot x_i)+\sum_{i,j}\alpha_i\alpha_j(x_i\cdot x_j)\le \mathrm{R}^2 \tag{7.5}$$

they are the target samples, whereas non-target samples.

Kernel function, which is $K(\boldsymbol{x_i}\ .\ \boldsymbol{x_j})$, is introduced to transform the non-linear problem in low-dimensional space into a linear problem in high-dimensional space. The problem can be described as:

$$\max\colon L=\sum_i\alpha_i K(x_i\cdot x_j)-\sum_{i,j}\alpha_i\alpha_j K(x_i\cdot x_j) \tag{7.6}$$

If the test samples such as z are the target samples, they should meet:

$$\|z-a\|^2=K(z\cdot z)-2\sum_i\alpha_i K(z\cdot x_i)+\sum_{i,j}\alpha_i\alpha_j K(x_i\cdot x_j)\le R^2 \tag{7.7}$$

7.3 Fault Diagnosis Method Based on SVDD-SVM

In fault diagnosis, traditional classifiers have larger misjudgment with the case of missing samples. The main reason is that the classification hyperplane of traditional classifiers cannot reasonably divide the space [6]. Supposing that the solving function of hyperplane is Φ, the improvement is made as follows:

$$H=\phi(\bigcup_{i=1}^{k} S_i \cup N) \bigcup other \tag{7.8}$$

where S_i is the fault type of sample i, N is the normal data samples, *other* represents other types, and the missing samples belong to it.

Supposing that there is a missing type of fault samples, use the Eq. (7.8) to build the hyperplane as follows:

$$H^{*}=\phi(\bigcup_{i=1}^{k-1} S_i \bigcup N) \bigcup other \tag{7.9}$$

Although a type of fault samples is missed, it belongs to *other*. So there will be little influence on the classification hyperplane.

For the incomplete fault samples in fault diagnosis, a method of fault diagnosis based on the combination of SVDD and SVM is proposed, and it achieves good results. The diagnosis steps are as follows:

Step 1: Collect data and extract the feature, composing the sample set D which contains normal samples and fault samples;
Step 2: Use D to build the hyperplane of SVDD, and use normal samples to train and build SVM classifier;
Step 3: For the test sample z, use the hypersphere to do the first diagnosis, and discrimination function is:

$$\delta=\operatorname{sgn}\left(R^2-\|z-a\|^2\right)=\begin{cases}1, & z\in\Omega\\ -1, & z\notin\Omega\end{cases} \tag{7.10}$$

If δ is equal to 1, z belongs to D and jump to Step 4. If δ is equal to -1, z belongs to unknown fault types and the diagnostic process is over;
Step 4: If δ is equal to 1 in Step 3, use SVM classifier created by Step 2 to do the second diagnosis and judge the concrete type of z. The diagnostic process is over.

7.4 Test Analyses

In this paper, the transformer fault data of Lv Ganyun et al. [7] are used to verify the effectiveness of the method. There are four kinds of data including the normal data (9), excessive heating data (38), low power data (11), high power data (17). N, P1,

P2 and P3 are used to represent these four kinds of data in sequence. The simulation experiment is build by MATLAB 2011b, LibSVM [8] and LibSVM-svdd-3.17. Here, the SVM type is C-SVC, and the kernel function is radial basis kernel function. The hypothesis is that P2, P3 are missing, structuring incomplete environment of fault samples.

With two types of fault samples missed, SVDD is used in the diagnosis in the first step. The training samples only include N and P1. And test samples include N, P1, P2, P3. The training set label of N and P1 is set to 1, and that of P2, P3 is set to −1 (unknown fault type). The hypersphere of SVDD is constructed by the training set, and the diagnosis performance is detected by the test set. The data are normalized and the range is [0, 1]. 5 groups of N, 20 groups of P1 are randomly chosen to as the common training set of SVDD and SVM. The SVDD test set is made up of the remaining 4 groups of N, 18 groups of P1 and all of P2 and P3 in a total of 50 groups of data. The SVM test set is made up of the data which the type is "1" diagnosed by SVDD.

In the SVDD diagnosis, C is set to 1 and g is set to 0.01 (g is the controlling factor of radial basis kernel function). Use the training set to train and build SVDD model, and test set to test it. The accuracy (ACC) is measured as 88 %. The testing result and the actual result are shown in Fig. 7.2.

After the first diagnosis by SVDD, the SVM classifier is trained by the training set, and it is tested by the data whose type are "1" diagnosed by SVDD in the first step (the label of six fault samples wrongfully diagnosed in first step is set to 3). In this step, C is set to 1 and g is set to 15. ACC is measured as 67.7 %. The test result and the actual result are shown in Fig. 7.3.

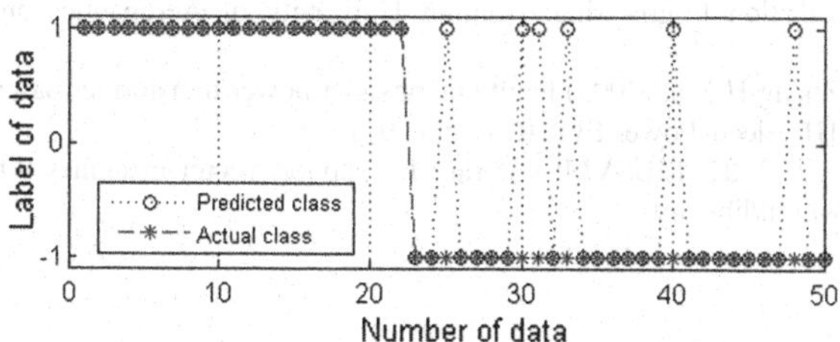

Fig. 7.2 Comparison between the actual values and SVDD test results in SVDD-SVM model

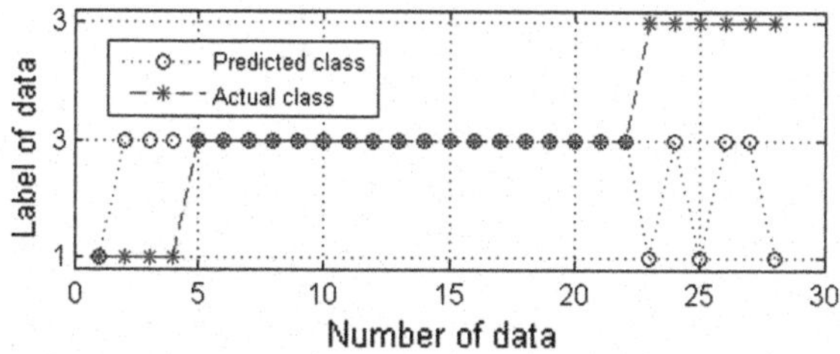

Fig. 7.3 Comparison between the actual values and SVM test result in SVDD-SVM model

7.5 Conclusions

In this paper, the feasibility of the proposed method of fault diagnosis based on SVDD and SVM is verified. The method has a better effect to the condition of incomplete fault samples. Although actual types of the missing samples can't be judged concretely, the types which are out of normal and known fault types can be diagnosed. It can inform the maintenance staffs that equipment has been in fault state, and it is the unknown fault type. It presents that this method has an applicable value for actual engineering.

Acknowledgement This work is supported by National Natural Science Foundation of China under Grant (61175059), and the nature science foundation of Hebei under contract (F2014205115).

References

1. Cortes C, Vapnik V (1995) Support-vector networks [J]. Mach Learn 20(3):273–297
2. Yu D, Yang Y, Cheng J (2005) Fault diagnosis approach for gears based on EMD and SVM [J]. Chin J Mech Eng 41(1):140–144
3. Lian K, Wang H, Long B (2007) Study on SVM based analog electronic system multiple fault diagnosis [J]. Chin J Sci Instrum 28(6):1029–1034
4. Tax DMJ, Duin RPW (2004) Support vector data description [J]. Mach Learn 54(1):45–66
5. Li L, Han J, Wang K, Hao W (2008) Mechanical fault diagnosis based on one-class classification [J]. J Mech Strength 30(5):697–701
6. Yi H (2011) A study of support vector machines based fault diagnosis and its applications [D]. College of Automation Engineering, Nanjing University of Aeronautics and Astronautics, Nanjing
7. Lv G, Chen H, Zhang H et al (2005) Fault diagnosis of power transformer based on multi-layer SVM classifier [J]. Electr Power Syst Res 75(1):9–15
8. Chang CC, Lin CJ (2002) LIBSVM: a library for support vector machines [OL]. http://www.csie.ntu.edu.tw/~cjlin/libsvm/

Chapter 8
Second-Order Sliding Mode Control for BUCK Converters

Jiadian Wang and Shihong Ding

Abstract This paper presents a second-order sliding mode (SOSM) control method for DC-DC Buck converters. First of all, by taking a subtraction between the output voltage and the desired voltage, the sliding mode variable with freedom of degree two can be constructed. Secondly, by using adding a power integrator method, the second-order sliding mode controller can be developed for the Buck converter. Under the proposed controller, it can be shown that the output voltage will track the desired voltage in a finite time. Finally, the theoretical considerations have been verified by simulations.

Keywords Buck converter · Finite-time control · Sliding mode control · Lyapunov stability

8.1 Introduction

The Buck converters are widely used in applications, such as mobile power supply equipment, photovoltaic system, DC supply system, etc. Generally speaking, the control design problems are based on a linear mathematical model of the Buck converters, and the linear controllers like conventional PID control play a dominated role. However, they can't obtain a satisfactory control performance under some large signal operating conditions, such as the huge variations of input voltage. Under this situation, the effect of disturbance, which is from the Buck converters' nonlinear behavior characteristics, can't be eliminated by conventional linear controllers. To this end, various nonlinear control strategies have been applied for Buck converters under large signal operating conditions, such as adaptive control [1], fuzzy logic control [2], artificial neural network control [3], etc. Among these nonlinear control strategies, the sliding mode control strategy has been paid much attention.

J. Wang · S. Ding (✉)
School of Electrical and Information Engineering, Jiangsu University,
Zhenjiang, Jiangsu 212013, China
e-mail: dsh@mail.ujs.edu.cn

© Springer-Verlag Berlin Heidelberg 2016

Y. Jia et al. (eds.), *Proceedings of the 2015 Chinese Intelligent Systems Conference*, Lecture Notes in Electrical Engineering 359,
DOI 10.1007/978-3-662-48386-2_8

Due to the switching operation, the Buck converters are inherently variable structure systems. Therefore, the SMC is particularly suit for the Buck converters. On this basis, a lot of works have introduced the application of SMC to Buck converters, such as [4–6]. The general design issues of sliding mode controllers in Buck converters is introduced in [4], where the basic design principles for sliding mode controllers are discussed. Later, an optimal sliding mode controller for Buck converters is proposed in [5], and a simple and efficient approach for choosing sliding mode coefficient is given. Meanwhile, to achieve a better performance of the closed loop system, a terminal sliding mode (TSM) controller is developed in [6], where the TSM manifold employs a nonlinear function which ensures that the output voltage error converges to zero in a finite time. Unfortunately, there may be a singular problem in [6]. To solve this problem, the non-singular TSM controller designed in [7] eliminates the singularity problem which arises in the terminal sliding mode due to the fractional power.

Different with the aforementioned SMC methods, we will propose a SOSM controller in the paper for controlling the Buck converter. Under the new SOSM controller, the finite-time Laypunov stability of Buck converter system rather than its finite-time convergence can be eventually proved. Simulation results show the effectiveness of the proposed method.

8.2 Modeling the Buck Converter

The circuit diagram of a Buck converter is shown as in Fig. 8.1.

It can be clearly seen from Fig. 8.1 that when the switch Sw is turned on, the operation of the Buck converter can be described as

$$\begin{cases} \frac{di_L}{dt} = \frac{1}{L}(V_{in} - v_O) \\ \frac{dv_O}{dt} = \frac{1}{C}(i_L - \frac{v_O}{R}). \end{cases} \tag{8.1}$$

When the switch Sw is turned off, the operation of the Buck converter can be described as

$$\begin{cases} \frac{di_L}{dt} = -\frac{v_O}{L} \\ \frac{dv_O}{dt} = \frac{1}{C}(i_L - \frac{v_O}{R}). \end{cases} \tag{8.2}$$

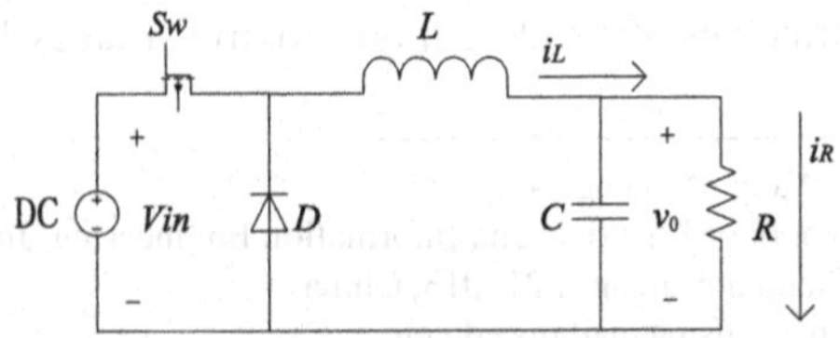

Fig. 8.1 The circuit diagram of Buck converter

Combine (8.1)–(8.2), we obtain the averaged model as follows

$$\begin{cases} \frac{di_L}{dt} = \frac{1}{L}(\mu V_{in} - v_0) \\ \frac{dv_0}{dt} = \frac{1}{C}(i_L - \frac{v_0}{R}), \end{cases} \tag{8.3}$$

where μ is the switch which takes "1" and "0" for the switch state ON and OFF, respectively. The switch μ is determined by a control scheme U, which will be designed later.

Define the voltage error s (i.e., the sliding variable) as

$$s = v_0 - V_{ref}, \tag{8.4}$$

where V_{ref} denotes the DC reference output voltage. By (8.3), the dynamics of the sliding variable s can be expressed as

$$\ddot{s} = a(t,x) + b(t,x)U, \tag{8.5}$$

with $a(t,x) = (\frac{1}{(RC)^2} - \frac{1}{LC})v_0 - \frac{i_L}{RC^2}$ and $b(t,x) = \frac{v_{in}}{LC}$.

Note that $0 \leq v_0 \leq v_{in}$, $0 \leq i_L \leq \frac{v_{in}}{R}$. It can be concluded that there exists a constant $\bar{a}$ such that

$$|a(t,x)| \leq \bar{a}. \tag{8.6}$$

The task here is to design a second-order sliding mode controller U such that the output voltage v_0 will track the reference voltage V_{ref}.

At the end of this section, we will list three lemmas which will be constantly used in proving the main results.

Lemma 1 ([11]) *If* $p_1 > 0$ *and* $0 < p_2 \leq 1$, *then* $\forall x \in R, \forall y \in R$,

$$|\lceil x \rfloor^{p_1 p_2} - \lceil y \rfloor^{p_1 p_2}| \leq 2^{1-p_2} |\lceil x \rfloor^{p_1} - \lceil y \rfloor^{p_1}|^{p_2}.$$

Lemma 2 ([8]) *Let c and d be positive constants. Given any positive number* $\gamma > 0$, *the following inequality holds:*

$$|x|^c |y|^d \leq \frac{c}{c+d}\gamma |x|^{c+d} + \frac{d}{c+d}\gamma^{-\frac{c}{d}} |y|^{c+d}, \forall x > 0, \forall y < 0. \tag{8.7}$$

Lemma 3 ([12]) *Let p be a real number with* $0 < p < 1$. *Then the following inequality holds:*

$$(|x_1| + \cdots |x_n|)^p \leq |x_1|^p + \cdots + |x_n|^p, \forall x_i > 0, i = 1, \ldots, n. \tag{8.8}$$

8.3 Second-Order Sliding Mode Controller Design

To simplify the expression, we first denote

$$\lceil x \rfloor^{\alpha} = |x|^{\alpha}\mathrm{sign}(x).$$

For system (8.5), the second-order sliding mode controller is constructed as

$$U = -\beta_2\ \mathrm{sign}(\lceil \dot{s} \rfloor^2 + \beta_1 s), \tag{8.9}$$

with proper chosen positive constants β_1 and β_2. Then, we have the following result.

Theorem 1 *Considering the second-order sliding mode dynamics (8.3), the second order sliding mode controller (8.9) provides for the finite-time establishment of second-order sliding mode* $s \equiv \dot{s} \equiv 0$*, which implies that the output voltage* v_0 *will track the reference voltage* V_{ref} *in a finite time.*

Proof Let $y_1 = s, y_2 = \dot{s}$. Then Eqs. (8.4), (8.5) and controller (8.9) can be rewritten as

$$\begin{aligned} \dot{y}_1 &= y_2 \\ \dot{y}_2 &= a(t,x) + b(t,x)U \end{aligned} \tag{8.10}$$

and

$$U = -\beta_2\ \mathrm{sign}(\lceil y_2 \rfloor^2 + \beta_1 y_1) \tag{8.11}$$

respectively. In the following, we will prove the finite-time stability of the closed-loop system (8.10) and (8.11) by using the adding a power integrator method proposed in [8, 9]. The proof will be carried out by two steps.

Step 1. We choose the following function

$$V_1(y_1) = \frac{2|y_1|^{5/2}}{5}.$$

Then taking derivative of $V_1(y_1)$ produces

$$\dot{V}_1(y_1) = -\beta_1^{1/2} y_1^2 + \lceil y_1 \rfloor^{3/2}(y_2 - y_2^*), \tag{8.12}$$

where y_2^* is a virtual control law. Design y_2^* as

$$y_2^* = -\beta_1^{1/2} \lceil y_1 \rfloor^{1/2}. \tag{8.13}$$

Step 2. Choose a function as

$$V_2(y_1, y_2) = V_1(y_1) + W(y_1, y_2),$$

with

$$W(y_1, y_2) = \int_{y_2^*}^{y_2} \lceil\lceil \kappa \rfloor^2 - \lceil y_2^* \rfloor^2 \rfloor^2 d\kappa.$$

Then we can take derivative of $V_2(y_1, y_2)$ as

$$\dot{V}_2(y_1, y_2) \leq -\beta_1^{1/2} y_1^2 + \lceil y_1 \rfloor^{3/2}(y_2 - y_2^*) + \frac{\partial W(y_1,y_2)}{\partial y_1}\dot{y}_1 + \lceil \xi \rfloor^2 \dot{y}_2, \tag{8.14}$$

with $\xi = \lceil y_2 \rfloor^2 - \lceil y_2^* \rfloor^2$. Next, we estimate each term in the right hand side of (8.14).

According to Lemma 1, we can obtain

$$\lceil y_1 \rfloor^{3/2}(y_2 - y_2^*) \leq \frac{\beta_1^{1/2}}{4} y_1^2 + \left(\frac{3}{\beta_1^{1/2}}\right)^3 \xi^2. \tag{8.15}$$

Noting that $\frac{\partial \lceil y_2^* \rfloor^2}{\partial y_1} = -\beta_1$, it can be concluded from Lemma 1 that

$$\frac{\partial W(y_1,y_2)}{\partial y_1}\dot{y}_1 \leq |y_2 - y_2^*||\xi||\frac{\partial \lceil y_2^* \rfloor^2}{\partial y_1} y_2| \leq 2^{1/2}\beta_1 |\xi|^{3/2}|y_2|. \tag{8.16}$$

It follows from Lemma 3 that $|y_2| \leq |\xi|^{1/2} + |y_2^*|$. Consequently, (8.16) can be rewritten as

$$\frac{\partial W(y_1,y_2)}{\partial y_1}\dot{y}_1 \leq 2^{1/2}\beta_1 |\xi|^{3/2}(|\xi|^{1/2} + |y_2^*|) \leq 2^{1/2}\beta_1 \xi^2 + 2^{1/2}\beta_1^{3/2}|\xi|^{3/2}|y_1|^{1/2}. \tag{8.17}$$

By using Lemma 2 again (let $\gamma = \frac{2^{1/2}}{\beta_1}, c = \frac{1}{2}, d = \frac{3}{2}$), one has

$$\frac{\partial W(y_1,y_2)}{\partial y_1}\dot{y}_1 \leq \frac{1}{2}\beta_1^{1/2} y_1^2 + \left(2^{1/2}\beta_1 + \beta_1^{11/6}\right)\xi^2.$$

It can be concluded from (8.10) and (8.14) that

$$\dot{V}_2(y_1, y_2) \leq -\frac{\beta_1^{1/2}}{4} y_1^2 + \left(\frac{27}{\beta_1^{3/2}} + 2^{1/2}\beta_1 + \beta_1^{11/6}\right)\xi^2 + \lceil \xi \rfloor^2 (a(t,x) + b(t,x)U). \tag{8.18}$$

Substituting (8.11) into (8.18), we can obtain

$$\begin{aligned}\dot{V}_2(y_1, y_2) \leq &-\frac{\beta_1^{1/2}}{4} y_1^2 + \left(\frac{27}{\beta_1^{3/2}} + 2^{1/2}\beta_1 + \beta_1^{11/6}\right)\xi^2 \\ &+ \lceil \xi \rfloor^2 (a(t,x) - b(t,x)\beta_2 \text{sign}(\lceil y_2 \rfloor^2 + \beta_1 y_1)).\end{aligned} \tag{8.19}$$

Note that $\lceil y_2 \rfloor^2 - \lceil y_2^* \rfloor^2 = \lceil y_2 \rfloor^2 + \beta_1 y_1 = \xi$ and $b(t,x) = \frac{v_{in}}{LC}$. It follows from (8.19) that

$$\dot{V}_2(y_1, y_2) \le -\frac{\beta_1^{1/2}}{4} y_1^2 + (\frac{27}{\beta_1^{3/2}} + 2^{1/2}\beta_1 + \beta_1^{11/6})\xi^2 + \lceil \xi \rfloor^2 a(t,x) - |\xi|^2 \frac{v_{in}}{LC}\beta_2. \tag{8.20}$$

Letting β_2 satisfies the following condition:

$$\frac{v_{in}}{LC}\beta_2 > \bar{a} + \frac{27}{\beta_1^{3/2}} + 2^{1/2}\beta_1 + \beta_1^{11/6} + \frac{1}{4}\beta_1^{1/2}. \tag{8.21}$$

Then, we have the following estimate

$$\dot{V}_2(y_1, y_2) \le -\frac{\beta_1^{1/2}}{4}(y_1^2 + \xi^2).$$

By the fact

$$\int_{y_2^*}^{y_2} \lceil \lceil \kappa \rfloor^2 - \lceil y_2^* \rfloor^2 \rfloor^2 d\kappa \le |y_2 - y_2^*||\xi|^2 \le 2^{\frac{1}{2}} |\xi|^{\frac{5}{2}},$$

we obtain

$$V_2(y_1, y_2) \le 2(|y_1|^{\frac{5}{2}} + |\xi|^{\frac{5}{2}}).$$

Letting $c = 2^{-\frac{14}{5}}\beta_1^{1/2}$, $\alpha = \frac{4}{5}$, by using Lemma 2, we can obtain

$$\dot{V}_2(y_1, y_2) + cV_2^{\alpha}(y_1, y_2) \le 0.$$

Note that $0 < \alpha < 1$. It follows from the finite-time Lyapunov theory give in [10] that system (8.10) can be globally stabilized by controller (8.11).

Next, we will design the Actuating Mechanism. By (8.11), we know that the second-order sliding mode controller includes two modes: $U = \beta_2$ or $U = -\beta_2$. It is obvious that the switch μ can be determined by the following relation

$$\mu = \begin{cases} 1 & when \quad U = \beta_2 \\ 0 & when \quad U = -\beta_2. \end{cases} \tag{8.22}$$

The control scheme can be illustrated as follows:

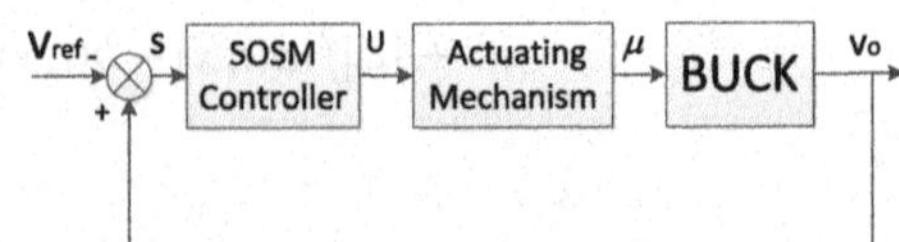

Fig. 8.2 Block diagram of control scheme for Buck Converter

8.4 Simulation Results

In order to demonstrate the performance of the SOSM approach, the closed loop of Buck converter system has been tested in Matlab/Simulink. The sampling time takes $0.001s$. The parameters of the controller (8.11) are $\beta_2 = 1, \beta_1 = 5$, and thus the controller can be expressed as (Fig. 8.2)

$$U = -\text{sign}(\lceil \dot{s} \rfloor^2 + 5s). \tag{8.23}$$

Under controller (8.23), the simulation results are shown in Figs. 8.3 and 8.4. Figure 8.3 shows the output voltage v_0. Figure 8.4 shows the switching state μ.

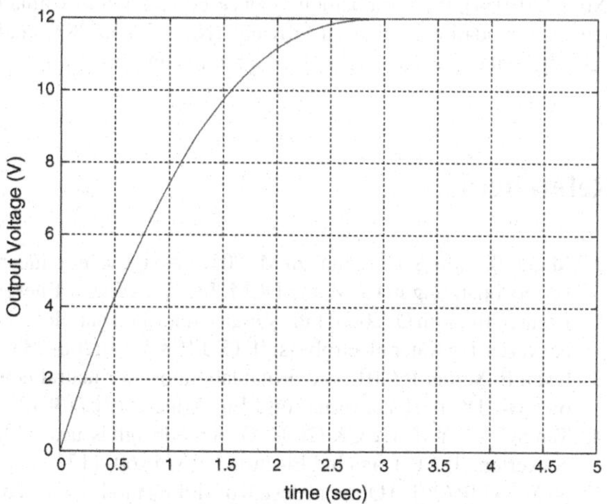

Fig. 8.3 Time history of the output voltage v_0 under SOSM controller (8.23)

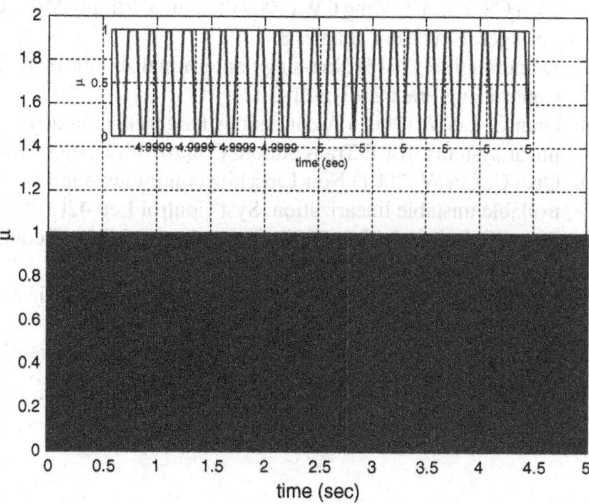

Fig. 8.4 Time history of switching state μ under SOSM controller (8.23)

8.5 Conclusion

A SMC control method has been presented for Buck converters in this paper. The advantages of the propose sliding mode controller are three folds. First of all, the design of sliding mode manifold which is frequently used in the conventional sliding mode control can be avoided. Instead, an error variable of freedom of degree two can be considered as the sliding variable directly. Secondly, the Lyapunov stability for the sliding variables has been tested, while only convergence can be guaranteed in conventional sliding mode control. Thirdly, the proposed controller has a simple structure and is easy to be implemented. Simulation results show the effectiveness of the SOSM controller.

Acknowledgments This work was supported by the National Natural Science Foundation of China (No. 61203014), the Postdoctoral Science Foundation of China (No. 2015M571687), the Natural Science Foundation of Jiangsu Province (No. BK2012283), and the Priority Academic Program Development of Jiangsu Higher Education Institutions.

References

1. Mahdi S, Jafar S, Gholamreza M (2013) Adaptive nonlinear control of the DC-DC buck converters operating in CCM and DCM. Int Trans Electr Energy Syst 23(8):1536–1547
2. Ergin S, Ibrahim O (2013) Fuzzy logic controlled parallel connected synchronous buck DC-DC converter for water electrolysis. IETE J Res 59(3):280–288
3. Okan B, Serdar P (2012) A virtual laboratory for neural network controlled DC motors based on a DC-DC buck converter. Int J Eng Educ 28(2):713–723
4. Tan SC, Lai YM, Tse CK (2008) General design issues of sliding-mode controllers in DC-DC converters. IEEE Trans Ind Electron 55(3):1160–1174
5. Ni Y, XU JP (2013) Optimal design of sliding mode control buck converter with bounded input. ACTA Electronica Sinica 41(3):555–560
6. Chiu CS, Lee YT, Yang CW (2009) Terminal Sliding Mode Control of DC-DC Buck Converter. Springer, New York chapter 8
7. Komurcugil H (2013) Non-singular terminal sliding-mode control of DC-DC buck converters. Control Eng Pract 21(3):321–332
8. Qian C, Lin W (2001) A continuous feedback approach to global strong stabilization of nonlinear systems. IEEE Trans Autom Control 46(7):1061–1079
9. Qian C, Lin W (2001) Non-Lipschitz continious stabilizers for nonlinear systems with uncontrollable unstable linearization. Syst Control Lett 42(3):185–200
10. Bhat SP, Bernstein DS (2000) Finite-time stability of continuous autonomous system. SIAM J Control Opimization 38(3):751–766
11. Ding SH, Li SH, Zheng WX (2012) Nonsmooth stabilization of a class of nonlinear cascaded systems. Automatica 48(10):2597–2606
12. Hardy GH, Littlewood JE, Polya G (1952) Inequalities. Cambridge University Press, Cambridge

Chapter 9
Open-World Planning Algorithm Based on Logic

Jie Gao, Ya-song Liu and Rui Bian

Abstract Existence of certain objects or fluent is often unknown before planning in many domains. Plan synthesis in such open worlds is challenging since we have to take various scenarios into account before searching plans. One way to do this is to employ sensors to observe unknown objects or fluent, assuming the sensors are capable of correctly capturing all information needed for planning. We aims at solving automated planning problem in open world, and call goal state with variables as query-goal. Instead of using sensors, we proposed a novel algorithm PQG (Planner with query-goal) to solve automated planning problem with query-goal, by encoding the planning problem into a planning logic problem, and then applying planning inference method to solve it. Finally, inference result is transformed into a planning solution. We empirically exhibit that our approach is effective in several planning domains.

Keywords Automated planning · Open world · Logic

9.1 Introduction

Automated Planning [1] is a popular branch of artificial intelligence in recent years. By analyzing surrounding environment according to its goal, planning agent can reason from available actions and limited resources, then comprehensively obtain a planning solution, namely action sequences. Currently, automated planning research focuses on the assumption of the close world, namely with complete initial state and goal state. However, the planning problems with incomplete initial state or

J. Gao (✉) · Y. Liu
Zhuhai College, Jilin University, Zhuhai 519000, Guangdong, China
e-mail: jiegao26@163.com

R. Bian
School of Public Administration, University of Business Studies, Guangzhou 510320, Guang Dong, China

© Springer-Verlag Berlin Heidelberg 2016

Y. Jia et al. (eds.), *Proceedings of the 2015 Chinese Intelligent Systems Conference*, Lecture Notes in Electrical Engineering 359,
DOI 10.1007/978-3-662-48386-2_9

goal state are more common, which are summarized as planning problems in open world [2]. Plan synthesis in an open world is challenging since we have to consider a wide variety of unknown scenarios before planning. One way to handle unknown scenarios is to equip planners with sensing devices and search plans by replanning and monitoring with the sensing devices [3, 4]. This online planning method assumes that sensors can sense everything needed for planning. It is often, however, difficult to determine how many and what types of sensors are needed before the planning tasks are provided.

In this paper, we aim at solving planning problem in an open world, without using sensors. Take blocks world domain as an example. If a goal state includes *on (A,B)*, we call such a goal state with concrete objects as a "certain goal". If a goal state includes *on(?x,B)*, *?x* representing a variable with block type, which means some unknown block is on the block *B*, we call such a goal state with variables as a "query goal". In this paper, for solving planning problems with query-goals, we propose a novel algorithm called PQG, namely Planning with query-goals. Since planning problems are logically explainable, we encode planning problem into logic formulas, and then apply logic programming technology (Prolog) to solve it. Finally we transform solution results into planning solution.

The paper is organized as follows: in Sect. 9.2, we introduce related work, including automated planning and logic programming technology (Prolog); in Sect. 9.3, we define planning problem with query-goals formally, and introduce main steps of algorithm PQG in detail; in Sect. 9.4, we evaluate PQG in three planning domains with query goals and analyze experimental results; finally, we summarize the paper and discuss research direction in future.

9.2 Related Work

Our work is firstly related to open-world planning [2]. Classical planning presupposes that complete and correct information about the world is available at any point of planning (by having a completely specified initial situation, and deterministic actions). However, in a more realistic setting, the knowledge about the initial state may be incomplete, the effects of actions may be not deterministic, or there may be other agents acting in the world. These are some sources of uncertainty in planning. Open world planning is the planning problem without the assumption of complete knowledge of initial state of the world. It is challenging to adapt those close-world planners to handle open-world problems, since it is ultimately flawed to assume that the world is close, or closing the world deliberately by acquiring all the missing knowledge before planning. A knowledge representation scheme called PSIPLAN [5] was proposed to efficiently handle domains in an open world. A number of systems [3, 6] have functioned by performing execution monitoring and subsequent plan repair or replanning upon the discovery of an inconsistent execution state with sensing capability. A planning model [6] was proposed to use arbitrary plan structures to handle open world planning, based on structural constraint satisfaction. Zhuo [7] presented an action-model acquisition system called CAMA to acquire background knowledge, including information about preconditions and effects of actions, from the crowd. Talamadupula et al. [8] showed that the teaming problem presents the need to handle open-world problems in the USAR scenario. They investigated the notion of conditional goals which foreground the trade-offs between goal reward and sensing cost. Assumption based planning [9], which computed a set of assumptions about aspects of the world to support plan generation, was also proposed to deal with uncertain scenarios of initial states based on preset assumptions. These systems rely on either sensing capability or preset assumptions.

9.3 Problem Formulation

A normal planning problem is defined by a triple <s_0, g, *O*>, where s_0 represents an initial state and g represents a goal state, and both of them are composed of a set of propositions. *O* denotes a set of action models, each of which is composed of a quadruple <*a, PRE, ADD, DEL*>, where *a* indicates an action schema, composed of an action name with zero or more parameters, *PRE* indicates a list of precondition of *a*, specifying the conditions that should be satisfied before applying action *a*, *ADD* indicates a list of adding effects, specifying the new effects created after applying action *a*, *DEL* indicates a list of deleting effects, specifying the set of effects deleted after applying action *a*. Note that in this paper we consider STRIPS action models. A solution to a planning problem is a plan, i.e., an action sequence that transits the initial state s_0 to the goal g.

We formulate our open planning problem as a quadruple $\langle \tilde{s}_0, \tilde{g}, \mathrm{O}, U \rangle$, where $\tilde{s}_0$ is an open initial state which is composed of a set of open propositions. A proposition is called open if there exist some variables in the parameter list of the proposition. For example, on(A,?x) is an open proposition since ?x is a variable in the parameter list of proposition on. An open initial state can be incomplete, i.e., some propositions are missing. $\tilde{g}$ is a an open goal which is likewise composed of a set of open propositions. The set of variables in both $\tilde{s}_0$ and $\tilde{g}$ is denoted by V. O is a set of possible objects that can be selected and assigned to variables in V. We assume O can be easily collected based on historical applications. A is a set of STRIPS action models as defined above. A solution to an open planning problem is an action sequence, as well as an assignment of variables in $\tilde{s}_0$ and $\tilde{g}$. Note that an assignment θ is defined by $\theta = \{(v, o) | v \in V \wedge o \in \mathrm{O}\}$.

9.4 Planning Algorithm with Query-Goals

Since logic language is applied to describe states in STRIPS, it has obvious advantage to build up STRIPS planning system by Prolog. Variables can be used as input and output in Prolog by unification technology. In the progress of reasoning, when information is incomplete, unification technology can be used to acquire the condition satisfying results. In addition, Prolog is based on first-order logic, therefore it inherits the first-order logic expression ability and abstract ability. By considering the above features, we can construct intelligent planning system by logic programming technology, and solve planning problem with query goals. We call the algorithm of planning problem with query goals based on logic programming technology as PQG. The framework of algorithm PQG is shown in Table 9.1. We will describe the steps of algorithm PQG in the following subsections.

9.4.1 *Step 1: Encode an Open Planning Problem into a Planning Logic Problem*

We encode an open planning problem into a planning logic problem (called PLP), which includes the set of facts and rules. It should be pointed out that there exists obvious difference between a planning logic problem and a classical logic problem. In classical logic problem, facts should be fully instantiated, but in planning logic problem, facts can contain some variables which are used to represent the states with incomplete information. Firstly, we encode the set of initial states and the set of goal states into two sets of facts in PLP respectively. Secondly, we encode each action model as a rule in PLP, denoted as

$$DEL(T_1), \ldots, DEL(T_m), ADD(K_1), \ldots, ADD(K_n) \colon - L_1, \ldots L_s,$$

where $L_1, \ldots, L_s$ correspond to the preconditions of action model; $DEL(T_i)$, $1 \leq i \leq m$ correspond to delete list of action model; $ADD(K_j)$, $1 \leq j \leq n$ correspond to add list of action model. For example, the action model of unstack(X)

Table 9.1 Framework of PQG algorithm

Input: The set of initial states and goal states, action models of planning problem
Output: Solution of planning problem
Steps:
1. Encoding planning problem into logic formulas
2. Reasoning logic formulas by logic programming technology
3. Converting results into planning solution

Table 9.2 The algorithm of planing inference

Input: The set of initial states, called Stateinitial; the set of goal states, called Stategoal; and the set of rules, called Actions
Output: A sequence of rules applied
Steps:
1. Let states=Stateinitial, newstates:={};
2. For each rule in Actions, denoted by
$DEL(T_1), \ldots, DEL(T_m), ADD(K_1), \ldots, ADD(K_n){:} -L_1, \ldots L_s$
we apply planning inference method (PIM), to acquire renewed set of facts, denoted as newstates=*PIM(R,States)*
If newstates is not empty, then states=states ∪ newstates, and keep record the applied rules as sequence
3. If states ⊃ Stategoal, then output the sequence of applied rules
Otherwise, return step 2

can be encoded as ADD(onblock(Y,noblock)),ADD(inhand(X)),DEL(onblock(Y, X)),DEL(onblock(X,noblock)),DEL(inhand(noblock))

$$: -\ (\text{onblock}(Y, X)), \text{onblock}(X, \text{noblock}), \text{inhand}(\text{noblock})$$

9.4.2 Step 2: Reasoning Logic Formulas by Logic Programming Technology

To solve planning logic problem (PLP), we apply rules to infer from the set of initial states to the set of goal states. The solution of PLP, is the sequence of rules applied in the process. Since there exists significant difference between planning logic problem and classical logic problems, inference method in classical logics can not be used in PLP, therefore a distinctive inference method is necessary for PLP. We define planning inference method, called PIM, as follows.

Definition Let $DEL(T_1), \ldots, DEL(T_m), ADD(K_1), \ldots, ADD(K_n){:} -L_1, \ldots L_s$ be a rule. Let the set of facts be $\boldsymbol{\phi} = \{F_1, \ldots, F_s\}$. If there exists a replacement $\boldsymbol{\theta}$,such that $L_t\boldsymbol{\theta} = F_t, 1 \le t \le s$, then facts $T_i\boldsymbol{\theta}$, $1 \le i \le m$ will be deleted in $\boldsymbol{\phi}$, and facts $K_j\boldsymbol{\theta}$, $1 \le j \le n$ will be added in $\boldsymbol{\phi}$. Let the renewed set of facts be *PIM*(*R*, $\boldsymbol{\phi}$).

By applying planning inference method PIM, we propose the algorithm of planing inference to acquire solution of PLP, as shown in Table 9.2.

9.4.3 Converting Results into Planning Solution

Since logic language is applied to describe states, it has obvious advantage to realize the algorithm of planning inference by Prolog. Variables can be used as input and output in Prolog by unification technology. In addition, Prolog is based on first-order logic, therefore it inherits the first-order logic expression ability and abstract ability. We use SWI-Prolog 6.5.2 [10] to realize the algorithm of planning inference and map the required sequence of rules into a sequence of actions, namely a planning solution.

9.5 Test and Analysis

We test effects of algorithm PQG in three planning domains including blocks world, logistics-strips and zeno travel. We test ability of algorithm PQG to obtain planning solution, comparing with other planners in the domains with certain states and query goals respectively. Firstly, we compare FF planner [11] (version 2.3) with PQG in the planning domains only with certain rules. FF is an excellent open-source planner with high efficiency, and makes great success in international intelligent planning competitions. We will list the most complicated twenty test problems as experimental data. Testing platform is Windows 7.0, RAM (4.0 G), CPU (2.30 GHZ). We set the maximum of planning time as 30 min, according to international planning competitions.

Firstly, we compare quality of PQG and FF in two aspects, namely planning time, length of planning solutions. Figure 9.1 shows the difference of planning time in three domains. Since problem 1-problem 8 are more simple with fewer predicates and actions, planning times of PQG and FF are very close in blocksworld domain. But problem 9-problem 20 are more complicated, FF solves planning problems faster than PQG. In logistics-strips and zenotravel domains, FF also solves planning problems faster than PQG. It is not difficult to find that FF is more efficient than PQG in solving planning problem, since FF is a forward heuristic planner in states space, and defines favorable actions as effective pruning strategy. Therefore, we will add heuristic strategy into present PQG algorithm, which is our future work.

Secondly, existing planner can not solve planning problem with query goals. We still choose the most complicated twenty test problems in blocks world, logistics-strips and zeno travel domains. For each test problem, we randomly generate three planning problems with variables in initial states and goal states to form a problem set to test PQG. We show average planning time and length of planning solution in Fig. 9.2.

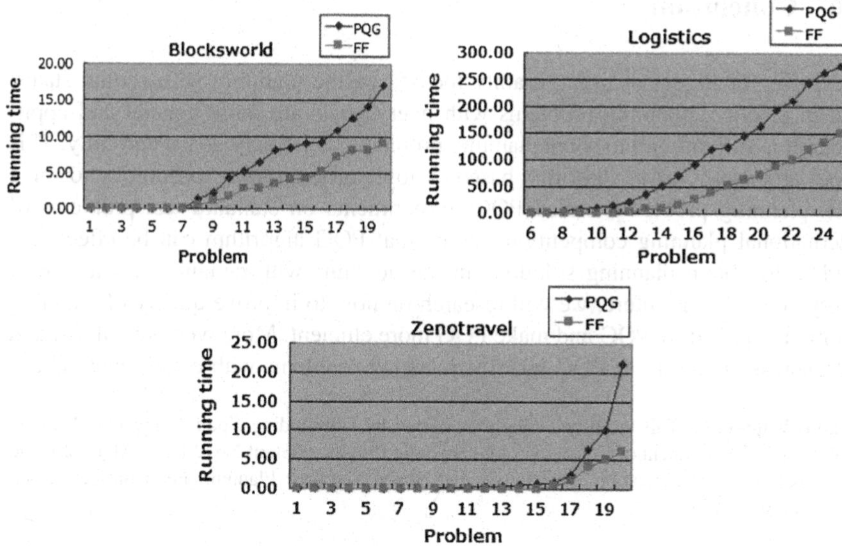

Fig. 9.1 Comparing running time of PQG and FF in three domains

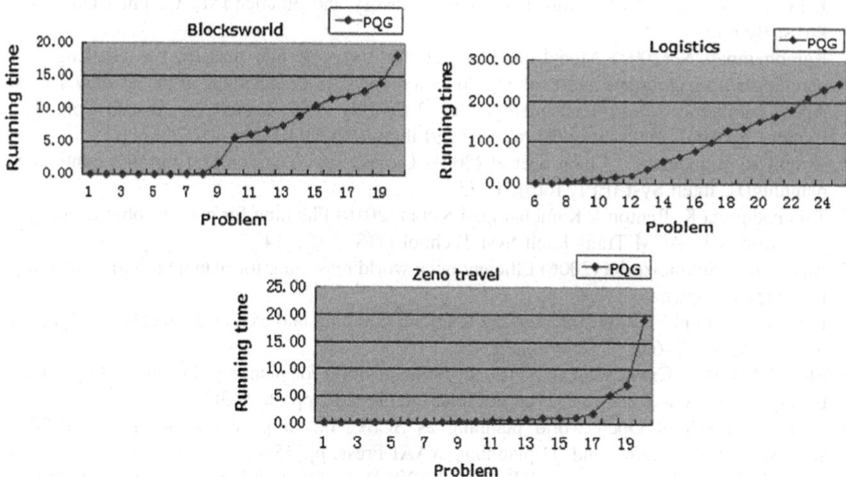

Fig. 9.2 Comparing running time of PQG in three domains

9.6 Conclusion

Presently, intelligent planning mainly focuses on the planning with certain states. But in practice, planning problems with query goals are more general and applicable. It is still difficult to solve planning problems with query goals, currently. This paper proposes a novel algorithm based on logic programming technology to solve such planning problems, called PQG. Experiments on standard test problems of international planning competitions, show that PQG algorithm can be effectively applied to obtain planning solutions in the domains with certain states and with query goals. In the future, we will research on how to improve quality of planning solution obtained by PQG and make PQG more efficient. Moreover, we will expand the expression ability of PQG to solve planning problems with conditional effects.

Acknowledgments This work was supported in part by Outstanding Youth Innovation Training Project in the Department of Education of Guangdong Province (Grant No. 2012LYM_0065), and by the Ministry of Education, Humanities and Social Sciences Planning Fund project (Grant No. 12YJA630157).

References

1. Jiang Y, Yang Q (2006) Automated planning: theory and practice [M]. Qinghua University Press, Beijing
2. Kambhampati S (2007) Model-lite planning for the web age masses: the challenges of planning with incomplete and evolving domain models. In Proceedings of the Twenty-second AAAI conference on artificial intelligence, 22–26 July 2007, Vancouver, British Columbia, Canada, pp 1601–1605. http://www.aaai.org/Library/AAAI/2007/aaai07-254.php
3. Knight R, Rabideau G, Chien S et al (2001) Casper: space exploration through continuous planning[J]. Intell Syst IEEE 16(5):70–75
4. Talamadupula K, Benton J, Kambhampati S et al (2010) Planning for human-robot teaming in open worlds[J]. ACM Trans Intell Syst Technol (TIST) 1(2):14
5. Babarian T, Schmolze JG (2006) Efficient open world reasoning for planning. logical methods in computer. Science 2:1–39
6. Lemai S, Ingrand F (2004) Interleaving temporal planning and execution in robotics domains [C]. AAAI 4:617–622
7. Zhuo HH (2015) Crowdsourced action model acquisition for planning. In: Proceedings of the twenty-eighth AAAI conference on artificial intelligence, pp 3004–3009
8. Nareyek A (2000) Open world planning as SCSP. In: In papers from the AAAI-2000 workshop on constraints and AI planning. AAAI Press, pp 35–46
9. Davis-Mendelow S, Baier JA, McIlraith SA (2013) Assumption-based planning: generating plans and explanations under incomplete knowledge [J]. AAAI 1101–1109
10. Wielemaker J, Schrijvers T, Triska M, Lager T (2012) SWI-prolog [J]. Theory Pract Log Program 12:67–96
11. Hoffmann J, Nebel B (2001) The FF planning system: fast plan generation through heuristic search. J Artif Intell Res 14:253–302

Chapter 10
Apple Nighttime Images Enhancement Algorithm for Harvesting Robot

Xingqin Lv, Bo Xu, Wei Ji, Gang Tong and Dean Zhao

Abstract In order to enhance the applicability and efficiency of harvesting robot to ensure that people can timely pick ripe fruit, the robot need to have an ability of continuous recognition and harvest at night. For some disadvantages of night vision images, Retinex algorithm for image enhancement based on bilateral filter is presented. Bilateral filter which has a function of edge preservation is adopted to improve the smooth, evaluate the illumination and remove unfavorable illumination effects from the original image. Then the reflectance of the image from above that contains just the characteristics of the object itself can be obtained. Finally, apple nighttime image enhancement is implemented. The experimental results show that the above method can more accurately evaluate the illumination of high-contrast edge regions, to suppress noise, enhance image contrast and improve overall visual effects of the image.

Keywords Nighttime image · Image enhancement · Retinex · Bilateral filter

10.1 Introduction

In china, apple harvesting largely rely on manual operations which contain high labor intensity, long-time consuming, low efficiency, and a certain risk to complete [1]. As the market demands increase, people grow wide areas of apples. For promptly picking ripe apples, the robot not only needs to work during the daytime, but also has to work at night. The primary task of night job is still accurately recognition for the targets. For the lack of light, the images acquired at night are

X. Lv (✉) · B. Xu · W. Ji · G. Tong · D. Zhao
School of Electrical and Information Engineering, Jiangsu University,
Zhenjiang 212013, China
e-mail: lvxingqin02204921@126.com

B. Xu
e-mail: xubo@ujs.edu.cn

© Springer-Verlag Berlin Heidelberg 2016

Y. Jia et al. (eds.), *Proceedings of the 2015 Chinese Intelligent Systems Conference*, Lecture Notes in Electrical Engineering 359,
DOI 10.1007/978-3-662-48386-2_10

fuzzy, dim, low-contrast and unclear [2–4]. So people need to propose a different approach from the daytime. Currently, few researchers have conducted studies on fruit recognition at night. A. Payne collected mango images at the 'stone hardening' stage under artificial lighting at night, and combined with the color feature of YCbCr and the texture characteristics to calculate the number of mangos and estimated mango crop yield [5]. D. Font used artificial lighting to gather RGB images of ripe grapes in the vineyard at night and calculated the number of grapes by detecting the peak of spherical reflection on their surface to estimate the production [6]. They both do not precisely recognize nighttime fruit to realize the robot harvesting.

To accurately recognize the night vision images, it takes to improve the brightness, increase dynamic range, suppress noise and express clearly the details of the darker by image enhancement, especially highlight edge of the fruit and eliminate mistiness. Histogram equalization etc. can only enhance the partial characteristics of the degraded image, such as suppressing noise and improving brightness or contrast, but it's difficult to obtain satisfactory results on keeping the color, clearing the details and so on. So based on the perceptual characteristics of the human eye on an object, this paper adopts Retinex algorithm to enhance the night vision images, which realizes color constancy, detail enhancement and other fine features. The purpose of Retinex algorithm is to remove the influence of the illumination from the image, and get the actual reflection component. It evaluates the illumination of the image to enhance the dark information while maintaining its brightness. Retinex algorithm has a unique enhancement effect in dealing with low-light images, shade images and foggy images, even has broad application prospects in the aeronautics and astronautics, remote sensing images, aerial images and video monitoring etc. [7]. However, using traditional Retinex algorithm may enlarge noise and bring about a certain color distortion.

Based on the above analysis, this paper proposes Retinex algorithm for image enhancement based on bilateral filter to recognize the night vision images. Using the bilateral filter to evaluate the illumination, it can be seen that this method more accurately evaluates the illumination of high-contrast edge regions, highlights the details and avoids color distortion. The proposed algorithm can not only make the enhanced images more natural and satisfy human observation, but also realize accurately recognition of the night vision images, greatly increase the accuracy of picking apples at night and improve the picking efficiency and picking quality. Meanwhile, the stability of the proposed algorithm ensures that the robot continuously work at night. In addition, related experimental operations provide a new research method and direction for the recognition of the night vision images.

10.2 Analysis of Night Vision Image

The change of illumination determines the nighttime images having characteristics different from daytime images. Affected by illumination, it's so difficult to capture images at night that it needs to use artificial lighting. This experiment uses a white fluorescent lighting whose power is 32 W and color temperature is over 5000 K. Apple nighttime images are collected by external camera whose type is M216 and pixel is 5 million from the apple orchard in DaShahe of Fengxian in Xuzhou. To match harvesting robot camera pixels, the resolution of these images is set to 640 × 480. And color information can be better hold in a format of jpg. Experimental computer configuration information is as follows: processor with Intel Core i3 CPU with 2.66 GHz, RAM with 2 GB and hard disk with 500 GB. This system developed based on the platform of MATLAB R2010b. Two images respectively taken during the day and at night assisted with a white fluorescent lighting are shown in Fig. 10.1. Their histograms of the images are shown in Fig. 10.2.

Comparing Fig. 10.1a with Fig. 10.1b, it can be seen that there are a lot of dark regions in the background of the nighttime image and its resolution is not high enough. Despite the target fruits are bright in the nighttime image, there are some highlight reflective areas, shadows and fuzzy edges. Meanwhile, in Fig. 10.2b, the histogram of the nighttime image is concentrated in the left part of coordinate system, indicating dark areas occupy a major part of this image, and the contrast is low. Thus, different from the daytime images, it needs to use image enhancement technology to make apple fruits in the darker stand out, clear their edges, and detail clearly. So it will lay foundation for the subsequent image segmentation and the feature extraction.

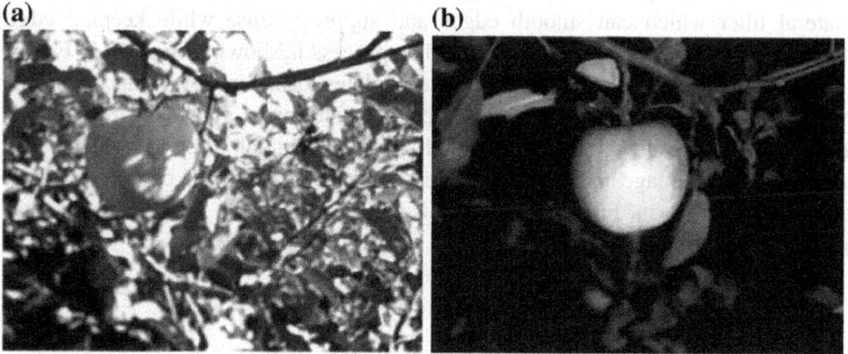

Fig. 10.1 Daytime image and nighttime image. **a** Daytime image. **b** Nighttime image

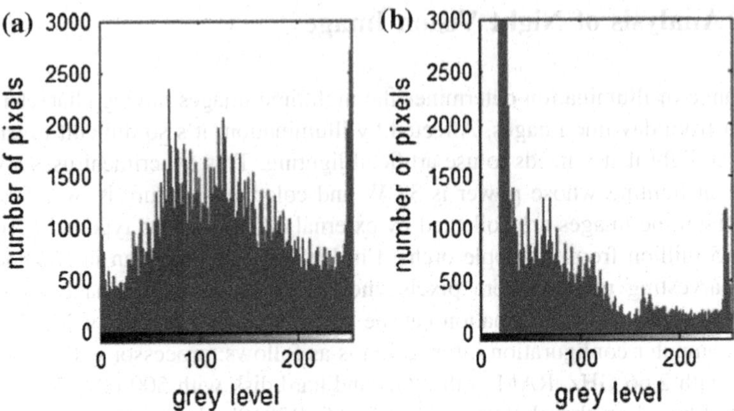

Fig. 10.2 Histograms of daytime and nighttime images. **a** Histogram of daytime image. **b** Histogram of nighttime image

10.3 Retinex Algorithm for Apple Nighttime Image Enhancement Based on Bilateral Filter

Traditional Retinex algorithm can enhance the dark information while maintaining its brightness, and improving its contrast. The obtained images have the advantages of sharpening, large compression of dynamic range and color constancy. But there are some shortcomings that the contrast of light and darkness parts is easy to produce serious halo. The bilateral filter can keep edge information and effectively smooth any slightest change in the images.

Considering the complex features of nighttime images, Retinex algorithm based on bilateral filter is presented for image enhancement in this paper. It mainly uses bilateral filter which can smooth edges and suppress noise while keeping edge information to evaluate the illumination. The process is shown in Fig. 10.3. Firstly it takes the logarithm of a image and then completes local contrast enhancement for the logarithm-taken image to effectively compensate losses generated from evaluating the brightness. On the other hand, it uses bilateral filter to evaluate the brightness of the image and suppress noise, then gets the reflection component from logarithmic transform. Finally, it synthesizes these two partial images for the final enhanced image.

10.3.1 Retinex Theory

Retinex theory was proposed by Land et al. [8]. It has many advantages including sharpening, color constancy, large compression of dynamic range, color fidelity and high computational efficiency. This theory mainly includes two aspects: the color of

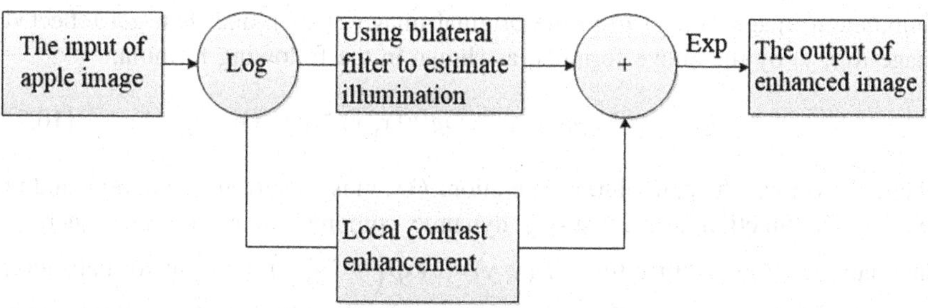

Fig. 10.3 The flow chart of image enhancement method

an object is determined by the reflectivity of long wave light, medium wave light and short wave light rather than the absolute value of reflected light intensity; the other one is that the color is not affected by non-uniformity of light source. According to Retinex theory, a given image $S(x, y)$ can be decomposed into two different images: a reflected image $R(x, y)$ and a brightness $L(x, y)$ of the image (it was also called the incident image), the principle is shown in Fig. 10.4:

As shown above, the image can be viewed as the product of an illumination image and a reflection image, as shown in the following formula

$$S(x, y) = R(x, y) * L(x, y) \tag{10.1}$$

There $S(x, y)$ is the original image, $R(x, y)$ is the reflection image, as is to say, the essential attribute of the image, $L(x, y)$ is the illumination image, which directly determines the pixels' dynamic range. In the traditional Retinex calculations, it firstly uses logarithmic transformation to convert the product to summation, as shown in the following equation:

$$log(S(x, y)) = log(R(x, y)) + log(L(x, y)) \tag{10.2}$$

However, it's not very realistic to get the reflected image in the actual process. Under the normal circumstances, Retinex processing is that, firstly it will evaluate

Fig. 10.4 Retinex algorithm schematics

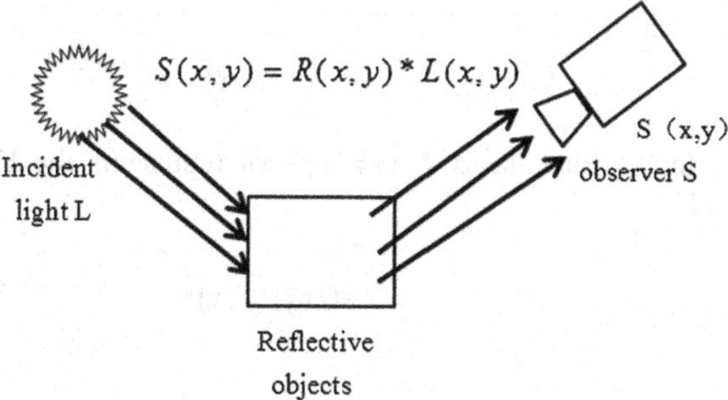

illumination image $L(x, y)$ from the original image $S(x, y)$, and then get reflected image $R(x, y)$ by the above formula, as shown in the following formula:

$$R(x, y) = logS(x, y) - log[F(x, y) * S(x, y)] \quad (10.3)$$

where '*' denotes the convolution operation; $F(x, y)$ is an surround function, and in general, the function use an strong dynamic compression capabilities, such as Gaussian function, at this time, $F(x, y) = \lambda\, exp\left(-\frac{x^2+y^2}{n^2}\right)$, λ is a normalization constant, $\int\int F(x, y)\, dxdy = 1$; n is a scale constant which controls the field range.

10.3.2 Bilateral Filter

Bilateral filter [9, 10] is a nonlinear filtering method which is combined space of proximity with similarity of pixel values during an image processing, at the same time considering the airspace information and gray similarity to reach the purpose of edge preserving and noise removing.

Bilateral filter can be expressed as

$$h(x) = k^{-1} \int_{-\infty}^{\infty} \int_{-\infty}^{\infty} f(\xi)c(\xi, x)s(f(\xi), f(x))d\xi \quad (10.4)$$

The normalized parameters can be listed as

$$k(x) = \int_{-\infty}^{\infty} \int_{-\infty}^{\infty} c(\xi, x)s(f(\xi), f(x))d\xi \quad (10.5)$$

In the formula, $c(\xi, x)$ shows the distance similarity between neighboring point ξ and central point x, $s(f(\xi), f(x))$ indicates brightness similarity between neighboring point ξ and central point x, $f(x)$ represents the brightness value of the point x in the input image, $h(x)$ represents the brightness value of the point x in the output image, $k(x)$ is a normalization constant and its value is independent of the image content, so its value is constant in the same geometric position.

The bilateral filter is extended to the Gauss kernel, as follows

$$c(\xi, x) = e^{-\frac{1}{2}\left(\frac{d(\xi, x)}{\sigma}\right)^2} \quad (10.6)$$

In the formula, $d(\xi, x) = \|\xi - x\|$ represents the Euclidean distance between ξ and x.

$$s(f(\xi), f(x)) = e^{-\frac{1}{2}\left(\frac{\delta(f(\xi), f(x))}{\sigma}\right)^2} \quad (10.7)$$

In the formula, $\delta(\phi, f) = \|\phi - f\|$ represents the difference between the brightness value ϕ and the brightness value f.

10.4 Experimental Results and Analysis

Using the above algorithm to test an apple nighttime image, and the effect will be compared with other methods. The results are showed in Fig. 10.5. This figure shows original image, Histogram equalization, traditional Retinex algorithm and proposed algorithm from left to right. It can be seen that the image's brightness is greatly improved while the reflective parts of apples are magnified, and there is some larger image distortion. In contrast, after using proposed algorithm to improve the image, some details in darker areas become visible, giving more prominence to the edge. Then the visual effects are significantly improved, while the bright areas are well suppressed. Figure 10.6 presents their corresponding histograms. Compared to the traditional enhancement algorithms, the gray level of the proposed algorithm histogram has also been stretched to some extent, and the distribution of pixels is more uniform in the entire dynamic range. It shows that after using this method, the image has clearer details, richer grayscale and larger dynamic range, and remarkable contrast. Moreover, it highlights the target apple in the image, so the details of their outlines become more clearly.

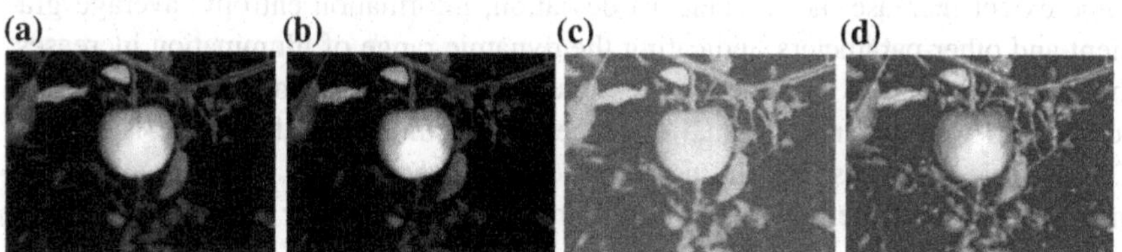

Fig. 10.5 Contrast maps of results. **a** Original image. **b** Histogram equalization. **c** Traditional Retinex. **d** Proposed algorithm

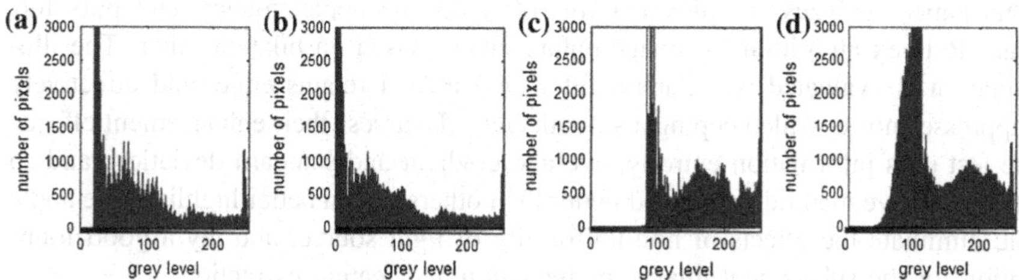

Fig. 10.6 The corresponding histograms. **a** Original image. **b** Histogram equalization. **c** Traditional Retinex. **d** Proposed algorithm

Table 10.1 The evaluation parameters of the visual quality of the first image

	Mean	Standard deviation	Information entropy	Average gradient
Original image	46.66471	41.07926	7.24861	1.97385
Contrast conversion	34.07745	63.84031	5.91147	2.37039
Traditional Retinex algorithm	80.50617	49.96841	8.78628	2.68896
Proposed algorithm	122.87591	53.62097	10.23761	4.19653

In the image processing, evaluation parameters of the enhanced image quality include average gradient, mean, entropy, standard deviation [11]. There the average gradient that is image clarity which reflects degree of improvement in image quality and change rate of any tiny details contrast. The larger the average gradient being, the more levels an image has, and the image becomes clearer. Image entropy represents how much information contained in a digital image, and by comparing the changes in the amount of information, the quality of the image enhancement is determined. The change of mean reflects a change in the dynamic image and a change in standard deviation reflects the level of contrast.

To further evaluate the advantages of the enhancement algorithm researched in this article, mean, standard deviation, information entropy and average gradient are used to evaluate the image quality after using three different image enhancement algorithms. The main parameters are shown in Table 10.1.

As can be seen from the above table, the above-mentioned three methods to some extent increase mean, standard deviation, information entropy, average gradient and other parameters, indicating the dynamic range of illumination increases, the contrast enhances and the clarity increases. And this method has the obvious advantages over other traditional enhancement methods. So the image is clearer. The average increases, but the noise is not enhanced. It also improves the information entropy of the image so that the image contains richer color information.

10.5 Conclusions

This paper analyzes complex features of nighttime apple images, and puts forward Retinex algorithm for image enhancement based on bilateral filter. The illumination is evaluated by bilateral filter, and then it retains edge, and effectively suppresses noise while keeping a smooth edge. To assess their enhancement effects, the test uses information entropy, average gradient and standard deviation, and so on. The above method performed better than others. It can better highlight the apple and eliminate the effects of non-uniformity of light source, and lay a good foundation for the subsequent image segmentation and feature extraction.

Acknowledgements This work was supported by a project funded by the Priority Academic Program Development of JiangSu Higher Education Institutions and Research Fund for the Doctoral Program of Higher Education of China under Grant 20133227110024.

References

1. Fengwu Z, Fenghua Y et al (2013) Research status and development trend of agricultural robot [J]. Agric Eng 2(6):10–13
2. Rao Y, Hou L et al (2014) Illumination-based nighttime video contrast enhancement using genetical gorithm [J]. Multimed Tools Appl 70(3):2235–2254
3. Martínez Cañada P, Morillas C et al (2013) Embedded system for contrast enhancement in low-vision [J]. J Syst Archit 59(1):30–38
4. Zhigang Z et al (2014) Global brightness and local contrast adaptive enhancement for low illumination color image [J]. J Light Electron Opt 125(6):1795–1799
5. Payne A et al (2014) Estimating mango crop yield using image analysis using fruit at 'stone hardening' stage and night time imaging [J]. Comput Electron Agric 100:160–167
6. Font D et al (2014) Counting red grapes in vineyards by detecting specular spherical reflection peaks in RGB images obtained at night with artificial illumination [J]. Comput Electron Agric 108:105–111
7. Xujia Qing et al (2013) Structured light image enhancement algorithm based on Retinex in HSV color space [J]. Comput Aided Des Comput Graph 25(4):488–493
8. Chuangbai X et al (2013) Rapid Retinex algorithm for night color image enhancement based on guided filtering [J]. J Beijing Univ Technol 39(12):1869–1873
9. Qian Y, Ying L et al (2014) Research on fog-degraded image restoration based on bilateral filter of RGB channel [J]. Comput Eng Appl 50(6):157–160
10. Weiwei H, Ronggui W, Shuai F et al (2010) Retinex algorithm for image enhancement based on Bilateral filtering [J]. J Eng Graph 2:104–109
11. Yuwei Zu, Qiang C et al (2014) Remote sensing image enhancement based on dark channel prior and bilateral filtering [J]. J Image Graph 19(2):313–321

Chapter 11
Research on Grasping Planning for Apple Picking Robot's End-Effector

Feiyu Liu, Wei Ji, Wei Tang, Bo Xu and Dean Zhao

Abstract Aiming at the lack of all-purpose and effective planning study for apple picking robot's end-effector during the grasping process, which has caused great inconvenient in the accuracy of fruit picking process and design of end-effector. This paper studies the contact process of three-finger end-effector with apples. Taking contact of apples with fingers as point contact of friction between hard objects, and based on the fact that the contact force is decomposed into the orthogonal operating force component and internal force component, the regulation of internal force on contact stability is discussed. Stability would be attributed to the existence of the internal force of concurrent polygon and the position of the internal force concurrent node within the concurrent polygon. Regarding the circle center of concurrent polygon of the maximum inscribed circle as the intersection of three internal force action lines, we get the size of each internal force, and calculate the internal force meeting the friction cone constraints to avoid complex operation such as matrix operation. Eventually, a numerical example shows the feasibility of the method.

Keywords Picking robot · End-effector · Grasping stability · Internal force

11.1 Introduction

In apple production operation, the harvesting task accounts for about 40 % of the whole work labor [1]. Apple harvesting robot instead of labor could support the fruit industry in the face of a decreasing labor force and global market competition.

F. Liu · W. Ji (✉) · W. Tang · B. Xu · D. Zhao
School of Electrical and Information Engineering, Jiangsu University, Zhenjiang 212013, China
e-mail: jwhxb@163.com

F. Liu
e-mail: 524781646@qq.com

© Springer-Verlag Berlin Heidelberg 2016

Y. Jia et al. (eds.), *Proceedings of the 2015 Chinese Intelligent Systems Conference*, Lecture Notes in Electrical Engineering 359,
DOI 10.1007/978-3-662-48386-2_11

However, due to the fragile appearance of fruit and the complexity of shape and growth condition, the end-effector is regarded as one of the critical technology of agricultural harvesting robots [2]. At present, the researches on end-effector almost focus on the configuration design and grasping control. And few researchers have conducted studies on fruit grasping planning of end-effector. However, the grasping planning of end-effector plays an important role for design of end-effector as well as the choice of the fruit grasping control technology in fruit harvesting robot.

When the end-effector of robot manipulator grasps apples, the grasping stability, namely no relative motion between the fingers and apples, is an important index. Too small force may cause the relative slide, but excessive force may cause apples' extrusion damage. Therefore, we needs to obtain a suitable grasping force to guarantee the stability of grasping and to ensure the quality of apple picking. At present, in order to get the appropriate grasping force, researchers mostly adopt the method of physical experiment to examine. Li Zhiguo, etc. [3] developed the finite element contact model for harvesting robot fingers and tomato fruit by investigating mechanical properties of tomato fruits. Qian Shaoming, etc. [4] analyzed the holding model of cucumber with the static mechanical analysis method, established relationship between the pressure value of compressed air in the pneumatic actuator and picking capacity. However, the method of physical experiment is only applicable to the specific en-effector and contact points, but easily causes large contact force error on different size and quality of fruit.

In other research areas, manipulator grasping force planning more adopts the method of analysis and calculation to be completed, which is reasonably determine needed contact force balancing external force on the objects in the premise of satisfying certain constraints. Jiang Li, etc. [5] described multiple fingers contact force planning problems as smooth manifold optimization problems corresponding to linear constraint positive definite matrix, and calculated to get grasping force by adopting the method of linear constraint gradient flow. Nakamura, etc. [6] adopted nonlinear planning method based on Lagrange multiplier method to solve the problem of force distribution, obtained the minimum contact force meeting friction constraints. By analyzing relationship between the internal forces of the dexterous hand and the stability of grasping, Liu Qingyun, etc. [7] constructed the optimization model for the best intersection point of internal forces and planning algorithm for contact force. Contact force could be obtained by calculating the operating force and the internal force components.

Based on the fact that the contact force is usually decomposed into operating force component and internal force component, this paper investigates influence of internal force components on the grasping stability, by taking grasping layout planning and contact force distribution into a unified process, and studies the planning problem of contact force of three-finger hand with the apple. By the method which overcomes the shortage of physical experiment and reflects the advantages of analysis and calculation, contact force of greater adaptability and flexibility can be worked out, which is applied to various fruit with different size and quality.

11.2 The Contact Model of Finger with Apple

As the shape of apples is regular and approximates the ball, the apple is regarded as a sphere in this study. Because the apple's hardness is larger, the deformation of apple is very little when pressure imposed on apple is small [8]. Force to normally grasp the apple is relatively smaller and it deformation is very small, so the contact of the finger with apples is considered as point contact with friction between hard objects. According to the experience of person's fingers grasping a spherical object, when contact points of fingers with the object are distributed on a circular surface through its center of a sphere, grasping is easier to reach a steady state in the process of grasping a spherical object. It is also easy to be achieved from the aspect of control. So contact points of three-finger hand with the apple are distributed on a circular surface through its center of a sphere.

11.3 Contact Force Equilibrium Equation and Contact Force Decomposition

In apples' spherical model, the center of a sphere is taken as the origin of coordinates to establish object coordinate system. When grasping an object, in object coordinate system, generalized force on object is $F_e \in R^6$, contact force of three-finger hand with the apple is $f_C = (f_{C1}^T, f_{C2}^T, f_{C3}^T)^T \in R^9$. Where $f_{ci}(i = 1, 2, 3)$ is contact force of finger i with the apple. When three-finger hand grasps the apple stably, force equilibrium equation is as follows

$$G\,(\mathrm{f}_{C1}^T, \mathrm{f}_{C2}^T, \mathrm{f}_{C3}^T)^T = -F_e \quad (11.1)$$

where $G \in R^{6\times 9}$ is the grasping matrix, which is only related to the distribution of contact point on the object. Contact force f_{ci} can be decomposed into two parts of orthogonal to each other [9], which is as follows

$$f_{ci} = f_M + f_N = -G^T (GG^T)^- F_e + f_N \quad (11.2)$$

where the function of operating force component $f_M = (f_{M1}^T, f_{M2}^T, f_{M3}^T)^T \in R^9$ is to balance external force on objects or make objects produce required movement. Internal force component $f_N = (f_{N1}^T, f_{N2}^T, f_{N3}^T)^T \in R^9$ is a collection of that resultant force on the object is zero, belonging to zero space of grab matrix **G**, which is

$$Gf_N = 0 \quad (11.3)$$

By Eq. (11.3), it indicates that when grasping stably, three internal force formed by three-finger hand acting on the apple have its own balanced force system. This three-force balance can only cause two cases, one is that three-force is coplanar and

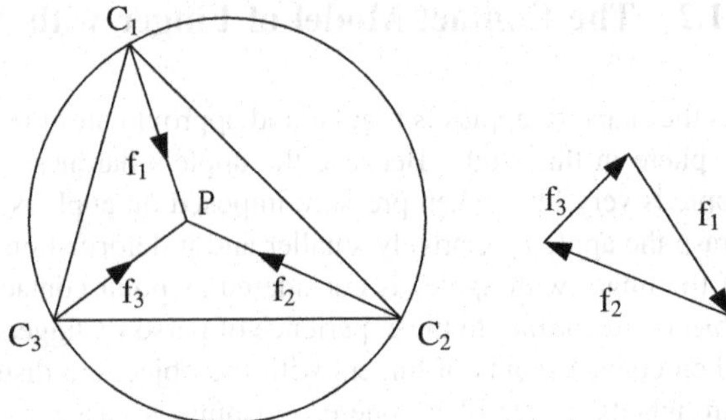

Fig. 11.1 Internal force intersecting and action lines of closed force

intersects at one point and the other is that three-force is coplanar and parallel. So contact of the three-finger hand with the apple is obvious the former one. Assuming that three contact points of three fingers with the object are C_1, C_2, C_3 respectively, then $\Delta C_1C_2C_3$ is constituted. It sets that the plane is **S**. When stably contacting, internal force f_{N1},f_{N2},f_{N3} of the contact point C_1, C_2, C_3 needs to meet force equilibrium condition, so action lines of f_{N1},f_{N2},f_{N3} must be located in the plane of $\Delta C_1C_2C_3$, and intersect at one point P within the plane and can form a force-closure triangle. Schematic diagram is shown in Fig. 11.1.

11.4 Stability Conditions and Influence of Internal Force on Stability

For contact of three-finger hand with the apple, fulfilling for grasp force-closure condition is a prerequisite for stably grasping. The existence of internal force is sufficient and necessary conditions for force-closure, so it is necessary to judge the existence of internal force before grasping. According to the Ref. [10], internal force existence condition of three-finger force-closure grasp is summarized as:

(1) Three contact points all meet existence conditions of friction fan;
(2) Existence conditions of concurrent polygon;
(3) Constraints conditions of internal force equilibrium.

Among them, the friction fan which located in the objects inside is taken as positive friction fan. Concurrent polygon refers to the polygon which is constituted by public area of positive friction fan of three contact points. The internal force equilibrium constraint refers to that action lines of three internal force intersect at one point within concurrent polygon, and direction of the force is pressed to the object.

When grasping planning can cause the internal force, grasping internal force has an important role in adjusting the contact force. Whatever the size and direction of operating force f_M changes, the internal force f_N of suitable size and direction always can be found to keep the contact force locating inside friction cone and grasping stably. Therefore, how to reasonably determine the internal force imposed on the apple is the important condition of stably grasping objects. In the contact force, internal force component is the extrusion pressure on the apple's by fingers, and operating force component is friction that fingers impose on the apple. They are two important factors in the contact force caused by three-finger hands with apples. A suitable extrusion pressure will cause no damage to apples and keeps stable even if disturb emerges. It is necessary to keep the internal force as gentle as possible under the premise of existence. But it does not mean the smaller the better, because the min force under the constraints is easy to make operating force close to the constrain boundary of friction cone, which increases possibility of contact sliding. Therefore, in the following research, the internal force component of contact force is examined through concurrent polygon.

11.5 Concurrent Polygon and Its Relationship with Grasping Stability

On the basis of a variety of constraint conditions of stably grasping, Liu Qingyun studied the existence of internal force, the relationship between internal force and stability of grasping and the calculation of internal force combining with the concept of concurrent polygon [7, 10, 11].

Under the existence condition of internal force, internal force plays an important role in regulating operating force [12], and a higher grasping stability is achieved. But the location of concurrent node of internal force within concurrent polygon has a great influence on stability. When concurrent node **P** is relatively close to boundary of concurrent polygon, the concurrent node of internal force may move outside of concurrent polygon when it is disturbed. This increases the probability of contact point sliding and affects the overall grasping stability. Conversely, when concurrent node **P** is far away from boundary of concurrent polygon and it is disturbed, the probability of concurrent node of internal force moves outside of the concurrent polygon is small, so the three-finger hand can continue to keep high stability. That's why center of the maximum inscribed circle of concurrent polygon can be chosen as such concurrent node. At this point, it can show that internal force which is most close to friction cone boundary and friction cone boundary reaches a relative maximum, the concurrent node of internal force is least likely to move to outside of the concurrent polygon.

11.6 Contact Planning of Three-Finger Hand with the Apple

The purpose of planning is to find stable contact point between three-finger hand with the apple, and to calculate the appropriate contact force. In practical grasping, the outline of the apple can be get according to its image, and three points fulfilling the existence conditions of internal force is found. The calculation process of contact force can be divided into the following steps. (1) Finding three contact points meeting the existence condition of internal force. (2) Calculating the operating force component of contact force according to the Eq. 11.2. (3) Determining the concurrent polygon. (4) Determining the position of concurrent node of internal force, calculating the interior angle of triangle of internal forces and relative size of internal force, and determining the amplitude of each internal force based on other conditions. (5) Determining the value of the contact force.

11.7 The Actual Example

As the following Fig. 11.2, it is taken a ball of 3 cm radius as an example to calculate the contact force of an apple. Assuming that its quality is 0.3 KG, friction coefficient of epidermis and mechanical finger is 0.5, and external force on the object is the gravity of 3 N and that contact points of three-finger with the object are distributed in the equator of the sphere. The Fig. 11.3 is the sectional drawing of the sphere, and rectangular coordinate is established taking the center of a circle as the origin. Two contact points C_1, C_2 of three-finger with the object are distributed on each side of Y axis, so angle of radius of two points and Y axis is 50°, angle of C_3's radius and radius of C_1, C_2 is 130°.

In rectangular coordinate taking center of a circle as the origin, coordinates of C_1, C_2, C_3 are (3sin(50°), −3cos(50°), 0), (−3sin(50°), −3cos(50°), 0), (0, 3, 0) respectively, so grab matrix is:

$$G=\begin{pmatrix} 1 & 0 & 0 & 1 & 0 & 0 & 1 & 0 & 0 \\ 0 & 1 & 0 & 0 & 1 & 0 & 0 & 1 & 0 \\ 0 & 0 & 1 & 0 & 0 & 1 & 0 & 0 & 1 \\ 0 & 0 & -3\cos 50^\circ & 0 & 0 & -3\cos 50^\circ & 0 & 0 & 3 \\ 0 & 0 & -3\sin 50^\circ & 0 & 0 & 3\sin 50^\circ & 0 & 0 & 0 \\ 3\cos 50^\circ & 3\sin 50^\circ & 0 & 3\cos 50^\circ & -3\sin 50^\circ & 0 & -3 & 0 & 0 \end{pmatrix} \tag{11.4}$$

According to the Eq. (11.2), three-finger's operating force that can be obtained is $f_{M1}=(0\quad 0\quad 0.9131)^T$, $f_{M2}=(0\quad 0\quad 0.9131)^T$, $f_{M3}(0\quad 0\quad 1.1738)^T$ respectively.

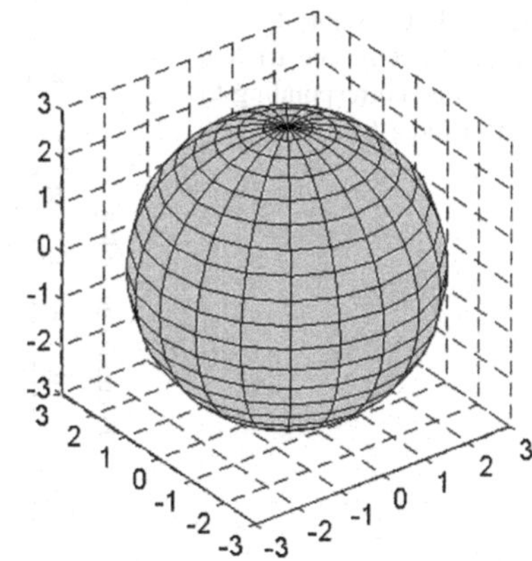

Fig. 11.2 Apple three-dimensional coordinate

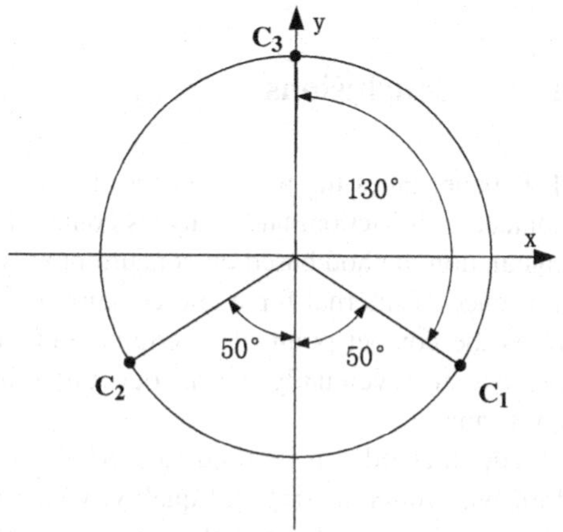

Fig. 11.3 Three contact points coordinate

The calculation method of grasping internal force is as follow: firstly it sets that friction coefficient of the apple and three-finger is $\mu = 0.5$, to get contact point's friction cone angle $\alpha = \arctan\mu = 26.57°$. To get each point friction fan and intersection of friction fan boundary fulfills the existence condition of concurrent polygon, which is that there is point of boundary located in the strict interior of third friction fan and concurrent polygon exists. After computing it obtains that concurrent polygon is hexagon and six vertex are (0, −1.3796), (−1.8671, −0.9338), (−1.6512, −0.4356), (0, 3), (1.6512, −0.4356), (1.8671, −0.9338) respectively. When the circle center is (0, 0.0323), the maximum of R is 1.3104 cm by MATLAB. Three interior angles of closed triangular which consists of action lines of internal force are respectively: $\beta_1 = \beta_2 = 50.4704°$, $\beta_3 = 70.0592°$, to obtain $f_{N1} = f_{N2}$, $f_{N3} = 1.2731 f_{N1}$. The position relationship between the contact force f_c corresponding to different sizes of internal force and friction cone angle is shown in Fig. 11.4. When $|f_{N3}|$ is more than 2.3476, three contact forces all meet friction cone constraints.

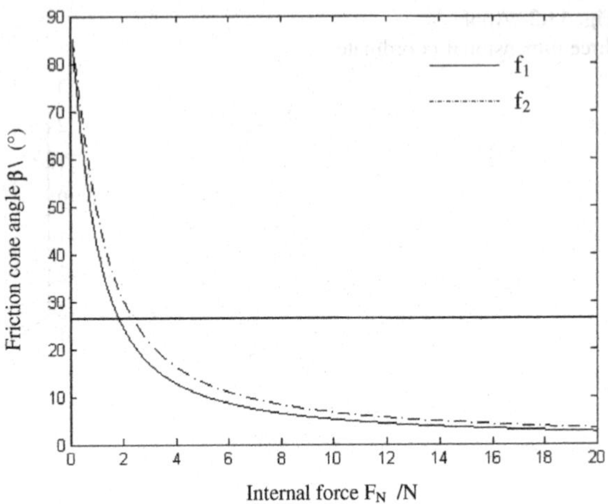

Fig. 11.4 Diagram of internal force f_N and friction angle cone corresponding to contact force f_c

11.8 Conclusions

This paper takes the point contact of three-finger hand with the apple as the point contact with friction, and analyzes contact force of fingers with the apple. It adopts calculation method based on concurrent polygon. Taking constraints problem as the existence of internal force and concurrent polygon, calculation is much simpler. It fuses the contact point of determine and calculation of contact force as a unified process to eventually work out three-finger contact force fulfilling friction constraint.

This method can be used to work out different stable contact force of various fruit with different size and quality, which has greater adaptability and flexibility. Compared with adopting the same contact force on the same fruit in different individuals, it is of greater significance for nondestructive picking robots.

Acknowledgments This work was supported by a project funded by the Priority Academic Program Development of JiangSu Higher Education Institutions and Research Fund for the Doctoral Program of Higher Education of China under Grant 20133227110024.

References

1. Jian S, Tiezhong Z, Liming X et al (2006) Research actuality and prospect of picking robot for fruits and vegetables [J]. Trans CSAM 37(5):158–162
2. Qinchuan L, Ting H, Chuanyu W et al (2008) Review of end-effectors in fruit and vegetable harvesting robot [J]. Trans CSAM 39(3):175–179
3. Zhiguo L (2012) Study on the grasp damage of harvesting robot based on the biomechanical properties of tomato fruit [D]. Jiangsu University, Zhenjiang
4. Shaoming Q, Qinghua Y, Zhiheng W et al (2010) Research on holding characteristics of cucumber and end-effetor of cucumber picking [J]. Trans CSAE 26(7):107–112

5. Li J, Hong L (2007) Real time force optimization algorithm of multi-fingered grasp [J]. Chin J Mech Eng 43(12):144–149
6. Nakamura Y, Nagai K, Yoshikawa T (1989) Dynamics and stability in coordination of multiple robotic mechanisms [J]. Int J Robot Res 8(2):44–61
7. Qingyun L, Wan CY, Nenggang X et al (2010) Optimization algorithm for contact forces based on grasping stability [J]. J Mech Eng 46(7):57–62
8. Xiaoyu L, Wei W (1998) A study on compressive properties of apple [J]. J NW Sci Tech Univ Agric For 26(2):107–110
9. Kerr J, Roth B (1986) Analysis of multi-fingered hands [J]. Int J Robot Res 4(3):3–17
10. Qingyun L, Ruiming Q, Jingping Y (2007) On internal force existence of three-fingered hand force-closure grasping [J]. China Mech Eng 18(3):265–268
11. Qingyun L (2009) Algorithm for internal force concurrent polygon existence of three-fingered robot hand [J]. J Mech Eng 45(3):119–123
12. Jiawei L, Haiying H, Bin W et al (2005) Simple algorithm of multi-fingered grasping internal force [J]. Prog Nat Sci 6(15):733–738

Chapter 12
Extended State Observer Based Sliding Mode Control for Mechanical Servo System with Friction Compensation

Chenhang Li and Qiang Chen

Abstract This paper proposes a tracking control method based on the extended state observer for the nonlinear mechanical servo system with friction compensation. The friction nonlinearity is described by a continuously differentiable LuGre model and compensated by using neural network (NN). Then, an extended state observer (ESO) is employed to estimate the system states and uncertainties including friction compensation error. A sliding mode control (SMC) scheme is developed based on ESO estimation to guarantee the convergence of the tracking error. Comparative simulations are conducted to show the superior performance of the proposed method.

Keywords Servo system · Sliding model control · Neural network · Extended state observer

12.1 Introduction

Over the past decades, the mechanical servo system has been widely used in instrumentation, flight control, industrial production, etc. It becomes more and more important to develop the performance of mechanical servo system in modern industries. However, many factors can influence the precision of the system, such as friction, dead zone, and disturbances [1–3].

The friction is the main factor in the system. Therefore, in order to eliminate the effects of friction and improve the performance of system, many efforts have been devoted to build friction models, such as Armstrong's model, Dahl model, and

C. Li · Q. Chen (✉)
College of Information Engineering, Zhejiang University of Technology,
Hangzhou 310023, China
e-mail: sdnjchq@zjut.edu.cn

C. Li
e-mail: lich0571@outlook.com

© Springer-Verlag Berlin Heidelberg 2016
Y. Jia et al. (eds.), *Proceedings of the 2015 Chinese Intelligent Systems Conference*, Lecture Notes in Electrical Engineering 359,
DOI 10.1007/978-3-662-48386-2_12

LuGre model [4–6]. In those friction models abovementioned, LuGre model is widely used to accurately describe the friction phenomenon in the mechanical system. However, it is difficult to identify all parameters in the LuGre model accurately. Since neural network can learning and approximating nonlinear function, it can be employed to design friction compensators for the dynamical model [7].

For the high precision control in the system, many control schemes have been proposed, in which the sliding model control (SMC) is regarded as an effective robust control scheme to control nonlinear uncertain systems. However, the SMC needs the information of the complete state vectors. In [8–10], the extended state observer (ESO) was developed to estimate the state vectors and the uncertainties including external disturbances.

In this paper, a sliding mode control scheme is presented by combining neural network compensation and extended state observer (ESO). LuGre model is employed to describe the friction dynamics, and a back propagation neural network (BPNN) is utilized to approximate and compensate the friction. In order to estimate the unknown states and uncertainties with friction compensation errors, the ESO and sliding mode technique are combined to design a composite controller for mechanical servo system with NN friction compensation.

12.2 System Description

The mechanical system under investigation is described as

$$\begin{cases} \frac{d\theta_m}{dt} = \omega_m \\ J\frac{d\omega_m}{dt} = K_t u - D\omega_m - F - T_l \end{cases} \tag{12.1}$$

where θ_m, ω_m are the state variables, denoting the output shaft of the motor position and velocity, respectively; D is the damping coefficient, K_t is the motor torque constant. J is the effective mass, T_l is the load torque, and F is the friction force, which can be described by the following LuGre model:

$$F = \sigma_0 z + \sigma_1 \dot{z} + \sigma_2 \dot{x} \tag{12.2}$$

where σ_o, σ_1 and σ_2 denote the stiffness, the damping coefficient, the viscous friction coefficient, respectively, and $x = \theta_m$.

Consider the contact surface effects which are lumped into an average asperity deflection z that is given by

$$\dot{z} = \dot{x} - \frac{|\dot{x}|}{h(\dot{x})} z \tag{12.3}$$

When $\dot{x}$ is constant, the deflection z approaches a steady and bonded state value z_s expressed as

$$z_s = h(\dot{x})\mathrm{sgn}(\dot{x}) \tag{12.4}$$

where the function $h(\dot{x})$ is

$$h(\dot{x}) = \frac{F_c + (F_s - F_c)e^{-(\dot{x}/x_s)^2}}{\sigma_0} \tag{12.5}$$

where F_c, and F_s are both unknown constants. F_c is the Coulomb friction coefficient, and F_s is the Stribeck friction coefficient. x_s is the Stribeck velocity.

Substituting (12.3) and (12.4) into (12.2) yields:

$$\begin{aligned} F = \sigma_2\dot{x} &+ \left[F_c + (F_s - F_c)e^{-(\dot{x}/x_s)^2}\right]\mathrm{sgn}(\dot{x}) \\ &+ \sigma_o\varepsilon\left[1 - \frac{\sigma_1}{F_c + (F_s - F_c)e^{-(\dot{x}/x_s)^2}}|\dot{x}|\right] \end{aligned} \tag{12.6}$$

with $\varepsilon = z - z_s$.

Remark 1 Due to the discontinuity of the sign function in (12.6), the friction F may not be directly estimated by using neural network. Therefore, we employ a continuous function $\tanh(\dot{x})$ instead of $\mathrm{sgn}(\dot{x})$ in the simulation part, and then (12.6) can be rewritten as

$$\begin{aligned} F = \sigma_2\dot{x} &+ \left[F_c + (F_s - F_c)e^{-(\dot{x}/x_s)^2}\right]\tanh(\dot{x}) \\ &+ \sigma_o\varepsilon\left[1 - \frac{\sigma_1}{F_c + (F_s - F_c)e^{-(\dot{x}/x_s)^2}}|\dot{x}|\right] \end{aligned} \tag{12.7}$$

12.3 Neural Network Compensation

In this section, a simple Back Propagation neural network (BPNN) is utilized to compensate the friction function. The whole structure of BPNN is shown in Fig. 12.1. There are two input nodes θ_m and ω_m, and one output node $\hat{F}$, which is the estimation of the friction F.

The hidden layer node value is:

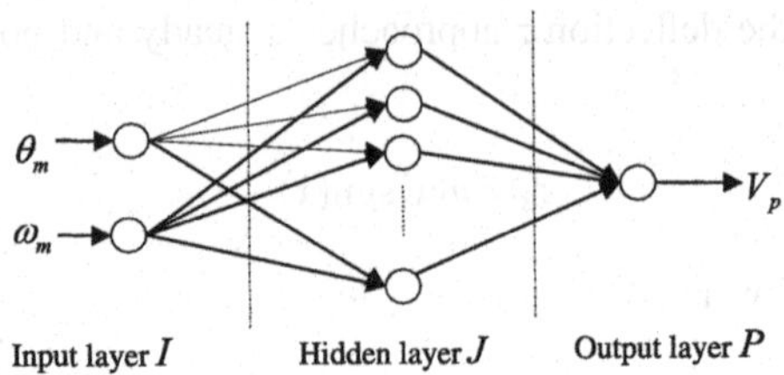

Fig. 12.1 The structure of BPNN

$$V_j = f(W_{ij} \cdot X_k + \theta_0) \tag{12.8}$$

The output layer node value is:

$$V_p = f(W_{jp} \cdot V_j + \theta_1) \tag{12.9}$$

where $f(x)$ is the sigmoid function, $f(x) = \frac{1}{1+e^{-ax}}$, a is the adjustment coefficient. W_{ij}, W_{jp} and θ_0, θ_1 are the weights and threshold matrix between the input layer and output layer.

The partial derivative of the output layer and the hidden layer δ_p, δ_j are

$$\delta_p = V_p(1 - V_p)(F - V_p) \tag{12.10}$$

$$\delta_j = V_j(1 - V_j) \sum \delta_p W_{jp} \tag{12.11}$$

The weight updating laws are given as

$$W_{jp}(n+1) = W_{jp}(n) + \eta \delta_p(n) V_j \tag{12.12}$$

$$W_{ij}(n+1) = W_{ij}(n) + \eta \delta_j(n) V_j \tag{12.13}$$

where η is the learning rate.

BPNN repeats learning of samples until meeting the requirements. After the end of algorithm, the output value V_p is the estimation of friction F, i.e., $V_p = \hat{F}$.

12.4 Controller Design

Let $x_1 = \theta_m$, $x_2 = \omega_m = \dot{x}_1$, and then system (12.1) can be rewritten as

$$\begin{cases} \dot{x}_1 = x_2 \\ \dot{x}_2 = \frac{K_t}{J} u - \frac{D}{J} x_2 - \frac{F}{J} - \frac{T_l}{J} \end{cases} \tag{12.14}$$

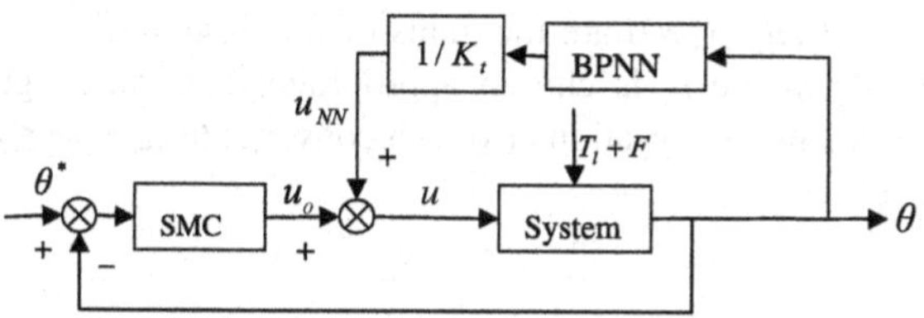

Fig. 12.2 The schematic of control system with BPNN compensation

The schematic of control system with BPNN compensation is shown in Fig. 12.2. The control signal is designed as

$$u = u_{NN} + u_o \tag{12.15}$$

where $u_{NN} = \hat{F}/K_t$ is the BPNN compensation term and u_o is the equivalent control signal.

Then (12.14) can be rewritten as

$$\begin{cases} \dot{x}_1 = x_2 \\ \dot{x}_2 = a(x) + bu_o \end{cases} \tag{12.16}$$

with $a(x) = -\frac{D}{J}x_2 - \frac{F}{J} - \frac{T_l}{J} + \frac{\hat{F}}{J}$, $b = \frac{K_t}{J}$;

Define $d = a(x) + \Delta bu_o$, $\Delta b = b - b_o$, where b_o is the estimate of b, which can be given by experience. Now, we define $x_3 = d$, and then (12.16) can be rewritten as the following equivalent form:

$$\begin{cases} \dot{x}_1 = x_2 \\ \dot{x}_2 = x_3 + b_o u_o \\ \dot{x}_3 = h \end{cases} \tag{12.17}$$

with $h = \dot{d}$.

12.4.1 Extended State Observer

Define z_i, $i = 1, 2, 3$, as the estimates of the state variables x_i, and then the linear extended state observer (LESO) is given by

$$\begin{cases} \dot{z}_1 = z_2 - \beta_1 e_{o1} \\ \dot{z}_2 = z_3 - \beta_2 e_{o1} + b_o u_o \\ \dot{z}_3 = -\beta_3 e_{o1} \end{cases} \tag{12.18}$$

where $e_{o1} = z_1 - x_1, \beta_1, \beta_2, \beta_3 > 0$ are the gains of the observer.

As stated in [11], if the β_i is chosen appropriately, it can be guaranteed that $z_i \to x_i,\ i = 1, 2, 3$ i.e., the observation error can converge to $|z_i - x_i| \le d_i$, where d_i is a very small figure.

12.4.2 Sliding Mode Control

In order to guarantee the convergence of system errors, the sliding mode controller is designed and the sliding surface is given as

$$s = e_2 + \lambda_1 e_1 \tag{12.19}$$

where $e_1 = x_1^* - x_1, e_2 = x_2^* - x_2 = \dot{e}_1$, with x_1^*, x_2^* being the desired position and velocity, respectively, and satisfying $\dot{x}_1^* = x_2^*$, and $\lambda_1 > 0$ is a control parameter.

Then, the derivative of s is

$$\begin{aligned} \dot{s} &= \dot{e}_2 + \lambda_1 \dot{e}' \\ &= \ddot{x}_1^* - x_3 - b_o u_o + \lambda_1 \left(x_1^* - x_2\right) \end{aligned} \tag{12.20}$$

The equivalent controller based on LESO (12.18) can be designed as

$$u_o = \frac{1}{b_o}\left(\ddot{x}_1^* - z_3 + \lambda_1 (\dot{x}_1^* - z_2) + k \cdot sign(s)\right) \tag{12.21}$$

where $k > 0$ satisfies $k \ge d_3 + \lambda_1 d_2$.

12.5 Stability Analysis

Theorem 1 *Consider the system* (12.15) *with equivalent controller* (12.21), *sliding surface* (12.19) *and neural network compensation* u_{NN}, *and then the tracking error* e_1 *will asymptotically converge to zero.*

Proof Define a Lyapunov candidate function as:

$$V = \frac{1}{2}s^2 \tag{12.22}$$

Then according to (12.20), the derivative of V is

$$\begin{aligned} \dot{V} &= s\dot{s} \\ &= s\left[\ddot{x}_1^* - x_3 - bu_o + \lambda_1\left(\dot{x}_1^* - x_2\right)\right] \end{aligned} \tag{12.23}$$

Combining (12.21) and (12.23), and we have

$$\begin{aligned} \dot{V} &= s[z_3 - x_3 + \lambda_1(z_2 - x_2) - ksign(s)] \\ &\le |s|(|z_3 - x_3| + \lambda_1|z_2 - x_2|) - k|s| \\ &\le |s|(d_3 + \lambda_1 d_2) - k|s| \\ &= |s|(d_3 + \lambda_1 d_2 - k) \end{aligned} \tag{12.24}$$

Since $k > 0$ and $k \ge d_3 + \lambda_1 d_2$, we can conclude that $|s|(d_3 + \lambda_1 d_2 - k) \le 0$, therefore, it is guaranteed that s can converge to zero asymptotically.

It is not hard to see that $\dot{e}_1 + \lambda_1 e_1 = 0$ in the sliding surface $s = 0$. By solving the first-order differential equation, we have $e_1 = e^{-\lambda_1 t}$, which means the system error e_1 will asymptotically converge to zero.

12.6 Simulations

In order to show the superior performance of the proposed method, the conventional sliding mode control based on extended state observer(ESO + SMC) in [10] is performed for the comparison with the proposed ESO + SMC with friction compensation (ESO + SMC + FC) scheme.

For the fair comparison, the same parameters of both schemes are set same. The nominal values of the servo system parameters are $J = 0.5$, $K_t = 1$, $D = 0.3$, $T_l = 0.5$. LuGre friction parameters are $\sigma_o = 0.5$, $\sigma_1 = 0.3$, $\sigma_2 = 0.1$, $F_s = 0.335$, $F_c = 0.285$, $V_s = 1$. In controller we set $\lambda_1 = 5$, $k = 1.5$. In BPNN, the initial weights of neural network are $W_{ij}(0) = W_{jp}(0) = 0$, the Hidden layer N = 25, and $a = 1, \eta = 1$, and the NN weight update laws are given by (12.12)–(12.13).

In this part, $y_d = \sin(t) + 0.5\cos(0.5t)$ is employed as the reference signal. The control parameters are $\beta_1 = 10, \beta_2 = 30, \beta_3 = 55$. Comparative tracking performance

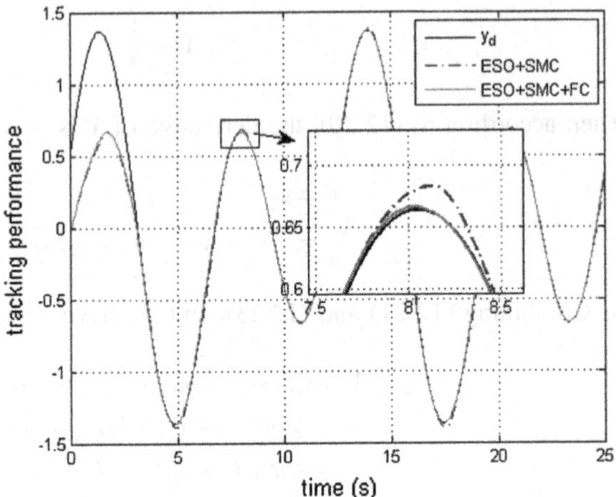

Fig. 12.3 Tracking performance

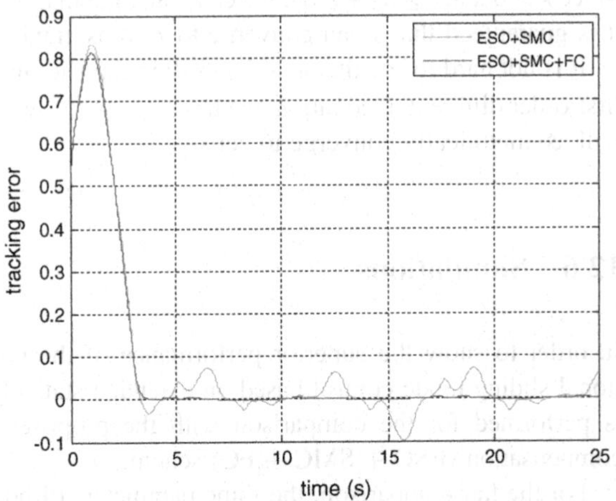

Fig. 12.4 Tracking errors

and corresponding tracking errors are shown in Fig. 12.3 and Fig. 12.4, respectively. It becomes more obvious that our method has a smaller error than ESO + SMC. Figure 12.5 shows the observation errors, we can see that ESO + SMC + FC also has a smaller observation error than ESO + SMC. The BPNN compensation performance is shown in Fig. 12.6.

From all the aforementioned simulation results, it is not hard to conclude that the proposed ESO + SMC with friction compensation scheme has a better tracking performance and a smaller tracking error.

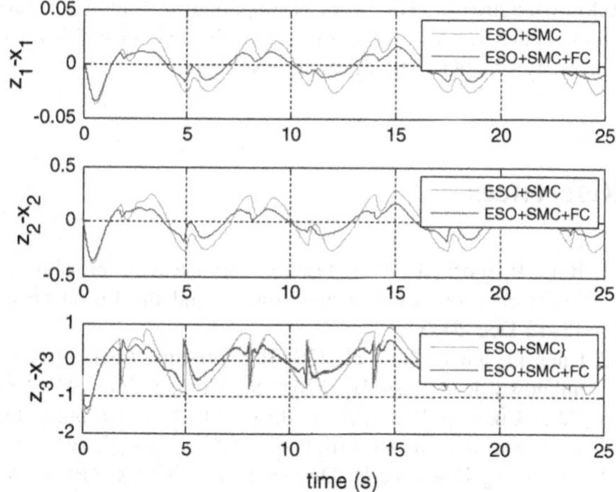

Fig. 12.5 Observation errors of LESO

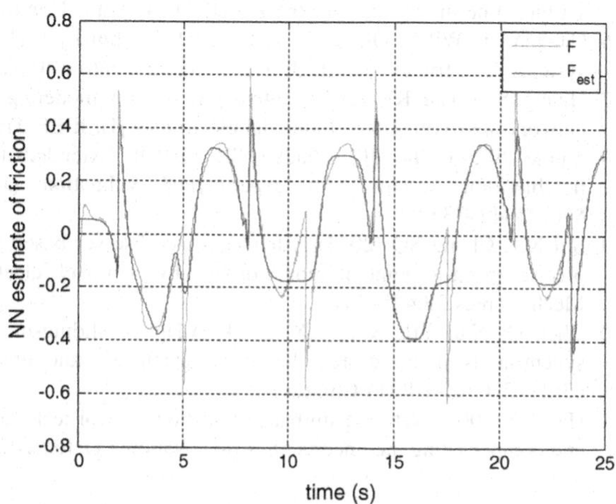

Fig. 12.6 BPNN compensation performance

12.7 Conclusion

A tracking control for the mechanical servo system with friction compensation is proposed in this paper. The difficulty from the unavoidable friction is circumvented by adopting a new continuously differentiable LuGre model, which is lumped into the BPNN for approximating unknown dynamics. Then, ESO is used to estimate the system states and uncertainties with friction compensation error. Based on the ESO estimation, the controller is designed by a sliding model principle to guarantee the convergence of the tracking error. It is shown in the simulation that the proposed method can improve the tracking performance of the mechanical system.

Acknowledgments This work is supported by National Natural Science Foundation (NNSF) of China under Grant 61403343 and 12th Five-Year Plan Construction Project of Emerging University Characteristic Specialty (No. 080601).

References

1. Bona B, Indri M (2005) Friction compensation in robotics: an overview [C]. In: 44th IEEE Conference on decision and control, and the European control conference, Seville, Spain, 12–15 Dec 2005
2. Chen Q, Yu L, Nan YR (2013) Finite-time tracking control for motor servo systems with unknown dead-zones [J]. J Syst Sci Comput 26(6):940–956
3. Na J, Chen Q, Ren XM, Yu G (2014) Adaptive prescribed performance motion control of servo mechanisms with friction compensation [J]. IEEE Trans Ind Electron 61(1):486–494
4. Armstrong-Helouvry B, Dupont P, De Wit CC (1994) A survey of models, analysis tools and compensation methods for the control of machines with friction [J]. Automatica 30(7):1083–1138
5. Bucci BA, Cole DG, Ludwick SJ et al (2013) Nonlinear control algorithm for improving settling time in system with friction [J]. IEEE Trans Control Syst Technol 21(4):1365–1373
6. Canudas de Wit C, Olsson H, Astrom KJ, Lischinsky P (1995) A new model for control of systems with friction [J]. IEEE Trans Autom Control 40(3):419–425
7. Huang SN, Tan KK (2012) Intelligent friction modeling and compensation using neural network approximations. Industrial electronics [J]. IEEE Trans 59(8):3342–3349
8. Talole SE, Kolhe JP, Phadke SB (2010) Extended-state- observer-based control of flexible-joint system with experimental validation [J]. IEEE Trans Ind Electron 57(4):3342–3349
9. Shi XX, Chang SQ (2013) Extended state observer-based time-optimal control for fast and precise point-to-point motions driven by a novel electromagnetic linear actuator [J]. Mechatronics 23:445–451
10. Chen Q, Nan YR, Xing KX (2014) Adaptive sliding-mode control of chaotic permanent synchronous motor system based on extended state observer [C]. Acta Physica Sinica 63(22):220506-1-8 (In Chinese)
11. Han JQ (2008) Active disturbance rejection control technique—the technique for estimating and compensating the uncertainties [M]. National Defense Industry Press, Bejing (In Chinese)

Chapter 13
Local Linear Discriminant Analysis Using ℓ_2-Graph

Ya Gu, Caikou Chen, Yu Wang and Rong Wang

Abstract A recently proposed method, called Local Fisher Linear Discriminant Analysis (LLDA), the experiment showed that compared with the traditional Fisher Linear Discriminant Analysis, it has a better result. However, it uses Euclidean distance selecting nearest neighbor samples which has some flaws in the way, such as the robustness is not good and not sparse, and so on. The paper presents an improved approach, called Local Linear Discriminant Analysis Using ℓ_2-Graph (L2G_LLDA). It remains reconstructed coefficient of samples to select the nearest samples, which enhances the robustness of the algorithms and makes it sparse. The extensive experimental results over several standard face databases have demonstrated the effectiveness of the proposed algorithm

Keywords Facial feature extraction · Sparse representation · Local linear discriminant analysis · Target classification

13.1 Introduction

Manifold learning algorithms has been widely used in computer vision processing. There are two main ways for manifold learning algorithm to constructing a graph, one of which is to select K nearest neighbor samples; another is the ε-ball based method. Recently, Xu et al. [1] proposed a new feature extraction method based on manifold Local Linear Discriminant Analysis Framework Using Sample Neighbors (LLDA), experimental results show the effectiveness of the algorithm. However, this algorithm is calculated by the Euclidean distance to select K nearest neighbor samples, there are some shortcomings when we construct the graph, such as robustness to noise, or not sparse.

Y. Gu · C. Chen (✉) · Y. Wang · R. Wang
College of Information Engineering,
Yangzhou University, Yangzhou 225127, China
e-mail: yzcck@126.com

© Springer-Verlag Berlin Heidelberg 2016

Y. Jia et al. (eds.), *Proceedings of the 2015 Chinese Intelligent Systems Conference*, Lecture Notes in Electrical Engineering 359,
DOI 10.1007/978-3-662-48386-2_13

Recent years sparse representation theory is widely used in image compression, feature extraction and the signal analysis and other fields. After sparse representation is introduced into the field of face recognition by Wright [2], many scholars are committed to research in this area. ℓ_1 norm is introduced into the manifold learning algorithm by Yan et al. [3], using ℓ_1-graph algorithm to construct adjacency matrix. By experimental analysis, with respect to the use of K-nearest-neighbor method or ε-ball based method, ℓ_1-graph algorithm has the following advantages, first, to enhance the robustness to noise, the second is sparse. Zhang et al. [4] proposed ℓ_2-graph algorithm (Constructing the ℓ_2-Graph for Subspace Learning and Subspace Clustering), In his view, with respect to the ℓ_1 norm, ℓ_2 norm is better, more efficient.

Based on the above discussion, we combine the advantages of ℓ_1-graph, ℓ_2-graph and some shortcomings of LLDA, this paper make some improvements on the basis of LLDA. Using ℓ_2-graph method to calculate the reconstructed coefficient of the sample, we choose K samples which are the former largest reconstructed coefficient, use them as K nearest neighbor of the given sample. We conducted experiments on common ORL face database, YALE face database and AR face database, and verify the effectiveness of the algorithm.

13.2 Local Linear Discriminant Analysis (LLDA)

We assume that there are C classes of samples and let $\mathbf{A} = [\mathbf{x}_1, \mathbf{x}_2, \ldots, \mathbf{x}_N]$. From the total training set, we can determine the K nearest neighbors $\mathbf{x}_i'(i = 1, 2, 3, \ldots, K)$ for a given sample which from c' classes, and let $c_i'(i = 1, 2, \ldots, c')$ donate their classes which are a part of the total classes. l_i' indicates the number of the nearest neighbor samples in each class that satisfies

$$\sum\nolimits_{i=1}^{c'} l_i' = K \tag{13.1}$$

The mean image of all determined neighbors is denoted by m' and the mean of the determined neighbors in class c_i' is denoted by m_i' $(i = 1, 2, \ldots, c')$, $\mathbf{x}_{ij}'$) denotes the jth determined neighbor of the given sample which in class c_i'.

The between-class scatter matrix $\mathbf{S}_b$ as follows:

$$\mathbf{S}_b = \frac{1}{K} \sum\nolimits_{i=1}^{c'} l_i' (m_i' - m')(m_i' - m')^{\mathrm{T}} \tag{13.2}$$

The within-class scatter matrix $\mathbf{S}_w$ as follows:

$$\mathbf{S}_w = \frac{1}{K} \sum\nolimits_{i=1}^{c'} \sum\nolimits_{j=1}^{l_i'} (\mathbf{x}_{ij}' - m_i')(\mathbf{x}_{ij}' - m_i')^{\mathrm{T}} \tag{13.3}$$

We can get the objective function as follows:

$$J_{LLDA}(\mathbf{w}) = \frac{\mathbf{w}^{\mathrm{T}}\mathbf{S}_b\mathbf{w}}{\mathbf{w}^{\mathrm{T}}\mathbf{S}_w\mathbf{w}} \tag{13.4}$$

For the given sample, if there is only one nearest neighbor in each of the c' classes, the within-class scatter matrix of these nearest neighbors is a zero matrix. The Eq. (13.4) can be rewritten as the following equation:

$$J_{LLDA}(\mathbf{w}) = \mathbf{w}^{\mathrm{T}}\mathbf{S}_b\mathbf{w} \tag{13.5}$$

For the given sample, if all of the K nearest neighbors are from only one class, then the between-class matrix is a zero matrix. The Eq. (13.4) can be rewritten as the following equation:

$$J_{LLDA}(\mathbf{w}) = \mathbf{w}^{\mathrm{T}}\mathbf{S}_w\mathbf{w} \tag{13.6}$$

13.3 Algorithm for Construction the ℓ_2-Graph

Let $\mathbf{D}$ donate the data set, and each column of $\mathbf{D}_i (i = 1, 2, \ldots, C)$ is the sample of class i, $\mathbf{X} = \{\mathbf{x}_1, \mathbf{x}_2, \ldots, \mathbf{x}_N\}$ be a collection of data located in the set of the $\{\mathbf{D}_i\}_{i=1}^{c}$, and $\mathbf{X}_i = [\mathbf{x}_1, \mathbf{x}_2, \ldots, \mathbf{x}_{i-1}, 0, \mathbf{x}_{i+1}, \ldots, \mathbf{x}_N]$ be the pointed dictionary for the data $\mathbf{x}_i$.

The steps of ℓ_2-Graph algorithm are summarized as follows:

Step 1 Normalize the columns of $\mathbf{D}$.

Step 2 Get the optimal solution of the problem which is given by

$$J(\mathbf{c}) = \arg\min_{\mathbf{c}_i} \frac{1}{2}\|\mathbf{x}_i - \mathbf{X}_i\mathbf{c}_i\|_2^2 + \gamma\|\mathbf{c}_i\|_2^2$$

Step 3 Construct a similarity graph $\mathbf{W}$. Denote $w_{ij} = |\widehat{c}_{ij}| + |\widehat{c}_{ji}|$ as an element of $\mathbf{W}$.

Step 4 Let $\boldsymbol{\Theta}$ denote the projection matrix. We can get the solution by solving $\min_{\boldsymbol{\Theta}} \|\boldsymbol{\Theta}^{\mathrm{T}}\mathbf{X} - \boldsymbol{\Theta}^{\mathrm{T}}\mathbf{X}\mathbf{W}\|_F^2$, s.t. $\boldsymbol{\Theta}^{\mathrm{T}}\mathbf{X}\mathbf{X}^{\mathrm{T}}\boldsymbol{\Theta} = \mathbf{I}$

13.4 Local Linear Discriminant Analysis Using ℓ_2-Graph

We suppose that there are C classes of samples and let $\mathbf{A} = [\mathbf{x}_1, \mathbf{x}_2, \ldots, \mathbf{x}_N]$, and we take N as the total number of samples. Let $\mathbf{x}_i$ denote the given sample, we can get the coefficients by using

$$J(\mathbf{b}) = \arg\min_{\mathbf{b}_i} \frac{1}{2} \|\mathbf{x}_i - \mathbf{X}_i \mathbf{b}_i\|_2^2 + \gamma \|\mathbf{b}_i\|_2^2 \tag{13.7}$$

We can solve the optimization problem (13.7) by

$$\mathbf{b}_i = (\mathbf{X}^{\mathrm{T}}\mathbf{X} + \gamma \mathbf{I})^{-1}[\mathbf{X}^{\mathrm{T}}\mathbf{x}_i - \frac{e_i^{\mathrm{T}}(\mathbf{X}^{\mathrm{T}}\mathbf{X} + \gamma \mathbf{I})^{-1}\mathbf{X}^{\mathrm{T}}\mathbf{x}_i}{e_i^{\mathrm{T}}(\mathbf{X}^{\mathrm{T}}\mathbf{X} + \gamma \mathbf{I})^{-1} e_i} e_i] \tag{13.8}$$

Let $\mathbf{P} = (\mathbf{X}^{\mathrm{T}}\mathbf{X} + \gamma \mathbf{I})^{-1}$, The Eq. (13.8) can be rewritten as the following equation:

$$\mathbf{b}_i = \mathbf{P}[\mathbf{X}^{\mathrm{T}}\mathbf{x}_i - \frac{e_i^{\mathrm{T}}\mathbf{P}\mathbf{X}^{\mathrm{T}}\mathbf{x}_i}{e_i^{\mathrm{T}}\mathbf{P}e_i} e_i]$$

We can obtain K samples that whose coefficients are the top K largest as the nearest neighbor samples for the given sample, they come from c' classes, l_i' indicates the number of the nearest neighbor samples in each class that satisfies:

$$\Sigma_{i=1}^{c'} l_i' = K \tag{13.9}$$

The mean image of all determined neighbors is denoted by m' and the mean of the determined neighbors in class c_i' is denoted by $m_i'(i = 1, 2, \ldots, c')$, $\mathbf{x}_{ij}'$ denotes the jth determined neighbor of the given sample which in class c_i'.

The between-class scatter matrix $\mathbf{S}_b^{L2G}$ as follows:

$$\mathbf{S}_b^{L2G} = \frac{1}{K} \Sigma_{i=1}^{c'} l_i'(m_i' - m')(m_i' - m')^{\mathrm{T}} \tag{13.10}$$

The within-class scatter matrix $\mathbf{S}_w^{L2G}$ as follows:

$$\mathbf{S}_w^{L2G} = \frac{1}{K} \Sigma_{i=1}^{c'} \Sigma_{j=1}^{l_i'} (\mathbf{x}_{ij}' - m_i')(\mathbf{x}_{ij}' - m_i')^{\mathrm{T}} \tag{13.11}$$

We can get the objective function as follows:

$$J_{L2G\_VLLDA}(\mathbf{w}) = \frac{\mathbf{w}^{\mathrm{T}}\mathbf{S}_b^{L2G}\mathbf{w}}{\mathbf{w}^{\mathrm{T}}\mathbf{S}_w^{L2G}\mathbf{w}} \tag{13.12}$$

For the given sample, if there is only one nearest neighbor in each of the c' classes, the within-class scatter matrix of these nearest neighbors is a zero matrix. The Eq. (13.12) can be rewritten as the following equation:

$$J_{L2G\_VLLDA}(\mathbf{w}) = \mathbf{w}^{\mathrm{T}}\mathbf{S}_b^{L2G}\mathbf{w} \tag{13.13}$$

For the given sample, if all of the K nearest neighbors are from only one class, then the between-class matrix is a zero matrix. The Eq. (13.12) can be rewritten as the following equation:

$$J_{L2G\_VLLDA}(\mathbf{w}) = \mathbf{w}^{\mathrm{T}}\mathbf{S}_w^{L2G}\mathbf{w} \tag{13.14}$$

Finally, we can get a projection matrix whose column are the eigen-vectors connected with the top eigen-values of the eigen-equation $\mathbf{S}_b^{L2G}\mathbf{w} = \lambda\mathbf{S}_w^{L2G}\mathbf{w}$, or the top eigen-values of the matrix $\mathbf{S}_b^{L2G}$ or $\mathbf{S}_w^{L2G}$.

13.5 Analysis of Algorithm

In this section, we will discuss the effect of the ℓ_2-graph which's applied to LLDA framework. Our method has some advantages as follows, firstly, Using ℓ_2-graph method to calculate the reconstructed coefficient of the sample, we choose K samples which are the top largest reconstructed coefficient, use them as K nearest neighbor of the given sample, if we do so, there will be two benefits, first is to enhance the robustness to noise, the second is sparse. Secondly, when solving the eigen projection matrix, we only train a small part of the training samples. The proposed algorithm is also flawed, similar to the traditional LDA; the algorithm also is faced with the problem of Small Sample Size problem, in order to avoid the Small Sample Size problem, we use PCA to reduce the dimension. Our algorithm is effective to improve the performance of LLDA, in summary, the main steps of Local Linear Discriminant Analysis Using ℓ_2-Graph are as follows:

Step 1 PCA pretreatment. We used PCA to reduce the dimension of each image.

Step 2 Use ℓ_2-graph method to calculate the reconstructed coefficient of the sample, we choose K samples which are the top largest reconstructed coefficient, use them as K-nearest-neighbor of the given sample.

Step 3 Use (13.10) to construct the between-class scatter matrix $\mathbf{S}_b^{L2G}$ of these nearest neighbors, use (13.11) to construct the within-class scatter matrix $\mathbf{S}_w^{L2G}$ of these nearest neighbors.

Step 4 If there is only one nearest neighbor in each of the c' classes, the within-class scatter matrix of these nearest neighbors is a zero matrix. The Eq. (13.12) can be rewritten as (13.13), then solve the eigen-Eq. (13.13).

Step 5 If all of the K nearest neighbors are from only one class, then the between-class matrix is a zero matrix. The Eq. (13.12) can be rewritten as (13.14), then solve the eigen-Eq. (13.14).

Step 6 If the condition is not satisfied in step (4) or in step (5), get the solution by solving $\mathbf{S}_b^{L2G}\mathbf{w} = \lambda\mathbf{S}_w^{L2G}\mathbf{w}$.

Step 7 Get the feature matrix and then to classify.

13.6 Experiments on ORL Database

In this section, we used the ORL face database to test our method and the other methods. The ORL database includes 400 face images taken from 40 subjects, with each subject providing 10 face images. We divide the samples of each subject into two parts; we take one part as the training samples, and take the remaining as the test samples. Number of training samples can be respectively selected 4, 5, 6, 7, and 8 which make up the five subsets of the training data. The sample images of some person from ORL face database are shown in Fig. 13.1.

In order to reduce the amount of calculation, in the experiment, we use PCA to reduce the dimension of each image to 40, and then we verify the performance of the algorithm in several experiments, and compare with the LLDA and other algorithms.

In this experiment, each type of algorithms is randomly conducted 15 times; Table 13.1 shows that our method is better than others. We can observe that as the number of training samples increases, the rate of recognition algorithm is increasingly high.

13.6.1 Experiments on YALE Database

There are 165 face images in the YALE face database; those face images comes from 15 individuals each providing 11 images. We divide the samples of each

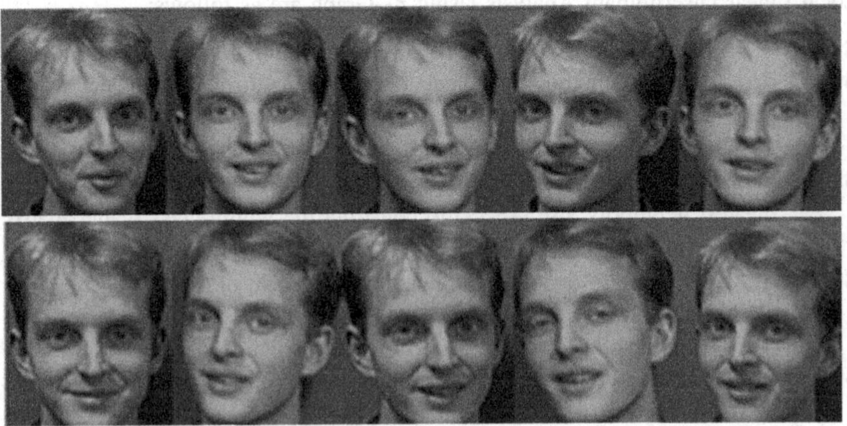

Fig. 13.1 Some faces of a subject of the ORL database

Table 13.1 Classification accuracies (mean ± std-dev percent) of algorithms of ORL data set

	N = 4	N = 5	N = 6	N = 7	N = 8
VLLDA	92.58 ± 2.0	95.15 ± 1.3	96.32 ± 1.6	97.58 ± 1.8	98.00 ± 1.3
L2G_LLDA	95.39 ± 1.6	96.30 ± 1.1	97.46 ± 0.7	98.28 ± 0.7	98.33 ± 0.8

Fig. 13.2 Some faces of a subject of the YALE database

Table 13.2 Classification accuracies (mean ± std-dev percent) of algorithms of YALE data set

	N = 4	N = 5	N = 6	N = 7	N = 8
VLLDA	84.48 ± 3.0	86.78 ± 3.1	86.95 ± 2.9	87.00 ± 3.1	87.77 ± 3.1
L2G_LLDA	90.54 ± 2.6	92.82 ± 2.9	92.62 ± 3.0	92.56 ± 3.1	92.89 ± 2.8

subject into two parts; we take one part as the training samples, and take the remaining as the test samples. Number of training samples can be respectively selected 4, 5, 6, 7, and 8 which make up the five subsets of the training data. The sample images of some person from YALE face database are shown in Fig. 13.2.

In order to reduce the amount of calculation, in the experiment, we use PCA to reduce the dimension of each image to 40, and then we verify the performance of the algorithm in several experiments, and compare with the LLDA and other algorithms.

In our experiment, each type of algorithms is randomly executed 15 times; the data in Table 13.2 is the average recognition rate of 15 groups of experiments. We observed that as the number of training samples increases, the recognition rate is also increasing, and our algorithm at least 5–6 % higher than the other methods, in other words, our algorithm is better than others.

13.6.2 Experiments on AR Database

The AR face database has 3120 images taken from 120 subjects, which were taken at different times, with different facial expression, pose, or lighting. On AR face database, we divide the samples of each subject into two parts; we take one part as the training samples, and take the remaining as the test samples. Number of training samples can be respectively selected 10, 11, 12, 13, and 14 which make up the five subsets of the training data. The sample images of some person from AR face database are shown in Fig. 13.3.

Fig. 13.3 some faces of a subject of the AR database

Table 13.3 Classification accuracies (mean ± std-dev percent) of algorithms of AR data set

	N = 10	N = 11	N = 12	N = 13	N = 14
VLLDA	92.72 ± 0.8	94.18 ± 0.6	95.57 ± 0.7	96.10 ± 0.4	96.93 ± 0.5
L2G_LLDA	97.28 ± 1.2	97.98 ± 1.4	97.65 ± 1.7	99.05 ± 0.6	99.06 ± 0.8

In order to reduce the amount of calculation, in the experiment, we use PCA to reduce the dimension of each image to 40, and then we verify the performance of the algorithm in several experiments, and compare with the LLDA and other algorithms.

In our experiment, each type of algorithms is randomly executed 15 times; the data in Table 13.3 is the average recognition rate of 15 groups of experiments. We observed that as the number of training samples increases, the recognition rate is also increasing. Overall, the classification performance of the algorithm is superior to other algorithms.

13.7 Conclusions

Combining the advantages of ℓ_2-graph and LLDA, we propose a method that uses the size of the reconstructed coefficient to choose the nearest neighbor samples, we call it Local Linear Discriminant Analysis Using ℓ_2-Graph. It uses local information data effectively, and retains the internal structure among samples. In YALE, ORL and AR face database show the effectiveness of the method.

References

1. Fan Z, Xu Y, Zhang D (2011) Local linear discriminant analysis framework using sample neighbors. IEEE Trans Neural Netw 22(7):1119–1132
2. Wright J, Yang AY, Ganesh A et al (2009) Robust face recognition via sparse representation. IEEE Trans Pattern Anal Mach Intell 31(2):210–227
3. Cheng B, Yang J, Yan S et al (2010) Learning with-l1graph for image analysis. IEEE Trans Image Process 19(4):858–866
4. Peng X, Zhang L, Yi Z (2012) Constructing l2-graph for subspace learning and segmentation. arXiv preprint arXiv:1209.0841

Chapter 14
A New Detecting and Tracking Method in Driver Fatigue Detection

Yang Yu, Xiaobin Li and Haiyan Sun

Abstract In order to reduce the traffic accidents caused by driver fatigue, this paper extends the TLD tracking algorithm to the case of detecting and tracking driver's face and local areas which is based on the machine vision. We address problems of detecting driver's face and local areas, which contain the driver's fatigue information, via adaboost cascade classifier based on haar-like features and track the target areas by the TLD method. The main results are given in terms of the accuracy of detecting driver's face and local areas. Lay a foundation for the driver fatigue detection.

Keywords Driver fatigue detection · Detecting and tracking · Haar-like features · Adaboost cascade classifier · TLD

14.1 Introduction

In traffic accidents, fatigue driving has a high proportion. In China, only in 2008, there are 2568 traffic accidents, in which 1353 people were killed and 3219 people were injured. That caused 57.38 million yuan direct economic losses [1]. To reduce the huge losses caused by fatigue driving, a lot of research has been done. The method of driver fatigue detection based on machine vision technology because of

Y. Yu · X. Li (✉)
School of Electrical and Electronic Engineering, Shanghai Institute of Technology, Shanghai 201418, China
e-mail: lixiaobinauto@163.com

Y. Yu
e-mail: 704403329@qq.com

H. Sun
School of Ecological Technology and Engineering, Shanghai Institute of Technology, Shanghai 201418, China
e-mail: sj-shy@163.com

© Springer-Verlag Berlin Heidelberg 2016

Y. Jia et al. (eds.), *Proceedings of the 2015 Chinese Intelligent Systems Conference*, Lecture Notes in Electrical Engineering 359,
DOI 10.1007/978-3-662-48386-2_14

its completeness of theory and nonaggression of detection properties has quickly become the research highlights.

The key about the fatigue driving detection based on machine vision is how to detect the areas accurately which include the driver's fatigue information in the driver monitoring video. Now, the main method is using adaboost cascade classifier based on the haar-like feature to detect the face and local areas in the static images (frames) in the video. Then extract the information of driver fatigue, such as the perclos parameter [2] and the yawn parameter, to detect the driver's fatigue. So the method has not use the implied dependency information between frames of video sequence. That makes the calculation amount large. And it will be larger with the video quality improving. To meet a better real-time property and reduce the hardware requirements, this paper presents a new method that combines adaboost cascade classifier with TLD (Tracking-Learning-Detection) tracking algorithm. Thus, it uses the correlation between frames of video sequence sufficiently, in the situation that the driver usually do not move a lot when driving, to reduce the calculated amount.

14.2 Algorithm Flow

In order to locate the driver's face and local areas precisely with a lower computational complexity, first of all, detect the driver's face and local areas by the adaboost cascade classifier in the first few frames of the video. Then track the target area (face and local areas) by the TLD tracking method. The algorithm flow chart is shown in Fig. 14.1.

Detecting driver's face and the local areas:

Haar-like features originally has been presented by Papageorgiou et al. [3, 4] and used in face detection. After that, Viola, et al. [5, 6] expanded the haar-like features to improve the detection accuracy. Finally, Lienhart et al. [7] developed haar-like features to 15 kinds. They are shown in the Fig. 14.2.

Aim to reduce the calculation amount of haar-like features, Viola et al. presented the concept of image integration. It changes the calculation of haar-like features from complex numerical integration to simple table lookup computation. That makes the quick detection possible.

Adaboost algorithm [8] namely the adaptive boosting, was presented by Freund and Schapire, in 1995. The main idea is to construct a learning framework that cascades the weak classifier which based on the haar-like features to get the strong classifier which can achieve any degree of accuracy.

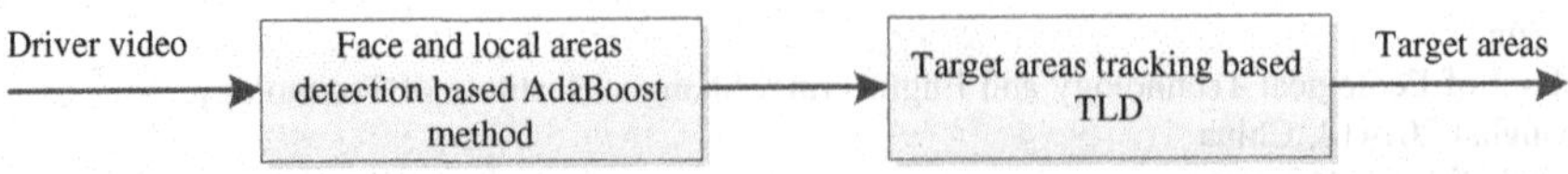

Fig. 14.1 Detection and tracking flowchart

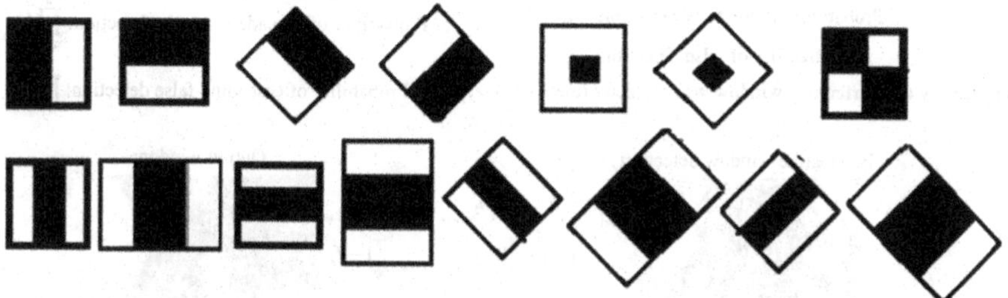

Fig. 14.2 Haar-like features developed by Lienhart

Construct the weak classifier: construct the weak classifier based on the haar-like feature of the face and local areas. The equation of weak classifier construction is as (1). Where, $f_j(x)$ is the haar-like feature. p_j is the category orientation parameter. θ_j is the threshold.

$$h_j(x) = \begin{cases} +1 & \textit{if } p_j f_j(x) < \theta_j \\ -1 & \textit{otherwise} \end{cases} \tag{1}$$

Construct the strong classifier: construct the strong classifier by cascading several weak classifiers. The main idea is dynamic adjusting the classifier by iteration. In each iteration, update the sample weights through the analysis of the classification results, then generate a new weak classifier to compensate the lack of the preceding classifiers. After that, the strong classifier can be gained.

Detection experimental results:

Use the adaboost cascade classifier got from the above to detect the driver's face and local areas. The detection flow chart is shown in Fig. 14.3.

And the adaboost cascade classifier is used to detect the driver's face and local areas in the first frames of the video as the Fig. 14.4 shown. The detection result is shown in Fig. 14.5.

Tracking driver's face and the local areas:

Comparing with the detecting driver's face and local areas frame by frame, using effective tracking algorithm can reduce the calculated amount. Consequently, use

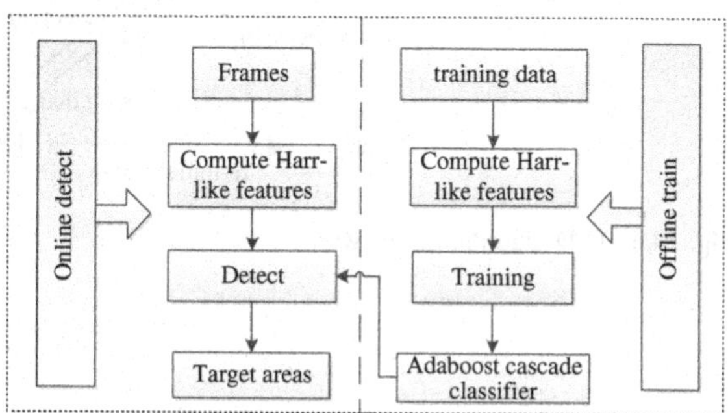

Fig. 14.3 Detection by Adaboost cascade classifier flowchart

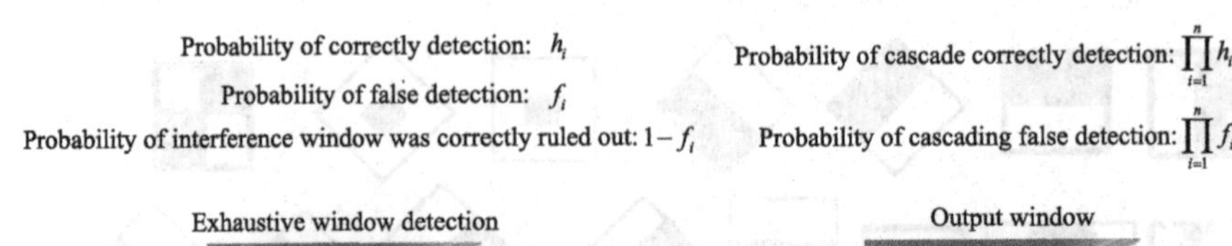

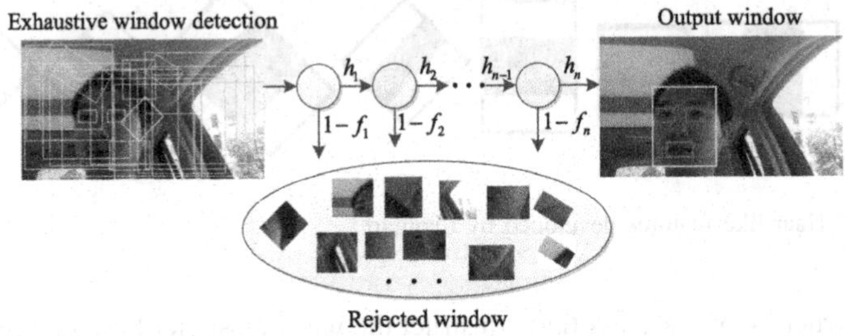

Fig. 14.4 Adaboost classifier cascade detecting structure diagram

Fig. 14.5 Adaboost classifier cascade detecting chart

TLD (Tracking-Learning-Detection) [9] proposed in 2012 by Kalal to track face and the local areas. The main feature of TLD is the learning mechanism, which extends the self-learning to the simple tracking method to deal with the problem of long time tracking. The algorithm framework is shown in Fig. 14.6.

The TLD algorithm can update the detector by the online learning mechanism and use the detector to reinitialize the tracker when the tracker lost the target. So, TLD realizes the long time tracking precisely and rapidly. The module structure is shown in Fig. 14.7.

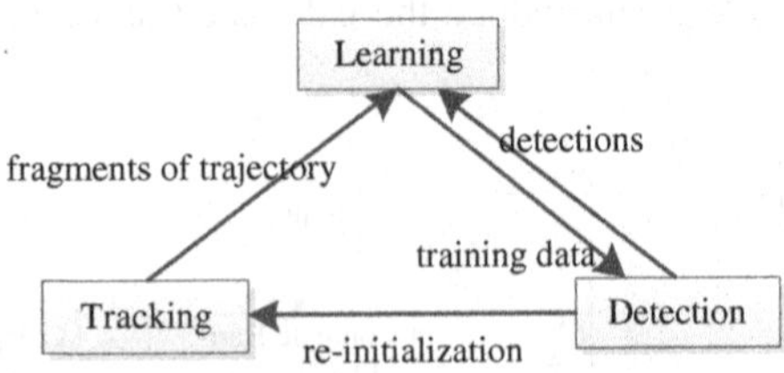

Fig. 14.6 TLD algorithm framework

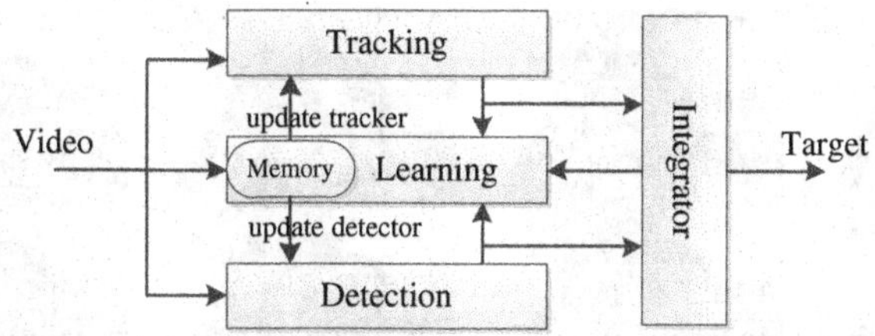

Fig. 14.7 TLD module diagram

TLD process:

Step 1 Define the tracking target in the first frames of the input video
Step 2 Initialize the detector and the tracker
Step 3 For the new frames, use detector to detect the target. Then use tracker to predict the location of the target
Step 4 According to the comprehensive results of the detector and tracker, determine the location of the target
Step 5 Update the detector by learning mechanism
Step 6 Terminate or skip to Step 3 to continue

Merits and demerits of TLD are that the Learning mechanism makes it can long time tracking the target with satisfactory precision and speed. Even lost the tracking target for some interference, it can be recovered by the detector. But it needs the definition of the target.

14.3 Simulation Experiment

This paper uses the adaboost cascade classifier to define the tracking target for the TLD. Then use the TLD to track the driver's face and local areas to lay a foundation for the driver fatigue detection.

The above algorithm is applied to the driver fatigue detection. The simulation experiment results are shown in Fig. 14.8 and Table 14.1.

This method can quickly and accurately capture driver's face and the local areas for extracting fatigue information to detect the driver fatigue.

14.4 Analysis of Experimental Results

This paper uses adaboost cascade classifier based on the haar-like to detect the driver's face and local areas. It sets the tracking target for TLD. Then use TLD to tracking the target areas to reduce the computational complexity by making use of the correlation between frames of video sequence sufficiently. From the

Fig. 14.8 Driver's face and local areas detection and tracking results

Table 14.1 Experimental results

Video no.	Frames	False frames in face detection	Face localization accuracy (%)	False frames in local areas detection (%)	Local areas localization accuracy (%)
1	506	2	99.6	66	86.96
2	295	0	100	0	100
3	876	0	100	15	98.21
4	634	0	100	36	94.35
5	671	0	100	18	97.23
Mean accuracy			99.92		95.35

From the simulation experiment results, the method of this paper has a satisfactory accuracy in detecting and tracking the driver's face and local areas.

experimental results, the mean accuracy of detecting driver's face localization is 99.92 %. And the accuracy of local areas localization is 95.35 %. The precision meets the requirement of driver fatigue detection. Lay a foundation for extracting driver fatigue information and fatigue driving discrimination.

Acknowledgments This work is partially supported by Shanghai key scientific research project No. 11510502700, and science and technology innovation focus of SHMEC No. 12ZZ189.

References

1. Duhou L, Qun L, Wei Y et al (2010) Relationship between fatigue driving and traffic accident [J]. J Traffic Transport Eng 10(2):104–109
2. Wierwille WW, Ellsworth LA, Wreggit SS et al (1994) Research on vehicle-based driver status/performance monitoring; development, validation, and refinement of algorithms for detection of driver drowsiness [R]. National Highway Traffic Safety Administration. Report No. DOT HS 808247
3. Papageorgiou CP, Oren M, Poggio T (1998) A general framework for object detection [C]. In: The 6th International Conference on Computer Vision (ICCV 1998), pp 555–562
4. Mohan A, Papageorgiou CP, Poggio T (2001) Example-based object detection in images by components [J]. IEEE Trans Pattern Anal Mach Intell 23(4):349–361
5. Viola P, Jones M (2001) Rapid object detection using a boosted cascade of simple features [C]. IEEE Computer Society Conference on Computer Vision and Pattern Recognition (CVPR 2001). Kauai, Hawaii, USA, 2001, vol 1, pp 511–518
6. Viola P, Jones MJ (2004) Robust Real-time Face Detection [J]. Int J Comput Vision 57 (2):137–154
7. Lienhart R, Maydt J (2002) An extended set of haar-like features for rapid object detection [C]. In: IEEE International Conference on Image Processing (ICIP 2002), pp 900–903. Rochester, New York, USA
8. Freund Y, Schapire RE (1996) Experiments with a new boosting algorithm [C]. In: Proceedings of the 13rd international conference on machine learning, pp 148–156
9. Kalal Z, Mikolajczyk K, Matas J (2012) Tracking-learning-detection [J]. IEEE Trans Pattern Anal Mach Intell 34(7):1409–1422

Chapter 15
A Multi-objective Dual-Resource Shop Scheduling Model Considering the Differences Between Operational Efficiency

Weizhong Wang, Jialian Wang, Chencheng Ma, Zhenqiang Bao and Richao Yin

Abstract Multiple resource constraints exist widely in the actual job shop under the complicate manufacturing environment, so a multi-objective dual-resource shop scheduling model considering the differences between operational efficiency is established, which contains two resource constraints: the machines and workers. An improved SPEA2 is proposed for the model which achieves three goals: the shortest completion time, the lowest cost and the minimum total tardiness. Since the evolutionary operation in the traditional algorithm for single-resource model has not been able to meet requirements of the model in this paper, the chromosome coding, chromosome decoding, initialization of the population, crossover operation and mutation operation are especially improved so that the machines and workers can be assigned to the process reasonably. Finally, the feasibility and efficiency of the model are proved by the simulation experiment.

Keywords Operational efficiency · Dual-resource · SPEA2 · Multi-objective · Scheduling model

---

W. Wang (✉) · J. Wang · C. Ma · Z. Bao · R. Yin
Information Engineering College, Yangzhou University, Jiangsu 225127, Peoples's Republic of China
e-mail: 1126829265@qq.com

J. Wang
e-mail: 381474594@qq.com

C. Ma
e-mail: 1014436010@qq.com

Z. Bao
e-mail: yzbzq@163.com

R. Yin
e-mail: 1013683495@qq.com

© Springer-Verlag Berlin Heidelberg 2016
Y. Jia et al. (eds.), *Proceedings of the 2015 Chinese Intelligent Systems Conference*, Lecture Notes in Electrical Engineering 359,
DOI 10.1007/978-3-662-48386-2_15

15.1 Introduction

The job-shop scheduling problem is one of the most basic and famous scheduling problem. It is also the core part of the production management in the manufacturing enterprises. When the vast majority of scholars solve the problem, they consider machines the only resource constraint. At present, research about dual-resource shop scheduling have already been done at home and abroad, but not too much. Xu et al. studied the development of dual-resource constrained system. Considering many uncertain factors in the actual job-shop scheduling system, Ju Quanyong and Zhu Jianying established the mathematical model of fuzzy scheduling based on the study of multi-routing dual-resource job-shop scheduling. Based on learning amnesia, John et al. solved the dual-resource shop scheduling model considering the factors of task-type. Bernd Scholz-Reiter et al. studied the dual-resource shop scheduling based on the priority rules. Literature [5–11, 14] utilized optimization algorithm to optimize the dual-resource shop scheduling. Chen Xi et al. studied the optimization of dual-resource double-objective job-shop scheduling in which routing is variable. Liu Xiaoxia et al. proposed a new calculation method to calculate the production cost of the dual-resource job-shop scheduling. But problems still exist: (1) While considering the two resources of machines and workers simultaneously, we usually regarded them as two independent resources. (2) In the single part small batch production mode, the workers' operational efficiency is generally changeable when various workers operate the same machine or the same worker operates various machines. Most literature generally considered the production cost as the single goal when studying the single-objective dual-resource scheduling. But in actual production, the differences between operational efficiency would directly affect multiple goals: completion time, due date, cost, quality and so on. That is to say, the multiple conflicting goals which are required to be optimized at the same time generally existed in actual production. Thus, the multi-objective dual-resource scheduling model considering the differences between operational efficiency is established, which can achieve three goals: the shortest completion time, the lowest cost and the minimum total tardiness. This model will be solved by the improved SPEA2.

15.2 Establishment of the Model

15.2.1 Problem Description

In the multi-objective dual-resource shop scheduling model, there are n jobs need to be processed in the workshop consisted of w workers and m machines. Each job contains one or several working procedures, sequence constraints exist in different working procedures, each working procedure can be processed on one or more machines, the processing cost per unit time of each machine is known, but performance varies in different machines, so the actual processing time of each working procedure differs with machine performance. There are w workers, each worker can operate one or more machines, and $w < m$. The same worker has different efficiency due to different machines, and different workers have different operational efficiency when operating the same machine. Therefore, the differences between operational efficiency cause effects on the actual processing time of the working procedure.

The goal of scheduling is to determine the best processing sequence of jobs on the machines, as well as the proper machines and operating workers, so the resource can be allocated optimally. Meanwhile the model employs the completion time, total cost and total tardiness as optimization objectives to optimize them overall.

In addition, this article assumes that:

(1) All machines and workers are available at time zero, all working procedures (the first working procedure of job) at time zero can be processed immediately.
(2) At the same time one machine can only process one working procedure and the machine can be only operated by one worker.
(3) There is strict sequence between different working procedures of the same job, but there are no sequence constraints in working procedures of different jobs. Different jobs have the same priority level.

15.2.2 Objective Functions

The objective functions need to be optimized simultaneously are as follows:

(1) Total cost

$$
\begin{aligned}
\mathrm{f}_1 = & I\sum_{i=1}^{n} C_i\left(TS_{i(Np_i+1)} - TR_{i1}\right) + \sum_{i=1}^{n}\sum_{j=1}^{Np_i}\left(Cm_k \times t'_{ijk} + Cw_k \times t_p\right) \\
& + I\sum_{i=1}^{n}\sum_{j=1}^{Np_i}\left\{\left(TS_{i(Np_i+1)} - TS_{ij}\right)\left(Cm_k \times t'_{ijk} + Cm_k \times t_p\right)\right\} \\
& i = 1,2,\ldots,n \quad j = 1,2,\ldots,Np_i \quad k = 1,2,\ldots,m \quad p = 1,2,\ldots,w \quad t'_{ijk} = t_{ijk}/e_{w_k}.
\end{aligned}
$$

Since the annual interest rate is relatively small, so it can be ignored. The function of total cost can be as follows:

$$
\begin{aligned}
\mathrm{f}_1 = & \sum_{i=1}^{n}\sum_{j=1}^{Np_i}\left(Cm_k \times t'_{ijk} + Cw_k \times t_p\right) \\
& i = 1,2,\ldots,n \quad j = 1,2,\ldots,Np_i \quad k = 1,2,\ldots,m \quad p = 1,2,\ldots,w
\end{aligned}
$$

(2) Completion time

$$f_2 = \max\{TS_{i(Np_i+1)} | i = 1, 2, \ldots, n\}$$

$$TS_{i(Np_i+1)} = \begin{cases} D_i, & \text{if } TE_{iNp_i} \le D_i \\ TE_{iNp_i}, & \text{if } TE_{iNp_i} > D_i \end{cases}$$

(3) Total tardiness

$$f_3 = \max\left\{\sum_{i=1}^{Np_i} (TE_{iNp_i} - D_i),\ 0\right\}$$

The follows are the meanings of each letter: $J_i(i \in \{1,2,3,\ldots,n\})$ means the job i, Np_i means the total number of working procedures for job i, $O_{ij}(j \in \{1,2,3,\ldots,Np_i\})$ means the number j working procedure of job i. $M_k(k \in \{1,2,3,\ldots,m\})$ means the machine k, $W_p(p \in \{1,2,3,\ldots,w\})$ means the worker p. C_i means the cost of raw material of job i, I means the annual interest rate. TR_{ij} means the ready moment of O_{ij}, TS_{ij} means the starting moment of O_{ij}, TE_{ij} means the completion moment of O_{ij}. Cm_k means the hourly rate of machine k, Cm_k means the hourly rate of a worker operating machine k. t_{ijk} means the processing time in theory for O_{ij} operating on machine k, t'_{ijk} means the actual processing time for O_{ij} operating on machine k with the effect of workers' operational efficiency. t_p means the total time starting from workers operate the first machine to the last completed. e_{w_k} means the impact factor of a worker operating machine k.

15.3 The Improved SPEA2 for the Model

In this section, we propose an algorithm based on the improved SPEA2. In this algorithm, we improve the coding, decoding, selection operation, crossover operation and mutation operation of chromosome, so that the algorithm can assign machines and workers to process reasonably and solve the multi-objective dual-resource shop scheduling model effectively.

15.3.1 The Algorithm Program

(1) Generate the initial population P_0 of size N randomly, the archive set Q_0 of size M is empty, T means the evolution algebra, and set t = 0;
(2) Calculate the fitness of each individual in P_t and Q_t, denoted by F(i);
(3) Choose the non-dominant individuals from P_t and Q_t, then put them into Q_{t+1}. If $|Q_{t+1}| > M$, then cut the needless individual by the pruning process.

If $|Q_{t+1}| < M$, then choose the dominant individuals of number $M - |Q_{t+1}|$ with smaller fitness value into Q_{t+1};

(4) If $t > T$, end the evolution, all of the non-dominant solutions in Q_{t+1} are the final optimization solutions, else back to step (5);

(5) Use the tournament method to choose the pairs for Q_{t+1};

(6) Execute the crossover and mutation operation on the paired chromosomes, choose the excellent individuals into the next generation, $t = t + 1$, then back to step (2).

15.3.2 Coding and Decoding

Coding: according to the characteristic of the multi-objective dual-resource shop scheduling model in this paper, we designed a three dimensional coding scheme. As shown in Fig. 15.1, a complete chromosome is consisted of three parts: process coding, machine coding and worker coding.

Process coding uses the following method: the same job expresses in the same real number, working procedure order is the appearing order of corresponding real numbers. As shown in Fig. 15.1, the first 1 represents the first working procedure of job 1, the second 1 represents the second working procedure of job 1, the third 1 represents the third working procedure of job 1, and so on.

Machine coding adopts the method based on real number. According to the simulation in this paper, the 1–3 position on the machine coding chromosome represent machines that respectively work for the first, the second, the third working procedure of job 1, and so on.

Worker coding is similar to machine coding.

Decoding: correspond the sequence of process chromosome, machine chromosome and worker chromosome, the top three position corresponding with the process coding are decoded as follows: (O_{21}, M_4, W_3), (O_{22}, M_1, W_1), (O_{11}, M_7, W_6).

15.3.3 Selection Operation

When the algorithm constructs new populations, environment selection starts firstly, then mating selection starts. The detailed selection program is shown in the program 3.1.

process coding	2	2	**1**	3	4	2	4	5	3	3	**1**	5	**2**	3	4	**1**	2	2
machine coding	7	6	7	4	1	4	7	5	2	2	2	8	5	6	4	4	4	8
worker coding	6	5	6	3	1	3	6	4	2	2	2	6	4	5	3	3	3	6
	J_1			J_2						J_3				J_4			J_5	

Fig. 15.1 Three dimensional coding scheme

15.3.4 Crossover Operation

Process crossover operation: adopt the IPOX crossover operation based on process coding:

As shown in Fig. 15.2, all the jobs are randomly divided into two non-empty set *J*1 and *J*2, copy the jobs of *P*1 included in *J*1 into *C*1, copy the jobs of *P*2 included in *J*2 into *C*2, retain their gene-bits in the original chromosome, copy the jobs of *P*2 included in *J*2 into *C*1, copy the jobs of *P*1 included in *J*1 into *C*2, retain their order in the original chromosome simultaneously.

Machine crossover operation: first generate a crossover points randomly (a is a real number not greater than the length of the chromosome), select the machines corresponding with the crossover points above in M1 and M2 orderly, then exchange them, other machines in M1 and M2 are reserved to the children C1 and C2. As shown in Fig. 15.3, the number of random crossover points is 3, respectively located in position 2, 5, 8.

Worker crossover operation: similar to machine crossover operation.

15.3.5 Mutation Operation

Process mutation operation: using the method of random insertion mutation. As shown in Fig. 15.4, randomly select one working procedure first, and then randomly generate an insertion point, insert the working procedure into the position of the insertion point.

Machine mutation operation: as shown in Fig. 15.5, select a gene randomly from machine coding chromosome, replace it by a random gene in the alternative processing sets.

Worker mutation operation: similar to machine mutation operation.

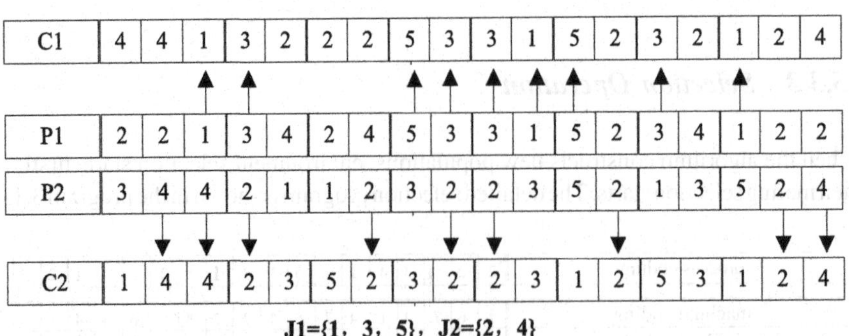

Fig. 15.2 Process crossover operation

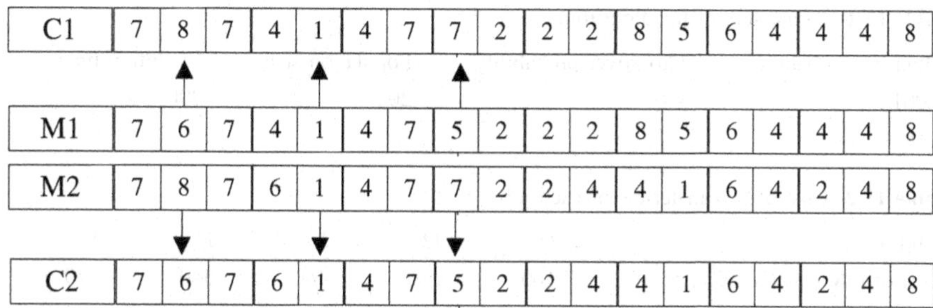

Fig. 15.3 Machine crossover operation

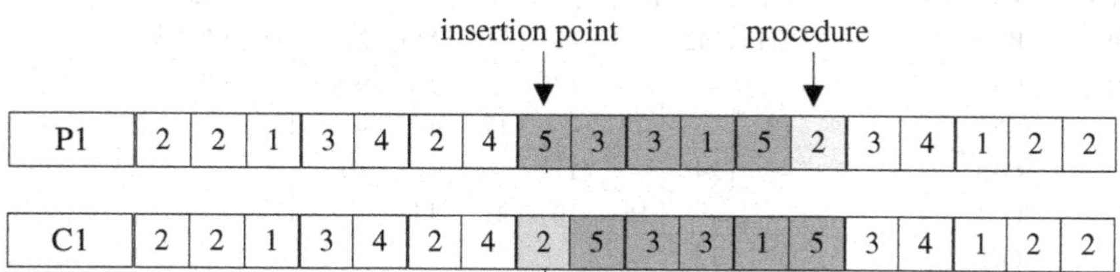

Fig. 15.4 Process mutation operation

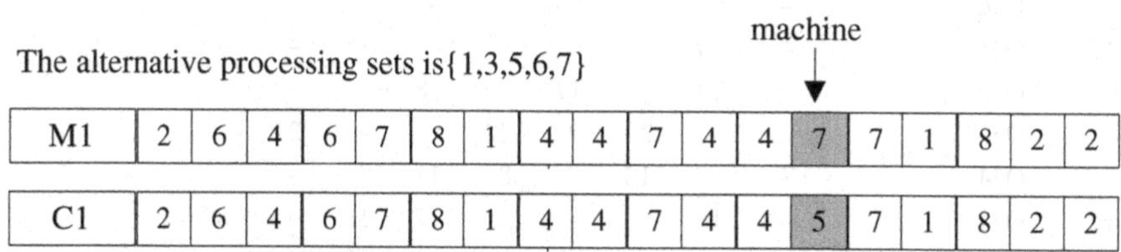

Fig. 15.5 Machine mutation operation

15.4 Simulation and Analysis

The proposed algorithm is coded with Java, the running environment for the program is as follows: P4 CPU, 2.8 GHz, and 1.25G RAM.

The execution parameters of the multi-objective evolutionary algorithm are shown in Table 15.1:

The efficiency of the model and the algorithm is proved by the application of a mechanical processing workshop. This workshop plans to use 8 machines and 6 workers to process five kinds of jobs with multiple working procedures respectively. Detailed data are shown as follows: Table 15.2 shows the cost of raw materials of each job. Table 15.3 shows the detailed processing parameters of each job. Table 15.4 shows the worker set corresponding to each machine. Table 15.5 shows the processing fees per unit time of each worker.

Table 15.1 Parameters of the algorithm

Mutation probability	Crossover probability	Population size	Evolution times
0.001	0.6	20	200

Table 15.2 Cost of raw materials of each job

Job no.	J1	J2	J3	J4	J5
Cost of raw materials C_m(yuan)	160	210	440	340	570

Table 15.3 Detailed processing parameters of each job

Job no.	Working procedure	Processing time t (hour)								Due date D_i (hour)
		M1	M2	M3	M4	M5	M6	M7	M8	
J1	O11	–	–	–	12	–	10	9	–	60
	O12	17	–	–	–	17	10	–	15	
	O13	–	24	–	11	–	–	10	–	
J2	O21	–	–	16	10	21	14	–	17	–
	O22	8	12	–	–	19	11	–	–	
	O23	–	–	–	15	–	21	25	–	
	O24	–	–	–	–	18	–	9	–	
	O25	12	15	–	14	9	–	10	–	
	O26	–	9	–	7	10	8	–	–	
J3	O31	–	14	–	–	–	–	17	–	–
	O32	–	23	23	17	–	18	–	–	
	O33	–	–	20	9	22	–	–	21	
	O34	7	–	10	–	8	11	9	–	
J4	O41	–	18	–	–	–	17	18	–	80
	O42	–	–	–	10	–	–	12	–	
	O43	–	8	11	8	–	–	9	20	
J5	O51	–	–	24	10	16	–	–	–	60
	O52	–	–	–	17	–	–	–	8	
Processing fees per unit time a/(yuan h^{-1})		8	5	10	9	7	6	9	4	

A set of solutions generated after the program finished, parts of the Pareto optimal solutions as shown in Table 15.6 are chosen from the set of solutions according to our personal preference on the premise of giving priority to the due date. In order to analyse comparatively, the dual-resource shop scheduling model considering the average operational efficiency is simulated by the improved SPEA2 simultaneously, parts of the Pareto optimal solutions are shown in Table 15.7. Through the analysis of the data in the two tables, the Pareto optimal solutions

Table 15.4 Worker set corresponding to each machine

Worker no.	M1	M2	M3	M4	M5	M6	M7	M8
W1	1.2	1						
W2		0.8	0.9					
W3			0.9	0.9				
W4					1.1	0.9		
W5						0.9	0.7	
W6							0.9	1.1

Table 15.5 Processing fees per unit time of each worker

Worker no.	W1	W2	W3	W4	W5	W6
Wage (yuan/hour)	20	15	15	20	15	20

Table 15.6 Pareto optimal solutions of workers with different operational efficiency

No.	Optimization goals		
	Processing cost	Completion time	Total tardiness
1	34063	114	0
2	27051	134	0
3	27089	104	0
4	25685	112	0

Table 15.7 Pareto optimal solutions of workers with average operational efficiency

No.	Optimization goals		
	Processing cost	Completion time	Total tardiness
1	34820	133	52
2	33209	147	0
3	30778	108	35
4	31138	116	0

considering the difference between operational efficiency are generally better than the solutions considering the average operational efficiency.

Due to the limited space, according to personal preference, Gantt charts corresponding with the fourth set of results chosen from Tables 15.6 and 15.7 respectively are shown in Figs. 15.6 and 15.7. The horizontal axis means the completion time of jobs, the ordinate means the machine sequence. Take the top green square in Figs. 15.6 as an example, 3-3-6 means the third working procedure of the job 3 is processed on the machine 8 by worker 6.

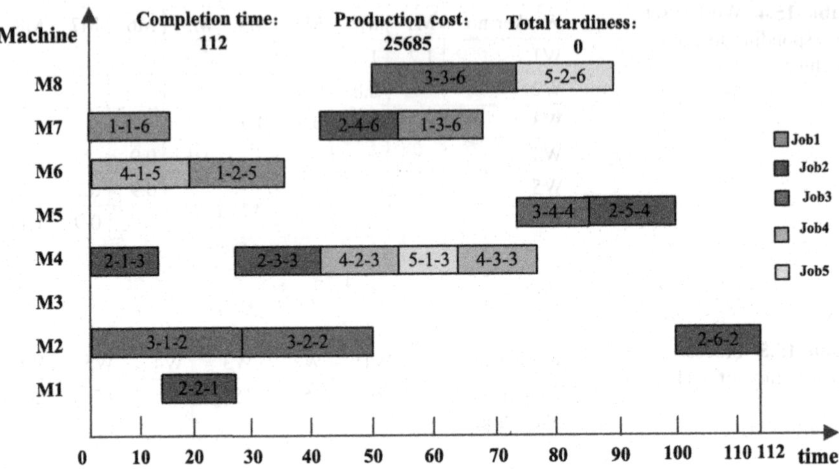

Fig. 15.6 Gantt for workers with different operational efficiency

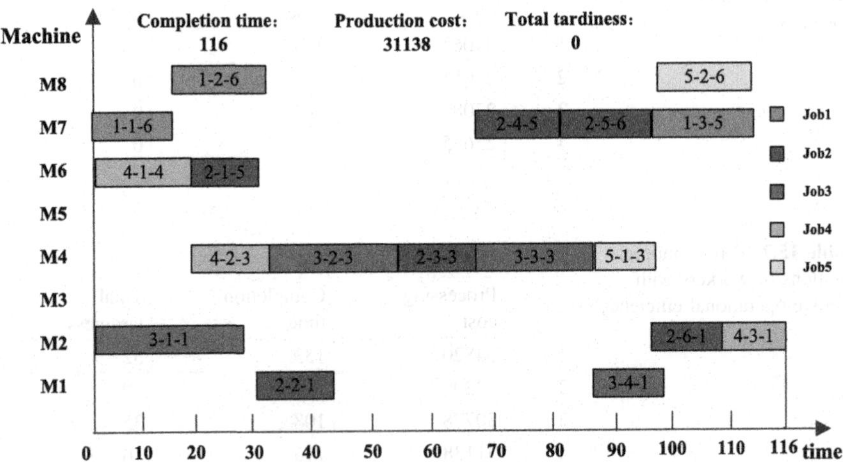

Fig. 15.7 Gantt for workers with average operational efficiency

15.5 Conclusion

In this paper, based on the study of the dual-resource shop scheduling, a multi-objective dual-resource shop scheduling model considering the differences between operational efficiency was established. According to the characteristic of this model, we put forward an improved SPEA2. In this algorithm, we designed a three-dimensional coding theme consisted of process coding, machine coding and worker coding, which was suitable for the requirements of the dual-resource, and then improved the crossover operation and mutation operation of chromosome, repaired the illegal chromosome possibly appeared in the process. Finally, we solved the corresponding model through the contrast experiment. The results showed that the reasonable distribution of workers could not only shorten the production cycle and reduce the production cost, also could improve the comprehensive performance of enterprises. So, the study of multi-objective dual-resource shop scheduling model considering the differences between operational efficiency could reflect the actual production accurately. Finally, the model is correct and the improved algorithm is effective.

Acknowledgements This research was financially supported by the National Natural Science Foundation of China (Grant NO. 61170201) and Research and Innovation Project for College Graduates of High Education of Jiangsu Province of China No. CXZZ13_0919.

References

1. Xu J, Xu X, Xie SQ (2011) Recent developments in Dual Resource Constrained (DRC) system research [J]. Eur J Oper Res 215:309–318
2. Quangyong J, Jianying Z (2006) Study of fuzzy scheduling problems with dual resource and multi process routes [J]. Mech Sci Technol 25(12):1424–1427
3. Zamiska JR, Mohamad Jaber Y, Kher Hemant V (2007) Worker deployment in dual resource constrained systems with a task-type factor [J]. Eur J Oper Res 177(3):1507–1519
4. Scholz-Reiter B, Heger J, Hildebrandt T (2010) Analysis of priority rule-based scheduling in dual-resource-constrained shop-floor scenarios [J]. Mach Learn Syst Eng 68:269–281
5. Melnyk SA, Ragatz G (1996) Information and scheduling in a dual resource constrained job shop [J]. Int J Prod Res 10(34):2783–2802
6. Jingyao L, Shudong S, Yuan H, Ning W (2011) Solving dual resource constrained job-shop scheduling problem (DRCJSP) based on hybrid ant colony algorithm with self-adaptive parameters [J]. J Northw Polytech Univ 29(1):54–61
7. Yinghui D, Hongyan C, Renyuan F, Juan F (2009) The algorithm research on dual-resource and multi-objective job shop scheduling [J]. Mech Des Manuf 12:15–17
8. Di L, Liyang X, Tianzhong S, Ze T (2006) Scheduling optimization based on hybrid genetic-tabu search algorithm for dual-resource constrained job shop [J]. J Northe Univ (Natural Science) 27(8):895–898
9. Zuhong L, Qingxin C, Ning M, Jiayu L (2013) Dual-resource near-optimal configuration method of manufacturing systems with production capability constraints [J]. Comput Integ Manufact Syst 19(6):1263–1271
10. Hoda E, Vishvas P, Imed Ben A (2000) Scheduling of manufacturing systems under dual-resource constraints using genetic algorithms [J]. J Manuf Syst 19(3):186–201
11. Binghai Z, Shuyu J, Ping H, Lifeng X (2008) Scheduling algorithm of flexible production system based on dual resource [J]. Huanan Ligong Daxue Xuebao/J South China Univ Technol (Natural Science) 36(4):45–49
12. Xi C, Quanke P, Ningsheng W (2003) Intelligent optimization of bi-objective job-shop scheduling using genetic algorithms [J]. Mech Sci Technol 22(3):398–401
13. Xiaoxia L, Gangyi C, Liyang X (2009) Research on bi-objective scheduling optimization for DRC Job Shop [J]. Mod Mach Tool Autom Manuf Techn 10:107–112

14. Huiyuan R, Lili J, Xiaoying X, Heping P (2009) Algorithm based on efficiency algorithm for dual-resource constrained job-shop scheduling [J]. Mech Electr Eng Technol 38(6):67–68
15. Chaoyong Z, Yunqing R, Peigeng L (2004) POX crossover operation-based GA for job-shop scheduling problems [J]. China Mech Eng 15(23):2149–2153

Chapter 16
A Vision-Based Traffic Flow Detection Approach

Hongpeng Yin, Kun Zhang and Yi Chai

Abstract Traffic flow detection plays an important role in Intelligent Transportation System (ITS). However, the conventional traffic flow detection approaches are high cost or complex installation. In this paper, a reliably vision-based traffic flow detection approach is proposed. In this approach, Gaussian mixture model (GMM) is employed to model the dynamic background of traffic scene. Then, the binary foreground contours are extracted by image subtraction. Comparing the binary vehicle contours' location and the current frame, the real and complete vehicles are obtained for detecting and monitoring. In the part of vehicle counting, to gather the vehicle flow parameter in each lane of the road and avoid the trouble of counting vehicles repeatedly, a discriminative method is presented to classify vehicles into different lanes. Experiment shows that a desired result can be achieved in the traffic flow detection system by the vision-based approach.

Keywords ITS · GMM · Traffic flow detection · Background subtraction

16.1 Introduction

In last decades, Intelligent Transportation System (ITS) [1] emphasizing the integrated optimization in views of system, control and information science, has got comprehensive development. As a key component of ITS, the traffic flow detection

H. Yin (✉)
National Mountain Highway Engineering and Technology Research Center, Chongqing, China
e-mail: yinhongpeng@gmail.com

H. Yin · K. Zhang · Y. Chai
College of Automation, Chongqing University, Chongqing 400044, China

Y. Chai
Key Laboratory of Power Transmission Equipment and System Security, Chongqing 400044, China

© Springer-Verlag Berlin Heidelberg 2016

Y. Jia et al. (eds.), *Proceedings of the 2015 Chinese Intelligent Systems Conference*, Lecture Notes in Electrical Engineering 359,
DOI 10.1007/978-3-662-48386-2_16

system can distinguish the vehicle targets from the background and provide relative traffic flow information about the urban roads. According to these information, ITS can quickly generate traffic induced control schemes to increase traffic efficiency and safety, reduce the probability of traffic accidents.

The common methods for vehicle flow detection mainly base on annular induction coils, ultrasonic sensors, pressure switch, radar scanner, pneumatic tubes, photoelectric devices and computer vision [2]. With the progress in information technology, the video camera as a promising and low-cost sensor is applied widely in the traffic flow detection system [3]. Thus, vision-based traffic flow detector plays an increasingly important role as the subsystem of ITS.

At present, various vision-based methods of object detection have been introduced in the literature [4–11], and the traditional vehicle detection methods mainly include optical flow filed analysis, temporal difference, virtual loop method and background subtraction. The method of optical flow analysis [4] can obtain the speed of moving objects or segment objects even in moving camera. However, the contour edge of objects segmented by this method is not clear enough. In addition, the high computation complexity is resistive to the implementation of real-time system. Temporal differencing [5] computes the difference between consecutive frames reflecting the features of moving objects. The algorithm is easy to implement and meet the real-time requirements. However, it is easy to be interfered by noise in traffic scene. The virtual loop method [6] can detect the vehicles passed according to the pixel difference between consecutive frames within the virtual-loop area. But it usually is hard to obtain the whole contour edge features for the reconstruction of complete vehicle information and vehicle tracking further in vehicle monitor system. Background subtraction is commonly based on background model technologies [7]. In traffic flow detection, background-updating models are mostly used to distinguish and segment foreground vehicles from the traffic scene, including frame average method, statistical background modeling method [8], codebook model method [9], single Gaussian, and Gaussian Mixture Model (GMM) [10]. Among these methods, GMM can make full use of pixels to model background temporally in the whole analysis process for all video frames, and update the relevant parameters in time online. So it has been widely used in vehicle detection system. More comprehensive methods and development about vehicle detection can be found in [11].

In the traffic flow detection, it is necessary to exact complete vehicles and tackle the problem that the same vehicle is counted repeatedly in order to gather the vehicle flow parameter of each lane in real time. Inspired by the fact that the relationship between any dynamic point and the reference line can be defined in each frame, because of the invariant viewing angle feature under stationary camera, we propose a discriminative method to classify vehicles into different lanes and avoid counting vehicles repeatedly in this paper. Combining the background subtraction method and the discriminative method, a vision-based approach is proposed in this paper to exact complete vehicles and detect the traffic flow in each lane of the road in real time.

The remainder of this paper is organized as follows. In Sect. 16.2, our approach for traffic flow is described. In Sect. 16.3, some experiments are conducted and results are analyzed. Finally, Sect. 16.4 concludes this paper.

16.2 Our Approach for Traffic Flow Detection System

16.2.1 Framework Designed for the System

In this paper, a vision-based approach is proposed for detecting traffic flow in each lane of the road. The flowchart for the detection system is briefly shown in Fig. 16.1. In the traffic scene, many changeable factors exist in the environment related to the background, such as, the shadows, trees, road lights, thus methods well to model the background for traffic scene are needed. Given that, the updating background method based on GMM is employed to segment foreground objects in the framework. Thus, the binary foreground contours of vehicles can be extracted by image subtraction after modelling the background. Comparing the binary vehicle contours' location and the current frame, the real and complete vehicles can be obtained for detecting and monitoring. In the stage of current traffic flow detection, the extracted vehicle targets are classified into different lanes by the *Lane Discriminant Module* based on the invariant viewing angle feature. Then, the number of vehicle in each lane is accumulated and displayed in monitor terminal.

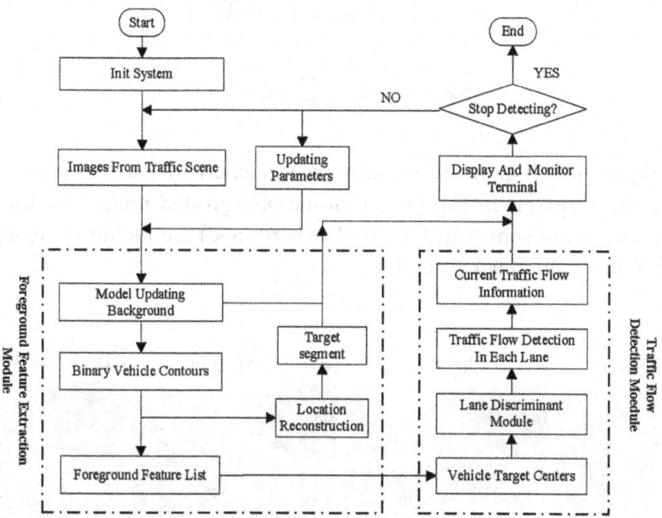

Fig. 16.1 The brief framework for the detection system

16.2.2 GMM-Based Background Model

Background modeling based on Gaussian model is a pixel based process. The pixels' probability density of image can be made into the representation of single Gaussian or the group of several Gaussian functions. In modeling updating background, the most widely used algorithm is Gaussian Mixture Mode (GMM) [10]. In this method, the recent historical values of each pixel from video scene are modeled by a mixture of K Gaussian distributions. The current pixel value X_t can be described by Eq. (16.1).

$$P(X_t) = \sum_{i=1}^{K} \omega_{i,t} * \eta(X_t, \mu_{i,t}, \Sigma_{i,t}) \tag{16.1}$$

where $\omega_{i,t}$ is the weight of the ith Gaussian distribution at time t, η represents the Gaussian probability density function, μ is the mean and Σ is the covariance. In the application, every Gaussian model of each pixel should be updated by learning from recent frames in the traffic scene. The new coming pixel at time t is compared with every Gaussian in the background model until a matched Gaussian is found. If a match is found, the mean and the variance of matched Gaussian are updated accordingly. Otherwise, one new Gaussian with the current mean value and an initially variance, is introduced into the mixture to replace the least probable distribution. The prior weights for new Gaussian at time t are obtained by Eq. (16.2). The background is estimated using these modeled Gaussian models. Firstly, the Gaussians are ordered by the value of ω/σ, where σ is the standard deviation value of Gaussian. The first Bg in ranked distributions is selected as background from Eq. (16.3), here, T is the measure of minimum portion of data accounted for by the background.

$$\omega_{k,t} = (1 - \alpha)\omega_{k,t-1} + \alpha(M_{k,t}) \tag{16.2}$$

$$Bg = \arg\min_b \left(\sum_{k=1}^{b} \omega_k > T \right) \tag{16.3}$$

The background of traffic scene can be modelled in current frame after the comparison for all pixels in Eq. (16.3). Some background images by this method under traffic scene are shown in Fig. 16.2, where (a–c) are the unprocessed frames and (d–f) are the traffic background.

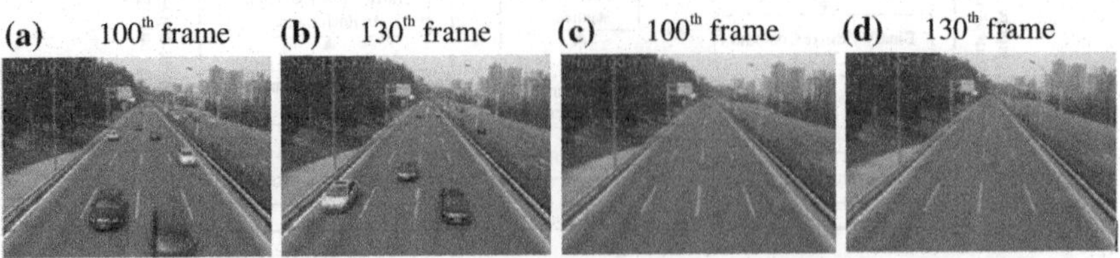

Fig. 16.2 **a**, **b** Are the unprocessed frames and **c**, **d** are the obtained background images

16.2.3 Foreground Feature Extraction

In order to track the vehicles and get the traffic flow, segmenting foreground objects and extracting vehicles' contour feature are needed after modelling the background of traffic scene. For the purpose, the process module for foreground extraction is designed.

In this module, to reduce random noise, Gaussian filter is used in smoothing the current image. Then, every pixel from the new image is processed by the GMM algorithm and the parameters of GMM are updated as described above. After modelling the background of the current scene, image subtraction is adopted to get foreground information, through subtracting the gray background image and the current gray frame image obtained behind smoothing filter. Before extracting contour of vehicle targets, the diff-image is transformed to the binary image. Simultaneously, the dilate operator are used to avoid the discontinuity of partial points in the binary image. After these operations, the object contours in the current diff-image are detected and are saved into the contour feature list firstly. If the area of one counter is smaller than the threshold determined by the average of all contours' area and the location of contours, the counter will be removed from the contour feature list. The satisfying contours will be considered as vehicle contours and be processed as the new feature list further. In the Location Reconstruction, the vehicles under the real traffic scene can be segmented completely by comparing the input frame image with the vehicle contours' location from the new feature list. Thus, the 3-dim vehicles' information can be obtained and sent to the system terminal for displaying and monitoring in ITS. Figure 16.3a, b are the extracted contours for 130th and 200th frame, Fig. 16.3c, d are the detected foreground vehicles under real traffic scene during the process for traffic video.

16.2.4 Vehicle Flow Detection

In the part of vehicle flow detection, it is necessary to gather the vehicle flow parameter from each lane of the road for monitoring the road in real time. It is also needed to improve the accuracy of system by tackling the problem well that the same vehicle is counted repeatedly.

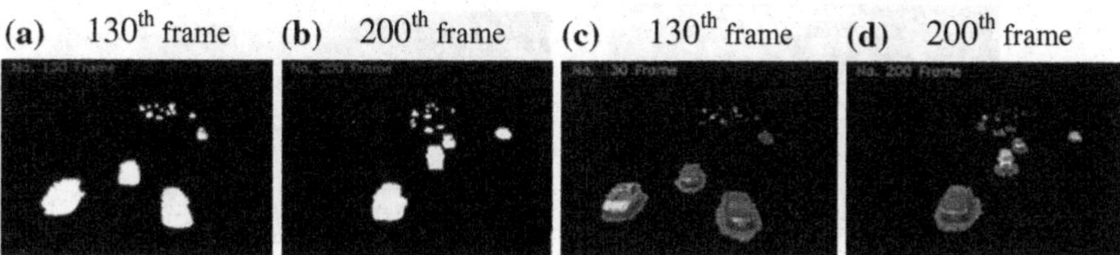

Fig. 16.3 **a**, **b** Are the extracted contours, **c**, **d** are the detected vehicles

In our approach, the traffic video is obtained from the stationary camera, so the viewing angle for every frame can be considered as an invariance factor relative to the ground. T the relationships can be defined between any dynamic point and the reference line given in the coordinate system of image. Moreover, the centers of detected vehicles can be classified into different lanes quickly according to the relationships between the point and the section consisted of different lines. In the same lane, the sorted position of different vehicles is usually not changed in the traffic scene. Thus, the approach can quickly find the new-appearing vehicle and count the number of vehicles by accessing and comparing the position lists of the same lane at two consecutive frames, so it can be helpful to avoid counting vehicles repeatedly and makes the foundation for tracking the new-appearing vehicles quickly.

The coordinate system of image is shown in Fig. 16.4a, which is divided into *part I* and *part II* by the line L defined by two given points A and B. We define the function as shown in Eq. (16.4).

$$f(x,y) = (x - x_A) - (x_B - x_A)(y - y_A)/(y_B - y_A) \quad (16.4)$$

where (x_A, y_A) is the location of A, (x_B, y_B) is the B's location, (x, y) is the location of any point and y_A, y_B satisfies $y_A \neq y_B$. Then the equation of L is described as $f(x,y) = 0$. For any point M (x_M, y_M), the relationships between M and L are described as Eqs. (16.5) and (16.6).

$$f(x_M, y_M) < 0, \text{ then M is in } \textit{part I} \quad (16.5)$$

$$f(x_M, y_M) > 0, \text{ then M is in } \textit{part II} \quad (16.6)$$

Based on the principle above, the relationship between any point and the section consisted of the basic lines can also be definite in the image coordinate system. If $f_L(x,y) = 0$ describes the left boundary of one lane, and $f_R(x,y) = 0$ is the right boundary, the point $C(x_c, y_c)$ can be considered to locate in the lane when it satisfies the discriminative relationship described by Eq. (16.7).

$$f_L(x_c, y_c) > 0, f_R(x_c, y_c) \leq 0 \quad (16.7)$$

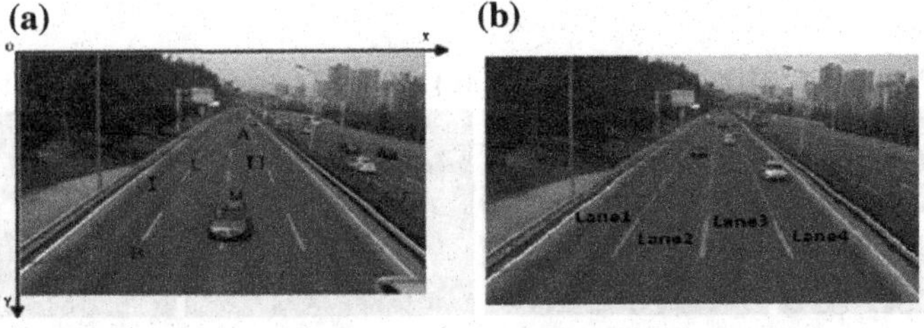

Fig. 16.4 **a** The coordinate system of images, **b** the extracted traffic *lane lines*

According the discriminative relationship, all lane lines need be extracted when detecting traffic flow for one road, as shown in Fig. 16.4b. The section surrounded by these traffic lane lines extracted is called *Detecting Section*, where it is necessary to count the number of vehicles passed in each lane. The procedure for vehicle flow detection based on the theory above is shown in Fig. 16.5.

After the vehicle contour feature list is obtained from the foreground extraction module, every contour will be converted to the approximate rectangle to extract the contour's center. Once the center is in the corresponding lane according to Eq. (16.7), the contour will be recorded in the corresponding lane information list, where the elements are arranged in the descending order according to the y value of the contour centers' location.

After these operations, it can be decided whether one center represents the new vehicle in the corresponding lane through the condition in Eq. (16.8).

$$Cx\_cur - Cx\_last > Cx\_thred, Cy\_cur - Cy\_last > Cy\_thred \quad (16.8)$$

where (Cx_cur, Cy_cur) is one contour's center in current frame, (Cx_last, Cy_last) is last frame's contour center with the same sequence number in the order list recording the vehicles' location, (Cx_thred, Cy_thred) is the minimum difference value of the same vehicle's locations in two consecutive frames. The new vehicle

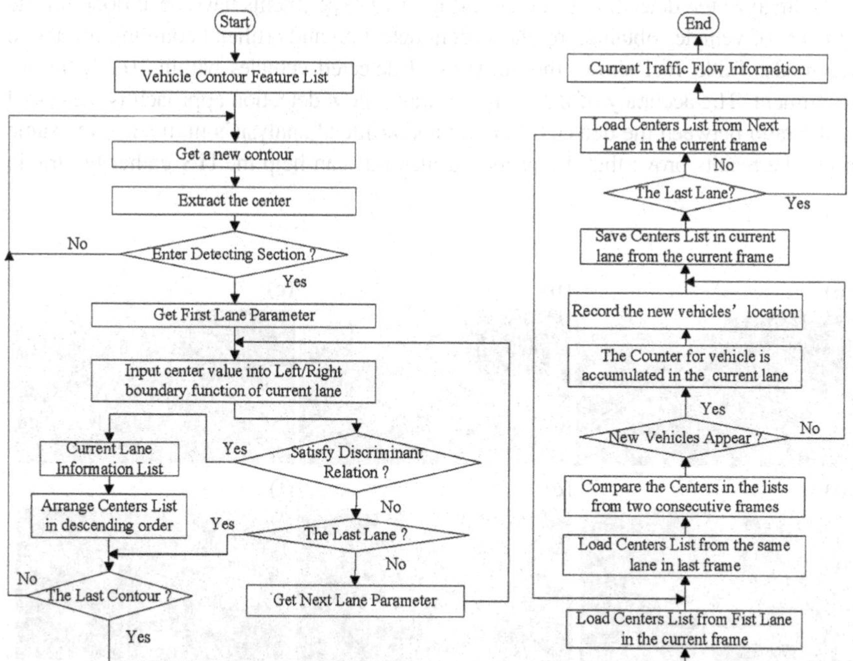

Fig. 16.5 The vehicle flow detection module

appears in the corresponding lane when the condition is satisfied or the center's sequence number in current frame exceeds the total number of vehicle contours in last frame. Then the new vehicle will be recorded into the sequence list of vehicle targets in the relevant lane, simultaneously, the counter for vehicles in the corresponding lane will be accumulated in order to complete the task of traffic flow detection.

16.3 Experiments and Results Analysis

The proposed approach has been implemented on the PCM-3362, an embedded microcontroller with Intel Atom N450 processor and 2 GB DDR2 SDRAM. Figure 16.6 show some image results processed toward the traffic scenes from two roads under different weather conditions in Chongqing, China. Figure 16.6a–c is related to the traffic scene in a cloudy day, and Fig. 16.6d–f shows the traffic scene in a rainy day. The resolution of all the images processed is 320*240 pixels. The results of vehicle flow in each lane and the total number are displayed for the current frame in real time. The experimental image results show that the rate of process can be around 20 frames per second in the cloudy day, and it is around 17 frames per second in the rainy day. It shows that the proposed approach can meet the real-time requirement for the detection system.

To analyze the detection accuracy further, two experiments have been done for the numbers of vehicles obtained by the system detection and artificial counting for above two roads. Table 16.1 shows the numbers of detected vehicles within 10 min in our experiment. The accuracy of the proposed traffic flow detection approach is measured by the ratio between the detected number and artificial analysis's number in the same time. The results prove that the proposed methods can help the ITS gather the traffic

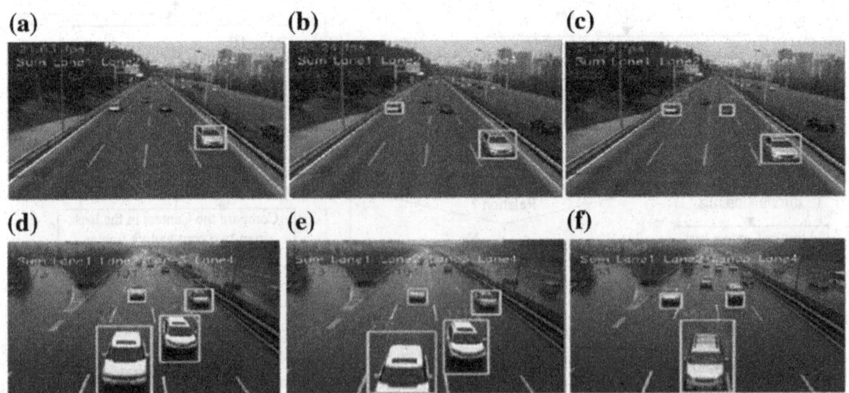

Fig. 16.6 Traffic flow detection result: **a–c** in a cloudy day, **d–f** in a rainy day

Table 16.1 The traffic flow detection result in two roads under different weather conditions

Cloudy day	Detected number	Artificial number	Accuracy (%)	Rainy day	Detected number	Artificial number	Accuracy (%)
Lane1	161	163	98.8	Lane1	88	93	94.6
Lane2	173	177	97.7	Lane2	132	142	93.0
Lane3	106	109	97.2	Lane3	149	161	92.5
Lane4	184	187	98.4	Lane4	134	142	94.4
Sum	624	636	98.1	Sum	503	538	93.5

flow parameters from different lanes in the same road and provide high accuracy for the real-time traffic flow detection under different weather conditions.

16.4 Conclusions

In this paper, a vision-based approach is proposed for traffic flow detection using the invariant viewing angle feature and the updating background model. The proposed approach can extract the contour features of vehicle and segment the vehicle objects well. Besides, the vehicles in each lane can be counted based on the proposed discriminative condition. Experimental result shows that the approach can detect vehicle flow accurately under different weather conditions. But, this approach can't define the lanes adaptively and doesn't perform well when the camera is moving. These are the future work we need to study.

Acknowledgments We would like to thank the supports by Fond National Engineering and Research Center For Mountainous Highways (GSGZJ-2014-07).

References

1. Mo GL, Zhang SY (2010) Vehicles detection in traffic flow [C]. In: Proceedings of 2010 6th international conference on natural computation, INIC2010, pp 751–754
2. Yang MT, Jhang RK, Hou JS (2013) Traffic flow estimation and vehicle-type classification using vision-based spatial-temporal profile analysis. IET Comput Vision 7(5):394–404
3. Tian B, Yao QM, Gu Y, Wang KF, Li Y (2011) Video processing techniques for traffic flow monitoring: a survey [C]. In: Proceedings of 2011 14th international IEEE conference on intelligent transportation systems, ITSC, pp 1103–1108
4. Antonio FC, Carlos CJ, Javier MC, Rafael MT (2010) Optical flow or image subtraction in human detection from infrared camera on mobile robot. Rob Auton Syst 58(12):1273–1281
5. Joshi KA, Thakore DG (2012) A survey on moving object detection and tracking in video surveillances system. Int J Soft Comput Eng 2(3):44–48
6. Yin ZZ, Cao L (2002) Traffic parameters detection based on video virtual-loop technologies [C]. In: Proceedings of the international conference on applications of advanced technologies in transportation engineering, pp 885–893

7. Tsai WK, Lin CC, Sheu MH (2012) High-accuracy background model for real-time video foreground object detection. Opt Eng 51(2)
8. Lu XF, Izumi TS, Teng L, Horie T, Wang L (2012) A novel background subtraction method for moving vehicle detection. IEEJ Trans on Fundam Mater 132(10):857–863
9. Li XY, Fang XZ, Lu QC (2013) On-road vehicle and pedestrian detection using improved codebook model [C]. In: Proceedings of 2013 IEEE international conference on vehicular electronics and safety, ICVES2013, pp 1–4
10. Stauffer C, Grimson WEL (1999) Adaptive background mixture models for real-time tracking. Proc IEEE Comput Soc Conf Comput Vision Pattern Recogn 2:246–252
11. Buch N, Velastin SA, Orwell J (2011) A review of computer vision techniques for the analysis of urban traffic. IEEE Trans Intell Trans Syst 12(3):920–939

Chapter 17
The Research of High-Definition Video Processing System Based on SOC

Jian Gao, Xiaofu Zou, Yue Zhang and Fei Tao

Abstract As people puts forward more requirements for high-definition video industry, capture, transcoding, storage and display of high-definition video then become progressively imperative. This paper presents a parallel video SoC system for faster and flexible 1080P video processing. The key elements of this system, including parallel RGB to YUV transcoder, two-level pipeline and high speed memory controller, are elaborated. 0.028 um CMOS technology node is applied for implementing the SoC architecture. It is competitive in providing high quality images and is proved to be well supportive for DVI and HDMI signal.

Keywords SOC · Video processing · Parallel transcoding · Two-level pipeline

17.1 Introduction

High definition video processing systems have been widely applied in many industrial fields such as weather forecasting, robot vision, medical imaging, precision-guided. In recent years, with the development of mobile portable devices and high-definition video, people have higher demand for picture quality, picture detail, and the portability of video processing system [1]. But by far the most common high-definition video system, i.e. PC capture card model, not only lacks portability but also fails to meet the requirements in real-time video processing [2, 3].

To overcome these problems, researches on SoC video processing mainly focuses on the architecture design of SoC processing, encoding ways of processing data and parallelization of the procedure and so forth. Specifically, Liu et al. [4]

J. Gao (✉) · X. Zou · Y. Zhang · F. Tao
School of Automation Science and Electrical Engineering,
Beihang University, 100191 Beijing, China
e-mail: 15210794508@163.com

© Springer-Verlag Berlin Heidelberg 2016

Y. Jia et al. (eds.), *Proceedings of the 2015 Chinese Intelligent Systems Conference*, Lecture Notes in Electrical Engineering 359,
DOI 10.1007/978-3-662-48386-2_17

firstly presented a SoC architecture for digital HDTV and focused on the design of spatio-temporal adaptive TV decoder, square-nonlinear interpolation scaler and memory controller module. After that, Sun et al. [5] established multi-port AXI bus controller, robust film-mode detection and edge-directed content adaptive image interpolation for the display of Full HD LCD TV. Focusing on the encoder/decoder function, Kangas et al. [6] designed an encoder for SoC video processing system based on homogeneous master-slave architecture, in which each slave handles a part in Single Program Multiple Data (SPMD) data model. For further accelerating the processing speed, Chen et al. [7] achieved AAC and H.264/AVC decoders on parallel architecture core with the consideration of low-power DSP computations and the dynamic voltage and frequency scaling (DVFS) capability on heterogeneous SoC. Due to the rising SoC design complexity, Gupta then concentrated on the functional verification methods for both Video and Audio SoC systems.

On the whole, there are still various problems which hindered the processing speed of the video processing. Firstly, the transcoding module in these architectures is not well designed. Most of them considered only encoder methods without the introduction of decoder. Then, the storage rate is not fast enough which delayed the real-time display for users. At last, capture module is not considered in most of the systems.

Therefore, this paper presented a new SoC video processing architecture, which including a parallel RGB to YUV transcoder, new two-level pipeline and high speed memory controller. We implement the system by ZYNQ SoC. It is designed to provide the required processing and computing capability for video surveillance, automotive driver assistance and other high-end embedded applications [8]. This series is supported by tools and IP provider and integrates ARM ® Cortex-A9 MPCore with 28 nm low-power Programmable Logic closely [9]. Its HDMI transmitter can quickly and efficiently achieve stable transmission of high-definition video. The video processing SoC system proposed in this paper can complete high-definition video capture and processing quickly and efficiently. By using parallel transcoding and two-level pipeline, it can not only meet real-time requirements, but also reduce the system design and development costs.

17.2 System Architecture

This design of high-definition video processing system mainly consists of DVI capture, video transcoding, memory/AXI/VDMA controller and HDMI display, as shown in Fig. 17.1.

DVI capture: This part of the system is based on ZYNQ. Its PMOD interface can support multiple concurrent video capture. In the video capture terminal, ZYNQ SOC configures the video sensor module through SCCB bus and transmits video streaming to ZYNQ through DVI interface. SCCB bus can configure high-definition cameras in real time. During video capture period, the system aligns video streaming by horizontal sync, vertical sync and pixel synchronization signal.

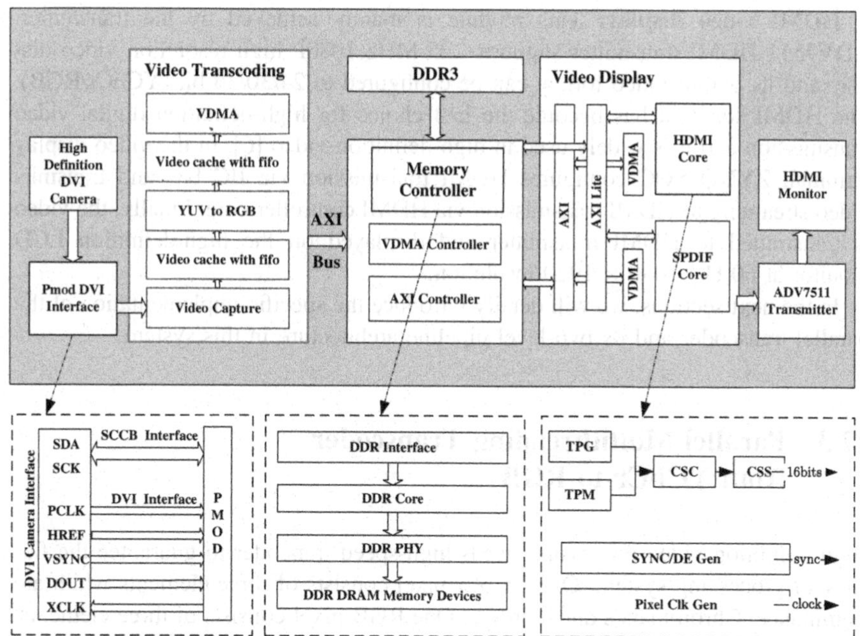

Fig. 17.1 System block diagram

Video transcoding: It is mainly achieved by Programmable Logic subsystem. In order to verify the performance of video processing, the system transcodes the raw video from YCbCr to RGB and stores it in the DDR3 temporarily. Then we will read the video for transcoding from RGB to YCbCr. In order to communicate between the asynchronous clock domains, two-level pipeline is designed which resolves data communication problem between asynchronous clock domains. Meanwhile, in order to improve the video processing speed, six sets of data of two pixels are processed concurrently to form a new parallelization scheme.

Memory/AXI bus/VDMA controllers: The DDR memory controller supports DDR2, DDR3, DDR3L and LPDDR2 devices [10]. It consists of three major blocks: an AXI memory port interface (DDRI), a core controller with transaction scheduler (DDRC) and a controller with digital PHY (DDRP).The system can complete 1080P high-definition video real-time storage by AXI HP interface. AXI bus interaction controller core and VDMA controller core are mainly used for video communication control and scheduling between ARM processors and Programmable Logic. AXI bus protocol is a high-performance, high-bandwidth, low-latency on-chip bus which can meet the design needs of high-performance SOC. Programmable Logic communicates with ARM through AXI bus in order to complete chip configuration and data transmission. As video cache, DDR3 is mainly responsible for coordinating the matching rate of asynchronous clock domains. Memory management, data communication and IP-core configuration are achieved by ZYNQ corresponding controller.

HDMI video display: This module is mainly achieved by the transmitter. ADV7511 HDMI transmitter supports 225 MHz 1080P high resolution video display and its output video format can be configured to 24/30/36 bit (YCbCr/RGB). The HDMI standard has become the first choice for high-definition digital video transmission and it is widely used in high-definition video [6]. In the video display terminal, ZYNQ SOC configures HDMI transmission via IIC bus and transmits video streaming to HDMI transmission via HDMI controller core. Finally, the video is transmitted to HDMI transmitter and displayed on the high-definition LCD monitor at 60 HZ 1080 × 1920 resolution.

In the next sections, we will detaily introduce the specific implementation of the parallel transcoder and its two-level pipeline architecture in this system.

17.3 Parallel Multithreading Transcoder from YCbCr to RGB

High definition video processing needs high-speed trancoder to guarantee the fluency of processing system. One YCbCr pixel consists of three elements whith are Luminance, Chrominance and Chroma. One RGB pixel consists of three elements, i.e., red, green and blue. The transcoding need to change these elements from YCbCr to RGB. In order to improve the video processing speed, parallel transcoding is used. The six sets of data of two pixels are processed in parallel by the system. The processing speed increases sixfold which meets the requirement of high-definition video transmission. The original video image format is 16 bit YCbCr4: 2: 2. The system transcodes it to 24 bit RGB 4: 4: 4. The following equations are used in transcoding process:

$$R = Y + 1.371*(Cr - 128) \tag{17.1}$$

$$G = Y - 0.698*(Cr - 128) - 0.336*(Cb - 128) \tag{17.2}$$

$$B = Y + 1.732*(Cb - 128) \tag{17.3}$$

Since the calculation process is completed in Programmable Logic, decimal in formula need be converted to integer. This can be achieved by shift operation. After shifting and amplifying the above formula, the following three equations can be obtained:

$$R = Y + (351*(Cr - 128)) \gg 8. \tag{17.4}$$

$$G = Y - (179*(Cr - 128) + 86*(Cb - 128)) \gg 8 \tag{17.5}$$

$$B = Y + (443*(Cb - 128)) \gg 8 \tag{17.6}$$

Transcoding one pixel from YCbCr to RGB needs three different calculations. If the system calculates one equation at one clock cycle, the processing speed will be largely delayed. In order to increase the speed of the whole HDMI display procedure, the transcoding architecture is changed to process six equations of two pixels at one clock cycle. The speed of the transcoder increases sixfold. In this way transcoding can be processed by Programmable Logic. Finally, we only need to truncate the data overflow within 0–255 to maintain the correctness of the whole transcoding.

17.4 Two-Level Pipeline Architecture Between Asynchronous Clock Domains

The speed among video data stream, parallel transcoder and bus transfer is different. In order to make data communication between asynchronous clock domains, we need to establish a higher efficient architecture, i.e., two-level pipeline, as shown in Fig. 17.2.

It is mainly composed of video data stream, two-stage buffer queue, parallel transcoding (as mentioned in the above section) and bus transfer. Because the speeds of video caputure, video transcoding and bus transfer are all different, data communication between asynchronous clock domains is required to be processed in the system. Hence, two asynchronous FIFOs which resolves data communication problem between asynchronous clock domains are established in this paper.

The FIFOs are configured with independent clock domains to write and read operations. The independent clock configuration of the FIFO enables us to implement unique clock domains on the wirte and read ports. Buffer No. 1 is used

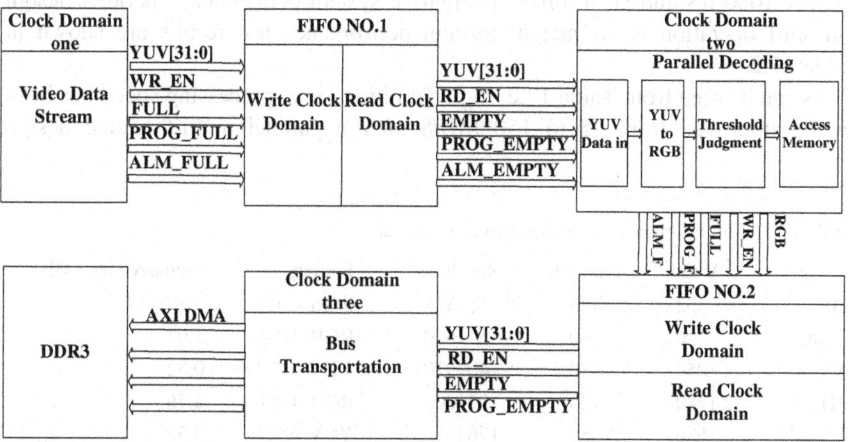

Fig. 17.2 Two-level pipeline architecture

to connect video data stream and parallel transcoding, while Buffer No. 2 is applied to connect parallel transcoding and bus transfer. Because the buffer is asynchronous, the read and write operations are independent. It is permitted to write and read data at the same time. To prevent the congestion of data channel, the speed of bus transfer is higher than parallel transcoding and the speed of parallel transcoding is higher than video data stream.

Thus it can be seen that during the process of video capture, pipelined architecture is designed to optimize the whole transcoding process and save system resources. Being different with traditional full image storing and processing, we use a new way that capturing and processing are performed simultaneously. Without reducing the processing speed of the system, resources are considerably saved, the real-time performance of the system is improved and the system process time delay is reduced significantly.

17.5 System Evaluation and Experimental Results

High definition video processing systems have become the development direction of security monitoring technology. Urban road monitoring, highway monitoring, airport monitoring and bank monitoring make high demands for high definition video in order to monitor license plate, vehicle, face, figure, luggage and other details. High definition video surveillance has become the best solution. In display resolution and fluency, high definition is much better than standard definition. The information of different video formats is listed in Table 17.1.

As shown in the Table 17.1, the resolution of 1080P is twenty times than the resolution of CIF. In the same display environment, high definition will be much clearer. With DVI high-definition camera, ZYNQ and HDMI display as mentioned above, high-definition video is displayed at a rate of 60 frames per second at 1920 × 1080 resolution. In order to improve system performance, parallel design and shift operation is optimized. System performance test results are shown in Table 17.2.

As can be seen from Table 17.2, high definition video processing speed has been greatly improved from 25 to 150 MB/S after a parallel multi-threaded design

Table 17.1 The information of different video formats

	Width	Height	Resolution	Format	Memory size (MB)
CIF	352	288	101376	RGB 4:4:4	0.30
NTSC	720	480	345600	RGB 4:4:4	1.04
WVGA	854	480	409920	YUV 4:2:2	0.82
HD 720	1280	720	921600	RGB 4:4:4	2.76
WSXGA	1680	1050	1764000	YUV 4:2:2	3.53
HD 1080	1920	1080	2073600	RGB 4:4:4	6.22

Table 17.2 HD video processing speed with different number of thread

	HD video processing speed (MB/S)	HD video capture speed (MB/S)	HD video display speed (MB/S)
Single thread	25	83.7	355.9
Dual threads	50	83.7	355.9
Triple threads	75	83.7	355.9
Five threads	100	83.7	355.9
Six threads	150	83.7	355.9

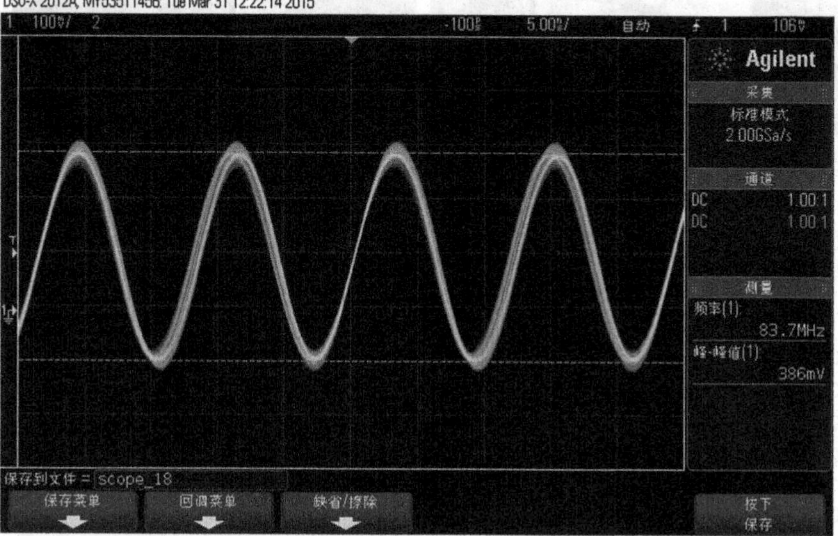

Fig. 17.3 High definition video capture speed

optimization. High definition video capture speed is 83.7 MB/S which is less than 150 MB/S to a great extent, as demonstrated in Fig. 17.3.

Besides, when the video is processed and displayed by single thread, the display is shown as the top of Fig. 17.4. That is because the processing speed of single tread is not high enough. They are full of messy code. The two pictures given in the bottom of Fig. 17.4 are processed and displayed by six threads. It can be seen that the display are clearer and have no messy code.

The optimization based on pipeline structure greatly reduces system processing delay and compile time. The test results are listed in Table 17.3 and Fig. 17.5.

As can be seen from Fig. 17.5, caching sixteen pixels (the figure on the right) consumes much less BRAM, LUT and FF compared with caching ten lines (the figure on the left). Resource consumption are greatly reduced by the optimization based on pipeline structure.

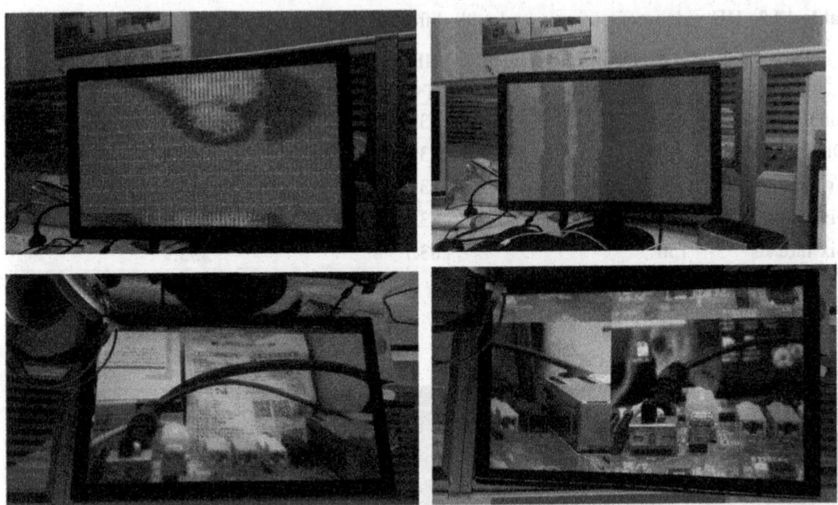

Fig. 17.4 The display of high definition video with different number of thread

Table 17.3 System real-time parameters and resource consumption

	System processing delay (ms)	Compile time (min)	RAM resource consumption (%)
Caching one image	133.0	Failure	1750
Caching ten lines	1.230	Failure	86
Caching two lines	0.246	120	23
Caching thirty-two pixels	0.002	30	6
Caching sixteen pixels	0.001	15	4

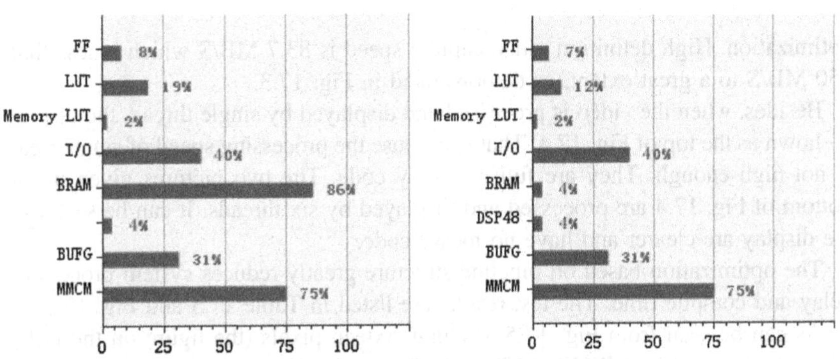

Fig. 17.5 The utilization (%) of system resources

17.6 Conclusion

This paper designed and implemented a high-definition video processing system. The system had achieved DVI high-definition video processing and HDMI display with ZYNQ SOC as the core. The algorithm parallelism and reconfigurability of the system are excellent since users can easily customize the video processing algorithms. The system can meet the requirements of portability, high definition and highly real-time capability for video processing. Its transmission performance is stable and reliable. The research of the paper is conducive to the further development of high-definition video processing.

References

1. Santanu Dutta, Rune Jensen, Alf Rieckmann (2001) Viper: a multiprocessor SOC for advanced set-top box and digital TV systems [J]. IEEE Des Test Comput 18(5):21–31
2. Schultz Richard R, Stevenson Robert L (1996) Extraction of high-resolution frames from video sequences [J]. IEEE Trans Image Process 5(6): 996–1011
3. van der Wolf P, Tomas H (2008) Video processing requirements on SoC Infrastructures [C]. Des Autom Test Europe 1124–1125
4. Longjun L, Hongbin S, Whenzhe Z et al (2011) A high performance and low cost video processing SoC for digital HDTV systems [C]. In: IEEE 9th International conference on ASIC, pp. 188–191
5. Hongbin S, Longjun L, Qiubo C et al (2012) Design and implementation of a video display processing SoC for full HD LCD TV [C]. In: 2012 International SoC design conference, pp. 297–300
6. Tero K, Hamalainen Timo D (2006) Scalable architecture for SoC video encoders. J VLSI Signal Process [C], 44:79–95
7. Jiaming Chen, Chunnan Liu, Jenkuei Yang (2011) Parallel architecture core (PAC)—the first multicore application processor SoC in Taiwan part II: application programming [J]. J Signal Process Syst 62:383–402
8. Sandeep G (2009) Hybrid functional verification methodology for video/audio SoC [C]. In: The 1st Asia Symposium on Quality Electronic Design, pp. 264–265
9. Jonathan Starck, Adrian Hilton (2007) Surface capture for performance-based animation [J]. IEEE Comput Graphics Appl 27(3):21–31
10. Shi Yun Q, Huifang S (1999) Image and video compression for multimedia engineering: fundamentals, algorithms, and standards [M]. CRC press, Boca Raton

Chapter 18
A Study on Feature Extraction of Surface Defect Images of Cold Steel

Yingjun Guo, Xiaoye Ge, Hexu Sun and Xueling Song

Abstract Feature extraction is one of the important characteristics used in classifying images. But the extracted features have big numbers and high dimension easily due to various type defects, complicated features and diverse methods offeature. Big numbers and high dimension of features are adverse for feature extraction. The effect of feature extraction decides the effect of image classification directly. According to these problems, experimental investigations are carried out on computer aiming at three typical surface defect images of cold steel strip, and this paper choose the gray features, textural features and Hu invariant moment features as the basis of classification finally. Experimental results demonstrated that features in this paper can be classification basis correctly.

Keywords Defect images of cold steel strip · Feature extraction · Gray features · Texture features · Hu invariant moment features

18.1 Introduction

With the growing demands of surface quality of cold steel strip, many specializes start to study the automatic surface quality inspection system for cold steel strip more widely and deeply, the methods of nondestructive testing based on machine vision has being a research hotspot in that field, and pattern recognition is a key step in surface defect inspection of cold steel strip [1–6]. Feature extraction is precondition of recognition and classification for defect images, and it also greatly influences the design and performance of classifier [7]. Methods and species of

Y. Guo · X. Ge (✉) · H. Sun · X. Song
College of Electrical Engineering, Hebei University of Science and Technology, 050018 Shijiazhuang, China
e-mail: m15533910949@163.com

Y. Guo · H. Sun
School of Control Science and Engineering, Hebei University of Technology, 300401 Tianjin, China

© Springer-Verlag Berlin Heidelberg 2016

Y. Jia et al. (eds.), *Proceedings of the 2015 Chinese Intelligent Systems Conference*, Lecture Notes in Electrical Engineering 359,
DOI 10.1007/978-3-662-48386-2_18

feature extraction keep on increasing because of the various type defects. If we do not effectively select the methods extraction but blindly adopt various methods or species of feature, it will only result in large types and huge numbers of features and it is prone to cause the curse of dimensionality, which has a big and adverse impact on the subsequent recognition and classification. In this paper we are concerned with the task of selecting a set of features for classifying. Experimental investigations are carried out on computer aimed at three typical surface defect images of cold steel strip, such as scratch defect, inclusion defect and hole defect, and the results are satisfactory.

18.2 Feature Extraction of Surface Defect Images of Cold Steel Strip

There are many types of surface defect images of cold steel strip, they are different in shape or locations, or they are same in shape but different in angles or locations, so effective feature extraction of defect images is the key of correctly identifying the type of defect images [8]. Different surface defects of steel strip have different reflects for texture features or grayscale features, the difference of identification for some surface defect images of steel strip can up to nearly 20 % if using different type of features to classify the defects, therefore, it is unscientific if we use one type of feature alone to recognize the surface defect images of cold steel strip, but too many types of feature may also cause the curse of dimensionality and reduce the separability of features. In this paper, we select the gray features, texture features and Hu invariant moment features as effective features for classification, which is based on extensive literature survey and experiments.

Gray histogram is a function of grayscale. It describes the number of each pixel gray levels and iterates through all pixels in the image, but it can only reflect image gray distribution rather than the position image pixels [9, 10]. Gray histogram is a two-dimensional figure, the horizontal axis represents gray level of each pixel of the image and the ordinate represents frequency or probability of each pixel appearing in the image. We can quantized the grayscale of the image to j gradation, so i = 0, 1, 2, … , j−1. The change of surface color of steel strip usually presented as the abnormality of gray, the gray distribution of zero-defect histogram is more uniform, which is similar to a Gaussian distribution [11]. The features of gray histogram include gray mean, variance, distortions, kurtosis, energy and entropy. The features we consider are

$$h_1 = \sum_{i=0}^{L-1} i \times p(i) \tag{18.1}$$

$$h_2 = \sum_{i=0}^{L-1} (i-\mu)^2 \times p(i) \tag{18.2}$$

$$h_3 = \frac{1}{\sigma^3} \sum_{i=0}^{L-1} (i-\mu)^3 \times p(i) \tag{18.3}$$

$$h_4 = \frac{1}{\sigma^4} \sum_{i=0}^{L-1} (i-\mu)^4 \times p(i) \tag{18.4}$$

$$h_5 = \sum_{i=0}^{L-1} (p(i))^2 \tag{18.5}$$

$$h_6 = -\sum_{i=0}^{L-1} p(i) \times \log p(i) \tag{18.6}$$

Because each pixel gray level is processed independently in the gray histogram, it cannot reflect the law of relativity of gray level well in texture, but the Gray Level Co-occurrence Matrix (GLCM) can describe the characteristics of spatial distribution and correlation of gray at the same time, reflecting the comprehensive information about the direction and neighbor interval and other aspects of gray [12, 13]. The GLCM is a second order statistical feature which is related to the change of intensity in the image and it is defined by joint probability density of pixels in two locations. In practice, the textural features are determined by some features which got by GLCM rather than GLCM itself [14]. The typical features include angular second moment, variance, inverse difference moment, entropy, contrast and correlation. The features we consider are

$$h_1 = \sum_{i=0}^{L-1} \sum_{j=0}^{L-1} \hat{P}^2(i,j) \tag{18.7}$$

$$h_2 = \sum_{i=0}^{L-1} \sum_{j=0}^{L-1} (i-\mu)^2 \hat{P}^2(i,j) \tag{18.8}$$

$$h_3 = \sum_{i=0}^{L-1} \sum_{j=0}^{L-1} \frac{\hat{p}(i,j)}{1+(i-j)^2} \tag{18.9}$$

$$h_4 = -\sum_{i=0}^{L-1} \sum_{j=0}^{L-1} \hat{P}(i,j) \log \hat{p}(i,j) \tag{18.10}$$

$$h_5 = \sum_{n=0}^{L-1} n^2 \left\{ \sum_{i=0}^{L-1} \sum_{j=0}^{L-1} \hat{P}(i,j) \right\} \tag{18.11}$$

$$h_6 = \frac{\sum_{i=0}^{L-1} \sum_{j=0}^{L-1} ij\hat{P}(i,j) - \mu_1\mu_2}{\sigma_1^2 \sigma_2^2} \tag{18.12}$$

Invariant moment is a highly condensed feature of images, which has invariant properties for translation, rotation, scaling and other geometric transformation. Two dimensional (p + q) step moments of a digital image which expressed by f(x, y) can be defined by

$$m_{pq} = \sum_{x}\sum_{y} x^p y^q f(x,y) \tag{18.13}$$

where

$$p,q = 0, 1, 2\ldots \tag{18.14}$$

seven invariant moments which insensitive for translation and scaling usually can be derived. The seven invariant moments we consider are

$$h_1 = \eta_{20} + \eta_{02} \tag{18.15}$$

$$h_2 = (\eta_{20} - \eta_{02})^2 + 4\eta_{11}^2 \tag{18.16}$$

$$h_3 = (\eta_{30} - 3\eta_{12})^2 + (\eta_{03} - 3\eta_{21})^2 \tag{18.17}$$

$$h_4 = (\eta_{30} + \eta_{12})^2 + (\eta_{21} + \eta_{03})^2 \tag{18.18}$$

$$\begin{aligned} h_5 = {} & (\eta_{30} - 3\eta_{12})(\eta_{30} + \eta_{12})[(\eta_{30} + \eta_{12})^2 - 3(\eta_{21} + \eta_{03})^2] \\ & + (3\eta_{21} - \eta_{03})(\eta_{21} + \eta_{03})[3(\eta_{30} + \eta_{12})^2 - (\eta_{21} + \eta_{03})^2] \end{aligned} \tag{18.19}$$

$$\begin{aligned} h_6 = {} & (\eta_{20} - \eta_{02})[(\eta_{30} + \eta_{12})^2 - (\eta_{21} + \eta_{03})^2] \\ & + 4\eta_{11}(\eta_{30} + \eta_{12})(\eta_{21} + \eta_{03}) \end{aligned} \tag{18.20}$$

$$\begin{aligned} h_7 = {} & (3\eta_{21} - \eta_{03})(\eta_{30} + \eta_{12})[(\eta_{30} + \eta_{12})^2 - 3(\eta_{21} + \eta_{03})^2] \\ & + (3\eta_{12} - \eta_{30})(\eta_{21} + \eta_{03})[3(\eta_{30} + \eta_{12})^2 - (\eta_{21} + \eta_{03})^2] \end{aligned} \tag{18.21}$$

where

$$\eta_{pq} = \frac{\mu_{pq}}{\mu_{00}^r} \tag{18.22}$$

$$\mu_{pq} = \sum_{x}\sum_{y} (x - \bar{x})^p (y - \bar{y})^q f(x,y) \tag{18.23}$$

$$\bar{x} = \frac{m_{10}}{m_{00}}, \bar{y} = \frac{m_{01}}{m_{00}} \tag{18.24}$$

$$r = \frac{p+q}{2} + 1 \tag{18.25}$$

18.3 Experimental Results and Analysis

Experimental investigations are carried out on computer aimed at three typical surface defect images of cold steel strip in this paper, such as scratch defect, inclusion defect and hole defect.

Characteristic parameters obtained from the gray histogram are shown in Table 18.1. From the data (Table 18.1), we can see that the change of intra-class eigenvalues of h2 is larger and the change of inter-class eigenvalues of h6 is less, so we can reject those parameters when gray features used to classify defects. The characteristic parameters of h1, h3, h4 and h5 have big difference in inter-class eigenvalues and slight difference in intra-class eigenvalues, so they are more suitable for classification. Further more, there is a big difference in the variance of gray histogram of zero-defect image and defective image due to the presence of defects, especially obvious defects. Variance of gray histogram may also be used as a threshold to distinguish whether the image is defective image in practice. The gray histogram of scratch defect image, inclusion defect image and hole defect image are shown in Figs. 18.1, 18.2 and 18.3 respectively.

The GLCM is an effective method for the extraction of texture feature, which is based on grayscale spatial correlation matrix. Characteristic parameters obtained by GLCM are given in Table 18.2. The angular second moment also known as energy, reflecting the degree of uniformity of the image gray distribution. Entropy is a measure of the amount of information of the image. Texture information also belongs to the image information, the GLCM almost is zero matrix and entropy is almost zero if the image does not have texture. Contrast can also be understood as the sharpness of the image, and the contrast becomes greater if the grooves of texture become deeper. Correlation is a measure of the degree of similarity in the direction of row or column for the element of GLCM. From the data (Table 18.2), we can see that the change of intra-class eigenvalues of h4 is larger and the change of inter-class eigenvalues of h6 is less, so we can reject those parameters.

Table 18.1 Gray histogram features

Samples	Features	h1	h2	h3	h4	h5	h6
Scratch	Sample 1	88.0685	53.9574	0.0185	2.2571	0.0180	4.5333
	Sample 2	81.5031	63.1348	0.0258	2.7621	0.0194	4.5540
	Sample 3	93.8029	37.0392	0.0270	3.5628	0.0193	4.7505
Inclusion	Sample 1	155.3854	34.3089	0.0191	1.3520	0.0086	4.8696
	Sample 2	154.1128	33.2145	0.0101	1.9726	0.0088	4.7854
	Sample 3	157.0459	40.3472	0.0148	1.1577	0.0075	4.9724
Hole	Sample 1	118.9985	51.8527	0.0393	4.6688	0.1129	4.4974
	Sample 2	112.6101	20.7288	0.0482	4.3958	0.1262	3.8304
	Sample 3	123.5608	11.7503	0.0369	4.7382	0.1081	4.9449

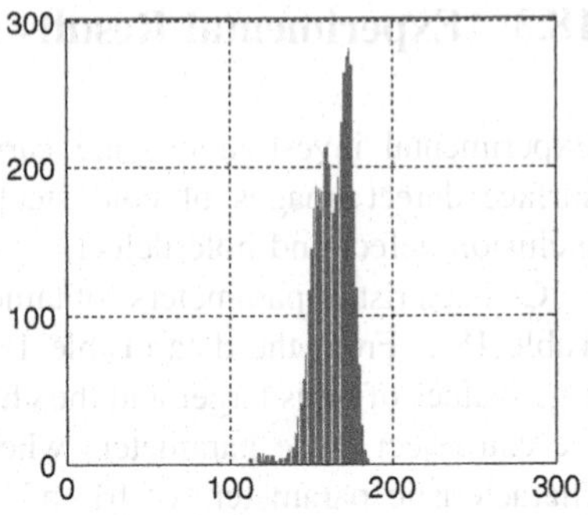

Fig. 18.1 Histogram of scratch

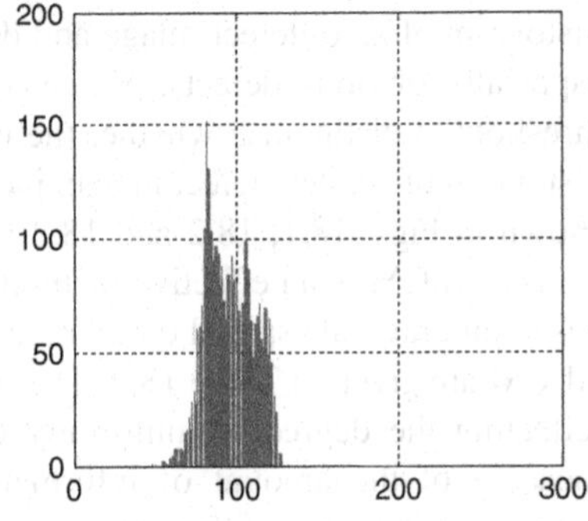

Fig. 18.2 Histogram of inclusion

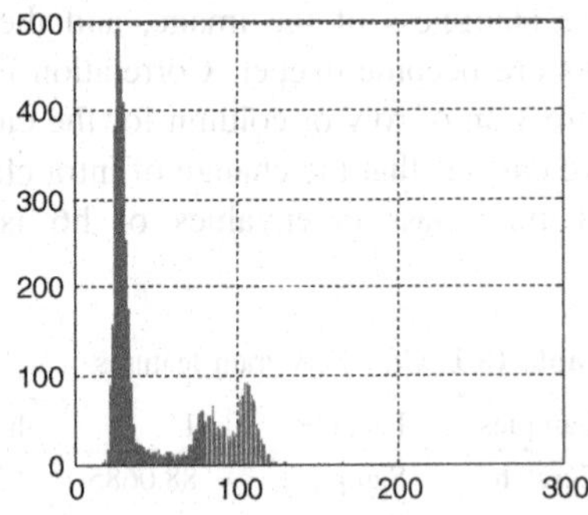

Fig. 18.3 Histogram of hole

The characteristic parameters of h1, h2, h5 and h6 are more suitable for classification because they have big difference in inter-class eigenvalues and slight difference in intra-class eigenvalues.

Hu invariant moments have invariance for rotation, scaling and other geometric transformation. Characteristic parameters obtained by Hu invariant moments for one original sample, the sample of adding Gaussian noise, the sample of rotating thirty degrees and the sample of halving the image are given in Table 4. From the data (Table 18.3), we can see that the characteristic parameters basically unchanged

Table 18.2 Textural features

Samples	Features	h1	h2	h3	h4	h5	h6
Scratch	Sample 1	91.0	10.2180	8.846	17.280	12.676	13.182
	Sample 2	85.3	10.4433	8.812	28.420	14.820	11.274
	Sample 3	85.8	9.9730	8.096	31.704	12.084	11.653
Inclusion	Sample 1	121.3	7.4012	8.242	30.958	3.097	2.008
	Sample 2	123.5	7.4561	8.811	22.819	3.715	3.565
	Sample 3	127.5	7.0408	8.721	25.413	4.293	4.223
Hole	Sample 1	243.5	5.5328	8.203	29.956	6.091	9.180
	Sample 2	271.6	5.1547	9.246	16.846	6.761	7.752
	Sample 3	252.6	4.1727	8.712	25.170	7.705	7.325

Table 18.3 Hu invariant moment features

Samples	Features	h1	h2	h3	h4	h5	h6	h7
Scratch	Origigal	5.8762	11.9168	20.4290	64.3023	56.0392	29.4364	48.7375
	Adding noise	5.8770	11.9184	20.4270	64.3362	56.0574	29.4500	48.7662
	Ratating	5.5685	11.3092	18.5231	62.2442	54.3771	28.3202	46.5840
	Halving	5.8770	11.9185	20.4310	64.3105	56.0435	29.4385	48.7438
Inclusion	Origigal	2.6770	5.6750	9.6209	9.6254	19.2554	12.4629	18.9455
	Adding noise	2.6757	5.6720	9.6208	9.6255	19.2556	12.4615	18.9595
	Ratating	2.6769	5.6750	9.2005	9.6256	19.0952	12.4631	19.2595
	Halving	2.6816	5.6930	9.6412	9.6458	19.2963	12.4923	18.9954
Hole	Origigal	5.2502	10.6992	10.9688	51.0223	41.6109	16.6880	46.6540
	Adding noise	5.2532	10.7053	10.9739	50.8403	41.6378	16.7073	46.5327
	Ratating	4.9450	10.0840	10.2899	48.0139	40.1132	15.6070	45.1907
	Halving	5.2518	10.7024	10.9743	51.0378	41.6126	16.6949	46.6676

after adding noise to the image, indicating Hu invariant moments have a certain immunity for noise. In addition, the parameters change little in the magnitude after rotating and halving expect individual parameters. Therefore, Hu invariant moment features are more suitable for classification. We also can see that any characteristic parameter of inclusion defect is smaller than the other defects, which can be better basis for classification. Due to big difference in inter-class eigenvalues of h3, h4, h5 and h6 of scratch defect and hole defect, they can be used as a valid basis for classification.

18.4 Conclusion

Aiming at the importance and existing questions of feature extraction in pattern recognition of surface defect images of cold steel strip, we select the gray features, textural features and Hu moment invariant features as the basis for subsequent classification, and experimental investigations are carried out on computer aiming at three typical surface defect images of cold steel strip, the results show that those features can be a valid basis for subsequent classification of the image. Any type of features or method has its advantages and disadvantages, combination of the features is an effective way for recognition system. Feature extraction is only a part of the entire automatic surface quality inspection system of cold steel strip, and follow-up works also include the feature screening, selection and design for classifier, we should take the appropriate program based on the actual situation in order to achieve the best results of classification.

Acknowledgments This paper was supported by the National Natural Science Foundation of China grant (NSFC61203275).

References

1. Guoyi Zhang, Zheng Hu, Ting Xu et al (2010) Classification method for defect Tmage based on feature extraction [J]. J Beijing Univ Technol 36(4):450–456
2. Chengming Wang, Yunhui Yan, Yingli Han et al (2007) New pattern recognition method of surface defect images of cold steel strip [J]. Comput Eng Appl 43(13):207–209
3. Pingchuan W, Tongjun L, Yan W (2000) Development and perspective of automatic strip surface inspection system [J]. Iron Steel 35(6):70–75
4. Huijun Hu, Yuanxiang Li, Maofu Liu et al (2014) Steel strip surface defects classification based on machine learning [J]. Comput Eng Design 35(2):621–624
5. Ranaee Vahid, Ebrahimzadeh Ata, Ghaderi Reza (2010) Application of the PSO-SVM model for recognition of control chart patterns [J]. ISA Trans 49:577–586
6. David Amit, Lerner Boaz (2005) Support vector machine-based image classification for genetic syndrome diagnosis [J]. Pattern Recogn Lett 26:1029–1038
7. Xu Ke Xu, Jinwu Liang Zhiguo et al (2005) Molecular dynamics simulation of interaction between calcite crystal and water-soluble polymers [J]. Acta Phys Chim Sin 21 (11):1198–1204
8. Junhai Zhai, Wenxiu Zhao, Xizhao Wang (2009) Research on the image feature extraction [J]. J Hebei Univ (Natural Science Edition) 29(1):106–109
9. Zhaozhun Zhong, Guangwei Xie, Shengkui Zhong (2014) Design of online surface defect inspection system for steel strips [J]. J Beijing Univ Technol 40(7):961–966
10. David Amit, Lerner Boaz (2005) Support vector machine-based image classification for genetic syndrome diagnosis [J]. Pattern Recogn Lett 26:1029–1038
11. Yang Xiang, Li Chen, Xiaolong Zhang (2012) Research on recognition of steel surface defect based on support vector machine [J]. Ind Control Comput 25(8):99–101
12. Hongying Yang, Xiangyang Wang, Zhongkai Fu (2012) A new image denoising scheme using support vector machine classification in shiftable complex directional pyramid domain [J]. Appl Soft Comput 12:872–886

13. Xanthopoulos Petros, Razzaghi Talayeh (2014) A weighted support vector machine method for control chart pattern recognition [J]. Comput Ind Eng 70:134–149
14. Yen Lin S, Shiang Guh R, Ren Shiue Y (2011) Effective recognition of control chart patterns in autocorrelated data using a support vector machine based approach [J]. Comput Ind Eng (61):1123–1134

Chapter 19
Consensus of Multi-agent System with Singular Dynamics and Time Delay

Hua Geng, Zengqiang Chen, Zhongxin Liu and Qing Zhang

Abstract In this paper, the consensus problem of high-order multi-agent systems with singular dynamics and time-varying time delay is investigated. By the restricted equivalent transformation, a differential-algebraic system is introduced which is equivalent to this singular system on consensus. Based on dynamic state information and time delay, a consensus protocol is proposed. A sufficient condition of consensus is given in terms of LMI. Furthermore, the consensus state is also obtained by calculating the solution of the subsystem. Finally, an example is presented to illustrate the theoretical results.

Keywords Consensus · Multi-agent system · Singular dynamics · Time delay · LMI

19.1 Introduction

Nowadays, more and more researchers have put themselves into the consensus field due to its widespread applications, such as flocking phenomena, formation control of multiple robots or unmanned aerial vehicles, synchronization phenomenon of networks, and attitude alignment of clusters of satellites [1–6]. Consensus, which is fundamental to the multi-agent system, means that all the agents in the multi-agent systems reach an agreement on a common value by following some control rules and negotiating with their neighbors [7].

At the beginning of studying consensus problem, the agents were simply described as first-order integrator agents. In [8], Olfati-Saber and Murray proposed

H. Geng (✉) · Z. Chen and Z. Liu
College of Computer and Control Engineering, Nankai University, Tianjin 300071, China
e-mail: huahua27102710@163.com

H. Geng · Z. Chen · Z. Liu
Tianjin Key Laboratory of Intelligent Robotics, Nankai University, Tianjin 300071, China

Z. Chen · Q. Zhang
College of Science, Civil Aviation University of China, Tianjin 300300, China

© Springer-Verlag Berlin Heidelberg 2016

Y. Jia et al. (eds.), *Proceedings of the 2015 Chinese Intelligent Systems Conference*, Lecture Notes in Electrical Engineering 359,
DOI 10.1007/978-3-662-48386-2_19

a classical research framework of consensus problem under fixed and switching networks. The proposed analysis method and consensus protocols were very useful for latter researchers. Based on [8], Hong investigated the consensus problem under leader-following network [9]. The designed neighbor-based local controller and neighbor-based state-estimation could guarantee the follower agents tracking the leader. Moreover, in a multi-agent system with first-order integrator agents, many results on consensus problem have been obtained in different focuses, such as consensus with time delay [10], finite-time consensus [11], consensus in uncertain communication environments [12], etc. Unlike the literature [8–12], Lin considered a class of second-order integrator multi-agent systems with time delay and obtained the sufficient condition of consensus by LMI tools [13]. Furthermore, the result showed that the communication time-delay could not affect consensus. Leader-following consensus problem with second-order dynamics were dealt with in [14].

In [8–14], the agents were described by first-order integrator or second-order integrator, but some complex systems are of high-order dynamics in real world. In [15], He and Cao proposed a high-order consensus protocol based on feedback controller and information from the neighbours. A sufficient and necessary condition for consensus was obtained and the relationship between feedback gain and system parameters was established. The consensus of high-order multi-agent systems with switching networks and unknown communication delays were investi-gated, respectively [16, 17].

In the aforementioned literatures, the agents are described by first-order agents, second-order agents or high-order agents, but the practical system may be the singular system which is also named as descriptor system, generalized system, implicit system and differential algebraic system. It is well known that consensus problem of singular multi-agent system are more challenging since these systems contain more complex dynamics and the normal systems are the special cases of singular systems. In recent years, more and more researchers have studied the consensus of singular multi-agent systems. By the restricted equivalence transformation, the admissible consensus condition was obtained under continuous-time model and discrete-time model, respectively [18]. Based on a linear matrix inequality and a modified Riccati equation, the proposed two protocols could guarantee all the agents reach consensus. In [19], Xi considered the consensualizing controller design problems for high-order singular multi-agent systems. By projecting the state of singular systems method, the necessary and sufficient was obtained in terms of LMI. Xi constructed a quadratic cost function for achieving a trade-off between consensus regulation performances and control energy consumptions [20]. Based on linear matrix inequality techniques, the condition of consensus, the cost function and explicit expressions of consensus functions were obtained. As we known, information delays naturally in the process of information transmission among agents. In [21], Xi investigated consensus problem of singular swarm systems with time delay. By constructing a system with singular dynamics, the admissible consensus problem was transformed into admissible problems of multiple singular subsystems with lower dimensions. The condition of consensus and consensus function were obtained.

In this paper, we mainly studied the differential-algebraic system induced by the high-order singular multi-agent systems. The current study covered the following two novel features. First, the time-varying time delay is considered, and the consensus protocol based on dynamic state information and time delay is proposed. The sufficient condition of consensus is established in terms of LMI. Second, after decomposing the state of system, the consensus state is also obtained by calculating the solution of the subsystem.

The remainder of the paper is organized as follows. In Sect. 19.2, some preliminaries are briefly outlined and the problem description is given. In Sect. 19.3, the sufficient condition of consensus is presented in terms of LMI. An example is given in Sect. 19.4. Finally, the conclusion is stated in Sect. 19.5.

Notations: $\mathbb{R}^{N\times N}$ and $\mathbb{C}^{N\times N}$ denote the set of $n \times n$ real and complex matrices. $\mathbf{1}_N$ is an N dimensional column vector with all components 1. I_N denotes the identity matrix. For a square matrix X, $X > 0$ means it is positive definite. $||\bullet||$ represents Euclidean norm.

19.2 Preliminaries

In this section, some basic concepts and results about graph theory are shown, and the problem description is given.

19.2.1 Graph Theory

Algebraic graph theory is very useful for our study. Now we introduce some fundamental knowledge on graph theory [22]. Let $\mathcal{G} = (\mathcal{V}, \mathcal{E}, \mathcal{A})$ be a weighted directed graph of order n with the set of nodes $\mathcal{V} = \{v_1, v_2, \ldots, v_n\}$, set of edges $\mathcal{E} \subseteq \mathcal{V} \times \mathcal{V}$, and an adjacency matrix $\mathcal{A} = [a_{ij}] \in \mathbb{R}^{N\times N}$ with nonnegative elements a_{ij}. The node indexes set is denoted by $\Gamma = \{1, 2, \ldots, n\}$. $e_{ij} = (v_i, v_j)$ denotes an edge from node v_j to v_i. The set of neighbors of node v_i is denoted by $\mathcal{N}_i = \{v_j \in \mathcal{V} : (v_i, v_j) \in \mathcal{E}\}$. There exists a directed path between node v_i and v_j if and only if the information flow begin with node v_i and end at node v_j (without self-loop and backflow). Furthermore, the concept of directed spanning tree is very important. It requires only one node (without parent node) can transmit information to every other node (only one parent node) via the existed directed path. The degree matrix $\mathcal{D} = \{d_1, d_2, \ldots, d_n\} \in \mathbb{R}^{N\times N}$ of graph $\mathcal{G}$ is a diagonal matrix, where diagonal $d_i = \sum_{j\in N_i} a_{ij}$ for $i = 1, 2, \ldots, n$. Then the Laplacian matrix of $\mathcal{G}$ is defined as $\mathcal{L} = \mathcal{D} - \mathcal{A}$.

Lemma 1 ([23]) *Let $\mathcal{L} \in \mathbb{R}^{N\times N}$ be the Laplacian matrix of a directed graph $\mathcal{G}$ and $\mathit{1} = [1, 1, \ldots, 1]^T \in \mathbb{R}^N$ then*

1. *$\mathcal{L}$ at least has a zero eigenvalue, and $\mathbf{1}$ is an associated eigenvector, that is,* $\mathcal{L}\mathbf{1} = 0$;
2. *If $\mathcal{G}$ has a spanning tree, then* 0 *is a simple eigenvalue of $\mathcal{L}$, and all the other $n-1$ eigenvalues have positive real part.*

19.2.2 Problem Description

Consider a multi-agent system composed of N agents with singular dynamics, each agent is denoted by

$$E\dot{x}_i = Ax_i + Bu_i, i = 1, 2, \ldots, N \tag{19.1}$$

where $E, A \in \mathbb{R}^{n \times n}$, $B \in \mathbb{R}^{n \times m}$ are constant matrices. $x_i \in \mathbb{R}^n$, $u_i \in \mathbb{R}^m$ are the state and input, respectively. $0 < rank(E) = r < n$.

Next, some basic concepts of singular system is given as follows

Definition 1 ([24]) The pair (E, A) in (19.1) is called

1. regular if $det(sE - A)$ is not identically zero for some $s \in \mathbb{C}$;
2. impulse-free if $deg(det(sE - A)) = rank(E)$ for $\forall s \in \mathbb{C}$;
3. admissible if it is regular, impulse-free and asymptotically stable.

Based on the singular system theory, there exist two nonsingular matrices Q, P, such that the system (19.1) is restricted equivalent to the following differential-algebraic system:

$$\begin{cases} \dot{x}_{i1} = A_{11}x_{i1} + A_{12}x_{i2} + B_1u_i \\ 0 = A_{21}x_{i1} + A_{22}x_{i2} + B_2u_i \end{cases} i = 1, 2, \ldots, N \tag{19.2}$$

where

$$QEP = \begin{bmatrix} I_r & 0 \\ 0 & 0 \end{bmatrix}, QAP = \begin{bmatrix} A_{11} & A_{12} \\ A_{21} & A_{22} \end{bmatrix}, QB = \begin{bmatrix} B_1 \\ B_2 \end{bmatrix}, P^{-1}x_i = \bar{x}_i = \begin{bmatrix} x_{i1} \\ x_{i2} \end{bmatrix} \tag{19.3}$$

and $A_{11} \in \mathbb{R}^{r \times r}$, $A_{12} \in \mathbb{R}^{r \times (n-r)}$, $A_{21} \in \mathbb{R}^{(n-r) \times r}$, $A_{22} \in \mathbb{R}^{(n-r) \times (n-r)}$, $B_1 \in \mathbb{R}^{r \times m}$, $B_2 \in \mathbb{R}^{(n-r) \times m}$, $x_{i1} \in \mathbb{R}^r$, $x_{i2} \in \mathbb{R}^{(n-r)}$.

For system (2), a useful lemma for latter studying is given as follows.

Lemma 2 ([24]) *For singular system* (19.1)*, the pencil* (E, A) *is regular and impulse-free if and only if* A_{22} *in* (19.2) *is invertible.*

As we known, if system (19.2) reach consensus, that is $\lim_{t \to \infty} ||\bar{x}_j - \bar{x}_i|| = 0$. From (19.3), we know system (19.1) reach consensus, so next we will investigate consensus of system (19.2). The definition of consensus as follows.

Definition 2 System (19.2) is said to reach consensus if it is regular, impulse-free and satisfies the following equation

$$\lim_{t\to\infty} ||x_{jm} - x_{im}|| = 0, m = 1, 2.i, j = 1, 2, \ldots, N \tag{19.4}$$

In the following, we introduce a useful lemma for latter studying.

Lemma 3 ([25]) *For any real differentiable vector function $x(t) \in \mathbb{R}^n$ and any $n \times n$ constant matrix $W = W^T > 0$, we have the following inequality*
$-h^{-1}[x - x(t-\tau(t))]^T W[x - x(t-\tau(t))] \le \int_{t-\tau(t)}^{t} \dot{x}^T(s)Wx(s)ds, t \ge 0$

19.3 Main Reasult

In this section, we propose a consensus protocol based on dynamic state information and time-varying time delay, the sufficient condition of consensus and consensus state are obtained.

It is well known that information delay appear naturally in the process of information transmission, so we propose the following consensus protocol with time-delay.

$$u_i = K \sum_{j=1}^{N} a_{ij}(x_{j1}(t-\tau(t)) - x_{i1}(t-\tau(t))), i, j = 1, 2, \ldots, N \tag{19.5}$$

where $K \in \mathbb{R}^{m\times r}$ is the constant gain, $0 \le \tau(t) \le h$, $h > 0$, $t \ge 0$.

Based on the protocol (19.5), we obtain the sufficient condition of consensus as follows.

Theorem 1 *Suppose that the digraph G of system* (19.2) *has a spanning tree, the consensus problem is solved if there exist two appropriate positive definite matrices R_1, R_2 and the following inequation hold*

$$\Xi = \begin{bmatrix} \Xi_{11} & \Xi_{12} \\ \Xi_{12}^T & \Xi_{22} \end{bmatrix} < 0$$

where $\Xi_{11} = R_1H_1 + H_1^TR_1 + hH_1^TR_2H_1 - h^{-1}R_2$, $\Xi_{12} = R_1H_2 + hH_1^TR_2H_2 + h^{-1}R_2$, $\Xi_{22} = hH_2^TR_2H_2 - h^{-1}R_2$.

Proof We define the error variables

$$\delta_{i1} = x_{i1} - x_{11}, \delta_{i2} = x_{i2} - x_{12}, i = 2, 3, \ldots, N$$

From (19.2), we have

$$\begin{cases} \dot{\delta}_{i1} = A_{11}\delta_{i1} + A_{12}\delta_{i2} + B_1(u_i - u_1) \\ 0 \quad = A_{21}\delta_{i1} + A_{22}\delta_{i2} + B_2(u_i - u_1) \end{cases} i = 2, \ldots, N \tag{19.6}$$

Let $\delta_1 = [\delta_{21}^T, \delta_{31}^T, \ldots, \delta_{N1}^T]^T, \delta_2 = [\delta_{22}^T, \delta_{32}^T, \ldots, \delta_{N2}^T]^T$. Based on (19.5), the error equation (19.6) can be rewritten as follows

$$\begin{cases} \dot{\delta}_1 = (I_{N-1} \otimes A_{11})\delta_1 + (I_{N-1} \otimes A_{12})\delta_2 - (L_{22} + 1_{N-1} \bullet \alpha^T) \otimes B_1 K \delta_1(t - \tau(t)) \\ 0 \;= (I_{N-1} \otimes A_{21})\delta_1 + (I_{N-1} \otimes A_{22})\delta_2 - (L_{22} + 1_{N-1} \bullet \alpha^T) \otimes B_1 K \delta_1(t - \tau(t)) \end{cases} \tag{19.7}$$

where $\alpha = [a_{12}, a_{13}, \ldots, a_{1N}]^T$, and $L_{22} = \begin{bmatrix} deg_{out}(2) & -a_{23} & \ldots & -a_{2N} \\ -a_{32} & deg_{out}(3) & \ldots & -a_{3N} \\ \vdots & \vdots & \vdots & \vdots \\ -a_{N2} & -a_{N3} & \ldots & deg_{out}(N) \end{bmatrix}$

From Lemma 2, A_{22} is invertible, so we have

$$\dot{\delta}_1 = H_1\delta_1 + H_2\delta_1(t - \tau(t)) \tag{19.8}$$

where $H_1 = I_{N-1} \otimes (A_{11} - A_{12}A_{22}^{-1}A_{21})$, $H_2 = (L_{22} + 1_{N-1} \bullet \alpha^T) \otimes (-B_1K + A_{12} A_{22}^{-1}B_2K)$

For system (19.8), select a Lyapunov candidate as follows

$$V = \delta_1^T R_1 \delta_1 + \int_{-h}^{0} \int_{t+\theta}^{t} \dot{\delta}_1^T(s) R_2 \dot{\delta}_1(s) ds d\theta \tag{19.9}$$

Taking the derivative of V, so we have

$$\dot{V} = 2\delta_1^T R_1 \dot{\delta}_1 + h\dot{\delta}_1^T R_2 \dot{\delta}_1 - \int_{t-h}^{t} \dot{\delta}_1^T(s) R_2 \dot{\delta}_1(s) ds \tag{19.10}$$

Since $0 \le \tau(t) \le h$, $h > 0$, $t \ge 0$, then

$$-\int_{t-h}^{t} \dot{\delta}_1^T(s) R_2 \dot{\delta}_1(s) ds \le -\int_{t-\tau(t)}^{t} \dot{\delta}_1^T(s) R_2 \dot{\delta}_1(s) ds \tag{19.11}$$

From Lemma 3, we can obtain

$$-\int_{t-\tau(t)}^{t} \dot{\delta}_1^T(s) R_2 \dot{\delta}_1(s) ds \le -h^{-1}[\delta_1 - \delta_1(t - \tau(t))]^T R_2 [\delta_1 - \delta_1(t - \tau(t))] \tag{19.12}$$

Based on (19.10)–(19.12), we have

$$\begin{aligned} \dot{V} \le\; & \delta_1^T(R_1H_1 + H_1^TR_1)\delta_1 + h\delta_1^T H_1^T R_2 H_1 \delta_1 - h^{-1}\delta_1^T R_2 \delta_1 \\ & + 2\delta_1^T R_1 H_2 \delta_1(t - \tau(t)) + 2h\delta_1^T H_1^T R_2 H_2 \delta_1(t - \tau(t)) + 2h^{-1}\delta_1^T R_2 \delta_1(t - \tau(t)) \\ & + h\delta_1^T(t - \tau(t)) H_2^T R_2 H_2 \delta_1(t - \tau(t)) - h^{-1}\delta_1^T(t - \tau(t)) R_2 \delta_1(t - \tau(t)) \end{aligned}$$

Let $y^T = [\delta_1^T, \delta_1^T(t - \tau(t))]$, we have $\dot{V} \le y^T \Xi y$. Since $\Xi < 0$, the system (19.8) is asymptotic stability, that is, $\delta_1 \to 0$, when $t \to \infty$. From (19.7), we can obtain $\delta_2 \rightarrow 0$, when $t \rightarrow \infty$. So system (19.2) reach consensus. The proof is complete. □

In the following, by analyzing the solution of system (19.2), we obtain the consensus state and Theorem 2 is described as follows

Theorem 2 *For the system* (19.2), *if it can reach consensus via protocol* (19.5), *the state will converge to*

$\lim_{t\to\infty} X_1(t) = \lim_{t\to\infty} U \bullet 1_N \otimes e^{\Xi_1 t} x_{11}(0)$

$\lim_{t\to\infty} X_2(t) = \lim_{t\to\infty} (-U \bullet 1_N \otimes A_{22}^{-1} A_{21} e^{\Xi_1 t} + LU \bullet 1_N \otimes A_{22}^{-1} B_2 K e^{\Xi_1 t}) x_{11}(0)$

respectively.

Proof Via (19.5), system (19.2) can be rewritten as

$$\begin{cases} \dot{X}_1 = (I_N \otimes A_{11})X_1 + (I_N \otimes A_{12})X_2 - L \otimes B_1 K X_1(t-\tau(t)) \\ 0 \;= (I_N \otimes A_{21})X_1 + (I_N \otimes A_{22})X_2 - L \otimes B_2 K X_1(t-\tau(t)) \end{cases} \tag{19.13}$$

where $X_1 = [x_{11}^T, x_{21}^T, \ldots, x_{N1}^T]^T$, $X_2 = [x_{12}^T, x_{22}^T, \ldots, x_{N2}^T]^T$.

There exist a invertible matrix $U \in \mathbb{R}^{N\times N}$, then

$$U^{-1} L U = \begin{bmatrix} 0 & 0 \\ 0 & J \end{bmatrix} \tag{19.14}$$

where J is upper triangular Jordan block, whose principal diagonal elements are nonzero eigenvalues $\lambda_i, i = 2, \ldots, N$ of L.

Taking the transformation $\overline{X}_1 = (U \otimes I_r)^{-1} X_1$, we have

$$\begin{aligned} \dot{\overline{X}}_1 &= I_N \otimes (A_{11} - A_{12}A_{22}^{-1}A_{21})\overline{X}_1 \\ &\quad + \begin{bmatrix} 0 & 0 \\ 0 & J \end{bmatrix} \otimes (-B_1 K + A_{12}A_{22}^{-1}B_2 K)\overline{X}_1(t-\tau(t)) \end{aligned} \tag{19.15}$$

From (19.15), we have

$$\dot{\overline{x}}_{11} = (A_{11} - A_{12}A_{22}^{-1}A_{21})\overline{x}_{11} \tag{19.16}$$

Further, (19.16) means

$$\overline{x}_{11}(t) = e^{\Xi_1 t}\overline{x}_{11}(0) \tag{19.17}$$

where $\Xi_1 = A_{11} - A_{12}A_{22}^{-1}A_{21}$. We can obtain the consensus state of $\overline{X}_1$ as follows $\lim_{t\to\infty} (I_N \otimes e^{\Xi_1 t})\overline{x}_{11}(0)$

From (19.13) and (19.14), we can obtain consensus state of X_1 and X_2 as follows

$\lim_{t\to\infty} X_1(t) = \lim_{t\to\infty} (U \bullet 1_N \otimes e^{\Xi_1 t}) x_{11}(0)$

$\lim_{t\to\infty} X_2(t) = \lim_{t\to\infty} (-U \bullet 1_N \otimes A_{22}^{-1} A_{21} e^{\Xi_1 t} + LU \bullet 1_N \otimes A_{22}^{-1} B_2 K e^{\Xi_1 t}) x_{11}(0)$

The proof is complete. □

19.4 Simulation

Consider a multi-agent system composed of 4 agents with singular dynamics, each agent is described by 4-order dynamics. The network is described as $A = \begin{bmatrix} 0\,0\,0\,1 \\ 1\,0\,0\,0 \\ 0\,1\,0\,0 \\ 0\,0\,1\,0 \end{bmatrix}$.

Each agent is described by $\begin{bmatrix} 1 & 0 \\ 0 & 1 \end{bmatrix} \dot{x}_{i1} = \begin{bmatrix} -1 & 0 \\ 0 & -2 \end{bmatrix} x_{i1} + \begin{bmatrix} -2 & 3 \\ 4 & 1 \end{bmatrix} x_{i2} + \begin{bmatrix} 1 \\ 1 \end{bmatrix} u_i, 0 = \begin{bmatrix} 1 & 1 \\ 2 & 2 \end{bmatrix} x_{i1} + \begin{bmatrix} 1 & -2 \\ 0 & 5 \end{bmatrix} x_{i2} + \begin{bmatrix} 1 \\ 1 \end{bmatrix} u_i$. The initial states of agents are $x_{11} = [1, 2], x_{21} = [5, 6], x_{31} = [-3, -2], x_{41} = [-5, -6], x_{12} = [-5.4, -1.2], x_{22} = [-19.8, -4.4], x_{32} = [9, 2], x_{42} = [19.8, 4.4]$. The time-varying time delay is given by $\tau(t) = 0.3|sin 10t|$, and it has an upper 0.3. The state trajectories of system (19.2) is shown as Fig. 19.1.

From Fig. 19.1, all states of each agent will reach consensus, respectively. Moreover, the states will tend to zero when $t \to \infty$.

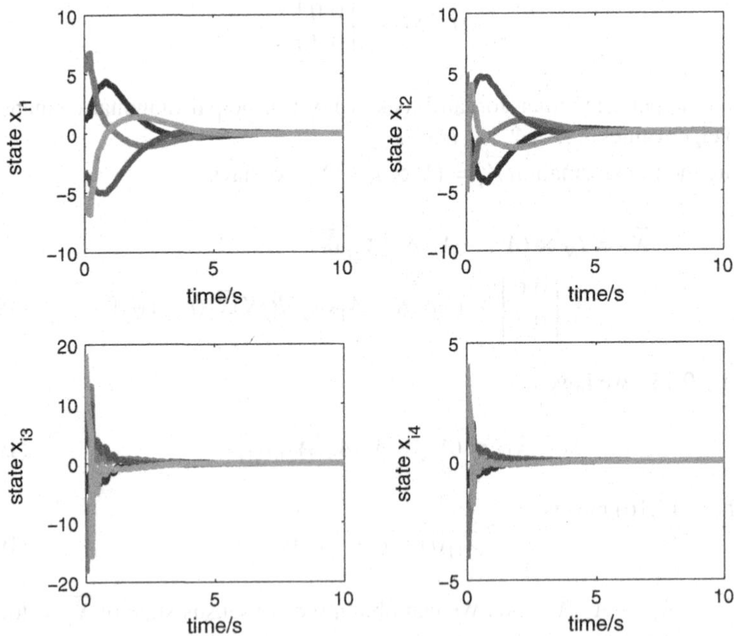

Fig. 19.1 State trajectories

19.5 Conclusions

In this paper, we consider the consensus problem of high-order multi-agent systems with singular dynamics and time delay. By applying the method of decomposing states and LMI tool, consensus problem is solved. Moreover, the consensus state is also obtained. Finally, an example is presented to illustrate the theoretical results.

Acknowledgments This research is supported by the National Natural Science Foundation of China (Grant No.61174094,61273138), and the Tianjin Natural Science Foundation of China (Grant No.14JCYBJC18700, 14JCZDJC39300).

References

1. Olfati-Saber R (2006) Flocking for multi-agent dynamic systems: algorithms and theory. IEEE Trans Autom Control 51(3):401–420
2. Xiao F, Wang L, Chen J (2009) Finite-time formation control for multi-agent systems. Automatica 45(11):2605–2611
3. Amarjeet S, Andreas K, Carlos G et al (2009) Efficient informative sensing using multiple robots. J Artif Intell Res 34:707–755
4. Xiang H, Tian L (2011) Development of a low-cost agricultural remote sensing system based on an autonomous unmanned aerial vehicle. Biosyst Eng 108(2):174–190
5. Alex A, Albert D-G, Jurgen K et al (2008) Synchronization in complex networks. Phys Rep-Rev Sect Phys Lett 469(3):93–153
6. Absessameud A, Tayebi A (2009) Attitude synchronization of a group of spacecraft without velocity measurement. IEEE Trans Autom Control 54(11):2642–2648
7. Olfati-Saber R, Fax JA et al (2007) Consensus and cooperation in networked multi-agent systems. Proc IEEE 95(1):215–233
8. Olfati-Saber R, Murray RM (2004) Consensus problems in networks of agents with switching topology and time-delays. IEEE Trans Autom Control 49(9):1520–1533
9. Hong YG, JiangPing H, Gao LL (2006) Tracking control for multi-agent consensus with an active leader and variable topology. Automatica 42(7):1177–1182
10. Lin P, Jia YM (2008) Average consensus in networks of multi-agents with both switch topology and coupling time delay. Phys A-Stat Mech Appl 387(2):303–313
11. Shang YL (2012) Finite-time consensus for multi-agent systems with fixed topologies. Int J Syst Sci 43(3):499–506
12. Li T, Zhang JF (2010) Consensus condition of multi-agent systems with time-varying topologies and stochastic communication noises. IEEE Trans Autom Control 55(9):2043–2057
13. Lin P, Jia YM (2010) Consensus of a class of second-order multi-agent systems with time-delay and jointly connected topologies. IEEE Trans Autom Control 53(3):778–784
14. Zhu W, Cheng DZ (2010) Leader-following consensus of second-order agents with multiple time-varying delays. Automatica 46(12):1994–1999
15. He W, Cao J (2011) Consensus control for high-order multi-agent systems. IET Control Theory Appl 5(1):231–238
16. Jiang FC, Wang L (2010) Consensus seeking of high-order dynamic multi-agent systems with fixed and switching topologies. Int J Control 83(2):404–420
17. Tian YP, Zhang Y (2012) High-order consensus of heterogeneous multi-agent systems with unknown communication delays. Automatica 48(6):1205–1212
18. Li M, Li QQ (2014) Admissible consensus of multi-agent singular systems. Asian J Control 16(4):1169–1178

19. Xi JX, Shi ZY, Zhong YS (2012) Admissible consensus and consensualization of high-order linear time-invariant singular swarm systems. Phys A 391:5839–5849
20. Xi JX, Yao Y, Liu GB et al (2014) Guaranteed-cost consensus for singular multi-agent systems with switching topologies. IEEE Trans Circuits Syst-I: Regul Pap 61(5):1531–1541
21. JianXiang X, FanLin M, ZongYing S et al (2014) Delay-dependent admissible consensualization for singular time-delayed swarm systems. Syst Control Lett 61(11):1089–1096
22. Godsil C, Royle G (2001) Algebraic graph theory. Springer, NewYork
23. Ren W, Beard RW (2005) Consensus seeking in multi-agent systems under dynamically changing interaction topologies. IEEE Trans Autom Control 50(5):655–661
24. Dai L (1989) Singular control systems. Springer, NewYork
25. YuanGong S, Long W, GuangMing X (2008) Average consen-sus in networks of dynamic agents with switching topologies and multiple time-varying de-lays. Syst Control Lett 57(2):175–183

Chapter 20
Data-Driven Filter Design for Linear Networked Systems with Bounded Noise

Yuanqing Xia, Li Dai, Wen Xie and Yulong Gao

Abstract Considering the case that the mathematical model of control plant is unavailable, this paper is concerned with the problem of data-driven filtering for linear networked systems with bounded noise and transmission data dropouts. One merit of the design is that the filter can be directly employed without identifying the model. To overcome the effect of data dropouts during the transmission, an output predictor is designed based only on the received output and input of the system. By utilizing the predicted output, a direct worst-case almost-optimal filter within the set membership framework is presented.

Keywords Networked system · Data-driven filter · Set membership filter · Data dropout

20.1 Introduction

Estimation for dynamical systems plays a crucial role in control system and signal processing and different kinds of estimation approaches are investigated [1–4]. In recent years, communication networks have been increasingly used to transmit data between the system components. However, the introduction of communication networks brings some challenging issues, such as data dropouts [5] and external disturbances. To solve these problems, many literatures have been developed on network-based filtering [6, 7]. For instance, network-based $\mathcal{H}_\infty$ filtering for discrete-time systems was addressed in [6] where a Markov jumping model approach was adopted.

Most of the above-mentioned methodologies relied on the exact mathematical model of the system, which are substantially model-based approaches. However, when the evolution process of the plant under consideration is not completely known, the model-based approach becomes no longer applicable. What we can obtain are

---

Y. Xia (✉) · L. Dai · W. Xie · Y. Gao
School of Automation, Beijing Institute of Technology, Beijing 100081, China
e-mail: xia_yuanqing@bit.edu.cn

© Springer-Verlag Berlin Heidelberg 2016

Y. Jia et al. (eds.), *Proceedings of the 2015 Chinese Intelligent Systems Conference*, Lecture Notes in Electrical Engineering 359,
DOI 10.1007/978-3-662-48386-2_20

only the measurements but not a known model. Motivated by these facts, data-driven filter design method is provided as a new tool to construct the desirable filters by using only the input and output measurements. In recent years, a great deal of literature on how to design direct data-driven filters has been emerged. One of the typical methods is the set membership approach [8–10], which established a set membership-based design framework on filtering. The direct data-driven linear filter with guaranteed worst-case performances was designed in [8]. In [9], the direct design approach was investigated in a nonlinear set membership setting, where the case of full observability and partial observability were discussed respectively to derive the desired filter design. However, the filters and error bounds in [9] are sub-optimal. Thus, the work of [10] developed some relevant improvements, and yielded optimal filters and optimal error bounds on the estimated variable.

This paper is motivated by the aforementioned works, especially, the work in [8]. However, for the system in the networked environment with the data dropout, the direct data-driven filter proposed in [8] may cease to be effective since the filter in [8] depends on the measurements at every time instant. Therefore, a direct filter design approach for the networked linear systems with data dropout is further investigated in the present paper. In this paper, to overcome the drawback of data dropouts in the network, an adaptive on-line output predictor is developed by using received output and input of the system, of which the convergence is guaranteed and the rate of convergence can be adjusted by parameters. Then, the networked data-driven filter is designed based on the predicted output instead of the received output.

20.2 Preliminaries and Problem Formulation

Consider a linear time-invariant (LTI) dynamic system S:

$$x(t+1) = Ax(t) + Bu(t),\ y(t) = C_1x(t) + Du(t),\ z(t) = C_2x(t), \tag{20.1}$$

where $x(t) \in \mathbb{R}^n$ is the unknown system state; $y(t) \in \mathbb{R}^{n_y}$ is the system output; $u(t) \in \mathbb{R}^{n_u}$ is the system input; $z(t) \in \mathbb{R}^{n_z}$ is the variable to be estimated; A, B, C_1, C_2 and D are constant real matrices with suitable finite dimensions.

As depicted in Fig. 20.1, the system matrices of system S are supposed to be unknown and the filtering problem for system S over a communication network is considered. Here the introduction of $u(t)$ is used to compensate the insufficient information of the system, which do not affect the filter design in essence. In the present framework we focus on the case that there exists a communication link from the sensors to the filter, and the filter knows the exact input at every time instant [11, 12]. At each time instant t, noised measurements are transmitted from the sensor through the network to the remote filter. The information we can employ are the input $u(t)$, and the measurements $\tilde{y}(t) = y(t) + w(t)$, $\tilde{z}(t) = z(t) + v(t)$, which can be collected from noisy sensors, where $w(t)$ and $v(t)$ are zero-mean white measurement noise. In the following, it is assumed that there may exist missing measurements of $\tilde{y}(t)$ and

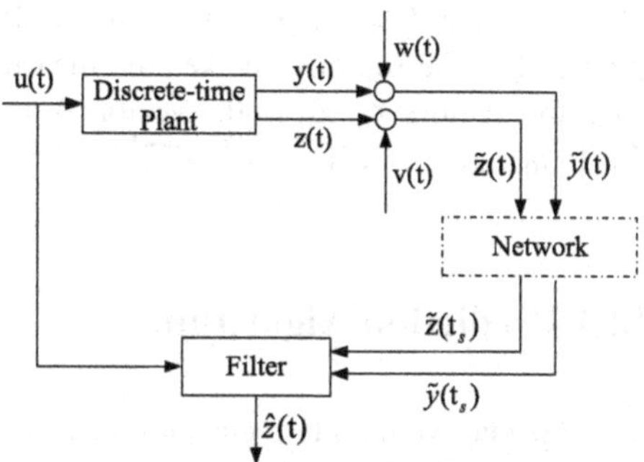

Fig. 20.1 Networked filter framework for LTI system

$\tilde{z}(t)$. Hence, the inputs of filter are $u(t)$, $\tilde{y}(t_s)$ and $\tilde{z}(t_s)$, which means the data between time t_{s-1} and t_s are missing. The objective of this paper is to design a direct filter that provides an estimation of z with minimizing the worst-case gain, which involves bounded w given in l_p-norm and the estimation error measured in l_q-norm.

For this purpose, recall the definition of l_p-norm for a one-sided discrete-time signal $s = \{s(0), s(1), \ldots\}, s(t) \in \mathbb{R}^{n_s}$: $\|s\|_p \doteq [\sum_{t=0}^{\infty} \sum_{i=0}^{n_s} |s_i(t)|^p]^{\frac{1}{p}}, p \in \mathbb{N}, 1 \leq p < \infty$, $\|s\|_\infty \doteq \max_{t=0,\ldots,\infty} \max_{i=0,\ldots,n_s} |s_i(t)|$ and the (l_q, l_p)-induced norm of a linear operator T: $\|T\|_{q,p} = \sup_{\|s\|_p=1} \|T(s)\|_q,\ p, q \in \mathbb{N}$.

Without loss of generality, we assume $n_z = 1$. In the following, the system S is supposed to be initially at rest (i.e., $x(0) = 0, u(t) = w(t) = 0\ \forall t < 0, u(0) \neq 0$, $w(0) \neq 0$) and the pair $[A, C_1]$ is supposed to be detectable. The inputs of system S are generated by digital computers and denoted by $u = \{u(0), u(1), \ldots, u(t), \ldots\}$. The measurements of system S are collected by noisy sensors, which can be represented by $\tilde{y} = \{\tilde{y}(0), \tilde{y}(1), \ldots, \tilde{y}(t), \ldots\}$ and $\tilde{z} = \{\tilde{z}(0), \tilde{z}(1), \ldots, \tilde{z}(t), \ldots\}$.

Denote the received data as $\tilde{y}(t_s), \tilde{z}(t_s)$, where t_s is the time instant when remote filter receives data, $s = 0, 1, \ldots$. Define the operator $\mathbf{T}$ maping the time series $\tilde{y}$, $\tilde{z}$ to $\{\tilde{y}(t_0), \tilde{y}(t_1), \ldots\}$, $\{\tilde{z}(t_0), \tilde{z}(t_1), \ldots\}$, respectively. We can easily obtain that $\mathbf{T}\tilde{y} \subseteq \tilde{y}$ and $\mathbf{T}\tilde{z} \subseteq \tilde{z}$ are the inputs of the remote filter.

In this paper, a filter composed of a predictor and a direct filter is designed, as depicted in Fig. 20.2. To derive the main results, the following assumption is given.

Assumption 1 The system S is asymptotically stable.

To achieve the purpose of filtering, an initial step is first carried out to collect a set of data with finite length $t_{N-1} + 1$, denoted by $\tilde{Y}_0 = \left[\tilde{y}(0), \tilde{y}(1), \ldots,\ \tilde{y}(t_{N-1})\right]^T \in$

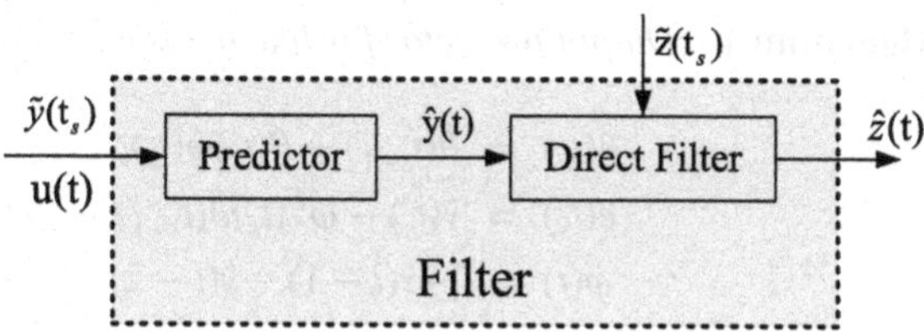

Fig. 20.2 Networked filter structure layout

$\mathbb{R}^{(t_{N-1}+1)n_y}$ and $\tilde{Z}_0 = [\tilde{z}(0), \tilde{z}(1), \ldots, \tilde{z}(t_{N-1})]^T \in \mathbb{R}^{(t_{N-1}+1)n_z}$. Similarly, assuming that at time t_{N-1} the remote receives the measurements, define the operator $\mathbf{T_N}$ which maps the vectors $\tilde{Y}_0$, $\tilde{Z}_0$ to the vectors $\tilde{Y} = [\tilde{y}(t_0), \tilde{y}(t_1), \ldots, \tilde{y}(t_{N-1})]^T \in \mathbb{R}^{Nn_y}$ and $\tilde{Z} = [\tilde{z}(t_0), \tilde{z}(t_1), \ldots, \tilde{z}(t_{N-1})]^T \in \mathbb{R}^{Nn_z}$.

20.3 Prediction Algorithm

In this paper, we mainly focus on how to solve the optimal filtering problem (OFP) in the presence of data dropouts. Without loss of generality, it is assumed that $\tilde{y}(t)$ and $\tilde{z}(t)$, i.e., samples between t_s and t_{s+1}, will be lost/dropped irregularly during the transmission. In this section, an efficient method for reconstruction of the missing measurements will be introduced.

Define the backward shift operator as ϱ^{-1}, which means $\varrho^{-1}x(t) = x(t-1)$. Equation (20.1) can be transformed into Deterministic AutoRegressive Moving Average (DARMA) model

$$A(\varrho^{-1})y(t) = B(\varrho^{-1})u(t), \tag{20.2}$$

where $A(\varrho^{-1}) = 1 + a_1\varrho^{-1} + \cdots + a_n\varrho^{-n}$, $B(\varrho^{-1}) = b_0 + b_1\varrho^{-1} + \cdots + b_n\varrho^{-n}$. Define ϑ and $\varphi(t)$ as $\vartheta = [a_1, \ldots, a_n, b_0, b_1, \ldots, b_n]^T$ and $\varphi(t) = [-y(t-1), \ldots, -y(t-n), u(t), u(t-1), \ldots, u(t-n)]^T$, respectively. Then, the DARMA model (20.2) can be expressed as $y(t) = \varphi(t)^T\vartheta$. Thus, we have $\tilde{y}(t) = y(t) + w(t) = \varphi(t)^T\vartheta + w(t)$.

To obtain the estimation of $y(t)$, the difficulty is that $\varphi(t)$ contains the unknown inner variables $y(t-i)$ and ϑ is unknown. The solution is to replace $y(t-i)$ with its estimate $\hat{y}(t-i)$. Furthermore, define $\hat{\varphi}(t) = [-\hat{y}(t-1), -\hat{y}(t-2), \ldots, -\hat{y}(t-n)$ $u(t), u(t-1), u(t-2), \ldots, u(t-n)]^T$. Let $\hat{\vartheta}(t)$ be the estimate of ϑ at time t. The value of $\hat{\vartheta}(t)$ updates only at time $t = t_s$ using the received data at time t_s, $s = 0, 1, 2, \ldots, N-1$, i.e., $\hat{\vartheta}(t_s + i) = \hat{\vartheta}(t_s)$, $i = 1, 2, \ldots, t_{s+1} - t_s - 1$. If the unknown $\varphi(t)$ and ϑ are replaced with $\hat{\varphi}(t)$ and $\hat{\vartheta}(t_s)$, respectively, then the estimate/prediction of $y(t_s - i)$ can be computed by $\hat{y}(t_s - i) = \hat{\varphi}(t_s - i)^T\hat{\vartheta}(t_{s-1})$.

According to the least square principle, minimizing the cost function

$$J(\vartheta) = \sum_{i=1}^{s} \lambda^{s-i}[\tilde{y}(t_i) - \hat{\varphi}^T(t_i)\vartheta]^2, \tag{20.3}$$

results in the following recursive least-squares algorithm with a forgetting factor λ for missing data systems [13].

Algorithm 1 (*Output predictor for linear system with missing measurements*)

$$\begin{aligned}
\hat{\vartheta}(t_s) &= \hat{\vartheta}(t_{s-1}) + P(t_s)\hat{\varphi}(t_s), && (20.4)\\
e(t_s) &= \tilde{y}(t_s) - \hat{\varphi}^T(t_s)\hat{\vartheta}(t_{s-1}), && (20.5)\\
\hat{\varphi}(t) &= [-\hat{y}(t-1), -\hat{y}(t-2), \ldots, -\hat{y}(t-n)
\end{aligned}$$

$$u(t), u(t-1), u(t-2), \ldots, u(t-n)]^T, \tag{20.6}$$

$$L(t_s) = P(t_s)\varphi(t_s) = \frac{P(t_{s-1})\varphi(t_s)}{\lambda + \varphi^T(t_s)P(t_{s-1})\varphi(t_s)}, \tag{20.7}$$

$$P(t_s) = \frac{1}{\lambda}[I - L(t_s)\varphi^T(t_s)]P(t_{s-1}), \quad 0 < \lambda \leq 1, \tag{20.8}$$

$$P(0) = p_0 I, \tag{20.9}$$

$$\hat{y}(t_s - i) = \hat{\varphi}(t_s - i)^T \hat{\vartheta}(t_{s-1}), \tag{20.10}$$

$$\hat{y}(t_s) = \hat{\varphi}(t_s)^T \hat{\vartheta}(t_s), \tag{20.11}$$

where p_0 is a large positive number. To initialize the algorithm, we take $\hat{\vartheta}(t_0)$ as a real vector with small entries, e.g., $\hat{\vartheta}(0) = \mathbf{1}_n/q_0$. The parameter λ is the forgetting factor to improve the rate of convergence. The convergence analysis of Algorithm 1 is presented in Theorems 1 and 3 of [13].

20.4 Direct Data-Driven Filtering with Bounded Disturbances

Assume that the convergence conditions in [13] hold in the rest of the paper, and apply Algorithm 1 on the received output $\tilde{y}$ and input u. Hence, the estimations of y can be obtained, denoted by $\hat{y}$, and collect them in the following column vector: $\hat{Y}_0 = [\hat{y}(0), \hat{y}(1), \hat{y}(2), \ldots, \hat{y}(t_{N-1})]^T \in \mathbb{R}^{(t_{N-1}+1)n_y}$.

Furthermore, consider the following $\mathscr{H}_\infty$ subset containing filters with bounded and exponentially decaying impulse response: $\mathcal{K}(L, \rho, \mu) = \{G \in \mathcal{H}_\infty : \|h_G^t\|_\infty \leq L, \forall t \in [0, \mu], \|h_G^t\|_\infty \leq L\rho^{t-\mu}, \forall t \geq \mu, t \in \mathbb{N}\}$, where the triplet (L, ρ, μ) is a parameter will be designed later, with $L > 0, 0 < \rho < 1, \mu \in N$, and $h_G = \{h_G^0, h_G^1, \ldots\}$ is the filter impulse response with $h_G^t \in R^{n_y}$.

The optimal worst-case filtering problem can be formulated as follows:

Optimal worst-case filtering problem (OFP): Given scalars $L > 0, 0 < \rho < 1$ and integers μ, p and q, design an optimal filter $G_o \in \mathcal{K}(L, \rho, \mu)$ such that the estimate $\hat{z}_{G_o} = G_o(\hat{y})$ achieves a finite gain $\gamma_o = \inf_{G_o \in \mathcal{K}(L,\rho,\mu)} \sup_{\|w\|_p=1} \|z - \hat{z}_{G_o}\|_q$.

The set of all the solutions to OFP is given by: $\mathcal{G}_o(L, \rho, \mu) = \{G \in \mathcal{K}(L, \rho, \mu) : \sup_{\|w\|_p=1} \|z - \hat{z}_G\|_q = \gamma_o\}$.

Assume the measurement noises are unknown with known bounds: $\|W\|_p \leq \delta$ and $\|V\|_q \leq \epsilon$. In the following, we assume $\delta = 1$. The purposes of this section are to determine a tight approximation of the optimal filter set $\mathcal{G}_o$ considering finite experiment length, and to find a filter with guaranteed worst-case performances, by suitably exploiting the information provided by the dataset $(\hat{Y}_0, \tilde{Z})$ and the noise bound.

Let $\hat{Z}_G^0 = [\hat{z}_G(0), \hat{z}_G(1), \ldots, \hat{z}_G(t_{N-1})]^T$ be the estimate vector provided by G when applied to received estimated output $\hat{Y}_0$. Then, applying the operator $\mathbf{T_N}$ to $\hat{Y}_0$ and

$\hat{Z}_G^0$, we have $\hat{Y} = [\hat{y}(t_0), \hat{y}(t_1), \ldots, \hat{y}(t_{N-1})]^T$ and $\hat{Z}_G = [\hat{z}_G(t_0), \hat{z}_G(t_1), \ldots, \hat{z}_G(t_{N-1})]^T$. Consider the following filter set: $FS(\tilde{\varepsilon}) = \{G \in \mathcal{K}(L, \rho, \mu) : \|\tilde{Z} - \hat{Z}_G\|_q \leq \tilde{\varepsilon}\}$. Next, a tight approximation to the set $\mathcal{G}_o(L, \rho, \mu)$ is provided considering an initial experiment of finite length $t_{N-1} + 1$.

Lemma 1 *Let the dataset $(\hat{Y}_0, \tilde{Z})$, scalars L, ρ, ϵ and integers μ, p, q be given.*
(i) $\mathcal{G}_o(L, \rho, \mu) \subseteq FS(\tilde{\varepsilon}), \forall \tilde{\varepsilon} \geq \gamma_o + \epsilon$
(ii) $FS(\gamma_o + \epsilon)$ is a tight outer approximation of $\mathcal{G}_o(L, \rho, \mu)$ as $N \to \infty$.

Proof (i) For any given filter $G \in \mathcal{G}_o(L, \rho, \mu)$, it follows that $\|z - \hat{z}_G\| \leq \gamma_o < \infty$, $\forall w \in \mathbb{R}^\infty$ such that $\|w\|_p \leq 1$. Since noisy measurements $\tilde{z} = z + v$ are available only, it turns out that $\|\tilde{z} - \hat{z}_G\|_q = \|z + v - \hat{z}_G\|_q \leq \|z - \hat{z}_G\|_q + \|v\|_q \leq \gamma_o + \|v\|_q$, $\forall w \in \mathbb{R}^\infty$ such that $\|w\|_p \leq 1$. To focus the attention on the received samples at time instant t_s, $s = 0, 1, \ldots, N-1$, we have $\|\tilde{Z} - \hat{Z}_G\|_q = \|\mathbf{T_N}(\tilde{Z}_0 - \hat{Z}_G^0)\|_q = \|\mathbf{T_N}\tilde{Z}_0 - \mathbf{T_N}\hat{Z}_G^0\|_q \leq \gamma_o + \|V\|_q \leq \gamma_o + \epsilon$, $\forall W \in \mathbb{R}^\infty$ such that $\|W\|_p \leq 1$. Thus, $G \in FS(\tilde{\varepsilon})$, $\forall \tilde{\varepsilon} \geq \gamma_o + \epsilon$.

(ii) The tightness property follows from the inclusion property (i) and the fact that, if the noise realization $v = \epsilon(z - \hat{z}_G)/\|z - \hat{z}_G\|_q$ occurs, then the inequality $\|\tilde{z} - \hat{z}_G\|_q = \|z + v - \hat{z}_G\|_q \leq \|z - \hat{z}_G\|_q + \|v\|_q \leq \gamma_o + \epsilon$ becomes tight.

The above lemma leads to an easier-to-operate solution set, and as in [8], we give the following definition.

Definition 1 Feasible Filter Set: $FFS = \{G \in \mathcal{K}(L, \rho, \mu) : \|\tilde{Z} - \hat{Z}_G\|_q \leq \gamma_o + \epsilon\}$.

In order to choose a suitable value of γ_o, a hypothesis validation problem is needed to be solved and the assumption of γ_o should guarantee that FFS is non-empty. Similar to [8], the following definition is given.

Definition 2 Let the dataset $(\hat{Y}_0, \tilde{Z})$, scalars L, ρ, ϵ and integers μ, p, q be given. Prior assumption on γ_o is validated if $FFS \neq \emptyset$.

When a filter $F \in \mathcal{K}(L, \rho, \mu)$ has been obtained by a design algorithm, for any estimated output $\hat{y}$, the difference between the estimate $\hat{z}_F$ provided by F and the estimate $\hat{z}_G$ can be measured by the term $\sup_{\|\hat{y}\|_q = 1} \|\hat{z}_G - \hat{z}_F\|_q = \|G - F\|_{q,q}$ where $G - F$ is the LTI dynamic system with input $\hat{y}$ and output $\hat{z}_G - \hat{z}_F$.

Since the induced norm $\|G - F\|_{q,q}$ depends on the particular G, it cannot be computed exactly. Fortunately, from Lemma 1 we know that $G \in \mathcal{G}_o(L, \rho, \mu) \subseteq FFS$. Thus, its tightest upper bound is given by: $\|G - F\|_{q,q} \leq \sup_{G \in \mathcal{G}_o(L,\rho,\mu)} \|G - F\|_{q,q} \leq \sup_{G \in FFS} \|G - F\|_{q,q}$, which leads to the following definition.

Definition 3 ([8]) Worst-case filtering error of a given $F \in \mathcal{K}(L, \rho, \mu)$: $E(F) = \sup_{G \in FFS} \|G - F\|_{q,q}$.

Then, the following optimality criterion can be defined.

Definition 4 ([8]) A filter $F_o \in \mathcal{K}(L, \rho, \mu)$ is optimal if $E(F_o) = \inf\limits_{F \in \mathcal{K}(L,\rho,\mu)} E(F)$.

It is known that the optimal filter can be computed by the Chebyshev center of *FFS*:

$$F_o = arg \inf_{F\in\mathcal{K}(L,\rho,\mu)} \sup_{G\in FFS} \|G - F\|_{q,q}. \tag{20.12}$$

However, it is very difficult to obtain the optimal solution of Eq. (20.12). This leads to the almost-optimal filter design.

Definition 5 ([8]) A filter F_a is almost-optimal if $F_a \in FFS$.

If the worst-case filtering error $E(F)$ can be computed, the following lemma gives a bound on the worst-case estimation error for any possible disturbance.

Lemma 2 *Under Assumption 1, for any given filter* $F \in \mathcal{K}(L, \rho, \mu)$*, the estimate* $\hat{z}_F = F(\hat{y})$ *guarantees* $\sup_{\|w\|_p=1} \|z - \hat{z}_F\|_q \le \gamma_o + E(F)\|S_{\hat{y}}\|_{q,p}$*, where* $S_{\hat{y}}$ *is the LTI dynamic subsystem of S such that* $\hat{y} = S_{\hat{y}}(w)$*.*

Proof If any filter $G \in \mathcal{G}_o(L, \rho, \mu)$ is considered, providing an estimate $\hat{z}_G = G(\hat{y})$, the worst-case gain from w to $z - \hat{z}_F$ is bounded as $\sup_{\|w\|_p=1} \|z - \hat{z}_F\|_q \le \sup_{\|w\|_p=1} \|z - \hat{z}_G\|_q + \sup_{\|w\|_p=1} \|\hat{z}_G - \hat{z}_F\|_q = \gamma_o + \sup_{\|w\|_p=1} \|\hat{z}_G - \hat{z}_F\|_q$, while $\sup_{\|w\|_p=1} \|\hat{z}_G - \hat{z}_F\|_q \le \sup_{\hat{y}\neq 0} \frac{\|\hat{z}_G-\hat{z}_F\|_q}{\|\hat{y}\|_q} \sup_{\|w\|_p=1} \|\hat{y}\|_q = \|G - F\|_{q,q}\|S_{\hat{y}}\|_{q,p} \le E(F)\|S_{\hat{y}}\|_{q,p}$.

20.5 Direct Data-Driven Design of Almost-Optimal Filters

Since the *FFS* is an infinite dimensional set, it is a difficult task to look for filters in the *FFS*. Finite impulse response (FIR) filters are used hereafter to approximate any filter $F \in \mathcal{K}(L, \rho, \mu)$ and the search in the *FFS* is transformed into a search in a finite dimensional space.

Hence, define $\mathcal{K}^m(L, \rho, \mu) = \{F \in \mathcal{K}(L, \rho, \mu) : h_F^t = 0, \forall t > m\}$ where $L > 0$, $0 < \rho < 1$, $\mu \in \mathbb{N}$, $m \in \mathbb{N}$, $m \ge \mu$ and $h_F^t \in \mathbb{R}^{n_y}$. It is obvious that $\mathcal{K}^m(L, \rho, \mu) \subset \mathcal{K}(L, \rho, \mu)$, $\forall m \in \mathbb{N}$.

Given a filter G, let G^m be its truncation of order m, i.e., the FIR filter having the same first $m + 1$ impulse response samples of G : $h_G^m = \{h_G^0, \dots, h_G^m, 0, 0, \dots\}$. The following standard result states that any filter $G \in \mathcal{K}(L, \rho, \mu)$ can be approximated with the required precision by the FIR filter $G^m \in \mathcal{K}^m(L, \rho, \mu)$ with suitable m.

Lemma 3 *Assume* $\mu \in \mathbb{N}$ *such that* $m \ge \mu$*. For any given filter* $G \in \mathcal{K}(L, \rho, \mu)$*, its truncation* $G^m \in \mathcal{K}^m(L, \rho, \mu)$ *guarantees* $\|G - G^m\|_{q,q} \le \|h_G - h_{G^m}\|_1 \le \eta_m$ *where* $1 \le q \le \infty$ *and* $\eta_m = n_y \frac{L\rho^{m+1-\mu}}{1-\rho}$ *is the truncation error of G through* G^m*.*

Proof For any given filter $G \in \mathcal{K}(L, \rho, \mu)$, we have $\|G - G^m\|_{q,q} = \sup_{\|\hat{y}\|_q=1} \|\hat{z}_G - \hat{z}_{G^m}\|_q = \sup \|h_G - h_{G^m}\|_q = \|\{0, \dots, 0, h_G^{m+1}, h_G^{m+2}, \dots\}\|_1 \le \sum_{t=m+1}^{\infty} n_y\|h_G^m\|_\infty \le \sum_{t=m+1}^{\infty} n_y L\rho^{t-\mu} = n_y \frac{L\rho^{m+1-\mu}}{1-\rho}$.

To determine the tightest set of FIR filters consistent with the overall available information (i.e., γ_o, L, μ, ρ, $\hat{Y}_0$, $\tilde{Z}$ and ϵ) and the truncation error η_m, let us define

$$\underline{FFS^m} = \{F \in \mathcal{K}^m(L, \rho, \mu) : \|\tilde{Z} - \mathbf{T}_{\hat{\mathbf{y}}}^{\mathbf{N}} H_F\|_q \leq \gamma_o + \epsilon\} \tag{20.13}$$

$$\overline{FFS^m} = \{F \in \mathcal{K}^m(L, \rho, \mu) : \|\tilde{Z} - \mathbf{T}_{\hat{\mathbf{y}}}^{\mathbf{N}} H_F\|_q \leq \gamma_o + \epsilon + \eta_m \|\hat{Y}\|_q\} \tag{20.14}$$

where $\mathbf{T}_{\hat{\mathbf{y}}}^{\mathbf{N}} H_F$ is the estimate of Z at the time t_s when the data are received, $H_F = [h_F^0, h_F^1, \ldots, h_F^m]^T \in \mathbb{R}^{(m+1)n_y}$ is the column vector of the first $m+1$ coefficients of the FIR filter F and $T_{\hat{y}}^N \in \mathbb{R}^{N\times(m+1)n_y}$ is defined as follows:

- if $m < N$, then $T_{\hat{y}}^N = T_{\hat{y}}^{N,m} = \begin{bmatrix} \hat{y}(t_0) & 0 & \cdots & 0 \\ \hat{y}(t_1) & \hat{y}(t_1 - 1) & \cdots & \hat{y}(t_1 - m) \\ \hat{y}(t_2) & \hat{y}(t_2 - 1) & \cdots & \hat{y}(t_2 - m) \\ \vdots & \vdots & \ddots & \vdots \\ \hat{y}(t_{N-1}) & \hat{y}(t_{N-1} - 1) & \cdots & \hat{y}(t_{N-1} - m) \end{bmatrix}$
- if $m \geq N$, then $T_{\hat{y}}^N = [T_{\hat{y}}^{N,m} \;\; 0_{N\times(m+1-N)n_y}]$.

Lemma 4 *Assume $m \in \mathbb{N}$ such that $m \geq \mu$.*
(i) $\underline{FFS^m} \subseteq FFS$.
(ii) For any given filter $G \in FFS$, its truncation $G^m \in \overline{FFS^m}$.

Proof (i) As $\mathcal{K}^m(L, \rho, \mu) \subseteq \mathcal{K}(L, \rho, \mu)$, the claim (i) follows from the definition of $\underline{FFS^m}$ and FFS.

(ii) Let $\hat{z}_G = G(\hat{y})$ and $\hat{z}_{G^m} = G^m(\hat{y})$ be the estimates provided by G and G^m, respectively. By applying the triangular inequality, the difference between $\tilde{z}$ and $\hat{z}_{G^m}$ can be norm-bounded as $\|\tilde{z} - \hat{z}_{G^m}\|_q \leq \|\tilde{z} - \hat{z}_G\|_q + \|\hat{z}_G - \hat{z}_{G^m}\|_q$. When the first N measurements are received, from Definition 1, it follows that $\|\tilde{Z} - \hat{Z}_G\|_q \leq \gamma_o + \epsilon$, $\forall G \in FFS$. Moreover, Lemma 3 states that $\|G - G^m\|_{q,q} = \sup_{\hat{y}\neq 0} \|\hat{z}_G - \hat{z}_{G^m}\|_q / \|\hat{y}\|_q \leq \eta_m$ and then $\|\hat{Z}_G - \hat{Z}_{G^m}\|_q \leq \eta_m \|\hat{Y}\|_q$ and $\forall G \in \mathcal{K}(L, \rho, \mu) \supset FFS$, with $\hat{Z}_{G^m} = T_{\hat{y}}^N H_{G^m}$. This implies that $\|\tilde{Z} - T_{\hat{y}}^N H_{G^m}\|_q \leq \gamma_o + \epsilon + \eta_m \|\hat{Y}\|_q$ for any G^m obtained as truncation of a filter $G \in FFS$, i.e., $G^m \in \overline{FFS^m}$.

The problem of checking the prior assumption validity is now considered.

Theorem 1 *Let the dataset $(\hat{Y}_0, \tilde{Z})$, scalars L, ρ, ϵ and integers μ, p, q, m be given, with $m \geq \mu$. Let ν^* be the solution to the optimization problem:*

$$\nu^* = \min_{F \in \mathcal{K}^m(L,\rho,\mu)} \|\tilde{Z} - T_{\hat{y}}^N H_F\|_q. \tag{20.15}$$

(i) A sufficient condition for prior assumption being validated is $\nu^* \leq \gamma_o + \epsilon$.
(ii) A necessary condition for prior assumption being validated is $\nu^* \leq \gamma_o + \epsilon + \eta_m \|\hat{Y}\|_q$.
(iii) If $m \geq t_{N-1}$ is chosen, a necessary and sufficient condition for prior assumption being validated is $\nu^* \leq \gamma_o + \epsilon$.

Proof For (i), if $v^* \leq \gamma_o + \epsilon$, then $\underline{FFS^m} \neq \emptyset$. Since $\underline{FFS^m} \subseteq FFS$, $FFS \neq \emptyset$,

For (ii), if $v^* > \gamma_o + \epsilon + \eta_m \|\hat{Y}\|_q$, then $\overline{FFS^m} = \emptyset$. From Lemma 4, it turns out that no filter G exists whose truncation G^m belongs to $\overline{FFS^m}$ and then $FFS = \emptyset$, i.e., the prior assumption is invalidated by data.

To prove condition (iii), at any time t, the estimate $\hat{z}_G(t)$ provided by any filter G can be written as $\hat{z}_G(t) = \sum_{k=-\infty}^{-1} (\hat{y}(k))^T h_G^{t-k} + \sum_{k=0}^{t} (\hat{y}(k))^T h_G^{t-k}$.

Taking into account that $\hat{y} = 0$, $\forall t < 0$, it follows that $\sum_{k=-\infty}^{-1} (\hat{y}(k))^T h_G^{t-k} = 0$, $\forall t < 0$, and then the estimate $\hat{z}_G(t)$ does not depend on h_G^k for $k > t$. When the first N measurements are received, the estimate vector $\hat{Z}_G$ does not depend on h_G^k for $k > t_{N-1}$ and then, if $m \geq t_{N-1}$ is chosen, it follows that: $\min_{G \in \mathcal{K}(L,\rho,\mu)} \|\tilde{Z} - \hat{Z}_G\|_q = \min_{F \in \mathcal{K}(L,\rho,\mu)} \|\tilde{Z} - T_{\hat{y}}^N H_F\|_q = v^*$. The condition $v^* \leq \gamma_o + \epsilon$ turns out to be necessary and sufficient for guaranteeing FFS non-empty since $\min_{G \in \mathcal{K}(L,\rho,\mu)} \|\tilde{Z} - \hat{Z}_G\|_q \leq \gamma_o + \epsilon$.

Let us consider the FIR filter F^* whose coefficients are given by the following algorithm:

$$H_{F^*} = \arg \min_{H_F \in \mathbb{R}^{(m+1)n_y}} \|\tilde{Z} - T_{\hat{y}}^N H_F\|_q, \tag{20.16}$$

such that $|h_{F,i}^t| \leq L, t = 0, \ldots, \mu, i = 1, \ldots, n_y$ and $|h_{F,i}^t| \leq L\rho^{t-\mu}, t = \mu + 1, \ldots, m, i = 1, \ldots, n_y$ where $h_{F,i}^t \in \mathbb{R}$ denotes the ith row element of h_F^t. Note that F^* is the filter class element that provides v^* as the solution to the optimization problem (20.15). The following theorem shows the properties of F^*, that hold for any l_p- and l_q-norms.

Theorem 2 *(i) If $v^* \leq \gamma_o + \epsilon$, the filter $F^* \in FFS$ and then is almost-optimal.*
(ii) If Assumption 1 is satisfied, then the estimate $\hat{z}_{F^*} = F^*(\hat{y})$ guarantees

$$\sup_{\|\omega\|_p = 1} \|z - \hat{z}_{F^*}\|_q \leq \gamma_o + E(F^*)\|S_{\hat{y}}\|_{q,p}. \tag{20.17}$$

Proof The claim (i) follows from the fact that, if $v^* \leq \gamma_o + \epsilon$, from Eqs. (20.13) and (20.15), then $F^* \in \underline{FFS^m} \subseteq FFS$, i.e., F^* is almost-optimal.

The claim (ii) directly follows from Lemma 2, Assumption 1, and the almost-optimality property of *FFS*.

Then, we can obtain the filter design algorithm as follows.

Algorithm 2 (*Direct data-driven almost-optimal filter design*)

- Step 1: Obtain the estimated output column vector $\hat{Y}_0$ by utilizing Algorithm 1 and collect the data column vector $\tilde{Z}$.
- Step 2: Estimate an appropriate γ_o by designing untuned filters G.

- Step 3: Build surface ν^* as the solution to problem (20.15) and adjust the estimation of γ_o. Appropriate parameters L, ρ, μ and m are chosen according to Theorem 1. Hence, the filter class $\mathcal{K}^m(L, \rho, \mu)$ is determined.
- Step 4: Design F^* using the algorithm (20.16). According to Theorem 2, F^* is an almost-optimal filter.

20.6 Conclusion

In this paper, a direct networked data-driven filter composed of an output predictor and a direct data-driven filter, is designed to give an accurate estimation of interests. The proposed approach can handle the filtering problem when the measurements of the system are transmitted through networks and the mathematical model is hard or even impossible to obtain. Moreover, the filter designed in this paper is almost-optimal.

Acknowledgments The work was supported by the National Basic Research Program of China (973 Program) (2012CB720000), the National Natural Science Foundation of China (61225015, 61105092), Foundation for Innovative Research Groups of the National Natural Science Foundation of China (Grant No.61321002).

References

1. Kalman RE (1960) A new approach to linear filtering and prediction problems. J Basic Eng 82(1):35–45
2. Zhang JH, Xia YQ, Shi P (2009) Parameter-dependent robust H_∞ filtering for uncertain discrete-time systems. Automatica 45(2):560–565
3. Zhang JH, Xia YQ, Tao R (2009) New results on H_∞ filtering for fuzzy time-delay systems. IEEE Trans Fuzzy Syst 17(1):128–137
4. Mahmoud MS, Xia YQ (2010) Robust filter design for piecewise discrete-time systems with time-varying delays. Int J Robust Nonlinear Control 20(5):544–560
5. Xia YQ, Fu MY, Liu GP (2011) Analysis and Synthesis of Networked Control Systems. Springer, Berlin
6. Zhang XM, Han QL (2012) Network-based H_∞ filtering for discrete-time systems. IEEE Trans Signal Process 60(2):956–961
7. Song HB, Yu L, Zhang WA (2011) Networked H_∞ filtering for linear discrete-time systems. Inf Sci 181(3):686–696
8. Milanese M, Ruiz F, Taragna M (2010) Direct data-driven filter design for uncertain LTI systems with bounded noise. Automatica 46(11):1773–1784
9. Milanese M, Novara C, Hsu K, Poolla K (2009) The filter design from data (FD2) problem: nonlinear set membership approach. Automatica 45(10):2350–2357
10. Novara C, Ruiz F, Milanese M (2013) Direct filtering: a new approach to optimal filter design for nonlinear systems. IEEE Trans Autom Control 58(1):86–99
11. Montestruque LA, Antsaklis PJ (2003) On the model-based control of networked systems. Automatica 39(10):1837–1843

12. Gupta V, Dana AF, Hespanha JP, Murray RM, Hassibi B (2009) Data transmission over networks for estimation and control. IEEE Trans Autom Control 54(8):1807–1819
13. Ding F, Ding J (2010) Least-squares parameter estimation for systems with irregularly missing data. Int J Adapt Control Signal Process 24(7):540–553

Chapter 21
Distributed Finite-Time Coordination Control for 6DOF Spacecraft Formation Using Nonsingular Terminal Sliding Mode

Fangya Gao and Yingmin Jia

Abstract This paper is devoted to the finite-time coordination control problem of 6DOF spacecraft formation with directed networks in the presence of external disturbances, and a distributed control algorithm using nonsingular terminal sliding mode (NTSM) is proposed. Based on Lyapunov methods, it is proved that all the spacecrafts achieve formation flying in finite time, which means the states of all the followers simultaneously converge to the states of the virtual leader. Simulation results are provided to validate the effectiveness of the theoretical analysis.

Keywords 6DOF spacecraft formation · Distributed · Coordination control · Finite-time · NTSM

21.1 Introduction

Recently, spacecraft formation flying (SFF) has attracted much attention due to its many advantages such as lower costs, shorter period, and more flexibility compared with one single spacecraft [1]. Therefore, there are comprehensive applications of SFF, for example, earth monitoring and deep space exploration [2]. To ensure that the formation flying are implemented successfully, spacecrafts in the formation must be coordinated with each other. In the coordination control, consensus is one of the most important and fundamental problems, that is, all spacecraft can reach an agreement on certain quantities of interest. In the past years, the consensus control problem for SFF has been widely studied [2–6].

F. Gao (✉) · Y. Jia
The Seventh Research Division and the Center for Information and Control, Beihang University (BUAA), Beijing 100191, China
e-mail: gaofybuaa@163.com

Y. Jia
e-mail: ymjia@buaa.edu.cn

© Springer-Verlag Berlin Heidelberg 2016

Y. Jia et al. (eds.), *Proceedings of the 2015 Chinese Intelligent Systems Conference*, Lecture Notes in Electrical Engineering 359,
DOI 10.1007/978-3-662-48386-2_21

It is important that the attitude of the spacecrafts should be synchronised while keeping a particular position in some applications, which means the rotational control is closely related to the translational control, thus 6DOF coordination control should be studied. Recently, more attention has been paid to 6DOF SFF [7–9]. In [7], a model of 6DOF is built up, whereas the consensus control problems are not involved. Chung et al. [8] propose decentralized tracking control laws that synchronize the attitude with global exponential convergence.

In the analysis of consensus problems, convergence rate acts as an important performance indicator, and now there are plenty of studies about asymptotical convergence. However, for practical systems, the control process needs to be terminated in finite time. Obviously, finite-time control are more desirable. Zhao et al. [10] design two time-varying TSM algorithms for attitude tracking control system, which ensure the tracking errors reach to zero in finite time. Li et al. [12] discuss the finite-time consensus problems for multi-agent systems with double-integrator dynamics, however, the linear model does not have generality.

Motivated by the above work, in this paper, we consider the finite-time coordination control for 6DOF leader-follower spacecraft formation in the presence of external disturbance. By defining consensus errors using local information, the error equations are derived. Hence, a distributed control law based on the NTSM theory is designed for each follower such that all the followers track the leader in finite time. Specifically, the communication topology among the spacecrafts is directed and the control law is robust for bounded external disturbance compared with much existing research, which is more challenging and practical.

The remainder of this paper is organized as follows. In Sect. 21.2, spacecraft 6DOF models and some preliminaries are given. In Sect. 21.3, the NTSM is defined and the finite-time control law is proposed. And simulation results are presented in Sect. 21.4. Finally, this paper is concluded in Sect. 21.5.

21.2 Preliminaries

21.2.1 Spacecraft 6DOF models

We employ modified Rodrigues parameters (MRPs) to describe the attitude of the spacecraft. The attitude dynamics of the ith spacecraft are given by [11]

$$\begin{cases} \dot{\boldsymbol{\sigma}}_i = G(\boldsymbol{\sigma}_i)\boldsymbol{\omega}_i \\ J_i\dot{\boldsymbol{\omega}}_i = -S(\boldsymbol{\omega}_i)J_i\boldsymbol{\omega}_i + \boldsymbol{\tau}_i + \boldsymbol{\tau}_{di} \end{cases} \tag{21.1}$$

where $i = 1, 2, \ldots, n$. The MRP vector $\boldsymbol{\sigma}_i = \boldsymbol{e}_i tan\frac{\phi}{4} \in \mathbb{R}^3$ represents the ith spacecraft attitude, where $\boldsymbol{e}_i \in \mathbb{R}^3$ is the eigenaxis unit vector and ϕ the eigenangle corresponding to the given orientation, and $G(\boldsymbol{\sigma}_i) = \frac{1}{2}[\frac{1-\boldsymbol{\sigma}_i^T\boldsymbol{\sigma}_i}{2}I_{3\times3} + S(\boldsymbol{\sigma}_i) + \boldsymbol{\sigma}_i\boldsymbol{\sigma}_i^T]$ with $S(\boldsymbol{\sigma}_i)$ is the skew-symmetric matrix. $J_i \in \mathbb{R}^3$ is the inertia matrix, $\boldsymbol{\omega}_i \in \mathbb{R}^3$ is the angular velocity, $\boldsymbol{\tau}_i \in \mathbb{R}^3$ and $\boldsymbol{\tau}_{di} \in \mathbb{R}^3$ denotes the control torque and the external disturbance torque, respectively.

Remark 1 The particular MRPs set will cause singularity as ϕ approaches 2π. Define the mapping set of the original MRPs as $\boldsymbol{\sigma}_i^s = -\frac{\boldsymbol{\sigma}_i}{\boldsymbol{\sigma}_i^T\boldsymbol{\sigma}_i}$, calling its corresponding shadow counterpart, and by switching between $\boldsymbol{\sigma}_i$ and $\boldsymbol{\sigma}_i^s$, any attitude rotation without singularity can be described. Generally, we switch the MRPs when $\boldsymbol{\sigma}_i^T\boldsymbol{\sigma}_i > 1$, which can not only ensure the module value of $\boldsymbol{\sigma}_i$ or $\boldsymbol{\sigma}_i^s$ is not more than 1, but also can avoid singularity problem [10].

Then the attitude dynamics for the *i*th spacecraft can be expressed by Euler-Lagrangian form as

$$M_i^{\sigma}\ddot{\boldsymbol{\sigma}}_i + C_i^{\sigma}\dot{\boldsymbol{\sigma}}_i = G^{-T}(\boldsymbol{\sigma}_i)(\boldsymbol{\tau}_i + \boldsymbol{\tau}_{di}) \tag{21.2}$$

where $M_i^{\sigma} = G^{-T}(\boldsymbol{\sigma}_i)J_iG^{-1}(\boldsymbol{\sigma}_i)$, $C_i^{\sigma} = -G^{-T}(\boldsymbol{\sigma}_i)J_iG^{-1}(\boldsymbol{\sigma}_i)\dot{G}(\boldsymbol{\sigma}_i)G^{-1}(\boldsymbol{\sigma}_i) - G^{-T}(\boldsymbol{\sigma}_i)$ $S(J_iG^{-1}(\boldsymbol{\sigma}_i)\dot{\boldsymbol{\sigma}}_i)G^{-1}(\boldsymbol{\sigma}_i)$.

In the orbital frame, the relative dynamics of the *i*th spacecraft with respect to the reference point are [9]

$$\begin{cases} \ddot{x}_i = 2\dot{\theta}\dot{y}_i + \ddot{\theta}y_i + \dot{\theta}^2 - \dfrac{\mu(x_i + R_c)}{R_i^3} + \dfrac{\mu}{R_c^2} + \dfrac{1}{m_i}(f_{xi} + f_{dxi}) \\ \ddot{y}_i = -2\dot{\theta}\dot{x}_i - \ddot{\theta}x_i + \dot{\theta}^2y_i - \dfrac{\mu y_i}{R_i^3} + \dfrac{1}{m_i}(f_{yi} + f_{dyi}) \\ \ddot{z}_i = -\dfrac{\mu z_i}{R_i^3} + \dfrac{1}{m_i}(f_{zi} + f_{dzi}) \end{cases} \tag{21.3}$$

where $\boldsymbol{\rho}_i = \left[x_i\ y_i\ z_i\right]^T \in \mathbb{R}^{3\times1}$ is the *i*th spacecraft position vector with respect to the reference point, $\boldsymbol{F}_i = \left[f_{xi}\ f_{yi}\ f_{zi}\right]^T \in \mathbb{R}^{3\times1}$ and $\boldsymbol{F}_{di} = \left[f_{dxi}\ f_{dyi}\ f_{dzi}\right]^T \in \mathbb{R}^{3\times1}$ represent the control forces and the external disturbance forces, m_i is the mass of the *i*th spacecraft, μ, R_c and θ are the gravitational constant of the earth, the distance from the earth's center to the reference point, and the true anomaly of the reference orbit respectively, $R_i = \sqrt{(x_i + R_c)^2 + y_i^2 + z_i^2}$ is the distance between the earth's center and the *i*th spacecraft.

Equation (21.3) can be written in a Euler-Lagrange form as

$$M_i^{\rho}\ddot{\rho}_i + C_i^{\rho}\dot{\rho}_i + \boldsymbol{g}_i^{\rho} = \boldsymbol{F}_i + \boldsymbol{F}_{di} \tag{21.4}$$

where $M_i^\rho = m_i I_3$, $C_i^\rho = 2m_i\begin{bmatrix}0 & -\dot{\theta} & 0\\ \dot{\theta} & 0 & 0\\ 0 & 0 & 0\end{bmatrix}$, $g_i^\rho = m_i\begin{bmatrix}-\ddot{\theta}y_i - \dot{\theta}^2 + \frac{\mu(x_i+R_c)}{R_i} - \frac{\mu}{R_c^2}\\ \ddot{\theta}x_i - \dot{\theta}^2 y_i + \frac{\mu y_i}{R_i^3}\\ \frac{\mu z_i}{R_i^3}\end{bmatrix}$.

Combining (21.2) and (21.4), we get the 6DOF model of rotation and translation

$$\hat{M}_i\ddot{\boldsymbol{p}}_i + \hat{C}_i\dot{\boldsymbol{p}}_i + \hat{\boldsymbol{g}}_i = \hat{\boldsymbol{\delta}}_i + \hat{B}_i\boldsymbol{u}_i \tag{21.5}$$

where $\boldsymbol{p}_i = \begin{bmatrix}\boldsymbol{\sigma}_i\\ \boldsymbol{\rho}_i\end{bmatrix} \in \mathbb{R}^{6\times1}$, $\hat{M}_i = \begin{bmatrix}M_i^\sigma & 0_{3\times3}\\ 0_{3\times3} & M_i^\rho\end{bmatrix}$, $\hat{C}_i = \begin{bmatrix}C_i^\sigma & 0_{3\times3}\\ 0_{3\times3} & C_i^\rho\end{bmatrix}$, $\hat{\boldsymbol{g}}_i = \begin{bmatrix}\boldsymbol{0}_{3\times1}\\ \boldsymbol{g}_i^\rho\end{bmatrix}$, $\hat{\boldsymbol{\delta}}_i = \begin{bmatrix}-G^{-T}(\boldsymbol{\sigma}_i)\boldsymbol{\tau}_{di}\\ -\boldsymbol{F}_{di}\end{bmatrix}$, $\hat{B}_i = \begin{bmatrix}G^{-T}(\boldsymbol{\sigma}_i) & 0_{3\times3}\\ 0_{3\times3} & I_3\end{bmatrix}$, $\boldsymbol{u}_i = \begin{bmatrix}\boldsymbol{\tau}_i\\ \boldsymbol{F}_i\end{bmatrix}$.

Denote $C_i = \hat{M}_i^{-1}\hat{C}_i$, $\boldsymbol{g}_i = \hat{M}_i^{-1}\hat{\boldsymbol{g}}_i$, $\boldsymbol{\delta}_i = \hat{M}_i^{-1}\hat{\boldsymbol{\delta}}_i$, and $\tilde{\boldsymbol{u}}_i = \hat{M}_i^{-1}\hat{B}_i\boldsymbol{u}_i$, then a brief 6DOF model can be derived as follows:

$$\ddot{\boldsymbol{p}}_i + C_i\dot{\boldsymbol{p}}_i + \boldsymbol{g}_i = \boldsymbol{\delta}_i + \tilde{\boldsymbol{u}}_i \tag{21.6}$$

Assumption 1 *The external disturbance $\boldsymbol{\delta}_i$ is bounded, and satisfies $\|\boldsymbol{\delta}_i\|_\infty \le \bar{\delta}$, where $\bar{\delta}$ is a positive constant.*

21.2.2 Algebraic Graph Theory

Let $G = \{N, E\}$ be a directed graph of order n, where $N = \{n_1, n_2, \dots, n_n\}$ is the set of nodes, $E \subseteq N \times N$ is the set of edges. An edge of G is denoted by an ordered pair (n_i, n_j), and the node indexes belong to a finite index set $\Pi = \{1, 2, \dots, n\}$. $A = [a_{ij}]$ is a weighted adjacency matrix of G. $(n_j, n_i) \in E$ means there is an edge from n_j to n_i, then $a_{ij} > 0$. Moreover, we define $a_{ii} = 0$ for all $i \in \Pi$. If a directed graph has the property that $a_{ij} = a_{ji}$ for any $i, j \in \Pi$, then the graph is called undirected. The in-degree of node n_i is $d_i = \Sigma_{j=1}^n a_{ij}$, and the in-degree matrix is $D = diag\{d_1, d_2, \dots, d_n\}$, then the Laplacian matrix of G is defined as $L = D - A$. The set of neighbors of node v_i is denoted by $N_i = \{j | (n_j, n_i) \in E\}$. A directed path is a sequence of successive ordered edges in the form $(n_{i_1}, n_{i_2}), (n_{i_2}, n_{i_3}), \dots$, where the subscript $i_j \in \Pi$. The graph G is said to be strongly connected if any two nodes have a directed path. Moreover, if there is a vertex (the root node) such that there is a directed path from this node to every other node, the graph contains a directed spanning tree.

In this paper, we consider the spacecraft formation system consisting of one virtual leader denoted as 0, and n followers represented by $1, 2, \dots, n$. Define $B = diag\{b_1, b_2, \dots, b_n\} \in \mathbb{R}^{n\times n}$, where $b_i \ge 0 (i \in \Pi)$ is the weighted adjacency element between i and 0, and $b_i > 0$ if and only if there is an edge from 0 to i. The graph topology among the leader and the followers is defined as $\bar{G} = \{\bar{N}, \bar{E}\}$, where $\bar{N} = \{0, 1, 2, \dots, n\}$, and $\bar{E} \subseteq \bar{N} \times \bar{N}$.

Assumption 2 *There is a directed path from the leader to any follower, that is, the graph* $\bar{G}$ *contains a directed spanning tree.*

Lemma 1 ([12]) *All the eigenvalues of the matrix* $L + B$ *have positive real parts if and only if Assumption 2 holds.*

21.2.3 General Theory for Finite-Time Stability

Lemma 2 ([11]) *Consider the following continuous nonlinear system*

$$\dot{x} = f(x) \tag{21.7}$$

where $f(0) = 0$, $x \in \mathbb{R}^n$. *Suppose there exist a positive definite and continuously differentiable function* $V(x)$ *defined on a neighborhood of the origin, and* $k > 0$ *and* $\alpha \in (0, 1)$ *such that* $\dot{V}(x) + k(V(x))^{\alpha} \leq 0$, *then the origin of the system is finite-time stable, and the converage time* $T \leq \frac{V(x(0))^{1-\alpha}}{k(1-\alpha)}$.

Lemma 3 ([13]) *A vector* $\boldsymbol{x} = [x_1, x_2, \ldots, x_n]^T$ *is given, if* $0 < p < 2$, *then* $\|\boldsymbol{x}\|^p \leq \Sigma_{i=1}^{n} |x_i|^p$ *holds.*

Lemma 4 ([9]) *If* $x_1, \ldots, x_n \geq 0$ *and* $0 < q \leq 1$, *then* $(x_1 + \cdots + x_n)^q \leq x_1^q + \cdots + x_n^q$.

21.3 Main Results

21.3.1 Dynamic Error Equations

The consensus errors are defined as

$$\boldsymbol{e}_i^1 = \sum_{j=1}^{n} a_{ij}(\boldsymbol{p}_i - \boldsymbol{p}_j) + b_i(\boldsymbol{p}_i - \boldsymbol{p}_0) \in \mathbb{R}^6 \tag{21.8}$$

$$\boldsymbol{e}_i^2 = \sum_{j=1}^{n} a_{ij}(\dot{\boldsymbol{p}}_i - \dot{\boldsymbol{p}}_j) + b_i(\dot{\boldsymbol{p}}_i - \dot{\boldsymbol{p}}_0) \in \mathbb{R}^6 \tag{21.9}$$

Denote $\boldsymbol{e}_1 = [(e_1^1)^T, (e_2^1)^T, \ldots, (e_n^1)^T]^T$, $\boldsymbol{e}_2 = [(e_1^2)^T, (e_2^2)^T, \ldots, (e_n^2)^T]^T$. Without loss of generality, we assume that the states of the leader are time-invariant, so the error equations can be expressed as

$$\begin{cases} \dot{\boldsymbol{e}}_1 = \boldsymbol{e}_2 \\ \dot{\boldsymbol{e}}_2 = ((L + B) \otimes I_6)(-C\dot{\boldsymbol{p}} - \boldsymbol{g} + \boldsymbol{\delta} + \tilde{\boldsymbol{u}}) \end{cases} \tag{21.10}$$

where $C = \begin{bmatrix} C_1 & & \\ & \ddots & \\ & & C_n \end{bmatrix}, p = \begin{bmatrix} p_1 \\ \vdots \\ p_n \end{bmatrix}, g = \begin{bmatrix} g_1 \\ \vdots \\ g_n \end{bmatrix}, \delta = \begin{bmatrix} \delta_1 \\ \vdots \\ \delta_n \end{bmatrix}, \tilde{u} = \begin{bmatrix} \tilde{u}_1 \\ \vdots \\ \tilde{u}_n \end{bmatrix}$.

21.3.2 Finite-Time Consensus Control Law

In this section, based on NTSM theory, we design a distributed finite-time consensus algorithm for the ith follower that utilizes its own information and its neighbors' information to ensure each follower track the leader in finite time, that is, for any initial states, there is a constant $T_0 \in [0, +\infty)$ such that $\lim_{t \to T_0} \boldsymbol{p}_i(t) = \boldsymbol{p}_0(T_0)$.

Theorem 1 *Consider the 6DOF multiple spacecraft system (21.6). If Assumptions 1 and 2 hold, the NTSM is chosen as*

$$s_i = e_i^1 + \beta (e_i^2)^\alpha \tag{21.11}$$

and the control law is designed as

$$\tilde{u}_i = C_i \dot{p}_i + g_i + [(\sum_{j=1}^{n} a_{ij} + b_i)^{-1} \otimes I_6][-\frac{(e_i^2)^{2-\alpha}}{\alpha\beta} + \sum_{j=1}^{n} (a_{ij} \otimes I_6)$$
$$(\tilde{u}_j - C_j \dot{p}_j - g_j) - (\bar{\delta} \| (L + B) \otimes I_6 \|_\infty + \lambda) sign(s_i)] \tag{21.12}$$

where $i = 1, \ldots, n$, $\alpha = \frac{q}{p} \in (1, 2)$, p and q are positive odd integers, and $\beta, \lambda > 0$, then consensus can be reached in finite time, that is the followers can converge to the desired states in finite time.

Proof Consider the following Lyapunov function defined as

$$V = \frac{1}{2} S^T S \tag{21.13}$$

Take the derivative of V with respect to time

$$\dot{V} = S^T [e_2 + \alpha\beta diag(e_2^{\alpha-1})((L + B) \otimes I_6)(-C\dot{p} - g + \delta + \tilde{u})] \tag{21.14}$$

The control input $\tilde{u}$ can be written as

$$\tilde{u} = C\dot{p} + g - [(L+B)^{-1} \otimes I_6][\frac{e_2^{2-\alpha}}{\alpha\beta} + diag(\bar{\delta} \| (L+B) \otimes I_6 \|_\infty + \lambda) sign(S)] \tag{21.15}$$

Substituting (21.15) into (21.14) yields

$$\begin{aligned}\dot{V} &= \alpha\beta S^T diag(e_2^{\alpha-1})((L+B)\otimes I_6)\delta \\ &\quad -\alpha\beta S^T diag(e_2^{\alpha-1}) diag(\bar{\delta}\|(L+B)\otimes I_6\|_\infty + \lambda) sign(S) \\ &\le \alpha\beta\|S^T diag(e_2^{\alpha-1})\|_1 \|((L+B)\otimes I_6)\delta\|_\infty \\ &\quad -\alpha\beta S^T diag(e_2^{\alpha-1}) diag(\bar{\delta}\|(L+B)\otimes I_6\|_\infty + \lambda) sign(S) \\ &\le \alpha\beta[\bar{\delta}\|(L+B)\otimes I_6\|_\infty - (\bar{\delta}\|(L+B)\otimes I_6\|_\infty + \lambda)] \sum_{i=1}^{n}\sum_{t=1}^{6} |s_{it}|(e_{it}^2)^{\alpha-1} \\ &= -\alpha\beta\lambda \sum_{i=1}^{n}\sum_{t=1}^{6} |s_{it}|(e_{it}^2)^{\alpha-1} \end{aligned} \tag{21.16}$$

When $e_{it}^2 \neq 0$, $i = 1, 2, \ldots, n$, $t = 1, 2, \ldots, 6$, denote $\xi = min\{\alpha\beta\lambda(e_{it}^2)^{\alpha-1}\}$, and $\xi > 0$. Then we have

$$\dot{V} \le -\xi \sum_{i=1}^{n}\sum_{t=1}^{6} |s_{it}| \le -\sqrt{2}\xi V^{\frac{1}{2}} \tag{21.17}$$

Therefore, according to Lemma 2, the variable vector $\boldsymbol{S}$ will converge to the NTSM surface $\boldsymbol{S} = \boldsymbol{0}$ in finite time.

When $e_{it}^2 = 0$, from (21.10) and (21.15), we have

$$\dot{e_2} = ((L+B)\otimes I_6)\delta - diag(\bar{\delta}\|(L+B)\otimes I_6\|_\infty + \lambda) sign(S) \tag{21.18}$$

In the case of $s_{it} > 0$, we have $\dot{e}_{it}^2 \le -\lambda$, and thus $\dot{s}_{it}(t) = e_{it}^2(t) \le -\lambda t$. Since $s_{it}(0) > 0$, so the reaching time $T = \sqrt{2s_{it}(0)/\lambda}$. The same way, in the case of $s_{it} < 0$, it can be seen that $\dot{e}_{it}^2 \ge \lambda$, and the reaching time $T = \sqrt{-2s_{it}(0)/\lambda}$. So it can be concluded that $\boldsymbol{S} = \boldsymbol{0}$ is achieved in finite time.

On the NTSM surface $\boldsymbol{S} = \boldsymbol{0}$, we have $\boldsymbol{e}_2 = -(\boldsymbol{e}_1/\beta)^{\frac{1}{\alpha}}$, and it can be concluded that $\boldsymbol{e}_1 = \boldsymbol{e}_2 = \boldsymbol{0}$ can be reached in finite time. In order to prove this term, let the Lyapunov function be $V_1 = \frac{1}{2}\boldsymbol{e}_1^T\boldsymbol{e}_1$, and

$$\dot{V}_1 = \boldsymbol{e}_1^T\boldsymbol{e}_2 \le -\frac{1}{\beta^{1/\alpha}} 2^{\frac{\alpha+1}{2\alpha}} V_1^{\frac{\alpha+1}{2\alpha}} \tag{21.19}$$

From Lemma 2, the consensus error $\boldsymbol{e}_1$ and $\boldsymbol{e}_2$ defined in (21.10) will converge to zero in finite time. Therefore, the proof is completed.

Remark 2 The theorem above is proved under the assumption that the states of the leader are time-invariant. Similarly, the followers can track the specific dynamic leader if the control law is altered suitably. The proof here is omitted.

21.4 Simulations

In this section, simulation results are presented to illustrate the effectiveness of our algorithm. Figure 21.1 denotes the spacecraft formation consisting of one leader indexed by 0 and four followers indexed from 1 to 4 respectively and communication topology among the spacecrafts.

The reference orbit is assumed to be an ellipse, and the mass of each follower is identical, that is $m_i = 10$. Next, simulation parameters are given in Table 21.1, where a denotes the semi-major axis of the orbit and e represents the eccentricity, respectively. For simplicity, we suppose that

$$a_{ij} = \begin{cases} 1, & \text{if } (j, i) \in \bar{E} \\ 0, & \text{otherwise} \end{cases} \tag{21.20}$$

Based on the distributed finite-time consensus control law (21.12), the state tracking errors of the four followers are shown in Fig. 21.2, where the left sub-figure illustrates attitude tracking errors and the right illustrates position tracking errors, respectively. From the two sub-figures, it can be seen that the state errors converge to zero in finite time, which validates the effectiveness of the proposed algorithm.

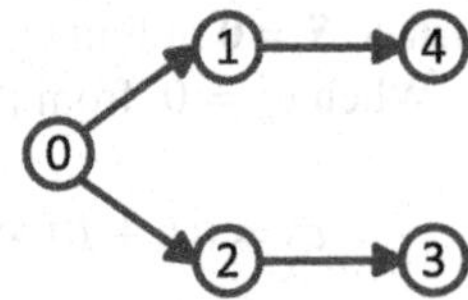

Fig. 21.1 Communication topology

Table 21.1 Simulation parameters

Parameters	Values
Inertia matrix	$J_1 = [10.10.1; 0.10.40.1; 0.10.10.9]$
	$J_2 = [0.80.20.2; 0.210.5; 0.20.51.3]$
	$J_3 = [10.10.1; 0.10.40.1; 0.10.10.9]$
	$J_4 = [1.20.20.4; 0.20.80.1; 0.40.11]$
Orbit parameters	$\mu = 3.98645 \times 10^{14}$, $a = 4.224 \times 10^7$, $e = 0.1$, $\theta = \pi/9$
Initial states	$\boldsymbol{p}_1 = [0.01\ -0.01\ 0\ -100\ 50\ 0]^T$
	$\boldsymbol{p}_2 = [0.1\ 0.04\ -0.01\ 20\ 80\ 100]^T$
	$\boldsymbol{p}_3 = [0.03\ 0\ -0.02\ 30\ 40\ 90]^T$
	$\boldsymbol{p}_4 = [0.1\ -0.2\ 0.05\ 70\ -50\ 60]^T$
External disturbance	$\boldsymbol{\tau}_{di} = [0.02sin(100t)\ 0.01cos(200t)\ -0.04sin(100t)]^T$
	$\boldsymbol{F}_{di} = [0.02sin(100t)\ 0.01cos(200t)\ -0.04sin(100t)]^T$
Desired states	$\boldsymbol{p}_d = [0.1\ 0 - 0.1\ 100 - 50\ 25]^T$
Controller parameters	$\alpha = 5/3$, $\beta = 1$, $\lambda = 0.5$

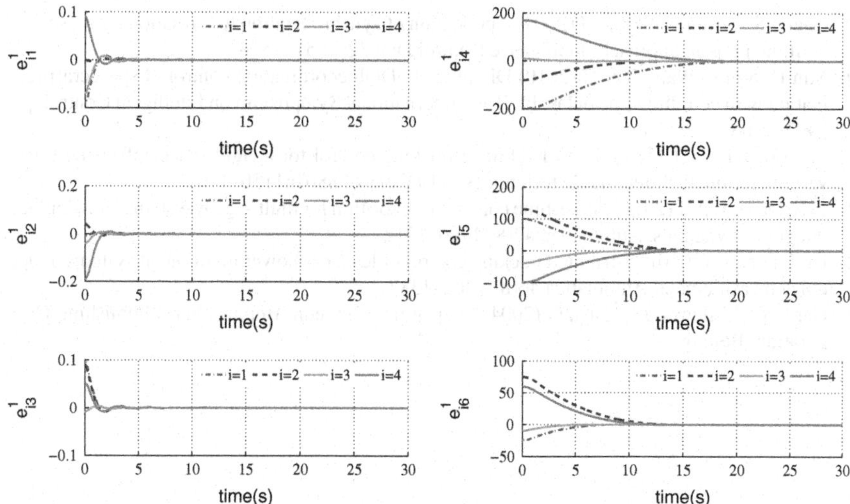

Fig. 21.2 State tracking errors

21.5 Conclusions

In this paper, a distributed algorithm is proposed via NTSM for 6DOF spacecraft formation with directed networks when there exists external disturbance. Based on algebraic graph theory and Lyapunov methods, we proved that the consensus can be achieved in finite time under the algorithm we proposed.

Acknowledgments This work was supported by the National Basic Research Program of China (973 Program: 2012CB821200, 2012CB821201) and the NSFC (61134005, 61221061, 61327807).

References

1. Zhang Y, Zeng G, Wang Z et al (2008) Thoery and application of distributed satellite system. Sience Press, Beijing (in Chinese)
2. Scharf D, Hadaegh F, Ploen S (2004) A survey of spacecraft formation flying guidance and control (Part II). In: American Control Conference, pp 2976–2985
3. Kapila V, Sparks A, Buffington J et al (2000) Spacecraft formation flying: dynamics and control. J Guidance Control Dyn 23(3):561–564
4. De Queiroz M, Kapila V, Yan Q (2000) Adaptive nonlinear control of multiple spacecraft formation flying. J Guidance Control Dyn 23(3):385–390
5. Li S, Du H, Shi P (2014) Distributed attitude control for multiple spacecraft with communication delays. IEEE Trans Aerosp Electron Syst 50(3):1765–1773
6. Zhao L, Jia Y (2014) Decentralized adaptive attitude synchronization control for spacecraft formation using nonsingular fast terminal sliding mode. Nonlinear Dyn 78(4):2279–2294
7. Kristiansen R, Nicklasson P, Gravdahl J (2007) Formation modelling and 6DOF spacecraft coordination control. In: American Control Conference, pp 4690–4696

8. Chung S, Ahsun U, Slotine J (2009) Application of synchronization to formation flying spacecraft: lagrangian approach. J Guidance Control Dyn 32(2):512–526
9. Min H, Sun F, Wang S et al (2010) Distributed 6DOF coordination control of spacecraft formation with coupling time delay. In: IEEE International Symposium on Intelligent Control, pp 2403–2408
10. Zhao L, Jia Y (2015) Finite-time attitude tracking control for a rigid spacecraft using time-varying terminal sliding mode techniques. Int J Control 88(6):1150–1162
11. Li S, Du H, Lin X (2011) Finite-time consensus algorithm for multi-agent systems with double-integrator dynamics. Automatica 47(8):1706–1712
12. Hu J, Feng G (2010) Distributed tracking control of leader-follower multi-agent systems under noisy measurement. Automatica 46(8):1382–1387
13. Hardy G, Littlewood J, Polya G (2004) Inequalities, 2nd edn. Beijing World Publishing Corporation, Beijing

Chapter 22
A Comparative Study on Optimal Transient Control of Aircraft Engines

Yang Gao, Jiqiang Wang, Xin Wu, Zhifeng Ye, Zhongzhi Hu and Georgi Dimirovski

Abstract Sequential quadratic programming has been widely used in the optimization of transient control of aircraft engine. On the other hand, the study of nonlinear optimization theory and method has been made great advance in recent years, of which active set method is one of the effective measures. In this article, the two algorithms are used to optimize the inputs based on one turbofan engine model. Results from the simulation comparisons of the two algorithms, advantages and disadvantages are given.

Keywords Optimization algorithm · SQP · ASM · Turbofan engine model

22.1 Introduction

Optimal control problem is a significant issue of aircraft engine control study. In recent years, a lot of research work on the engine of transition control has been done [1–5]. Numerous studies is based on the algorithm of sequential quadratic programming to optimize the transition control variables. However, the research on the algorithm of active set method that used to optimize the engine control is scant. To get some experience on engine optimization design, this paper describes the application of the two methods—active set method (ASM) [6] and sequential quadratic programming (SQP) [7] to optimize the acceleration control. All the study has been based on a module of two-spool mixed exhaust turbofan engine.

Y. Gao · J. Wang (✉) · Z. Ye · Z. Hu
Jiangsu Province Key Laboratory of Aerospace Power Systems,
Nanjing University of Aeronautics and Astronautics, Nanjing 210016, China
e-mail: jiqiang.wang@nuaa.edu.cn

J. Wang · X. Wu
AVIC Shenyang Engine Design and Research Institute, Shenyang 110015, China

G. Dimirovski
Dogus University of Istanbul, 34722 Istanbul, Turkey

© Springer-Verlag Berlin Heidelberg 2016

Y. Jia et al. (eds.), *Proceedings of the 2015 Chinese Intelligent Systems Conference*, Lecture Notes in Electrical Engineering 359,
DOI 10.1007/978-3-662-48386-2_22

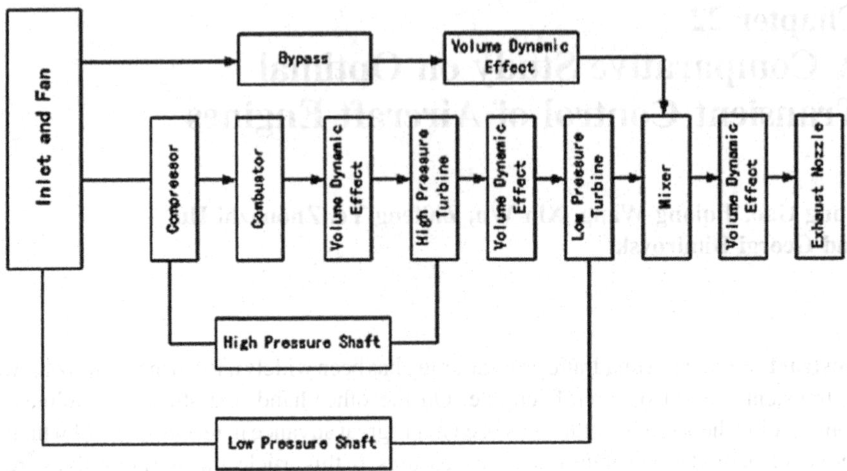

Fig. 22.1 The turbofan engine system block diagram

22.2 Engine Model and Algorithm

22.2.1 Engine Model

This paper uses some Component-Level model for a Turbofan Engine [8] which base on the platform of Simulink to compare the two algorithms. Volume dynamic effect has been introduced to realize the dynamic simulation of engine. Therefore, the study of the transient and the set-point control can be taken. As shown in Fig. 22.1 is the system block diagram. This paper selects fuel flow and exhaust nozzle area as control inputs.

As we known, a turbofan engine must perform over a wide range of operating conditions. It is worth mentioning that both the Height and the Mach number are 0 to consider the problem of comparison in this paper. According to the similarity principle, it will convert to other operating point of the engine within the flight envelope and doesn't consider it here.

22.2.2 Nonlinear Optimal Algorithm

As aircraft engine is a strongly nonlinear system with complex operating environment and high reliability requirements, it gets many constraints on it. Consequently, it is obvious that nonlinear optimization theory and method is introduced to solve the problem. Furthermore, the quadratic programming is the simplest aspect that has been first studied by people in the theory. The existence of the optimal solution can be verified by the limited amount of computation steps, it can also be obtained by means of numerical methods in a finite number of steps. Thus, to get the better method in this paper SQP and ASM (both of them are quadratic programming algorithms) are compared. Abundant of research on the Nonlinear Optimization Algorithm has been taken. So calculation steps are shown but not the proof. The problem can be abstracted as the following functions:

$$\min Q(x) = \frac{1}{2}x^T Gx + g^T x$$
$$s.t. \begin{cases} a_i^T x = b_i, i \in \varepsilon, \\ a_i^T x \geq b_i, i \in I. \end{cases} \tag{22.1}$$

where $G \in R^{n \times n}$ is symmetric positive definite, and its feasible region is Ω.

Framework of SQP is shown as the following:

Step 1: Set starting point $x_0 \in R^n$. Set positive definite matrix B_0, and let $k = 0$.
Step 2: Solve the following sub-problems to get d_k

$$\min f_k + \nabla f(x_k)^T d + \frac{1}{2} d^T B_K d$$
$$s.t. \begin{cases} c_i(x_k) + \nabla c_i(x_k)^T d = 0, i \in \varepsilon \\ c_i(x_k) + \nabla c_i(x_k)^T d \geq 0, i \in \mathrm{I}. \end{cases} \tag{22.2}$$

Step 3: If $d_k = 0$, end. x_k is the K-T point of the original planning problem; Else set $x_{k+1} = x_k + \alpha_k d_k$ where step length $\alpha_k \geq 0$ is based on some linear search.
Step 4: Evaluate B_k to get B_{k+1} positive definite, set $k = k + 1$, go to Step 2.

Framework of ASM is shown as following:

Step 1: Set starting point $x_0 \in R^n$, let $S_0 = A(x_0)$, $k = 0$
Step 2: Calculate the following questions:

$$\min Q(x) = \frac{1}{2}(x_k + d)^T G(x_k + d) + g^T(x_k + d)$$
$$s.t.: a_i^T d = 0, i \in S_k \tag{22.3}$$

Where $S_k \subset A(x_k)$ and S_k is an estimate of x_k for effective constraint index set $A(x_k)$.
Get the K-T point (d_k, λ^k). If $d_k \neq 0$, go to Step 3; Else, verify if x_k is the K-T point of the following question:

$$\min Q(x) = \frac{1}{2}x^T Gx + g^T x$$
$$s.t.: a_i^T d = b_i, i \in S_k \tag{22.4}$$

If any $i \in S_k \cap I(x_k)$ gets $\lambda_i^k \geq 0$, then end the steps; Else set $i_k = \arg \min_{i \in S_k \cap I(x_k)} \{\lambda_i^k | \lambda_i^k < 0\}$, $S_{k+1} = S_k / \{i_k\}$ and let $x_{k+1} = x_k$, $k = k+1$ then go to Step 3.

Step 3: Set $x_{k+1} = x_k + \alpha_k d_k$

Step 4: If $\alpha_k \neq \hat{\alpha}_k$ let $S_{k+1} = S_k k = k+1$ and go to Step 2; Else, if $i_k \notin S_k$ meet $a_{ik}^T(x_k + \alpha_k d_k) = b_{ik}$ let $S_{k+1} = S_k \cup \{i_k\} k = k+1$ and go to Step 2, where $\hat{\alpha}_k$ is the special estimates of α_k. Not described in detail here.

22.3 Objective Function and Constraint Function

22.3.1 *Nomenclature*

N_h: Rotational speed of high-pressure spool
N_l: Rotational speed of low-pressure spool
$N_{h_{\max}}$: Max speed of N_h
$N_{l_{\max}}$: Max speed of N_l
T_4: Turbine forward temperature
W_f: Fuel flow
A_8: Nozzle area
far: Fuel to air mass flow ratio
smf: Surge margin of fan
smc: Surge margin of compressor

22.3.2 *Objective Function*

Because only the process of the turbofan engine acceleration is discussed in this article. Moreover, the best process of acceleration will limit the parameters maximum, so the process is reliable and short [9]. N_h reflects the engine working state directly and T_4 guarantee the safety of the engine. That's the reason we set the following objective function. In summary, we set

$$\min J = \omega_1 \int_{t_0}^{t_f} (N_h - N_{h_{\max}})^2 dt + \omega_2 \int_{t_0}^{t_f} (T_4 - T_{4_{\max}})^2 dt \tag{22.5}$$

to be the objective function, where $x = (N_h, T_4)$ is set as the controlled variable and $u = (W_f, A_8)$ is set as the controlled variable. t_o is the starting time of the acceleration and t_f is the ending time. Then discrete the formula to get

$$\min J = \alpha_1 (1 - \frac{N_h(k)}{N_{h_{\max}}})^2 \Delta t + \alpha_2 (1 - \frac{T_4(k)}{T_{4_{\max}}})^2 \Delta t, k = 1, 2, \ldots n \tag{22.6}$$

where α_1, α_2 is the weight coefficient of either part.

More specifically,

$$\begin{cases} \alpha_1 = \dfrac{(1 - \frac{N_h(k)}{N_{h\max}})^2}{(1 - \frac{N_h(k)}{N_{h\max}})^2 + (1 - \frac{T_4(k)}{T_{4\max}})^2} \\ \alpha_2 = \dfrac{(1 - \frac{T_4(k)}{T_{4\max}})^2}{(1 - \frac{N_h(k)}{N_{h\max}})^2 + (1 - \frac{T_4(k)}{T_{4\max}})^2} \end{cases} (k = 1, 2, \ldots n) \tag{22.7}$$

Where $\alpha_1 + \alpha_2 = 1. \alpha_1 \geq 0.08, \alpha_2 \geq 0.08$. Δt is step length and set it 0.02 s.

22.3.3 Constraint Function

The process of acceleration requires the physical limit of the shaft speed, the maximum temperature for turbine blades, the maximum limit of the burner (or combustor) pressure and the surge (or stall) limit of the compression system. Here we consider the following several limitations with the component-level model.

$$\begin{array}{cc} N_h \leq 12674\,\text{rad/min}, & N_l \leq 9907\,\text{rad/min} \\ T_4 \leq 1604\,\text{k}, & far \leq 0.022 \\ smc \geq 0.1, & smf \geq 0.1 \end{array}$$

Meanwhile it exists some physical factors that limits the calculation:

Flue flow limit: $0.3\,\text{kg/s} \leq W_f \leq 1.144\,\text{kg/s}$

Nozzle area limit: $0.2642\,\text{m}^2 \leq A_8 \leq 0.2935\,\text{m}^2$

The maximum rate of flue flow changing: $\Delta W_f \leq (0.6\,\text{kg/s})/\text{s}$

The maximum rate of nozzle area changing: $\Delta A_8 \leq 0.2\,\text{m}^2/\text{s}$

22.4 Simulation and Analysis

Based on the existed engine model, we write the two optimization programs (ASM and SQP) in accordance with the objective function and the constraint function. The Matlab program implementation of the optimization is shown by different color as the following figures.

As the results shown in Fig. 22.2, it is obvious that the SQP is faster than the ASM to get the expected engine working state from the labeled points. The specific time consuming: SQP costs 2.46 s and ASM costs 2.54 s. It can be seen that SQP saves 0.08 s.

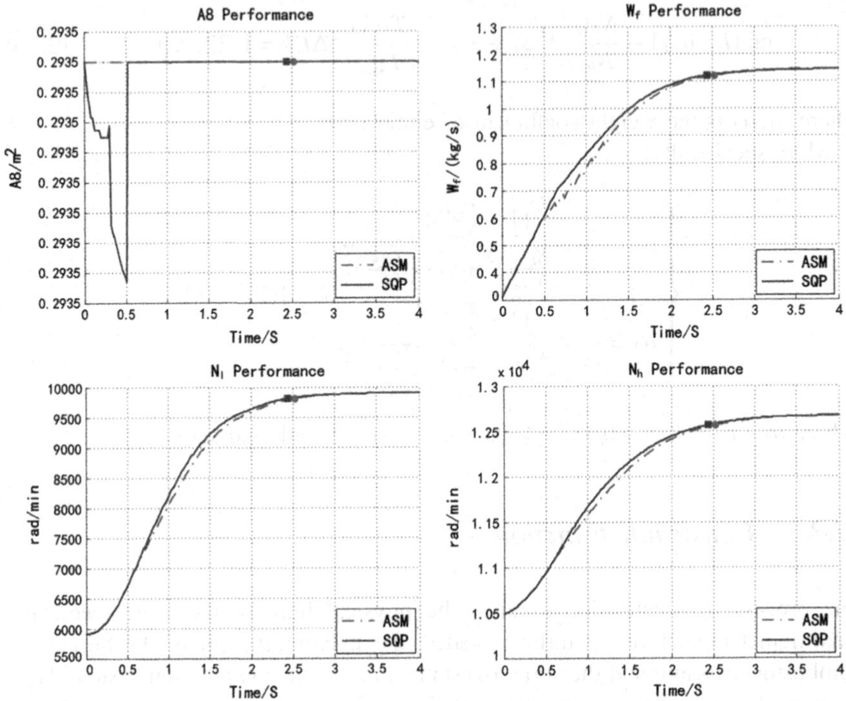

Fig. 22.2 The simulation comparison of A_8 $W_f N_l$ and N_h

It can be seen that there is no significant change between ASM and SQP from the comparison chart of A_8 in Fig. 22.2. The A_8 optimization of SQP gradually decreases in the initial 0.5 s, and regains after that. It is very closely with the model we used here. In the contrary, ASM gets constant. It's related to the optimization program realization in detail. The W_f rise of ASM is a bit slower than the SQP and it gets less fuel flow at the same time. From the comparison picture of N_l and N_h, SQP responses faster than ASM. Tracking for the objective N_l and N_h is still good, with no steady-state error.

It can be seen the results of T_4 *smc smf* and *far* in Fig. 22.3. The ASM gets slight oscillation between 0.5 to 1.4 s, and the curve of optimization is smooth. Turbine forward temperature of ASM gets lower than the SQP at the same time. T_4 has a strong relationship with the operating life of engine. We see ASM gets slight oscillation but the ASM gets distinct advantage than SQP based on the principle of engine operating life.

On the other hand, the surge margin of fan and compressor limit the engine operating normally hard, the values of surge margin gets higher, the more reliability of engine operating. It's crucial to the extremely high reliability requirements of aircraft engines. Likewise, it can be seen from the comparison between 0.5 to 1.4 s ASM maintain a relatively higher compressor (or fan) surge margin value than SQP. The same situation is also maintained at a lower *far* which saves some fuel and it has some significance of economics.

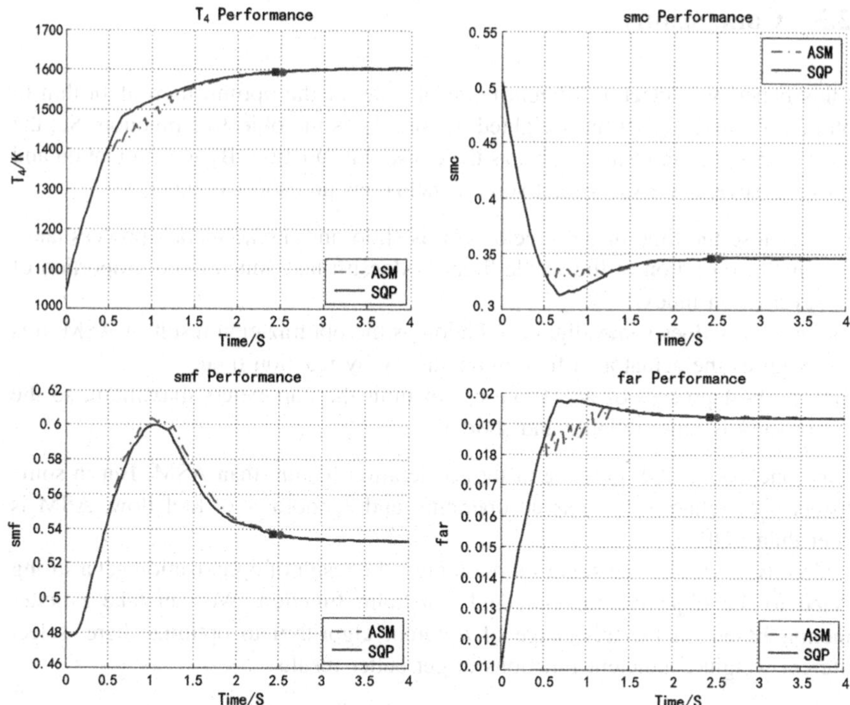

Fig. 22.3 The simulation comparison of T_4 *smc smf* and *far*

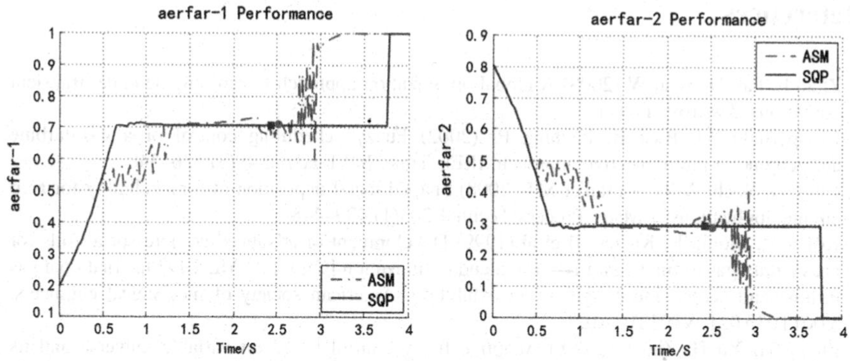

Fig. 22.4 The simulation comparison of α_1, α_2

The Fig. 22.4 shows the weight value α_1, α_2 that we set in this paper. It is obvious that ASM gets strong oscillation and the frequency is higher. In the contrast, SQP gets smooth and almost has no oscillation.

22.5 Conclusions

In this paper, we select a different point of view of the optimization algorithm to optimize W_f and A_8, set the weighted N_h and T_4 as the objective function. Set the actual working limit of the engine as the constraint function. By the simulation and following several views, we get some conclusions and shortcomings.

(1) Because the time of the acceleration is short and the accelerate process takes small proportion of the whole flight, it doesn't take much more proportion of evaluation index.
(2) Fuel flow does some slight oscillation as the optimization result of ASM, this requires the actuator of fuel pump strictly by reaction time.
(3) The best process of acceleration will limit the parameters maximum, so the process is more reliable and shorter.

In conclusion, SQP gets the faster acceleration feature than ASM. But in some aspects, like reliability of engine operating and economics of fuel flow, ASM is better than SQP.

Moreover, we also get some shortcomings: the result of optimization get a strong related to the objective function and constraint function. We can take a better function based on the advantage of different algorithm to optimize here, select different weight coefficient functions to get better results.

Acknowledgments We are grateful for the support of the Natural Science Foundation of Jiangsu Province (No. BK20140829); Jiangsu Postdoctoral Science Foundation (No.1401017B).

References

1. Zhao L, Fan D, Shan W (2008) A global optimization approach to aero-engine under transient conditions. J Aerosp Power
2. Chipperfield AJ, Bica B, Fleming PJ (2002) Fuzzy scheduling control of a gas turbine aero-engine: a multiobjective approach. IEEE Trans Ind Electron 49(3):536–548
3. Clark D, Bache MR, Whittaker MT (2008) Shaped metal deposition of a nickel alloy for aero engine applications. J Mater Process Technol 203(1):439–448
4. Köller U, Mönig R, Küsters B et al (1999) Development of advanced compressor airfoils for heavy-duty gas turbines: Part I—design and optimization [C]. In: ASME 1999 international gas turbine and aeroengine congress and exhibition. American society of mechanical engineers, V001T03A021–V001T03A021
5. Zheng H, Xu H, Zhu G (2008) Adaptive fuzzy control based on variable universe and its application to aero engine turbine power generator. Control Theory Appl 2:253–256
6. Nettleton D, Torra V (2001) A comparison of active set method and genetic algorithm approaches for learning weighting vectors in some aggregation operators. Int J Intell Syst 16(9):1069–1083
7. Ceng G, Fan D (1999) Application of SQP to acceleration control of turbofan Engine. J Aerosp Power 14(3):313–316
8. Xia C, Wang J, Shang G, Zhou M (2012) Component-level modeling and analysis of aeroengine based on matlab/simulink. Aeroengine
9. Jaw LC, Mattingly JD (2011) Aircraft engine controls design, system analysis and health monitoring. Aviation Ind Press, Beijing

Chapter 23
Region-Based Shape Control for Multi-Robot Systems with Uncertain Kinematics and Dynamics

Jia Yu, Yingmin Jia and Changqing Chen

Abstract This paper is devoted to the shape control problem of multi-robot systems with uncertain kinematics and dynamics. An adaptive control method is proposed, based on which the closed-loop systems satisfy the region shape and collision avoidance requirements in task space. Using the Lyapunov-like function and the Barbalat's lemma, a theoretical analysis on the motion of the robot end-effectors is proposed. Simulation results illustrate the feasibility and effectiveness of our controller.

Keywords Uncertain kinematics · Shape control · Adaptive control · Task space

23.1 Introduction

Cooperative control of multi-manipulator systems is a hot topic in recent research, due to its wide range of applications such as automatic parking system, satellites and space station maintenance and so on. Adaptive control is a basic method to achieve this goal, and lots of researchers have spent many years to improve the method [1–4]. Among these, the synchronization strategies in [1] and [2] were both based on the passivity of the system, and respectively solved the synchronization

J. Yu (✉) · Y. Jia
The Seventh Research Division and the Center for Information and Control, Beihang University (BUAA), Beijing 100191, China
e-mail: yujiabuaa@126.com

Y. Jia
e-mail: ymjia@buaa.edu.cn

C. Chen
The Beijing Institute of Control Engineering and National Key Laboratory of Science and Technology on Space Intelligent Control, Beijing 100191, China
e-mail: changqingchen@hotmail.com

© Springer-Verlag Berlin Heidelberg 2016

Y. Jia et al. (eds.), *Proceedings of the 2015 Chinese Intelligent Systems Conference*, Lecture Notes in Electrical Engineering 359,
DOI 10.1007/978-3-662-48386-2_23

problems in balanced graphs and strongly connected graphs. Papers [3] and [4] extended the result in [1] to the task space but only considering about the dynamic uncertainties. In case that the end-effector of robot changes, such as picking up a new tool or putting down an object, both the dynamics and kinematics of the robots change, so it is important and essential to consider about the kinematics uncertainties. From [5–8], we get several synchronization schemes to achieve task-space synchronization with both the dynamic and kinematic uncertainties. But all the schemes above can only keep the robots tracking with a desired path, and the formation of system is out of control.

The multi-robot systems formation problems usually have three approaches, the leader-following approach, the virtual structure approach and the graph theoretic approach. In leader-following method, as [9, 10], the leaders are identified, and the followers need to track with the leaders. Obviously, the failure of one leader must lead to the failure of the whole system. In the virtual structure method, as [11, 12], the whole system is considered as a virtual entity, and the relationship among the robots is rigid. When the formation moves, every robot needs to track the virtual point. When the system is larger, the control turns to be much more complicated, so the virtual method cannot deal with the large system. The graph theoretic method is a way to control a large number of robots system, as [13–15]. But the formation shape is out of control, which only depends on the number of robots. From [16], we get a region-based shape control method, which can keep a large number of robots in a desired shape and moving along with the shape maintaining a safe distance between each other. But this control method can only deal with the joint space problem with only dynamic uncertainties.

In this paper, we propose an adaptive shape controller, not only considering about the dynamic and kinematic uncertainties, but also controlling the formation as we want. In our control method, all end-effectors of the closed-loop systems under our controller satisfy the global objective and the local objective proposed in [16], that is to say, all end-effectors stay inside the desired shape and avoid collision among themselves. Using the Lyapunov-like function and the Barbalat's lemma, a theoretical analysis on the motion of the robot end-effectors is proposed. Simulation results illustrate the feasibility and effectiveness of our controller.

23.2 Robot Kinematics and Dynamics

We consider a group of n fully actuated robots with joint spaces and task spaces. The kinematical equation of ith robot is described as [17],

$$\dot{x}_i = J_i(q_i)\dot{q}_i \tag{23.1}$$

where $x_i \in R^m$ is the position vector of ith robot in task space, $q_i \in R^m$ is the joint variable speed vector of ith robot in joint space, and $J_i(q_i) \in R^{m\times m}$ is the Jacobian matrix. The dynamical equation of ith robot is described as [18],

$$M_i(q_i)\ddot{q}_i + C_i(q_i,\dot{q}_i)\dot{q}_i + g_i(q_i) = u_i \tag{23.2}$$

where $M_i(q_i) \in R^{m\times m}$ is an initial matrix which is symmetric and positive definite, $C_i(q_i,\dot{q}_i) \in R^{m\times m}$ is the Coriolis and centrifugal matrix, which can be appropriately chosen such that $\dot{M}_i - 2C_i$ is skew-symmetric, $g_i(q_i) \in R^{m\times m}$ is the gravitational torque vector, and $u_i \in R^m$ is the control input.

The kinematics (23.1) is linear with respect to a constant kinematic parameter vector θ_i, and can be rewritten as

$$\dot{x}_i = J_i(q_i)\dot{q}_i = Z_i(q_i,\dot{q}_i)\theta_i \tag{23.3}$$

where $Z_i(q_i,\dot{q}_i)$ is the kinematic regressor matrix.

The dynamics (23.2) is linear with respect to a constant dynamic parameter vector a_i, and can be rewritten as

$$M_i(q_i)\dot{x}_i + C_i(q_i,\dot{q}_i)x_i + g_i(q_i) = Y_i(q_i,\dot{q}_i,x,\dot{x})a_i \tag{23.4}$$

where $Y_i(q_i,\dot{q}_i,x,\dot{x})$ is the dynamic regressor matrix, and $x \in R^m$ is a differentiable vector.

23.3 Region-Based Control Conditions

In this section, we present a region-based method according to paper [16]. First, we define a global objective to keep all robots in the given region shape and go along with the shape. Second, we define a local objective to keep robots maintaining a safe distance among each other.

Define

$$\Delta\xi_i \triangleq \frac{\partial P_{Gi}(\Delta x_{iol})}{\partial \Delta x_{iol}} = \sum_{l=1}^{M} k_l \max(0, f_{Gl}(\Delta x_{iol}))\left(\frac{\partial f_{Gl}(\Delta x_{iol})}{\partial \Delta x_{iol}}\right)^T \tag{23.5}$$

where $l = 1, 2, \ldots M$, M is the number of the desired region functions, $k_l > 0$ is a positive constant, Δx_{iol} is the distance between ith robot and the reference point in the lth desired region, $f_{Gl}(\Delta x_{iol})$ is a continuous differentiable function, and $P_{Gi}(\Delta x_{iol})$ is the global potential energy function.

Define

$$\Delta\rho_i \triangleq \frac{\partial (Q_{Lij}(\Delta x_{ij}))}{\partial \Delta x_{ij}} = \sum_{j\in N_i} k_{ij} \max(0, g_{Lij}(\Delta x_{ij}))\left(\frac{\partial (g_{Lij}(\Delta x_{ij}))}{\partial \Delta x_{ij}}\right)^T \tag{23.6}$$

where $\Delta x_{ij} = x_i - x_j$ is the distance between ith robot and jth robot, $k_{ij} = k_{ji} > 0$ are positive constants, N_i is the neighborhood of ith robot, $g_{Lij}(\Delta x_{ij}) = r_0^2 - \|\Delta x_{ij}\|^2$ with r_0 being the minimum distance between each other, and $Q_{Lij}(\Delta x_{ij})$ is the local potential energy function.

Define

$$\Delta\varepsilon_i = \alpha_i \Delta\xi_i + \gamma \Delta\rho_i \tag{23.7}$$

where $\alpha_i > 0$ and $\gamma > 0$ are positive constants.

23.4 Adaptive Task-Space Tracking Controller

In this section, we consider the region based task-space tracking control with uncertain kinematics and dynamics. Our objective is to design adaptive control protocols such that the robots can track the desired region and maintain a safe distance among each other.

First, let us define some variables in task-space:

$$\dot{x}_{r,i} = \dot{x}_0 - \Delta\varepsilon_i \tag{23.8}$$

$$s_{X,i} = \dot{x}_i - \dot{x}_{r,i} = \Delta\dot{x}_i + \Delta\varepsilon_i \tag{23.9}$$

where $\Delta x_i = x_i - x_0$ represents the tracking error of ith robot.

Next, define some variables in joint-space based on (23.1):

$$\dot{q}_{r,i} = \hat{J}_i^{-1}(q_i)\dot{x}_{r,i} = \hat{J}_i^{-1}(q_i)(\dot{x}_0 - \Delta\varepsilon_i) = \hat{J}_i^{-1}(q_i)(\dot{x}_i - s_{X,i}) \tag{23.10}$$

$$s_i = \dot{q}_i - \dot{q}_{r,i} \tag{23.11}$$

where $\hat{J}_i(q_i)$ is the estimate of $J_i(q_i)$ with the estimate parameters θ_i instead of θ_i. The following equation reveals the relationship between joint-space and task-space.

$$s_{X,i} = \dot{x}_i - \dot{x}_{r,i} = J_i(q_i)\dot{q}_i - \hat{J}_i(q_i)\dot{q}_{r,i} = \hat{J}_i(q_i)s_i - Z_i\Delta\theta_i \tag{23.12}$$

where $\Delta\theta_i = \theta_i - \theta_i$ is the kinematic parameter estimation error of ith robot.

Now we consider the region-based task-space tracking controller for a swarm of robots, and the control input of ith robot is supposed as,

$$u_i = Y_i(q_i, \dot{q}_i, q_{r,i}, \dot{q}_{r,i})\hat{a}_i - \hat{J}_i^T k_i s_{X,i} - \hat{J}_i^T k_p \Delta\varepsilon_i \tag{23.13}$$

where $\hat{a}_i$ is the estimate of a_i, $k_i > 0$ and $k_p > 0$ are positive constants, and $Y_i(q_i, \dot{q}_i, q_{r,i}, \dot{q}_{r,i})$ is a known dynamic regressor matrix.

The estimated parameters $\hat{a}_i$ and θ_i are updated by

$$\begin{cases} \dot{\hat{a}}_i = -\Gamma_i Y_i^T(q_i, \dot{q}_i, q_{r,i}, \dot{q}_{r,i}) s_i \\ \dot{\theta}_i = \Lambda_i Z_i^T(q_i, \dot{q}_i)\left(k_p \Delta\varepsilon_i + k_i s_{X,i}\right) \end{cases} \tag{23.14}$$

Substituting (23.4) and (23.14) into (23.2), we can get the closed-loop dynamics as follows,

$$M_i(q_i)\dot{s}_i + C_i(q_i, \dot{q}_i)s_i = Y_i(q_i, \dot{q}_i, q_{r,i}, \dot{q}_{r,i})\hat{a}_i - \hat{J}_i^T k_i s_{X,i} - \hat{J}_i^T k_p \Delta\varepsilon_i \tag{23.15}$$

Now, we are ready to give the following theorem:

Theorem 1 *Consider a group of N robots described by* (23.1) *and* (23.2) *with both kinematics and dynamics uncertainties. If the control input is chosen as* (23.13) *and the adaptive updating law is chosen as* (23.14)*, then the region-based shape control problem is achieved.*

Proof We define a Lyapunov-like function V consisting of V_1 and V_2 for the multi-robot systems as follows,

$$\begin{aligned} V = & \underbrace{\sum_{i=1}^{N} \frac{1}{2} s_i^T \mathrm{M}_i(q_i) s_i + \sum_{i=1}^{N} \frac{1}{2} \Delta a_i^T \Gamma_i^{-1} \Delta a_i + \sum_{i=1}^{N} \frac{1}{2} \Delta\theta_i^T \Lambda_i^{-1} \Delta\theta_i}_{V_1} \\ & + \underbrace{\sum_{i=1}^{N} \alpha_i k_p P_{Gi}(\Delta x_{iol}) + \frac{1}{2} \sum_{i=1}^{N} \gamma k_p Q_{Lij}(\Delta x_{ij})}_{V_2} \geq 0 \end{aligned} \tag{23.16}$$

□

Referring to the result of paper [6], the derivative of V_1 is

$$\begin{aligned} \dot{V}_1 = & -\sum_{i=1}^{N} s_i^T \hat{J}_i^T(q_i) k_i s_{X,i} - \sum_{i=1}^{N} s_i^T \hat{J}_i^T(q_i) k_p \Delta\varepsilon_i \\ & + \sum_{i=1}^{N} \Delta\theta_i^T Z_i^T(q_i, \dot{q}_i) k_p \Delta\varepsilon_i + \sum_{i=1}^{N} \Delta\theta_i^T Z_i^T(q_i, \dot{q}_i) k_i s_{X,i} \\ = & -\sum_{i=1}^{N} s_{X,i}^T k_p \Delta\varepsilon_i - \sum_{i=1}^{N} s_{X,i}^T k_i s_{X,i} \end{aligned} \tag{23.17}$$

Next, taking the derivative of V_2 with respect to time, we have

$$\begin{aligned}\dot{V}_2 &= \sum_{i=1}^{N} \alpha_i k_p \Delta\dot{x}_i^T \Delta\xi_i + \frac{1}{2}\sum_{i=1}^{N} \gamma k_p \Delta\dot{x}_i^T \Delta\rho_i \\ &\quad - \frac{1}{2}\sum_{i=1}^{N} \gamma k_p \sum_{j\in N_i} k_{ij} \max\left(0, g_{Lij}(\Delta x_{ij})\right)\left(\frac{\partial\left(g_{Lij}(\Delta x_{ij})\right)}{\partial \Delta x_{ij}}\right)^T \\ &= \sum_{i=1}^{N} k_p s_{X,i}^T \Delta\varepsilon_i - \sum_{i=1}^{N} k_p \Delta\varepsilon_i^T \Delta\varepsilon_i\end{aligned} \tag{23.18}$$

From (23.17) and (23.18), we get

$$\dot{V} = \dot{V}_1 + \dot{V}_2 = -\sum_{i=1}^{N} s_{X,i}^T k_i s_{X,i} - \sum_{i=1}^{N} k_p \Delta\varepsilon_i^T \Delta\varepsilon_i \le 0 \tag{23.19}$$

Now we have proved that the derivative of the Lyapunov-like function is negative semi-definite.

In the following statement, we shall use Barbalat's lemma to prove the convergence of the swarm system. From (23.16) and (23.19), we can obtain that V is bounded, and s_i, Δa_i, $\Delta\theta_i$ are bounded as well. From (23.19), we get $s_{X,i}, \Delta\varepsilon_i \in L_2$. Differentiating (23.5) and (23.6), we find that $\Delta\xi_i$ and $\Delta\dot{\rho}_{ij}$ are bounded, therefore $\Delta\dot{\varepsilon}_i = \alpha_i \Delta\xi_i + \gamma\Delta\dot{\rho}_{ij}$ is bounded. From (23.10), we obtain $\dot{q}_{r,i}$ is bounded because the estimated Jacobian matrix $\hat{J}_i(q_i)$ has full rank, it follows that $\dot{q}_i = s_i + \dot{q}_{r,i}$ is bounded. From the closed-loop Eq. (23.15), we can conclude that $\dot{s}_i$ is bounded. Differentiating (23.10), we get $\ddot{q}_{r,i} = \dot{\hat{J}}_i^{-1}(q_i)(\dot{x}_0 - \Delta\varepsilon_i) + \hat{J}_i^{-1}(q_i)\left(\ddot{x}_0 - \Delta\dot{\varepsilon}_i\right)$ is bounded, which implies $\ddot{q}_i = \dot{s}_i + \ddot{q}_{r,i}$ and $\ddot{x}_i = \dot{J}_i(q_i)\dot{q}_i + J_i(q_i)\ddot{q}_i$ are both bounded. So $\dot{s}_{X,i}$ is bounded. It turns out that $\ddot{V} = -2\sum_{i=1}^{N} s_{X,i}^T k_i \dot{s}_{X,i} - 2\sum_{i=1}^{N} k_p \Delta\varepsilon_i^T \Delta\dot{\varepsilon}_i$ is bounded. Using the Barbalat's lemma [18], we have $\dot{V} \to 0$ as $t \to \infty$, that is to say, $\Delta\varepsilon_i \to 0$ and $s_{X,i} \to 0$ as $t \to \infty$. Since $s_{X,i} = \Delta\dot{x}_i + \Delta\varepsilon_i$, we have $\Delta\dot{x}_i \to 0$ as $t \to \infty$, which implies the velocity consistency of end-effectors with the center of the desired region.

Note that

$$\Delta\varepsilon_i = \alpha_i \Delta\xi_i + \gamma\Delta\rho_i = 0 \tag{23.20}$$

as $t \to \infty$. Summing two sides of the equation with respect to $i = 1, \ldots N$, we have

$$\sum_{i=1}^{N} \Delta\varepsilon_i = \sum_{i=1}^{N} \alpha_i \Delta\xi_i + \sum_{i=1}^{N} \gamma\Delta\rho_i = 0 \tag{23.21}$$

For the interactive forces of local constrains between each robots are bi-directional and can cancel each other out, the summation of all the forces turns out to be zero. So the second item in (23.21) is zero ($\sum_{i=1}^{N} \gamma \Delta\rho_i = 0$), therefore $\sum_{i=1}^{N} \alpha_i \Delta\xi_i = 0$. There are two situations: The first situation is that all robots are in the desired region shape, that is to say, for all i, $\Delta\xi_i = 0$. From (23.20), we get $\Delta\rho_i = 0$, which means all robots will stay in the desired region shape all the time and maintaining the safe distance between each other. The second situation is that some of the robots are outside the region shape, that is to say, there exists i such that $\Delta\xi_i \neq 0$, in which situation the robots outside the region must be on the opposite sides of the desired region, so that $\Delta\xi_i, i = 1, \ldots N$ can cancel out each other. Generally speaking, the desired region is so large that the robots outside and opposite cannot communicate with each other, and we can destroy the balance by adjusting the weighing factor α_i. Finally, it can change to the situation one.

In conclusion, the end-effectors of all the robots in system can converge to the same velocity and stay in the desired region maintaining the safe distance among themselves.

23.5 Simulation Results

In this section, we give a simulation of the proposed synchronization scheme. Six robotic agents are considered and they are all two-DOF manipulators, whose physical parameters are listed in Table 23.1, and the initial configurations are listed in Table 23.2. The desired region shape is a unit circle with radius $r = 0.3$. The desired safe distance is set to $r_0 = 0.1$. Simulation results are shown in Fig. 23.1, which presents the end-effectors moving converge to the desired region.

Table 23.1 The physical parameters of the six robots

*i*th robot	l_{i1}, l_{i2} (m)		m_{i1}, m_{i2} (kg)		I_{i1}, I_{i2} (kg m^2)		l_{ci1}, l_{ci2} (m)	
1	2.0	1.8	1.6	1.7	0.52	0.41	1.00	0.90
2	1.9	1.8	1.4	1.5	0.37	0.46	0.95	0.90
3	2.0	1.7	1.3	1.2	0.32	0.51	1.00	0.85
4	1.7	1.6	1.6	1.4	0.50	0.51	0.85	0.80
5	1.8	1.8	1.5	1.7	0.51	0.40	0.90	0.90
6	1.7	1.7	1.4	1.5	0.46	0.49	0.80	0.85

Table 23.2 The initial configurations of the six robots

ith robot	q_{i1}, q_{i2} (π rad)	
1	0.43	−0.52
2	0.36	−0.71
3	0.37	−0.78
4	0.40	−0.60
5	0.38	-0.70
6	0.35	-0.65

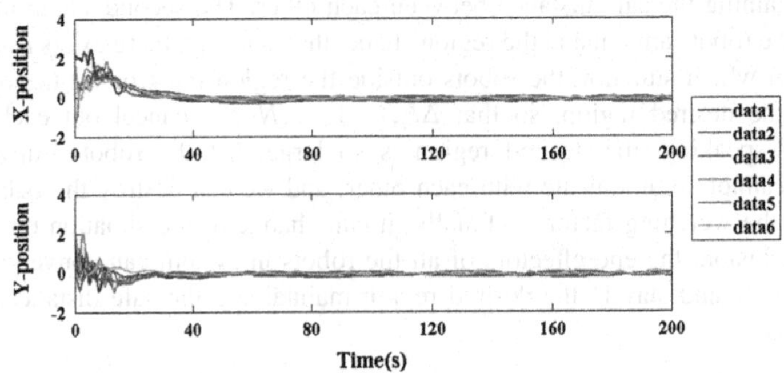

Fig. 23.1 Task-space tracking positions of six robots

23.6 Conclusion

In this paper, we have proposed a region-based adaptive control method to solve the problem of synchronization for a swarm of robots in task-space with uncertain kinematics and dynamics. The controller can keep all robots in the desired region shape moving along with the shape as a group, and maintaining the safe distance among each other. Our future research will extend our result to time-delay systems which is more reasonable and closer to the reality.

Acknowledgments This work was supported by the National Basic Research Program of China (973 Program: 2012CB821200, 2012CB821201, 2013CB733100) and the NSFC (61134005, 60921001, 90916024).

References

1. Chopra N, Spong MW (2006) Passivity-based control of multi-agent systems. In: Advances in robot control, pp. 107–134. Springer, Berlin
2. Liu YC, Chopra N (2010) Synchronization of networked robotic systems on strongly connected graphs. In: 2010 49th IEEE Conference on decision and control (CDC), IEEE, pp. 3194–3199

3. Liu Y, Chopra N (2009) Controlled synchronization of robotic manipulators in the task space. In: ASME 2009 dynamic systems and control conference, American society of mechanical engineers, pp. 443–450
4. Chopra N (2012) Output synchronization on strongly connected graphs. IEEE Trans Autom Control 57(11):2896–2901
5. Cheng L, Hou ZG, Tan M, Liu D, Zou AM (2008) Multi-agent based adaptive consensus control for multiple manipulators with kinematic uncertainties. In: ISIC 2008 IEEE international symposium on intelligent control, IEEE, pp. 189–194
6. Wang H (2013) Passivity based synchronization for networked robotic systems with uncertain kinematics and dynamics. Automatica 49(3):755–761
7. Liu X, Tavakoli M, Huang Q (2010) Nonlinear adaptive bilateral control of teleoperation systems with uncertain dynamics and kinematics. In: 2010 IEEE/RSJ International conference on intelligent robots and systems (IROS), IEEE, pp. 4244–4249
8. Dongya Z, Shaoyuan L, Quanmin Z (2011). Adaptive Jacobian synchronized tracking control for multiple robotic manipulators. In: 2011 30th Chinese control conference (CCC), IEEE, pp. 3705–3710
9. Consolini L, Morbidi F, Prattichizzo D, Tosques M (2008) Leader–follower formation control of nonholonomic mobile robots with input constraints. Automatica 44(5):1343–1349
10. Dimarogonas DV, Egerstedt M, Kyriakopoulos KJ (2006). A leader-based containment control strategy for multiple unicycles. In: 2006 45th IEEE Conference on decision and control, IEEE, pp. 5968–5973
11. Egerstedt MB, Hu X (2001). Formation constrained multi-agent control
12. Ren W, Beard RW (2004). Formation feedback control for multiple spacecraft via virtual structures. In: IEE Proceedings-control theory and applications, IET 151(3):357–368
13. Olfati-Saber R (2006) Flocking for multi-agent dynamic systems: algorithms and theory. IEEE Trans Autom Control 51(3):401–420
14. Pereira AR, Hsu L (2008). Adaptive formation control using artificial potentials for Euler-Lagrange agents. In: Proceedings of the 17th IFAC world congress pp. 10788–10793
15. Belta C, Kumar V (2004) Abstraction and control for groups of robots. IEEE Trans Robot 20 (5):865–875
16. Cheah CC, Hou SP, Slotine JJE (2009) Region-based shape control for a swarm of robots. Automatica 45(10):2406–2411
17. Spong MW, Vidyasagar M (2008). Robot dynamics and control. Wiley, New York
18. Slotine JJE, Li W (1991) Applied nonlinear control, vol 60. Prentice-Hall, Englewood Cliffs, NJ

Chapter 24
Image Threshold Processing Based on Simulated Annealing and OTSU Method

Yue Zhang, Hong Yan, Xiaofu Zou, Fei Tao and Lin Zhang

Abstract This chapter analyzes Maximum between-Cluster Variance method to conduct image threshold, coming up with an optimizing searching method of image segmentation with simulated annealing optimization algorithm. This algorithm determines the optimal threshold adaptively, and has strong adaptability and good effect of image segmentation, and it can greatly reduce the computational complexity. And it is optimized by multi-threading, which improves the parallel algorithm, and speeds up the efficiency of the algorithm.

Keywords Image threshold · Maximum between-Cluster variance method · Simulated annealing algorithm

24.1 Introduction

In image processing, the storage space required which two value image is small, its calculation speed is fast, and it can reflect the important information of image. First it should be converted into gray image. Then a threshold should be selected and conduct process for each pixel in the picture which is to be processed. Compare the gray value of each pixel in the image with the threshold value, for the pixels which gray value lower than the threshold, set them to 0, and for the pixels which gray value higher than the threshold, set them to 1.

In the image threshold segmentation, the most important topic is the selection of threshold value. The unreasonable selection of threshold value will result in the bad effect of threshold processing. OTSU use the between cluster variance as criteria in

Y. Zhang (✉) · X. Zou · F. Tao · L. Zhang
School of Automation Science and Electrical Engineering,
Beihang University, Beijing 100191, China
e-mail: xiaohuyyyy@163.com

H. Yan
Beijing Aerospace Automatic Control Institute, Beijing 100854, China

© Springer-Verlag Berlin Heidelberg 2016
Y. Jia et al. (eds.), *Proceedings of the 2015 Chinese Intelligent Systems Conference*, Lecture Notes in Electrical Engineering 359,
DOI 10.1007/978-3-662-48386-2_24

threshold segmentation, and proposed the between-cluster variance method. Pun introduced the concept of entropy in information theory into the image segmentation. Pal introduced the fuzzy theory into image segmentation, and proposed fuzzy threshold method. Kapur proposed the maximum entropy method. Cheng proposed the method combining the fuzzy division and the maximum entropy. The OTSU method is the most used method of threshold segmentation [1].

An important characteristic of image processing is its high computational complexity. Using intelligent optimization algorithm in image threshold segmentation can highly improve its calculation time. This chapter will introduce a method of using simulating annealing optimization method to search threshold value in threshold segmentation process.

24.2 Maximum Between-Cluster Variance Method

Maximum between-Cluster variance method is also called OTSU Method which is a kind of dynamic image threshold segmentation method. It is proposed by a Japanese OTSU. OTSU method is based on the gray characteristic, using the threshold value to divide the picture into two parts: the background and the objective. The background is divided into one class and the objective is divided into another class. And eventually make the calculation value of the class variance between the background and the target.

The basis of this method is, the bigger the class variance between the background and the objective, the more different it is between the two parts of the image. When the background is divided into the target wrongly, or the target is divided to the background wrongly, the class variance becomes smaller [3]. Using the maximum between-cluster variance method can ensure that the background and the target will not be divided wrongly, and after the threshold processing, the difference between the target and the background will be the biggest.

The main idea of maximum between-cluster variance method is:

(1) Establish an image gray histogram, there are L level in the gray level, each level's appearance frequency is p [2].

$$N = \sum_{i=0}^{L-1} n_i \tag{1}$$

$$p_i = \frac{n_i}{N} \tag{2}$$

(2) Calculate the frequency of background and the target, the formula is as follows [2]:

$$p_A = \sum_{i=0}^{t} p_i, \quad p_B = \sum_{i=t+1}^{l-1} p_i \tag{3}$$

In the formula, t represents the selected threshold, and A represents the background (gray level for 0 ~ N), P_A represents the probability of background. And B represents the target, P_B is the probability of the target [3].

(3) Calculate the between-cluster variance of A and B:

$$\omega_A = \sum_{i=0}^{t} ip_i/p_A \tag{4}$$

$$\omega_B = \sum_{i=t+1}^{l-1} ip_i/p_B \tag{5}$$

$$\omega_0 = p_A\omega_A + p_B\omega_B = \sum_{i=0}^{L-1} ip_i \tag{6}$$

$$\sigma^2 = p_A(\omega_A - \omega_0)^2 + p_B(\omega_B - \omega_0)^2 \tag{7}$$

The expression (4) and (5) calculate the average gray value the of cluster A and B. Expression (6) calculates the global gray average value;
Expression (7) calculates the between-cluster variance of region A and region B.

(4) The above steps calculated the between-cluster variance on a certain gray value. The goal is to get a threshold value t, which make the between-cluster variance value of class A and B become the biggest. This need to conduct the optimization in the entire gray value space [3].

24.3 Simulated Annealing Algorithm

The traditional maximum between-cluster variance method needs to search every point in the entire gray value space by computing the between-cluster variance value one by one, which needs a large amount of calculation and consumes a lot of time. It limits the performance of maximum between-cluster variance method. Therefore, on the basis of maximum between-cluster variance method to conduct image threshold segmentation, the simulated annealing algorithm method is added.

Simulated annealing algorithm is derived from the principle of solid annealing, the solid is heated to a sufficient temperature, and cooled slowly (annealing), which make it achieve the lowest point of energy. During the heating, the internal particles in the solid enter into a disordered state with the rising temperature, internal energy increases. During the cooling, the particles become ordered, it reaches equilibrium at each temperature, and finally reached the ground state at room temperature, and the internal energy is reduced to a minimum [13]. According to Metropolis criterion, the probability of the particle tends to equilibrium at a temperature of T is

$$P = e^{-\frac{E}{KT}} \tag{8}$$

Where E is the internal temperature of T, E for the change, K is Boltzmann constant. With the solid simulated annealing optimization, the objective function value simulates the internal energy E and temperature T simulates control parameters t. This is the simulated annealing algorithm: From the initial solution I and initial control parameters t, repeat the step of "Creating new solutions, calculating the objective function difference, to accept or discard" on the current solution [14]. And gradually decrease the value of t, when the algorithm terminates, the current solution is the optimal solution. It is a heuristic random process based on Monte Carlo iterative development.

Pseudo code of simulated annealing algorithm

```
Begin
    t := 0;
    P(t) := InitPopulation();
    Evaluate(P(t));
    While (stop criteria unsatisfied OR T>T_min)
            P' (t) = NeighborSearch(P(t));
            Fori = 1 to N
                    e:= I_i'(t) –I_i(t);
              If (e> 0)
                    I_i(t + 1) = I_i'(t);
              Else if (exp(e / kT) > random(0, 1))
                    I_i(t + 1) = I_i'(t);
              Else
                    I_i(t + 1) = I_i(t);
                    T = r * T;
          t = t + 1;
End
```

Its main disadvantage is that it required higher initial temperature, slow cooling rate temperature and lower ending temperature. Its performance is related to the initial value, and it is parameter sensible.

Use the threshold value as the solution of simulated annealing, and the between-cluster variance as the objective function. Get the best segmentation through the iteration of simulated annealing optimization. The objective function is:

$$f(x) = \sigma^2(t) = \omega_0(t)\omega_1(t)(\mu_0(t) - \mu_1(t))^2 \tag{9}$$

T is the threshold value, gray value range: *{0, 1,…,k-1}*. Objective probability is $\omega_0(t) = \sum_{i=0}^{t} p_i$, background part probability is $\omega_1(t) = \sum_{i=t+1}^{k-1} p_i$, the average value of the target part is $\mu_0(t) = \sum_{i=0}^{t} ip_i/\omega_0$,the average value of background part is $\mu_1(t) = \sum_{i=t+1}^{k-1} ip_i/\omega_1$. The target is to get the best threshold value which makes f(x) has the biggest value.

In the function input, t represents the threshold value, nHistogram[] represents the gray level histogram, W0 represents the overall average gray value. The output of the function is the between cluster variance value in threshold value t.

In the simulated annealing algorithm, set the Markoff chain length MarkovLength = 20, step factor is set as StepFactor = 0.02, the initial temperature is set as Temperature = 100, the decay scale is set as DecayScale = 0.50.

24.4 Multi-thread Optimization

In order to accelerate the process of parameter searching in simulated annealing algorithm, the multi-thread programming technology is used. Thread is the smallest unit of program execution. Multithreading is the technology to implement the concurrent execution of multiple thread task through software and hardware. A single thread program can only complete a task within a certain time, and the multi thread program can handle their independent tasks at the same time. In the multi-core processor, multiple threads concurrently executes at the same time, which can improve the overall efficiency treatment.

In order to pass parameters, the definite structure:

```
struct OtsuArgs
{
    Mat pGray;
    int XMIN;
    int XMAX;
} ;
```

Mat is the image to be processed, XMIN is the upper bound of search, and XMAX is the lower bound of search.

Set the parameters of the two threads. The first thread search in (0 ~ 127) and the second thread search in (128 ~ 256):

Start two thread:

```
    int err1, err2;
        pthread_t ntid1, ntid2;
        err1 = pthread_create(&ntid1, NULL, simu_otsu,
(void*)&OtsuArgs1);
        if (err1!=0)
            printf("can't create thread");//thread 1
        err2 = pthread_create(&ntid2, NULL, simu_otsu,
(void*)&OtsuArgs2);
        if (err2!=0)
            printf("can't create thread");//thread 2
        void *tret1, *tret2;
        pthread_join(ntid1, &tret1);
        pthread_join(ntid2, &tret2);//wait for thread 1 and thread
2 to exit
```

The ID numbers of the two threads are *ntid1* and *ntid2*. A new thread can be created through function *pthread_create()*. The prototype of *pthread_create()* is:

int pthread_create(pthread_t *restrict tidp, const pthread_attr_t *restrict attr, void *(*start_rtn)(void *), void *restrict arg);

When *pthread_create()* returns successfully, the new thread will be set to the memory unit pointed to by *tidp*. The newly created thread will start to run at the address of function *simu_otsu()*. This function has a void type pointer parameter *arg*, which use structure *OtsuArgs1* and *OtsuArgs2* as *arg* to pass parameters.

If *pthread_create()* fails, it will return error ID number to *err1* and *err2*. If it succeeds, it will return 0. So if the value of *err1* and *err2* does not equal to zero, we can determine that the call failed, make the thread exit and print error information.

pthread_join() is used to block the thread, and realize the synchronization between threads. *pthread_join(ntid1, &tret1), pthread_join(ntid2, &tret2)* will block the program, until the two thread *ntid1* and *nitd2* return from the function. This will achieve the synchronization of the data.

24.5 Results

Run tests on PC, the operating environment is as following: processor Intel Core i3-2130 CPU @ 2.53 GHz. Memory (RAM): 2.00 GB. Operating system: Ubuntu 12.04 LTS.

The processing object of the algorithm is Pepper image 512 * 512, its gray level is 256. The experiment results is shown as the picture following (Figs. 24.1, 24.2 and 24.3), (Table 24.1).

The best threshold value of the three algorithms is the same, the value is 119. And the max between class variance is 2129.03. The OTSU method takes time

Fig. 24.1 Original image of pepper

Fig. 24.2 256 gray image

Fig. 24.3 Two value image

Table 24.1 Comparison between OTSU and simulated annealing

Algorithm	Best threshold value	Max between class variance	Runtime/s
OTSU method	119	2129.03	0.1415
OTSU with simulated annealing	119	2129.03	0.0169
Multi-thread OTSU with simulated annealing	119	2129.03	0.0079

0.1415 s. When the simulated annealing method is added to OTSU, the runtime decreases greatly, the run time becomes 0.0169 s. And the Multi-thread OTSU with simulated annealing takes time of 0.0079 s. By using the multi-thread programming method, the time is the half of the original, which reflects the effect of the multi-CPU conducting multi-thread algorithm.

From the above experiment results, it can be concluded that OTSU method with simulated annealing can find the best value the threshold value in image threshold segmentation. The result is stable and accurate, and can largely shorten the time to get the best value. It effectively solves the problem of large calculation volume, and time-consuming in threshold value selection, which is conductive to the following image processing. And with the multi-thread method which further improved the algorithm, the calculation is assigned to the two core on PC computer by the multi-thread, which make the algorithm realizes parallelization, and shorten the calculating time by half.

References

1. Hai-kun Z, Wei-can Z (2007) Image segmentation based on an improved OTSU algorithm. J Chongqing Inst Technol
2. Zhang J, Hu J (2008) Image segmentation based on 2D OTSU method with histogram analysis. In: International conference on computer science and software engineering, pp 105–108
3. Wang HY, Pan DL, Xia DS (2007) A fast algorithm for two-dimensional OTSU adaptive threshold algorithm. Acta Automatica Sinica 9:968–971
4. Mei-yan C, Qing-xian W, Chang-sheng J (2007) Target image segmentation based on modified OTSU algorithm. Electron Opt Control
5. Yu J (2009) OTSU method and K-means. Ninth Int Conf Hybrid Intell Syst 2009:344–349
6. Sthitpattanapongsa p, Srinark T (2012) An equivalent 3D OTSU's thresholding method. Adv Image Video Technol doi:10.1007/978-3-642-25367-6_32
7. Xiang-yang X, En-min S, Liang-hai J (2009) Characteristic analysis of threshold based on OTSU criterion. Acta Electronica Sinica 37(12):2716–2719
8. Shi J, Malik J (2000) Normalized cuts and image segmentation. Ranaon on Arn Analy and Mahn Nllgn 22(8):888–905
9. Cheng HD, Jiang XH, Sun Y et al (2001) Color image segmentation: advances and prospects. Pattern Recognit 34:2259–2281
10. Fu KS, Mui JK (1981) A survey on image segmentation. Pattern Recognit 13(81):3–16
11. Felzenszwalb PF, Huttenlocher DP (1998) Image segmentation using local variation. In: Proceedings of IEEE conference on computer vision and pattern recognition, pp 98–104
12. Sharon E, al E (2000) Fast multiscale image segmentation. Proc IEEE Conf Comput Vis Pattern Recognit 1:70–77
13. Dowsland KA, Thompson JM (2012) Handbook of natural computing. Springer, Berlin
14. Sorkin GB (1991) Efficient simulated annealing on fractal energy landscapes. Algorithmica 6 (1–6):367–418

Chapter 25
Fault Tolerant Tracking Control for a Team of Non-Minimum Phase VTOL Aircrafts Based on Virtual Leader Structure

Shuyao Cheng, Hao Yang and Bin Jiang

Abstract This paper develops a fault tolerant approach for a team of thrust vector vertical take-off and landing (VTOL) aircrafts based on virtual leader structure. First, for each VTOL aircraft, a cascaded observer system with two extended state fault diagnosis observers (ESFDO) is designed. Then, local fault tolerant tracking controllers are developed. Finally, simulation results for a team of three VTOL aircrafts are presented to demonstrate the effectiveness of our proposed method.

Keywords Vertical take-off and landing aircraft · Non-minimum phase system · Virtual leader structure · Fault tolerant control · Integral sliding mode

25.1 Introduction

In recent years, vertical take-off and landing (VTOL) aircrafts based on thrust vector technology have attracted a great deal of interest in both civilian and military applications. Such aircrafts can take off in short distance and are not limited to conventional airport runways [1].

The dynamics of a VTOL aircraft involves some specific characteristics including multiple inputs and multiple outputs (MIMO), non-minimum phase, under-actuated, and coupling, which make it difficult to control and prone to anomalies and faults. The existing approaches to control VTOL aircrafts include sliding mode control [2, 3], feedback linearization [4–6] and stable inversion [7]. However, few works have been done on the fault tolerant control for VTOL aircrafts.

Moreover, as the missions become more complex and diverse, cooperative VTOL aircrafts are necessary to meet the requirements of the assigned missions and

S. Cheng · H. Yang (✉) · B. Jiang
College of Automation Engineering, Nanjing University of Aeronautics and Astronautics, 210016 Nanjing, China
e-mail: haoyang@nuaa.edu.cn

© Springer-Verlag Berlin Heidelberg 2016

Y. Jia et al. (eds.), *Proceedings of the 2015 Chinese Intelligent Systems Conference*, Lecture Notes in Electrical Engineering 359,
DOI 10.1007/978-3-662-48386-2_25

various methods have been proposed for the formation control such as the virtual leader structure [8], the leader-follower structure [9], distributed control [10] and the behavior based control [11]. Each method has its own advantages and disadvantages. For example, the virtual leader structure is based on some virtual points or virtual agents and agents follow them in a rigid body formation that is maintained during the missions. The merit of this structure is that the virtual leader never fails. A number of fault tolerant cooperative control strategies have been proposed in the literature, such as optimal cooperation [12].

In this work, a novel fault tolerant tracking control (FTTC) method based on the concept of the virtual leader structure is proposed. A virtual leader is utilized to track the desired output vector and it provides an external reference signal for each team member. Then, local fault tolerant controller is developed. First, a novel cascade observer system consisting of two extended observers is designed to realize the estimation of unknown faults. Then, by utilizing the coordinate transformation and feedback linearization, the dynamic system of each aircraft is decoupled and the FTTC problem is tackled by using optimal control and integral sliding mode schemes.

25.2 System Description

Consider a team of N VTOL aircrafts in the vertical plane with identical dynamics. The model of a single VTOL aircraft is shown in Fig. 25.1 and the dynamics of a VTOL aircraft is governed by the equations as follows

$$\begin{aligned} \ddot{x} &= -(u_{1f} - f_{a1})\sin\theta + \varepsilon(u_{2f} - f_{a2})\cos\theta \\ \ddot{y} &= -(u_{1f} - f_{a1})\cos\theta + \varepsilon(u_{2f} - f_{a2})\sin\theta - g \\ \ddot{\theta} &= u_2 \end{aligned} \tag{25.1}$$

where x and y represent the horizontal and vertical axes, g is the acceleration of gravity, θ is the roll angle, control inputs u_{1f} and u_{2f} are respectively the thrust and the rolling moment, ε represents the coupling coefficient between the rolling moment and the lateral acceleration, and f_{a1} and f_{a2} are the additional faults occur in u_1 channel and u_2 channel, respectively.

The formation control method we utilize in this paper, that is, the virtual leader structure method, involves the synthesis of a virtual leader $z(t) \in R^m$, which provides information on the reference trajectory for each team member.

For simplicity, we model the virtual leader as follows which can describe most linear reference path such as sinusoidal path:

$$\dot{z}(t) = Fz(t), \tag{25.2}$$

where $F \in R^{m \times m}$ is a constant matrix with proper dimension.

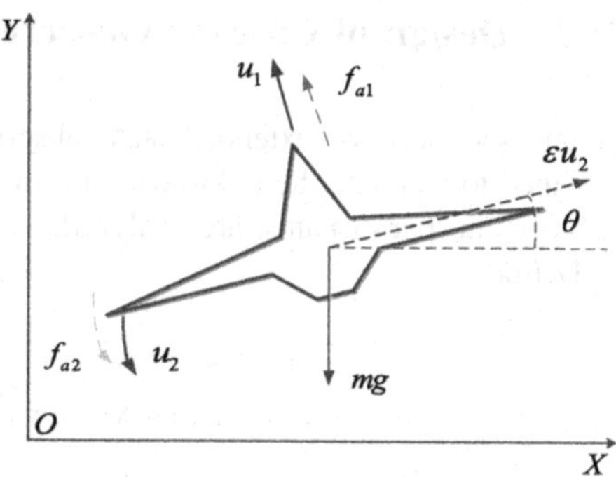

Fig. 25.1 VTOL aircraft

Each corresponding agent is specified by a vector d_i whose entries represent the desired reference output offsets from the virtual leader states. Therefore, the desired reference output takes the form:

$$\bar{y}(t) = Hz(t) + d \tag{25.3}$$

where $\bar{y}(t) = (\bar{y}_1(t), \bar{y}_2(t), \bar{y}_3(t))^T$ denotes the desired reference output vector, vector $d = (d_1, d_2, d_3)^T$ represents relative position on the virtual leader z(t), and H is a constant matrix of appropriate dimension.

Assumption 1 There exists a positive constant M such that

$$|\dot{f}_{a1}| \le M, |\dot{f}_{a2}| \le M. \tag{25.4}$$

Define fault signals

$$f_{1a} = \sin x_5 f_{a1}, f_{2a} = -\varepsilon \cos x_5 f_{a2}, f_{3a} = \cos x_5 f_{a1}, f_{4a} = -\varepsilon \sin x_5 f_{a2}. \tag{25.5}$$

By utilizing the following coordinate transformation

$$x_1 = x, x_2 = \dot{x}, x_3 = y, x_4 = \dot{y}, x_5 = \theta, x_6 = \theta \tag{25.6}$$

we can rewrite the system model (25.1) and divide it into two subsystems as follows:

$$\begin{cases} \dot{x}_5 = x_6 \\ \dot{x}_6 = u_{2f} - f_{a2} \end{cases} \text{and} \begin{cases} \dot{x}_1 = x_2 \\ \dot{x}_2 = -u_{1f}\sin x_5 + \varepsilon u_{2f}\cos x_5 + f_{1a} + f_{2a} \\ \dot{x}_3 = x_4 \\ \dot{x}_4 = u_{1f}\cos x_5 + \varepsilon u_{2f}\sin x_5 + f_{3a} + f_{4a} - g \end{cases} \tag{25.7}$$

One can see that one of the subsystems is affected by fault f_{a2} only.

25.3 Design of Cascade Observers

In this section, two extended state observers (ESO) with cascade structure are designed to estimate the unknown fault information for each VTOL aircraft. Their stability and performance are analyzed.

Define

$$\begin{aligned} & z_1 = x_5, z_2 = x_6, z_3 = -f_{a2}, \dot{z}_3 = -\rho_1(t), \\ & z_4 = x_1, z_5 = x_2, z_6 = f_{1a}, \dot{z}_6 = \rho_2(t). \end{aligned} \tag{25.8}$$

For two subsystems (25.7), two extended state observers with cascade structure are constructed respectively

$$\begin{aligned} & \begin{cases} \dot{\hat{z}}_1 = \hat{z}_2 + \alpha_1\omega_1(z_1 - \hat{z}_1) \\ \dot{\hat{z}}_2 = u_{2f} + \hat{z}_3 + \alpha_2\omega_1^2(z_1 - \hat{z}_1) \\ \dot{\hat{z}}_3 = \alpha_3\omega_1^3(z_1 - \hat{z}_1) \end{cases} \\ & \begin{cases} \dot{\hat{z}}_4 = \hat{z}_5 + \beta_1\omega_2(z_4 - \hat{z}_4) \\ \dot{\hat{z}}_5 = -u_{1f}\sin z_1 + \varepsilon u_{2f}\cos z_1 + \hat{z}_6 + \beta_2\omega_2^2(z_4 - \hat{z}_4) \\ \dot{\hat{z}}_6 = \beta_3\omega_2^3(z_4 - \hat{z}_4). \end{cases} \end{aligned} \tag{25.9}$$

Lemma 1 [13] *Under Assumption 1, there exist constants* $\lambda_1 > 0, \lambda_2 > 0, \mu_1 > 0, \mu_2 > 0, \mu_2 > 0, M_1 > 0, M_2 > 0$ *and a finite time constant T > 0 such that*

$$\lim_{t\to T}\|z_1(t)\| \le \frac{\mu_1 M_1}{w_1^3\lambda_1} = O\left(\frac{1}{w_1^3}\right),\ \lim_{t\to T}\|z_2(t)\| \le \frac{\mu_2 M_2}{w_2^3\lambda_2} = O\left(\frac{1}{w_2^3}\right) \tag{25.10}$$

where $\mathcal{O}\left(\frac{1}{w_1^3}\right)$ and $\mathcal{O}\left(\frac{1}{w_2^3}\right)$ *are respectively the third order infinitesimal of the observer gains* ω_1 *and* ω_2 .

25.4 Local Fault Tolerant Controller

The decomposition approach proposed in [4] is adopted here. For the subsystems (25.7), the new control inputs are chosen as follows:

$$\begin{bmatrix} w_1 \\ w_2 \end{bmatrix} = R(x)\begin{bmatrix} u_{1f} \\ u_{2f} \end{bmatrix} - \begin{bmatrix} 0 \\ g \end{bmatrix} \tag{25.11}$$

where R(x) is the nonsingular decoupling matrix:

$$R(x) = \begin{bmatrix} -\sin x_5 & \varepsilon\cos x_5 \\ \cos x_5 & \varepsilon\sin x_5 \end{bmatrix} \tag{25.12}$$

And by selecting the following coordinate transformation

$$\begin{aligned} &e_1 = x_1 - x_d, e_2 = x_2 - \dot{x}_d, e_3 = x_3 - y_d, e_4 = x_4 - \dot{y}_d, w_{s1} = w_1 - \ddot{x}_d, \\ &w_{s2} = w_2 - \ddot{y}_d, \eta_1 = x_5, \eta_2 = \varepsilon x_6 - e_2\cos x_5 - e_4\sin x_5 \end{aligned}$$

one obtains the tracking error system:

$$\begin{aligned} &\dot{e}_1 = e_2, \dot{e}_2 = w_{s1} + f_{1a} + f_{2a}, \dot{e}_3 = e_4, \dot{e}_4 = w_{s2} + f_{3a} + f_{4a}, \\ &\dot{\eta}_1 = \frac{1}{\varepsilon}(\dot{\eta}_2 + e_2\cos\dot{\eta}_1 + e_4\sin\dot{\eta}_1), \\ &\dot{\eta}_2 = \frac{1}{\varepsilon}(\dot{\eta}_2 + e_2\cos\eta_1 + e_4\sin\eta_1)(e_2\sin\eta_1 - e_4\cos\eta_1) \\ &\quad + (\ddot{x}_d - f_{1a} - f_{2a})\cos\eta_1 + (\ddot{y}_d - f_{3a} - f_{4a} + g)\sin\eta_1 - \varepsilon f_{a2} \end{aligned} \tag{25.13}$$

The zero dynamics is constituted by x_5 and x_6 and evidently unstable [3, 6]. Therefore, a novel fault tolerant controller is proposed in the following to stabilize the zero dynamics. The last two equations of (25.13) represent the zero dynamics and can be rewritten in the form of η-dynamics:

$$\dot{\eta} = q(\eta, E, F_a, \ddot{z}_d) \tag{25.14}$$

where $\eta = (\eta_1, \eta_2)^T$, $E = (e_1, e_2, e_3, e_4)^T$, $F_a = (f_{1a}, f_{2a}, f_{3a}, f_{4a}, f_{a2})^T$ and $\ddot{z}_d = (\ddot{x}_d, \ddot{y}_d)^T$, which is the second derivative of the reference trajectory.

Note that

$$\left.\frac{\partial q(\eta, E, F_a, \ddot{z}_d)}{\partial(e_1, e_2)}\right|_{(0,0)} \neq O_{2\times 2}, \left.\frac{\partial q(\eta, E, F_a, \ddot{z}_d)}{\partial(e_3, e_4)}\right|_{(0,0)} = O_{2\times 2} \tag{25.15}$$

This implies that the η-dynamics is linearly dependent on (e_3, e_4) while linearly independent of (e_1, e_2). Denote $e_m = [e_3, e_4]^T$. Thus, we can decompose the system (25.13) into two subsystems. One is a minimum phase subsystem which is independent of the zero dynamics:

$$\dot{e}_m = \begin{bmatrix} \dot{e}_3 \\ \dot{e}_4 \end{bmatrix} = A_1\begin{bmatrix} e_3 \\ e_4 \end{bmatrix} + B_1 w_{s2} \tag{25.16}$$

where $A_1 = \begin{bmatrix} 0 & 1 \\ 0 & 0 \end{bmatrix}, B_1 = \begin{bmatrix} 0 \\ 1 \end{bmatrix}$ and for the minimum phase part, the optimal control law is designed:

$$w_{s2} = -K_1 e_m \tag{25.17}$$

where $k_1 = (R_1)^{-1}(B_1)^T P_1$ is the optimal feedback gain matrix. P_1 satisfies the Riccati equation:

$$P_1A_1 + A^TP_1 - P_1B_1(P_1)^{-1}(B_1)^TP_1 = -Q_1 \tag{25.18}$$

where P_1 and Q_1 are symmetric and positive definite matrices with appropriate dimension.

Define new state variables:$z_1 \triangleq e_1, z_2 \triangleq (e_1, \eta^T)^T$, and the other subsystem, which is non-minimum phase, takes the form:

$$\dot{z}_1 = w_{s1} + f_{1a} + f_{2a}, \dot{z}_2 = A_{z_1} z_1 + A_{z_2} z_2 + O(\eta, E, F_a, \ddot{z}_d) \tag{25.19}$$

where $A_{z_1} = \begin{bmatrix} 1 \\ \frac{1}{\varepsilon} \\ 0 \end{bmatrix}, A_{z_2} = \begin{bmatrix} 0 & 0 & 0 \\ 0 & 0 & \frac{1}{\varepsilon} \\ 0 & g & 0 \end{bmatrix}$, and $O(\eta, E, F_a\ddot{z}_d) = \dot{z}_2 - A_{z_1} z_1 - A_{z_2} z_2$, which represents the higher order term remaining after linearization.

Denote $e_n \triangleq (z_1, z_2^T)^T$, the system (25.19) can be described in the following form

$$\dot{e}_n = A_2 e_n + B_2 w_{s1} + \Delta(\eta, E, F_a, \ddot{z}_d) \tag{25.20}$$

where $A_2 = \begin{bmatrix} 0 & 0 & 0 & 0 \\ 1 & 0 & 0 & 0 \\ \frac{1}{\varepsilon} & 0 & 0 & \frac{1}{\varepsilon} \\ 0 & 0 & g & 0 \end{bmatrix}, B_2 = \begin{bmatrix} 1 \\ 0 \\ 0 \\ 0 \end{bmatrix}$, and $\Delta(\eta, E, F_a, \ddot{z}_d) = \begin{bmatrix} 0 \\ O(\eta, E, F_a\ddot{z}_d) \end{bmatrix}$.

For the non-minimum phase subsystem (25.19), in order to stabilize the linear term, an optimal control law is designed:

$$w_{s1a} = -(K_2 e_n + \hat{f}_{1a} + \hat{f}_{2a}) \tag{25.21}$$

where $K_2 = (R_2)^{-1}(B_2)^T P_2$ is the optimal feedback gain matrix and P_2 satisfies the Riccati equation:

$$P_2A_2 + A_2^TP_2 - P_2B_2(P_2)^{-1}(B_2)^TP_2 = -Q_2 \tag{25.22}$$

where P_2 and Q_2 are positive definite symmetric matrices.

The optimal integral sliding manifold is selected to be:

$$s = z_1 + Mz_2 - M\left(W(0) + \int_0^t W(\tau)d\tau\right) \tag{25.23}$$

where $M \in R^{1\times 3}$ is a constant vector to be chosen, $W(t) = A_{z_1} z_1(t) + A_{z_2} z_2(t) - M^+ K_2 e_n(t)$ and W(0) is the initial value. Thus, the control law for the nonlinear higher order term is designed to be:

$$w_{s1b} = -M\widehat{O}(\cdot) - \sigma \mathrm{sgn}(\hat{s}) \tag{25.24}$$

where σ is the gain to be selected and $\widehat{O}(\cdot)$ is the estimation of the nonlinear higher order term $O(\eta, E, F_a, \ddot{z}_d)$.

Thus, the control law for the non-minimum phase subsystem takes the form:

$$w_{s1} = w_{s1a} + w_{s1b} = -(K_2 e_n + \hat{f}_{1a} + \hat{f}_{2a}) - M\widehat{O}(\cdot) - \sigma \mathrm{sgn}(\hat{s}) \tag{25.25}$$

Theorem 1 *If the gain σ satisfies the following inequality (25.26), using the the control law (25), the closed-loop system of the non-minimum phase subsystem (25.20) is asymptotically stable at origin:*

$$\sigma > |\tilde{f}_{1a}| + |\tilde{f}_{2a}| + \|M\| \left\| O(\cdot) - \widehat{O}(\cdot) \right\| \tag{25.26}$$

Proof Select the Lyapunov function:$V(s,t) = \frac{s^2}{2}$ and define $\tilde{f}_{1a} \triangleq f_{1a} - \hat{f}_{1a}$ and $\tilde{f}_{1a} \triangleq f_{1a} - \hat{f}_{1a}$, we have

$$\begin{aligned} \dot{V}(s,t) &= s\dot{s} = s[\dot{z}_1 + M\dot{z}_2 - M(A_{z_1} z_1 + A_{z_2} z_2 + M^+ K_2 e_n)] \\ &= (\hat{s} + s - \hat{s})\left\{ -\tilde{f}_{1a} - \tilde{f}_{2a} + M\left[O(\cdot) - \widehat{O}(\cdot)\right] - \sigma \mathrm{sgn}(\hat{s}) \right\} \\ &\le |\hat{s}|\left(|\tilde{f}_{1a}| + |\tilde{f}_{2a}| + \|M\| \left\| O(\cdot) - \widehat{O}(\cdot) \right\| - \sigma \right) \\ &\quad + |s - \hat{s}|(|\tilde{f}_{1a}| + |\tilde{f}_{2a}| + \|M\| \left\| O(\cdot) - \widehat{O}(\cdot) \right\| - \sigma) \end{aligned} \tag{25.27}$$

Apparently, for any chosen gain σ such that $\sigma > |\tilde{f}_{1a}| + |\tilde{f}_{2a}| + \|M\| \| O(\cdot) - \widehat{O}(\cdot) \|$, we have $\dot{V}(s,t) \le 0$. This completes the proof.

25.5 Numerical Results

In this section, simulation results are presented for faults that occur in both input channels in one of the VTOL aircraft in a team of three VTOL aircrafts. It should be pointed out that our proposed method is not limited to this and faults can occur in more than just one vehicle.

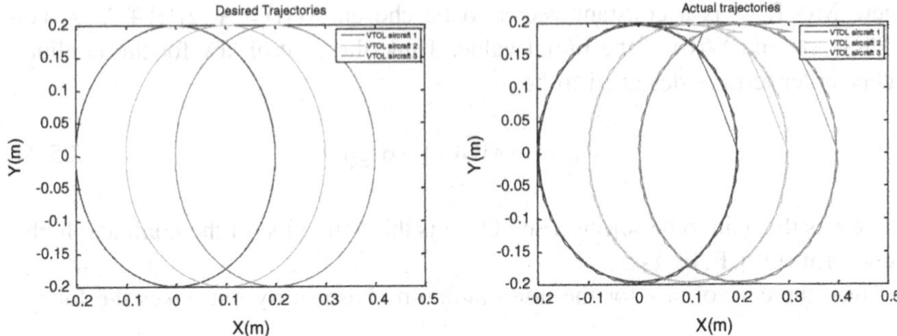

Fig. 25.2 The desired trajectories and the actual trajectories

According to (25.2) and (25.3), the reference output path is considered to be

$$X_d(t) = \begin{bmatrix} X_{d1}(t) \\ X_{d2}(t) \\ X_{d3}(t) \end{bmatrix} = \begin{bmatrix} 0.2\sin(0.2t) \\ 0.2\sin(0.2t)+0.1 \\ 0.2\sin(0.2t)+0.2 \end{bmatrix}, Y_d(t) = \begin{bmatrix} Y_{d1}(t) \\ Y_{d2}(t) \\ Y_{d3}(t) \end{bmatrix} = \begin{bmatrix} 0.2\cos(0.2t) \\ 0.2\cos(0.2t) \\ 0.2\cos(0.2t) \end{bmatrix}$$

For simplicity, the parameters are selected to be: $\varepsilon = 0.5$, $g = 1$, the initial states of the agent 1 are $(0.5, 0, 0.1, 0, 0, 0)^T$, the gains of optimal control are computed as follows

$$K_1 = [1.4142 \quad 2.1974], K_2 = [3.8143 \quad -0.4082 \quad 3.9150 \quad 5.5215]$$

The faults are selected as $f_{a1} = \begin{cases} 0, & 0\,s < t < 25\,s \\ 0.8, & 25\,s \le t < 40\,s \\ 0.4, & 40\,s \le t \le 50\,s \end{cases}$ and

$$f_{a2} = \begin{cases} 0, & t < 11\,s \\ 0.1\sin[0.18(t-11)] + 0.04\cos[0.43(t-11)], & t \ge 11\,s \end{cases}$$

As depicted in Fig. 25.2, desired formation is maintained and all the VTOL aircrafts can arrive at the desired values.

25.6 Conclusions

A novel fault tolerant tracking control (FTTC) method for a team of VTOL aircrafts is developed and introduced in this work. Good simulation performances have demonstrated the effectiveness of the proposed method. In this work, formation control is considered based on virtual leader structure. Future work will focus on more formation control method such us leader-follower structure method and modified leader-follower structure method in order to minimize the effect of the faulty agent on the team level or even dealing with faults that can't be handled on the agent level.

Acknowledgements This work is supported by National Natural Science Foundation of China (61273171, 61473143), Fundamental Research Funds for the Central Universities (NE2014202, NE2015002,NE2015103) and State Key Laboratory of Synthetical Automation for Process Industries.

References

1. Oishi M, Tomlin C (1999) Switched nonlinear control of a VSTOL aircraft [C]. In: Proceedings of the 38th IEEE conference on decision and control, vol 3(3), pp 2685–2690
2. Shkolnikov IA, Shtessel YB (2002) Tracking in a class of non-minimum phase systems with nonlinear internal dynamics via sliding mode control using method of system center [J]. Automatica 38(5):837–842
3. Wang XH (2013) Takeoff/landing control based on acceleration measurements for VTOL aircraft [J]. J Franklin Inst 350(10):3045–3063
4. Al-Hiddabi SA, McClamroch NH (1998) Output tracking for nonlinear non-minimum phase VTOL aircraft [C]. In: Proceedings of the 37th IEEE conference on decision and control, vol 4, pp 4573–4578
5. Cai KY, Liu JK, Wang XH (2009) Tracking control for a velocity sensor-less vertical take-off and landing aircraft with delayed outputs. Automatica 45(12):2876–2882
6. Hauser J, Meyer G, Sastry S (1992) Nonlinear control design for slightly non-minimum phase systems: application to V/STOL aircraft. Automatica 28(4):665–679
7. Devasia S, Zou Q (2007) Precision preview based stable inversion for nonlinear non-minimum phase systems: The VTOL example. Automatica 43(1):117–127
8. Essghaier A, Beji L, El. Kamel MA, Abichou A, Lerbet J (2011) Co-leaders and a flexible virtual structure based formation motion control [J]. Int J Veh Autom Syst pp 108–125
9. Consolini L, Morbidi F, Prattichizzo D, Tosques M (2008) Leader-follower formation control of nonholonomic mobile robots with input constraints [J]. Automatica 44(5):1343–1349
10. Hong YG, Chen GR, Bushnell L (2008) Distributed observers design for leader-following control of multi-agent networks [J]. Automatica 44(3):846–850
11. Wolley BG, Prterson GL, Kresge JT (2011) Real-time behavior-based robot control [J]. Autonom Robots 30(3):233–242
12. Semsar E, Khorasani K (2007) Optimal control and game theoretic approaches to cooperative control of a team of multi-vehicle unmanned systems [C]. In: Proceedings of IEEE international conference on networking, sensing and control, pp 628–633
13. Jiangyong Y, Zongxia J, Dawei M (2014) Adaptive robust control of dc motors with extended state observer [J]. IEEE Trans Industr Electron 61(7):3630–3637

Chapter 26
Sliding Mode Fault-Tolerant Control for Air-Breathing Hypersonic Vehicles with External Disturbances

Wei Huang and Yingmin Jia

Abstract This paper studies the fault-tolerant control for air-breathing hypersonic flight vehicles(AHFVs) subject to both actuator faults and external disturbances. The overall procedure includes the feedback linearization of the model and the adaptive sliding mode control design. Specifically, a feedback linearized model of AHFVs is firstly obtained with the actuator faults and external disturbances. Then, two adaptive laws are designed to approximately estimate their unknown bounds. Based on the Lyapunov analysis method, it is proved that adaptive sliding mode control law can achieve the asymptotic tracking of all commands signals and guarantee the stability of the system. Simulations are conducted to confirm the efficacy of the proposed control method.

Keywords AHFVs · Adaptive sliding mode control · Feedback linearization · Fault-tolerant control

26.1 Introduction

AHFVs, as a new kind of low-cost, high-speed and time-saving aircraft, allow sharp reduction in flight times for both commercial and military applications. The scramjet engine powered AHFVs are capable of obtaining oxygen from air while transporting at more than 5 mach number, which increases the carry payload. Direct access to earth orbit without the use of separate boosting stages also become possible as

W. Huang (✉) · Y. Jia
The Seventh Research Division and the Department of Systems and Control,
Beihang University (BUAA), Beijing 100191, China
e-mail: xbweier@163.com

Y. Jia
e-mail: ymjia@buaa.edu.cn

© Springer-Verlag Berlin Heidelberg 2016

Y. Jia et al. (eds.), *Proceedings of the 2015 Chinese Intelligent Systems Conference*, Lecture Notes in Electrical Engineering 359,
DOI 10.1007/978-3-662-48386-2_26

scramjet engine powered aircraft enter service [1]. In 2006, the successful record-breaking X-43A flight suggests that AHFVs constitute an important role in the next generation of aviation.

Although AHFVs have excellent performance and huge advantages, there are still numerous technical challenges in the flight control system. The controller of AHFVs is supposed to effectively handle the intricate interactions between the propulsive, structure and aerodynamic effects, as well as the significant flexibility, parameters uncertainties and other modeling errors. In the last decades, the design of tracking controllers for AHFVs concentrated on tracking the velocity and altitude reference command signals by using the simplified longitudinal dynamics of hypersonic vehicles [2–4]. Moreover, a precise linearized model of hypersonic vehicle has been developed by applying the input-output feedback linearization technology [1, 5], based on which various contol laws have been developed to fulfill the desired properties [3, 6, 7].

Despite the fact that various control methods have been employed in AHFVs flight contol system, such as robust control [2], T-S Fuzzy control [3], the researches on the control issue against actuator faults for AHFVs are relatively less and present great challenges. It is proved that the failure on the actuators may lead to performance degradation and even instability of the system [7]. Hence, it's practical to design fault-tolerant controller considering possible actuator faults.By applying model reference adaptive method, Gao [6] developed an adaptive controller for AHFVs subject to actuator faults associated with a sliding observer by Filippov's construction of the equivalent dynamics(FCED). However, the proposed controller can only track a specific class of signals. Besides, Li [4] developed a guaranteed cost fault-tolerant output tracking control for the flexible AHFVs system with regional pole constraints, while the controller may be invalid if the states are far away from the trim condition.

To this end, this paper is devoted to fault-tolerant control for the longitudinal model of AHFVs subject to actuator faults and external disturbances. The objective is to design an controller such that the asymptotic tracking of velocity and altitude can be achieved. The rest of this paper is organized as follows: The nonlinear longitudinal AHVFs model and the linearized AHVFs model with actuator faults and external disturbances are presented in Sects. 26.2 and 26.3, respectively. Then, a fault-tolerant adaptive sliding mode control is proposed in Sect. 26.4. In Sect. 26.5, simulations are conducted to confirm the efficacy of the proposed controller. Conclusions are given in Sect. 26.6.

26.2 Problem Formulation

In this section, the nonlinear longitudinal equations of motion for AHFVs [8] are presented as

$$
\begin{aligned}
\dot{V} &= \frac{T\cos\alpha - D}{m} - \frac{\mu\sin\gamma}{r^2} \\
\dot{\gamma} &= \frac{L + T\sin\alpha}{mV} - \frac{(\mu - V^2 r)\cos\gamma}{Vr^2} \\
\dot{h} &= V\sin\gamma \\
\dot{\alpha} &= q - \dot{\gamma} \\
\dot{q} &= \frac{M_{yy}}{I_{yy}}
\end{aligned} \tag{26.1}
$$

where the variables V, γ, α, h, q are velocity, flight path angle, angle of attack, altitude, and pitch rate, respectively. The aerodynamic forces and moment are given by

$$
\begin{aligned}
L(h, V, \alpha, \delta_e) &= \frac{1}{2}\rho V^2 S C_L(\alpha, \delta_e) \\
D(h, V, \alpha, \delta_e) &= \frac{1}{2}\rho V^2 S C_D(\alpha, \delta_e) \\
T(h, V, \alpha, \beta) &= \frac{1}{2}\rho V^2 S C_T(\alpha, \beta) \\
M(h, V, \alpha, \delta_e, \beta_c) &= \frac{1}{2}\rho V^2 S\bar{c}(C_{M,\alpha}(\alpha) + C_{M,\delta_e}(\delta_e) + C_M(q))
\end{aligned}
$$

The curve-fitted approximations for the lift, drag, moment and thrust coefficients [1] can be expressed as

$$
\begin{aligned}
C_L(\alpha, \delta e) &= C_L^{\alpha}\alpha + C_L^{\delta e}\delta_e + C_L^0 \\
C_D(\alpha, \delta e) &= C_D^{\alpha^2}\alpha^2 + C_D^{\alpha}\alpha + C_D^{\delta e^2}\delta_e^2 + C_D^{\delta e}\delta_e + C_D^0 \\
C_{M,\alpha}(\alpha) &= C_{M,\alpha}^{\alpha^2}\alpha^2 + C_{M,\alpha}^{\alpha}\alpha + C_{M,\alpha}^0 \\
C_{M,\delta_e}(\delta_e) &= \bar{c}(\delta_e + d_2(t) - \alpha) \\
C_T(\alpha, \beta) &= C_T^{\alpha^3}(\beta)\alpha^3 + C_T^{\alpha^2}(\beta)\alpha^2 + C_T^{\alpha}(\beta)\alpha + C_T^0(\beta)
\end{aligned}
$$

and the engine dynamics is modeled by a second-order system

$$
\ddot{\beta} = -2\xi\omega_n\dot{\beta} - \omega_n^2\beta + \omega_n^2(\beta_c + d_1(t)) \tag{26.2}
$$

where the throttle setting β_c and elevator deflection δ_e are control inputs, the velocity V and the altitude h are control outputs, $d_1(t)$ and $d_2(t)$ represent external disturbances impacting on throttle setting and elevator deflection, respectively. It's assumed that $d_i(t)$ and $\dot{d}_i(t)$ are bounded by constants [7], i.e., $|d_i(t)| \leq c_i$, $|\dot{d}_i(t)| \leq \bar{c}_i$, $i = 1, 2$. The controller is designed to regulate the outputs to track the desired command signals of velocity $V_d(t)$ and altitude $h_d(t)$.

26.3 Feedback Linearization with Actuator Failures and External Disturbances

In this section, a feedback linearized model of the AHFVs associated with actuator failures and external disturbances is developed, neglecting the weak elevator couplings by setting the coefficients $C_L^{\delta_e}$, $C_D^{\delta_e}$ and $C_D^{\delta_e^2}$ to zero. The inputs are explicitly shown respect to velocity and attitude by differentiating V three times and h four times [7]. Therefore, the relative degree of the system equals to the order of the system. The input-output feedback linearization model of AHFVs can be described as

$$\begin{bmatrix} \dddot{V} \\ h^{(4)} \end{bmatrix} = \begin{bmatrix} \dddot{V}_0 \\ h_0^{(4)} \end{bmatrix} + \boldsymbol{B} \begin{bmatrix} \beta_c \\ \delta e \end{bmatrix} + \boldsymbol{B} \begin{bmatrix} d_1(t) \\ d_2(t) \end{bmatrix} \tag{26.3}$$

where the specific expressions for $\dddot{V}_0, h^{(4)}$ and $\boldsymbol{B}$ can be seen in Ref. [7]. Besides, it can be easily calculated that $\boldsymbol{B}$ is irreversible if and only if $\gamma = \pm\frac{\pi}{2}$. Normally, γ is relatively small in the hypersonic cruise flight condition, it's reasonable to assume $\boldsymbol{B}$ is nonsingular.

Control law based on healthy actuators may have drawbacks and adverse effects on the performance and even stability of the vehicle system when possible actuators faults occur. Therefore, it's necessary to utilize fault-tolerant strategies to deal with unexpected actuator faults. Generally, actuator faults include loss of effectiveness and partially uncontrollable action [6]. In this paper, the uncontrollable action can be integrated with the external disturbances as an unknown part of the dynamic, the actuator faults are given as

$$\begin{bmatrix} \beta_c \\ \delta_e \end{bmatrix} = \begin{bmatrix} \rho_1(t) & 0 \\ 0 & \rho_2(t) \end{bmatrix} \begin{bmatrix} u_1 \\ u_2 \end{bmatrix} \tag{26.4}$$

where $0 < \underline{\rho}_1 \leq \rho_1(t) \leq 1, 0 < \underline{\rho}_2 \leq \rho_2(t) \leq 1$. Substituting (26.4) into (26.3), the feedback linearized model of AHFVs with actuator faults and external disturbances can be rewritten as

$$\begin{bmatrix} \dddot{V} \\ h^{(4)} \end{bmatrix} = \begin{bmatrix} \dddot{V}_0 \\ h_0^{(4)} \end{bmatrix} + \boldsymbol{B}\rho(t) \begin{bmatrix} u_1 \\ u_2 \end{bmatrix} + \boldsymbol{B} \begin{bmatrix} d_1(t) \\ d_2(t) \end{bmatrix} \tag{26.5}$$

26.4 Adaptive Sliding Mode Controller Design

A sliding mode control law combined with two adaptive laws estimating the bounds of the actuator faults and external disturbances is designed in this section. The integrated control law can make the velocity V and altitude h to track the command signals $V_d(t)$ and $h_d(t)$, respectively. The sliding surfaces are chosen as

$$s_1 = (\frac{d}{dt} + \lambda_1)^3 \int_0^t e_1(\tau)d\tau, \quad e_1(t) = V - V_d$$
$$s_2 = (\frac{d}{dt} + \lambda_2)^4 \int_0^t e_2(\tau)d\tau, \quad e_2(t) = h - h_d \tag{26.6}$$

where λ_1, λ_2 are strictly positive constants, $\int_0^t e(\tau)d\tau$ is introduced to eliminate the steady-state errors. Differentiating s_1 and s_2, we obtain

$$\begin{bmatrix} \dot{s}_1 \\ \dot{s}_2 \end{bmatrix} = \begin{bmatrix} v_1 \\ v_2 \end{bmatrix} + \boldsymbol{B} \begin{bmatrix} \rho_1(t) & 0 \\ 0 & \rho_2(t) \end{bmatrix} \begin{bmatrix} u_1 \\ u_2 \end{bmatrix} + \boldsymbol{B} \begin{bmatrix} d_1(t) \\ d_2(t) \end{bmatrix} \tag{26.7}$$

where the expressions of v_1 and v_2 are

$$v_1 = -\dddot{V}_d + \dddot{V}_0 + 3\lambda\ddot{e}_1 + 3\lambda_1^2\dot{e} + \lambda_1^3 e_1 \tag{26.8}$$

$$v_2 = -h_d^{(4)} + h_0^{(4)} + 4\lambda\dddot{e}_2 + 6\lambda_2^2\ddot{e} + 4\lambda^3 2\dot{e}_2 + \lambda_2^4 e_2 \tag{26.9}$$

Conventionally, if the bounds of external disturbances and actuator faults are known, the control law can be given as

$$\boldsymbol{u} = -\boldsymbol{B}^{-1}\rho^{-1}(\phi\psi(t)\frac{\boldsymbol{s}}{\|\boldsymbol{s}\|} + k\boldsymbol{s}) \tag{26.10}$$

where $\psi(t) = \|v\| + \|\boldsymbol{B}\|$, $\rho = \min\{\rho_1(t), \rho_2(t)\}$, $\psi = \max\{1, \sqrt{c_1^2 + c_2^2}\}$, k is a strictly positive constant. Consider the following Lyapunov candidate

$$V_1 = \frac{1}{2}\boldsymbol{s}^T\boldsymbol{s} \tag{26.11}$$

Differentiate V_1 along system (26.7), we have

$$\begin{aligned} \dot{V}_1 &= \boldsymbol{s}^T(\boldsymbol{v} + \boldsymbol{B}\rho\boldsymbol{u} + \boldsymbol{B}\boldsymbol{d}) \\ &\le \|\boldsymbol{s}\|\phi\psi(t) - \boldsymbol{s}^T\boldsymbol{B}\rho\boldsymbol{B}^{-1}\rho^{-1}(\phi\psi(t)\frac{\boldsymbol{s}}{\|\boldsymbol{s}\|} + k\boldsymbol{s}) \\ &\le \|\boldsymbol{s}\|\phi\psi(t) - \|\boldsymbol{s}\|\phi\psi(t) - k\boldsymbol{s}^T\boldsymbol{s} \\ &\le -k\boldsymbol{s}^T\boldsymbol{s} \end{aligned} \tag{26.12}$$

then $\dot{V}_1 \le -2kV_1$, the system (26.7) is exponentially stable with the controller (26.10), i.e., the outputs track the reference command signals at an exponential convergence rate. However, the bounds of actuator faults and external disturbances are usually unknown in practice. Hence, the estimation of the unknown bounds is essential in the process of controller design. Before the adaptive sliding mode controller designing, a corollary of Barbalat's lemma which will be utilized in the convergence analysis is presented here.

Lemma 1 (Sastry and Bodson [9]) *if* $f(t), \dot{f}(t) \in \mathcal{L}_\infty$ *and* $f(t) \in \mathcal{L}_p$ *for some* $p \in [1, +\infty)$, *then* $\lim_{t\to\infty} f(t) = 0$.

Theorem 1 *The AHFVs system (26.3) in the presence of external disturbances and actuator faults can asymptotically track the reference trajectories* $V_d(t)$ *and* $h_d(t)$ *with the proposed sliding mode control law (26.13) and adaptive laws (26.14). The overall control laws are designed as*

$$\boldsymbol{u} = -\hat{\rho}\hat{\phi}\psi(t)\frac{\boldsymbol{B}^{-1}}{\|\boldsymbol{s}\|}\boldsymbol{s} - k\boldsymbol{B}^{-1}\boldsymbol{s} \tag{26.13}$$

$$\dot{\hat{\rho}} = \Gamma_1 \hat{\rho}^3 \psi(t)\|\boldsymbol{s}\|, \quad \dot{\hat{\phi}} = \Gamma_2 \psi(t)\|\boldsymbol{s}\| \tag{26.14}$$

where $\psi(t) = \|\boldsymbol{v}\| + \|\boldsymbol{B}\|$, $\phi = \max\{\min\{c_1, c_2\}, 1\}$, $\rho = min\{\rho_{11}, \rho_{12}\}$, Γ_1, Γ_2 *and* k *are strictly positive constants,* $\hat{\rho}$ *and* $\hat{\phi}$ *are the estimation of the bounds of actuator faults and external disturbances,* $\hat{\rho}(0) > 0$, $\hat{\phi}(0) > 0$.

Proof The Lyapunov candidate is defined as

$$V_2 = \frac{1}{2}\boldsymbol{s}^T\boldsymbol{s} + \frac{1}{2\Gamma_1}\tilde{\rho}^2 + \frac{1}{2\Gamma_2}\tilde{\phi}^2 \tag{26.15}$$

where $\boldsymbol{s}^T = [s_1, s_2]$, $\tilde{\rho} = \hat{\rho}^{-1} - \rho$, $\tilde{\phi} = \hat{\phi} - \phi$. Differentiate V_2 along system (26.7), we have

$$\begin{aligned}
\dot{V}_2 &= \boldsymbol{s}^T(\boldsymbol{v} + \boldsymbol{B}\boldsymbol{\rho}(t)\boldsymbol{u} + \boldsymbol{B}\boldsymbol{d}) + \frac{1}{\Gamma_1}\tilde{\rho}\dot{\tilde{\rho}} + \frac{1}{\Gamma_2}\tilde{\phi}\dot{\tilde{\phi}} \\
&\le \|\boldsymbol{s}\|\,\|\boldsymbol{v} + \boldsymbol{B}\boldsymbol{d}\| + \boldsymbol{s}^T\boldsymbol{B}\boldsymbol{\rho}(t)\boldsymbol{u} + \frac{1}{\Gamma_1}\tilde{\rho}\dot{\tilde{\rho}} + \frac{1}{\Gamma_2}\tilde{\phi}\dot{\tilde{\phi}} \\
&\le \|\boldsymbol{s}\|\phi\psi(t) - \hat{\phi}\psi(t)\boldsymbol{s}^T\boldsymbol{B}\rho\hat{\rho}\boldsymbol{B}^{-1}\frac{\boldsymbol{s}}{\|\boldsymbol{s}\|} - k\rho\boldsymbol{s}^T\boldsymbol{s} - \frac{1}{\Gamma_1}\hat{\rho}^{-2}\dot{\hat{\rho}}\tilde{\rho} + \frac{1}{\Gamma_2}\tilde{\phi}\dot{\hat{\phi}} \\
&\le -k\rho\boldsymbol{s}^T\boldsymbol{s} + \|\boldsymbol{s}\|\phi\psi(t) - \hat{\phi}\rho\hat{\rho}\psi(t)\|\boldsymbol{s}\| - \hat{\phi}\hat{\rho}\tilde{\rho}\psi(t)\|\boldsymbol{s}\| + \frac{1}{\Gamma_2}\tilde{\phi}\dot{\hat{\phi}} \\
&\le -k\rho\boldsymbol{s}^T\boldsymbol{s} - \tilde{\phi}\psi(t)\|\boldsymbol{s}\| + \tilde{\phi}\psi(t)\|\boldsymbol{s}\| \\
&= -k\rho\boldsymbol{s}^T\boldsymbol{s}
\end{aligned} \tag{26.16}$$

Therefore, V_2 is bounded and $\boldsymbol{s}, \hat{\rho}, \hat{\phi} \in \mathcal{L}_\infty$. According to Eq. (26.7) we have $\dot{\boldsymbol{s}} \in \mathcal{L}_\infty$. Moreover, integrating inequality (26.16) from 0 to ∞ we can obtain that $s \in \mathcal{L}_2$. Then based on the Lemma, it can be concluded that $\lim_{t\to\infty} \boldsymbol{s} = 0$. Consequently, e_1 and e_2 converge to zero when $t \to \infty$, i.e., the outputs asymptotically track the respective reference trajectories in the present of actuator faults and external disturbances. □

Remark 1 In the control law (26.13), the part $-\hat{\rho}\hat{\phi}\psi(t)\frac{\boldsymbol{B}^{-1}}{\|s\|}\boldsymbol{s}$ may cause unexpected chattering. Thus a saturation function $-\hat{\rho}\hat{\phi}\psi(t)\boldsymbol{B}^{-1}\boldsymbol{sat}(\frac{s}{\boldsymbol{\Phi}})$ is used to eliminate the chattering. Consequently, the control law (26.13) can be rewritten as

$$\boldsymbol{u} = -\hat{\rho}\hat{\phi}\psi(t)\boldsymbol{B}^{-1}\boldsymbol{sat}(\frac{\boldsymbol{s}}{\boldsymbol{\Phi}}) - k\boldsymbol{B}^{-1}\boldsymbol{s} \tag{26.17}$$

where $\boldsymbol{sat}(\frac{s}{\boldsymbol{\Phi}}) = [sat(\frac{s_1}{\Phi_1}), sat(\frac{s_2}{\Phi_2})]$. But (26.17) is just a approximation of (26.13), the system (26.7) with (26.17) is uniform ultimate bounded of the switching sufaces [10].

26.5 Simulations

In the simulation, the parameters $R_E, S, I_{yy}, \mu, \rho, \bar{c}, c_e$ and the coefficients of the forces and moment in Ref. [7] are adopted. The command signals are $V_d(t) = 4700m/s$, $h_d(t) = 33100m$. In addition, the flexible body modes [5] are expressed as

$$\ddot{\eta} = -2\xi_i\omega_i\dot{\eta}_i - \omega_i^2\eta_i + N_i, i = 1, 2 \tag{26.18}$$

$$N_1 \approx N_1^{\alpha^2}\alpha^2 + N_1^{\alpha}\alpha + N_1^0, \quad N_2 \approx N_1^{\alpha^2}\alpha^2 + N_1^{\alpha}\alpha + N^{\delta_1}\delta_e + N_1^0$$

The faults and external disturbances are assumed as $\rho_1(t) = 0.75 + 0.05\sin(0.005t)$, $\rho_2(t) = 0.85 + 0.14\sin(0.03t)$, $d_1(t) = 0.001\sin(t)$, $d_2(t) = 0.001\sin(t)$.

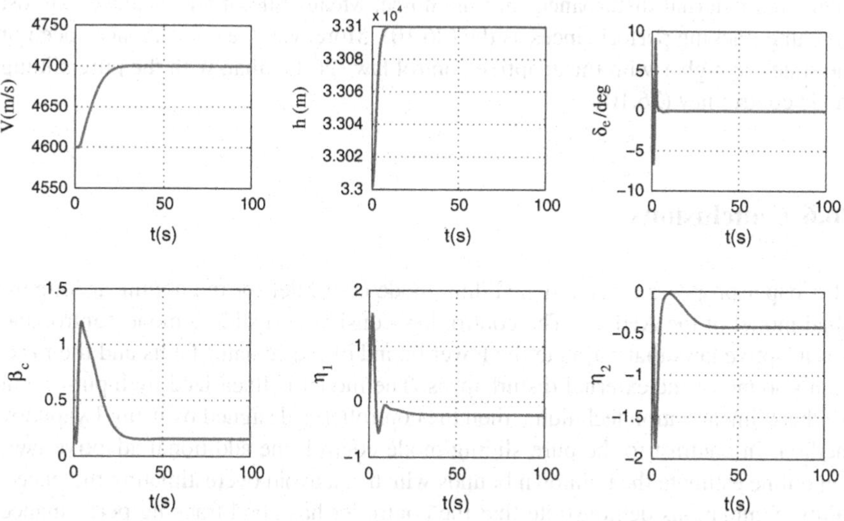

Fig. 26.1 Response curves of sliding mode controller

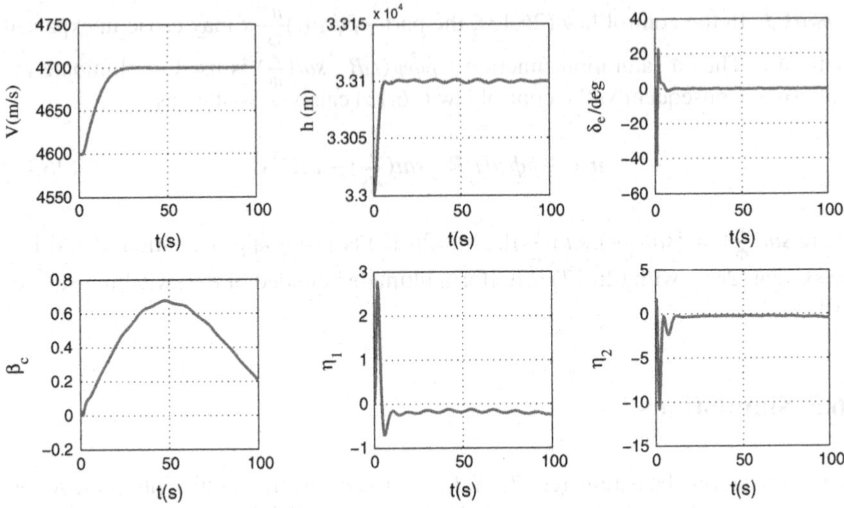

Fig. 26.2 Response curves of adaptive sliding mode controller

The controller parameters are chosen as $k = 5, \Gamma_1 = 0.001, \Gamma_2 = 0.002, \Phi_1 = 0.1, \Phi_2 = 0.1, \lambda_1 = 0.3, \lambda_2 = 1.2$. The initial values of the system are chosen as $\hat{\rho}(0) = 0.001, \hat{\phi}(0) = 0.002, V(0) = 4600\,\mathrm{m/s}$ and $h(0) = 33000\,\mathrm{m}$.

Figures 26.1 and 26.2 show the tracking performance of velocity and altitude under control law (26.10) and (26.17), respectively. It can be observed that the adaptive sliding mode controller has high tracking performance and good robust to actuator faults and unknown disturbances, even though the bounds of the actuator faults and external disturbances are unknown. Meanwhile, it still achieves almost the same tracking performances as the (26.10). Moreover, the convergence speed of the state are higher with the adaptive control law (26.17) than with the pure sliding mode control law (26.10).

26.6 Conclusions

This paper proposes an adaptive sliding mode controller for the nonlinear longitudinal model of the AHFVs. The control law consists of a sliding mode control and two adaptive laws that estimate the lower bound of the actuator faults and the maximum bound of the external disturbances. The model is linearized by input-output feedback linearization technique, then the controller is designed by using Lyapunov method. In contrast to the pure sliding mode control, the additional adaptive laws can online estimate the unknown bounds which can avoid overestimating the uncertainty. Simulations demonstrate that the controller has good tracking performance and robustness with respect to actuator faults and external disturbances.

Acknowledgments This work was supported by the National Basic Research Program of China (973 Program: 2012CB821200, 2012CB821201) and the NSFC (61134005, 61221061, 61327807).

References

1. Parker JT, Serrani A, Yurkovich S, Bolender M, Doman D (2006) Approximate feedback linearization of anair-breathing hypersonic vehicle. J Guidance Navig Control Confe Exhib 1–20
2. Sigthorsson D, Jankovsky P, Serrani A, Yurkovich S, Bolender M, Doman DB (2008) Robust linear output feedback control of an airbreathing hypersonic vehicle. J Guidance Control Dyn 31(4):1052–1066
3. Shen Q, Jiang B, Cocquempot V (2012) Fault-tolerant control for T-S fuzzy systems with application to near-space hypersonic vehicle with actuator faults. IEEE Trans Fuzzy Syst 20(4): 652–665
4. Li H, Wu L, Si Y, Gao H, Hu X (2010) Multi-objective fault-tolerant output tracking control of a flexible air-breathing hypersonic vehicle. J Syst Control Eng. In:Proceedings of the Institution of Mechanical Engineers, Part I: 224(6):647–667
5. Parker JT, Serrani A, Yurkovich S, Bolender MA, Doman DB (2007) Control-oriented modeling of an air-breathing hypersonic vehicle. J Guidance Control Dyn 30(3):856–869
6. Gao G, Wang J (2014) Observer-based fault-tolerant control for an air-breathing hypersonic vehicle model. Nonlinear Dyn 76(1):409–430
7. Gao G, Wang J (2014) Finite time integral sliding mode control of hypersonic vehicles. Nonlinear Dyn 73(1–2):229–244
8. Bolender MA, Doman DB (2006) Nonlinear longitudinal dynamical model of an air-breathing hypersonic vehicle. J Spacecr Rocket 44(2):374–387
9. Sastry S, Bodson M (2011) Adaptive control: stability, convergence, and robustness. Cour Corp
10. Tao G (2003) Adaptive control design and analysis. Wiley, New York

Chapter 27
On Indirect Model Reference Adaptive Learning Control

Wen Du

Abstract This chapter deals with tracking problem of single-input single-output linear time-invariant (SISO LTI) system with parametric uncertainties by using a model reference adaptive iterative learning control (MRAILC) scheme. The tracking error converges to zero pointwisely after infinite iterations. A major feature of this proposed method is that system output could track the desired trajectory whether the plant system is minimum-phrase or not.

Keywords Iterative learning control · Model reference adaptive control

27.1 Introduction

Iterative learning control (ILC) is one of the most desirable control methods when a perfect tracking performance is required during a finite time interval [1–4]. One reason is that this method is independent of system model, and it can usually handle nonlinear and strongly coupled systems. However, a limitation of ILC is that most existing ILC methods can only be applicable to plants without parametric uncertainties. That is, the system plant must be a determined plant. This limitation draws back its application.

As we all know, adaptive control (AC) is a powerful approach to dispose parameter uncertainties by estimating and updating the unknown parameters. So many scholars combine the ILC with AC to solve a series of uncertain parametric problems, such as iteration-varying problems [5, 6], random initial condition problems [7], time-varying parametric uncertainties problems [8, 9] and so on.

W. Du (✉)
School of Mathematics and Systems Science, Beihang University (BUAA), Beijing 100191, China
e-mail: duwen1992@sohu.com

© Springer-Verlag Berlin Heidelberg 2016

Y. Jia et al. (eds.), *Proceedings of the 2015 Chinese Intelligent Systems Conference*, Lecture Notes in Electrical Engineering 359,
DOI 10.1007/978-3-662-48386-2_27

It is worth mentioning that one of the scholars combine the ILC with AC to solve MRC problems [10]. Consequently, he concludes that the tracking error converges to zero pointwisely by using the direct method of AC. But one obvious shortage of the direct method is that it can't deal with minimum-phase plant. The reason is it is impossible for nonminimum-phase plant to parameterize as the parameters of desired controller. Inspired by his work, we also consider the same system with relative degree one but use an indirect method of AC. That is, we do not get the updating law directly from the parametric vector but from the unknown system coefficients. Then we use matching equations to connect the estimated coefficients with the system input. Thus, we can cope with the uncertain parameter plant. On the one hand, compared to the model reference adaptive control (MRAC), the transient performance of MRAILC is improved a lot. On the other hand, compared to the ILC, MRAILC can dispose a class of system with unknown parameters.

The passage is organized as follows: in Sect. 27.2, we give the background and the basic assumptions of this problem; in Sect. 27.3, we prove the tracking error convergence; finally, we summarize the whole passage and give the perspective of the further work in Sect. 27.4.

27.2 Problem Formulation

Consider the SISO-LTI system, expressed by the transfer function

$$y_k(t) = G_p(s)[u_k(t)] = k_p \frac{Z_p(s)}{R_p(s)}[u_k(t)] \tag{27.1}$$

where $Z_p(s)$ is a monic Hurwitz polynomials of degree m_p, the degree of $R_p(s)$ equals to n, $t \in [0, T]$ stands for the system operating time, and the integral number k represents the iteration number, the relative degree n^* of $G_p(s)$ equals to 1, $u_k(t)$, $y_k(t)$ are the system input and output respectively at the kth iteration, and the high frequency gain k_p is a constant and its sign is known.

The reference model, which describes the desired features of the plant, is described in the same form of (27.1)

$$y_d(t) = G_m(s)[r_f(t)] = k_m \frac{Z_m(s)}{R_m(s)}[r_f(t)] \tag{27.2}$$

where $Z_m(s)$, $R_m(s)$ are monic Hurwitz polynomials, the degree of $R_m(s)$ is p_m ($p_m \leq n$), the relative degree of $G_m(s)$ is identical to $G_p(s)$, $r_f(t) \in R^1$ is the reference input which is uniformly bounded and piecewisely continuous, and k_m is a constant.

Our goal is to design the system input $u_k(t)$ such that after infinite iterations, the system output $y_k(t)$ can track the reference output $y_d(t)$ as much as possible for any reference input $r_f(t)$. In the following part, we omit t on the condition that this action will not cause ambiguity.

Since we consider the same plant and goal referred in [10], some analysis process of control law designing can be omitted here. We propose the same control law

$$\begin{aligned} \dot{w}_{1,k} &= Fw_{1,k} + gu_k, \quad w_{1,k}(0) = 0 \\ \dot{w}_{2,k} &= Fw_{2,k} + gy_k, \quad w_{2,k}(0) = 0 \\ u_p &= \theta^{*T} w \end{aligned} \tag{27.3}$$

where

$$w_{1,k}, w_{2,k} \in R^{n-1},$$
$$\theta^* = [\theta_1^{*T}, \theta_2^{*T}, \theta_3^*, c_0^*]^T,$$
$$w = [w_{1,k}, w_{2,k}, y_p, r]^T,$$
$$F = \begin{bmatrix} -\lambda_{n-2} & -\lambda_{n-3} & -\lambda_{n-4} & \cdots & -\lambda_0 \\ 1 & 0 & 0 & \cdots & 0 \\ 0 & 1 & 0 & \cdots & 0 \\ \cdots & \cdots & \cdots & \cdots & \cdots \\ 0 & 0 & 1 & \cdots & 0 \end{bmatrix}, \quad g = \begin{bmatrix} 1 \\ 0 \\ \cdots \\ 0 \end{bmatrix}$$

λ_i are the coefficients of

$$\Lambda(s) = s^{n-1} + \lambda_{n-2}s^{n-2} + \cdots + \lambda_1 s + \lambda_0 = \det(sI - F).$$

Unlike the direct method [1], we estimate the coefficients of the plant at each time and then we use the estimated coefficients to calculate $\theta^*(t)$ in the control law.

According to [10], we can obtain the following matching equations:

$$c_0^* = \frac{k_m}{k_p} \tag{27.4}$$

$$\theta_1^{*T}\alpha(s) = \Lambda(s) - Z_p(s) \tag{27.5}$$

$$\theta_2^{*T}\alpha(s) + \theta_3^*\Lambda(s) = \frac{R_p(s) - \Lambda_0(s)R_m(s)}{k_p} \tag{27.6}$$

From the matching equations the relationship between unknown plant coefficient and $\theta^*(t)$ can be confirmed.

In order to simplify (27.4)–(27.6), we order

$$
\begin{aligned}
Z_p(s) &= s^{n-1} + p_1^T \alpha(s) \\
R_p(s) &= s^n + a_{n-1}s^{n-1} + p_2^T \alpha(s) \\
\Lambda(s) &= s^{n-1} + \lambda^T \alpha(s) \\
\Lambda_0(s)R_m(s) &= s^n + r_{n-1}s^{n-1} + \upsilon^T \alpha(s)
\end{aligned}
$$

where $p_1 \in R^{n-1}, p_2 \in R^{n-1}$, $a_{n-1} \in R$ are plant parameters, i.e., $\theta_p^* = [k_p, p_1^T, a_{n-1}, p_2^T]^T$; λ, $\upsilon \in R^{n-1}$ and $r_{n-1} \in R$ are the coefficients of the known polynomials $\Lambda(s)$, $\Lambda_0(s)R_m(s)$ and $\alpha(s) = [s^{n-2}, s^{n-3}, \ldots, s, 1]^T$

Using the equations above, we can obtain

$$
c_0^* = \frac{k_m}{k_p},\ \theta_1^* = \lambda - p_1,\ \theta_2^* = \frac{p_2 - a_{n-1}\lambda + r_{n-1}\lambda - \gamma}{k_p},\ \theta_3^* = \frac{a_{n-1} - r_{n-1}}{k_p}
$$

Assuming that the $\hat{k}_p(t), \hat{p}_1(t), \hat{p}_2(t), \hat{a}_{n-1}(t)$ are the estimates of the $k_p, p_1, p_2, a_{n\text{-}1}$ respectively at each time t, we can calculate the $\theta(t) = [\theta_1^T, \theta_2^T, \theta_3, c_0]^T$ by replacing the k_p, p_1, p_2, a_{n-1} with $\hat{k}_p(t), \hat{p}_1(t), \hat{p}_2(t), \hat{a}_{n-1}(t)$, respectively.

Using the closed-loop system information analyzed in [10], we can get:

$$
e_{1,k} = G_m(s)\rho^*(u_k - \theta^{*T} w) \tag{27.7}
$$

$$
\rho^* = \frac{1}{c_0^*} \tag{27.8}
$$

Rewrite (27.7) as

$$
e_{1,k} = G_m(s)\frac{1}{k_m}(k_p u_k - k_p\theta^{*T} w - \hat{k}_p u_p + \hat{k}_p \theta^T w) \tag{27.9}
$$

By substituting for $k_p\theta^*$, $\hat{k}_p\theta$, we can get:

$$
\begin{aligned}
e_{1,k} = G_m(s)\frac{1}{k_m}&[\tilde{k}_p(\lambda^T w_{1,k} - u_k) + \tilde{k}_2^T w_{2,k} + \tilde{a}_{n-1}(y_k - \lambda^T w_{2,k}) \\
&- \hat{k}_p \hat{p}_1 w_{1,k} + k_p p_1^T w_{1,k}]
\end{aligned}
$$

where the parameter errors are:

$$
\tilde{k}_p = \hat{k}_p - k_p,\quad \tilde{p}_2 = \hat{p}_2 - p_2,\quad \tilde{a}_{n-1} = \hat{a}_{n-1} - a_{n-1}
$$

Because

$$-\hat{k}_p\hat{p}_1^T w_{1,k} + k_p p_1^T w_{1,k} + k_p\hat{p}_1^T w_{1,k} - k_p\hat{p}_1^T w_{1,k} = -\tilde{k}_p\hat{p}_1^T w_{1,k} - k_p\tilde{p}_1^T w_{1,k}$$

we have

$$e_{1,k} = G_m(s)\frac{1}{k_m}[\tilde{k}_p\xi_{1,k} + \tilde{a}_{n-1}\xi_{2,k} + \tilde{p}_2^T w_{2,k} - k_p\tilde{p}_1^T w_{1,k}] \tag{27.10}$$

where

$$\xi_{1,k} = \lambda^T w_{1,k} - u_k - \hat{p}_1^T w_{1,k},\ \xi_{2,k} = y_k - \lambda^T w_{2,k},\ \tilde{p}_1 = \hat{p}_1 - p_1$$

A minimal state-space representation of (27.10) is given by

$$\begin{aligned}\dot{e}_k &= A_c e_k + B_c\left[\tilde{k}_p\xi_{1,k} + \tilde{a}_{n-1}\xi_{2,k} + \tilde{p}_2^T w_{2,k} - k_p\tilde{p}_1^T w_{1,k}\right]\\ e_{1,k} &= C_c^T e_k\end{aligned}$$

where

$$C_c^T(sI - A_c)^{-1}B_c = G_m(s)\frac{1}{k_m}$$

Define the Lyapunov-like function

$$V_k = \frac{e_k^T P_c e_k}{2} + \frac{\tilde{k}_p^2}{2\gamma_p} + \frac{\tilde{a}_{n-1}^2}{2\gamma_1} + \frac{\tilde{p}_1^T\Gamma_1^{-1}\tilde{p}_1}{2}|k_p| + \frac{\tilde{p}_2^T\Gamma_2^{-1}\tilde{p}_2}{2}$$

where P_c satisfies the algebraic equations of the Lefschetz-Kalman-Yakubovich (LKY) Lemma. $\gamma_1, \gamma_p > 0$ and Γ_1, Γ_2 are symmetric positive definite matrix. By designing the updating laws, we can get the following theorem.

27.3 Convergence Analysis and Conclusion

Theorem Assume that $G_m(s)$ is strictly positive real. Consider the system (27.1), with a relative degree $n^* = 1$, under the following control law

$$u_k(t) = \theta_k^T(t)\Omega_k(t) \quad for \ \ k \geq 0$$

where at the initial iteration, i.e., $k = 0$, the parametric adaption law is given by

$$\begin{aligned}\dot{\hat{a}}_{n-1,0}(t) &= -\gamma_1 e_{1,0}(t)\xi_{2,0}(t)\\ \dot{\hat{p}}_{1,0}(t) &= \Gamma_1 e_{1,0} w_{1,0}(t)\mathrm{sgn}(k_p)\\ \dot{\hat{p}}_{2,0}(t) &= -\Gamma_2 e_{1,0}(t) w_{2,0}(t)\end{aligned}$$

$$\dot{\hat{k}}_{p,0}(t) = \begin{cases} -\gamma_p e_{1,0}(t)\xi_{1,0}(t), & \text{if } \left|\hat{k}_{p,0}\right| > 0 \text{ or } \left|\hat{k}_{p,0}\right| = k_0 \text{ and } e_{1,k}\xi_{1,k}\mathrm{sgn}(k_p) \le 0 \\ 0, & \text{otherwise} \end{cases}$$

and for $k \ge 1$, the parametric adaption law is given by

$$\begin{aligned}\hat{a}_{n-1,k}(t) &= \hat{a}_{n-1,k-1}(t) - \gamma_1 e_{1,k}(t)\xi_{2,k}(t)\\ \hat{p}_{1,k}(t) &= \hat{p}_{1,k-1}(t) + \Gamma_1 e_{1,k} w_{1,k}(t)\mathrm{sgn}(k_p)\\ \hat{p}_{2,k}(t) &= \hat{p}_{2,k-1}(t) - \Gamma_2 e_{1,k}(t) w_{2,k}(t)\\ \hat{k}_{p,k}(t) &= \hat{k}_{p,k-1}(t) - \gamma_p e_{1,k}(t)\xi_{1,k}(t)\end{aligned}$$

Then,

(1) All the signals in the closed-loop system are bounded.
(2) $\lim_{k\to\infty} e_{1,k}(t) = 0$, for all $t \in [0, T]$.

Proof Let the state-space representation of (27.10) be

$$\begin{aligned}\dot{e}_k &= A_c e_k + B_c\left[\tilde{k}_p\xi_{1,k} + \tilde{a}_{n-1}\xi_{2,k} + \tilde{p}_2^T w_{2,k} - k_p\tilde{p}_1^T w_{1,k}\right],\ e_k(0) = 0\\ e_{1,k} &= C_c^T e_k\end{aligned}$$

Now let us consider the following Lyapunov-like functional candidate:

$$\begin{aligned}&W_k(e_k, \hat{a}_{n-1,k}, \hat{p}_{1,k}, \hat{p}_{2,k}, \hat{k}_{p,k})\\ &= V_k(e_k) + \int \frac{\hat{a}_{n-1,k}^T\gamma_1^{-1}\hat{a}_{n-1,k}}{2}d\tau + \left|k_p\right| \int \frac{\hat{p}_{1,k}^T\Gamma_1^{-1}\hat{p}_{1,k}}{2}d\tau + \int \frac{\hat{p}_{2,k}^T\Gamma_2^{-1}\hat{p}_{2,k}}{2}d\tau\\ &\quad + \int \frac{\hat{k}_{p,k}^T\gamma_p^{-1}\hat{k}_{p,k}}{2}d\tau\end{aligned}$$

where $V_k(e_k) = \frac{1}{2}e_k^T P e_k$ and the positive defined symmetric P satisfies the algebraic equations of the (LKY) Lemma, and the $V_k(e_k)$ can be written as follows:

$$\begin{aligned} V_k(e_k(t)) &= V_k(e_k(0)) + \int_0^t \dot{V}_k(e_k(\tau))d\tau \\ &= V_k(e_k(0)) + \frac{1}{2}\int_0^t \dot{e}_k^T(t)Pe_k(t) + e_k^T(t)P\dot{e}_k(t)d\tau \\ &= -\frac{1}{2}\int_0^t e_k^T(qq^T + \gamma L)e_k - 2e_{1,k}(\tilde{k}_p\xi_{1,k} + \tilde{a}_{n-1}\xi_{2,k} \\ &\quad + \tilde{p}_2^T w_{2,k} - k_p\tilde{p}_1^T w_{1,k})d\tau \end{aligned}$$

The difference of the Lyapunov-like function is given by:

$$\begin{aligned} \Delta W_k &= V_k - V_{k-1} + \frac{1}{2}\int \left(\hat{a}_{n-1,k}^T\gamma_1^{-1}\hat{a}_{n-1,k} - \hat{a}_{n-1,k-1}^T\gamma_1^{-1}\hat{a}_{n-1,k-1}\right)d\tau \\ &\quad + \frac{|k_p|}{2}\int \left(\hat{p}_{1,k}^T\Gamma_1^{-1}\hat{p}_{1,k} - \hat{p}_{1,k-1}^T\Gamma_1^{-1}\hat{p}_{1,k-1}\right)d\tau \\ &\quad + \frac{1}{2}\int \hat{p}_{2,k}^T\Gamma_2^{-1}\hat{p}_{2,k} - \hat{p}_{2,k-1}^T\Gamma_2^{-1}\hat{p}_{2,k-1}\,d\tau \\ &\quad + \frac{1}{2}\int \hat{k}_{p,k}^T\gamma_p^{-1}\hat{k}_{p,k} - \hat{k}_{p,k-1}^T\gamma_p^{-1}\hat{k}_{p,k-1}\,d\tau \\ &= V_k - V_{k-1} - \frac{1}{2}\int_0^t \left(\bar{\hat{a}}_{n-1,k}^T\gamma_1^{-1}\bar{\hat{a}}_{n-1,k} - 2\bar{\hat{a}}_{n-1,k}^T\gamma_1^{-1}\hat{a}_{n-1,k}\right)d\tau \\ &\quad - \frac{|k_p|}{2}\int_0^t \bar{\hat{p}}_{1,k}^T\Gamma_1^{-1}\bar{\hat{p}}_{1,k} - 2\bar{\hat{p}}_{1,k}^T\Gamma_1^{-1}\hat{p}_{1,k}\,d\tau \\ &\quad - \frac{1}{2}\int_0^t \left(\bar{\hat{p}}_{2,k}^T\Gamma_2^{-1}\bar{\hat{p}}_{2,k} - 2\bar{\hat{p}}_{2,k}^T\Gamma_2^{-1}\hat{p}_{2,k}\right)d\tau \\ &\quad - \frac{1}{2}\int_0^t \bar{\hat{k}}_{p,k}^T\gamma_p^{-1}\bar{\hat{k}}_{p,k} - 2\bar{\hat{k}}_{p,k}^T\gamma_p^{-1}\hat{k}_{p,k}\,d\tau \end{aligned}$$

where $\bar{\hat{a}}_{n-1,k} = \hat{a}_{n-1,k} - \hat{a}_{n-1,k-1}$ $\bar{\hat{p}}_{1,k} = \hat{p}_{1,k} - \hat{p}_{1,k-1}$ $\bar{\hat{p}}_{2,k} = \hat{p}_{2,k} - \hat{p}_{2,k-1}$ $\bar{\hat{k}}_{p,k} = \hat{k}_{p,k} - \hat{k}_{p,k-1}$. So we can obtain:

$$\begin{aligned} \Delta W_k &= -V_{k-1} - \frac{1}{2}\int_0^t \bar{\hat{a}}_{n-1,k}^T\gamma_1^{-1}\bar{\hat{a}}_{n-1,k}d\tau - \frac{|k_p|}{2}\int_0^t \bar{\hat{p}}_{1,k}^T\Gamma_1^{-1}\bar{\hat{p}}_{1,k}d\tau \\ &\quad - \frac{1}{2}\int_0^t \bar{\hat{p}}_{2,k}^T\Gamma_2^{-1}\bar{\hat{p}}_{2,k}d\tau - \frac{1}{2}\int_0^t \bar{\hat{k}}_{p,k}^T\gamma_p^{-1}\bar{\hat{k}}_{p,k}d\tau \\ &\quad + V_k + \int_0^t \bar{\hat{a}}_{n-1,k}^T\gamma_1^{-1}\hat{a}_{n-1,k}d\tau + \frac{|k_p|}{2}\int_0^t \bar{\hat{p}}_{1,k}^T\Gamma_1^{-1}\hat{p}_{1,k}d\tau \\ &\quad + \frac{1}{2}\int_0^t \bar{\hat{p}}_{2,k}^T\Gamma_2^{-1}\hat{p}_{2,k}d\tau + \frac{1}{2}\int_0^t \bar{\hat{k}}_{p,k}^T\gamma_p^{-1}\hat{k}_{p,k}d\tau \end{aligned}$$

$$
\begin{aligned}
&= -V_{k-1} - \frac{1}{2}\int_0^t \bar{\tilde{a}}_{n-1,k}^T \gamma_1^{-1} \bar{\tilde{a}}_{n-1,k} d\tau - \frac{|k_p|}{2}\int_0^t \bar{\tilde{p}}_{1,k}^T \Gamma_1^{-1} \bar{\tilde{p}}_{1,k}\, d\tau \\
&\quad - \frac{1}{2}\int_0^t \bar{\tilde{p}}_{2,k}^T \Gamma_2^{-1} \bar{\tilde{p}}_{2,k} d\tau - \frac{1}{2}\int_0^t \bar{\tilde{k}}_{p,k}^T \gamma_p^{-1} \bar{\tilde{k}}_{p,k}\, d\tau \\
&\quad - \frac{1}{2}\int_0^t e_k^T (qq^T + \gamma L) e_k - 2e_{1,k}^T (\tilde{k}_p \xi_{1,k} + \tilde{a}_{n-1} \xi_{2,k} + \tilde{p}_2^T w_{2,k} \\
&\quad - k_p \tilde{p}_1^T w_{1,k}) d\tau - \int_0^t e_{1,k}^T \xi_{2,k} \hat{a}_{n-1,k} d\tau + |k_p| \int_0^t e_{1,k}^T w_{1,k} \operatorname{sgn}(k_p) \hat{p}_{1,k}\, d\tau \\
&\quad - \int_0^t e_{1,k}^T w_{2,k} \hat{p}_{2,k} d\tau - \int_0^t e_{1,k}^T \xi_{1,k} \hat{k}_{p,k} d\tau \\
&= -V_{k-1} - \frac{1}{2}\int_0^t \bar{\tilde{a}}_{n-1,k}^T \gamma_1^{-1} \bar{\tilde{a}}_{n-1,k} d\tau - \frac{|k_p|}{2}\int_0^t \bar{\tilde{p}}_{1,k}^T \Gamma_1^{-1} \bar{\tilde{p}}_{1,k}\, d\tau \\
&\quad - \frac{1}{2}\int_0^t \bar{\tilde{p}}_{2,k}^T \Gamma_2^{-1} \bar{\tilde{p}}_{2,k} d\tau - \frac{1}{2}\int_0^t \bar{\tilde{k}}_{p,k}^T \gamma_p^{-1} \bar{\tilde{k}}_{p,k}\, d\tau \\
&\quad - \frac{1}{2}\int_0^t e_k^T (qq^T + \gamma L) e_k\, d\tau \\
&\leq 0
\end{aligned}
$$

Hence the $W_k(t)$ is non-increasing and consequently e_k^T, $e_{1,k}(t)$, $\int_0^t \bar{\tilde{a}}_{n-1,k}^T \gamma_1^{-1} \bar{\tilde{a}}_{n-1,k} d\tau$, $\int_0^t \bar{\tilde{p}}_{1,k}^T \Gamma_1^{-1} \bar{\tilde{p}}_{1,k} d\tau$, $\int_0^t \bar{\tilde{p}}_{2,k}^T \Gamma_2^{-1} \bar{\tilde{p}}_{2,k} d\tau$ and $\int_0^t \bar{\tilde{k}}_{p,k}^T \gamma_p^{-1} \bar{\tilde{k}}_{p,k} d\tau$ are bounded if $W_0(t)$ is bounded.

Now, let us proof the boundness of $W_0(0)$. Consider the following function:

$$
\begin{aligned}
S_0(e_0, \hat{a}_{n-1,0}, \hat{p}_{1,0}, \hat{p}_{2,0}, \hat{k}_{p,0}) &= \frac{1}{2} e_0^T P e_0 + \frac{1}{2} \hat{a}_{n-1,0}^T \gamma_1^{-1} \hat{a}_{n-1,0} \\
&\quad + \frac{|k_p|}{2} \hat{p}_{1,0}^T \Gamma_1^{-1} \hat{p}_{1,0} + \frac{1}{2} \hat{p}_{2,0}^T \Gamma_2^{-1} \hat{p}_{2,0} \\
&\quad + \frac{1}{2} \hat{k}_{p,0}^T \gamma_p^{-1} \hat{k}_{p,0}
\end{aligned}
$$

whose time derivative is given by

$$
\dot{S}_0 = \begin{cases} -\frac{1}{2} e_0^T (qq^T + vL) e_0, \\ \qquad \qquad if\ |\hat{k}_{p,0}| > 0\ or\ |\hat{k}_{p,0}| = k_0\ and\ e_{1,k}\xi_{1,k} \operatorname{sgn}(k_p) \leq 0 \\ -\frac{1}{2} e_0^T (qq^T + vL) e_0 + e_1 \xi_1 \tilde{k}_p, \\ \qquad \qquad if\ |\hat{k}_{p,0}| = k_0\ |\hat{k}_{p,0}| = k_0\ and\ e_{1,k}\xi_{1,k} \operatorname{sgn}(k_p) > 0 \end{cases}
$$

which means the $e_0(t)$, $\hat{a}_{n-1,0}$, $\hat{p}_{1,0}$, $\hat{p}_{2,0}$ and $\hat{k}_{p,0}$ are globally bounded. Hence $W_0(t)$ is bounded over the finite time interval $[0, T]$.

Let us rewrite $W_k(t)$ as follows:

$$W_k = W_0 + \sum_{j=0}^{j=k} \Delta W_j \leq W_0 - \sum_{j=0}^{j=k} V_j \leq W_0 - \frac{1}{2}\sum_{j=0}^{j=k} e_j^T P e_j$$

which leads to

$$\sum_{j=1}^{k} e_j^T P e_j \leq 2(W_0 - W_k) \leq 2W_0$$

Since $W_0(t)$ and $e_k(t)$ are bounded for all $k \in Z_+$ and $t \in [0, T]$, one can conclude that $\lim_{k \to \infty} e_{1,k} = 0, \quad \forall t \in [0, T]$. □

27.4 Conclusion

MRAILC method is presented for solving the tracking problem of a class of uncertain-parameter plants. Compared with the exiting MRAILC, the new indirect method could deal with both minimum- and nonminimum-phase plant for the reason that parameterizing the plant model as the desired controller parameters is convenient. Both theoretical analysis and simulation example verify the effectiveness of the MRAILC. MRAIC with normalized adaptive laws and the application to a higher order relative degree system would be our further work.

Acknowledgement This paper was supported by the National Natural Science Foundation of China (NSFC: 61473010).

References

1. Xiong ZH, Zhang J, Wang X, Xu YM (2007) Integrated tracking control strategy for batch processes using a batch-wise linear time-varying perturbation model. IET Control Theory Appl 1:179–188
2. Hou ZS, Xu JX, Zhong HW (2007) Freeway traffic control using iterative learning control based ramp metering and speed signaling. IEEE Trans Veh Technol 56:466–477
3. Wang YQ, Zhou DH, Gao FR (2008) Iterative learning reliable control of batch processes with sensor faults. Chem Eng Sci 63:1039–1051
4. Hahn B, Oldham K (2012) On-off iterative adaptive controller for low-power micro-robotic step regulation. Asian J Control 14:624–640
5. Chi R, Wang D, Hou Z, Jin S, Zhang D (2012) A discrete-time adaptive iterative learning from different reference trajectory for linear time-varying systems. In: Proceedings of the 31st Chinese control conference
6. Xu JX, Xu J (2004) On iterative learning from different tracking tasks in the presence of time-varying uncertainties. IEEE Trans Syst Man Cybern 34:589–597

7. Chi R, Hou Z, Wang D (2013) Discrete-time adaptive ILC for non-parametric uncertain nonlinear systems with iteration-varying trajectory and random initial condition. Asian J Control 15:562–570
8. Chi R, Sui S, Hou Z (2008) A new discrete-time adaptive ILC for nonlinear systems with time-varying parametric uncertainties. Acta Automatic Sinica 34:805–808
9. Xu J, Tan Y (2002) A composite energy function-based learning control approach for nonlinear systems with time-varying parametric uncertainties. IEEE Tran Autom Control 47:1940–1945
10. Tayebi A (2006) Model reference adaptive iterative learning control for linear systems. Int J Adapt Control Signal Process 20:475–489

Chapter 28
Stereo Vision Pose Estimation for Moving Objects by the Interacting Multiple Model Method

Ying Peng, Shihao Sun, Yingmin Jia and Changqing Chen

Abstract The stereo vision measurement system is very widely employed to obtain the 6 DOF pose information for the moving objects in space. However, the linear and angular velocities are impossible to estimate using these systems while the dynamic model is unknown and disturbances exist, and their applications is limited. To overcome this disadvantage, we propose an approach based on the IMM algorithm for moving objects. Our approach is verified in the feature points of a moving object. And the simulating results show its validity.

Keywords 6DOF · Pose estimation · IMM · Kalman filter

28.1 Introduction

Estimating the position and pose of multiple-degrees-of-freedom moving objectives is one of the principal goals in the actual applications such as intelligent navigation, objective tracking, space docking and so on [1]. To this end, a great number of estimating systems have been conducted in recent years which are: inertial navigation system, GPS system, laser system and stereo vision system.

Y. Peng (✉) · S. Sun · Y. Jia
The Seventh Research Division and the Center for Information and Control, Beihang University (BUAA), Beijing 100191, China
e-mail: pengying1202@126.com

S. Sun
e-mail: jxcrssh@126.com

Y. Jia
e-mail: ymjia@buaa.edu.cn

C. Chen
The Beijing Institute of Control Engineering and National Key Laboratory of Science and Technology on Space Intelligent Control, Beijing 100191, China
e-mail: changqingchen@hotmail.com

© Springer-Verlag Berlin Heidelberg 2016
Y. Jia et al. (eds.), *Proceedings of the 2015 Chinese Intelligent Systems Conference*, Lecture Notes in Electrical Engineering 359,
DOI 10.1007/978-3-662-48386-2_28

In these systems, the stereo vision system is a powerful and versatile system to measure the 6DOF pose information [2]. Based on the positions of the non-collinear feature points on left and right images [3] [4], the pose of the moving object in inertial coordinate can be obtained by 3D reconstruction method [5, 6, 7, 8]. As the measurement noise existing, it is difficult to obtain the accurate motion state by stereo vision measurement. Although, based on dynamic model of the object, the Kalman Filter is effective to estimate the noise, it is unable to estimate the motion state normally when the dynamic model is unknown. To overcome this disadvantage, the IMM estimator, which has been shown to be a cost-effective motion state estimation approach [9], is employed for the motion state estimation of the particle object when the dynamic model is unknown. With several behaviour modes, such as CV, CA, CT, which can be switched from one to another, the IMM method has the ability to estimate the motion state of a particle object. Recently, the IMM estimator is widely applied in target tracking [10] [11]. Tan [12] develops an algorithm of target tracking based on IMM-Singer model, which improved the accuracy. Xiao [13] proposes an adaptive IMM tracking algorithm of based on coordinated turn model.

In this paper, the position in inertial coordinate of the feature points is measured by stereo vision approach. And the IMM method is employed to estimate the motion state of the feature points. Then, according to the model obtained by IMM method, the position, velocity, attitude angle, and angular velocity are all accurately estimated. At last we show the validity of the method in a moving object.

28.2 Background Knowledge

28.2.1 Stereo Vision-Based Measurement Method

As shown in Fig. 28.1, $O_l - x_l y_l z_l$ and $O_r - x_r y_r z_r$ are left and right camera coordinate systems respectively. The vectors $[x_l, y_l, z_l]^T$, $[x_r, y_r, z_r] \in R^3$ denote the position of the space point P, resolved in the coordinate $O_l - x_l y_l z_l$ and $O_r - x_r y_r z_r$.

According to the principle of perspective transformation, we can get

$$s_l \begin{bmatrix} X_l \\ Y_l \\ 1 \end{bmatrix} = \begin{bmatrix} f_l & 0 & 0 \\ 0 & f_l & 0 \\ 0 & 0 & 1 \end{bmatrix} \begin{bmatrix} x_l \\ y_l \\ z_l \end{bmatrix} \tag{28.1}$$

$$s_r \begin{bmatrix} X_r \\ Y_r \\ 1 \end{bmatrix} = \begin{bmatrix} f_r & 0 & 0 \\ 0 & f_r & 0 \\ 0 & 0 & 1 \end{bmatrix} \begin{bmatrix} x_r \\ y_r \\ z_r \end{bmatrix} \tag{28.2}$$

where vectors $[X_l, Y_l]$, $[X_r, Y_r]$ denote the position of the space point P, resolved in the left and right image coordinate.

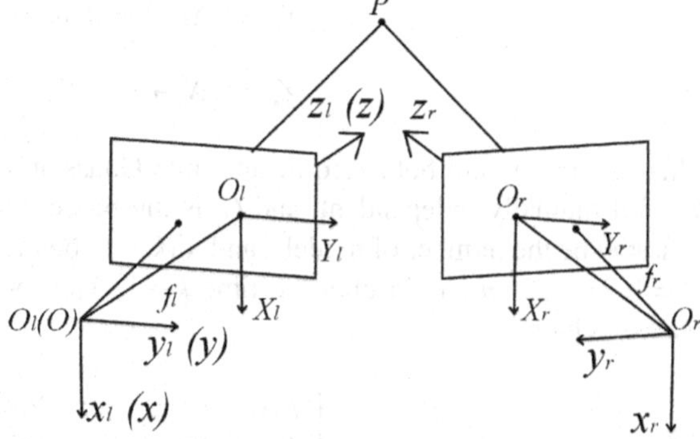

Fig. 28.1 Structure of the observing space points by double cameras

Let $M_{lr} = [R|T]$ be space transformation matrix, where $T = [\,t_x \quad t_y \quad t_z\,]^T$ and $R = [\,r_1 \quad r_2 \quad r_3; \quad r_4 \quad r_5 \quad r_6; \quad r_7 \quad r_8 \quad r_9;\,]$. Then,

$$\begin{bmatrix} x_r \\ y_r \\ z_r \end{bmatrix} = M_{lr} \begin{bmatrix} x_l \\ y_l \\ z_l \\ 1 \end{bmatrix} = \begin{bmatrix} r_1 & r_2 & r_3 & t_x \\ r_4 & r_5 & r_6 & t_y \\ r_7 & r_8 & r_9 & t_z \end{bmatrix} \begin{bmatrix} x_l \\ y_l \\ z_l \\ 1 \end{bmatrix} \tag{28.3}$$

Assume that the world coordinate coincides with the left camera coordinate, which means that $[x, y, z] = [x_l, y_l, z_l]$, the 3-D world coordinate of the space point P can be rewritten as

$$\begin{aligned} x &= \frac{zX_l}{f_l} \\ y &= \frac{zY_l}{f_l} \\ z &= \frac{f_l(f_r t_x - X_r t_z)}{X_r(r_7X_l + r_8Y_l + f_l r_9) - f_r(r_1X_l + r_2Y_l + f_l r_3)} \\ &= \frac{f_l(f_r t_y - X_r t_z)}{Y_r(r_7X_l + r_8Y_l + f_l r_9) - f_r(r_4X_l + r_5Y_l + f_l r_6)} \end{aligned} \tag{28.4}$$

28.2.2 The IMM Algorithm

Consider a typical discrete-time state-space model of target tracking, the process equation and measurement equation are given by

$$X_k = F_j X_{k-1} + G_j \omega_k \tag{28.5}$$

$$Z_k = H_j X_k + v_k \tag{28.6}$$

where ω_k and v_k are both zero-mean white Guassian noises with variances Q_k and R_k, and mutually independent, and G_j is the process noise gain matrix.

Let m be the number of models, and $\mu_i(k-1)$ be the mode probability that model $i(i=1,2,3,\ldots,m)$ is in effect at time $k-1$. The mode transition governed by a Markov chain

$$P = \begin{bmatrix} p_{11} & p_{12} & \cdots & p_{1m} \\ p_{21} & p_{22} & \cdots & p_{2m} \\ \vdots & \vdots & \cdots & \vdots \\ p_{m1} & p_{m2} & \cdots & p_{mm} \end{bmatrix} \tag{28.7}$$

where p_{ij} is the Markov transition probability from mode i to mode $j(i,j=1,2,3,\ldots m)$.

Assume that the state estimations and their associated covariance are denoted respectively by $\hat{X}_i(k)$ and $\hat{P}_i(k)$ at time k conditioned on the ith mode. A typical IMM algorithm is expanded from [14] [15].

Firstly, the mixing estimations and the associated mixing covariance are computed by

$$\hat{X}_{0j}(k-1) = \sum_{i=1}^{m} \hat{X}_i(k-1)\mu_{ij}(k-1) \tag{28.8}$$

$$\begin{aligned} \hat{P}_{0j}(k-1) = \sum_{i=1}^{m} \mu_{ij}(k-1)\{\hat{P}_i(k-1) + [\hat{X}_i(k-1) - \hat{X}_{0i}(k-1)] \\ [\hat{X}_i(k-1) - \hat{X}_{0i}(k-1)]'\} \end{aligned} \tag{28.9}$$

where

$$\mu_{ij}(k-1) = \frac{p_{ij}\mu_i(k-1)}{\bar{c}_j} \tag{28.10}$$

$$\bar{c}_j = \sum_{i=1}^{m} p_{ij}\mu_i(k-1) \tag{28.11}$$

Then, according to Kalman Filter or other filtering methods, with $\hat{X}_{0j}(k-1)$ and $\hat{P}_{0j}(k-1)$ as input on filter matched the jth mode, the state estimation $\hat{X}_j(k)$, the covariance $\hat{P}_i(k)$ and the likelihood function $\Lambda_i(k)$ are all calculated.

Thirdly, we update the mode probabilities as follows

$$\mu_j(k) = \frac{\Lambda_j(k)\bar{c}_j}{C} \tag{28.12}$$

where

$$C = \sum_{i=1}^{m} \Lambda_i(k)\bar{c}_i \tag{28.13}$$

Finally, combine all the state estimations and covariance based on the mode probabilities $\mu_i(k)$ as following:

$$\hat{X}(k) = \sum_{i=1}^{m} \hat{X}_i(k)\mu_i(k) \tag{28.14}$$

28.3 Kinematics Models of Feature Points

As shown in Fig. 28.2, $O - x_d y_d z_d$ is the moving coordinate. With pitch angle α, roll angle β and yaw angle θ, the rotation matrix R and translation vector T are

$$R = \begin{bmatrix} 1 & 0 & 0 \\ 0 & cos\alpha & sin\alpha \\ 0 & -sin\alpha & cos\alpha \end{bmatrix} \begin{bmatrix} cos\beta & 0 & -sin\beta \\ 0 & 1 & 0 \\ sin\beta & 0 & cos\beta \end{bmatrix} \begin{bmatrix} cos\theta & sin\theta & 0 \\ -sin\theta & cos\theta & 0 \\ 0 & 0 & 1 \end{bmatrix}$$
$$= \begin{bmatrix} cos\beta cos\theta & cos\beta sin\theta & -sin\beta \\ sin\alpha sin\beta cos\theta - cos\alpha sin\theta & cos\alpha cos\theta + sin\alpha sin\beta sin\theta & sin\alpha cos\beta \\ cos\alpha sin\beta cos\theta + sin\alpha sin\theta & sin\alpha cos\theta + cos\alpha sin\beta sin\theta & cos\alpha cos\beta \end{bmatrix}$$
$$T = [x, y, z]^T$$

The vectors $P_{js} = [x_{js} \quad y_{js} \quad z_{js}]^T$, $P_j = [x_j \quad y_j \quad z_j]^T$, $j = 1, 2, \ldots, 5$ denote the initial coordinate and the new coordinate after moving of the point P_j. The connection of P_{js} and P_j can be written as

$$P_j = RP_{js} + T, \ j = 1, 2, \ldots, 5 \tag{28.15}$$

then,

$$P_i - P_j = R(P_{is} - P_{js}) \tag{28.16}$$

With (28.15) and (28.16), the 6-DOF pose of spacecraft can be calculated. Differentiate Eq. (28.16) to yield

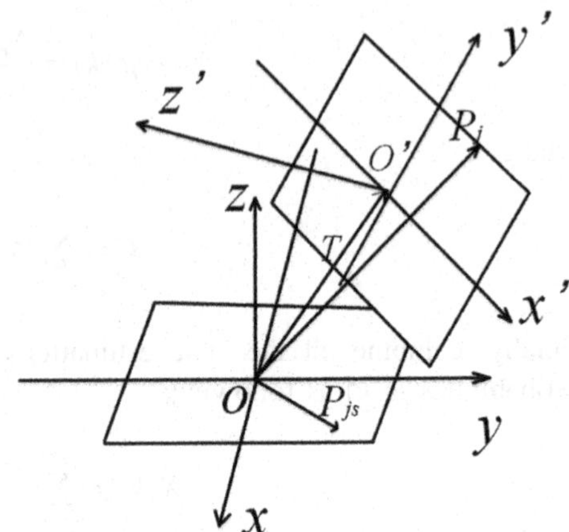

Fig. 28.2 Spacecraft's position and orientation

$$\dot{P}_i - \dot{P}_j = \dot{R}(P_{is} - P_{js}) \tag{28.17}$$

Through the equations obtained from (28.5), the unknown parameters of them (α, β, θ, $\dot{\alpha}$, $\dot{\beta}$, $\dot{\theta}$) can be solved on the condition that $\dot{P}_j = [\dot{x}_j \quad \dot{y}_j \quad \dot{z}_j]^T$, $j = 1, 2 \ldots, 5$ are known, which can be estimated by IMM method.

The matrix $[\dot{r}_{11} \quad \dot{r}_{12} \quad \dot{r}_{13}; \quad \dot{r}_{21} \quad \dot{r}_{22} \quad \dot{r}_{23}; \quad \dot{r}_{31} \quad \dot{r}_{32} \quad \dot{r}_{33};]$ denotes $\dot{R}$. Then, the angular velocities $\dot{\alpha}$, $\dot{\beta}$, $\dot{\theta}$ are

$$\begin{aligned} \dot{\beta} &= \frac{-\dot{r}_{13}}{cos\beta} \\ \dot{\theta} &= \frac{\dot{r}_{11} + \dot{\beta} sin\beta cos\theta}{cos\beta sin\theta} \\ \dot{\alpha} &= \frac{\dot{r}_{23} + \dot{\beta} sin\alpha sin\beta}{cos\alpha cos\beta} \end{aligned} \tag{28.18}$$

According to [16], $R = [e_1 \quad e_2 \quad e_3][o_1 \quad o_2 \quad o_3]^{-1}$, where $e_k = \frac{P_{i_k} - P_{j_k}}{||P_{i_k} - P_{j_k}||_2}$,

$o_k = \frac{P_{i_k s} - P_{j_k s}}{||P_{i_k s} - P_{j_k s}||_2}, k = 1, 2.$ $e_3 = e_1 \times e_2$, $o_3 = o_1 \times o_2$.

Thus,

$$\dot{R} = [\dot{e}_1 \quad \dot{e}_2 \quad \dot{e}_3][o_1 \quad o_2 \quad o_3]^{-1}$$

where

$$\dot{e}_k = \left(\frac{1}{||P_{i_k} - P_{j_k}||_2} - \frac{(P_{i_k} - P_{j_k})(\dot{P}_{i_k} - \dot{P}_{j_k})^T}{||P_{i_k} - P_{j_k}||_2} \right)(P_{i_k} - P_{j_k}), \ k = 1, 2.$$

$$\dot{e}_3 = \dot{e}_1 \times e_2 + e_1 \times \dot{e}_2$$

are easily obtained based on $\dot{\mathrm{R}}$, $\dot{\alpha}$, β, θ.

28.4 Simulations

Consider a state-space model of a moving object given by Eq. (28.7). The state vector is $X_k = [x_k \quad \dot{x}_k \quad \ddot{x}_k \quad y_k \quad \dot{y}_k \quad \ddot{y}_k \quad z_k \quad \dot{z}_k \quad \ddot{z}_k\,]^T$. The transition matrix is $F = diag\{F_1,\ F_1,\ F_1\}$. The process noise gain matrix is $G = diag\{G_1,\ G_1,\ G_1\}$.

For constant velocity model:

$$F_1 = \begin{bmatrix} 1 & T & 0 \\ 0 & 1 & T \\ 0 & 0 & 1 \end{bmatrix} \quad G_1 = \begin{bmatrix} 0.5T^2 \\ T \\ 0 \end{bmatrix}$$

For constant acceleration model:

$$F_1 = \begin{bmatrix} 1 & T & 0.5T^2 \\ 0 & 1 & T \\ 0 & 0 & 1 \end{bmatrix}, \quad G_1 = \begin{bmatrix} 0.5T^2 \\ T \\ 1 \end{bmatrix}$$

For 3D coordinate turn model:

$$F_1 = \begin{bmatrix} 1 & \frac{\sin(\omega T)}{\omega} & \frac{1-\cos(\omega T)}{\omega^2} \\ 0 & \cos(\omega T) & \frac{\sin(\omega T)}{\omega} \\ 0 & -\omega\sin(\omega T) & \cos(\omega T) \end{bmatrix}, \quad G_1 = \begin{bmatrix} 0.5T^2 \\ T \\ 1 \end{bmatrix}$$

where ω is the angular rate. In this paper, the model set contains 2 3D coordinate turn models with different angular rates.

Figure 28.3 shows the trajectory of the 3D maneuvering target, which starts from $(30, 30, 30\,\text{km})$. The concrete motion situation is showed as Table 28.1.

Assume that the standard deviation of the measured data, calculated by stereo vision-based measurement system, is 400 m on each coordinate orientation respectively. The mode transition matrix is

$$P = \begin{bmatrix} 0.97 & 0.01 & 0.01 & 0.01 \\ 0.01 & 0.97 & 0.01 & 0.01 \\ 0.01 & 0.01 & 0.97 & 0.01 \\ 0.01 & 0.01 & 0.01 & 0.97 \end{bmatrix}$$

Figure 28.4 shows the root mean square error (RMSE) in the estimation and the RMSE in the measurement of position for $x-y-z$. As shown in Fig. 28.3, the

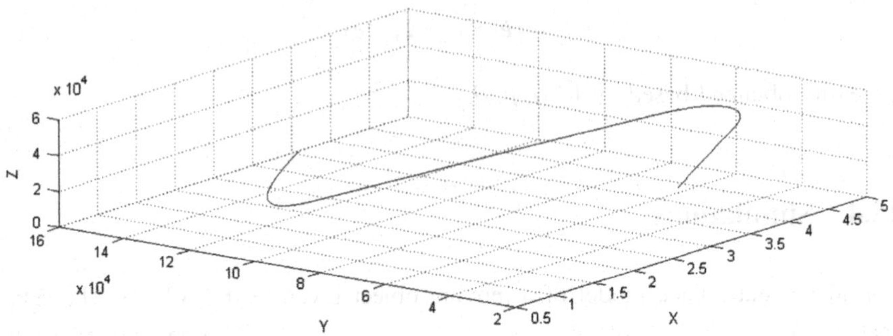

Fig. 28.3 3D maneuvering target trajectory

Table 28.1 The target movement pattern

Time	0–50 s	50–100 s	100–200 s	200–250 s
Motion situation	Uniform linear	Uniform circular	Uniform linear	Uniform acceleration

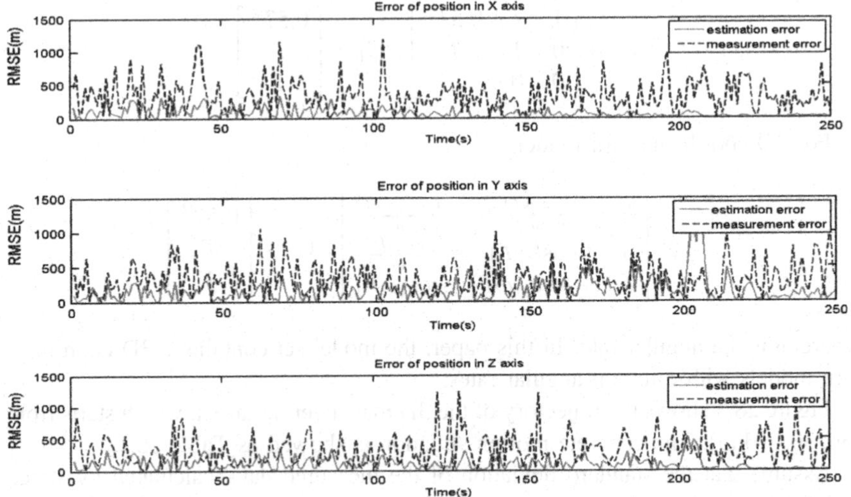

Fig. 28.4 Estimation error of position

accuracy of position\increased by IMM algorithm. At the target's maneuvering time (50, 100, 200 s), the IMM estimator maintains a fairly accuracy.

As shown in Fig. 28.5, the RMSE percentage of velocity estimation is basically within 2 %. The larger fluctuation in the error of velocity estimation appeared at the target's maneuvering time, especially at 200 s. However, the IMM algorithm has capability to make the velocity estimation accurate after several times of iteration.

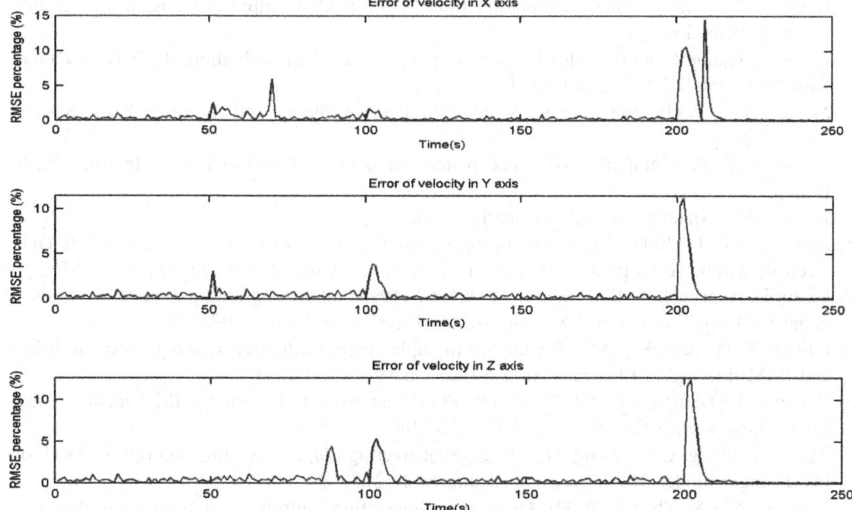

Fig. 28.5 Estimation error of velocity

The simulation results show that the high precision estimation of position and velocity can be obtained by IMM algorithm. According to Sect. 28.3, the angular velocity is acquired.

28.5 Conclusion

In this paper, a stereo vision-based state estimator using the IMM method which contains three different models is presented to estimate the 6 DOF pose and velocity of the simulated spacecraft. From the simulating results, the proposed algorithm has ability to obtain the translational velocity and angular velocity and improves the pose estimation accuracy significantly.

Acknowledgments This work was supported by the National Basic Research Program of China (973 Program: 2012CB821200, 2012CB821201) and the NSFC (61134005, 61221061, 61327807, 61304232).

References

1. Soatto S, Frezza R, Perona P (1996) Motion estimation via dynamic vision. IEEE Trans Autom Control 4(3):61–72
2. Ma T, Wei C (2004) An overview of the space rendezvous and docking. Aerosp China 7:33–34

3. Bouguet JY (2004) Camera calibration toolbox for matlab. http://www.vision.caltech.edu./bouguetj/calib_doc/
4. Zhang Z, Zhang Z (2000) A flexible new technique for camera calibration. IEEE Trans Pattern Anal Mach Intell 22(11):1330–1334
5. Ma S, Zhang Z (1998) Computer Vision: computation theory and algorithm basis. Science Press, Beijing
6. Zheng N (1998) Computer vision and pattern recognition. National Defend Industry Press, Beijing
7. Jia Y (2000) Machine vision. Science Press, Beijing
8. Zuo A, Wu J (2000) Measurement of position and orientat ion of a parallel 6–DOF electrohydraulic servo platform based on stereo vision. China Mech Eng 11(7):814–816
9. Mazor E, Averbuch A, Bar-Shalom Y, Dayan J (1998) Interacting multiple model methods in target tracking: a survey. IEEE Trans Aerosp Electron Syst 34(1):103–123
10. Puranik S, Tugnait JK (2007) Tracking of multiple maneuvering targets using multiscan JPDA and IMM filtering. IEEE Trans Aerosp Electron Syst 43(1):23–35
11. Rapoport I, Oshman Y (2007) Efficient fault tolerant estimation using the IMM methodology. IEEE Trans Aerosp Electron Syst 43(2):492–508
12. Tan SC, Wang GH, Wang N (2012) Maneuvering target tracking algorithm based on IMM-Singer model. Huoli yu Zhihui Kongzhi 37(2):32–34
13. Xiao W, Nie X, Zhen J (2009) An adaptive interacting multiple models tracking algorithm based on coordinated tum model. Command Control Simul 31(2):36–41
14. Ng G (2003) Intelligent systems: fusion, tracking and control CSI, control and signal image processing series. Philadelphia Research Studies Press, USA
15. Watson GA, Blair WD (1992) IMM algorithm for tracking targets that maneuver through coordinated turns. In: Aerospace Sensing, International society for optics and photonics pp. 236–247
16. Hao Y, Zhu F, Ou J (2001) 3D visual methods for object pose measurement. SPIE-Int Soc Opt Eng

Chapter 29
A Carrier Tracking Algorithm of Kalman Filter Based on Combined Maneuvering Target Model

Zhulin Xiong, Shijie Ren, Celun Liu and Jianping An

Abstract Since the traditional current statistical Kalman (CS-Kalman) filter doesn't perform well enough when used in high-dynamic carrier tracking, an carrier tracking algorithm of Kalman filter based on combined maneuvering target model composed of high dynamic CS-Kalman filter and steady-state self-adaptive CS-Kalman filter is presented. The proposed algorithm achieves stable and accurate carrier synchronization by adjusting the CS-Kalman filter type and parameter corresponding to the dynamic condition in real time. Simulation results show that compared with the traditional CS-Kalman algorithm, the proposed algorithm is more realistic, practical valuable and adaptable in high dynamic environment.

Keywords High dynamic · Carrier tracking · Current statistical kalman (CS-Kalman) filter · Maneuvering target model

29.1 Introduction

In the field of aerospace, all kinds of aircraft such as airplane and satellite have high dynamic character. Their communication signal contains high Doppler frequency shift and Doppler rate-of-change far beyond the tracking capabilities of traditional carrier synchronization loop. Therefore, the research on the stable synchronization algorithm in high dynamic environments has high theoretical and practical significance.

Z. Xiong · S. Ren · C. Liu (✉) · J. An
School of Information and Electronics, Beijing Institute of Technology,
100081 Beijing, China
e-mail: liucelun@bit.edu.cn

S. Ren
Institute of Physics Science and Information Engineering, Liaocheng University,
252000 Liaocheng, China

© Springer-Verlag Berlin Heidelberg 2016

Y. Jia et al. (eds.), *Proceedings of the 2015 Chinese Intelligent Systems Conference*, Lecture Notes in Electrical Engineering 359,
DOI 10.1007/978-3-662-48386-2_29

A lot of in-depth relevant research has been carried out in recent years. Many early algorithms [1], [2] are open loop and based on the maximum likelihood criterion. These algorithms reach high estimation precision at the cost of exorbitant calculation. The non-stationary parameter estimation algorithms based on the time and frequency domain orthogonal basis transform [3] or the Markov Monte-Carlo method [4] also have the disadvantage of high computational complexity. The traditional carrier tracking loops [5] such as phase-locked loop and frequency-locked loop have low complexity and high precision, but they can't work reliably in high dynamic environment because of the limited noise bandwidth and loop order. Some existing researches try to improve the tracking loops in three ways: raising loop order, reducing signal dynamic and utilizing nonlinear filtering technique. High order loop filter [6] can track high order Doppler component, but its multiple poles lead to poor stability. There are two kinds of methods to reduce signal dynamic: one is forecasting dynamic roughly by inertial navigation [7] and the other is reducing the order of Doppler component by self-conjugate difference [8]. Inertial navigation system can effectively reduce the complexity of the receiving terminal, but its application is seriously restricted by the high cost. The algorithm of self-conjugate difference performs well in high signal-to-noise ratio (SNR) environment, but it would bring large performance deterioration when SNR is low. The extended Kalman filter [9] (EKF) based on local linearization of Kalman filter has a simple construction and good performance in Gaussian and mild nonlinear environment. However, the given parameter recursive model of the classical EKF may not match the actual dynamic of communication signal and that will lead to a steady-state tracking error and low probability of lock in present of the high order frequency components. Reference [10] proposed a carrier tracking algorithm based on the CS-Kalman filter which combines current statistical model and Kalman filter. Compared with EKF, the CS-Kalman filter has a much wider tracking dynamic range and similar computational complexity. But in the low dynamic environment, CS-Kalman filter will produce additional noise due to the model mismatch. To solve the above problem, a carrier tracking algorithm of Kalman filter based on combined maneuvering target mode composed of high dynamic CS-Kalman filter and steady-state self-adaptive CS-Kalman filter is proposed in this paper. By choosing the appropriate CS-Kalman filter corresponding to the dynamic condition, the proposed algorithm achieves stable and accurate tracking of maneuvering target in a variety of dynamic conditions.

29.2 CS Model and Its Limitations

According to the CS model, when the target is maneuvering, the acceleration value of the next moment must be in the neighborhood of "current" acceleration. The CS model is a non-zero-mean time-correlated maneuvering acceleration model whose "current" probability follows the modified Rayleigh distribution. The stochastic acceleration belongs to the first order time-dependent process in the time domain [11]

$$\ddot{x}(t) = \bar{a}(t) + a(t) \tag{29.1}$$

$$\dot{a}(t) = -\alpha a(t) + \omega(t) \tag{29.2}$$

where $\bar{a}(t)$ is the "current" mean of the maneuvering acceleration, α is the reciprocal of maneuvering time constant named maneuvering frequency, $\omega(t)$ is zero mean Gaussian noise with a variance $\sigma_\omega^2 = 2\alpha\sigma_a^2$, σ_a^2 is the variance of the target acceleration.

Assuming $a_1(t) = \bar{a}(t) + a(t)$, rewrite Eqs. (29.1) and (29.2) as follows

$$\ddot{x}(t) = a_1(t) \tag{29.3}$$

$$\dot{a}_1(t) = -\alpha a_1(t) + \alpha\bar{a}(t) + \omega(t). \tag{29.4}$$

From discretizing Eqs. (29.3) and (29.4) we get the equation of state

$$\begin{bmatrix} \dot{x}(t) \\ \ddot{x}(t) \\ \dddot{x}(t) \end{bmatrix} = \begin{bmatrix} 0 & 1 & 0 \\ 0 & 0 & 1 \\ 0 & 0 & -\alpha \end{bmatrix} \begin{bmatrix} x(t) \\ \dot{x}(t) \\ \ddot{x}(t) \end{bmatrix} + \begin{bmatrix} 0 \\ 0 \\ \alpha \end{bmatrix} \bar{a}(t) + \begin{bmatrix} 0 \\ 0 \\ 1 \end{bmatrix} \omega(t) \tag{29.5}$$

where $[x(t)\ \dot{x}(t)\ \ddot{x}(t)]^T$ is the state vector which represents the displacement, velocity and acceleration of target. Equation (29.5) shows that the selection of maneuvering frequency α and acceleration variance σ_a^2 is very important for the tracking performance. The dynamic of high-speed aircraft such as satellite navigation is very complex. In order to guarantee the robustness of the tracking algorithm in such a dynamic environment, the tracking accuracy has to be sacrificed. The satellite navigation signal of the airplane in the take-off stage [12] is taken as an example, whose dynamic characteristics are shown in Fig. 29.1.

In Fig. 29.1, 0 ~ 100 s is the preparation stage (stage I) where the airplane is static and the dynamic characteristics of navigation signals are entirely determined by the satellite; 100 ~ 186 s is the take-off stage (stage II) where the aircraft pitches up by a sharp acceleration and the dynamic is mainly determined by the airplane; 186 ~ 400 s is the stable flying stage (stage III) where the airplane flies into the predetermined height and the dynamic is codetermined by the airplane and the

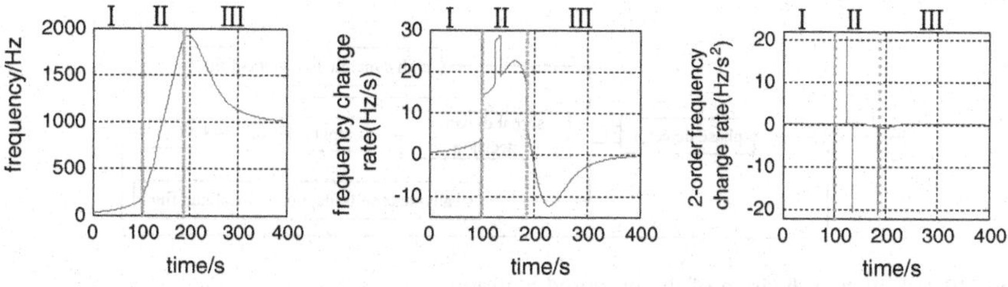

Fig. 29.1 Satellite navigation signal dynamic of the airplane in the take-off stage

satellite. In stage I and stage III the 2-order frequency changing rate of navigation signal is approximately zero and the difficulty of tracking signal is relatively low, while in stage II the 2-order frequency changing rate as well as the tracking difficulty increases greatly due to the sharp acceleration. In order to maintain the stability of CS-Kalman filter in the worst case, maneuvering frequency α and acceleration variance σ_a^2 should match the maximum dynamic of stage II which will greatly reduce the estimation accuracy of another stage.

29.3 Carrier Tracking Method

Assuming that the channel is Gaussian, the input signal of carrier tracking loop with white Gaussian noise and Doppler frequency shift is given by

$$s_{in}(k) = e^{j\left[\pi\left(f_1'' + f_2''\right)k^3T^3/-0pt3 + \pi\Delta f'k^2T^2 + 2\pi\Delta fkT + \theta_0\right]} + n(k) \tag{29.6}$$

where f_1'' and f_2'' are 2-order frequency changing rates determined by the satellite and the maneuvering target, T is the time interval of sampling, $\Delta f'$, Δf and θ_0 are the initial frequency changing rate, frequency shift and carrier phase, $n(k)$ is the complex zero mean Gaussian noise whose variance of real and imaginary component is $\sigma^2/-0pt2$.

Structure diagram of the proposed algorithm is shown in Fig. 29.2. Carrier tracking module receives the I, Q base band signal and utilizes the phase detector to calculate the real-time phase difference between the received signal and the local numerically controlled oscillator (NCO). The phase difference is taken as the input parameter of the high dynamic CS-Kalman filter and the steady-state self-adaptive CS-Kalman filter. These two filters track the input parameter respectively and the algorithm chooses the appropriate filter output according to the estimation result of the high dynamic CS-Kalman filter.

Assuming the state vector of CS-Kalman filter $X(k) = \left[\theta(k)\, f(k)\, f'(k)\right]^T$. From Eq. (29.5), the time update equation and the observation equation

$$X(k) = \boldsymbol{\Phi}(k|k-1)X(k-1) + \boldsymbol{\Lambda}(k-1)f'(k-1) + Q(k-1) \tag{29.7}$$

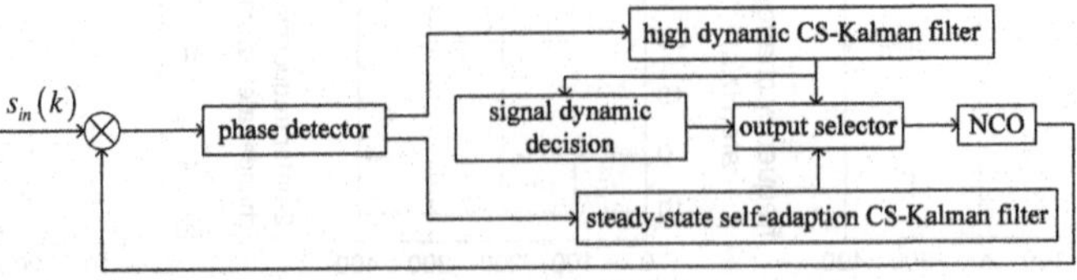

Fig. 29.2 Structure diagram of the proposed algorithm

$$z(k) = H(k)X(k) + V(k). \tag{29.8}$$

The state transfer matrix of Eq. (29.7) is

$$\boldsymbol{\Phi}(k|k-1) = e^{AT} = \begin{bmatrix} 1 & T & (-1+\alpha T+e^{-\alpha T})/-0pt\alpha^2 \\ 0 & 1 & (1-e^{-\alpha T})/-0pt\alpha \\ 0 & 0 & e^{-\alpha T} \end{bmatrix} \tag{29.9}$$

where $X(k) = \begin{bmatrix} 0 & 1 & 0 \\ 0 & 0 & 1 \\ 0 & 0 & -\alpha \end{bmatrix}$.

The control matrix of system is

$$\boldsymbol{\Lambda}(k-1) = \int_0^T eA(t)\begin{bmatrix} 0 \\ 0 \\ \alpha \end{bmatrix} dt = \begin{bmatrix} -T/-0pt\alpha + 0.5T^2 + (1-e^{-\alpha T})/-0pt\alpha^2 \\ T-(1-e^{-\alpha T})/-0pt\alpha \\ (1-e^{-\alpha T}) \end{bmatrix}. \tag{29.10}$$

The covariance matrix of the zero mean process noise is [9]

$$Q = qT\begin{bmatrix} T^4/-0pt20 & T^3/-0pt8 & T^2/-0pt6 \\ T^3/-0pt8 & T^2/-0pt3 & T/-0pt2 \\ T^2/-0pt6 & T/-0pt2 & 1 \end{bmatrix} \tag{29.11}$$

where q represents the variance of 2-order frequency changing rate. The observation matrix **H**(k) = [1 0 0] and **V**(k) is the zero mean Gaussian observation noise with variance $R(k) \approx \beta^2\theta^2(k) + 2\alpha\sigma_{f'}^2$ where β is the relative error coefficient.

The filtering process of CS-Kalman filter is as follows:

The time update equations are

$$\widehat{X}(k|k-1) = \boldsymbol{\Phi}(k|k-1)\widehat{X}(k-1) + \boldsymbol{\Lambda}(k-1)f'(k-1) \tag{29.12}$$

$$P(k|k-1) = \boldsymbol{\Phi}(k|k-1)P(k-1)\boldsymbol{\Phi}^T(k|k-1) + Q(k-1). \tag{29.13}$$

The state update equations are

$$K = P(k|k-1)H^T\left(HP(k|k-1)H^T + R\right)^{-1} \tag{29.14}$$

$$\widehat{X}(k) = \widehat{X}(k|k-1) + K\left(\arg(s_{in}(k)) - H\widehat{X}(k|k-1)\right) \tag{29.15}$$

$$P(k) = [I-KH]P(k|k-1). \tag{29.16}$$

There are two adjustable parameters in CS-Kalman filter: maneuvering frequency α and the variance of 2-order frequency changing rate q. In general, the estimation error is inversely proportional to α when T < 0.1 [11] which can be easily satisfied in the navigation system. Therefore, the dynamic tracking performance and tracking accuracy of CS-Kalman filter is mainly determined by parameter q. The selection of q is the major difference of high dynamic CS-Kalman filter and steady-state self-adaptive CS-Kalman filter.

The high dynamic CS-Kalman filter is designed for reliable tracking of maneuvering target in the high-speed motion. From Eq. (29.7) we know that q is determined by the maximum 2-order frequency disturbance increment of the target

$$q=\max\left(\left|\Delta f_1^{''}+\Delta f_2^{''}\right|^2\right). \tag{29.17}$$

The steady-state self-adaptive CS-Kalman filter is designed for the accurate tracking of maneuvering target in the steady motion, so the parameter q need to be adjusted according to the actual state of target. The 2-order frequency disturbance increment at time k can be approximated by the difference between the estimation of frequency changing rate at time k and the prediction of frequency changing rate at time k−1, which is

$$\Delta f^{''}(k)\approx\left[\widehat{f}^{'}(k)-\widehat{f}^{'}(k|k-1)\right]/T. \tag{29.18}$$

q is proportional to the square of disturbance increment

$$q(k)=\left|\widehat{f}^{'}(k)-\widehat{f}^{'}(k|k-1)\right|^2/T^2. \tag{29.19}$$

From Eq. (29.19) we know that the estimation $\widehat{f}^{'}(k)$ is close to the prediction $\widehat{f}^{'}(k|k-1)$ in low dynamic, the variance $q(k)$ as well as the Kalman filter gain is very small and the steady-state self-adaptive CS-Kalman filter has a slow but accurate tracking of maneuvering target. When the dynamic is high, $\widehat{f}^{'}(k)$ has a obviously deviation from $\widehat{f}^{'}(k|k-1)$, and the Kalman filter gain becomes much larger which makes the target tracking faster and inaccurate.

In order to realize the automatic switch of the high dynamic and the steady state tracking model, the threshold of 2-order s $f^{''}(k)$ should be set as follows

$$f_{th}^{''}=\max\left|f_1^{''}\right|. \tag{29.20}$$

Compared with the steady-state self-adaptive CS-Kalman filter, the high dynamic CS-Kalman filter has a better dynamic tracking ability. The estimation of $f^{''}(k)$ is

$$\widehat{f}''(k) = \left| \widehat{f}'(k) - \widehat{f}'(k-1) \right| \tag{29.21}$$

where $\widehat{f}'(k|k-1)$ and $\widehat{f}'(k-1)$ are the estimation values of frequency changing rate at time k and k-1 in high dynamic CS-Kalman filter. The appropriate filter output is selected by comparing $\widehat{f}''(k)$ and f''_{th}, if $\widehat{f}''(k) > f''_{th}$ the output of high dynamic CS-Kalman filter is selected as the estimation of target dynamic, otherwise the output of the steady-state self-adaptive CS-Kalman filter is selected.

The carrier NCO is adjusted by the estimation of phase at time k to compensate the input signal $s_{in}(k+1)$, which is

$$\theta_{nco}(k+1) = \theta_{nco}(k) + \widehat{\theta}(k) + \widehat{f}(k)T + \widehat{f}'(k)T^2/2. \tag{29.22}$$

29.4 Simulation Results

In order to verify the effectiveness of the algorithm, we choose the satellite navigation signal shown in Fig. 29.1 for simulation. Various performance indicators including probability of losing lock and root-mean-square error (RMSE) of phase estimation are simulated and compared with the traditional algorithm based on CS-Kalman filter [10]. The key parameters are as follows: $T = 0.01\text{s}$, $\alpha = 1$, $\beta = 0.01$, $q = 400\text{Hz}^2/-0pts^4$, $f''_{th} = 1\text{Hz}/-0pts^2$. The simulation results are shown in Figs. 29.3 and 29.4.

Figures 29.3 and 29.4 show that compared with traditional algorithm, the proposed algorithm has a lower probability of losing lock as well as RMSE of phase estimation. In the traditional algorithm, severe floor effect of phase estimation appears when the SNR is greater than 1.5 dB due to the fixed parameters of

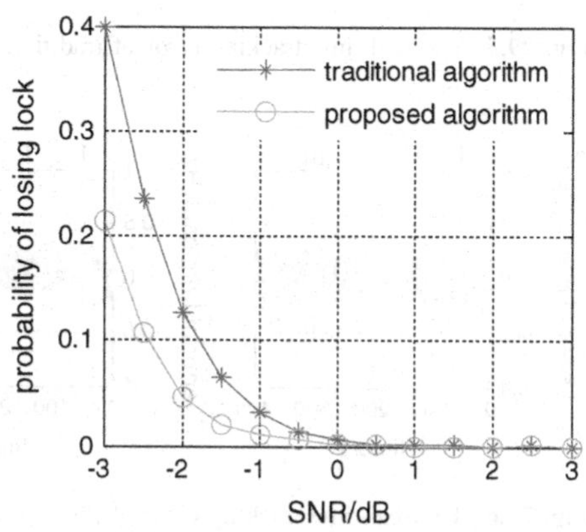

Fig. 29.3 Probability of losing lock under different SNRs

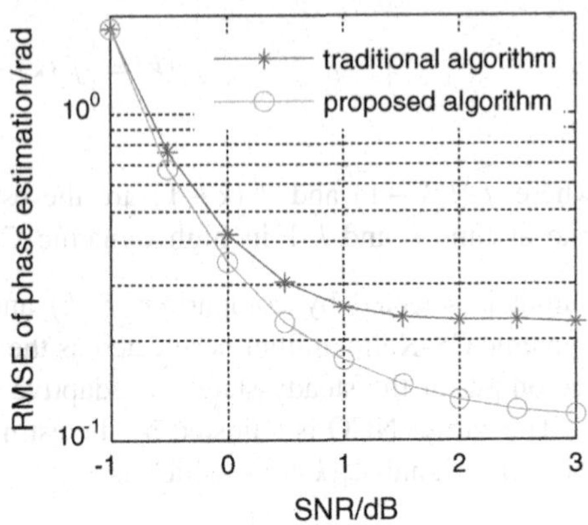

Fig. 29.4 RMSE of phase estimation under different SNRs

CS-Kalman filter. The proposed algorithm makes use of the steady-state self-adaptive CS-Kalman filter to match the dynamic characteristics of target, thus the floor effect can be effectively restrained.

The real-time tracking error of traditional algorithm and proposed algorithm at SNR = 10 dB is shown in Figs. 29.5 and 29.6. In stage II, while the tracking error suddenly increasing both algorithms respond quickly and reduce the tracking error within a short period of time. In stage I and III, the tracking error of proposed algorithm is much lower than that of traditional algorithm.

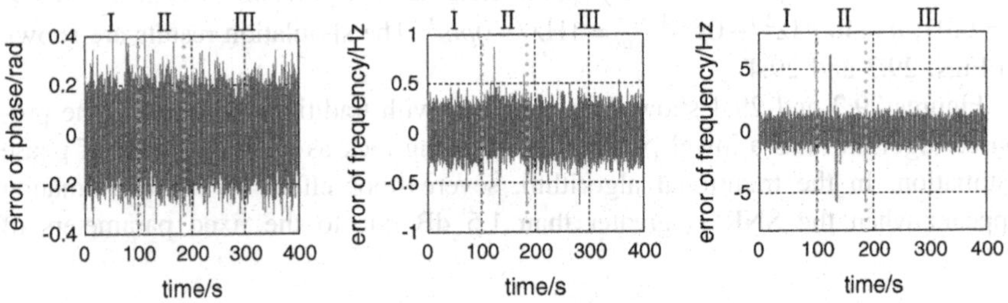

Fig. 29.5 The real-time tracking error of traditional algorithm when the SNR is 10 dB

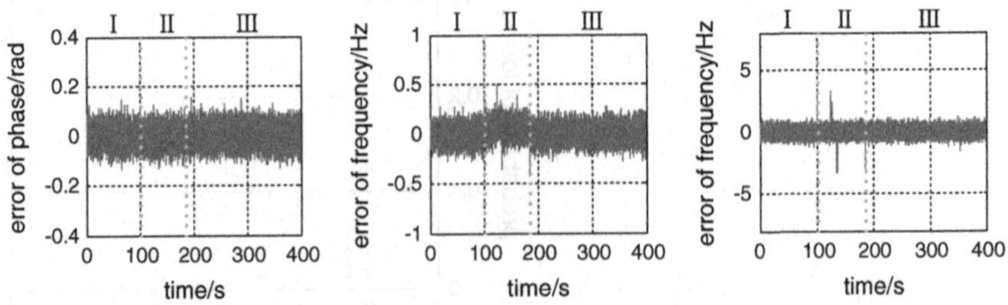

Fig. 29.6 The real-time tracking error of proposed algorithm when the SNR is 10 dB

In summary, the proposed algorithm is robust enough to realize stable and accurate carrier tracking in high dynamic environment.

29.5 Conclusion

A carrier tracking algorithm based on combined maneuvering target model Kalman filter is proposed for the accurate tracking of carrier in high dynamic environment. It utilizes the high dynamic CS-Kalman filter to ensure the dynamic characteristics of tracking, and takes advantage of steady-state self-adaptive CS-Kalman filter to improve the steady-state estimation accuracy. The simulation results show that the proposed algorithm has strong robustness, high precision and wide range of applications. Compared with the traditional algorithm based on CS-Kalman filter, the proposed algorithm has much better tracking performance and higher practical value.

References

1. Vilnrotter VA, Hinedi S, Kumar R (1989) Frequency estimation techniques for high dynamic trajectories[J]. IEEE Trans Aerosp Electron Syst 25(4):559–577
2. Jong-Hoon W, Pany T, Eissfeller B (2012) Iterative maximum likelihood estimators for high-dynamic GNSS signal tracking[J]. IEEE Trans Aerosp Electron Syst 48(4):2875–2893
3. Millioz F, Davies M (2012) Sparse detection in the chirplet transform: application to FMCW radar signals[J]. IEEE Trans Signal Process 60(6):2800–2813
4. Dhanoa JS, Hughes EJ, Ormondroyd RF (2003) Simultaneous detection and parameter estimation of multiple linear chirps[C]. In: Proceedings of international conference on acoustic, speech, and signal processing, Hong Kong, pp. 129–132
5. Ara P (1999) On phase-locked loops and Kalman filters[J]. IEEE Trans Commun 47 (5):670–672
6. Shu H, Simon EP, Ros L (2013) Third-order kalman filter: tuning and steady-state performance[J]. IEEE Signal Process Lett 20(11):1082–1085
7. Li CJ, Li XC (2012) Performance of INS-aided tracking loop for GPS high dynamic receiver [C]. In: International conference on computer science and network technology, Harbin, pp. 757–761
8. Gazor S, Rabiei AM, Pasupathy S (2002) Synchronized per survivor MLSD receiver using a differential Kalman filter[J]. IEEE Trans Commun 50(3):364–368
9. Shademan A, Janabi-Sharifi F (2005) Sensitivity analysis of EKF and iterated EKF pose estimation for position-based visual serving[C]. In: Proceedings of IEEE conference on control applications, pp. 755–760
10. Cui SQ, An JP, Wang AH (2014) An Kalman filter used for carrier tracking based on matched maneuvering target models[J]. Syst Eng Electron 36(2):376–381
11. Zhou H, Kumar KSP (1984) A "current" statistical model and adaptive algorithm for estimating maneuvering targets[J]. AIAA J Guidance 7(5):596–602
12. Lin KX, Cen GP, Li L et al (2012) Simulation and analysis for airplanes performance of takeoff and landing[J]. J Air Force Eng Univ 13(4):21–25

Chapter 30
Smooth Time-Varying Formation Control of Multiple Nonholonomic Agents

Chenghui Yang, Wenjing Xie, Chen Lei and Baoli Ma

Abstract In this paper, we investigate the position/orientation formation problem of multiple nonholonomic agents. Coordinate transformations are first presented to obtain the matrix form of formation error model, then a distributed smooth time-varying control law is designed based on a Lyapunov-like function. We prove that the closed-formation-system is globally asymptotically stable by Barbalat's lemma if the communication graph is undirected, time-invariant and connected. Simulation results verify the effectiveness of the proposed control scheme.

Keywords Asymptotic formation · Cooperative control · Nonholonomic agents · Distributed control

30.1 Introduction

Distributed control of dynamic multi-agents has catched many researchers' eyes in recent years, including consensus, rendezvous, flocking, cooperative control, and formation control. It has been applied both in networks of classical linear systems (e.g. integrator systems) and in networks of complicated nonlinear systems (e.g. nonholonomic systems), which may have communication delays and provide switching topologies.

The problem of formation control is one of distributed cooperative control objectives for networks of dynamic agents, that is, designing a distributed algorithm

---

C. Yang · C. Lei · B. Ma
The Seventh Research Division, School of Automation Science and Electrical Engineering, Beihang University, 100191 Beijing, China
e-mail: sandy198907@126.com

W. Xie (✉)
Faculty of Computer and Information Science, Southwest University, 400715 Chongqing, China
e-mail: x_wenjing@aliyun.com

© Springer-Verlag Berlin Heidelberg 2016

Y. Jia et al. (eds.), *Proceedings of the 2015 Chinese Intelligent Systems Conference*, Lecture Notes in Electrical Engineering 359,
DOI 10.1007/978-3-662-48386-2_30

such that all the agents converge to specific relative positions and maintain the patterns. Relatively, protocols of consensus, agreement or rendezvous are all special cases of formation problems. It has been extensively studied and applied in multiple agent systems with simple linear models [1–3], networks of continuous time first-order integrator agents [4–6], discrete-time first-order integrator agents [7–9], double-integrator agents [10], high-order linear agents [11] as well as other kinds of linear agents [12].

Besides the previous study of multi-agent with linear models, there are also some researches with emphasis on nonlinear multi-agents, especially the nonholonomic agents, due to the obvious fact that the methods proposed for the formation control problem of multiple general linear agents are not applicable for multiple nonholonomic agents [13, 14]. For the networked nonholonomic mobile robots, the formation problem is investigated in [15–23]. However, the formation scheme in [15] cannot be used for arbitrary formation, the ones in [16, 18, 22, 23] require the velocity nonzero, the ones in [16, 17, 19, 21] have singularities because of the vanishing denominators, and the ones in [18, 20] are both discontinuous and hence cannot be extended to obtain the dynamic control law by backstepping method.

Based on the statements above, the formation problem is still an open one. Therefore, this paper addresses the position/orientation formation control problem of multiple nonholonomic agents. Firstly, the matrix form for the multiple nonho-lonomic agents is derived, then a time-varying distributed control law is designed by the Lyapunov method and Barbalat's lemma. The stability analysis shows that the formation errors are globally asymptotically convergent to zero provided that the communication graph is connected, time-invariant and undirected. Finally, the simulation examples confirm the effectiveness of the proposed control strategy.

The rest of the paper is organized as follows. Section 30.2 describes the formation control problem, and introduces some basic concepts. Section 30.3 presents the smooth time-varying position/orientation formation protocol. Then, in Sect. 30.4, the simulation examples verify the effectiveness of the proposed algorithm, and Sect. 30.5 concludes this work finally.

30.2 Problem Formulation

In this section, we firstly review the graph tools and some relative basic concepts in graph and matrix theories. Let $G=(\mathcal{V},\mathcal{E},\mathcal{A})$ be an undirected graph of order n with the set of nodes $\mathcal{V}=(\nu_1,\nu_2,\ldots,\nu_n)$, set of edges $\mathcal{E}\subseteq\mathcal{V}\times\mathcal{V}$, and a adjacency matrix $\mathcal{A}=[a_{ij}]$. The node indices belong to a finite index set $\mathcal{L}=\{1,2,\cdots,n\}$, i.e., the order of the graph G is finite. An edge of G is denoted by $e_{ij}=(\nu_i,\nu_j)$. The adjacency matrix is the $n\times n$ matrix given by $a_{ij}=1\Leftrightarrow e_{ij}\in\mathcal{E};\quad a_{ij}=0\Leftrightarrow e_{ij}\notin\mathcal{E};\quad a_{ii}=0$.

The set of neighbors of node ν_i is denoted by $N_i = \{\nu_j \in \mathcal{V} | e_{ij} \in \mathcal{E}\}$. For undirected graph, it is obvious that $e_{ij} \in \mathcal{E} \Leftrightarrow e_{ji} \in \mathcal{E}$. A *path* from i to j is a sequence of distinct nodes starting from node ν_i and ending with node ν_j. An undirected graph is said to be *connected* if there is a path between any two nodes, otherwise, *disconnected*. For the undirected graph G, the degree of note i is equal to the number of its neighboring vertices, and the *degree matrix* of G is an $n \times n$ diagonal matrix $\Delta = \mathrm{diag}\{d_1, d_2, \ldots, d_n\}$, $d_i = \sum_{j=1}^{n} a_{ij}$. The *Laplacian* of G is $L = \Delta - \mathcal{A}$ which is symmetric. From the definition of L, we know that the sum of each row and column in L are both zero. Moreover, it can be shown that zero is a simple eigenvalue of L with the corresponding eigenvector $\mathbf{1} = [1, 1, \cdots, 1]^{\mathrm{T}} \in R^{n \times 1}$.

The considered agents are the classic unicycle-like nonholonomic robotic systems. In this paper, the kinematic model of the robot indexed by $i(i \in \mathcal{L})$ is represented by

$$x_i = v_i \cos\theta_i\ ,\ \ y_i = v_i \sin\theta_i\ ,\ \ \dot{\theta}_i = \omega_i\ , \tag{30.1}$$

where (x_i, y_i) is the position coordinates of robot i in the earth frame, θ_i the orientation of robot i in the body fixed frame, and (v_i, ω_i) the linear and angular velocity control inputs of robot i, respectively. Note that the robots described by (30.1) are nonholonomic due to the constraint on their velocities

$$x_i \sin\theta_i - y_i \cos\theta_i = 0, \tag{30.2}$$

meaning they cannot move side-away.

In our paper, each robotic agent locates in a communication network with undirected information flow, and naturally is considered as a node ν_i with the position and the orientation values, i.e., (x_i, y_i, θ_i). Given a desired geometric pattern $\mathcal{P}$ that has n vertices and is defined by orthogonal coordinates (p_{ix}, p_{iy}). Define $\bar{x}_i = x_i - p_{ix}, \bar{y}_i = y_i - p_{iy}$, and

$$\begin{aligned}
&\bar{x} = [\bar{x}_1, \bar{x}_2, \ldots, \bar{x}_n]^{\mathrm{T}} \in R^{n \times 1}, \bar{y} = [\bar{y}_1, \bar{y}_2, \ldots, \bar{y}_n]^{\mathrm{T}} \in R^{n \times 1},\\
&X = [\bar{x}_1, \bar{y}_1, \bar{x}_2, \bar{y}_2, \ldots, \bar{x}_n, \bar{y}_n]^{\mathrm{T}} \in R^{2n \times 1}, \Theta = [\theta_1, \theta_2, \ldots, \theta_n]^{\mathrm{T}} \in R^{n \times 1},\\
&L_{(2)} = L \otimes I_2 \in R^{2n \times 2n}, I_2 = \mathrm{diag}\{1, 1\},
\end{aligned}$$

where X and Θ represent the stack vectors of position and orientation coordinates, $\otimes$ is Kronecker product. Let $\bar{x}_{ij} = \bar{x}_i - \bar{x}_j, \bar{y}_{ij} = \bar{y}_i - \bar{y}_j$ denote the formation error, and $\theta_{ij} = \theta_i - \theta_j$ the direct orientation error between robot i and j.

Assumption 1 The communication graph $G = (\mathcal{V}, \mathcal{E}, \mathcal{A})$ is undirected, time-invariant and connected.

Under Assumption 1, we know that *rank* $(L) = n - 1$ and each row sum of L is zero, hence the null space of L is the spanning space of $\{\mathbf{1}\}$, meaning that the formation problem for multi-robots in undirected connected topology graph can be stated as: design distributed protocols $v_i(\cdot), \omega_i(\cdot)$ such that

$$x_i - x_j \to p_{ix} - p_{jx}, \quad y_i - y_j \to p_{iy} - p_{jy}, \quad \theta_i - \theta_j \to 0,$$

which are equal to

$$L\bar{x} \to 0, L\bar{y} \to 0, L\Theta \to 0; \quad \text{or} \quad L_{(2)}X \to 0, L\Theta \to 0. \tag{30.3}$$

30.3 Protocols for Multiple Nonholonomic Agents

In this section, we focus on the controller design for the formation problem of multiple nonholonomic robots. For this case, we solve the position/orientation formation problem by smooth time-varying scheme. We assume that robot i can only get the formation error $(\bar{x}_{ij}, \bar{y}_{ij})$ and its own orientation θ_i, and cannot known its position (x_i, y_i) or the other one's orientation $\theta_j (j \neq i)$.

First, we establish the model of all the robots in graph in form of compact matrix for simplicity. Define $U = [v_1, v_2, \ldots, v_n]^{\mathrm{T}} \in R^{n\times 1}, \Omega = [\omega_1, \omega_2, \ldots, \omega_n]^{\mathrm{T}} \in R^{n\times 1}$, and rewrite the unicycle-like model as

$$\dot{X} = H_1 U, \quad \dot{\Theta} = \Omega, \tag{30.4}$$

where X, Θ are defined in Sect. 30.2, and H_1 is a block diagonal matrix given by

$$H_1 = \begin{bmatrix} \alpha_1 & O_{2\times 1} & \cdots & O_{2\times 1} \\ O_{2\times 1} & \alpha_2 & \cdots & O_{2\times 1} \\ \vdots & \vdots & \ddots & \vdots \\ O_{2\times 1} & O_{2\times 1} & \cdots & \alpha_n \end{bmatrix} \in R^{2n\times n}, \tag{30.5}$$

where $\alpha_i = [\cos\theta_i, \sin\theta_i]^{\mathrm{T}} \in R^{2\times 1}, O_{2\times 1} = [0 \quad 0]^{\mathrm{T}}$. Differentiating H_1 yields

$$\dot{H}_1 = H_2 H_\omega, \tag{30.6}$$

where

$$H_\omega = \mathrm{diag}\{\omega_1, \omega_2, \cdots, \omega_{\mathrm{n}}\} \in R^{n\times n},$$

$$H_2 = \begin{bmatrix} \beta_1 & O_{2\times 1} & \vdots & O_{2\times 1} \\ O_{2\times 1} & \beta_2 & \vdots & O_{2\times 1} \\ \vdots & \vdots & \ddots & \vdots \\ O_{2\times 1} & O_{2\times 1} & \cdots & \beta_n \end{bmatrix} \in R^{2n\times n}, \tag{30.7}$$

$$\beta_i = \frac{d\alpha_i}{d\theta_i} = [-\sin\theta_i, \cos\theta_i]^{\mathrm{T}} \in R^{2\times 1}.$$

It is hence that

$$\Omega = H_{\omega}\mathbf{1}, \dot{H}_2 = -H_1 H_{\omega}. \tag{30.8}$$

For system (30.4), we have the next theorem.

Theorem 1 *Under Assumption* 1, *the control law*

$$U = -k_1 H_1^{\mathrm{T}} L_{(2)} X, \Omega = -k_3(\Theta - \alpha\mathbf{1}) + k_2 f(t) H_2^{\mathrm{T}} L_{(2)} X \tag{30.9}$$

can guarantee (30.3) *holds, where* $k_1 > 0$, $k_2 \neq 0$, $k_3 > 0, \alpha \in R$, *and* $f(t) \in R$ *a smooth bounded periodic function with* $\lim_{t\to\infty} \dot{f}(t) \neq 0$.

Proof To prove Theorem 1, we construct the following formation positive definite Lyapunov function as

$$V = X^{\mathrm{T}} L_{(2)} X = X^{\mathrm{T}} L_{(2)}^{\mathrm{T}} X = 0.5 \sum_i \sum_{j \in N_i} \left(\bar{x}_{ij}^2 + \bar{y}_{ij}^2\right) \geq 0 \tag{30.10}$$

whose derivative along (30.4) (30.9) is derived as

$$\dot{V} = 2X^{\mathrm{T}} L_{(2)}^{\mathrm{T}} H_1 U = -2k_1 \left[H_1^{\mathrm{T}} L_{(2)} X\right]^{\mathrm{T}} H_1^{\mathrm{T}} L_{(2)} X \leq 0 \tag{30.11}$$

For undirected connected graph, Eq. (30.10) implies that $2X^{\mathrm{T}} L_{(2)}^{\mathrm{T}} X$ represents the quadric sum of $\bar{x}_{ij}$ and $\bar{y}_{ij}$, which is similar in [4], and also can be found in [3, 5]. Control law (30.9) can be also rewritten in component form as

$$\begin{aligned} v_i &= -k_1 \left(\cos\theta_i \textstyle\sum_{j \in N_i} a_{ij}\bar{x}_{ij} + \sin\theta_i \sum_{j \in N_i} a_{ij}\bar{y}_{ij}\right) \\ \omega_i &= -k_3(\theta_i - \alpha) + k_2 f(t)\left(-\sin\theta_i \textstyle\sum_{j \in N_i} a_{ij}\bar{x}_{ij} + \cos\theta_i \sum_{j \in N_i} a_{ij}\bar{y}_{ij}\right) \end{aligned} \tag{30.12}$$

It follows from $V \geq 0, \dot{V} \leq 0$ that $V \in L_\infty, \bar{x}_{ij} \in L_\infty, \bar{y}_{ij} \in L_\infty$ for all $i \in \mathcal{L}$, $j \in N_i$, thus $v_i \in L_\infty, U \in L_\infty, \dot{X} \in L_\infty$ and so is the term $k_2 f(t) H_2^{\mathrm{T}} L_{(2)} X$ in (30.9) due to $f(t) \in L_\infty$, implying that $\theta_i \in L_\infty$ for all i. Therefore, $\Omega \in L_\infty$, $\dot{\Theta} \in L_\infty, \dot{H}_1 \in L_\infty$, and $\ddot{V} = -4k_1 [H_1^{\mathrm{T}} L_{(2)} X]^{\mathrm{T}} (\dot{H}_1^{\mathrm{T}} L_{(2)} X + H_1^{\mathrm{T}} L_{(2)} \dot{X}) \in L_\infty$. By $\ddot{V} \in L_\infty$, we can verify $\dot{V}$ is uniformly continuous. Using Barbalat's lemma and the forging analysis [24], we have $\lim_{t\to\infty} \dot{V} = 0$, i.e., $U = -k_1 H_1^{\mathrm{T}} L_{(2)} X \to 0, \dot{X} \to 0$ as $t \to \infty$.

To simplify the next analysis, we combine (30.8) (30.9) and calculate $\mathbf{1}^{\mathrm{T}}\dot{U}, \mathbf{1}^{\mathrm{T}}\ddot{U}, \dot{\Omega}$ as

$$
\begin{aligned}
\mathbf{1}^{\mathrm{T}}\dot{U} &= -k_1\mathbf{1}^{\mathrm{T}}\dot{H}_1^{\mathrm{T}}L_{(2)}X - k_1\mathbf{1}^{\mathrm{T}}H_1^{\mathrm{T}}L_{(2)}\dot{X} \\
&= -k_1\left(\Omega^{\mathrm{T}}H_2^{\mathrm{T}}L_{(2)}X + \mathbf{1}^{\mathrm{T}}H_1^{\mathrm{T}}L_{(2)}H_1U\right) \\
\mathbf{1}^{\mathrm{T}}\ddot{U} &= -k_1\left[\dot{\Omega}^{\mathrm{T}}H_2^{\mathrm{T}}L_{(2)}X + \Omega^{\mathrm{T}}\left(-H_\omega^{\mathrm{T}}H_1^{\mathrm{T}}\right)L_{(2)}X + \Omega^{\mathrm{T}}H_2^{\mathrm{T}}L_{(2)}\dot{X}\right] \\
&\quad -k_1\left(\mathbf{1}^{\mathrm{T}}2H_1^{\mathrm{T}}L_{(2)}\dot{H}_1U + \mathbf{1}^{\mathrm{T}}H_1^{\mathrm{T}}L_{(2)}H_1\dot{U}\right) \\
\dot{\Omega} &= -k_3\Omega + k_2\dot{f}(t)H_2^{\mathrm{T}}L_{(2)}X + k_2f(t)\left(-H_\omega^{\mathrm{T}}H_1^{\mathrm{T}}\right)L_{(2)}X + k_2f(t)H_2^{\mathrm{T}}L_{(2)}\dot{X}
\end{aligned}
$$

Applying Barbalat's lemma to signals U, $\dot{U}$, we have $\lim_{t\to\infty}U=0$ and $\lim_{t\to\infty}\ddot{U}=0$, where the former equality shows that $\lim_{t\to\infty}\mathbf{1}^{\mathrm{T}}U=0 \Rightarrow \lim_{t\to\infty}(\Omega^{\mathrm{T}}H_2^{\mathrm{T}}L_{(2)}X)=0$, and the later one shows that $\lim_{t\to\infty}(\dot{\Omega}^{\mathrm{T}}H_2^{\mathrm{T}}L_{(2)}X)=0 \Rightarrow \lim_{t\to\infty}[(-k_3\Omega + k_2f(t)H_2^{\mathrm{T}}L_{(2)}X)^{\mathrm{T}}H_2^{\mathrm{T}}L_{(2)}X]=0$.

Therefore, by $\lim_{t\to\infty}(\Omega^{\mathrm{T}}H_2^{\mathrm{T}}L_{(2)}X)=0$ and $\lim_{t\to\infty}f(t)\neq 0$, we can get $\lim_{t\to\infty}(H_2^{\mathrm{T}}L_{(2)}X)=0$.

So far, we have proven $\lim_{t\to\infty}(H_1^{\mathrm{T}}L_{(2)}X)=0$ and $\lim_{t\to\infty}(H_2^{\mathrm{T}}L_{(2)}X)=0$, which are equal to $\lim_{t\to\infty}(L_{(2)}X)=0$.

We now return to Θ-dynamics $\dot{\Theta}=-k_3(\Theta-\alpha\mathbf{1})+k_2f(t)H_2^{\mathrm{T}}L_{(2)}X$, and can easily obtain that Θ exponentially approaches $\alpha\mathbf{1}$ by using the proven fact $\lim_{t\to\infty}(H_2^{\mathrm{T}}L_{(2)}X)=0$. □

The difference of the convergence of X, Θ is that the former converges to a given geometric pattern with its centroid unknown, while the latter converges to a pre-specified vector $\alpha\mathbf{1}$. For the formation problem of multiple nonholonomic agents, distinct from the time-varying law in [23] that realizes position formation only, our control law can realize orientation consensus at the same time; distinct from the discontinuous laws in [18, 20] that cannot be extended to the dynamic level, our control law is smooth and can be extended to dynamic robot model.

30.4 Simulation

In this section, simulation examples are presented in order to demonstrate the effectiveness of the proposed control algorithm in Sect. 30.3. The initial states, geometric pattern and control parameters are chosen as:

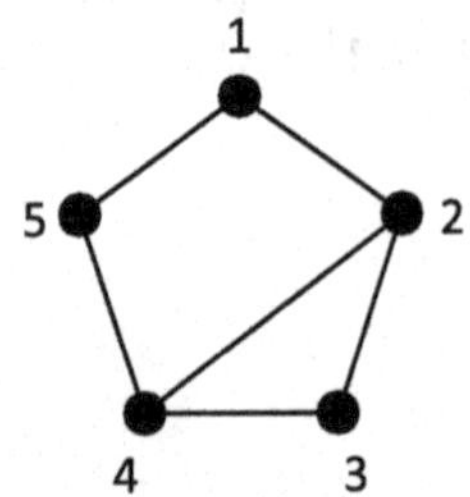

Fig. 30.1 Connected undirected graph

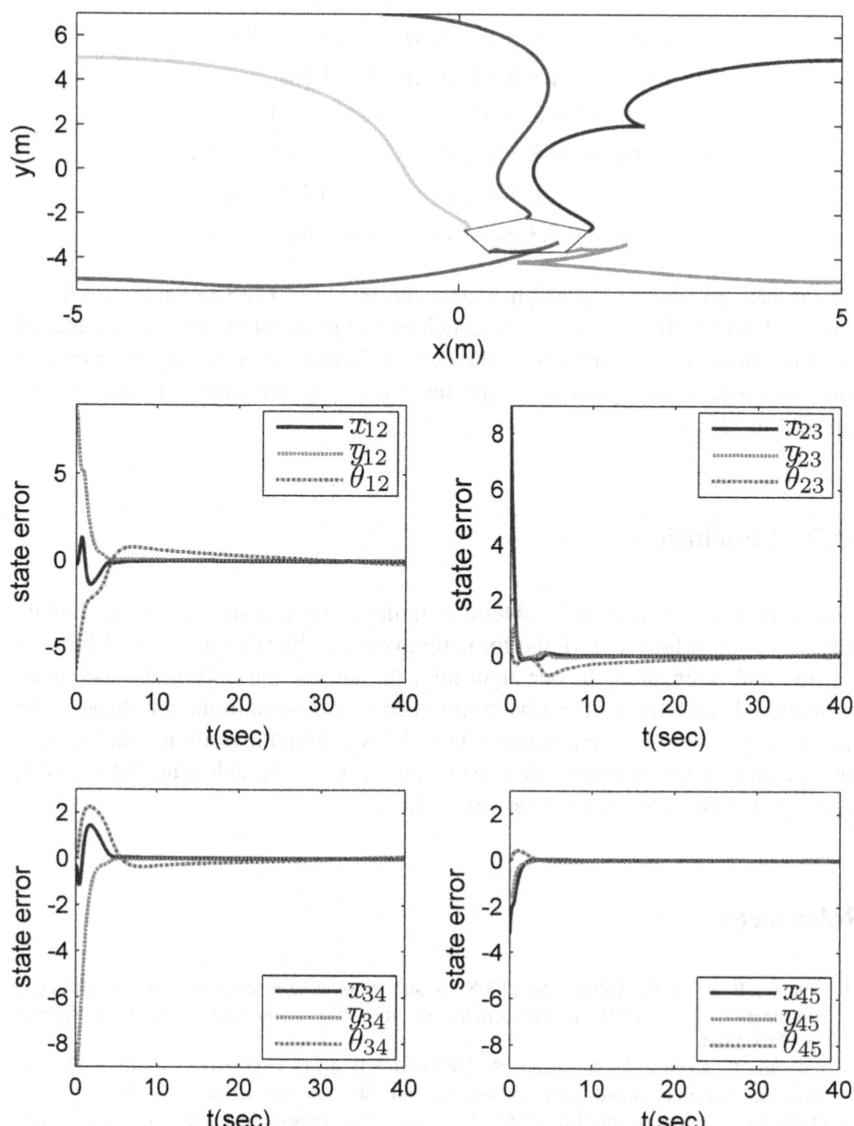

Fig. 30.2 Position/orientation formation of nonholonomic agents under (30.9)

$$
\begin{aligned}
&(x_1,y_1,\theta_1)=(5,5,-\pi),(p_{1x},p_{1y})=(1.31,0.95),\\
&(x_2,y_2,\theta_2)=(5,-5,\pi),(p_{2x},p_{2y})=(1,0),\\
&(x_3,y_3,\theta_3)=(-5,-5,\pi),(p_{3x},p_{3y})=(0,0),\\
&(x_4,y_4,\theta_4)=(-5,5,\pi),(p_{4x},p_{4y})=(-0.31,0.95),\\
&(x_5,y_5,\theta_5)=(-1,7,\pi),(p_{5x},p_{5y})=(0.5,1.54),\\
&\qquad k_1=0.5,k_2=3,k_3=1,f(t)=\sin(0.1t),\alpha=1,
\end{aligned}
$$

and the network undirected graph is depicted in Fig. 30.1. Simulation results are illustrated in Fig. 30.2, where we can conclude that protocol (30.9) can guarantee all the five robotic agents converge to the desired formation shape, and their orientations converge to 1. In summary, the main result in Theorem 1 holds, and our approach is effective.

30.5 Conclusion

This work solves formation problem of multiple nonholonomic robots, and the proposed control law achieve the formation control objective in terms of both the position and orientation. In our algorithm, the agents can only make use of the orientation themselves, and relative position error of desired formation shape, rather than direct position. Future researches include extending the result to other systems with parameter uncertainties, discussing the case of digraph with time-varying topology and considering the time-delay effect.

References

1. Ren W, Beard RW, Atkins EM (2005) A survey of consensus problems in multi-agent coordination [C]. In: IEEE Proceedings of the 2005 American Control Conference, pp 1859–1864
2. Kranakis E, Krizanc D, Rajsbaum S (2006) Mobile agent rendezvous: a survey [M]. In: Structural information and communication complexity, Springer Berlin, pp 1–9
3. Olfati SR, Fax JA, Murray RM (2007) Consensus and cooperation in networked multi-agent systems [J]. Proc IEEE 95(1):215–233
4. Olfati-Saber R, Murray RM (2004) Consensus problems in networks of agents with switching topology and time-delays [J]. IEEE Trans Autom Control 49(9):1520–1533
5. Ren W, Chao H, Bourgeous W et al (2008) Experimental validation of consensus algorithms for multivehicle cooperative control [J]. IEEE Trans Control Syst Technol 16(4):745–752
6. Hui Q (2011) Finite-time rendezvous algorithms for mobile autonomous agents [J]. IEEE Trans Autom Control 56(1):207–211
7. Cortés J, Martínez S, Bullo F (2006) Robust rendezvous for mobile autonomous agents via proximity graphs in arbitrary dimensions [J]. IEEE Trans Autom Control 51(8):1289–1298
8. Conte G, Pennesi P (2007) On convergence conditions for rendezvous [C]. In: The 46th IEEE Conference on decision and control, p 2375–2378

9. Fan Y, Feng G, Wang Y (2011) Combination framework of rendezvous algorithm for multi-agent systems with limited sensing ranges [J]. Asian J Control 13(2):283–294
10. Su H, Wang X, Chen G (2010) Rendezvous of multiple mobile agents with preserved network connectivity [J]. Syst Control Lett 59(5):313–322
11. Fax JA, Murray RM (2004) Information flow and cooperative control of vehicle formations [J]. IEEE Trans Autom Control 49(9):1465–1476
12. Smith SL, Broucke ME, Francis BA (2007) Curve shortening and the rendezvous problem for mobile autonomous robots [J]. IEEE Trans Autom Control 52(6):1154–1159
13. Dong W, Farrell JA (2008) Consensus of multiple nonholonomic systems [C]. In: The 47th IEEE conference on decision and control (CDC), p 2270–2275
14. Dong W, Farrell JA (2008) Cooperative control of multiple nonholonomic mobile agents [J]. IEEE Trans Autom Control 53(6):1434–1448
15. El-Hawwary MI, Maggiore M (2013) Distributed circular formation stabilization for dynamic unicycles [J]. IEEE Trans Autom Control 58(1):149–162
16. Liu T, Jiang ZP (2013) Distributed formation control of nonholonomic mobile robots without global position measurements [J]. Automatica 49(2):592–600
17. Wang P, Ding BC (2014) Distributed RHC for tracking and formation of nonholonomic multi-vehicle systems [J]. IEEE Trans Autom Control 59(6):1439–1453
18. Consolini L, Morbidi F, Prattichizzo D et al (2008) Leader–follower formation control of nonholonomic mobile robots with input constraints [J]. Automatica 44(5):1343–1349
19. Park BS, Park JB, Choi YH (2011) Robust adaptive formation control and collision avoidance for electrically driven non-holonomic mobile robots [J]. IET Control Theory Appl 5(3):514–522
20. López-Nicolás G, Aranda M, Mezouar Y et al (2012) Visual control for multirobot organized rendezvous [J]. IEEE Trans Syst Man Cybern B Cybern 42(4):1155–1168
21. Listmann KD, Masalawala MV, Adamy J (2009) Consensus for formation control of nonholono- mic mobile robots [C]. In: IEEE International conference on robotics and automation (ICRA), p 3886–3891
22. Yoshioka C, Namerikawa T (2008) Formation control of nonholonomic multi-vehicle systems based on virtual structure [C]. In: The 17th IFAC world congress, p 5149–5154
23. Lin Z, Francis B, Maggiore M (2005) Necessary and sufficient graphical conditions for formation control of unicycles [J]. IEEE Trans Autom Control 50(1):121–127
24. Khalil HK, Grizzle JW (1996) Nonlinear systems, vol 3 [M]. Prentice hall, New Jersey

Chapter 31
Robust Stability Analysis for Uncertain Time-Delay System

Wenfang Xin, Yaoyong Duan, Yi Luo and Qingmei Ji

Abstract In this paper, the robust stability problem of time-delay system is discussed by constructing a Lyapunov-Krasovskii function. And two different systems are given in this paper, one of them has not consider the interference and unmodeled dynamics components, the other one add this uncertainties into the system. Based on this two different systems model, several general and powerful algorithms can also be given as theorems.

Keywords Time-delay · Lyapunov-Krasovskii function · LMI · Uncertainties · H infinite

31.1 Introduction

It is generally known that time delay always causes destabilization and performance degradation. But it is well known that time delay is frequently encountered in many practical engineering systems, such as communication, electronics and chemical systems. So it is necessary to study the stability criteria of time-delay system. On the back of the urgency of robust stability analysis, people have study a lot of stability sufficient conditions [14–16]. Such as in 1961 V.M. Popov [13] improved Nyquist criterion by introduce analysis in frequency domain. But in fact, the Lyapunov theory is the important and fundamental one [12, 18]. On the other hand, with the occurrence of uncertain system, the classical theory could not afford the necessary. So in 1981, Zames [2] introduce a new objective function to complete optimal design. Since then, the H_∞ optimal control has becoming a fashion research targets of study-

W. Xin · Q. Ji
Special Police of China, Beijing 102211, China

Y. Duan (✉) · Y. Luo
The Chinese People's Armed Police Forced Academy, Langfang 065000, China
e-mail: duanyaoyong@126.com

© Springer-Verlag Berlin Heidelberg 2016

Y. Jia et al. (eds.), *Proceedings of the 2015 Chinese Intelligent Systems Conference*, Lecture Notes in Electrical Engineering 359,
DOI 10.1007/978-3-662-48386-2_31

ing time-delay system. Depending on H_∞ optimal control include the information of delay, the attention can be classified into delay-independent ones [3, 6] and delay-dependent ones [4, 5, 7]. Since delay-dependent results are usually less conservative than delay-independent ones, so much attention has been paid to the study of delay-dependent H_∞ for time-delay systems. Such as [1] solve the problem by constructing a Riccati differential equation, [8, 9] derive the problem by constructing a Lyapunov-Krasovskii function, and [10, 11] derive the problem by constructing a Lyapunov-Razumikhin function.

In this paper, we consider the problem of robust H_∞ control analysis for uncertain time-delay system by constructing a new Lyapunov functional. The distributed delays are assumed to appear in terms of interference and unmodeled dynamics components, and the parameter uncertainties are allowed to be time-varying but norm-bounded. Some sufficient conditions are obtained to guarantee the existence of H_∞ performance, which can be constructed by solving LMIs.

31.2 Preliminaries

In this section, before giving main result, some necessary lemmas are shown as follow:

Lemma 1 *In linear algebra and the theory of matrices, the Schur complement of a matrix block (i.e., a submatrix within a larger matrix) is defined as follows. Suppose* A, B, C, D *are respectively* $p \times p, p \times q, q \times p$ *and* $q \times q$ *matrices, and* D *is invertible. Let* $M = \begin{bmatrix} A & B \\ C & D \end{bmatrix}$ *so that* M *is a* $(p+q) \times (p+q)$ *matrix.*

Then the Schur complement of the block D *of the matrix* M *is the* $p \times p$ *matrix* $A - BD^{-1}C$.

Lemma 2 *In linear algebra, an* $R^{n \times n}$ *square matrix* A *is called invertible, and* $A = B + XCY$. *Suppose* B^{-1} *has been known, and* C *is a* $r \times r$ *invertible matrix, where* $r \le n$. *So if* $C^{-1} + YB^{-1}X$ *is invertible, then*

$$A^{-1} = B^{-1} - B^{-1}X(C^{-1} + YB^{-1}X)^{-1}YB^{-1}$$

Lemma 3 *In linear algebra, for any* $X, Y \in R^n$ *and positive-definite matrix* $P \in R^{n \times n}$,

$$2X^T Y \le X^T P^{-1} X + Y^T PY.$$

Lemma 4 *When a matrix inequality has some variables that appear in a certain form, we can derive an equivalent inequality without those variables. [17] Consider*

$$G(z) + U(z)XV(z)^T + V(z)X^T U(z)^T > 0,$$

where the vector z and the matrix X are(independent) variables, and $G(z) \in R^{n\times n}$, $U(z)$ and $V(z)$ do not depend on X. Suppose that for every z, $\tilde{U}(z)$, $\tilde{V}(z)$ are orthogonal complements of $U(z)$, $V(z)$ respectively. Then

$$\tilde{U}(z)^T G(z)\tilde{U}(z) > 0, \tilde{V}(z)^T G(z)\tilde{V}(z) > 0, \tag{31.1}$$

We can express (31.1) in another form using Finsler's lemma,

$$G(z) - \sigma U(z)U(z)^T > 0, G(z) - \sigma V(z)V(z)^T > 0. \tag{31.2}$$

31.3 Problem Formulation

We consider the below system as the first system.

$$\begin{cases} \dot{x}(t) = A_1 x(t) + A_2 x(t-h(t)) \\ z(t) = C_1 x(t) + C_2 x(t-h(t)) \end{cases} \tag{31.3}$$

where $x(t) = \phi(t)$ is the system state variable, $h(t)$ is the time delay of system, and we assume that $x \in R^n, t \in [-\tau, 0]$. The matrices $A_1 \in R^{n\times n}$, $A_2 \in R^{n\times n}$, $C_1 \in R^{n\times m}$, $C_2 \in R^{n\times m}$ are known as real constant matrices.

According to system (31.3), we choose the following Lyapunov-Krasovskii function candidate as

$$V(t) = x^T(t)Px(t) + \int_{t-h(t)}^{t} x^T(s)Qx(s)ds + \int_{-h_0}^{0}\int_{t+\beta}^{t} \dot{x}^T(\alpha)Z\dot{x}(\alpha)d\alpha d\beta. \tag{31.4}$$

Theorem 1 *The system (31.3) is asymptotically stable for any time-delay* $0 < h(t) \le h_0$, $\dot{h}(t) \le h_1 < 1$, $\varepsilon > 0$, $P > 0$, $Q > 0$, $Z > 0$, *when*

$$\Sigma_1 = \begin{bmatrix} PA_1 + A_1{}^T P + Q & PA_2 & 0 & A_1 \\ * & -(1-h_1)Q & 0 & A_2 \\ * & * & -h_0 Z & 0 \\ * & * & * & -\frac{1}{h_0{}^2}Z^{-1} \end{bmatrix} < 0$$

Proof If we consider the Lyapunov-Krasovskii function candidate as (31.4), and where

$$V_1 = x^T(t)Px(t),$$

$$V_2 = \int_{t-h(t)}^{t} x^T(s)Qx(s)ds,$$

$$V_3 = \int_{-h_0}^{0}\int_{t+\beta}^{t} \dot{x}^T(\alpha)Z\dot{x}(\alpha)d\alpha d\beta.$$

then the Lyapunov-Krasovskii function (31.4) is written as

$$V(t) = V_1(t) + V_2(t) + V_3(t).$$

With matrix P, Q, Z are known as real constant matrices, we can get:

$$\begin{aligned}
\dot{V}_1(t) &= x^T(t)(A_1{}^T P + PA_1)x(t) + x^T(t)PA_2 x(t-h(t)) + x^T(t-h(t))A_2^T Px(t) \\
\dot{V}_2(t) &= x^T(t)Qx(t) - (1-h_1)x^T(t-h(t))Qx(t-h(t)), \\
\dot{V}_3(t) &= h_0 \dot{x}^T(t)Z\dot{x}(t) - \int_{t-h_0}^{t} \dot{x}^T(\alpha)Z\dot{x}(\alpha)d\alpha.
\end{aligned}$$

Then the Lyapunov-Krasovskii function (31.4) can be derived

$$\dot{V}(t) = \dot{V}_1(t) + \dot{V}_2(t) + \dot{V}_3(t) = \frac{1}{h_0}\int_{t-h_0}^{t} y_1^T \Sigma_1 y_1 d\alpha \le 0.$$

where

$$y_1 = \left[x^T(t)\; x^T(t-h(t))\; \dot{x}^T(\alpha)\right],$$

$$\Sigma_1 = \Sigma_1' + \begin{bmatrix} A_1^T \\ A_2^T \\ 0 \end{bmatrix} h_0^2 Z \begin{bmatrix} A_1^T \\ A_2^T \\ 0 \end{bmatrix}^T$$

$$\Sigma_1' = \begin{bmatrix} PA_1 + A_1{}^T P + Q & PA_2 & 0 \\ * & -(1-h_1)Q & 0 \\ * & * & -h_0 Z \end{bmatrix}$$

Consider Schur Lemma (1), we can get if

$$\Sigma_1 = \begin{bmatrix} PA_1 + A_1{}^T P + Q & PA_2 & 0 & A_1 \\ * & -(1-h_1)Q & 0 & A_2 \\ * & * & -hZ & 0 \\ * & * & * & -\frac{1}{h_0^2}Z^{-1} \end{bmatrix} < 0$$

the system (31.3) is asymptotically stable.

31.4 Stability Analysis

Consider the upper system (31.3), we add interference and unmodeled dynamics components into the system (31.3). Thus we get the following system:

$$\begin{cases} \dot{x}(t) = A_1x(t) + A_2x(t-h(t)) + f(x(t), x(t-h(t)), t) + B_1w(t) \\ z(t) = C_1x(t) + C_2x(t-h(t)) + Dw(t) \end{cases} \tag{31.5}$$

where $x(t) = \phi(t)$ is the system state variable, $h(t)$ is the time delay of system, and we assume that $x \in R^n$, $u \in R^m$,$w \in R^m$, $t \in [-\tau, 0]$. The matrices $A_1 \in R^{n\times n}$, $A_2 \in R^{n\times n}$, $B_1 \in R^{n\times m}$, $C_1 \in R^{n\times m}$, $C_2 \in R^{n\times m}$, $D \in R^{n\times m}$ are known as real constant matrices. The nonlinearity function $f(x(t), x(t-h(t)), t)$ satisfies

$$f^T(x(t), x(t-h(t)), t)f(x(t), x(t-h(t)), t) \leq 1. \tag{31.6}$$

According to the stability analysis of system (31.3), we choose the Lyapunov-Krasovskii function candidate as Eq. (31.4).

By following a similar lines as does in Theorem 1, the following theorem gives the delay-dependent stability result for system (31.5).

Theorem 2 *Consider system (31.5), if* $w(t) = 0$, *for any time-delay* $0 < h(t) \leq h_0, \dot{h}(t) \leq h_1 < 1$, *and* $\varepsilon_1 > 0$, $\varepsilon_2 > 0$, $P > 0, Q > 0, Z > 0$, *the system (31.5) is asymptotically stable, when*

$$\Sigma_1 = \begin{bmatrix} PA_1 + {A_1}^TP + Q - \varepsilon_2P^TP & PA_2 & h_0 & P & A_1^T \\ * & -(1-h_1)Q & -h_0 & 0 & A_2^T \\ * & * & -h_0Z & 0 & 0 \\ * & * & * & -\varepsilon_1I & I \\ * & * & * & * & -\frac{1}{h_0^2}Z^{-1} \end{bmatrix} < 0$$

Proof Similar with Theorem 1, if we consider the Lyapunov-Krasovskii function candidate as (31.4), and where

$$V_1 = x^T(t)Px(t),$$
$$V_2 = \int_{t-h(t)}^{t} x^T(s)Qx(s)ds,$$
$$V_3 = \int_{-h_0}^{0}\int_{t+\beta}^{t} \dot{x}^T(\alpha)Z\dot{x}(\alpha)d\alpha d\beta.$$

then the Lyapunov-Krasovskii function (31.4) is written as

$$V(t) = V_1(t) + V_2(t) + V_3(t).$$

And with matrix $P > 0, Q > 0, Z > 0$,

$$\begin{aligned}\dot{V}_1(t) =& x^T(t)(A_1{}^T P + PA_1)x(t)\\ &+ x^T(t)PA_2x(t - h(t)) + x^T(t - h(t))A_2^T Px(t)\\ &+ f^T(x(t), x(t - h(t)), t)Px(t) + x^T(t)Pf(x(t), x(t - h(t)), t),\\ \dot{V}_2(t) =& x^T(t)Qx(t) - (1 - h_1)x^T(t - h(t))Qx(t - h(t)),\\ \dot{V}_3(t) =& h_0\dot{x}^T(t)Z\dot{x}(t) - \int_{t-h_0}^{t} \dot{x}^T(\alpha)Z\dot{x}(\alpha)d\alpha.\end{aligned}$$

For $\dot{V}_1(t)$,

$$\begin{aligned}&x^T(t)(A_1{}^T P + PA_1)x(t) + x^T(t)PA_2x(t - h(t)) + x^T(t - h(t))A_2^T Px(t)\\ &\quad + f^T(x(t), x(t - h(t)), t)Px(t) + x^T(t)Pf(x(t), x(t - h(t)), t)\\ &\leq 0\end{aligned}$$

based on S-Procedure, for $\varepsilon_1 > 0$, $\varepsilon_2 > 0$,

$$\begin{aligned}&x^T(t)(A_1{}^T P + PA_1)x(t) + x^T(t)PA_2x(t - h(t)) + x^T(t - h(t))A_2^T Px(t)\\ &\quad - \varepsilon_1 f^T(x(t), x(t - h(t)), t)f(x(t), x(t - h(t)), t) - \varepsilon_2 x^T(t)P^T Px(t)\\ &\leq 0\end{aligned}$$

Suppose $\varepsilon_1 > 0$, $\varepsilon_2 > 0$,

$$\dot{V}(t) = \dot{V}_1(t) + \dot{V}_2(t) + \dot{V}_3(t) \leq 0.$$

So if

$$y_2 = \left[x^T(t)\ x^T(t - h(t))\ \dot{x}^T(\alpha)\ f^T\right]^T,$$

it can be get

$$\dot{V}(t) \leq \frac{1}{h_0}\int_{t-h_0}^{t} y_2^T \Sigma_2 y_2 d\alpha$$

where

$$\Sigma_2 = \Sigma_2' + \begin{bmatrix} A_1^T \\ A_2^T \\ 0 \\ I \end{bmatrix} h_0^2 Z \begin{bmatrix} A_1^T \\ A_2^T \\ 0 \\ I \end{bmatrix}^T$$

$$\Sigma_2' = \begin{bmatrix} PA_1 + A_1{}^T P + Q - \varepsilon_2 P^T P & PA_2 & h_0 & 0 \\ * & -(1-h_1)Q & -h_0 & 0 \\ * & * & -h_0 Z & 0 \\ * & * & * & -\varepsilon_1 I \end{bmatrix}$$

Consider Lemma (1), we can get if

$$\Sigma_2 = \begin{bmatrix} PA_1 + A_1{}^T P + Q - \varepsilon_2 P^T P & PA_2 & h_0 & P & A_1^T \\ * & -(1-h_1)Q & -h_0 & 0 & A_2^T \\ * & * & -h_0 Z & 0 & 0 \\ * & * & * & -\varepsilon_1 I & I \\ * & * & * & * & -\frac{1}{h_0^2} Z^{-1} \end{bmatrix} < 0$$

the system (31.5) is asymptotically stable.

Theorem 3 *Consider Theorem 2, if* $w(t) \neq 0$, *for any time-delay* $0 < h(t) \leq h_0, \dot{h}(t) \leq h_1 < 1$, *and* $\varepsilon_1 > 0$, $\varepsilon_2 > 0$, $\gamma \geq \|D\|$, $P > 0, Q > 0, Z > 0$, *the system (31.5) satisfied* H_∞ *performance* $\|z(t)\| \leq \gamma \|\omega(t)\|$ *and asymptotically stable, when*

$$\Sigma_3 = \begin{bmatrix} PA_1 + A_1{}^T P + Q + C_1^T C_1 - \varepsilon_2 P^T P & PA_2 + C_1^T C_2 & h_0 & 0 & PB_1^T + C_1^T D & A_1^T \\ * & (h_1 - 1)Q + C_2^T C_2 & -h_0 & 0 & C_2^T D & A_2^T \\ * & * & -h_0 Z & 0 & 0 & 0 \\ * & * & * & -\varepsilon_1 I & 0 & I \\ * & * & * & * & D^T D - \gamma^2 I & B_1^T \\ * & * & * & * & * & -\frac{1}{h_0^2} Z^{-1} \end{bmatrix} \leq 0.$$

Proof For Lyapunov-Krasovskii function (31.4), we construct equation

$$\begin{aligned} J(t) &= \int_0^t z^T(t) z(t) - \gamma^2 \omega^T(t) \omega(t) dt \\ &= \int_0^t z^T(t) z(t) - \gamma^2 \omega^T(t) \omega(t) + \dot{V}(t) dt - V(t) + V(0) \end{aligned} \tag{31.7}$$

So similar with Theorem 2, we get

$$\Sigma_3 = \Sigma_3' + \begin{bmatrix} A_1^T \\ A_2^T \\ 0 \\ I \\ B_1^T \end{bmatrix} h_0^2 Z \begin{bmatrix} A_1^T \\ A_2^T \\ 0 \\ I \\ B_1^T \end{bmatrix}^T$$

where

$$\Sigma_3' = \begin{bmatrix} PA_1{+}A_1{}^TP{+}Q{+}C_1^TC_1{-}\varepsilon_2P^TP & PA_2{+}C_1^TC_2 & h_0 & P & PB_1^T{+}C_1^TD \\ * & (h_1-1)Q{+}C_2^TC_2 & -h_0 & 0 & C_2^TD \\ * & * & -h_0Z & 0 & 0 \\ * & * & * & -\varepsilon_1 I & 0 \\ * & * & * & * & D^TD-\gamma^2 I \end{bmatrix}$$

So consider Lemma (2), if

$$\Sigma_3 = \begin{bmatrix} PA_1{+}A_1{}^TP{+}Q{+}C_1^TC_1{-}\varepsilon_2P^TP & PA_2{+}C_1^TC_2 & h_0 & 0 & PB_1^T{+}C_1^TD & A_1^T \\ * & (h_1-1)Q{+}C_2^TC_2 & -h_0 & 0 & C_2^TD & A_2^T \\ * & * & -h_0Z & 0 & 0 & 0 \\ * & * & * & -\varepsilon_1 I & 0 & I \\ * & * & * & * & D^TD-\gamma^2 I & B_1^T \\ * & * & * & * & * & -\frac{1}{h_0^2}Z^{-1} \end{bmatrix} \leq 0.$$

then

$$J(t) = \int_0^t z^T(t)z(t) - \gamma^2\omega^T(t)\omega(t) + \dot{V}(t)dt \leq 0$$

With zero initial condition, it is easily to know $V(t) - V(0) = V(t) \geq 0$, so

$$\begin{aligned} J(t) &= \int_0^t z^T(t)z(t) - \gamma^2\omega^T(t)\omega(t) + \dot{V}(t)dt - V(t) + V(0) \\ &= \int_0^T z^T(t)z(t) - \gamma^2\omega^T(t)\omega(t)dt \\ &\leq 0 \end{aligned}$$

Meantime, if

$$\Sigma_3 = \begin{bmatrix} PA_1{+}A_1{}^TP{+}Q{+}C_1^TC_1{-}\varepsilon_2P^TP & PA_2{+}C_1^TC_2 & h_0 & 0 & PB_1^T{+}C_1^TD & A_1^T \\ * & (h_1-1)Q{+}C_2^TC_2 & -h_0 & 0 & C_2^TD & A_2^T \\ * & * & -h_0Z & 0 & 0 & 0 \\ * & * & * & -\varepsilon_1 I & 0 & I \\ * & * & * & * & D^TD-\gamma^2 I & B_1^T \\ * & * & * & * & * & -\frac{1}{h_0^2}Z^{-1} \end{bmatrix} \leq 0.$$

The system (31.5) satisfied H_∞ performance $z^T(t)z(t)-\gamma^2\omega^T(t)\omega(t) \leq 0$ ($\|z(t)\| \leq \gamma\,\|\omega(t)\|$) and asymptotically stable.

31.5 Conclusions

In this paper, a delay-dependent stabilization problem for continuous-time-delay system has been discussed. And two different systems have been established. One of them has not consider the interference and unmodeled dynamics components, the other one add this uncertainties into the system. Based on the constructed Lyapunov-Krasovskii function, an improved robust stability criterion has been given in terms of LMIs. And the result shows that it is possible to analysis the stability problem by this way. The next target of us is to design controller.

References

1. Pila AW, Shaked U, de Souza CE (1999) Hi filtering for continuous-time linear systems with delay. IEEE Trans Autom Control 44(7):1412–1417
2. Zames G (1981) Feedback and optimal sensitivity: model reference transformations, multiplicative seminorms, and approximate inverses. IEEE Trans Autom Control 26(2):301–320
3. Wang Z, Huang B, Unbehauen H (1999) Robust Hi observer design of linear state delayed systems with parametric uncertainty: the discrete-time case. Automatica 35(6):1161–1167
4. Wu ZG, Zhou WN (2007) Delay-dependent robust Hi control for uncertain singular time-delay systems. IET Control Theory Appl 1(5):1234–1241
5. Palhares RM, Campos CD, Ekel PY, Leles MCR, d'Angelo MFSV (2005) Delay-dependent robust Hi control of uncertain linear systems with lumped delays. IEE Proc-Control Theory Appl 152(1):27–33
6. Xu S, Chen T (2004) An LMI approach to the Hi filter design for uncertain systems with distributed delays. IEEE Trans Circuits Syst II: Express Briefs 51(4):195–201
7. Glover William, Lygeros John (2004) A stochastic hybrid model for air traffic control simulation. In: Alur Rajeev, Pappas George J (eds) HSCC 2004, vol 2993., LNCSSpringer, Heidelberg, pp 372–386
8. Zhang XM, Han QL (2008) Robust Hi filtering for a class of uncertain linear systems with time-varying delay. Automatica 44(1):157–166
9. Kharitonov VL, Zhabko AP (2003) Lyapunov CKrasovskii approach to the robust stability analysis of time-delay systems. Automatica 39(1):15–20
10. Dashkovskiy S, Naujok L (2010) Lyapunov-Razumikhin and Lyapunov-Krasovskii theorems for interconnected ISS time-delay systems. In: Proceedings of the 19th international symposium on mathematical theory of networks and systems (MTNS), pp 5–9
11. Jankovic M (2001) Control Lyapunov-Razumikhin functions and robust stabilization of time delay systems. IEEE Trans Autom Control 46(7):1048–1060
12. Gu K, Chen J, Kharitonov VL (2003) Stability of time-delay systems. Springer, Berlin
13. Popov VM (1962) Absolute stability of nonlinear systems of automatic control. Autom Remote Control 22(8):857–875
14. Anand DK (1984) Introduction to control systems. Elsevier Science Inc., London
15. Bistritz Y (1984) A new unit circle stability criterion. In: Mathematical theory of networks and systems, pp 69–87
16. Liu X, Wang J, Duan Z, Huang L (2010) New absolute stability criteria for time-delay Lur'e systems with sector-bounded nonlinearity. Int J Robust Nonlinear Control 20(6):659–672
17. Yaz EE (1997) Linear matrix inequalities in system and control theory (Book Review). Proc IEEE 85(5):798–799
18. Mukhija P, Kar IN, Bhatt RKP (2012) Delay-distribution-dependent robust stability analysis of uncertain Lurie systems with time-varying delay. Acta Autom Sinica 38(7):1100–1106

Chapter 32
Interconnection Power Converter for Multi-Rail DC Distribution System

Huiyang Lu, Yuru Zhu, Ying Hua, Wei Jiang and Nailu Li

Abstract This paper presents an interconnection bi-directional dc-dc converter for multi-rail DC distribution system. The principles of the energy conversion process, small signal modeling and controller design for the new topology are discussed in detail. The system parameters are carefully calculated for an intended application of automotive 14 V/42 V dual-bus DC distribution system. A double-zero-double-pole PI controller is designed to control the high rail voltage as well as bidirectional power. The simulation and experimental results show that the proposed topology has the advantages of power expansibility, smaller capacitance current stress and bidirectional operation. It is proved to be applicable in high current high temperature automotive DC distribution systems.

Keywords Multi-rail · DC distribution system · Interconnection converter

32.1 Introduction

DC power distribution provide a flexible solution to a variety application, Transmitting power in a dc form could save one stage of ac-dc conversion, such that the overall system efficiency can be improved. However, when the load increases, the conduction loss due to current in DC power system will increase in square. If the dc voltage level is low, the loss problem will be more stringent. A common solution will be the voltage boost. If the existing voltage level is increased, the transmission less due to the line resistance will be reduced significantly.

This paper presents an interconnection bi-directional dc-dc converter for multi-rail DC distribution system [1–5]. The principles of the energy conversion process, small signal modeling and controller design for the new topology are

H. Lu · Y. Zhu · Y. Hua · W. Jiang (✉) · N. Li
School of Hydraulic, Energy and Power Engineering, Yangzhou University, Yangzhou 225000, China
e-mail: jiangwei@yzu.edu.cn

© Springer-Verlag Berlin Heidelberg 2016

Y. Jia et al. (eds.), *Proceedings of the 2015 Chinese Intelligent Systems Conference*, Lecture Notes in Electrical Engineering 359,
DOI 10.1007/978-3-662-48386-2_32

discussed in detail. The system parameters are carefully calculated for an intended application of automotive 14 V/42 V dual-bus DC distribution system. A double-zero-double-pole PI controller is designed to control the high rail voltage as well as bidirectional power. The simulation results show that the proposed topology has the advantages of power expansibility, smaller capacitance current stress and bidirectional operation. It is proved to be applicable in high current high temperature automotive DC distribution systems.

32.2 Converter Topology Analysis

The proposed bidirectional dc-dc converter topology is shown in Fig. 32.1. The basic switching cell [6] is the bridge circuit. Input source V1 is connected to the middle point of the phase-leg via inductor L1, L2, and L3 respectively. The energy-transfer capacitor C2 is connected across the bridge. The input energy is processed and transferred to the output capacitor C3. The output voltage is the summation of the input voltage and the VC3.

To simplify the analysis, the switches S3-S6 and their anti-parallel diodes are removed to obtain a single-phase circuit. According to the switching status of the S2, there are two operational modes for this topology [7–9]. When power MOSFET S2 is closed, the input inductor L1 stores energy. The current in the inductor L1 increases linearly. Meanwhile, the energy-transfer capacitor C2 reverse biases the diode D1. The energy on C2 is transmitted to the output capacitor C3 and the load through the output inductor L4. When power MOSFET S2 is open, the voltage polarity in the inductor L1 is reversed. The diode D1 is forced to be forward-biased. The energy-transfer capacitor C2 is charged by the power supply voltage V1 and the input inductor L1. The energy then is transferred through output inductor L4 to output capacitor C3 and the load. The superposition of the voltage in output capacitor C3 and input voltage is yielded to be the high voltage output of the system.

In one switching cycle, energy-transfer capacitor C2 charges and discharges each once. The equivalent circuit diagram when the switch S1 is open and closed is

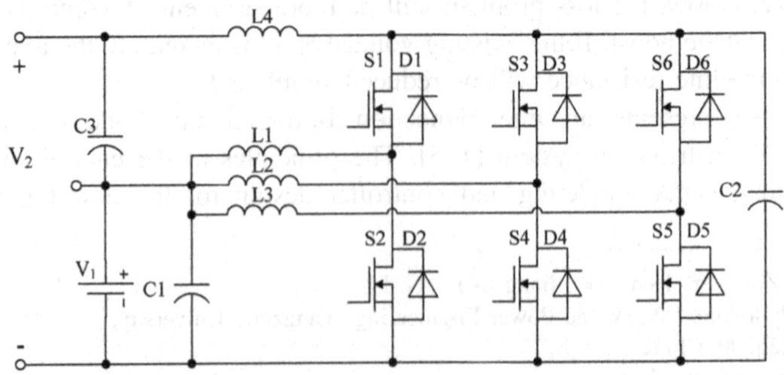

Fig. 32.1 Interconnection converter topology

shown in Fig. 32.2a, b respectively. It can be observed that when the switch is closed, the capacitor C2 discharges to the load through the output inductor L4. When the switch is open, the energy in L1 is transferred to C2. According to the principle of capacitor charge balance, it can be drawn that the charging current of capacitor is input current and the discharging current is output current. In Fig. 32.3a, one can find that the RMS value of capacitor current is equivalent to the output and input current level during the switch on and off interval respectively. As ESR exists in all capacitors, there could be considerable conduction losses with such large exchanged currents.

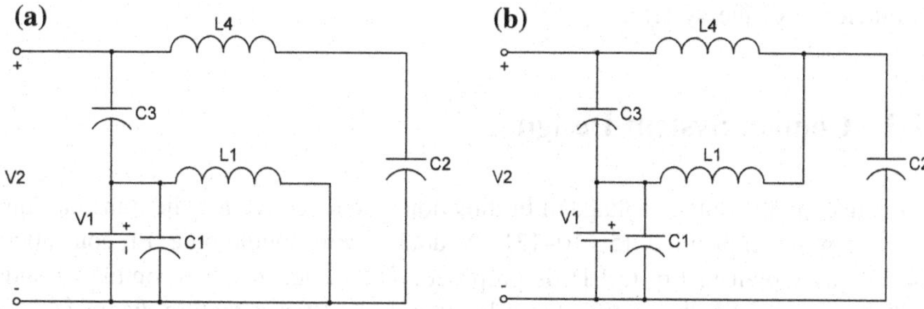

Fig. 32.2 Equivalent circuits, **a** the switch is on, **b** the switch is off

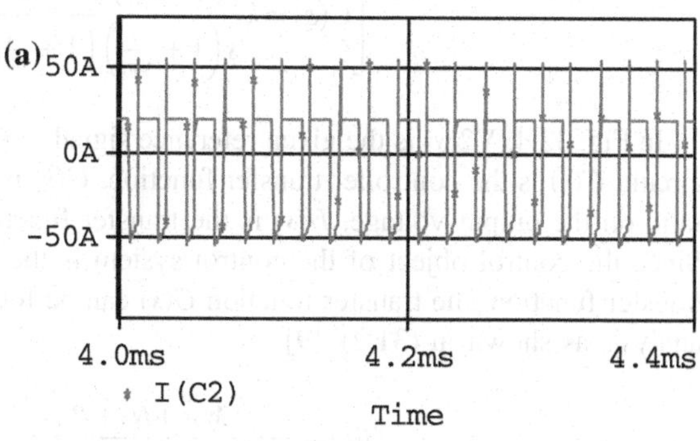

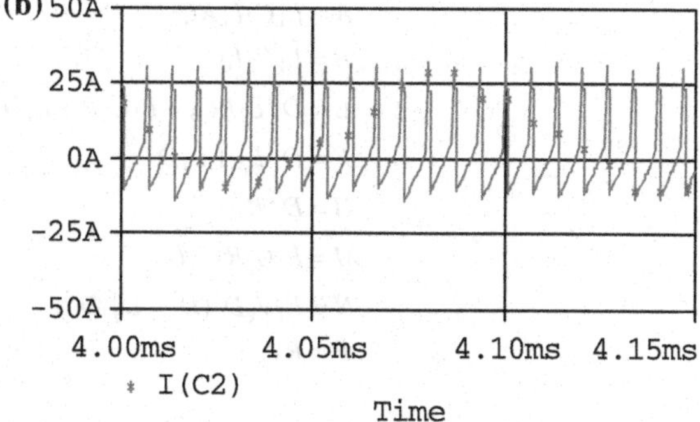

Fig. 32.3 Current in energy-transfer capacitor C2. **a** single-phase; **b** three-phase

Table 32.1 Comparison of the current impact amplitude in capacitor

Topology	Peak-peak current (A)
1 phase	72
3 phase	24

If the number of bridge arms is expanded and the phase shift in the drive signal is implemented, [6] the charging and discharging current in capacitor can be greatly reduced. Thus the current in capacitor I_{C2} is shown in Fig. 32.3b.

By comparing the magnitude of current impact in Fig. 32.3a, b and specific data shown in Table 32.1, it can be found that three-phase complementary circuit can reduce the amplitude of current impact in the energy-transfer capacitor to increase the efficiency of the system.

32.3 Control System Design

By controlling the output voltage of bi-directional dc-dc converter, the transmission power can be controlled [5, 10–13]. A double-zero-double-pole PI controller [14–17], as shown in Eq. (32.1), is proposed. The design result is simulated and verified. Figure 32.3 shows the closed-loop control system structure diagram.

$$\left[C(s)=k\frac{\left(1+\frac{s}{\omega_{z1}}\right)\left(1+\frac{s}{\omega_{z2}}\right)}{s\left(1+\frac{s}{\omega_{p1}}\right)\left(1+\frac{s}{\omega_{p2}}\right)}\right] \tag{32.1}$$

In Fig. 32.4, V2(s) is the given reference signal. $v_2(s)$ is the output voltage of the circuit. $C(s)$ is the controller transfer function. $G(s)$ is the transfer function of duty ratio on the output voltage. $H(s)$ is the transfer function of the feedback network. Since the control object of the control system is the output voltage to duty cycle transfer function, the transfer function $G(s)$ can be found according to small-signal analysis as shown in (31.2). [9]

$$\begin{aligned}
G(s) &= \frac{Ms^2+Ns+P}{As^4+Bs^3+Es^2+Fs+G}\\
A &= L_1C_2L_2RC_2\\
B &= L_1C_2L_2\\
E &= D'^2L_2RC_3+L_1C_2R+L_1D^2RC_3\\
F &= D'^2L_2+L_1D^2\\
G &= D'^2R\\
M &= L_1C_2RV_g/D'\\
N &= L_1V_gD^2(D'-D)/D'^2\\
P &= RV_g
\end{aligned} \tag{32.2}$$

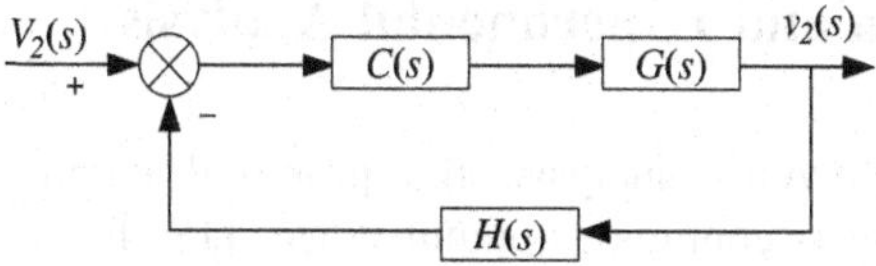

Fig. 32.4 Closed-loop control system block diagram

The parameters of the control object are shown in Table 32.2.

Figure 32.5 shows the bode plot of the system with close-loop compensation. The zeros are selected at 1.3 kHz; and the poles are selected at 20 kHz. After tuning, crossover frequency of the open loop transfer function is at 10 Hz. The low-frequency gain and intermediate frequency bandwidth are increased. High-frequency noise is attenuated. The anti-interference performance of the system is improved, and phase margin reaches 92.8°.

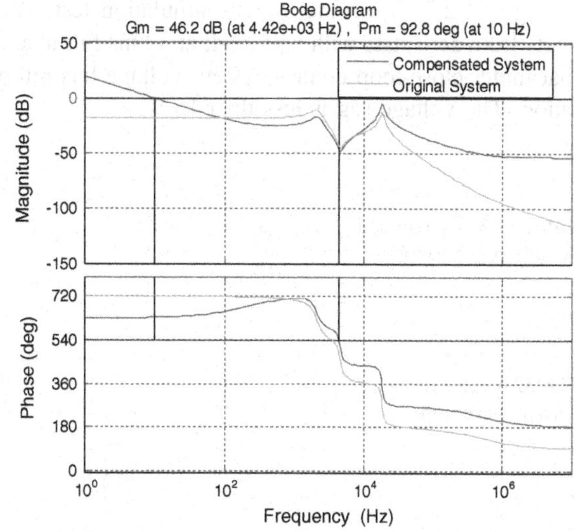

Fig. 32.5 System Bode plot after tuning

Table 32.2 Comparison of the current impact amplitude in capacitor

V_g	42 V
V_c	14 V
C_2	40 uF
C_3	60 uF
D	0.66
R	1.764
L_1, L_2, L_3	10 uH
L_4	2 uH

32.4 Simulation and Experimental Analysis

According to the previous analysis, the proposed circuit topology achieves continuous-input power- continuous-output power [18, 19]. Appropriate inductor current ripple factor should be selected to reach a reasonable inductor size and to promise that the value of energy transfer capacitance is small enough to be able to use conventional ceramic chip elements. According to the design specification, that the current ripple factor of the input inductor is less than 30 %; and the ripple factor of output voltage is less than 1 % [20].

After adjusting parameters and simulating, the ripple factors meet the requirement, and specific data are shown in Table 32.3.

The input voltage is selected at 14 V, and the rated power is selected for 1 kW at 42 V output voltage. Figure 32.6 shows the current waveforms when the system is under the steady—state operation. The first sub-figure shows the system input current and three-phase inductor currents, the input current ripple is reduced from 15 to 1.5 A. Because the energy transferred by output inductor L4 is that buffed by C2, the ripple of output current is very small which is under 0.2 A.

Figure 32.7 shows the system simulation test with the load-step. The system starts from zero state with half load; at 30 ms full load is switched in. It can be seen that under close-loop control system voltage bus returns to 42 V after 3 ms regulation. The voltage sag is less than 15 %.

Table 32.3 Current and voltage ripple factors

Δi_{L4}	5 %
Δv_{C3}	0.5 %

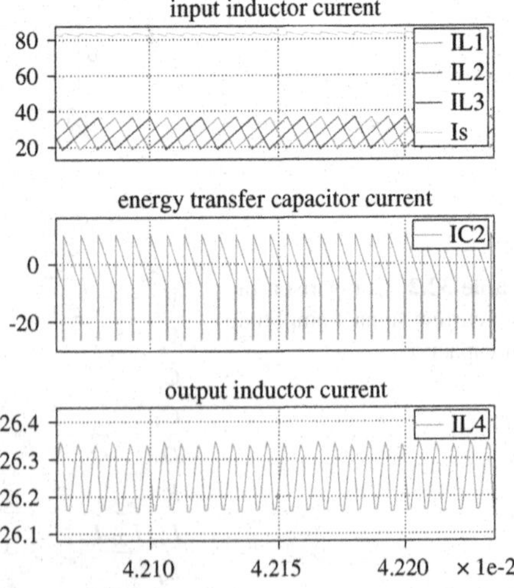

Fig. 32.6 Steady-state current waveform

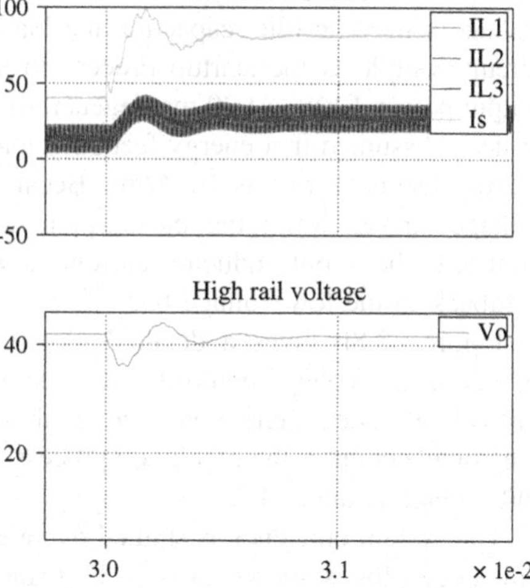

Fig. 32.7 Step load simulation

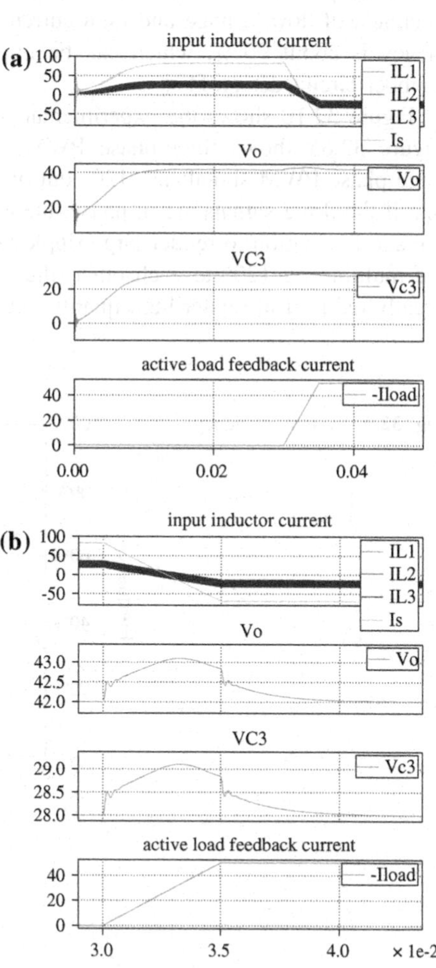

Fig. 32.8 System startup waveforms, **a** system start and bidirectional operation, **b** detailed view of the regenerative transient

Figure 32.8 shows the waveforms of input inductor current, output voltage, the voltage across the filter capacitor and load feedback current. From 0 to 10 ms, the circuit completes the startup process to achieve output voltage 42 V and stable output power 1 kW. At 30 ms, regenerative load starts to feed current back into the system. Assumed that energy feedback load is high-performance motor drive load, current feedback rate is 10 A/ms. Because of the energy feedback, the high rail voltage surges. With the closed-loop control, the duty cycle is automatically adjusted, the input inductor current is automatically reversed and the current feedback to the low voltage bus.

Figure 32.8b gives a detailed plot of the regeneration transient. As can be observed, the voltage controller feedbacks the current in order to control the high rail voltage. The regenerative current takes about 5 ms to rise from 0 to 50 A; under the voltage control, the regulation process takes about 15 ms. The overshoot in high rail voltage is under 1 V.

The system efficiency is studied by sweeping the load from 10 to 120 %.System efficiency plot is shown in Fig. 32.9. Indicating that the system with can stably operate with low voltage and high current at various power levels. Full load efficiency is about 90 %, which can be improved by device optimization and the snubber circuit design [17].

Figure 32.10 shows the experimental waveform for the steady state operation. Figure 32.8a shows three-phase PWMs are in phase while Fig. 32.8b shows three-phase PWM signals are 120° out of phase to each other. It can be observed that if the drive signals are in phase, the energy transfer capacitor C2 shows large voltage fluctuation to render large ripple output voltage. If the phase drive signals are multiplexed 120° to each other, the stress in the energy transfer capacitor is greatly reduced to render high quality output voltage.

Fig. 32.9 System efficiency

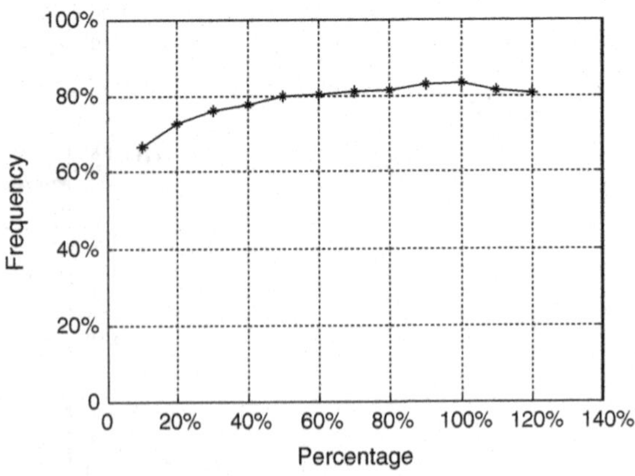

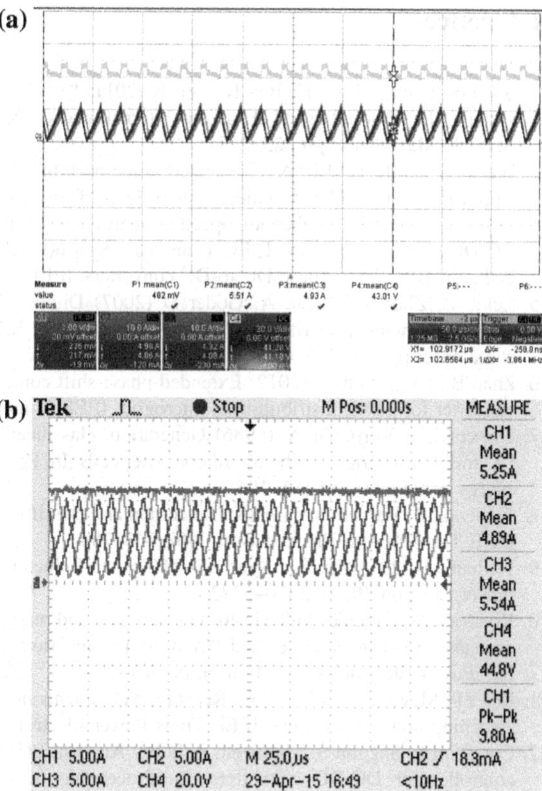

Fig. 32.10 The steady state inductor current and output voltage, **a** three-phase in phase, **b** three-phase 120° multiplexed

32.5 Conclusions

This paper proposes an interconnection bi-directional dc-dc converter for multi-rail DC distribution system. Principles and characteristics of the interconnected converter topology are analyzed. A double-zero-double-pole PI controller is designed to control the output voltage. Through the simulation test, when the system starts up and in steady state, the current and voltage characteristics are analyzed. The results show that the proposed topology has the advantages of power expansibility, smaller capacitance current stress and bidirectional operation. It is proved to be applicable in high current high temperature automotive DC distribution systems with multi-DC-bus structure.

Acknowledgment This work is sponsored by National Students' Innovation and entrepreneurship training program (201411117001), National Natural Science Foundation of China (grant number 51207135), Jiangsu Natural Science Foundation (grant number BK2012266), and YZU-Yangzhou City Joint Fund (grant number 2012038-10), CSC No. 201409300007.

References

1. Seo GS, Baek J, Choi K, Bae H, Cho B (2011) "Modeling and analysis of DC distribution systems. In: Proceeding of IEEE 8th International conference on power electronics ECCE Asia, 30 May–3 Jun, pp 223–227
2. Wang Z, Li H (2013) An integrated three-port bidirectional DC–DC converter for PV Application on a DC distribution system. IEEE Trans Power Electron 28(10):4612–4624
3. Mooney J (2013) Application-specific instruction-set processor for control of multi-rail DC-DC converter systems. IEEE Trans. Circ Syst 60(1):243–254
4. Williams BW (2008) Basic DC-to-DC converters. IEEE Trans Power Electron 23(1):387–401
5. Lukic Z, Zhao Z, Prodic A, Goder D (2007) Digital Controller for Multi-Phase DC-DC Converters with Logarithmic Current Sharing. In: IEEE Power electronics specialists conference, pp 119–123
6. Zhao B, Yu Q, Sun W (2012) Extended-phase-shift control of isolated bidirectional DC–DC converter for power distribution in microgrid. IEEE Trans Power Electron 27(11):4667–4680
7. Tymerski R, Vorperian V (1986) Generation, classification and analysis of switched-mode DC-to-DC converters by the use of converter cells. In: Proceeding of IEEE telecommunication energy conference, pp 181–195
8. Tymerski R, Vorperian V (1988) Generation and classification of PWM DC-to-DC converters. IEEE Trans Aerosp Electron Syst 24(6):743–754
9. Williams B (2014) Unified Synthesis of Tapped-Inductor DC-to-DC Converters. IEEE Trans Power Electron 29(10):5370–5383
10. Kondrath N, Kazimierczuk M (2011) Loop gain and margins of stability of inner-current loop of peak current-mode-controlled PWM dc-dc converters in continuous conduction mode. In: IET Power Electronics, vol 4, no 6, pp 701–707
11. Liu YF, Meyer E, Liu X (2009) Recent developments in digital control strategies for DC/DC switching power converters. IEEE Trans Power Electron 24(11):2567–2577
12. Corradini L, Saggini S, Mattavelli P (2008) Analysis of a high-bandwidth event-based digital controller for DC-DC converters. In: Proceeding of IEEE Power electronics specialists conference, (PESC), pp 4578–4584
13. Qahouq JAA, Huang L, Huard D, Hallberg A (2007) Novel current sharing schemes for multiphase converters with digital controller implementation. In: IEEE Applied power electronics conference, pp 148–156
14. Zeng Jianwu (2014) An interconnection and damping assignment passivity-based controller for a DC–DC boost converter with a constant power load. IEEE Trans Ind Appl 50 (4):2314–2322
15. Mooney J, Mahdi AE, Kelly A, Rinne K (2008) DSP-based controller for multi-output/multi-phase high switching frequency DCDC converters. In: IEEE Workshop on control and modeling for power electronics, pp 1–6
16. Mooney J (2010) DSP-Based control of multi-rail DC-DC converter systems with non-integer switching frequency ratios. In: 2010 1st International conference on energy, power and control (EPC-IQ), pp 203–207
17. Lukic Z, Rahman N, Prodic A (2007) Multibit sigma-delta pwm digital controller IC for DC-DC converters operating at switching frequencies beyond 10 MHz. IEEE Trans Power Electron 22:1693–1707
18. El Khateb AH (2015) DC-to-DC Converter with low input current ripple for maximum photovoltaic power extraction. IEEE Trans. Ind Electron. 62(4):2246–2256
19. Lukic Z, Zhao Z, Prodic A, Goder D (2007) Digital Controller for Multi-Phase DC-DC Converters with Logarithmic Current Sharing. In: IEEE Power electronics specialists conference, pp. 119–123
20. Williams B (2013) DC-to-DC converters with continuous input and output power. IEEE Trans Power Electron 28(5):2307–2316

Chapter 33
Sensitivity Analysis and Simulation of Performance for $M/G/1/K$ Queuing Systems

Liping Yin, Hongyan Zhang, Lihua Wang and Li Zhou

Abstract In this paper, the performance potential and performance derivatives are analyzed for $M/G/1/K$ queuing systems by the embedded Markov Chain. A computation algorithm is also given basing on a single sample path. To demonstrate the effectiveness of this algorithm, a special $M/G/1/K$ queuing systems is given in the simulation section, which indicates the estimation error is very small.

Keywords $M/G/1/K$ queuing systems · Performance potential · Sensitivity analysis · Simulation

33.1 Introduction

The network queueing system is one of the most commonly used mathematical models of the discrete event dynamic systems (DEDS). Due to the wide application of the network queueing systems in many actual systems, such as communication networks, flexible manufacturing and public services, more and more attention has been paid for the performance analysis and optimization problems of network queueing systems. For example, in [1], a new theory named *Markov performance potential theory* was proposed. In [2–6], the performance potential theory is used to study the sensitivity estimation and simulations of of one sample path based Markov closed queuing networks. In this kind of queueing systems, it is assumed that the systems are of infinite capacities, which means the customers can enter the system whenever he(she) arrives. However, actual queueing systems are all of of finite capacities

L. Yin (✉)
School of Automation Science and Electrical Engineering, Beijing University of Aeronautics & Astronautics, Beijing 100191, China
e-mail: lpyin@nuist.edu.cn

L. Yin · H. Zhang · L. Wang · L. Zhou
CICAEET, Nanjing University of Information Science & Technology, Nanjing 210014, China

© Springer-Verlag Berlin Heidelberg 2016

Y. Jia et al. (eds.), *Proceedings of the 2015 Chinese Intelligent Systems Conference*, Lecture Notes in Electrical Engineering 359,
DOI 10.1007/978-3-662-48386-2_33

[7, 8]. Taking $M/G/1/K$ as an example, the semi-Markov process, which is used to describe the states' changes shares the same steady-state probability with the embedded Markov chain [7]. Therefore, the embedded Markov chain can be used to analyze the sensitivity and the steady-state performance of the system. And it is more easier to formulate the sensitivity of the steady-state performance using embedded Markov chain.

In this paper, a simulation algorithm is also studied to compute the potential and performance derivative for the $M/G/1/K$ queueing system. This paper is organized as follows:

Firstly, a brief introduction of $M/G/1/K$ queueing system is be provided. Secondly, the one-step transition probability matrix and the steady-state probability are formulated, and then the formulas of the sensitivity and the steady-state performance are given. Finally, a simple simulation is carried out based on the one sample path simulation method.

33.2 $M/G/1/K$ Queuing System

It is supposed that there is only one service desk. The arrival process of the customers is assumed to be a homogeneous poisson process with intensity λ. The service time for each customer is independent with identical distribution G, which is also independent of the arrival process. The service rule is first come first served(FCFS). When the number of queuing customers reaches $K-1$, the new customers will be refused to enter into the system unless some customer leave. Denote Z_t as the number of the customers staying in the system before time t. It can be conclude that $Z = \{Z_t; t \geq 0\}$ is a semi-Markov process with the state space $\phi = \{0, 1, 2, \ldots, K-1\}$. Denote $X = \{x_n; n \geq 0\}$ as an embedded homogeneous Markov chain, that is to say, x_n is the number of customers waiting to be served when the nth customer leaves. To determine the transition probability, denote y_n as the number of customers arriving the system during the period when the nth customer was served.

Let $p_{ij} = P\{x_{n+1} = j | x_n = i\}$, then
when $i = 0, j = 0, 1, 2, \ldots, K-2$, $p_{ij} = P\{y_{n+1} = j\} = a_j$;
when $i = 0, j = K-1$, $p_{ij} = \sum_{j=K-1}^{\infty} P\{y_{n+1} = j\} = 1 - \sum_{j=0}^{k-1} a_j$;
while
when $1 \leq i \leq K-1, i-1 \leq j \leq K-2$, $p_{ij} = P\{y_{n+1} = j-i+1\} = a_{j-i+1}$;
when $1 \leq i \leq K-1, j = K-1$,

$$p_{ij} = \sum_{j=k-i}^{\infty} P\{y_{n+1} = j-i+1\} = 1 - \sum_{j=0}^{k-(i+1)} a_{j-i+1}.$$

Thus the one-step transition probability matrix is

$$P = \begin{pmatrix} a_0\ a_1\ \dots\ a_{K-2}\ 1 - \sum_{j=0}^{K-2} a_j \\ a_0\ a_1\ \dots\ a_{K-2}\ 1 - \sum_{j=0}^{K-2} a_j \\ 0\ a_0\ \dots\ a_{K-3}\ 1 - \sum_{j=0}^{K-3} a_j \\ \vdots\ \vdots\ \vdots\ \vdots\ \vdots \\ 0\ 0\ \dots\ 0\ 1 - a_0 \end{pmatrix} \tag{33.1}$$

where $a_j = \lim_{t\to\infty} \int_0^t \left[\frac{e^{-\lambda x}(\lambda x)^j}{j!}\right] dG(x), j = 0, 1, 2, \dots, K-2$.

Denote $\rho = \sum_{j=0}^{\infty} ja_j$ as the average number of arriving customers during the period when one customer is being served. When $\rho < 1$, there is only steady-state probability matrix for X, which can be written as

$$\pi = (\pi(0), \pi(1), \pi(2), \dots, \pi(K-1))$$

and satisfies:

$$\pi e = 1 \tag{33.2}$$

$$\pi(P - I) = 0 \tag{33.3}$$

$$\pi(i) \geq 0, i = 0, 1, 2, \dots, K-1.$$

where I is the $K-1$-dim unit matrix, e is an $K-1$-dim column vector whose components are all 1.

It is supposed the distribution G is differentiable with respect to θ within the range $J \subset R$, the performance function f is differentiable with respect to λ and θ, then the sensitivity of steady state performance η_f can be formulated as

$$\frac{\partial \eta_f}{\partial \theta} = \pi \frac{\partial P}{\partial \theta} x^{\{f\}} + \pi \frac{\partial f}{\partial \theta} \tag{33.4}$$

$$\frac{\partial \eta_f}{\partial} = \pi \frac{\partial P}{\partial \lambda} x^{\{f\}} + \pi \frac{\partial f}{\partial \lambda} \tag{33.5}$$

where $x^{\{f\}}$ is the potential vector of the embedded Markov chain X , that is to say, the following Poisson equation holds:

$$(P - I)x^{\{f\}} = -f + \eta_f e \tag{33.6}$$

Remark 33.1 Denote the one step transition probability matrix of $M/G/1$ queuing system as P^*, then when $\rho < 1$, there is only one steady state and let's just denote it as

$$\pi^* = \left(\pi^*(0), \pi^*(1), \pi^*(2), \dots\right),$$

which satisfies

$$\pi^*(P^* - I^*) = 0, \tag{33.7}$$

$$\pi^* e^* = 1 \tag{33.8}$$

where I^* is the infinite-dimensional unit matrix, e^* is the infinite-dimensional column vector whose elements are all 1. According to (8),

$$\sum_{i=1}^{K-1} \pi^*(i) = constant.$$

Comparing (33.3) with (33.7), it can be found that the first $K-1$ equations of (33.3) are the same as the first $K-1$ equations of (33.7). combining with (33.2), it can be concluded that

$$\pi(i) = c\pi^*(i), i = 0, 1, 2, \ldots, K-1.$$

33.3 Algorithm Analysis

In this section, the performance potential will be estimated by analyzing one sample path, and then the statistic of derivatives of the performance potential will be obtained. In the above analysis, the sensitivity analysis of the steady-state performance is carried out for $M/G/1/K$ queuing system by using the embedded Markov chain. The following algorithm is designed from the aspect of Markov Chain.

Denote T_n as the state transition time of the $M/G/1/K$ queuing system. Let $X_n = Z_{T_n^+}$, then $\{x_n\}_{n=0}^{\infty}$ is the embedded Markov chain of the queuing system. Denote $\{x_n^i\} = \{x_n : x_0 = i, n \geq 0\}$ as an embedded Markov chain of $\{x_n\}_{n=0}^{\infty}$ initiating from state i. Define

$$\varepsilon^j(X_n) = \begin{cases} 1, X_n = j; \\ 0, X_n \neq j. \end{cases}$$

According to Borel Theorem, the steady-state probability of state j is

$$\pi(j) = \lim_{c\to\infty} \frac{1}{c} \sum_{n=0}^{c-1} \varepsilon^j(X_n).$$

It is noted that when the Markov chain reaches $\{x_n^i\}$ reaches the steady state, the following equation holds:

$$E\{\varepsilon^j(X_n)\} = \pi_j$$

then the unbiased estimations for π_j and η_f can be represented by

$$\hat{\pi} = \frac{1}{c}\sum_{n=0}^{c-1} \varepsilon^j(X_n)$$

and $\hat{\eta}_f = \pi\hat{f}$ respectively.

Denote $L^{\{j\}}(i) = \inf\{n|X_n^{\{j\}} = i; n \geq 0\}$ as the first time period when X_n starts from state j and reaches state i ; Correspondingly, denote $L_s^{\{j\}}(i)$ as the sth time period when X_n starts from state j and reaches state i. Moreover, denote $t(i,j,s)$ as the time period when X_n arrives at state j from the sth arrival of state i. Obviously

$$L_s^{\{j\}} = t(j,i,s) - t(t,j,s-1).$$

According to the basic characteristics of Markov Chain, the samples $L_s^{\{j\}}(i), s = 1,2,\ldots$ are independent and yields identical distribution(i.i.d). Denote the performance in time period $[t(i,j,s-1), t(i,j,s)]$ as

$$\sum_{t(i,j,s-1)}^{t(i,j,s)} f(X_k) = R_s^{\{j\}}(i), s = 1,2,\ldots$$

then $R_s^{\{j\}}(i), s = 1,2,\ldots$ are i.i.d. and

$$E\{R_s^{\{j\}}(i)\} = \overline{R} < +\infty$$

According to the strong law of large numbers, we have

$$E\{L^{\{j\}}(i)\} = \lim_{N\to\infty}\frac{1}{N}\sum_{s=1}^{N} L_s^{\{j\}}(i),$$

$$E\left\{\sum_{n=0}^{L^{\{j\}}(i)-1} f(X_n^i)\right\} = \lim_{N\to\infty}\frac{1}{N}\sum_{s=1}^{N} R_s^{\{j\}}(i)$$

Statistically, they can be estimated as

$$\hat{E}\{L^{\{j\}}(i)\} = \frac{1}{H^{\{j\}}(i)}\sum_{s=1}^{H^{\{j\}}(i)} L_s^{\{j\}}(i),$$

$$\hat{E}\left\{\sum_{n=0}^{L^{\{j\}}(i)-1} f(X_n^i)\right\} = \lim_{N\to\infty}\frac{1}{N}\sum_{s=1}^{N} R_s^{\{j\}}(i) \tag{33.9}$$

where $H^{\{j\}}(i)$ is the number of transition from state j to state i. Combining (33.8) and (33.9) yields

$$\hat{d}_{i,j} = \hat{E}\left\{\sum_{n=0}^{L^{\{j\}}(i)-1} f(X_n^i)\right\} - \hat{E}\left\{\hat{L}^j(i) - 1\right\}\hat{\eta}_f,$$

$$\hat{x}^{(f)} = -\hat{D}\hat{\pi}, \hat{D} = \left[\hat{d}_{i,j}\right] \quad (33.10)$$

where $d_{i,j}$ is the realization factor of X with respect to f. Then the unbiased estimation of the steady-state derivative of the embedded Markov chain can be obtained as

$$\frac{\partial \hat{\eta}_f}{\partial \theta} = \hat{\pi}\frac{\partial P}{\partial \theta}\hat{x}^{\{f\}} + \hat{\pi}\frac{\partial f}{\partial \theta} \quad (33.11)$$

which converges to $\frac{\partial \eta_f}{\partial \theta}$ in probability 1.

The algorithm can be summarized as follows:

(1) Estimate $\hat{E}\left\{\sum_{n=0}^{L^{\{j\}}(i)-1} f(X_n^i)\right\}$, $E\{L^{\{j\}}(i)\}$, π and η_f through one sample path;
(2) Compute $d_{i,j}$ and D through (33.10);
(3) Compute the potential $\hat{x}^{(f)}$ basing on $\hat{x}^{(f)} = -\hat{D}\hat{\pi}$ in (33.10);
(4) Compute the derivative of steady-state potential in (33.11).

33.4 Numerical Simulation

Consider an $M/E_m/1/K$ queuing system. The arrival process yields the following mth order Erlang distribution:

$$f(t) = \frac{m\mu(m\mu t)^{m-1}e^{-m\mu t}}{(m-1)!}, t \geq 0.$$

Choose the parameters as $\lambda = 1, \mu = 2, m = 2, K = 5$, and the performance index as $f(i) = i, i = 1, 2, 3, 4, 5$.

The potential estimation is

$$-0.6230, 0.3857, 0.6029, 1.6019, 2.5963$$

The derivatives of the potential with respect to λ and μ are estimated as

$$\frac{\partial \hat{\eta}_f}{\partial \lambda} = 0.3588, \frac{\partial \hat{\eta}_f}{\partial \mu} = -0.1794$$

The theoretical value for $\frac{\partial \eta_f}{\partial \lambda}, \frac{\partial \eta_f}{\partial \mu}$ are

$$\frac{\partial \eta_f}{\partial \lambda} = 0.3591, \frac{\partial \hat{\eta}_f}{\partial \mu} = -0.1796$$

Obviously, the estimation error is very small.

33.5 Conclusion

An algorithm to compute the potential sensitivity for $M/G/1/K$ queuing system is given in this paper. This algorithm avoids compute the state transition matrix and thus reduce the computation load. It can be directly applied in the online potential analysis and optimization because it is designed in one sample path.

Acknowledgments The work reported here is jointly supported by NSFC under grant No. 61320106010, 61573190, 61571014, 61403207, 51405241, China Postdoctoral Science Foundation funded project (2012M520141), Jiangsu Outstanding Youth Fund(BK20140045) and the Practice Innovation Training Program Projects for the Jiangsu College students(201410300092X). These are gratefully acknowledged.

References

1. Cao XR, Chen HF (1997) Perturbation realization, potentials and sensitivity analysis of Markov processes. IEEE Trans Autom Control 42(10):1382–1393
2. Zhou YP, Xi HS, Yin BQ et al (2002) Optimality equations based performance potentials for a class of controlled closed queuing networks. Control Theory Appl 19(4):521–526
3. Yin BQ, Zhou YP, Xi HS et al (1999) Sensitivity formulas of performance in two- server cyclic queuing networks with phase-type distributed service times. Int Trans Oper Res 6(6):649–663
4. Dai GP, Yin BQ, Zhou YP et al (2003) Sensitivity estimate and simulation of performance for $M/G/1$ queuing systems. J Syst Simul 15(7):950–952
5. Yin BQ, Xi HS, Zhou YP (2000) Sensitivity analysis of performance for $M/G/1$ queuing systems. Appl Math J Chin Univ Ser A 16(2):235–242
6. Heidelberger P, Towsley D (1989) Sensitivity analysis from sample paths using likehoods. Manag Sci 35(12):1475–1488
7. Papoulis A, Pillai SU (2004) Probability, random variables and stochastic processe. Xian jiaotong University Press, Xian
8. Konstantopoulos P, Zazanis MA (1992) Sensitivity analysis for stationary and ergodic queues. Adv Appl Probab 24(3):738–750

Chapter 34
Bayesian Network Structure Learning Algorithms of Optimizing Fault Sample Set

Hongyang Guo, Rupei Zhang, Jiaqi Yong and Bin Jiang

Abstract As a representative heuristic search algorithm of structure learning algorithms for Bayesian network, K2 algorithm is easily trapped in local optimum in the process of Bayesian network structure optimization. This paper proposes a K2 algorithm based on chaotic perturbation. First of all, define the order of the node based on mutual information to construct a Bayesian network structure. Then introduce distribution function skew tent map combined with chaotic search. And ergodicity is used to jump out of local optimum and achieve global optimization. Finally, taking steering shaft control structure of B777 control system as an example, the proposed algorithm presents a good effect in Bayesian network structure learning.

Keywords Bayesian network · Chaotic perturbation · K2 algorithm · Mutual information

34.1 Introduction

During the past 20 years, Bayesian network (BN) technology which was widely applied to the engineering practice had been used to analyze uncertainties and had ability to describe logical relationship of the uncertain events by the graph pattern of variables' probability. Actually, network structure based on expert knowledge is time-consuming and laborious, sometimes even impossible [1].

For optimal Bayesian network structure, there are mainly two kinds of methods: one is scoring function and network structure searching; the other is correlation analysis [2]. Considering huge space of structure, heuristic search algorithm is mainly adopted. The K2 algorithm, proposed in 1992 by Cooper and Herskovits, is

---

H. Guo (✉) · R. Zhang · J. Yong · B. Jiang
Automation College, Nanjing University of Aeronautics and Astronautics, Nanjing 210016, China
e-mail: hongyanggnj@126.com

© Springer-Verlag Berlin Heidelberg 2016

Y. Jia et al. (eds.), *Proceedings of the 2015 Chinese Intelligent Systems Conference*, Lecture Notes in Electrical Engineering 359,
DOI 10.1007/978-3-662-48386-2_34

the most used algorithm. Then some authors utilized other metrics via K2 algorithm such as the minimum description length (MDL) metric to improve performance [3]. Confronting gigantic space of search, the other authors used this algorithm with other search strategies to avoid being trapped in local optimum. The K2 algorithm was combined with simulated annealing algorithm to get the approximate global solution in Ref. [4], but the efficiency was insufficient. Reference [5] proposed the scoring function of K2 with the Hill-Climbing strategy to solve converging to a local optimum, but the convergence rate was reduced.

In this paper, we propose the K2 algorithm combined with chaos search. In the framework of classical K2 algorithm, mutual information is applied because K2 algorithm nodes need to confirm the order. Meanwhile, uniform distribution function—skew tent map is introduced to increase global scope of the search. The hybrid optimization algorithm can search the global optimal solution with faster search capability. The successful application of B777 shows that the method is feasible and effective.

34.2 Bayesian Network

Combined with the probability theory and graph theory, Bayesian network is shown by the visual graph with the connected probability of the variables. There are two parts: the directed acyclic graph (G) and the node conditional probability table (P). BN topology structure represents the composition of edges between variables and connected nodes from parent node to child node [6]. Among them, the representative of the causal relationship between the variable is the edge. Conditional probability table P depends on the probability of the node and the parent node. For those nodes without any parent node, the prior probability can be directly used [7].

34.3 Learning Algorithm

34.3.1 K2 Algorithm

The core idea of K2 algorithm is to define the evaluation function scaling the network structure. Finally, the optimal structure (G_s) is selected because of bigger posteriori probability [8].

$$P(G_s|D) = \frac{P(D|G_s)P(G)}{P(D)} = \frac{P(G_s, D)}{P(D)} \tag{34.1}$$

Where D represents the data base, $P(D|G_s)$ is marginal likelihood function; $P(G)$ is priori probability distribution of the structure, usually supposed uniform distribution that is $P(G) = c$, and c is constant; $P(D)$ is independent of the structure G. Just find G_s to maximize $P(G_s, D)$. The following notation is used: n is node number, π_i is the set of parent node of X_i, $q_i = \varphi_i$ is number of possible configurations of π_i, r_i is the number of the possible value of X_i, N_{ijk} is a kind of frequency of X_i [9].

$$\max_{G_s} P(G_s, D) = c \prod_{i=1}^{n} \max_{\pi_i} \left[\prod_{j=1}^{q_i} \frac{(r_i - 1)\,!}{(N_{ij} + r_i - 1)\,!} \prod_{k=1}^{r_i} N_{ijk}! \right] \quad (34.2)$$

Formula 34.2 is separated into the product of n local structure which consist of X_i and π_i. The score of local structure is:

$$score(X_i, \pi_i) = \prod_{j=1}^{q_i} \frac{(r_i - 1)\,!}{(N_{ij} + r_i - 1)\,!} \prod_{k=1}^{r_i} N_{ijk}! \quad (34.3)$$

$$SCORE(G_s, D) = \prod_{i=1}^{n} score(X_i, \pi_i) \quad (34.4)$$

The concrete process of $score(X_i, \pi_i)$ increased is: from the sequence of nodes constantly find the node in front of X_i to join π_i until the score no longer increases. The limit of the number of parent node is η [10].

The pseudo code of K2 algorithm:

Input: A set of n nodes, a given order among them, an upper limit η on the number of parents for each node and a database D on the set
Output: A DAG with oriented arcs, parents of the node (Fig. 34.1).

```
for i := 1 to n do
    π_i := φ;
    S_old = SCORE(X_i, π_i);
    OKToProceed := true;
    while(OKToProceed and |π_i| < u do)
        let w be the node in (Pred(X_i) − π_i) that maximizes SCORE(X_i, π_i ∪ {w});
        S_new := SCORE(X_i, π_i ∪ {w});
        if S_new > S_old
        then S_new = S_old;
        π_i := π_i ∪ {w};
        else OKToProceed := false;
        end if
    end while
    write('Node:', X_i, 'Parent:', π_i)
end for
```

Fig. 34.1 Bayesian network learning of K2 algorithm

34.3.2 Mutual Information Define Variable Sequence

A node sequence set must be given in K2 algorithm although it is difficult to obtain by experts in practical problem. In order to determine the variable sequence, this paper introduces the mutual information (MI) that describes the dependence between two variables as $I(X;Y)$. If the interactional relationship exists between two variables, the state of a variable must contain another variable's. Considering each node as a variable in the network, calculate MI of every node by type 1 and find the order of the node

$$\begin{aligned} I(X;Y) &= H(X) - H(X|Y) \\ &= H(Y) - H(Y|X) \\ &= H(X) + H(Y) - H(X,Y) \end{aligned} \tag{34.5}$$

$H(X)$ shows entropy of X, $H(X|Y)$ shows entropy of X under the condition Y, $H(X,Y)$ shows joint entropy in the following equations.

$$H(X) = -\sum_{i=1}^{n} P(X_i) \cdot \log P(X_i) \tag{34.6}$$

$$H(X|Y) = -\sum_{i=1}^{n}\sum_{j=1}^{m} P(X=x_i, Y=y_i) \cdot \log P(X=x_i|Y=y_i) \tag{34.7}$$

$$H(X,Y) = -\sum_{i=1}^{n}\sum_{j=1}^{m} P(X=x_{ij}, Y=y_{ij}) \cdot \log P(X=x_{ij}|Y=y_{ij}) \tag{34.8}$$

where n *and* m is said the number of state variable X and Y, $P(X=x_i, Y=y_i)$ joint probability distribution of X and Y. The symmetry of $I(X;Y)$ is by MI. The greater value of $I(X;Y)$ shows the closer relationship between variables that leads up to the greater chance of connection between nodes. How to build node chain by the following:

1. Calculate the mutual information of each node, $I_i(X_i;Y_i)$
2. Descending order of mutual information, $I_i(X_i;Y_i)$
3. According to the order, take a pair of nodes in turn and calculate whether it is connection or circuit. If there were no, add an edge between nodes.
4. After take all nodes, form a node chain without direction.

The sequence as the input of K2 algorithm is still needed to confirm the direction of node chain. According to the forward η_1 and reverse η_2, structure learning uses small samples of the sequence extracted from data sets by K2 algorithm to get

SCORE_1 and SCORE_2. If SCORE_1 > SCORE_2, then take η_1 as input variable sequence of K2 algorithm, otherwise take η_2 as input [11].

34.3.3 Chaotic Perturbation

The trajectories of chaotic motion have a strong sensitivity to initial value of the change and separate exponentially over time, presenting ergodicity and stochastic characteristic. On the basis of these characteristic, optimization algorithm which main idea is to put the chaotic variable linear mapping to the value space of optimized variables is proposed.

This method has higher speed than random search or genetic algorithm, such as skew tent mapping system:

$$u_{n+1} = \begin{cases} u_n/a, & 0 < u_n \leq a \\ (1-u_n)/(1-a), & a < u_n \leq 1 \end{cases} \tag{34.9}$$

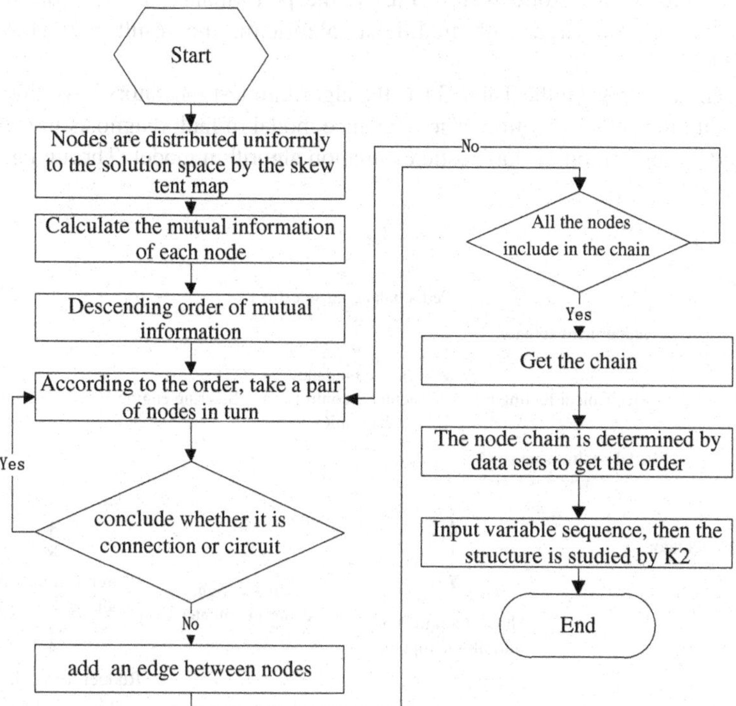

Fig. 34.2 A K2 algorithm based on chaotic search

When a = 0.6, the Skew Tent map produces more uniform distribution of chaotic variables to optimize quickly and efficiently [12].

34.3.4 Algorithmic Flow

The process of global algorithm as shown (Fig. 34.2):

34.4 Simulation Experiment

In order to validate the algorithm proposed in this paper, steering shaft control structure of B777 control system was taken as an example and only the normal mode of control surfaces was considered [13] (Fig. 34.3).

Artificial flight control system requires signal source for yaw control as shown in Table 34.1:

The rudder faults are omitted here.

Then the Bias network was established for verifying the performance of the K2 algorithm based on chaotic search. Lastly, the performance of the proposed K2 algorithm was compared with traditional algorithm, the result was shown in Table 34.2.

As can be seen from the Table 34.1, the algorithm test indicators have improved in line with the actual situation. The relevance model of fault diagnosis models can be found in application of diagnostic evaluation algorithm model. Through guiding

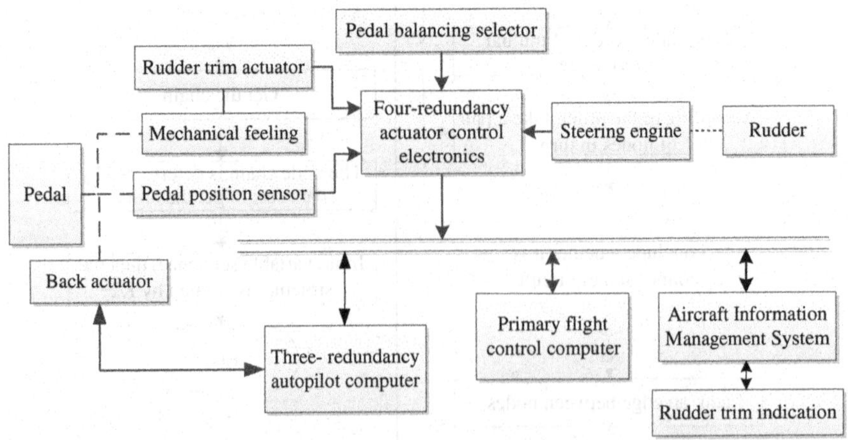

Fig. 34.3 B777 fly-by-wire flight control system yaw axis control structure

Table 34.1 The sensors of B777 yaw control

	Function	Desired signal	Corresponding sensor
Yaw control	Mode suppression	Mode-acceleration	Modal accelerometer
	Gust suppression	Gust pressure	Gust suppression pressure sensor
	Hydraulic pressure control	Height	ADIRU
	Yaw damping	Inertial data	ADIRU/SAARU
		Air data	ADIRU/SAARU
	Wheel/Rudder crosslinking	Steering wheel position	Steering wheel position sensor
	Command control of rudder	Pedal position	Pedal position sensor
	Asymmetric thrust compensation	Calculated air speed	ADIRU/SAARU

Table 34.2 The indicators of three algorithms

Test indicator (%)	This paper	K2 (orders given)	K2 (random)
Fault detect rate	100	99.87	98.35
The key fault detection rate	100	100	99.52
Fault isolation rate	83.1	80.15	81.12

the diagnosis model modified, the rate of fault diagnosis and fault isolation rate of diagnosis model would be improved.

Through comparing the conditional probability of nodes with random numbers in uniform distribution, sampling of the nodes can be realized. Based on the Bayesian network structure, sampling nodes in accordance with the order of the network structure, a group of random sampling data network can be generated. By using the sample_bnet function in Matlab toolboxes, a small data set containing N data sets will be produced. Then datas based on the set is expanded. Finally the structure is got by K2 algorithm. In order to reduce the accidental error, the experiment was repeated for 10 times, the average value is taken as the output for comparison.

In Table 34.3, compare our method with the previous method, **IE** refers to the multi edge, **LE** refers to the reduced edge, **RE** represents the reversed edge. The original sample number was set to N = 500.

Obviously, the learning algorithm determines the order of nodes by MI with traversing the solution space of the Skew Tent map to converge to global optimum. In the condition of small sample data set, the effect of Bayesian-chaotic network structure learning with prior sequence of nodes is better than K2 algorithm. With the data set reduced, the learning error increases, but the algorithm effects more superiority obviously.

In the condition of small sample data set, the effect of Bayesian-chaotic based on mutual information was flat to K2 algorithm in which information of orders of prior

Table 34.3 Comparison of the effect of the algorithm

Algorithm	IE	LE	RE
This paper	1.1	3.1	0
K2 (orders given)	0.2	4.7	0
K2 (random)	1.5	5.1	1.3

nodes was obtained, and it's better than K2 algorithm in random order. So it can be proved that the algorithm is effective to improve the performance when the sequence of nodes must be specified in advance.

Because Bayesian evaluation method has nothing to do with the specific application, the algorithm proposed by this paper can be applied to any decision support system based on Bayesian network. Therefore, this method has wide application value.

34.5 Conclusion

According to the defect that traditional K2 algorithm is easy to jump into local optimum in the process of optimization with Bayesian network structure learning, this paper proposes a new K2 algorithm based on chaos search, calculating node degree sequence using mutual information, and then form acquired Bayesian network based on the expanded sample datas. Through the simulation analysis, the proposed algorithm has good ability and high learning speed. So in the engineering background, it's a reliable and efficient method.

Acknowledgments Fund Projects: 1. Fund of National Engineering and Research Center for Commercial Aircraft Manufacturing (No.SAMC14-JS-15-053) 2. Fund of Graduate Innovation Center in NUAA (kfjj20150318), supported by "the Fundamental Research Funds for the Central Universities"

References

1. Zhongbao Z, Doudou D, Jinglun Z (2006) Application of Bayesian networks in reliability analysis[J]. Syst Eng Theory Pract 26(6):95–100
2. Tianmei L, Changhua H, xin Z (2011) Fault injection method resulting from inaccessible location fault based on fault propagation characteristics[J]. Acta Aeronautica et Astronautica Sinica 32(12):2277–2286
3. Bouckaert RR (1993) Probabilistic network construction using the minimum description length principle[M]. In: Symbolic and quantitative approaches to reasoning and uncertainty. Springer, Berlin, pp 41–48
4. Yan J, Yun'an H, Jin Z, Jun H (2012) Bayesian network structure learning combining K2 with simulated annealing[J]. J Southeast Univ Nat Sci 42:82–86
5. Yongguang W, Shichun P (2014) K2 & HC structure learning algorithm[J]. Comput Digital Eng 7:1137–1140

6. Haijun L, Dengwu M, Xiaom L et al (2009) Bayesian network theory in the application of the equipment fault diagnosis[M]. National Defence Industry Press, Beijing
7. Lijia X, Jianguo H, Houjun W et al (2009) Hybrid optimized algorithm for learning bayesian network structure[J]. J Comput Aided Des Comput Graphics 5:633–639
8. Cooper GF, Herskovits E (1992) A Bayesian method for the induction of probabilistic networks from data[J]. Mach Learn 9(4):309–347
9. Bouchaala L, Masmoudi A, Gargouri F et al (2010) Improving algorithms for structure learning in Bayesian Networks using a new implicit score[J]. Expert Syst Appl 37 (7):5470–5475
10. Ko S, Kim D (2014) An efficient node ordering method using the conditional frequency for the K2 algorithm[J]. Pattern Recogn Lett 40(4):80–87
11. Shaojin H, Jianxun L (2014) Structure learning algorithm for Bayesian network based on probability density kernel estimation[J]. Comput Eng Appl 50(15):107–112
12. Jiajie S, Feng L (2012) Structure learning of Bayesian network using adaptive hybrid Memetic algorithm[J]. Syst Eng Electron 34(6):1293–1298
13. Xiao X, Z J, Ping Z (2010) Analysis and research of fly-by-wire control system reliability modeling for large civil aircraft[J]. J Syst Simul 22:228–233

Chapter 35
The Face to Face and Teleoperation by Using Remote Control of Robots

Xue Wang and Chaoli Wang

Abstract In this paper, we propose a system capable of remotely controlling mobile robots based on the wireless. In this system, we can not only control mobile robots to monitor each corner of home when we are out, but also establish interactive videos among friends and family members when we are far-flung geographically. In addition, we can also teleoperate the robot to accomplish certain task at home in case some switches are not well cut-off when leaving. Meanwhile, we can also make use of sensor to ensure the robot security in the case of network delay.

Keywords Remote control · Service robot · Video connection · Sensor

35.1 Introduction

In recent years, the exponential growth in the mobile robot market and advances in communication technology make the intelligent robots be widely used [1]. Intelligent robots are increasingly applied in the risk industries, even in the other environmental barriers. For example: the application in search, rescue, maintenance of nuclear plants and space exploration [2]. Meanwhile, the remote control of mobile robot has also been used more diffusely, which not only plays an important role in these areas above, but also plays an indispensable role in the home service industry.

No matter what area the remote control of mobile robot is applied in, the remote control system contains three following modules:

Robot module that's server, which is mainly responsible for the surrounding environmental data acquisition of mobile robots and ensures its own security.

This paper was partially supported by The Scientific Innovation program(13ZZ115), National Natural Science Foundation (61374040, 61203143), Hujiang Foundation of China (C14002), Graduate Innovation Program of Shanghai(54-13-302-102).

X. Wang (✉) · C. Wang
Department of Control Science and Engineering, University of Shanghai for Science and Technology, Shanghai 200093, People's Republic of China
e-mail: xuewang_2015@163.com

© Springer-Verlag Berlin Heidelberg 2016

Y. Jia et al. (eds.), *Proceedings of the 2015 Chinese Intelligent Systems Conference*, Lecture Notes in Electrical Engineering 359,
DOI 10.1007/978-3-662-48386-2_35

Transmission module—it's mainly responsible for transmitting control commands from control module to robot and transmitting the environmental data between mobile robot and control module. Control module named as client which is mainly responsible for displaying the environmental data, sending the control commands to the mobile robot [3–5].

Implementation platform of the remote control robot has a wide range of methods. For example, [6, 7] realizes the mobile application for the Android operating system, which uses Bluetooth technology. Li and Dai [8] designs the remote navigate robot via a web browser. It also adopts the wireless to remote the robot. However, these measures above are based on Internet.

It's well-known, in a remote control system, the operator need remotely control the mobile robot. As the proverb says, seeing is believing. So, the client's grip of the remote environment plays a vital role to the success of the remote control. There're many methods to arrive at this goal, but video feeds are the most common form acting as the feedback in the remote control [9]. Based on visual feedback, [10] accomplishes teleoperation of a robot manipulator, [11] actualizes the robots remote real-time navigation. Li and Li [12] presents the remote monitoring system of robot.

In this paper, we accomplish to control robot based on the wireless. We can not only remotely control robot to monitor each corner of the home when nobody at home, and we can establish video connection between the client and the robot when we're not at home. In addition, the robot can ensure its own security by sensor when the network is delay.

35.2 System Overview

In this paper, our main purpose is to remote control service robot based on the wireless, clients can establish real-time video transmission and server, and the server avoids obstacles using the sensor fixed on it to ensure its own security. Accordingly, we know remote control of service robot consists mainly of two parts: the client and the server. In this experiment, the server is MT-R robot with a laptop and the client is a laptop. The system chart is as Fig. 35.1.

As Fig. 35.1, we can deduce the functions of the two parts:

The client—(1) It sends control commands and transmits video to the server, that's robot. (2) receives and displays the video containing the surrounding information of the robot, simultaneously, it receives the warning from the robot sensor.

The server—(1) it executes the control commands from the client and displays the video containing the surrounding information of the client. (2) sends the video to the client. (3) updates the sensor data and sends the data to the client when it comes across obstacles.

This system works as follows: After the server responses the requests sent by the client, the client can send control commands and transmits videos with its own surrounding information to the server. Meanwhile, the client receives the video information from the robot, and sensor data when the robot meets the obstacles.

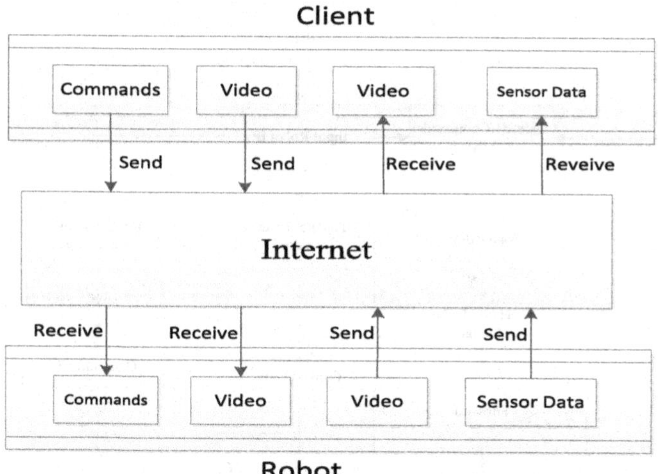

Fig. 35.1 System chart

35.3 System Function

The function of the client:

As shown in Fig. 35.2, we can roughly know how the client works. Meanwhile, we can infer the function of the client each plate by the flow chart.

The client control panel as Figs. 35.3 and 35.4. We can obtain the functions of each section from below figures.

Remark: Figs. 35.3 and 35.4 are both the control panel modules of the client, in order to make the figure of the module clearer, so they are separated.

(1) As shown in Fig. 35.3, the client establish connection with the server by the connect button. After both establish connection, the client sends control commands to the server by the button directions (forward, backward, left, right, stop). The robot always moves on the basis of the speed provided by the procedure except when the client changes the speed by inputting the number to the edit-control "LeftSpeed" and "RightSpeed".
(2) As shown in Fig. 35.4, the two video windows result from the experiment, one video window named video sending sends the real-time video information around the client to the server; the other video window named video received receives the real-time video information from the robot. By these, both sides establish video connection.
(3) As shown in Fig. 35.4, the status next to the video windows shows the real-time compressed ration.

The function of server:

As shown in Fig. 35.5, we can roughly know how the robot works. What is more, we can obtain the function of the server each panel.

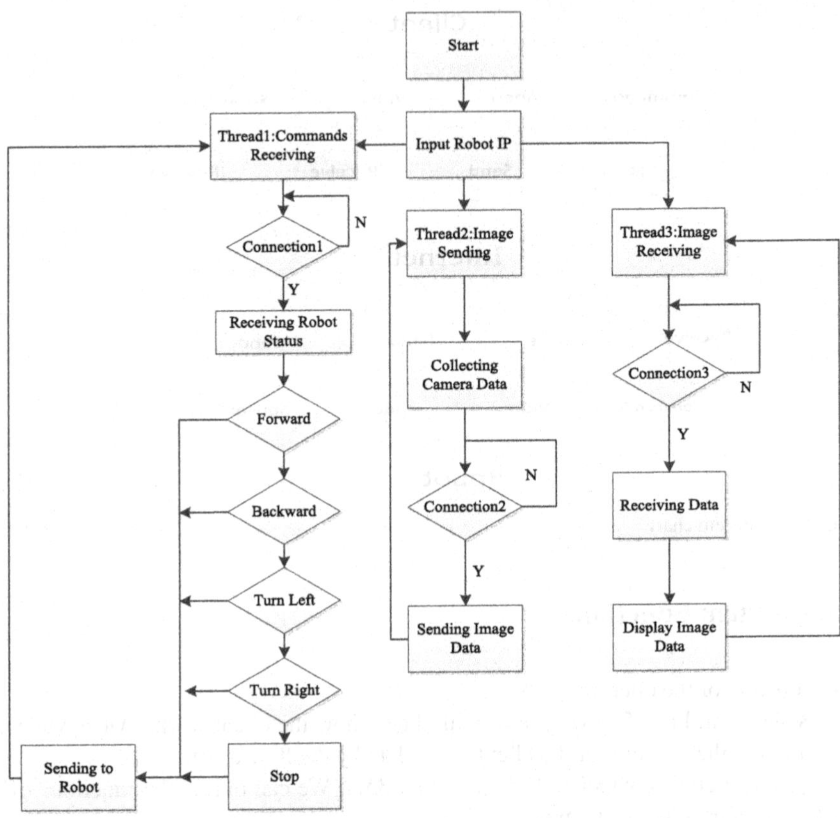

Fig. 35.2 The flow chart of the client

The server control panel is shown as Figs. 35.6a, b and 35.7. We can obtain the function of each section from the below figures.

Remark: The Figs. 35.6a, b and 35.7 are all the control panel modules of the server. In order to make the figure of the module clearer, so they are separated.

(1) As shown in Fig. 35.6a, After this module receives the connection requests from the client, it begins to receive the control commands. The window named "Command Received" shows real-time control command.
(2) As shown in Fig. 35.6b, the module shows the real-time sensor data. The twelve sensors are distributed in the robot, including six infrared sensors and six ultrasonic sensors. Meanwhile, the module under sensor is control command of the robot. We can also control the robot motion by this control panel easily.
(3) As shown in Fig. 35.7, the two windows result from the experiment, the server has two video windows similar to the client, one named "video received" receives the real-time video from the client. The other named "video sending" sends the real-time video to the client.

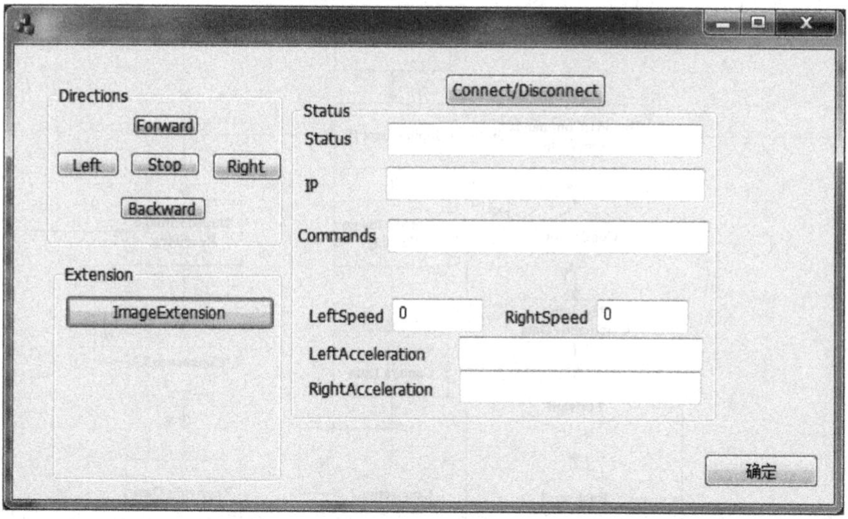

Fig. 35.3 Experimental set-up

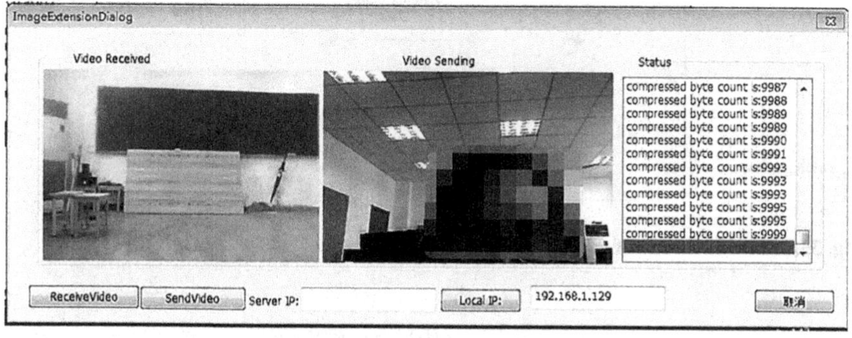

Fig. 35.4 The client control panel-2

35.4 Video Acquisition and Transmission

In this paper, we need to remote control the robot, and establish video between the client and the server, so it's critical to ensure the real-time of video acquisition and accuracy of video transmission. Both acquire video information by the camera of the laptop instantly, and get the compression sequence of the video information compressed by H.263 compression algorithm. Then, both transmit the video information based on the RTP/RTCP protocol. Finally, both decompress the compression sequence separately and display the video on their own screen. Meanwhile, the client remote controls the robot by the video information.

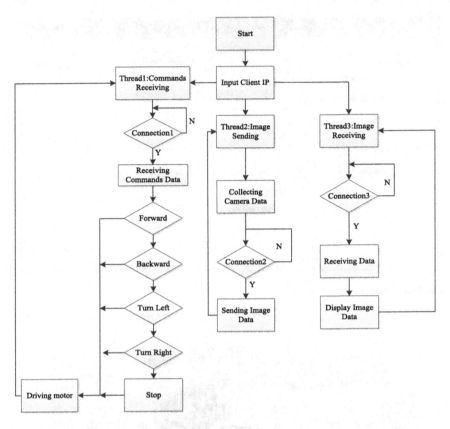

Fig. 35.5 The flow chart of the robot

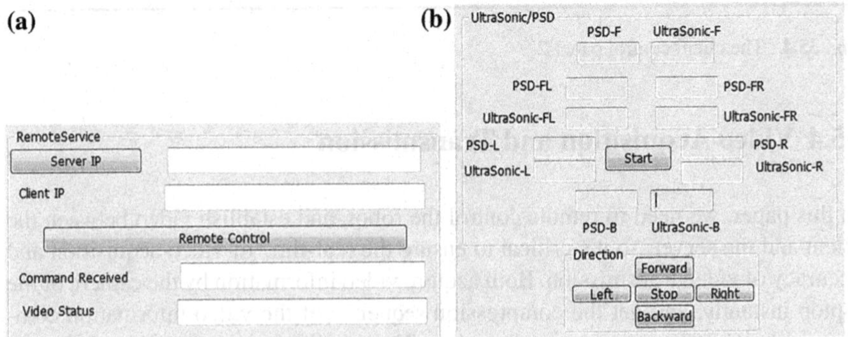

Fig. 35.6 The server control panel. **a** The server control panel-1. **b** The server control panel-2

Fig. 35.7 The server control panel-3

(1) Video acquisition—there're a variety of video acquisition, but in this experiment, we adopt the camera by the laptop. It's mainly because laptop screen is large enough to carry both sides of the video information and control templates.
(2) Video compression—because real-time video acquisition contains vast data. So, we adopt H.263 compression algorithm used widely to improve the efficiency and accuracy of transmission.
(3) Video transmission—during transmission of the video, we can stand with the errors caused by package loss, but the compression sequence can't be changed. Therefore, in this paper, we adopt RTP protocol based on UDP, and RTCP is always with RTP.
(4) Video decompression and display. In order to display video, we need to decompression the compression sequence, in this way, both establish video connection.

In this paper, our purpose is to remote control robot and establish video connection. Therefore, in order to improve the transmission ration of video and control commands, we utilization multi-thread technique. We apply three threads in the server, one thread is control command transmission, the other two threads are acquisition and transmission. The client is the same as the server. We can improve the efficiency and accuracy by multi-thread.

35.5 Motion of Robot

Command parsing—in the remote control system, first, we need to build a control command database on robot. The control commands from the client to the server is actually a string, including forward, backward, left, right and stop. When the robot receives the control string, it needs to match the string with the control command database so that the robot can acquire the exact motion command. When the robot confirms the corresponding command, it will drive the motor by communication interface. Then, the robot will move as the speed shown in the following table.

Table 35.1 Left and right speed of robot

Forward		Backward		Turn left		Turn right		Stop	
$v_l = c$	$v_r = c$	$v_l = -c$	$v_r = -c$	$v_l = c$	$v_r = 0$	$v_l = 0$	$v_r = c$	$v_l = 0$	$v_r = 0$

Among, c is constant, v_l and v_r are on behalf of the left and right wheel speed respectively (Table 35.1).

Sensor—in the remote control system, it's unreliable to remote control robot just by video transmission. Because we can't acquire real-time video information when the performance of network is poor, which could lead to a collision between the robot and an obstacle. In order to improve the robot's own safety and reduce the probability of a collision with an obstacle, we install twelve sensors on the robot, including six infrared sensors and six ultrasonic sensors. As the distance range of the two kinds of sensors is different, we apply up and down staggered distribution to these two kinds of sensors. The distribution is shown in Fig. 35.8.

The real-time data that sensors acquire is shown in the module of Fig. 35.8. In order to ensure the robot's safety, we set up a threshold. When the distance between the robot and the obstacle is less than the threshold, the robot will stop to move. Meanwhile, the robot can send a feedback signal to the client, here, the status on the client shows "obstacle" to remind the user that it come across the obstacle and stop. Even though the client send the control command to the robot, it still stops to move.

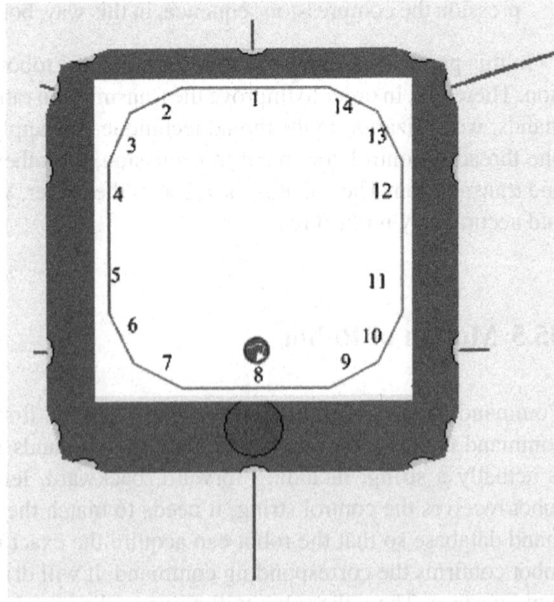

Fig. 35.8 The distribution of sensors

35.6 Conclusions

In this experiment, although we encountered a lot of problems in the beginning, finally we achieve the goal, that's we can remote control the robot. More importantly, we establish video connection between the client and the robot. In addition, we apply twelve sensors to detect the distance between the client and the robot. In these methods, the robot can ensure the robot's security.

References

1. Ryu JG, Shim HM, Kil SK et al (2006) Design and implementation of real-time security guard robot using CDMA networking [C]. In: Advanced Communication Technology. ICACT 2006. The 8th International Conference. IEEE, vol 3(6). pp 1906
2. Salmanipour S, Sirouspour S (2013) Teleoperation of a mobile robot with model-predictive obstacle avoidance control. In: Industrial Electronics Society, IECON 2013–39th Annual Conference of the IEEE, 4270 C 4275
3. Li En LZZTM, Yang S (2005) Inspection robot remote control based on image transmission. High Tech Lett 15
4. Petrovic I, Babic J, Budisic M (2007) Teleoperation of collaborative mobile robots with force feedback over internet. In: Proceedings of international conference on informatics in control, automation and robotics, robotics and automation, vol 2, 2007
5. Jung S, Song T, Jeon J (2008) Obstacle collision prevention of mobile robot using force feedback in remote sites. In: Proceedings of the 2nd international conference on Ubiquitous information management and communication. ACM, pp 515–519
6. Mook Jung S, Houn Song T, Wook Jeon J(2008) Obstacle collision prevention of a mobile robot using force feedback in remote sites. In: International conference on ubiquitous information management and communication—ICUIMC, pp 515–519
7. Nadvornik J, Smutny P (2014) Remote Control Robot Using Android Mobile Device. In: 15th international carpathian control conference (ICCC), 2014
8. Li HK, Dai ZD (2010) A semiautonomous sprawl robot based on remote wireless control. Robot Biom—ROBIO
9. Sheridan Thomas B (1992) Musings on telepresence and virtual presence. Teleoperators Virtual Environ—Presence 1(1):120–125
10. Kofman J, Xianghai W, Luu TJ, Verma S (2005) Teleoperation of a robot manipulator using a vision-based human-robot interface. IEEE Trans Ind Electr—IEEE TRANS IND ELECTR 52(5):1206–1219
11. Wang Q, Pan W, Li M (2014) Robot's remote real-time navigation controlled by smart phone. In: IEEE international conference on robotics and biomimetics (ROBIO), China, 2012
12. Li P, Li R (2012) The research of remote monitoring system for home partner robot[z]

Chapter 36
Robust Adaptive Control for Robotic Systems with Guaranteed Parameter Estimation

Baorui Jing, Jing Na, Guanbin Gao and Guoqing Sun

Abstract In this paper, we propose a novel adaptive control scheme for robotic systems by incorporating the parameter error into the adaptive law. By carrying out filter operations, the robotic system is linearly parameterized without using the measurements of acceleration. Then a new adaptive algorithm is introduced to guarantee that the parameter error and control error exponentially converge to zero. In particular, we provide an intuitive method to verify the standard PE condition for the parameter estimation. The robustness against disturbances is also studied and comparisons to several adaptive laws are provided. Simulations with a realistic robot arm are presented to validate the improved performance.

Keywords Adaptive control · Parameter estimation · Robotic system · PE condition

36.1 Introduction

Adaptive control [1, 2] has been widely used to achieve tracking control for systems with unknown parameters. In the classical framework, the adaptive laws are driven by the control errors to retain the tracking error convergence and the boundedness of the estimated parameters [2]. However, it is not trivial to guarantee that the estimated parameters converge to their true values. In particular, the required persistent excitation (PE) condition is difficult to verify. In [3, 4], a composite adaptive control has been developed, where the adaptive laws are designed by combining the

*This work was supported by the National Natural Science Foundation of China (No. 61203066).

B. Jing · J. Na (✉) · G. Gao · G. Sun
Faculty Of Mechanical & Electrical Engineering, Kunming University of Science and Technology, 650500 Kunming, China
e-mail: najing25@163.com

© Springer-Verlag Berlin Heidelberg 2016

Y. Jia et al. (eds.), *Proceedings of the 2015 Chinese Intelligent Systems Conference*, Lecture Notes in Electrical Engineering 359,
DOI 10.1007/978-3-662-48386-2_36

tracking error and the prediction error. However, the convergence performance heavily depends on the presented predictor because the adaptive laws are driven by the induced output control and predictor errors.

A recent work [5] proposed novel adaptive laws based on the information of parameter error to further address the estimation performance. In particular, in our previous work [6–8], several novel *direct* estimation schemes are proposed by introducing novel filter operations, so that the unstable integrator and the online calculation of a matrix inverse in [5] are all avoided. Moreover, exponential and/or finite-time error convergence is retained under a verifiable excitation condition.

On the other hand, adaptive control of robotic systems has also been developed (e.g., [3, 9, 10]) since 1980s when the linear parameterization of nonlinear robot dynamics was first introduced [3, 11]. In [10], the need of robotic joint acceleration measurements limits the practical applicability [12]. Moreover, in these methods only the convergence of the tracking error can be proved, while the parameter estimation convergence was not addressed. In our previous work [13], a terminal sliding mode (TSM) control was proposed for robotic systems to achieve finite-time convergence by incorporating the ideas of adaptive laws [7] into the adaptive control; the potential singularity problem of TSM was avoided and the joint accelerations were not used. However, this two-phase control leads to complexities in the analysis and practical control implementations.

In this paper, we further revisit the adaptive control for robotic systems with unknown parameters, where the parameter estimation convergence is considered. The new adaptive control strategy will incorporate the parameter estimation methods in [7] into the adaptive control to achieve exponential convergence of the tracking control and parameter estimation simultaneously. In particular, a new modification term with the estimation error is adopted as leakage term in the adaptive law. In contrary to [13], only one phase control is used to simplify the control implementation. The robustness of the proposed schemes against external disturbances is analyzed and comparisons to classical adaptive laws are provided. Specifically, we show that the required excitation condition is equivalent to the standard PE condition, and then suggested an intuitive scheme is to verify the standard PE condition. Finally, simulations based on a 6-DOF robot arm model are presented to validate the performance of the new method.

36.2 Problem Formulation

We consider an n-degree of freedom (DOF) robot arm modeled by:

$$M(q)\ddot{q} + C(q,\dot{q})\dot{q} + G(q) = \tau \tag{36.1}$$

where $q, \dot{q}, \ddot{q} \in \mathbb{R}^n$ are the joint position, velocity and acceleration, respectively; n is the number of the DOF, $\tau \in \mathbb{R}^n$ is the control input torque; $M(q) \in \mathbb{R}^{n\times n}$ is the inertia matrix, $C(q,\dot{q})\in\mathbb{R}^{n\times n}$ represents the Coriolis/centripetal torque, viscous and nonlinear damping, and $G(q)\in\mathbb{R}^{n}$ denotes the gravity torque.

The following properties will be used in the following control design [11]:

Property 1 *The matrix $\dot{M}(q)-2C(q,\dot{q})$ is skew-symmetric, then*

$$x^{T}\left[\dot{M}(q)-2C(q,\dot{q})\right]x=0,\qquad \forall x,q,\dot{q}\in\mathbb{R}^{n} \tag{36.2}$$

Property 2 *The dynamics of robotic system (36.1) can be represented as a linearly parameterized form*

$$M(q)\ddot{q}+C(q,\dot{q})q+G(q)=\phi(q,\dot{q},\ddot{q})\theta \tag{36.3}$$

where $\theta\in\mathbb{R}^{N}$ is a constant parameter vector which contains the parameters to be estimated, $\phi(q,\dot{q},\ddot{q})\in\mathbb{R}^{n\times N}$ is the known dynamic regressor matrix.

36.3 Adaptive Control with Guaranteed Parameter Estimation

To achieve tracking control of a given reference q_d and the estimation of unknown parameter θ, we first define a control error variable as

$$S=\dot{e}+\lambda e=\dot{q}_r-\dot{q} \tag{36.4}$$

where $e=q_d-q$; $\dot{q}_r=\dot{q}_d+\lambda e$ is the tracking error and its derivative $\ddot{q}_r=\ddot{q}_d+\lambda\dot{e}$ can be calculated based on $q,\dot{q},\dot{q}_d,\ddot{q}_d$.

Then based on Property 2, we have

$$R(q,\dot{q})=M(q)\ddot{q}_r+C(q,\dot{q})\dot{q}_r+G(q)=\Phi_{\mathrm{R}}(q,\dot{q})\theta \tag{36.5}$$

Thus the robotic system (36.1) can be reformulated as

$$M(q)\dot{S}+C(q,\dot{q})S-R(q,\dot{q})=-\tau \tag{36.6}$$

As shown in (36.5), the joint acceleration measurement $\ddot{q}$ does not appear in the regressor $\Phi_{\mathrm{R}}(q,\dot{q})$, thus $\Phi_{\mathrm{R}}(q,\dot{q})$ can be used in the control design. However, for the purpose of parameter estimation, the joint acceleration $\ddot{q}$ is involved in the derivative $\dot{S}$ in (36.6). To eliminate the requirements of $\dot{S}$, we define auxiliary functions $F(q,\dot{q})=M(q)S$ and $H(q,\dot{q})=-\dot{M}(q)S+C(q,\dot{q})S$ as [12] as

$$F(q,\dot{q})=M(q)S=\Phi_F(q,\dot{q})\theta,\qquad H(q,\dot{q})=-\dot{M}(q)S+C(q,\dot{q})S=\Phi_H(q,\dot{q})\theta \tag{36.7}$$

Then, the system (36.6) can be rewritten as

$$\dot{F}(q,\dot{q})+H(q,\dot{q})-R(q,\dot{q})=[\Phi_F(q,\dot{q})+\Phi_H(q,\dot{q})-\Phi_R(q,\dot{q})]\theta=\Phi(q,\dot{q},\ddot{q})\theta=-\tau \tag{36.8}$$

where $\dot{F}(q,\dot{q})=\frac{d}{dt}[M(q)S]=\Phi_F(q,\dot{q})\theta$ is the derivative of $M(q)S$ and $\Phi(q,\dot{q},\ddot{q})=[\Phi_F(q,\dot{q})+\Phi_H(q,\dot{q})-\Phi_R(q,\dot{q})]$ is the regressor matrix. However, because $\dot{F}(q,\dot{q})$ and $\Phi(q,\dot{q},\ddot{q})$ contain the joint acceleration $\ddot{q}$, Eq. (36.8) cannot be used for parameter estimation when the joint acceleration $\ddot{q}$ is not measurable.

To eliminate the use of $\ddot{q}$ in the parameter estimation, we introduce a stable filter operation on both sides of (36.8) as

$$\begin{cases} k\dot{\Phi}_{Ff}+\Phi_{Ff}=\Phi_F, & \Theta_{Ff}|_{t=0}=0\\ k\dot{\Phi}_{Hf}+\Phi_{Hf}=\Phi_H, & \Theta_{Hf}|_{t=0}=0\\ k\dot{\Phi}_{Rf}+\Phi_{Rf}=\Phi_R, & \Theta_{Rf}|_{t=0}=0\\ k\dot{\tau}_f+\tau_f=\tau, & \tau_f|_{t=0}=0 \end{cases} \tag{36.9}$$

where $k>0$ is a constant filter parameter, $\Phi_{Ff}(q,\dot{q}),\Phi_{Hf}(q,\dot{q}),\Phi_{Rf}(q,\dot{q})$ and $F_f(q,\dot{q}),H_f(q,\dot{q}),R_f(q,\dot{q}),\tau_f$ are the filtered form of $\Phi_F(q,\dot{q}),\Phi_H(q,\dot{q}),\Phi_R(q,\dot{q})$ and $F(q,\dot{q}),H(q,\dot{q}),R(q,\dot{q}),\tau$, respectively. Then from (36.8) and (36.9), we obtain

$$\dot{F}_f(q,\dot{q})+H_f(q,\dot{q})-R_f(q,\dot{q})=[\dot{\Phi}_{Ff}(q,\dot{q})+\Phi_{Hf}(q,\dot{q})-\Phi_{Rf}(q,\dot{q})]\theta=-\tau_f \tag{36.10}$$

According to (36.9), we get $\dot{\Phi}_{Ff}=\frac{\Phi_F-\Phi_{Ff}}{k}$, so that system (36.10) is rewritten as

$$\left[\frac{\Phi_F(q,\dot{q})-\Phi_{Ff}(q,\dot{q})}{k}+\Phi_{Hf}(q,\dot{q})-\Phi_{Rf}(q,\dot{q})\right]\theta=\Phi_f(q,\dot{q})\theta=-\tau_f \tag{36.11}$$

where $\Phi_f(q,\dot{q})=\frac{\Phi_F(q,\dot{q})-\Phi_{Ff}(q,\dot{q})}{k}+\Phi_{Hf}(q,\dot{q})-\Phi_{Rf}(q,\dot{q})$ is a new regressor. Thus, the acceleration $\ddot{q}$ is avoided and thus (36.11) can be used for estimation.

In order to accommodate the parameter estimation, we will introduce the auxiliary matrix $P(t)\in\mathbb{R}^{N\times N}$, vector $Q(t)\in\mathbb{R}^N$ and vector $W(t)\in\mathbb{R}^N$ as

$$\begin{cases} \dot{P}(t)=-\ell P(t)+\Phi_f^T\Phi_f, & P(0)=0\\ \dot{Q}(t)=-\ell Q(t)+\Phi_f^T\tau_f, & Q(0)=0\\ W(t)=P(t)\hat{\theta}-Q(t) \end{cases} \tag{36.12}$$

where $\ell>0$ is a designed parameter, $\hat{\theta}$ is the estimated parameter.

As shown in [8], we know $Q(t) = P(t)\theta$, and thus derive from (36.11) and (36.12) that

$$W(t) = P(t)\hat{\theta} - Q(t) = P(t)\hat{\theta} - P(t)\theta = -P(t)\tilde{\theta} \tag{36.13}$$

where $\tilde{\theta} = \theta - \hat{\theta}$ is the parameter estimation error.

Then we can design an adaptive control for system (36.1) as

$$\tau = \Phi_R(q, \dot{q})\hat{\theta} + KS \tag{36.14}$$

where $K > 0$ is a constant feedback gain matrix.

The adaptive law to update the parameter estimation $\hat{\theta}$ is provided by

$$\dot{\hat{\theta}} = \Gamma\left(\Phi_R^T(q, \dot{q})S - \kappa W\right) \tag{36.15}$$

where $\Gamma > 0$ and $\kappa > 0$ are the adaptive learning gains.

Substituting (36.14) into (36.6), we obtain the closed-loop error dynamics as

$$M(q)\dot{S} + C(q, \dot{q})S + KS = \Phi_R(q, \dot{q})\tilde{\theta} \tag{36.16}$$

Remark 1 The term W in (36.15) is a new leakage term used to achieve parameter estimation convergence. This is clearly different to classical e-modification and σ-modification [2]. Moreover, by introducing filter operation (36.9), system (36.8) is reformulated as (36.11), so that the joint acceleration $\ddot{q}$ is avoided in both of the adaptive control (36.14) and the adaptive law (36.15).

Lemma 1 *If the vector Φ_f in (36.8) is persistently excited (PE), then the matrix P in (36.12) is positive definite, i.e., $\lambda_{\min}(P) > \sigma > 0$ for a positive constant σ. On the other hand, the positive definiteness of P also implies that Φ_f is PE.*

Remark 2 In our previous work [6–8, 13], we have proved that the standard PE condition is *sufficient* to guarantee the positive definiteness of matrix $P(t)$, i.e., the PE condition of Φ_f implies $\lambda_{\min}(P) > \sigma > 0$. In this paper, the inverse of this claim (i.e., $\lambda_{\min}(P) > \sigma > 0$ implies the PE of Φ_f) can also be claimed. Thus Lemma 1 is a more complete result paving a way for online verifying the PE condition, i.e., calculating the minimum eigenvalue of P and testing for $\lambda_{\min}(P) > \sigma > 0$. The proof of this claim is omitted here due to the limited space.

Theorem 1 *For robotic system (36.1) with adaptive control (36.14) and adaptive law (36.15), if the regressor matrix Φ_f in (36.8) is PE, then the parameter error $\tilde{\theta}$ and the tracking error S converge to zero exponentially.*

Proof We define Lyapunov candidate function as

$$V = \frac{1}{2}S^T MS + \frac{1}{2}\tilde{\theta}^T \Gamma^{-1}\tilde{\theta} \tag{36.17}$$

Then the derivative $\dot{V}$ with respect to time t can be obtained along (36.15) and (36.16) as

$$\dot{V} = S^T\left[-C(q,\dot{q})S - KS + \Phi_R\tilde{\theta}\right] + \frac{1}{2}S^T\dot{M}(q)S + \tilde{\theta}^T\Gamma^{-1}\dot{\tilde{\theta}} = -S^T KS - \kappa\tilde{\theta}^T P\tilde{\theta} \le -\mu V \tag{36.18}$$

where $\mu = \min\{2\lambda_{\min}(K)/\lambda_{\max}(M), 2\kappa\sigma/\lambda_{\max}(\Gamma^{-1})\}$ is a positive constant. Then we conclude that S and $\tilde{\theta}$ converge to zero exponentially with the rate μ.

Remark 3 As shown in Theorem 1, by introducing the leakage term κW containing the parameter estimation error $P(t)\tilde{\theta}$ in the adaptive law (36.15), a quadratic term $\kappa\tilde{\theta}^T P\tilde{\theta}$ appears in the Lyapunov analysis (36.3), which can guarantee the exponential convergence of $\tilde{\theta}$ and S to zero simultaneously.

36.4 Robustness Analysis and Comparisons

To study the robustness of the proposed control and adaptive law, we introduce a bounded disturbance $\xi \in \mathbb{R}^n$ in the robotic system (36.1) such that

$$M(q)\ddot{q} + C(q,\dot{q})\dot{q} + G(q) = \tau + \xi \tag{36.19}$$

where $\|\xi\| \le \varepsilon_\xi, \varepsilon_\xi > 0$. Then similar to Sect. 36.3, system (36.4) can be rewritten as

$$\Phi_f(q,\dot{q})\theta = \tau_f + \xi_f \tag{36.20}$$

where ξ_f is the filtered version of ξ in terms by $k\dot{\xi}_f + \xi_f = \xi,\ \xi_f(0) = 0$.

Consequently, by defining the control error as (36.4) and using the same control (36.14), we can obtain the closed-loop error for (36.4) as

$$M(q)\dot{S} + C(q,\dot{q})S + KS = \Phi_R(q,\dot{q})\tilde{\theta} + \xi \tag{36.21}$$

In this case, the auxiliary variable W defined in (36.12) can be represented as

$$W = P\hat{\theta} - Q = -P\tilde{\theta} + \psi \tag{36.22}$$

where $\psi = -\int_0^t e^{-\ell(t-r)}\Phi_f^T(r)\xi_f(r)dr$ is bounded by $\|\psi\| \le \varepsilon_\psi$ for some $\varepsilon_\psi > 0$.

Corollary 1 *Consider system (36.4) with control (36.14) and adaptive law (36.15), if the regressor matrix Φ_f in (36.8) is PE, then the closed-loop system is stable, and the estimation error $\tilde{\theta}$ and the tracking error S converge to a small set around zero.*

Proof Consider the Lyapunov function $V_2 = V$ as (36.2), then $\dot{V}_2$ is calculated as

$$\begin{aligned}\dot{V}_2 &= -S^T KS + S^T \xi - \kappa\tilde{\theta}^T P\tilde{\theta} + \kappa\tilde{\theta}^T \psi \\ &= -\left(\lambda_{\min}(K) - \frac{1}{2\eta}\right)\|S\|^2 - (\kappa\sigma - \frac{1}{2\eta})\|\tilde{\theta}\|^2 + \frac{\eta\varepsilon_\xi^2}{2} + \frac{\eta\varepsilon_\psi^2}{2} \le -\mu_2 V_2 + \gamma_2\end{aligned} \tag{36.23}$$

where $\mu_2 = \min\{2(\lambda_{\min}(K) - 1/2\eta)/\lambda_{\max}(M), 2(\kappa\sigma - 1/2\eta)/\lambda_{\max}(\Gamma^{-1})\}$ and $\gamma_2 = \eta\varepsilon_\xi^2/2 + \eta\varepsilon_\psi^2/2$ are all positive constants for large $\eta > 0$. Thus we know that all signals in the closed-loop system are bounded, and the errors $\tilde{\theta}$ and S converges to a small residual set around zero.

Finally, we compare the proposed novel adaptive law (36.15) with the widely used gradient method and σ-modification.

(1) *Gradient method [1]*: The adaptive law for parameter estimation is solely driven by the tracking error S, i.e., $\kappa = 0$ in (36.15). Then the estimation error is

$$\dot{\tilde{\theta}} = -\Gamma\Phi_{\mathrm{R}}^T(q, \dot{q})S \tag{36.24}$$

A critical problem of the gradient adaptation is the potential bursting phenomenon of the estimation error $\tilde{\theta}$, i.e., the convergence of $\tilde{\theta}$ to zero cannot be claimed even the tracking error S converges to zero.

(2) *σ-modification [2]*: A modification term $\kappa\hat{\theta}$ is used to replace the term κW in (36.15) to give the σ-modification method

$$\dot{\hat{\theta}} = \Gamma\left(\Phi_{\mathrm{R}}^T(q, \dot{q})S - \kappa\hat{\theta}\right) \tag{36.25}$$

Then the estimation error can be obtained as

$$\dot{\tilde{\theta}} = -\Gamma\Phi_R^T(q, \dot{q})S + \Gamma\kappa\hat{\theta} = -\Gamma\kappa\tilde{\theta} - \Gamma\Phi_R^T(q, \dot{q})S + \Gamma\kappa\theta \tag{36.26}$$

The error Eq. (36.11) contains a damping term $\Gamma\kappa\tilde{\theta}$, thus the error dynamics in (36.11) are bounded-input-bounded-output (BIBO). However, this term makes the estimated parameter $\hat{\theta}$ stay in the neighborhood of the pre-selected value only. In fact, when the tracking error $S = 0$, the transfer function of (36.11) is $\tilde{\theta} = \frac{\Gamma\kappa\theta}{p + \Gamma\kappa}$ (p is the Laplace operation). Consequently, $\tilde{\theta}$ cannot be null.

(3) *Proposed method*: A new term κW is used in (36.15) leading to estimation error:

$$\dot{\tilde{\theta}} = -\Gamma \kappa P \tilde{\theta} - \Gamma \Phi_R^T(q, \dot{q}) S \tag{36.27}$$

The error Eq. (36.12) introduces a forgetting factor $\Gamma \kappa P \tilde{\theta}$ as σ-modification (36.11). Thus, the error $\tilde{\theta}$ in (36.12) is also BIBO stable. Consequently, the robustness of the proposed adaptive law is compatible to σ-modification (36.11). Moreover, the leakage term κW can update $\hat{\theta}$ towards its true value θ. In fact, the error Eq. (36.12) is represented as $\tilde{\theta} = \frac{\Gamma \Phi_R^T S}{p + \Gamma P \kappa}$, so that $\tilde{\theta} \to 0$ holds for $S \to 0$. Thus, the adaptive law (36.15) can obtain better convergence than (36.11).

36.5 Simulation

In this paper, we use a 6-DOF Robot Arm purchased from Reinovo Ltd for simulation. In order to simply the control design, only two joints of this robot arm are modeled in this paper. According to [14], we obtain the kinetic energy $\mathcal{K}$ and the potential energy $\mathcal{P}$ as

$$\begin{aligned} \mathcal{K} &= \frac{1}{2} m v^2 + \frac{1}{2} I \omega^2 = \frac{1}{2} m_1 l_1^2 \dot{q}_1^2 + \frac{1}{2} m_2 l_2^2 \dot{q}_1^2 + \frac{1}{2} m_2 l_2^2 \dot{q}_2^2 + m_2 l_2^2 \dot{q}_1 \dot{q}_2 \\ \mathcal{P} &= mgh = m_1 g (l_1 \sin q_1 + 320) + m_2 g [l_2 \sin(q_1 + q_2) + l_1 \sin(q_1) + 320] \end{aligned} \tag{36.28}$$

where v and ω are the velocity and angular velocity respectively. I is the moment of inertia and h is the height. The Lagrange's equation of a robotic system is given as $\frac{d}{dt}\frac{\partial \mathcal{L}}{\partial \Theta} - \frac{\partial \mathcal{L}}{\partial \Theta} = \tau$, where $\mathcal{L} = \mathcal{K} - \mathcal{P}$ is the Lagrangian.

Then the dynamics of the robotic arm is molded as

$$\begin{bmatrix} M_{11}(q) & M_{12}(q) \\ M_{21}(q) & M_{22}(q) \end{bmatrix} \begin{bmatrix} \ddot{q}_1 \\ \ddot{q}_2 \end{bmatrix} + \begin{bmatrix} C_{11}(q, \dot{q}) & C_{12}(q, \dot{q}) \\ C_{21}(q, \dot{q}) & C_{22}(q, \dot{q}) \end{bmatrix} \begin{bmatrix} \dot{q}_1 \\ \dot{q}_2 \end{bmatrix} + \begin{bmatrix} G_1(q) \\ G_2(q) \end{bmatrix} = \tau \tag{36.29}$$

with

$$\begin{aligned} & M_{11}(q) = m_2 l_2^2 + (m_1 + m_2) l_1^2 + 2 m_2 l_1 l_2 \cos(q_2), M_{12}(q) = M_{21}(q) = m_2 l_2^2 + m_2 l_1 l_2 \cos(q_2) \\ & M_{22}(q) = m_2 l_2^2, C_{11}(q, \dot{q}) = -2 m_2 l_1 l_2 \dot{q}_2 \sin(q_2), C_{12}(q, \dot{q}) = -m_2 l_1 l_2 \dot{q}_2 \sin(q_2) \\ & C_{21}(q, \dot{q}) = m_2 l_1 l_2 \dot{q}_1 \sin(q_2), C_{22}(q, \dot{q}) = 0, G_1(q) = m_2 g l_2 \cos(q_1 + q_2) + (m_1 + m_2) g l_1 \cos(q_1) \\ & G_2(q) = m_2 g l_2 \cos(q_1 + q_2) \end{aligned}$$

where m_1, m_2 are the mass of robot arm, l_1, l_2 are the length of each link, $g = 9.18$ is the gravity constant. In this study, the unknown parameters in system (36.14) is $\theta = [m_1, m_2]^T = [10, 5]^T$, so that $F(q, \dot{q}), H(q, \dot{q}), R(q, \dot{q})$ can be derived as

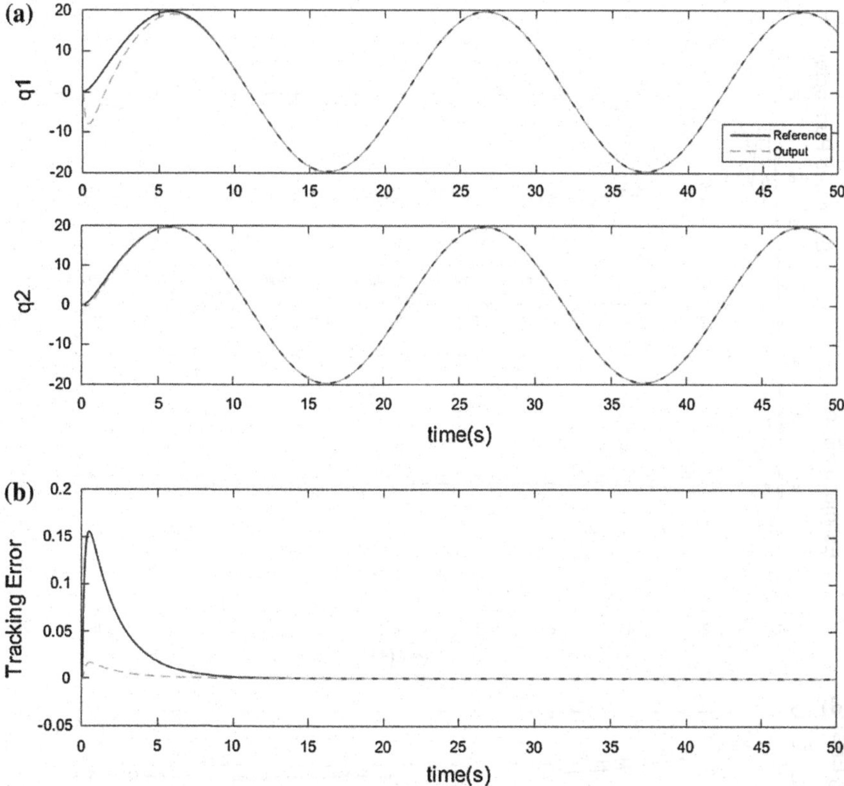

Fig. 36.1 Adaptive control and parameter estimation performance. **a** Tracking control performance. **b** Tracking errors. **c** Parameter estimation performance. **d** Tracking and estimation errors under disturbances

$$R(q,\dot{q}) = \underbrace{\begin{bmatrix} l_1^2 \ddot{q}_{r1} + l_1 g\cos(q_1) & \begin{array}{r} (l_1^2+l_2^2+2l_1l_2\cos(q_2))\,\ddot{q}_{r1} + (l_1l_2\cos(q_2)+l_2^2)\,\ddot{q}_{r2} + l_1 g\cos(q_1) \\ -2l_1l_2\sin(q_2)\dot{q}_2\dot{q}_{r1} - l_1l_2\sin(q_2)\dot{q}_2\dot{q}_{r2} + l_2 g\cos(q_1+q_2) \end{array} \\ 0 & \begin{array}{c} (l_1l_2\cos(q_2)+l_2^2)\,\ddot{q}_{r1} + l_2^2\,\ddot{q}_{r2} \\ + l_1l_2\sin(q_2)\dot{q}_1\dot{q}_{r1} + l_2 g\cos(q_1+q_2) \end{array} \end{bmatrix}}_{\Phi_R(q,\dot{q},\dot{q}_r,\ddot{q}_r)} \begin{bmatrix} m_1 \\ m_2 \end{bmatrix} \tag{36.30}$$

$$F(q,\dot{q}) = \underbrace{\begin{bmatrix} l_1^2 S_1 & (l_1^2+l_2^2+2l_1l_2\cos(q_2))S_1 + \frac{1}{12}(l_1l_2\cos(q_2)+l_2^2)S_2 \\ 0 & (l_1l_2\cos(q_2)+l_2^2)S_1 + l_2^2 S_2 \end{bmatrix}}_{\phi_F(q,\dot{q})} \begin{bmatrix} m_1 \\ m_2 \end{bmatrix} \tag{36.31}$$

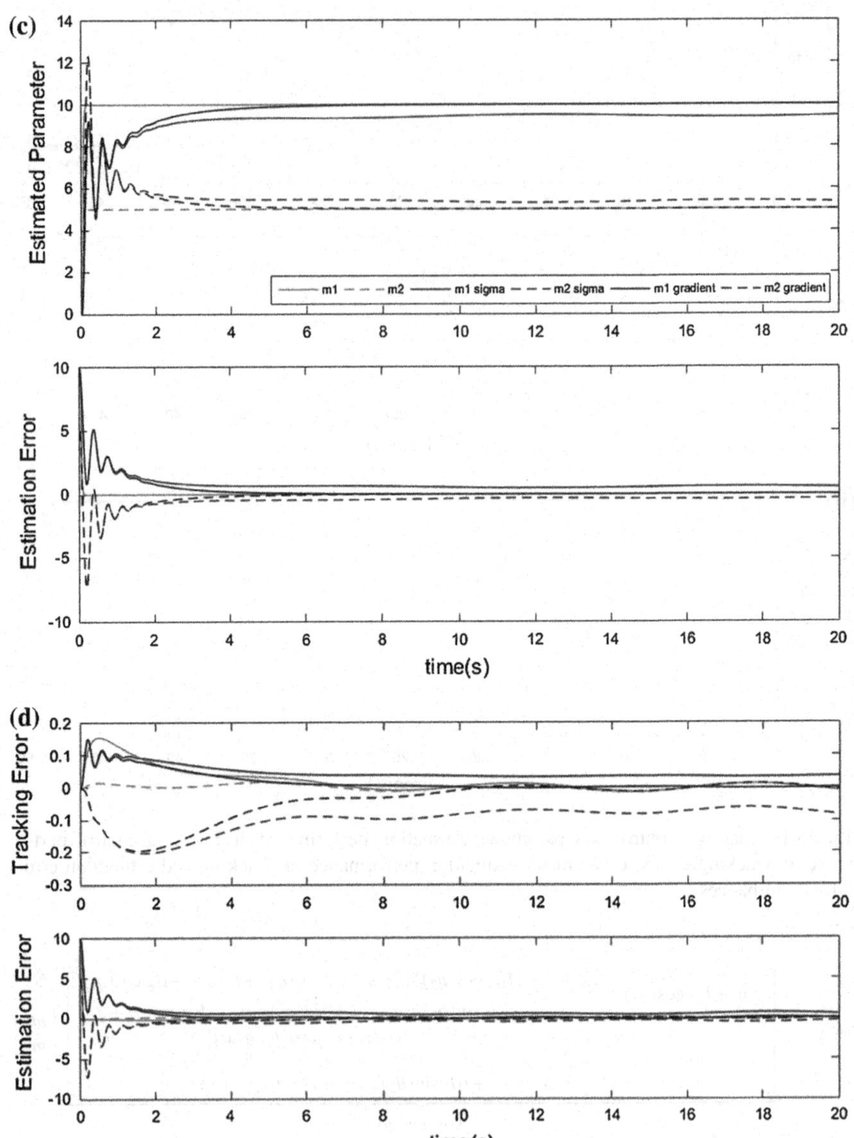

Fig. 36.1 (continued)

$$
\begin{aligned}
H(q,\dot{q}) &= -\begin{bmatrix} \dot{M}_{11}(q)S_1 + \dot{M}_{12}(q)S_2 \\ \dot{M}_{21}(q)S_1 + \dot{M}_{22}(q)S_2 \end{bmatrix} + \begin{bmatrix} C_{11}(q,\dot{q})S_1 \\ C_{21}(q,\dot{q})S_1 \end{bmatrix} \\
&= \underbrace{\begin{bmatrix} 0 & 0 \\ 0 & l_1 l_2 S_1(\dot{q}_1 + \dot{q}_2)\sin(q_2) \end{bmatrix}}_{\phi_H(q,\dot{q})} \begin{bmatrix} m_1 \\ m_2 \end{bmatrix}
\end{aligned} \tag{36.32}
$$

where $S = [S_1, S_2]^T$ is the control error with the parameter as $\lambda = diag([5,5])$.

To guarantee the required excitation, we choose reference as $q_d = 20\sin(0.3t)$, and set the control feedback gain as $K = diag([10,10])$. Other parameters are $\ell = 1, k = 0.001, \kappa = 50$ and $\Gamma = 20I$. Comparative simulation results are shown in Fig. 36.1, where the adaptive control (36.14) and the proposed adaptive law (36.15) and the gradient method (36.9) and σ-modification (36.10) are all simulated with the same conditions and parameters. The tracking profiles and the tracking error with (36.14) and (36.15) are shown in Fig. 36.1a, b. The evolutions of the estimated parameters with different adaptive laws are all depicted in Fig. 36.1c. It is shown that the estimated parameters with new adaptive law (36.15) converge to their true values very fast. However, the transient performance for the gradient scheme is sluggish. Nevertheless, the steady-state error for σ-modification (36.10) cannot converge to zero, which has been pointed out in Sect. 36.4. Finally, an external disturbance $\xi = 0.2\sin(t)$ is used to verify the robustness. As shown in Fig. 36.1d, the parameter and tracking errors of new methods converge to a very small set. However, the gradient method provides sluggish control and estimation performance though it performs slightly better than σ-modification method. All of simulation results show that the newly introduced leakage term in the adaptive law can improve the parameter estimation and thus the overall control performance.

36.6 Conclusion

This paper presents an alternative adaptive control method for robotic systems, which incorporates a new leakage term into the adaptive law. By introducing appropriate filter operations, the robotic acceleration measurements are avoided. Exponential convergence of the control error and parameter estimation error to zero can be achieved simultaneously. In particular, we prove that the required excitation condition is equivalent to the standard PE condition, and thus provide a numerically feasible and intuitive method to online verify the PE condition. The robustness and comparisons to other adaptive schemes are also provided and validated in terms of simulations based on a realistic robotic arm model.

References

1. Sastry S, Bodson M (1989) Adaptive control: stability, convergence, and robustness. Prentice Hall, New Jersey
2. Ioannou PA, Sun J (1996) Robust adaptive control. Prentice Hall, New Jersey
3. Slotine JJE, Li W (1989) Composite adaptive control of robot manipulators. Automatica 25 (4):509–519
4. Patre PM, MacKunis W, Johnson M, Dixon WE (2010) Composite adaptive control for Euler-Lagrange systems with additive disturbances. Automatica 46(1):140–147
5. Adetola V, Guay M (2010) Performance improvement in adaptive control of linearly parameterized nonlinear systems. IEEE Trans Autom Control 55(9):2182–2186
6. Na J, Herrmann G, Ren X, Mahyuddin MN, Barber P (2011) Robust adaptive finite-time parameter estimation and control of nonlinear systems. In: Proceeding of IEEE international symposium on intelligent control (ISIC), Denver, CO, USA, pp 1014–1019
7. Na J, Mahyuddin MN, Herrmann G, Ren X (2013) Robust adaptive finite-time parameter estimation for linearly parameterized nonlinear systems. In: Proceeding of 2013 32nd Chinese control conference (CCC), pp 1735–1741
8. Na J, Ren X, Xia Y (2014) Adaptive parameter identification of linear SISO systems with unknown time-delay. Syst Control Lett 66:43–50
9. Lewis FL, Liu K, Yesildirek A (2002) Neural net robot controller with guaranteed tracking performance. IEEE Trans Neural Netw 6(3):703–715
10. Wang L, Chai T, Zhai L (2009) Neural-network-based terminal sliding-mode control of robotic manipulators including actuator dynamics. IEEE Trans Ind Electron 56(9):3296–3304
11. Lewis FL, Dawson DM, Abdallah CT (2004) Robot manipulator control: theory and practice, vol 15. Marcel Dekker Inc, New York
12. Mahyuddin M, Herrmann G, Khan S (2012) A Novel adaptive control algorithm in application to a humanoid robot arm. Adv Auton Robot 7429:25–36
13. Na J, Mahyuddin MN, Herrmann G, Ren X, Barber P (2014) Robust adaptive finite time parameter estimation and control for robotic systems. Int J Robust Nonlinear Control, pp 1–27 (In press)
14. Craig JJ (2005) Introduction to robotics: mechanics and control vol 3. Pearson Prentice Hall, Upper Saddle River

Chapter 37
Cross-Media Big Data Tourism Perception Research Based on Multi-Agent

Dong Guan and Junping Du

Abstract In this paper, we introduced the design and implementation in detail the cross-media big tourist perception system based on agent. We used the technology based on agent to implement the parallel in the process of data collection. This system mainly consists of a set of cooperation agent, including the data collection agent; the URL Agent; the data update agent and the management agent. Compared with the ordinary distributed information collection system, we designed and implemented the URL agent and data update agent, we used the algorithm to analyze the page based on agent, and then improves the efficiency of the cross-media big tourist perception.

Keywords Cross-media · Big data sensing · Multi-agent · Distributed data collection

37.1 Introduction

With the development of the Internet, the mobile communication and the Internet of things, the experience of the travel will be more deeply. The change of the data collection is from centralization to distribution [1]. The emergence of the agent provides a new method for the distributed system, which is a major breakthrough in the software development. At present in the information management [2], intelligent database, data mining [3], network management and e-commerce, the agent technology has been widely used [4]. In the aspect of the big data collection,

D. Guan · J. Du (✉)
Beijing Key Laboratory of Intelligent Telecommunication Software and Multimedia, School of Computer Science Beijing University of Posts and Telecommunications Beijing, Beijing 100876, China
e-mail: junpingdu@126.com

© Springer-Verlag Berlin Heidelberg 2016

Y. Jia et al. (eds.), *Proceedings of the 2015 Chinese Intelligent Systems Conference*, Lecture Notes in Electrical Engineering 359,
DOI 10.1007/978-3-662-48386-2_37

although there are a lot of data collection systems and application models based on the agent [6], there is no prototype system or application system for the cross-media tourism big data collection. So in this paper, we study the cross media tourism perception, through the distributed data collection system based on Agent, to complete the data collection for cross-media tourism, and then to provide the data basis for further research.

37.2 Related Technologies

37.2.1 Nutch Technology

The Nutch system is made up of Nutch tool and a series of tools used to establish and maintain the data structure [8]. The principle and steps of the cross-media tourism big data perception system based on Nutch is shown in Fig. 37.1.

The process of distributed information collection system based on Nutch as follows:

(1) It generates an empty CrawlDB, and then adds the initial URL lists to the CrawlDB.
(2) According to certain rules, it takes out URLs from CrawlDB, and then creates a new segment to generate FetchList.
(3) According to the FetchList, it downloads web pages from the Internet.
(4) It parses the collected content into the text and data.
(5) It extracts the new web page link URL into the CrawlDB.
(6) It repeats steps 2–5 until reaching the designated depth or the fetchlist is empty.

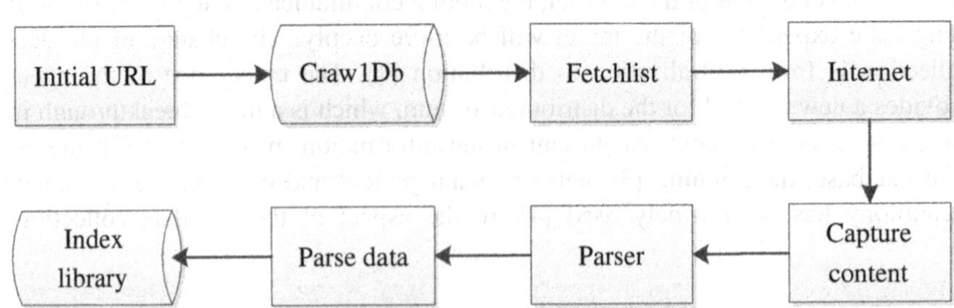

Fig. 37.1 Principle of the distributed data collection system based on Nutch

37.3 Cross-Media Tourism Big Data Perception System Based on Agent

37.3.1 Function Diagram of the Cross-Media Tourism Big Data System Based on Agent

In this paper, we used the Nutch tool to encapsulate the relatively independent modules of cross-media tourism big data perception function as agent. It is made up of data collection Agent, URL Agent, and data update agent. On the basis of it, we established a management Agent, which is responsible for the coordination and control of other three agents and sets different parameters and tasks for different Agent. The functional diagram of cross-media tourism big data perception system based on agent is shown in Fig. 37.2.

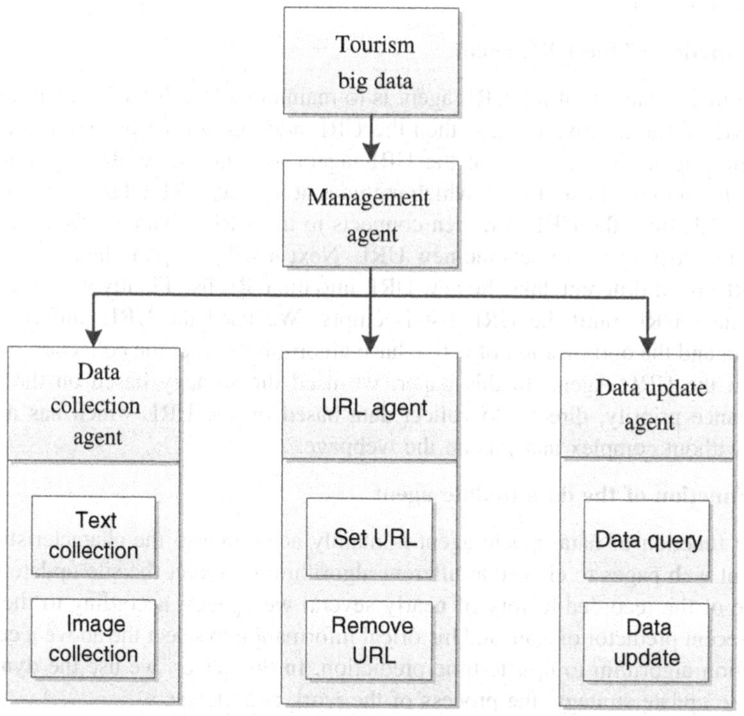

Fig. 37.2 Functional diagram of cross-media tourism big data perception system

37.3.2 Cross-Media Tourism Big Data Perception Agent Function Implementation

(1) **The function of the management agent**

The main function of the management agent is the organizer and leader for data collection. It is responsible for the management of the URL queue and assigns URL to the data collection agent, then products and recycles data collection agent according to the usage of server resources. The management agent regularly collects and backup to the server according to the system time.

(2) **Function of the data collection agent**

The main function of the data collection agent is to find web pages related to tourism and parses into URL format. It also collects and saves web pages related to tourism. When the data collection agent is generated by the management agent, firstly it will load the collection strategy and related parameters. Then data collection agent will adopt the way of Nutch combined with multi-thread to parallel collect data. In this way, it can fully use the system resources and shorten the time of data collection.

(3) **Function of the URL agent**

The main function of the URL agent is to maintain a URLlist which will access the URL of the information, and then the URL address will be passed to the data collection agent. The process of the URL agent as follows: At the beginning, it establishes an initial list of URL which includes at least an URL address. It retrieves a new URL from the URL list, then connects to the URL, obtains web document, parses the document, extracts the new URL. Next it will compare the new URL to the URL list, if different, take the new URL into the URL list. Finally it continues to select next URL until the URL list is empty. We used the URL uniform hash function and the optimization of active hash algorithm to filter the collection of web page in the URL Agent. In this paper, we used the strategy based on the URL importance priority, directly to collect data based on the URL which has a high score without complex analysis on the webpage.

(4) **Function of the data update agent**

The function of data update agent is mainly according to the characteristics of different web pages to choose a different algorithm to predict the site update time, change of the recorded history of nearly several web pages, according to the web site a recent predictor of state and historical information to select the above a certain prediction algorithm to update time prediction. In this paper, we use the dynamic web site update strategy, the process of the work as follows:

As to the new web page, the system gives an initial value of the next visit according to the type of the web page, and then it uses the adjacent method to predict, sets the state predictor to state one. When the visit the web page, the page is constantly changing, but the page don't change or visit the web page before, web page does not change, the web page changes, then sets the state predictor to state two.

As the state predictor in state two, it used the geometric interval method to predict. By using the method of average, it can get the change of web pages. According to the geometric interval method, we used it to access the web page until the page changed serial times, and then take the change of average.

As the state predictor in state two or three, the algorithm set the state predictor in state one according to the history of the change.

When predicting the effect on the change of web pages, we adopt three test items: the accuracy of the prediction, the time migration rate of the prediction, the efficiency of the prediction.

37.4 Results and Analysis of the Experiment

37.4.1 Experimental Result of the Collection Strategy Algorithm

In the URL agent, we used the main page coverage to judge the validity of the evaluation mechanism and the extended OPIC algorithm. Accordingly we selected 16 initial travel portals as the object of collection. First of all, we collected about 100,000 pages related to the tourism, then artificially selects 2000 pages, to the main page and then respectively by random sampling strategy and based on priority of URL rating value acquisition strategy for acquisition of the saved 100,000 page again to verify the important degree of priority to the URL sampling strategy is preferred to a more important pages. The experimental result is shown in Figs. 37.3 and 37.4.

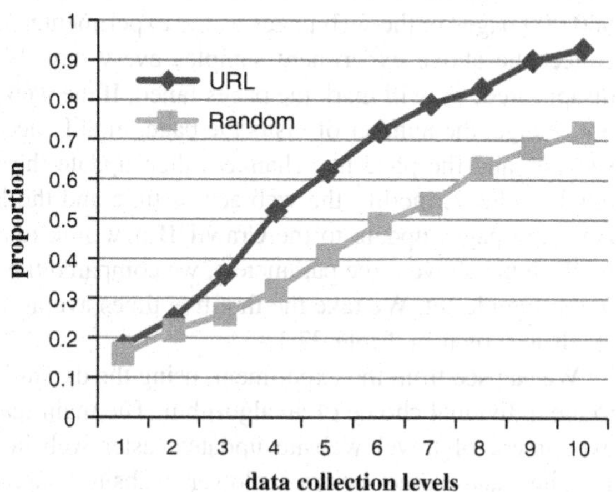

Fig. 37.3 Comparison effects of the two kinds of data collection strategy

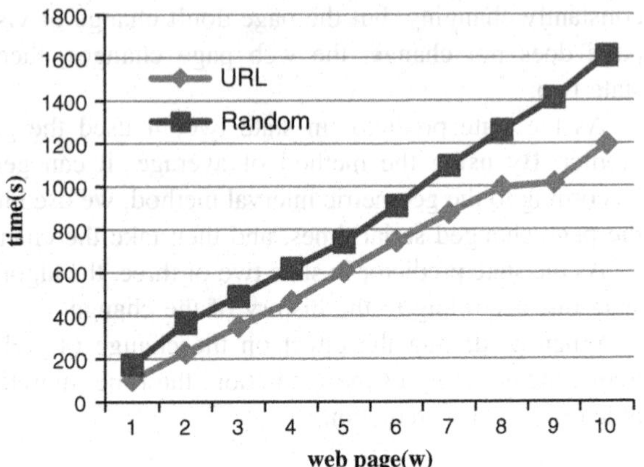

Fig. 37.4 Comparison time of the two kinds of data collection strategy

In Figs. 37.3 and 37.4, it is concluded that using random sampling strategy, sampling 8 times the cumulative only about 75 % of important web pages and cost more time, As using the URL priority algorithm to collect, collecting seven important web pages can be about 80 % of the total collection and cost less time, when collected 10 times, basically completed the collection of all the important web page. Thus it can be seen that give priority to your URL importance sampling strategy is able to complete the priority of the data collection.

37.4.2 *Experimental Results of the Predict Updating Algorithm*

Firstly we selected the home page of sixteen sites related to tourism, according to the breadth-first traversal scope of the web page, getting the collection of the start pages and inserting the web link into the CrawlDB. Then we randomly selected 500,000 pages in the web pages as the experimental sample. Finally we repeated to collect the above experiment samples every day. If founding the web pages has disappeared, we will mark the pages failed. If the web page success, but the web did not change, the number of visits the page, modify access time recently. Acquisition success, and the page has changed, then update the web fingerprints, web access number plus 1, modify the web access time and the last update time recently. The new web pages update to the CrawlDB new link relations.

With the above three parameters, we compared the difference of three algorithms in the sample set. We take the first five times average change of the web page. The result is shown in Table 37.1:

We can see from the experiment, using the dynamic selection method is superior to the individual choice of an algorithm. The main reason is that we selected in this experiment of travel website updates faster website (such as some portal travel website) and some updates slower website (such as some tourist attractions

Table 37.1 Efficiency of the three kinds of prediction algorithm

Algorithm	Accuracy	Efficiency	Migration rate
Adjacent	0.634	1.126	0.181
Interval	0.668	0.958	0.135
Dynamic selection	0.773	1.238	0.105

introduction website), and adjacent method and interval method are respectively applied to different frequency of site update, so choosing the dynamic selection method for the web update forecast result is better.

37.5 Conclusions

In this paper, we mainly studied the problem of the cross-media tourism big data perception under the Internet. Due to the need for the cross-media tourism big data collection, we choosed the Nutch distributed acquisition system to meet the perception of the big data. Considering the intelligent at the same time, we designed and implemented the distributed data acquisition system based on agent. In the system, we designed the data collection agent; URL agent; data update agent and management agent, and also adopted the URL importance priority data collection algorithm in the URL agent to avoid the blindness of data collection. We also used the dynamic selection algorithm in the URL agent to forecast the change of web pages, which can effectively improve the accuracy of the forecast of web pages. Through the experiment of the two algorithms in the paper, the result shows that the system can improve the efficiency of cross-media tourism big data collection.

Acknowledgments This work was supported by the National Basic Research Program of China (973 Program) 2012CB821200 (2012CB821206) and the National Natural Science Foundation of China (No. 61320106006).

References

1. Yanfang S (2014) The design and implement of the network forensics data acquisition system based on agent[D]. Shandong normal university, Jinan
2. Zhiqian Y (2013) The research and implement of the web data acquisition based on trust [D]. Donghua university, Shanghai
3. Ming M (2012) The data collection system based on agent[D]. Huazhong university of science and technology, Wuhan
4. Su W (2011) The research of the emergency information intelligent monitoring system based on multi-agent[D]. Beijing University of Posts and Telecommunications, Beijing
5. Futong Q, Yijun Z, Hao L, Zhiqiang L (2014) The network security data collection model based on semantic agent[J]. Ship Electron Eng 2:95–98

6. Hengfei Z, Yuexiang Y, Wang F (2011) The study and optimization of the distributed web crawler based on Nutch [J]. Comput Sci Explor 5(l):68–74
7. Song H-J Shen Z-Q, Miao C-Y, Tan A-H, Zhao G-P (2007) The multi-agentdata collection in HLA-Based simulation system[J]. In proceedings, 21st international workshop on principles of advanced and distributed simulation (PADS'07) 30:61–69
8. Suetsugu T , Matsunaga S, Torikai T, Furukawa H (2015) Effective data collection scheme by mobile agent over wireless sensor network[J]. IEEE 7057847:1–6

Chapter 38
Location and Navigation Study of Laser Base Station

Peng Wang, Weicun Zhang and Yuzhen Zhang

Abstract Underground navigation is one of critical underground intelligent mining technologies. On account of underground specialty and working conditions, this article presents a research on location mode based on laser ranging and laser-based active navigation mode. First, the hardware design of base station control system used in laser location navigation is introduced. Second, the design of base station motion controller and communication module are described. Third, based on the motion controller, a moving target acquisition algorithm, laser location tracking algorithm, and laser-based active navigation algorithm are presented. Finally, some experimental results conducted in base station are illustrated.

Keywords Motion control · Laser location · Active navigation · Target acquisition · Tracking

38.1 Introduction

At present, the automation and intelligent technology of underground metal mining in our country is far behind of the developed countries [1], which is a serious constraint to the efficient and the production security of underground metal mining and cannot meet the demand of the rapid mining development. Therefore, we need carry out a research on the intelligent mining technology and underground metal mining equipment with the goal of equipment intelligent operation and process intelligent monitor, to break the back of the key technology of underground metal mining and to enhance the market competitiveness of mining enterprises and equipment manufacturing in our country [2].

P. Wang (✉) · W. Zhang · Y. Zhang
School of Automation and Electrical Engineering,
University of Science and Technology Beijing, Beijing 100083, China
e-mail: wpby@vip.qq.com

© Springer-Verlag Berlin Heidelberg 2016

Y. Jia et al. (eds.), *Proceedings of the 2015 Chinese Intelligent Systems Conference*, Lecture Notes in Electrical Engineering 359,
DOI 10.1007/978-3-662-48386-2_38

The location method of related research abroad primarily focuses on RFID, radio frequency electromagnetic and wireless communication. Navigation means include absolute and relative navigation and SLAM technology with combination of them. And the relative navigation technology has been enough mature and been applied to some products. The mining industry in developed countries has been in the information age successively since 1990s. They have continuously improved the mining automation technology, for example, the technologies of remote control and shovel loading, unmanned equipment and automatic navigation are gradually or have been brought into operation phase [3]. The research of underground location technology and the system development started at the beginning of this century and mainly concentrated on the field of underground personnel location [4]. The development of underground location technology can be divided into three stages [5]. First, the underground location products were imported from abroad entirely. Second, our country researched and developed the underground personnel system independently [6]. Third, our country constantly perfected the active underground personnel location system of China's own research and development [7].

This article is mainly about the design of laser base station used in underground equipment location and navigation.

38.2 System Construction

To locate the moving target and actively correct the moving target position, as well as to acquire the feedback signal of action error, this article divides the base station system into four modules to design and implement: distance inspection module, motion control module, motion execution module and communication module. The entire system function block diagram shows in Fig. 38.1.

(1) Motion control module: It adopts the DSP 28335 of Texas Instruments Company as the processing core. This module is a core component of the whole base station.

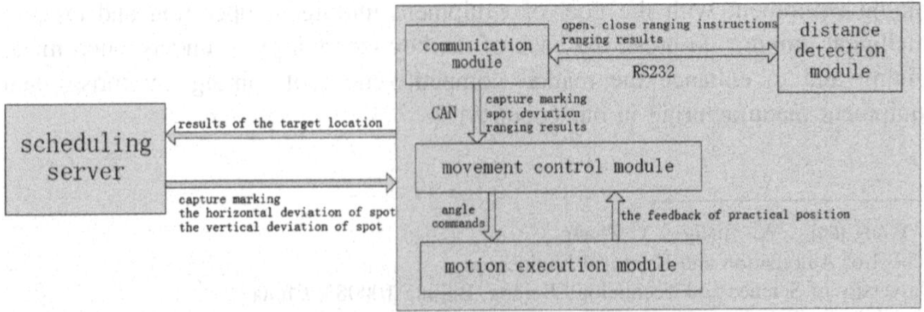

Fig. 38.1 Base station system function block diagram

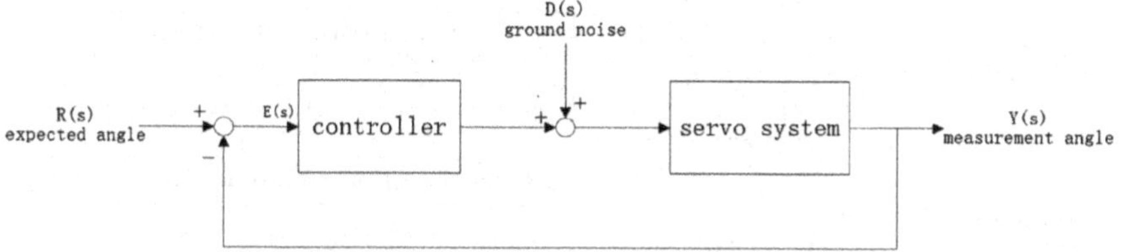

Fig. 38.2 Block diagram of base station following motion system

(2) Distance inspection module: It adopts the FTM-50 laser displacement sensor of Feituoxinda Company.
(3) Motion execution module: It adopts the precise alternating current servo system by Nikki Direct-drive Servo Motor Company.
(4) Communication module: It adopts the ARM kernel MCU stm32 of ST Microelectronics as the processor.

Figure 38.2 shows the system diagram in vertical direction. In this control module, Y(s) is the actual elevation (measured value), and interference effect of ground flatness is indicated as D(s). The object of the system design is to optimize the controller parameters and the selection of servo system. So it can make the response indicators of input angle meet the demand, and to make the errors caused by interference do not interrupt the normal working state of system.

38.3 The Selection and Design of System Hardware

This chapter introduces the hardware design of the motion controller and communication module.

38.3.1 The Circuit Design of Motion Controller

Motion control module is the core component in the system. The motion controller should complete the following three functions: (1) The output of the differential pulse signal; (2) The input of the quadrature encoder pulse signal; (3) The communicating function of CAN bus.

Based on the above demands, the design of this article applies the 28335 DSP chip by TI Company as the main control chip.

Motion control board is in charge of the comprehensive analysis of the target information and controlling the motion execution module according the processing results. Motion controller system diagram is showed in Fig. 38.3.

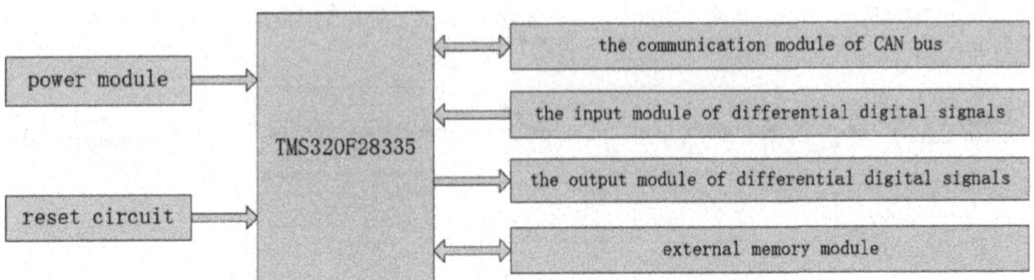

Fig. 38.3 Motion controller system function block diagram

(1) The operating voltage of the peripheral interface chip on main controller power module is 5 V, and to the DSP and ARM processor, the operating voltage is 3.3 V.
(2) The design of motor drive circuit. The Nikki direct-drive servo motor used in this system is controlled by the means of pulse sequence inputted by 5 V differential signal.
(3) The design of encoder pulse feedback. The encoder feedback pulse also applies the differential signal to output position information, so as to improve the system stability and anti-interference.
(4) Level switching circuit. Since the input port on 28335 chip only supports 3.3 V signal, before the 5 V TTL signal reaches into the DSP processor, the design should add a switching circuit that can convert 5 to 3.3 V, otherwise the 5 V signal will damage the pin circuit on main control chip [8].

38.3.2 Hardware Circuit Design of Communication Module

Since the system adopts multiple communication channels and communication protocols, such as CAN, RS232, wireless Ethernet communication, and in order to reduce the communication load of motion controller, a communication transmission module is designed that dedicated to information processing and transmission. This module uses the STM32F107VCT6, an interconnected embedded microprocessor, produced by ST Company. The STM32F107VCT6 has 100 pins with two communication modules of CAN bus. There are one Ethernet port and several serial communication ports that can configure as RS232 or 485 communication port within the selected type. The selection results meet the demand of communication module design.

Communication pin-board is mainly composed of minimum system, RS232 transceiver, CAN bus transceiver and RS485 bus transceiver. System function block diagram of communication module shows in Fig. 38.4.

(1) STM 32 micro processor needs 3.3 V power supply, and selects AS1117 steady voltage chip as the power.
(2) CAN bus is the key communication way between the base station and chips within base station. CAN (short for Controller Area Network) is one of the

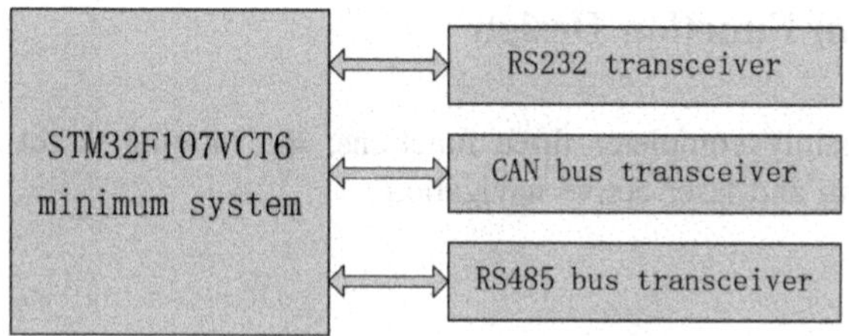

Fig. 38.4 System function block diagram of communication module

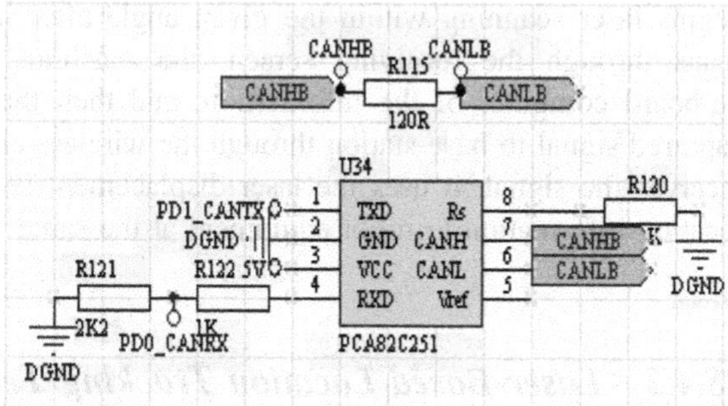

Fig. 38.5 CAN module circuit schematic diagram

most wildly used field buses around the world [9]. CAN communication transceiver is the PCA82C251CAN from Philips Company. Figure 38.5 shows the CAN module circuit schematic diagram.

(3) Ethernet module is used for the communication between the base station with server, and the server with intelligent devices. Although the STM32F107 chip has the Ethernet MAC controller, it does not provide the physical interface. So DP83848 is selected as the Ethernet interface chip. The circuit is connected according to the handbook as shown in Fig. 38.6.

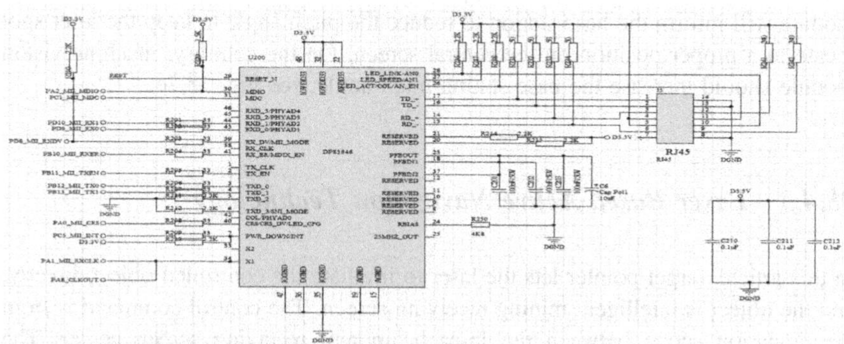

Fig. 38.6 Ethernet module circuit schematic diagram

38.4 System Function Design

This system mainly completes three functions: the moving target capture, laser location tracking and laser active navigation.

38.4.1 The Technology of Moving Target Capture

Base station's first capture to the intelligent equipment is that: the base station begins laser scanning within the given angle after initialization. When the laser scans through the receiving screen, the machine vision module informs the on-board computer of the information, and then the robot equipment sends the captured signal to base station through the wireless channel. After the base station receives the signal, it uses the laser displacement sensor to locate and sends the location information to robot equipment at the same time.

38.4.2 Laser-Based Location Tracking Technology

The base station conducts a real-time location by the spherical coordinate tracking measurement. During the location process, base station range finder and the servo system form a spherical coordinate. During the process of location measurement, the motion controller combines the laser displacement sensor measurement result with the heading angle and the elevation, and then calculates the target position in the local coordinate system. Combing the world coordinate of the base station, the system can calculate the intelligent mining equipment coordination in the world system through the transformation of coordinates.

Since the surface ground of the tunnel is uneven, the base station should control the upper and lower deviation of the laser on the screen. As it shown in Fig. 38.7a, if the laser spot was located on the upper half part of the screen, the machine vision module will inform the base station to reduce the pitch angle to keep the laser spot locate in a proper position on the central screen. On the contrary, machine vision module should increase the base station pitch angle, see Fig. 38.7c.

38.4.3 Laser-Based Active Navigation Technology

In this article, target pointer lets the laser to irradiate the controlled object directly, and the object is intelligent mining receiving screen. The control command is from the deviation scope between the laser beam and receiving screen center. The controlled object always tries to eliminate the geometry error by changing the

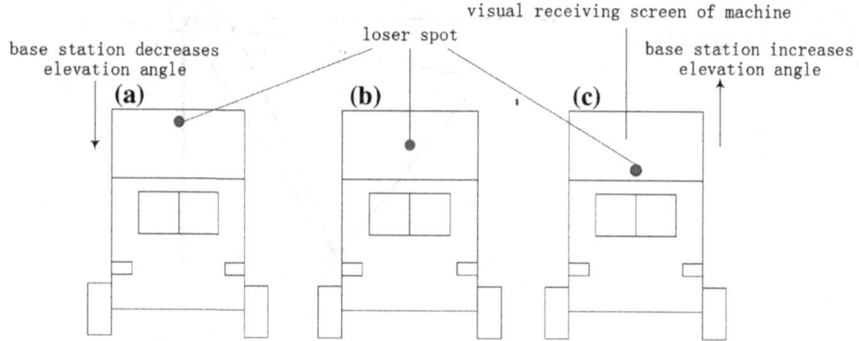

Fig. 38.7 Possible situation may occur on the laser receiving screen (pitching direction)

equipment splice angle to let the equipment move in the planned orbit. As it shown in Fig. 38.8b, when the laser spot is located on the screen center, the equipment moves within the planned orbit and it does not adjust the moving direction. In the Fig. 38.8a, when the laser spot is on the left of screen, the equipment will think that the moving direction is on the right of planned orbit, then it turns left and eliminates deviation between the laser and screen center to set the direction right. On the contrary, as it shown in Fig. 38.8c, Make the opposite adjustment to Fig. 38.8a.

The mathematical model of laser active navigation aims to definite the relationship between the equipment and moving target, then to obtain the next attitude according to the current attitude, and designs the moving position process by using interpolation algorithm. The specific working principle of linear interpolation algorithm shows in Fig. 38.9.

Bi(B-1, B0, B, B1, …) in the diagram is the predetermined point in the planned trajectory sequence, and the planned moving direction is from small to large. Point A is the equipment position, and point B is the planned target position that

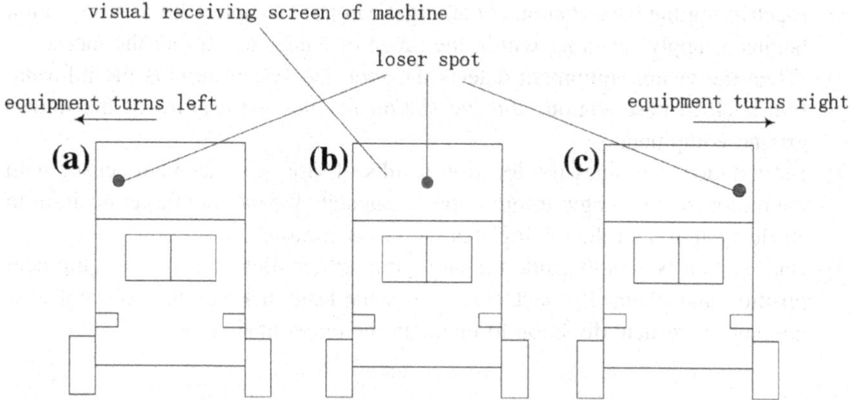

Fig. 38.8 Possible situation may occur on the laser receiving screen (vertical direction)

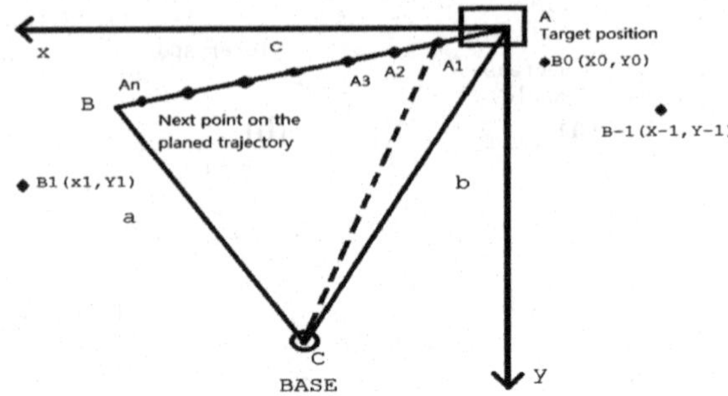

Fig. 38.9 Interpolation algorithm principle schematic diagram

will reach, and the point C is the base station position. The diagram can be explained that: the moving target is not on the planned trajectory because of the execution error or other reasons. At this time, base station have completed the capture and it is leading the target to point B (planned trajectory point).

Linear interpolation algorithm can be divided into following 5 procedures:

(1) Acquire the next guidance target coordinate B.
(2) Known three coordinates A, B and C, calculate the vector AB, AC direction as the target moving direction.
(3) According the information of the ABC point position, equipment's speed and the previous position of equipment, put them into the equipment kinetic model and get the horizontal swing motor deflection angle $\angle ACA1$.
(4) According to the result in (3), drive the horizontal swing motor to revolve specified angle.
(5) Finishing rotation, update the coordinate of point A, and return to procedure (1).

38.4.4 The Entire System Working Process

(1) Electrifying the base station, if the target can not be captured, the base station begins to apply scanning within the range of angles to explore the target.
(2) When the vision equipment detects the spot, the system reports the information back to base station, and the station finishes the capture of the underground equipment.
(3) Base station combines the location results of laser displacement sensor with the motor rotation angle information to calculate the moving target position in world system, and then completes the target location.
(4) Base station begins to guide the target on vertical direction by the equipment position and planned trajectory, at the same time, it keeps the spot not evanescent on vertical direction to maintain the target navigation.

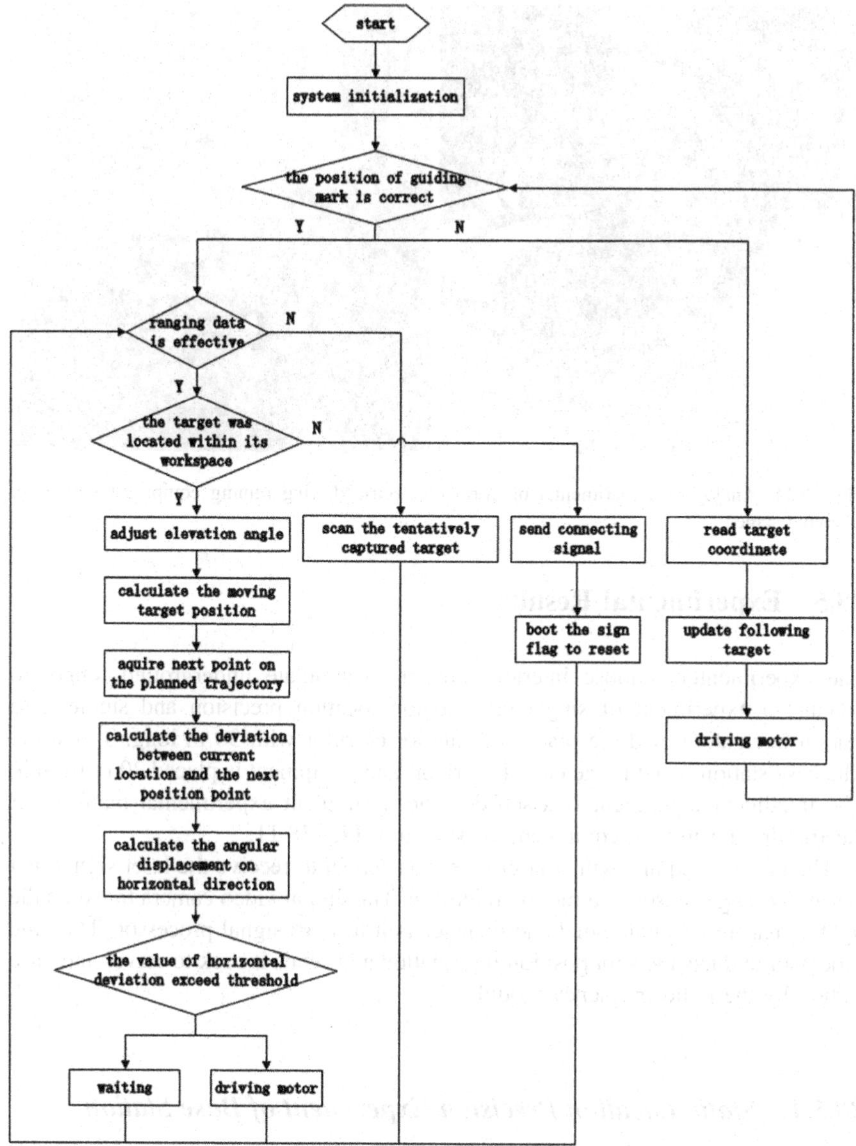

Fig. 38.10 System entire working process diagram

(5) Any unforeseen circumstances will cause the spot leave equipment screen, and the target informs the base station by wireless channel. Then the base station begins to recapture target according to memory location and direction, and step backs to (1).

Processes above can simplify a flow diagram, as shown in Fig. 38.10.

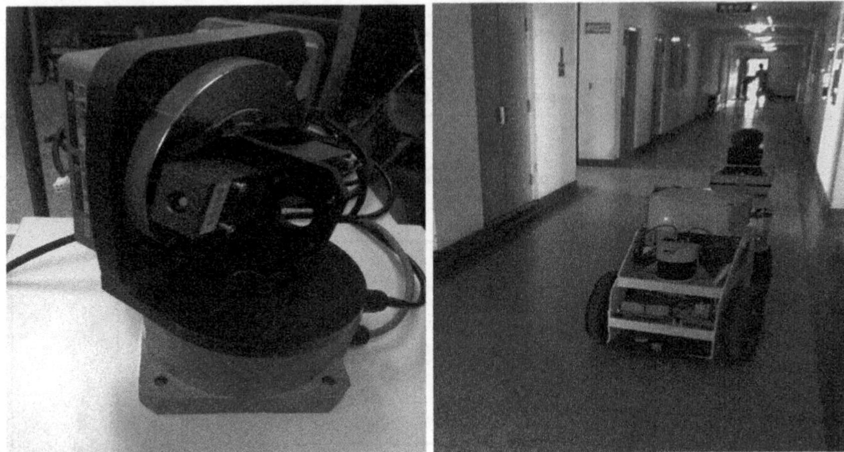

Fig. 38.11 Analogous experimental platform of electric driving mining equipment with laser receiving screen

38.5 Experimental Results

The experimenters choose interior corridor to simulate underground tunnel to conduct a experiment of single base station location precision and single base station navigation, and the place is a interior corridor with 30 m long, 3 m wide. The base station is put at the end of corridor, and equipment is placed 20 m from it. The intelligent equipment is a self-developed simulant experimental platform of electric driving mining equipment, as shown in Fig. 38.11.

The platform equips with a laser receiving screen to receive the laser signal that is sent by range sensor as a motion reference. The digital video camera transfers the light signal into digital signal and then sends it to DSP signal processor. Then the processor extracts the spot position information and sends it back to server and base station by the radio frequency module.

38.5.1 Static Location Precision Experiment of Base Station

In the experiment, the equipment is placed at specified position and the course angle is measured artificially. On the basis, the laser sensor is open, and reads the data when the target is captured, then compares the data with accurate measurement of total station. Part of the precision experiments is shown in Fig. 38.12.

Analyzing the static location data in above table, it can be inferred that the maximum static location error value is 97, and the minimum value is 1 mm, and the average location error value is 18.25 mm.

Data/ mm	X-value (location)	X-value (measurement)	measurement error	Y-value (location)	Y-value (measurement)	measurement error
	4499	4500	-1	9059	9076	-17
	6031	5995	36	10308	10255	53
	12420	12432	-12	1130	1149	-19
	15112	15089	23	5405	5409	-4
	17763	17843	-76	1881	1933	-52
	22767	22751	16	2608	2644	-36
	23233	23247	-14	3937	3967	-30
	27762	27758	4	3327	3375	-48
	33686	33638	48	4159	4221	-62
	38294	38391	-97	4833	4910	-77

Fig. 38.12 Partial data of indoor base station static position precision experiment

38.5.2 Dynamic Guidance Experiment of the Base Station

The planed path in the dynamic guidance experiment of the base station is straight line path, and a part of the experiment process is shown in Fig. 38.13. During the test process, base station transfers the current position to upper computer, and the upper computer displays and stores the position information. The finally result is shown in Fig. 38.14.

This article selects the distance from the actual moving tracking point $P(x_p, y_p)$ to the line path $Ax + By + C = 0$, to evaluate navigation precision.

Fig. 38.13 Base station dynamic guidance

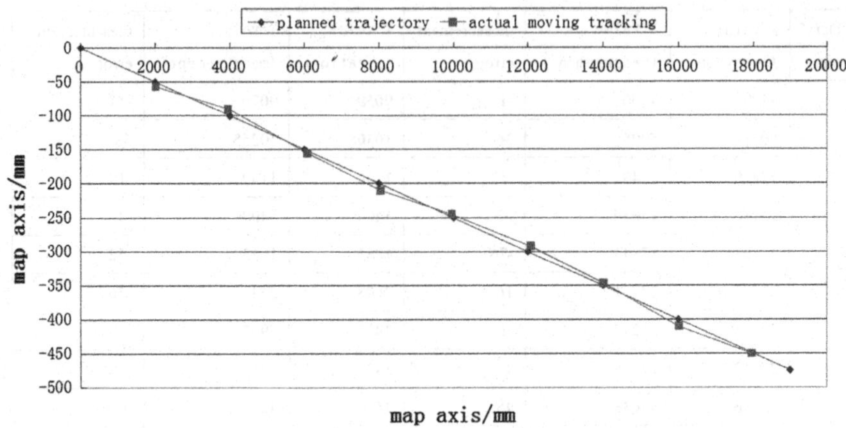

Fig. 38.14 Comparison diagram of straight moving path with practical moving path

The maximum deviation value between actual moving tracking point and planned tracking is less than ±9.75 mm, and the average deviation value is 6.11 mm.

38.6 Conclusion

This article designs a control system that is used to conduct laser location and active navigation to underground mining equipment, then a systematical study is performed on the crucial techniques of the design through using the system, such as, the moving target capture technology, laser location and laser-based active navigation and so on. Based on the above technologies, the article also verified the feasibility and practicability of the subject "The laser location navigation base station can assist the intelligent mining equipment and acquire location results with high precision under the mining tunnel that can't or is not suitable to use the traditional location technologies (GPS, Wifi, RFID and so on)".

References

1. Fang Z, Wang L, Huang W (2008) Present situation and advances of underground mining equipment of metal mine in China [J]. Express Inf Min Ind 24(11):2–4
2. Wang Y (2006) Development trend and science and technology development strategy of mining technology for metallic mines [J]. Met Mine 1:19
3. Wang S (2005) Research and implementation of 3D geological model [D]. Northeastern University, British Columbia

4. Zhu X (2007) With the implementation of the underground personnel positioning system hardware design [D]. University of Technology of Hubei, Hubei
5. Sun Y (2011) Design of Underground Personnel Positioning System Based on ZigBee Technology [D]. University of Science and Technology of Harbin
6. Shu L (2013) Design of downhole positioning communication terminal based on android [D]. University of Science and Technology of Shandong, Shandong
7. Feng L (2006) Research on application of wireless sensor network and ZigBee technology. [D]. University of Technology of Hefei, Hefei
8. Li J (2012) Study on the dust concentration measurement method based on DSP [D]. University Of Science and Technology of Shandong, Shandong
9. Ai X (2008) Design of CAN bus controller in SOC chip [D]. Southeast China University, Nanjing

Chapter 39
Sentiment Analysis Based on Evaluation of Tourist Attractions

Zhicheng Ma, Junping Du and Yipeng Zhou

Abstract Tourists satisfaction has become more and more an indicator of tourism development. Sentiment analysis on data of comments and hot discussion on travel site and Weibo can help judge real-time satisfaction trend of tourists with scenic spots according to the intensity of sentimental tendency. Considering the deficiencies in current sentiment analysis, in this paper, firstly the polarity value and strength value are used to calculate the sentimental intensity of the sentimental words. HIT-CIR Tongyici Cilin (extended) is used to expand the synonyms of the sentimental words in order to reduce the impact of words not in HowNet and some sentimental words with low frequency in the corpus. Then we improved the traditional semantic similarity method based on HowNet according to the characteristics of sentiment analysis, combining it with the method based on Point Mutual Information (PMI) and syntactic dependency relations. High accuracy is shown by the experimental results.

Keywords Sentiment analysis · HowNet · PMI · Syntactic dependency

39.1 Introduction

The rapid development of technology and application of the Internet has had a profound effect on consumer behavior patterns. Most tourists in front of the travel plan will refer to other travelers' review information and then choose the right

Z. Ma · J. Du (✉)
Beijing Key Laboratory of Intelligent Telecommunication Software and Multimedia School of Computer Science, Beijing University of Posts and Telecommunications, Beijing 100876, China
e-mail: junpingdu@126.com

Y. Zhou
School of Computer and Information Engineering, Beijing Technology and Business University, Beijing 100876, China

© Springer-Verlag Berlin Heidelberg 2016
Y. Jia et al. (eds.), *Proceedings of the 2015 Chinese Intelligent Systems Conference*, Lecture Notes in Electrical Engineering 359,
DOI 10.1007/978-3-662-48386-2_39

spots. Therefore, comment information on Weibo and travel sites affects the recommendation of tourism destinations, or the possibility of visiting again.

Travel notes and comment of attractions on tourism community or BBS are provided by the tourists themselves, which have certain truth and are generally preferred by the masses of tourists. Tourists satisfaction becomes more and more an indicator of tourism development. Sentiment analysis of Weibo and other tourist commentary not only can provide help for tourist destination choice, but also can timely understand the evaluation of tourist destination. Tourism enterprises can adopt corresponding measures to improve the products and services, and promote the long-term development of the tourism destination.

Firstly, the paper uses the polarity value and strength value to calculate the sentimental intensity of the sentimental words and HIT-CIR Tongyici Cilin (extended) to expand the synonyms of the sentimental words, then fuses point semantic analysis method (S0-PMI) [1] and improved semantic calculation method based on HowNet [2] to determine the words' sentimental polarity. Then uses syntactic dependency relation [3] to extract the sentimental dependency phrases and judges the sentimental tendencies and intensity of sentence.

39.2 Sentiment Analysis Technology

Chinese text sentiment analysis is generally divided into word, phrase, sentence and discourse level. The way of word level sentimental analysis is divided into method based on dictionary and corpus.

The method based on dictionary digs sentimental words by using correlation between words in dictionary. HowNet is commonly used in Chinese. It is a common sense knowledge base, which uses concept represented by Chinese and English words to describe object. The traditional calculation method of semantic similarity based on HowNet is proposed by Qun Liu, Su-jian Li [4]. The method based on corpus is mainly to collect enough corpus, using statistical characteristics to calculate words polarity. S0-PMI is commonly used. Its basic idea is that the statistical probability of two words appearing at the same time in the corpus, if the probability is larger, the relevance is closer and the similarity is higher.

The syntax rules of syntax analysis can be used to extract the combination evaluation units and can provide basis for the phrase level sentimental polarity analysis. Generally the shape of the structure and dynamic structure, verb-object combination, subject-predicate relation and "of" word structure include affective words which can be extracted to analyze the phrase sentimental polarity. If there is negative words or degree adverbs adv, the polarity of sentimental words can be opposite, enhanced or reduced through the Strength (adv) set in the dictionary.

According to the associated words, sentences relationship can be divided into three kinds: juxtaposition, progressive and adversative. Sentence sentimental polarity values of juxtaposition, progressive and adversative relationship can be computed respectively by addition, setting weight (sentence in the after has larger weight) and subtraction.

39.3 Sentiment Analysis Based on Sentimental Dictionary

39.3.1 Custom Sentimental Dictionary

At first, this paper builds a sentimental dictionary containing more than 35,000 words, whose strength is divided into 10 kinds (±1, ±3, ±5, ±7, ±9) with the negative words (3326), sentimental words (1330), positive evaluation words (3939) and sentimental words (890) in HowNet, HIT-CIR Tongyici Cilin (Extended) (7850), and Dalian university of technology sentimental vocabulary ontology library (27467). According to 219 degree adverbs in HowNet, the paper builds an adverb table and set strength for each degree adverbs. According to emoticons, this paper constructs the symbol table based on Weibo expression [5], at the same time sets sentimental polarity and the weights according to Chinese semantic expression.

39.3.2 Improvement of Word Semantic Similarity Method Based on HowNet

Semantic similarity method based on HowNet judges the similarity of words mainly by calculating sememe similarity. The semantital polarity of word is mainly related to the positive and negative property of words defined in HowNet. This property belongs to the other sememe. Therefore, this paper improves the other sememe similarity calculation method that Liu and Li proposed: if the other sememe of word contains positive and negative property, then we set the similarity of positive and negative property as the similarity of the other sememe, meanwhile raise the other sememe's weight.

39.3.3 Words Polarity Calculation Based on the Combining Method

While Judgment accuracy of low similarity words based on HowNet is low, it is effective to improve accuracy with method based on PMI to judge words polarity of low similarity by setting the similarity threshold proposed in reference [6]. However, in reference [6] certain words with similarity higher than the threshold with both positive and negative benchmark, are unable to judge correctly. So in this paper, the following improvements are made. For words whose similarity are higher than the threshold in both benchmarks, we can further determine the polarity through the difference of similarity between its synonym set and the two benchmarks. Meanwhile in order to gain better fusion with HowNet, for method based on PMI, we regard product of maximum probability of co-occurrence with polarity of benchmark words as polarity of words under test.

For word w_1, firstly we expand the synonyms of the w_1 to synonym set is $\{w_1, w_2, \ldots, w_n\}$, w_1 and its synonyms set has closer contact with positive (negative) seeds, the more the positive (negative) tendency of the word. Words similarity computation formula based on HowNet is as follows:

$$\mathrm{SO{-}Hownet}(w_1) = \frac{1}{N}\sum_{j=1}^{N}\left(\sum_{i=1}^{k} Sim(p_i, w_j) - \sum_{i=1}^{k} Sim(n_i, w_j)\right) \quad (39.1)$$

N is the number of synonyms expanded set. w_j is element of set. p_i, n_i are respectively positive and negative benchmark word.

The paper uses Baidu as corpus, regarding number of hits that search engine returns as frequency of words, then the word similarity computation formula based on So-PMI is as follows.

$$\mathrm{SO{-}PMI}(w_1) = \frac{1}{N}\sum_{i=1}^{N}\log_2\left(\frac{\prod_{pw \in Pws}(hits(w_i\&pw)+1) * \prod_{nw \in Nws}(hits(nw)+1)}{\prod_{pw \in Pws}(hits(pw)+1) * \prod_{nw \in Nws}(hits(w_i\&nw)+1)}\right) \quad (39.2)$$

N is the number of synonyms expanded set. w_j is element of set. p_i, n_i are respectively positive and negative benchmark word. Adding 1 is to smooth (39.2) in case that the denominator is zero. The algorithm process is as follows:

(1) For word w_1, firstly we expand the synonyms of the w_1 with HIT-CIR Tongyici Cilin (Extended) to synonym set $\{w_1, \ldots, w_n\}$.
(2) Calculating similarity of synonym set with all of the benchmarks words Sim (w_i, benchmark word j). Max is the highest value of similarity, θ_{sim} is predefined, If Max < θ_{sim}, go to 3). If Max > θ_{sim}, similar degree between w_1 (or its synonyms) and the benchmark words is high, calculating SO-Hownet(w_1) through formula (39.1), θ_{Hownet} is predefined (in paper, $\theta_{Hownet} = 0$), If the benchmark words are positive reference words, SO-Hownet (w_1) > θ_{Hownet} or benchmark words are negative reference words, SO-Hownet (w_1) < θ_{Hownet}, the sentimental tendency of w_1 and the benchmark words is tend to be the same: SO(w_1) = Sim(w_i, benchmark word j)*SO(benchmark word j), or go to 3).
(3) Calculating SO-PMI(w_1) through formula (39.2), a threshold θ_{PMI} is predefined, if SO-PMI(w_1) > θ_{PMI}, w_1 is judged to be positive, Pmax, the maximum of co-occurrence probability of w_1 and all its synonyms wirh benchmark

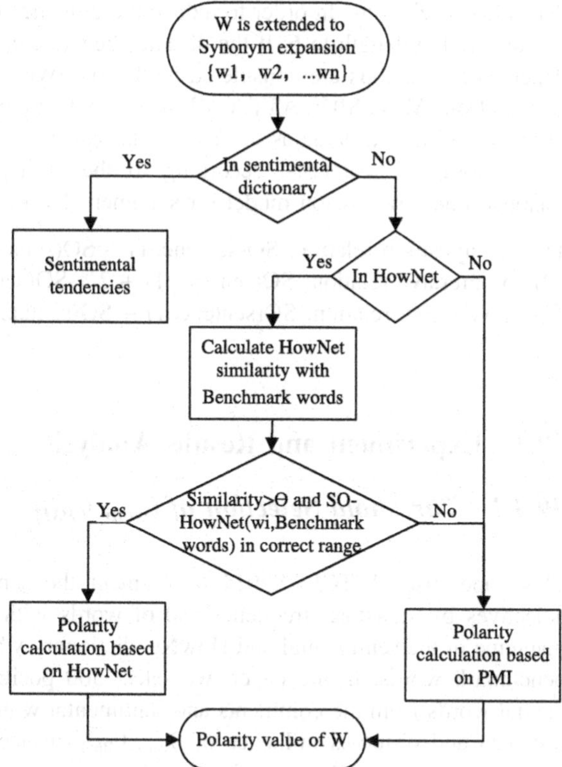

Fig. 39.1 Sentimental tendency calculation of words based on HowNet and PMI

words is to be calculated; If SO-PMI (w_1) < $-\theta_{PMI}$, w_1 is judged to be negative, then calculating the maximum probability of co-occurrence. Pmax = MAX (hits (w_i & benchmark word j)/hits (benchmark word j). Sentimental tendency of w_1: SO(w_1) = Pmax * SO (benchmark word). Otherwise, w_1 is judged to be neutral (Fig. 39.1).

39.3.4 Polarity Calculation of Sentimental Phrase and Sentence Based on Syntactic Dependency Relation

In this paper, the language of the Harbin industrial university technology platform LTP [7] is used to extract affective phrase in the scenic spot comments sentence. We analyze comments a total of more than 50,000 and Spot dynamic a total of more than one hundred thousand from Weibo about Beijing attractions with syntactic dependency relation. We select the syntactic dependency relations of large proportion containing sentimental words, which are ADV, SBV, ATT, CMP dependency relations. In order to reduce the influence of syntactic parser error, we set the window length to 6. If the distance between the two words is more than 6, dependency relations are considered ineffective. We extract candidate sentimental phrases from ADV, SBV, ATT, CMP structure. For simple sentences, we sum those phrases sentimental value as its sentimental value.

For complex sentences, according to the category of the correlatives, we establish linear regression model for sentimental value of simple sentences.

(1) juxtaposition relation: SO(sentence1) + SO(sentence2)
(2) progressive relation: SO(sentence1) + 2 * SO(sentence2)
(3) adversative relation: SO(sentence1) − SO(sentence2)

39.4 Experiment and Results Analysis

39.4.1 Threshold Selection of θ_{sim}, θ_{PMI}

This paper uses ICTCLAS2014 to segment the sentences and sort verbs and adjectives by statistical frequency. 80 of words with strong praise or blame semanteme in the sentimental and HowNet dictionary which rank top are selected as benchmark words. In the paper, we select 500 positive and 500 negative sentimental words from the comments and sentimental words issued by HowNet as the test data and observes influence of θ_{sim}, θ_{PMI} on accuracy respectively with the accuracy as the test indicators. The experimental results are shown in Figs. 39.2 and 39.3.

We set $\beta1 = 0.5$, $\beta2 = 0.28$, $\beta3 = 0.13$, $\beta4 = 0.09$. After many experiments, we select the threshold values: $\theta_{sim} = 0.82$, $\theta_{PMI} = 0.2$.

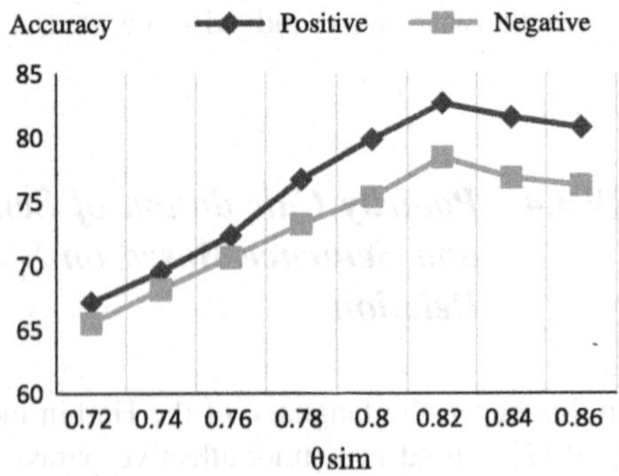

Fig. 39.2 Recognition accuracy of sentimental words varying with θ_{sim}

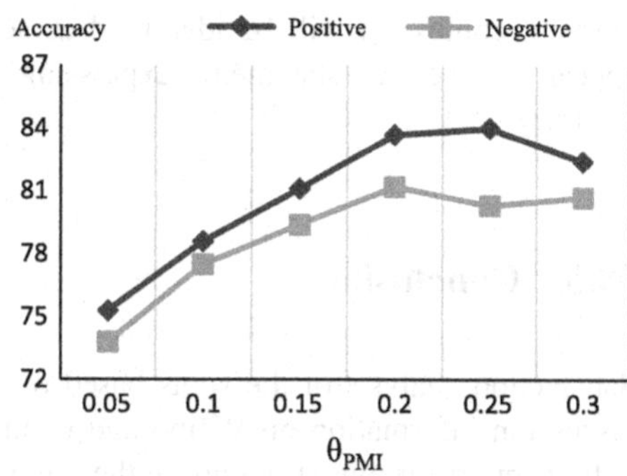

Fig. 39.3 Recognition accuracy of sentimental words varying with θ_{PMI}

39.4.2 Experimental Contrast of Three Kinds of Sentiment Analysis Method

This paper selects 9000 online comments and spot dynamic about Beijing attractions from the public comments and Sina, Weibo, in which 3000 are positive, 3000 are negative, 3000 are neutral. We made experiment with improved text sentimental polarity calculation method based on HowNet, method based on PMI and combination of these two methods proposed in this paper respectively, with accuracy P, recall rate R and F1 as the evaluation indicators. Experimental results of three methods are shown in Table 39.1.

From Table 39.1, we can see that the fusion method this paper proposed has raised accuracy and recall rate largely. The positive and negative emotion recognition accuracy has reached 80 %. Compared with the former two methods, the method proposed in this paper reduces the impact of words that are not in HowNet or have low frequency in corpus effectively. Meanwhile it solves the problem of wrong judgement of low or high similarity degree in positive and negative tendency in HowNet effectively. Accuracy of neutral emotion recognition is lower compared with the previous two. Firstly HowNet do not perform well in recognizing sentimental words which do not contain positive or negative attributes; Secondly as Chinese expressions are flexible and emotion of some words tend to be different in

Table 39.1 Experimental results

Method		Positive	Negative	Neutral
HowNet	P	0.754	0.736	0.705
	R	0.739	0.743	0.712
	F1	0.746	0.739	0.708
PMI	P	0.738	0.724	0.716
	R	0.727	0.750	0.701
	F1	0.732	0.737	0.708
Fusion	P	0.809	0.784	0.739
	R	0.782	0.795	0.753
	F1	0.795	0.789	0.746

different context, it will be identified as sentimental tendency when these words appear in objective statements expressing, which are neutral. That leads to low accuracy.

39.5 Conclusions

Satisfaction analysis of the spots based on the sentiment analysis of the real-time discussion information on Weibo and comments on travel sites can not only provides support for tourist to choose the scenic spots but also promotes the long-term development of tourist destination. This paper brings polarity value and strength value to calculate the sentimental intensity of the sentimental words, HIT-CIR Tongyici Cilin (extended) to expand the synonyms of the sentimental words in order to reduce the impact of words that are not in HowNet or have low frequency in corpus. Then we fused S0-PMI and improved semantic calculation method based on HowNet, meanwhile combined them with syntactic dependency relation to judge the sentimental tendencies and intensity of sentence, good results have been achieved so far.

Acknowledgement This work was supported by the National Basic Research Program of China (973 Program) 2012CB821200 (2012CB821206), the National Natural Science Foundation of China (No. 61320106006), Beijing Excellent Talent Founding Project (2013D005003000009).

References

1. wen-ying Z (2010) Sentiment analysis of travel destination reviews in Chinese. Harbin industrial University
2. Dong Z, Dong Q (2015) HowNet. http://www.keenage.com. Accessed 15 March 2015
3. Weiping W, Cuicui M (2011) Evaluation object extraction based on the analysis of the syntax and interdependence. Comput Appl Syst 20(8):52–57
4. Liu Q, Li S (2002) Lexical semantic similarity computation based on HowNet. Taipei
5. Li Y. Microblog Emotional Dictionary Built and Application on Sentiment Analysis of Microblog. Zhengzhou University, 2014
6. Wang ZY, Wu Z, Hu F (2012) Sementital polarity calculation of words based on HowNet and PMI. Comput Eng 38(15):187–189, 193
7. Che W, Li Z, Liu T (2010) LTP: a Chinese language technology platform. In: Proceedings of the 23rd international conference on computational linguistics: demonstraions. ACM, New York, pp 13–16

Chapter 40
Target Localization and Tracking of Unmanned Mining Equipment Based on Multi-sensor Information Fusion Technology

Xuan Li, Jiannan Chi and Weicun Zhang

Abstract The accurate localization is very important for mobile devices to make right decisions about autonomous path planning, avoiding obstacles and finishing other complex tasks. This paper presents a research on the localization and tracking technology of autonomous underground mining equipment. Two types of Kalman filters are considered as information fusion method: multi-sensor multi-model adaptive Kalman filtering and weighted adaptive multiple model Kalman filtering.

Keywords Multiple model adaptive kalman filtering · Weighted multiple model adaptive control · Information fusion

40.1 Introduction

The method of target tracking uses detectors, such as radar, sonar, infrared, to measure the moving objects and locate the targets' moving state. Filtering and prediction are two basic elements of the tracking system, at the same time are very important for estimate of object moving state parameters, such as position, velocity and acceleration [1], present or at future moment.

In the 60s of last century, Kalman filtering algorithm [2] (KF) was put forward for the first time, which not only got extensive application with deeper theoretic and practice research, but also established the foundation of modern filtering theory. Kalman filter is a software filtering method [3]. Its basic idea is to minimize the minimum-mean square error as the best estimation criterion, based on the state space model of signal and noise and utilization of previous estimation and current

X. Li (✉) · J. Chi · W. Zhang
School of Automation and Electrical Engineering, University of Science and Technology Beijing, Beijing 100083, China
e-mail: Lixuan_sky@163.com

© Springer-Verlag Berlin Heidelberg 2016

Y. Jia et al. (eds.), *Proceedings of the 2015 Chinese Intelligent Systems Conference*, Lecture Notes in Electrical Engineering 359,
DOI 10.1007/978-3-662-48386-2_40

measure value to update the estimation of state variables and to determine the current estimation.

Data fusion is a collaborative multilevel automatic message processing procedure that acquires the goal's state and feature estimation as well as the situation and threat assessment using the detection, correlation, estimation and synthesis of multi-classes, multi-aspects, multi-information. It uses different sources, mode, time, location and forms' information to conduct fusion, finally gets accurate description of the perceived objects [4]. Basically, data fusion comes from the redundancy of information. Based on the separate sensors' measure information to get more effective information, its ultimate aim is taking advantage of multiple sensors to enhance the effectiveness of the whole system.

40.2 Positioning Methods Description

Because of the complexity of underground mining environment, we can't use usual position ways such as GPS, in this paper, the main location methods are laser ranging cooperative with visual capture and UWB wireless positioning.

40.2.1 Laser Ranging Cooperative with Visual Capture Devices

Laser positioning base is a two-dimensional turntable fixed in the mine, and installs with laser transmitter. Using laser positioning information and 2D turntable's turning angle to calculate laser point's specific coordinate that irradiated on the screen. Using the Ethernet communication function to send the message to the computer of the articulated vehicle.

The vision receiving device is installed in the articulated vehicle, a device for recognizing laser ranging points, is mainly composed of a piece of screen, a camera and an image processing board. The camera will capture the image of light receiving screen and perform image processing. When the laser point hits the screen, the vision receiving device will identify the laser point's vertical and horizontal offset value relatively to the screen center, and send those value to the computer on the articulated vehicle by CAN bus, and the on-board computer will transfer it to laser positioning base by Ethernet.

When the laser base obtains the offset value, it will adjust 2D turntable to ensure that the laser point will always shine on the screen, thus realizes the tracking of laser point to the vehicle. As shown in Fig. 40.1.

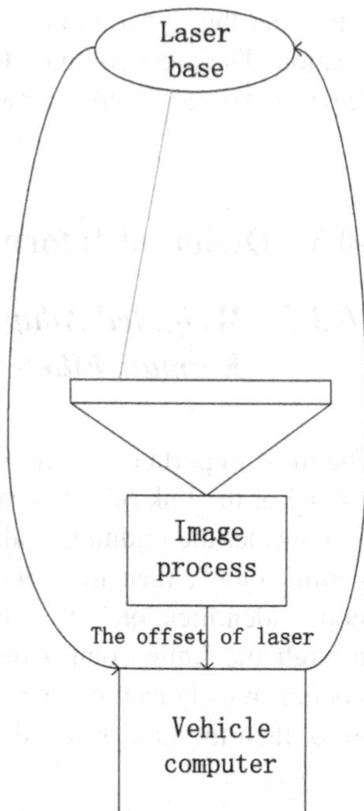

Fig. 40.1 The working process of laser base

When the on-board computer gets the laser point's coordinates and the offset value, it can calculate the vehicle's front axle center or hinge center coordinates information, thus realizes the position of articulated vehicle coordinates.

40.2.2 UWB Location Method

UWB wireless position module is a method using UWB (Ultra Wideband) technique, which is a high speed, low expense and low-power dissipation technology. It is a non carrier technology, using nanosecond pulse communication, carrying a high speed data transmission technique. Because the UWB system pulse duration is very short, and has strong time and space resolution, it can effectively combat multi-path fading caused by multi-path effect, which has a unique advantage in the field of wireless location. Related UWB products can achieve very high precision ranging and positioning.

Generally, underground vehicle precious position system of wireless method consists of two parts, one is fixed anchor nodes installed on the wall, this kind of nodes will be arranged with accurate coordinates in advance to conduct position calculation; the other is mobile nodes of system installed on the moving device to determine the dynamic node information through a variety of means of communication. The most common technique used in UWB are: Angle of Arrival (AOA), Time of Arrival (TOA), Time Difference of Arrival (TDOA) [5].

40.3 Design of Information Fusion Algorithm

40.3.1 Weighted Adaptive Multiple Model Control Kalman Filter

The most important feature of weighted adaptive multiple model WMMAC is that instead of the link of parameter estimation, it conducts the calculation of weight. If we consider the traditional adaptive control parameter estimation as "infinite model identification", then this kind of weight calculation can be considered as "finite model identification", that is, realizing the identification of finite local model through the finite control device's weight that is calculated off-line. When one model is mostly near to the true system and the weight is 1. While the other weights are 0, then the process of identification is successful [6].

40.3.2 Interacting Multiple Model-Probabilistic Data Association Filter

IMM-PDAF is the key point of data fusion and object tracking, combing with the research founding of Bar-Shalom and Houles [7], we study the feasibility of it through simulation experiment.

The contrast of the two information fusion theory (WMMAC and IMM-PDAF) showed in Figs. 40.2 and 40.3.

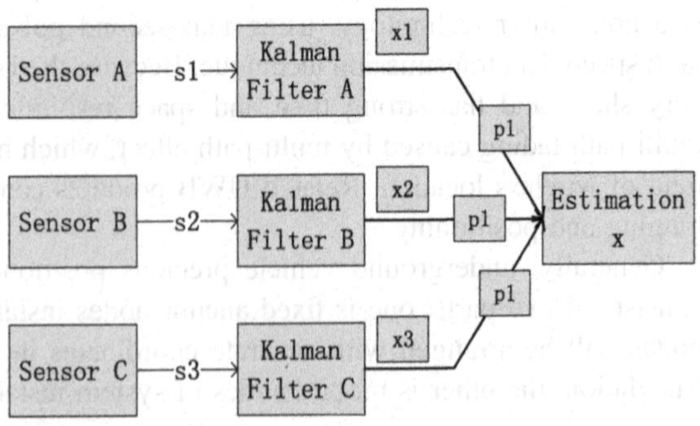

Fig. 40.2 The process of WMMAC

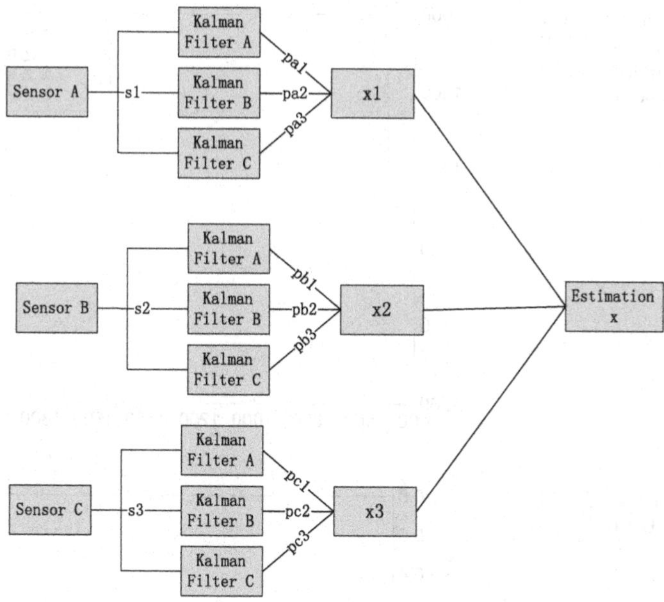

Fig. 40.3 The process of IMM-PDAF

40.4 Matlab Simulation

Assuming there are a laser base and UWB system to observe a target, the object is doing uniform velocity linear movement along side the y axis at t = 0–10 s, velocity is −1.5 m/s, the coordinate of start point is (10, 20 m); the object is tuning a 90° corner towards x axis, the acceleration speed is $u_x = u_y = 0.15$ m/s^2, when the turning is finished, the acceleration speed come down to 0. The laser measure error and UWB system measure 0.15 and 0.20 m through the experiment site gauge.

In Fig. 40.4, it shows the object's actual movement, laser and UWB observation trajectory, filtering estimation trajectory by the method of WMMAC. In Fig. 40.5, it shows the weight value of laser and UWB.

Under the same situation, the results from IMM-PDAF is shown in Fig. 40.6. The error between filtering estimation and true value shown in Fig. 40.7.

When the measure error is bigger, assuming the measurement noise standard deviation are 1 and 1.2 m for laser and UWB. In Fig. 40.8, it shows the object's actual trajectory, laser and UWB observation trajectory, filtering estimation trajectory using WMMAC. In Fig. 40.9, it shows the weight value of the two positioning strategies.

Under the same situation, the results from IMM-PDAF is shown in Fig. 40.10. The error between filtering estimation and true value shown in Fig. 40.11.

According to the simulation results, we can obtain the following conclusions:

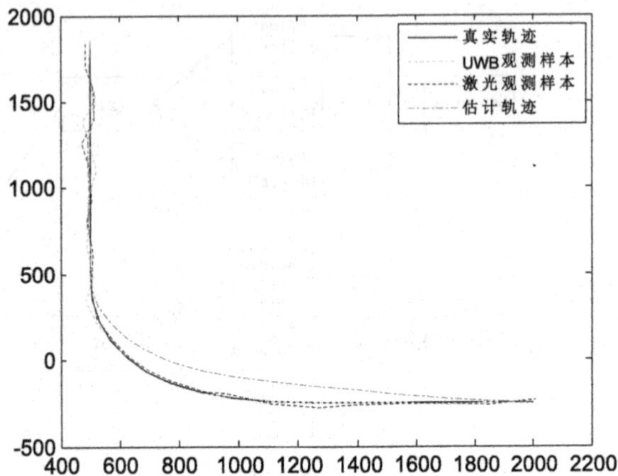

Fig. 40.4 Object's actual trajectory, laser and UWB observation trajectory and estimation trajectory

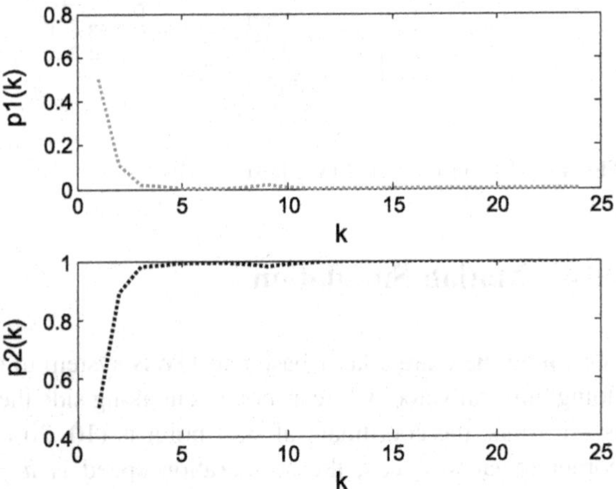

Fig. 40.5 The weight value of laser and UWB

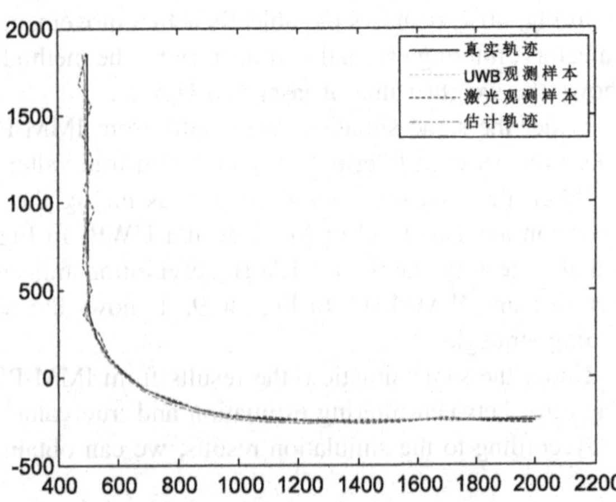

Fig. 40.6 Object's actual trajectory, laser and UWB observation trajectory and estimation trajectory

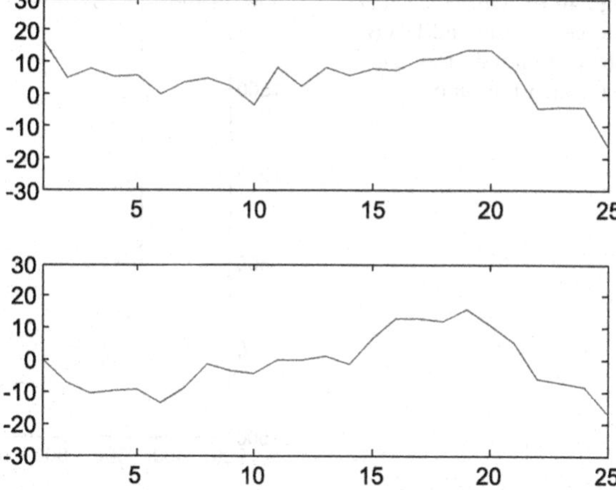

Fig. 40.7 The error between filtering estimation and true value

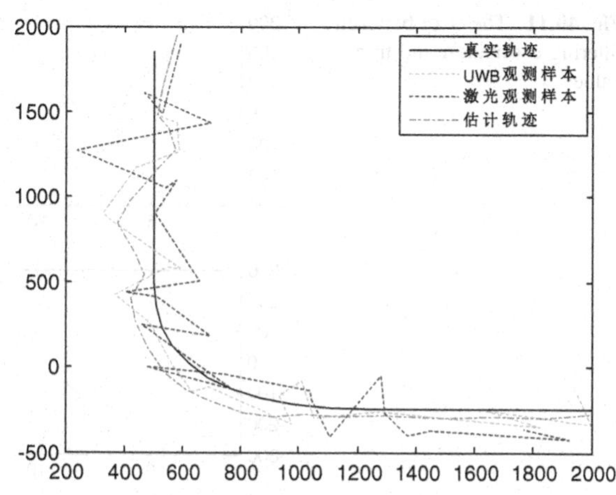

Fig. 40.8 Object's actual trajectory, laser and UWB observation trajectory and estimation trajectory

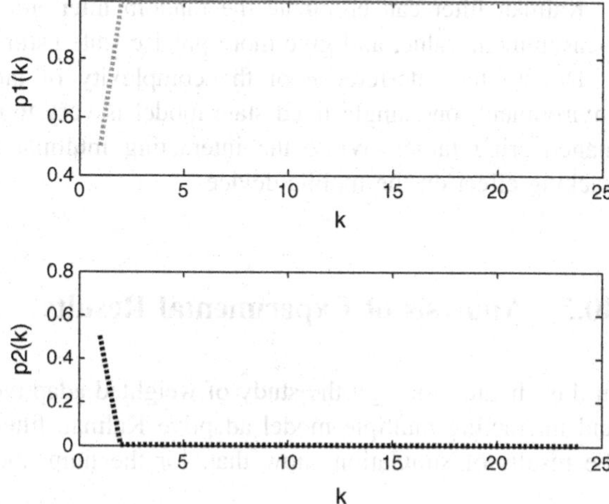

Fig. 40.9 The weight value of laser and UWB

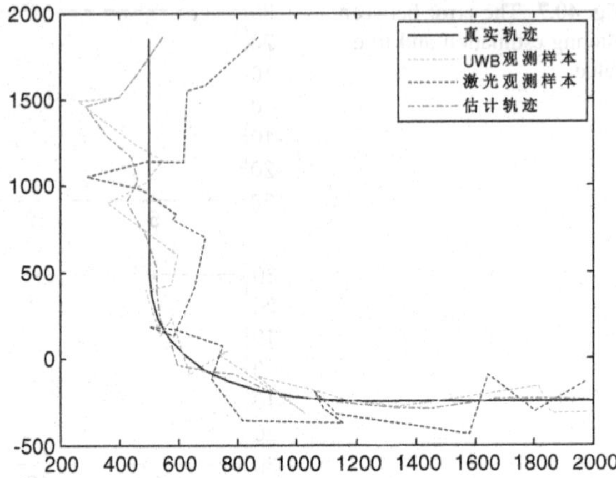

Fig. 40.10 Object's actual trajectory, laser and UWB observation trajectory and estimation trajectory

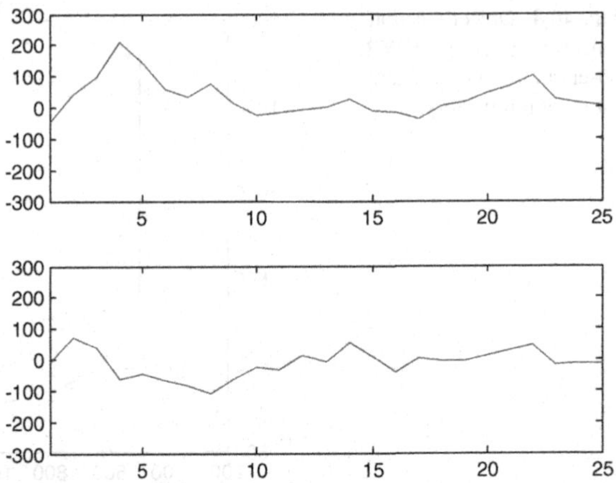

Fig. 40.11 The error between filtering estimation and true value

Kalman filter can eliminate the random interference based on the objection's measurement value, and give more precise state estimation;

Due to the interference of the complexity of target motion model and the environment, one single fixed state model is very tough to make good tracing of maneuvering target. While the interacting multiple model algorithm has better tracking effect on the mobile device.

40.5 Analysis of Experimental Results

In this chapter, through the study of weighted adaptive Kalman filtering algorithm and interacting multiple model adaptive Kalman filtering algorithm, analysis and the results of simulation show that, for the unmanned mining equipment, using

IMM-PDAF fusion algorithm of laser and UWB can get the optimal estimation, improve the accuracy of positioning.

References

1. Libin C, Mingan T (2000) Interacting multiple model adaptive filtering algorithm for maneuvering tracking. Fire Control Command 12:102–152
2. Kalman RE (1960) A new approach to linear filtering and prediction problem. IEEE Trans Aerosp Electron Syst Basic Eng 82(5):34–45
3. Dingcong P (2009) The basic principle and application of Kalman filtering. Softw Guide 8(11):34–45
4. Feng D, Qiuxi J, Nan Z (2007) The review and prospect of development of multi-sensor data fusion. Warsh Electron Warf 30(3):52–55
5. Fang H, Cui X, Liu Q (2005) Summary of the localization problem in wireless sensor networks. Comput Inf Technol (06)
6. Weicun Z (2012) Stability of the weighted multi-model adaptive control. Control Theory Appl 29(12):1657–1660
7. Blom HAP, Bar-shalom Y (1988) The interacting multiple model algorithm for systems with Markovian switching coefficients. IEEE Trans Autom Control AC-33(8):780–783

Chapter 41
Improved Robust Multiple Model Adaptive Control of the Two-Cart Mass-Spring-Damper System with Uncertainties

Ya Wang, Baoyong Zhao and Weicun Zhang

Abstract A new weighting algorithm called Posterior Possibility Generator (PPG) is proposed to replace PPE algorithm in robust multiple model adaptive control (RMMAC) architecture, resulting in the improved robust multiple model adaptive control (IRMMAC) architecture, and a two-cart mass-spring-damper system with uncertainties is used to illustrate the advantages of PPG against PPE.

Keywords RMMAC · IRMMAC · PPE · Kalman filter · PPG

41.1 Introduction

In recent years, Sajjad Fekri, Michael Athans and Antonio Pascoal put forward a new kind of weighted multiple model adaptive control (WMMAC), which is robust multiple adaptive control (RMMAC) [1, 2]. The architecture of so-called RMMAC blends the regular multiple model adaptive evaluation (MMAE) system and a bank of controllers designed by mixed μ-synthesis [8–10]. The details of the RMMAC system are in the references [5–7]. Because of the added controllers, the RMMAC architecture improves its robust stability and robust performance [11] compared with non-adaptive system [1]. From the original MMAE system to RMMAC system, posterior possibility evaluator (PPE) [3, 4] algorithm has been playing the key role among them.

The convergence properties of PPE algorithm highly depend on design of Kalman filters (KF) [1, 5–7] , which may lead to inaccurate identification if the design of KF doesn't satisfy the theoretical assumptions [16]. Unfortunately, there is a lack of studies in the KF design in the presence of unmodelled dynamics.

For the above reasons, a new type of weighting algorithm called posterior possibility generator (PPG) is proposed to replace PPE algorithm, bringing in the so-called

Y. Wang · B. Zhao · W. Zhang (✉)
University of Science and Technology Beijing, Beijing 100083, China
e-mail: weicunzhang@263.net

© Springer-Verlag Berlin Heidelberg 2016
Y. Jia et al. (eds.), *Proceedings of the 2015 Chinese Intelligent Systems Conference*, Lecture Notes in Electrical Engineering 359,
DOI 10.1007/978-3-662-48386-2_41

improved robust multiple model adaptive control (IRMMAC) architecture in this paper. PPG algorithm not only relaxes its convergence conditions, but also improves the convergence rate.

41.2 Improved Robust Multiple Model Adaptive Control Architecture

The IRMMAC architecture is shown in the Fig. 41.1, in which there are three subsystems. The first refers to the bank of KFs, based on the model set. The second subsystem is the calculation of posterior possibility using PPG algorithm, which is in charge of identification with KFs. The last subsystem consists of a bank of controllers designed by the mixed μ-synthesis.

With help of KFs, the requirements of robust stability and robust performance can be guaranteed at the start. However, unlike the RMMAC, where KF residuals, $r_k(t)$, $k = 1, 2, \ldots, N$, produced on-line, and residual covariance matrices, S_k, computed off-line, are both needed by the weighting algorithm to generate the posterior probability values $P_k(t), k = 1, 2, \ldots, N$, the IRMMAC only needs $r_k(t), k = 1, 2, \ldots, N$, to generate the posterior probability values $P_k(t)$.

The mixed μ-synthesis and Matlab Synthesis Toolbox determine the optimal compensator that guarantees the robust stability and robust performance of the system. μ is a mathematical object and very useful in analyzing the effect of parameter uncertainty and unmodeled dynamics on the robust stability and robust performance of the system. If the following sufficient condition is satisfied,

$$\mu_{ub}(\omega) \leq 1 \ \forall\omega \tag{41.1}$$

then the robust stability and robust performance can be thought as guaranteed.

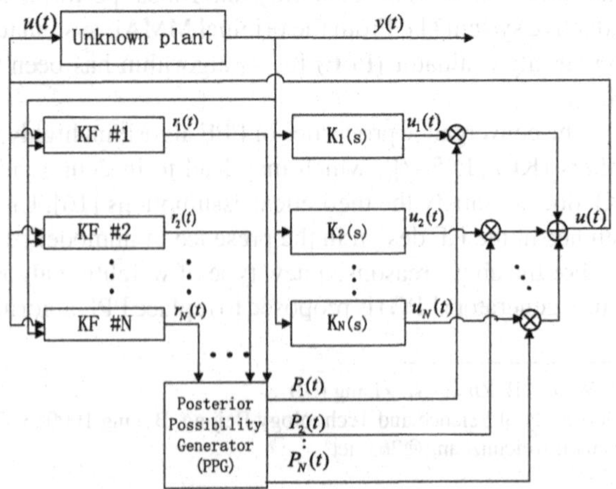

Fig. 41.1 The IRMMAC architecture

Each controller $K_j(s)$ generates a control signal $u_j(t)$ by itself. Then the ultimate control signal $u(t)$ is the sum of the each controller signal, $u_j(t), j = 1, 2, \ldots, N$, multiplied by the posterior probability, $P_j(t), j = 1, 2, \ldots, N$

$$u(t) = \sum_{j=1}^{N} P_j(t) u_j(t) \tag{41.2}$$

The calculation of $P_j(t), j = 1, 2, \ldots, N$, is introduced in Sect. 41.3 in detail.

41.3 Description of Posterior Possibility Generator

For each model $M_k, k = 1, 2, \ldots, N$, the discrete-time state-space is

$$\begin{aligned} x(t+1) &= A_k x(t) + B_k u(t) + L_k \varepsilon(t) \\ y(t+1) &= C_k x(t+1) + \theta(t+1);\ t = 0, 1, 2 \ldots \end{aligned} \tag{41.3}$$

where $\varepsilon(t)$ and $\theta(t+1)$ are both zero-mean white-noise sequence with constant intensity but independent to each other. Next, the bank of N steady-state KFs are associated with the set of models. For each model $M_k, k = 1, 2, \ldots, N$, its output is given by discrete-time steady-state Kalman Filter

Predict-cycle:

$$\begin{aligned} \widehat{x}_k(t+1\,|t) &= A_k \widehat{x}_k(t) + B_k u(t) \\ \widehat{y}_k(t+1\,|t) &= C_k \widehat{x}_k(t+1\,|t) \end{aligned} \tag{41.4}$$

Updated-cycle:

$$\widehat{x}_k(t+1\,|t+1) = \widehat{x}_k(t+1\,|t) + H_k r_k(t+1) \tag{41.5}$$

where the H_k is the constant steady-state KF gain-matrix. For details of designing Kalman filter, the reader is referred to [1, 5–7].

The residual $r_k(t)$ is defined below:

$$r_k(t+1) = y(t+1) - \widehat{y}_k(t+1\,|t) \tag{41.6}$$

In Fig. 41.1, each controller produces $u_k(t)$, the ultimate $u(t)$ is obtained by

$$u(t) = \sum_{k=1}^{N} P_k(t) u_k(t) \tag{41.7}$$

first

$$l_k(0) = \frac{1}{N} P_k(0) = l_k(0) \tag{41.8}$$

$$l_k'(t) = \alpha + \frac{1}{t}\sum_{i=1}^{t} r_k^2(i) \tag{41.9}$$

$$l'_{\min}(t) = \min_k\{l_k'(t)\} \tag{41.10}$$

$$V_k(t) = \frac{l'_{\min}(t)}{l_k'(t)} \tag{41.11}$$

$$l_k(t) = \begin{cases} l_k(t-1) & \text{if } V_k(t) = 1 \\ l_k(t-1)[V_k(t)]^{ceil(\frac{1}{1-V_k(t)})} & \text{if } V_k(t) < 1 \end{cases} \tag{41.12}$$

$$P_k(t) = \frac{l_k(t)}{\sum_{k=1}^{N} l_k(t)} \tag{41.13}$$

where $t = 1, 2, 3 \ldots$ and $\alpha > 0$ is a small constant to avoid $l_k'(t) = 0$

Withe the amount of measurements increasing, as $t \to \infty$, the PPG algorithm can make one of the possibilities converge to the nearest probabilistic neighbor, that is if kth model from the model set is mostly closet to the true model, that will cause

$$P_k(t) \to 1 \;\; a.s. \;\; as \;\; t \to \infty \tag{41.14}$$

The performance comparisons of PPG against PPE will be exhibited in numerical simulations in Sect. 41.4.

41.4 IRMMAC Simulations

41.4.1 Modeling of the Two-Cart Mass-spring-damper (MSD) System

A two-cart mass-spring-damper system is shown in Fig. 41.2. The mass m_1 is connected to the mass m_2 by the spring k_1 but the stiffness constant k_1 is uncertain. One end of the spring k_2 is tied to the wall and the other is connected to the mass m_2. b_1 and b_2 are the damping parameters of the spring k_1 and the spring k_2. $x_1(t)$ and $x_2(t)$ are the displacement of the mass m_1 and the mass m_2 from equilibrium. And the mass m_2 should be in an initial equilibrium state. The $u(t)$ is the control force, with a time delay, which is applied on the mass m_1 to minimize the displacement of the mass m_2. $d(t)$ is a disturbance force acting on the mass m_2, with continuous-time white noise $\xi(t)$, with zero mean and unit intensity, as follows:

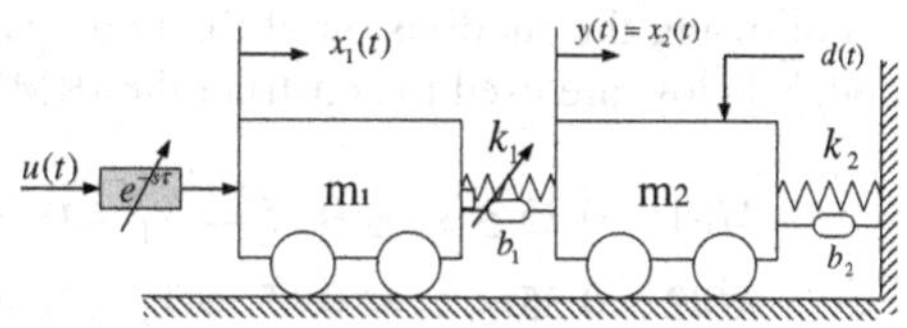

Fig. 41.2 The two-cart mass-spring-damper system

$$d(s) = \frac{\alpha}{s+\alpha}\xi(s) \tag{41.15}$$

The overall state-space representation is:

$$\begin{aligned} \dot{x}(t) &= Ax(t) + Bu(t) + L\xi(t) \\ y(t) &= Cx(t) + \theta(t) \end{aligned} \tag{41.16}$$

The state vector is

$$x^T(t) = \left[x_1(t)\; x_2(t)\; \dot{x}_1(t)\; \dot{x}_2(t)\; d(t) \right] \tag{41.17}$$

Then

$$A = \begin{bmatrix} 0 & 0 & 1 & 0 & 1 \\ 0 & 0 & 0 & 1 & 0 \\ -\frac{k_1}{m_1} & \frac{k_1}{m_1} & -\frac{b_1}{m_1} & \frac{b_1}{m_1} & 0 \\ \frac{k_1}{m_2} & -\frac{(k_1+k_2)}{m_2} & \frac{b_1}{m_2} & -\frac{(b_1+b_2)}{m_2} & \frac{1}{m_2} \\ 0 & 0 & 0 & 0 & -\alpha \end{bmatrix} \tag{41.18}$$

$$B^T = [0\; 0\; \tfrac{1}{m_1}\; 0\; 0],\; C = [0\; 1\; 0\; 0\; 0],\; L^T = [0\; 0\; 0\; 0\; \alpha] \tag{41.19}$$

$$m_1 = m_2 = 1,\; k_2 = 0.15,\; b_1 = b_2 = 0, \alpha = 0.1 \tag{41.20}$$

The range of the uncertain spring constant, k_1 is:

$$\Omega = \{k_1 : 0.25 \le k_1 \le 1.45\} \tag{41.21}$$

An unmodeled dynamic time-delay τ is assumed below

$$\tau \le 0.05s \tag{41.22}$$

The replacement of the mass m_2 is the output of the system,

$$y(t) = x_2(t) \tag{41.23}$$

Following the subdivision of the large parameter uncertainty in RMMAC, three models below are used to construct the IRMMAC

$$\begin{aligned} M\#1: \ 0.25 \le k_1 \le 0.35 \Rightarrow k_1 = 0.3(1+0.05\delta_{k_1}); \ |\delta_{k_1}| \le 1 \\ M\#2: \ 0.35 \le k_1 \le 0.65 \Rightarrow k_1 = 0.5(1+0.15\delta_{k_1}); \ |\delta_{k_1}| \le 1 \\ M\#3: \ 0.65 \le k_1 \le 1.45 \Rightarrow k_1 = 1.05(1+0.4\delta_{k_1}); \ |\delta_{k_1}| \le 1 \end{aligned} \quad (41.24)$$

41.4.2 Performance Evaluation

In this paper, we suppose that the model set of the two-cart MSD system includes three models and the uncertainty of the two-cart MSD system comes from k_1. In the model set, k_1 of the first model is 0.3, of the second model 0.5, and of the third model 1.05. Each model is complemented with a Kalman filter and a controller by mixed μ-synthesis. In order to compare the performance among the IRMMAC, RMMAC, non-adaptive control, we consider five cases in simulations (Fig. 41.3).
Case 1
If k_1 of the true model is set as 1.45, the output $y(t)$ of IRMMAC, RMMAC and Non-adaptive control is shown in Fig. 41.4 and the three weighting signals of RMMAC and IRMMAC in Fig. 41.5
Case 2
If k_1 of the true model is set as 0.6, the simulation results are shown in Figs. 41.6 and 41.7.
Case 3
If k_1 of the true model is set as 0.8, the simulation results are shown in Figs. 10 and 11.

In case 1, the true model is mostly close to the third model and both IRMMAC and RMMAC have good disturbance rejection performance. In case 2, the true model is obviously closest to the model 2 and both IRMMAC an RMMAC also have good disturbance rejection performance, although they both have bigger fluctuation at the

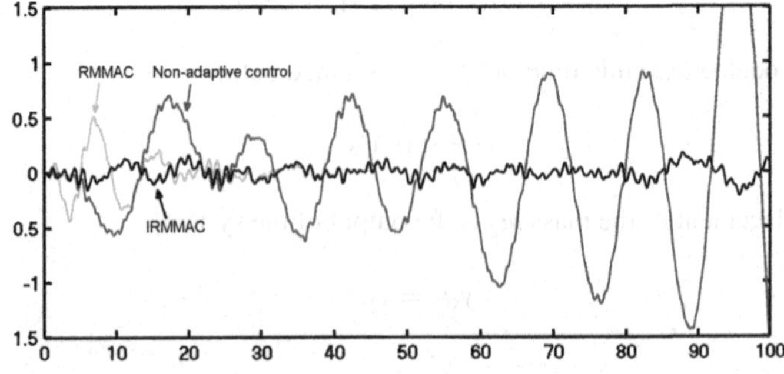

Fig. 41.3 Output performance when $k_1 = 1.45$

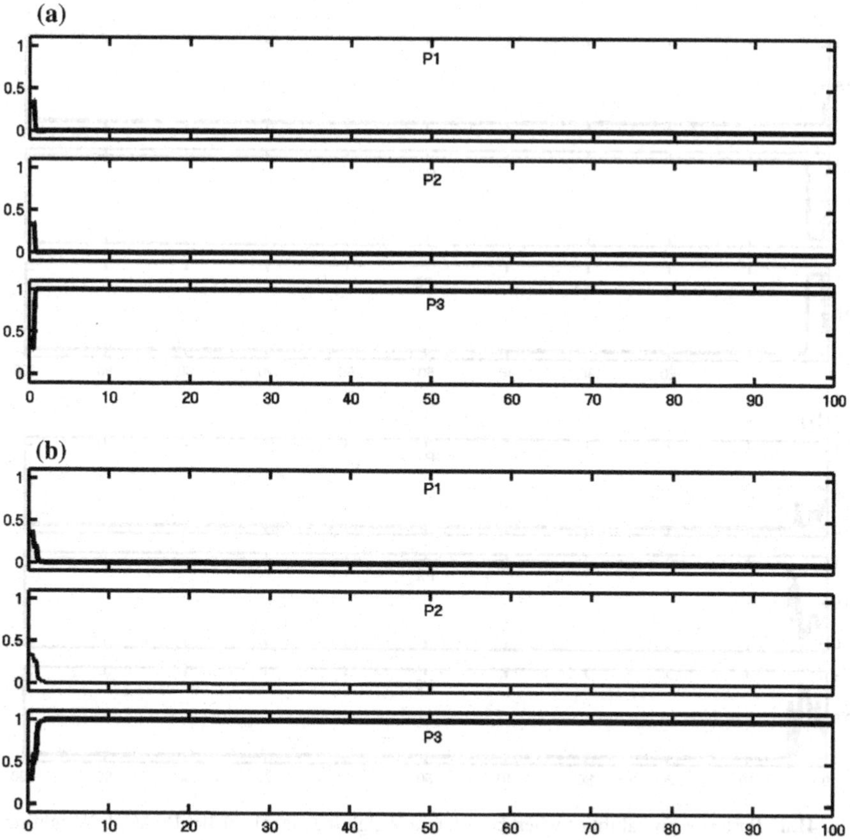

(a)

(b)

Fig. 41.4 Posterior probabilities when $k_1 = 1.45$. **a** Weighting signals of IRMMAC. **b** Weighting signals of RMMAC

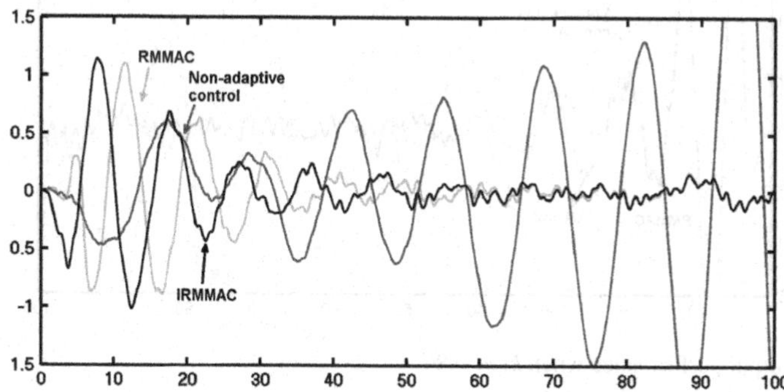

Fig. 41.5 Output performance when $k_1 = 0.6$

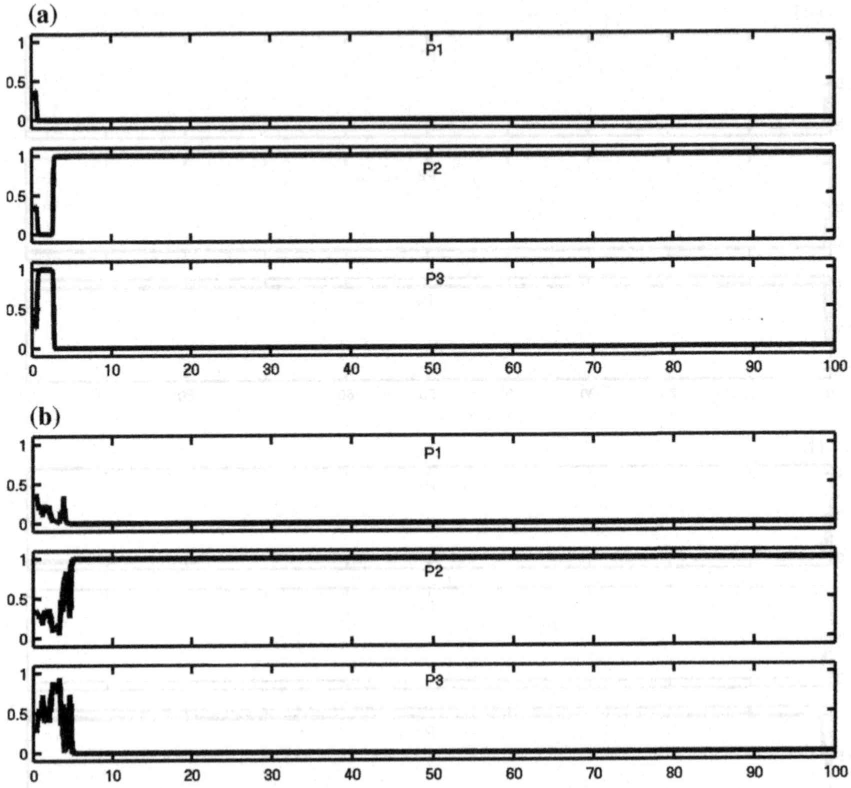

Fig. 41.6 Posterior probabilities when $k_1 = 0.6$. **a** Weighting signals of IRMMAC. **b** Weighting signals of RMMAC

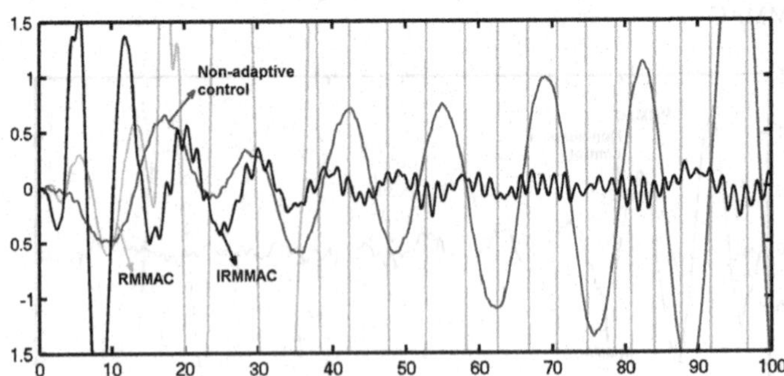

Fig. 41.7 Output performance when $k_1 = 0.8$

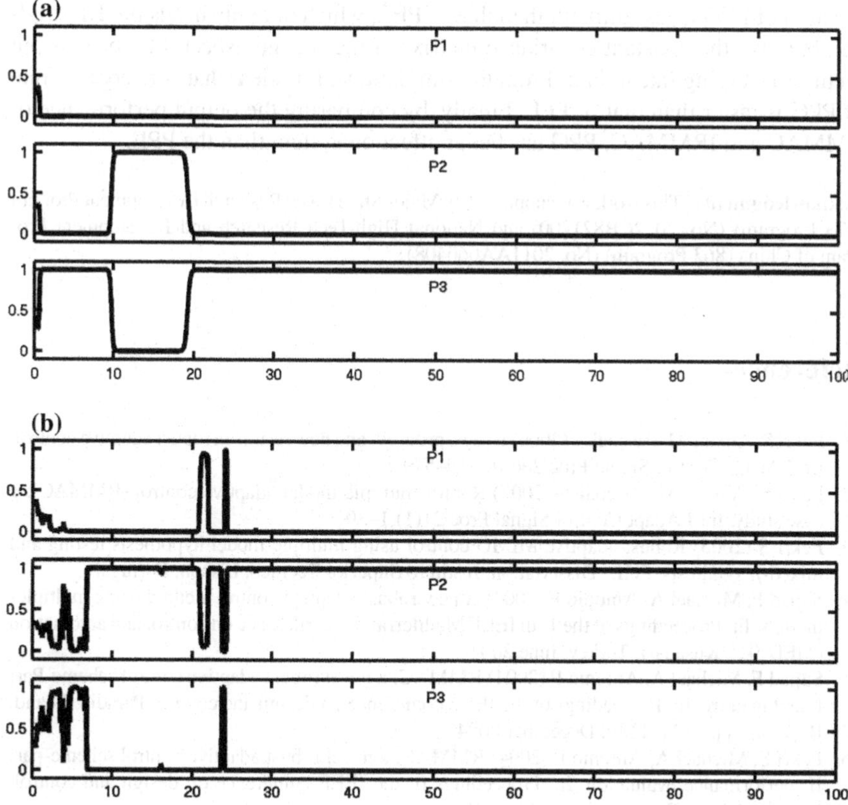

Fig. 41.8 Posterior probabilities when $k_1 = 0.8$. **a** Weighting signals of IRMMAC. **b** Weighting signals of RMMAC

start compared with the case 1, and converge to the proper model quickly. But as we adjust the value of k_1 between the model 1 and the model 2, IRMMAC can pick up the right model for it, but RMMAC chooses the model indecisively from the model set, so the output fluctuates more violently. After that, the true model's the value of k_1 is set between the model 2 and model 3, the output of the RMMAC worsens and can't converge to the proper model within the certain time; the IRMMAC's result deteriorates at the beginning but improves quickly (Fig. 41.8).

41.5 Conclusions

This paper is devoted to comparing the PPG algorithm with PPE algorithm in three aspects, with implementation of RMMAC and IRMMAC into the two-cart mass-spring-damper system. The first aspect is the ease of calculation. Obviously, the calculation of PPG is less difficult than that of PPE, which not only needs the KF residual, but also the constant covariance matrix. In the second aspect, the focus is put on the converging rate to unit. From the simulations, it is clear that convergence rate of PPG is faster than that of PPE. Finally, by comparing the output performance of RMMAC and IRMMAC, PPG has better effect on system than the PPE.

Acknowledgments This work was supported by Major State Basic Research Development Program (973 Program) (No. 2012CB821200) and National High-Tech Research and Development Program of China (863 Program) (No. 2011AA060408)

References

1. Fekri S, Athans M, Pascoal A (2006) Issues, progress and new results in robust adaptive control. Int J Adapt Control Signal Proc 20(10):519–579
2. Fekri S, Athans M, Pascoal A (2007) Robust multiple model adaptive control (RMMAC): a case study. Int J Adapt Control Signal Proc 21(1):1–30
3. Fekri S (2005) Robust adaptive MIMO control using multiple-model hypothesis testing and mixed-μ synthesis. Ph.D. Dissertation, Instituto Superior Tecnico, Lisbon, Portugal
4. Sajjad F, Michael A, Antonio P (2004) A new robust adaptive control method using multiple-models. In: Proceedings of the 12th IEEE Mediterranean conference on control and automation (MED'04), Kusadasi, Turkey, June 2004
5. Sajjad F, Michael A, Antonio P (2004) RMMAC: a novel robust adaptive control scheme-Part I: architecture. In: Proceedings of the IEEE conference on design and control, Paradise Island, Bahamas, pp 1134–1139, December 2004
6. Fekri S, Michael A, Antonio P (2004) RMMAC: a novel robust adaptive control scheme-part II: performance evaluation. In: Proceedings of the IEEE conference on design and control, Paradise Island, Bahamas, pp 1140–1145
7. Fekri S, Michael A, Antonio P (2006) A two-input two-output robust multiple model adaptive control (RMMAC) case study. In: Proceedings of american control conference, Minneapolis, June 2006
8. Young PM et al (1992) Practical computation of the mixed-μ problem. In: Proceedings of American control conference, Chicago, pp 2190–2194
9. Young PM et al (1994) Controller design with mixed-μ problem. In: Proceedings of American control conference, Baltimore, pp 2333–2337, June 1994
10. Young PM et al (1995) Computing bounds for the mixed-μ problem. Int J Robust Nonlinear Control 5:573–590

Chapter 42
An Improved Laplacian Eigenmaps Algorithm for Nonlinear Dimensionality Reduction

Wei Jiang, Nan Li, Hongpeng Yin and Yi Chai

Abstract Manifold learning is a popular recent approach to nonlinear dimensionality reduction. While conventional manifold learning methods are based on the assumption that the data distribution is uniform. They are hard to recover the manifold structure of data in low-dimension space when the data is distributed non-uniformly. This paper presents an improved Laplacian Eigenmaps algorithm, which improved the classical Laplacian Eigenmaps (LE) algorithm by introduce a novel neighbors selection method based on local density. This method can optimize the process of intrinsic structure discovery, and thus reducing the impact of data distribution variation. Several compared experiments between conventional manifold learning methods and improved LE are conducted on synthetic and real-world datasets. The experimental results demonstrate the effectiveness and robustness of our algorithm.

Keywords Dimensionality reduction · Manifold learning · Laplacian eigenmaps · Non-uniform distribution

42.1 Introduction

In many research fields, such as data mining, artificial intelligence and computer vision, people always need to deal with high-dimensional data. The high dimensionality causes a lot of difficulties in data analysis and classification, it also can

---

W. Jiang · N. Li · H. Yin (✉) · Y. Chai
School of Automation, Chongqing University, Chongqing 400044, China
e-mail: yinhongpeng@gmail.com

H. Yin
Key Laboratory of Dependable Service Computing in Cyber Physical Society
Ministry of Education, Chongqing 400030, China

Y. Chai
Key Laboratory of Power Transmission Equipment and System Security,
Chongqing 400044, China

© Springer-Verlag Berlin Heidelberg 2016

Y. Jia et al. (eds.), *Proceedings of the 2015 Chinese Intelligent Systems Conference*, Lecture Notes in Electrical Engineering 359,
DOI 10.1007/978-3-662-48386-2_42

slow the computational speed [1]. In order to handle these data effectively, their dimensionality need to be reduced. In essence, the goal of dimensionality reduction is to map the high-dimensional dataset into a low-dimensional space with discovering the meaningful information hidden in the data. Classical dimensionality reduction methods include Principal Component Analysis (PCA), Multidimensional Scaling (MDS), Linear Discriminant Analysis (LDA), etc. These methods usually work well when the data lie in a linear or almost linear space. Nevertheless, real-world data is generally highly nonlinear and contains much redundant information. In this situation, traditional linear dimensionality reduction methods are hard to discover nonlinear structures embedded in the set of data [2].

Recently, some new nonlinear dimensionality reduction techniques based on manifold learning have been proposed. They are based on the assumption that there exist a low-dimensional manifold embedded in the high-dimension space. In contrast to traditional linear methods, manifold learning methods can deal with complex nonlinear data and uncover the manifold structure in the data. These algorithms are generally divided into two categories: global methods and local methods. Global methods aim to preserve global properties of the original data in the low-dimensional representation. They include Isometric Mapping (ISOMAP) [3], Maximum Variance Unfolding (MVU) [4], Diffusion Maps [5], etc. Local methods attempt to preserve the local geometry of the original data in the low-dimensional representation. They include Locally Linear Embedding (LLE) [6], Hessian-based Locally Linear Embedding (HLLE) [7], Laplacian Eigenmaps (LE) [8], etc.

Although manifold learning methods are good at nonlinear dimensionality reduction, most of them don't take into account the variation of data distribution. As a result, a lot of conventional manifold learning algorithms fail to nicely deal with most real problems that are non-uniformly distributed. For this reason, Zhang [9] proposed an adaptive manifold learning based on LTSA, which dealt with non-uniformly distributed data by using an adaptive neighbors selection strategy. Wen [10] proposed adaptable neighborhood method to improve LLE algorithm. This approach is to divide the non-uniformly distributed manifold into some evenly distributed sub-manifolds, then chose different neighborhood size for each sub-manifold. In [11], authors modified locally linear embedding algorithm by using cam weighted distance, and obtained better performance on non-uniformly distributed data.

As to Laplacian Eigenmaps, it is a kind of manifold learning algorithm based on spectral graph theory. The goal of this algorithm is to construct a representation for data lying on a low-dimensional manifold embedded in a high dimensional space. The nearby data points in the high-dimensional space will still be neighborhoods in low-dimensional space after dimension reduction. However, LE algorithm is like other conventional manifold learning methods, which assumes the whole data manifold is uniformly distributed [8]. The non-uniform data distribution always leads to the performance of algorithm decline.

In this paper, we present an improved Laplacian Eigenmaps algorithm which aims at improving the performance of LE especially for the non-uniformly distributed dataset. It introduces a novel neighbors selection method based on local density. Moreover, it also makes LE algorithm more robust to the variation of neighborhood size. The remainder of this paper is organized as follows. Section 42.2 describes traditional Laplacian Eigenmaps algorithm. Section 42.3 describes how to improve LE. Section 42.4 shows the experiments on artificial datasets and real-world datasets, and then the experimental results are analyzed. Conclusion is drawn in Sect. 42.5.

42.2 Laplacian Eigenmaps Algorithm

Laplacian Eigenmaps, a popular manifold learning method, was presented by Belkin and Niyoki in 2001. It based on spectral graph theory. The goal of this algorithm is to find a set of points $y_1, y_2, \ldots, y_n$ in $\mathbb{R}^d$ to represents the set $x_1, x_2, \ldots, x_n$ in $\mathbb{R}^D$ $(d \ll D)$. During the procedure of dimensionality reduction, the local properties of the manifold are preserved. The algorithm procedure is formally stated below.

(1) Constructing the Adjacency Graph G
Nodes i and j are connected by an edge if j is among k nearest neighbors of i. The distance between two points is measured by the Euclidean distance between them.

(2) Defining the weights
If point i and j are connected by a edge, we define the weight of the edge

$$W_{ij} = e^{-\frac{\left\|x_i - x_j\right\|^2}{t}} \tag{42.1}$$

otherwise put $W_{ij} = 0$.
In that way, we can get a matrix $W = (W_{ij})_{n \times n}$,

(3) Eigenmaps
In order to get the low-dimensional representation $Y = \{y_1, y_2, \ldots, y_n \in \mathbb{R}^d\}$ with preserving the proximity relationship in low dimensional space and, we need to minimized the cost function

$$\varphi(Y) = \sum_{i,j} \left\|y_i - y_j\right\|^2 W_{ij} \tag{42.2}$$

The input of LE algorithm are $x_1, x_2, \ldots, x_n \in \mathbb{R}^D$, d, k, t, where d is the dimensionality, k is the neighborhood size parameter, t is the parameter of heat kernel. The output are $y_1, y_2, \ldots, y_n \in \mathbb{R}^d$.

42.3 Improved LE Algorithm

In traditional LE algorithm, each data point is regarded as the center of a probability distribution and its nearest neighbors are determined by Euclidean distance between them. It works well when whole data points are uniformly distributed. Nevertheless, real-world data distribution isn't always uniform due to different sample rate or other reasons. Neglecting the variation of data distribution and still only using Euclidean distance to describe the local structure is inappropriate. As a result, it will lead to the performance of algorithm decline.

As illustrated in Fig. 42.1, the data isn't uniformly distributed. If we select the neighbors by k-neighborhoods algorithm based on Euclidean distance, only the information on left side can be preserved. Meanwhile, no information in other directions are preserved. That means some local properties or intrinsic structure information will be lost after dimension reduction. What's more, the neighbor size for all data points are identical. So an appropriately chosen neighborhood size is the key to the success of LE algorithm. On the one hand, choosing a small neighborhood size in dense areas can avoid the appearance of "short circuit". On the other hand, choosing a large neighborhood size in sparse areas can avoid manifold be fragmented into disconnect regions. It is hard to select a neighborhood size parameter k suitable for different areas of data. If data distribution variation can be taken into account during the process of choosing neighbors, the algorithm will be more robust to distribution variation and parameter changes.

Based on this idea, in step (1) we introduce a novel neighbors selection method based on local density, the distance of point *i* and *j* is described as fellow:

$$D_{ij} = \frac{\left\| X_i - X_j \right\|}{\sqrt{M_i M_j}} \tag{42.3}$$

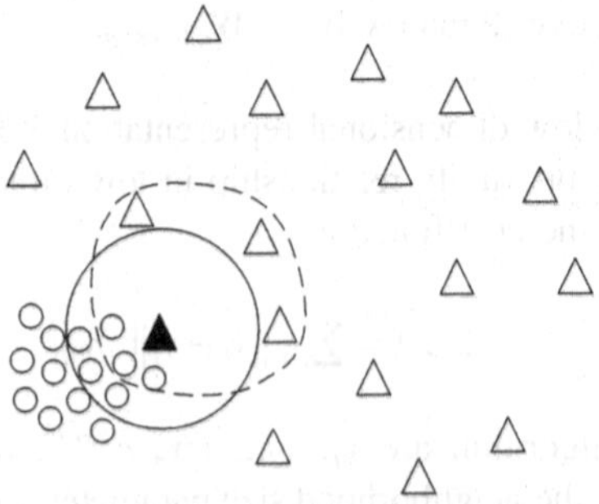

Fig. 42.1 Select nearest neighbors using k-neighborhoods algorithm by Euclidean distance (*solid line*) and measurement based on local density (*dash line*)

Here $\boldsymbol{M}_i$ is the mean Euclidean distance between point i and its k nearest neighbors.

$$M_i = \frac{\sqrt{\sum_{n=1}^{k} \|X_i - X_n\|^2}}{k} \tag{42.4}$$

where $\boldsymbol{M}_i$ represents the local density of point i. We search the k nearest neighbors by this new ways for every sample.

Accordingly in step (2), if point i and j are neighbors the weight of the edge is defined as fellow:

$$W_{ij} = e^{-\frac{D_{ij}^2}{t}} \tag{42.5}$$

otherwise put $W_{ij} = 0$.

The step (3) is same to traditional Laplacian Eigenmaps.

Through these new neighbors selection method, distribution variation will be taken into account in choosing the nearest neighbors. As a result the distance between two points in dense area will be shortened, and it in sparse area will be enlarged. It allows the intrinsic structure discovery more adaptable to the data characteristics, and make the distance measure more reasonable for representing the local distribution.

42.4 Experimental Evaluation

In order to show the limitation of traditional manifold learning algorithm and demonstrate the effectiveness of our algorithm, three datasets are chosen for experiments. One is synthetic dataset, the other two are real-world datasets.

42.4.1 Helix Dataset

The Helix dataset concludes 800 points, and the sampling rate can be changed. We chose five popular manifold learning methods to map the 3D dataset onto a 2D plane, so as to visualize the results. Then the sampling rate is changed to test if these manifold learning methods can handle non-uniform distribution too. Here a proper neighborhood size parameter has been chosen for each algorithm.

We can see from Fig. 42.2, ISOMAP, HLLE, LE and improved LE can unravel the dataset into a circle. But LLE cannot unfold the curved structure, a large number of data points are overlapping after been mapped onto 2D plane.

As shown in Fig. 42.3, when the data is distributed non-uniformly some manifold methods fail to keep good performance. That is because there exist much short-circuit edges in process of selecting neighbors. However the LE and its

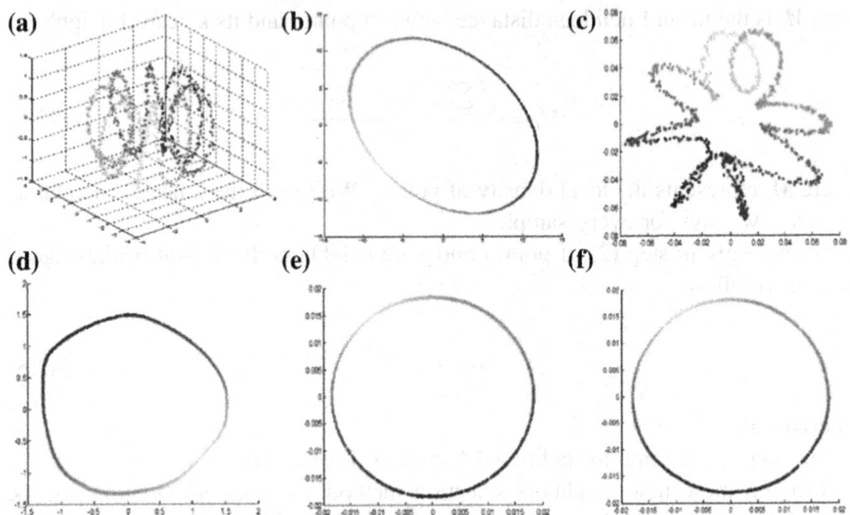

Fig. 42.2 Comparison results of uniform Helix dataset. **a** Helix dataset. **b** Result of ISOMAP. **c** Result of LLE. **d** Result of HLLE. **e** Result of LE. **f** Result of improved LE

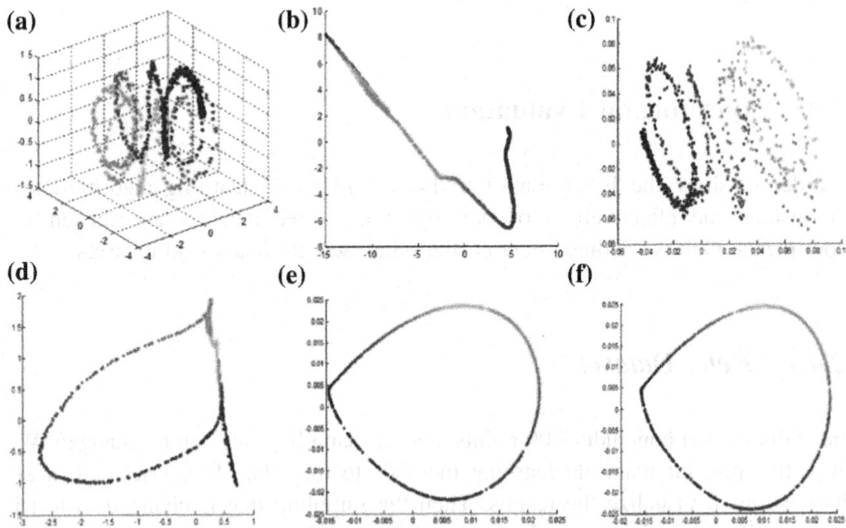

Fig. 42.3 Comparison results of non-uniform Helix dataset. **a** Non-uniform Helix dataset. **b** Result of ISOMAP. **c** Result of LLE. **d** Result of HLLE. **e** Result of LE. **f** Result of improved LE

improved algorithm still can unfold the 3D helix dataset. That means they are more robust to the variation of dataset distribution than other algorithms.

In general, most conventional manifold learning methods are sensitive to the choice of neighborhood size parameter k. To test whether LE algorithm still can keep a good performance when the value of k varies, some further experiments are conducted. In these experiments, several different values of k are selected to explore how it influences the performance of dimensionality reduction.

When k = 4, as is shown in Fig. 42.4b, the result of LE is bad. Some nearby points are disconnected after their dimensionality are reduced to 2D. While Improved LE algorithm still can recover the structure of manifold, preserve the neighborhood relationships among these data points (Fig. 42.5).

When k = 5, which is proper for LE and improved LE, the result of them are both good.

When k = 7, as shown in Fig. 42.6b, some points mapped by LE are overlapped. While the result of improved LE still keep good.

We can learn from these figures above when the parameter k is too large or too small, traditional LE algorithm can't recover the manifold structure of dataset in low-dimensional space. However, improved LE still can unfold the toroidal helix curve with preserving the local neighborhood relationship. In fact, if the dataset is distributed uniformly, LE algorithm works well when the parameter k varies in a wide range from 3 to 20. But when the data distribution is non-uniform, it can get a

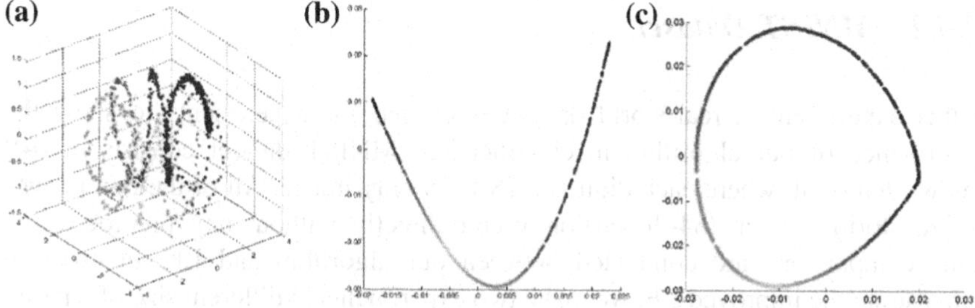

Fig. 42.4 k = 4, dimensionality reduction of non-uniform Helix dataset by LE and improved LE. **a** Non-uniform Helix dataset. **b** Result of LE. **c** Result of improved LE

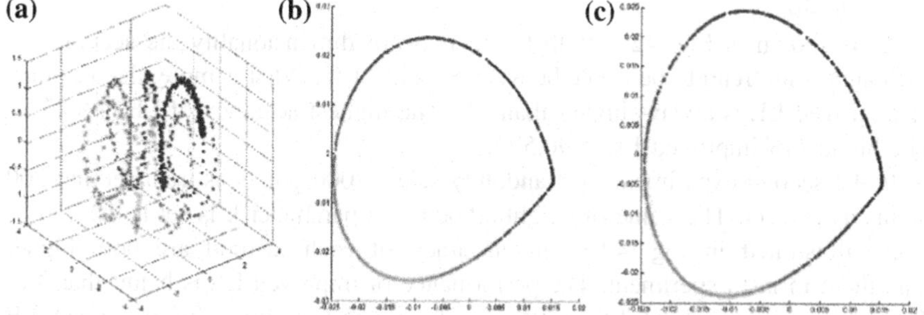

Fig. 42.5 k = 5, dimensionality reduction of non-uniform Helix dataset by LE and improved LE. **a** Non-uniform Helix dataset. **b** Result of LE. **c** Result of improved LE

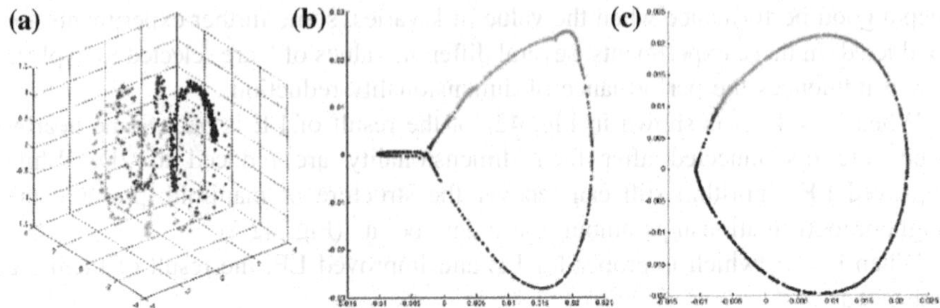

Fig. 42.6 k = 7, dimensionality reduction of non-uniform Helix dataset by LE and improved LE. **a** Non-uniform Helix dataset **b** Result of LE. **c** Result of improved LE

relatively good result only when the value of k is set in a narrow range (only 5, 6 in this experiment). To sum up, the improved LE is more robust to the variation of sample distribution. In addition, its alternative value range of neighborhood size parameter k is broader (from 4 to 13 in this experiment) than traditional LE.

42.4.2 MNIST Dataset

In this experiment, a real-world dataset is chosen for experiments to show the performance of our algorithm in classification. MNIST dataset consists of 0–9 handwritten digit, where each digit is a 28 × 28 gray image. All of the raw images are transformed to a 784-dimension vector directly without any preprocessing. Some comparisons are conducted between our algorithm and LE. In order to demonstrate the improved LE, we make two experiments in different size of dataset.

In the first experiment, we randomly select 1000 samples as the train set and 200 as the test set, and choose SVM as classifier. The neighborhood size k of each algorithm is set to 10, which is proper for each algorithm. Some compared experiment are conducted between two algorithms with the dimensionality range from 3 to 40.

As is shown in Fig. 42.7, with the increase of dimensionality the accuracy of each algorithm trend to be stable between 83 and 91 %. What is more, the accuracy of improved LE is always higher than LE. The highest accuracy of LE is 88.5 %. The highest of improved LE is 90.5 %.

In the second experiment, we randomly select 2000 points as train set and 200 points as test set. The value of neighborhood size parameter k is set to 10, too.

As illustrated in Fig. 42.8, the accuracy of each method are both higher than them in last experiment. The performance of improved LE is better than LE. The highest accuracy of LE is 94 %. The highest accuracy of improved LE is 97.5 %.

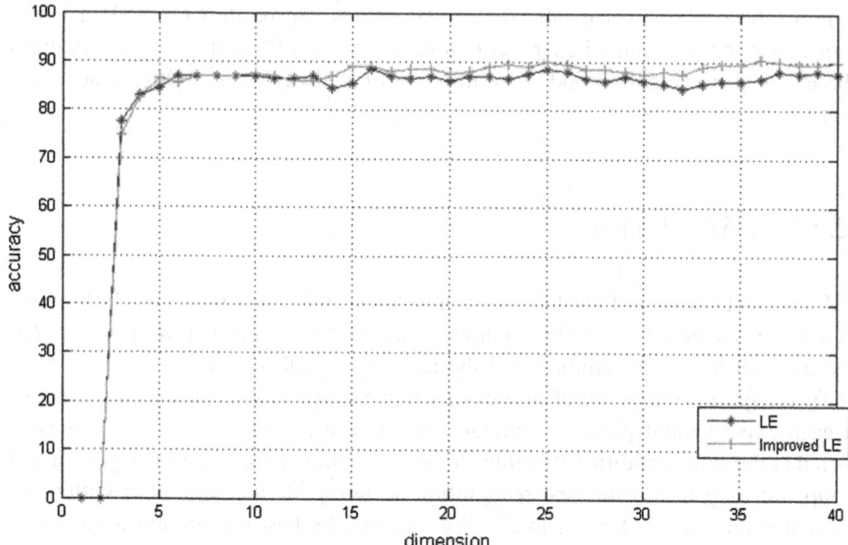

Fig. 42.7 Recognition accuracy for different dimensionality

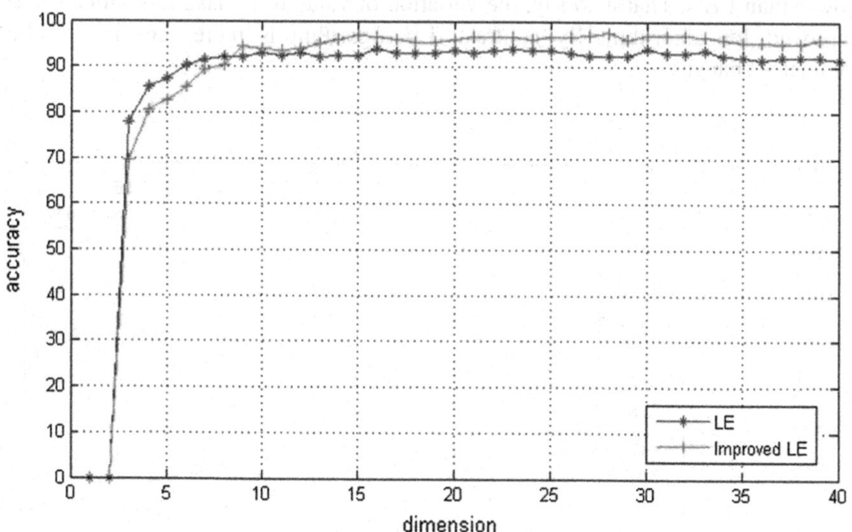

Fig. 42.8 Recognition accuracy for different dimensionality

From these two experiments, some conclusions can be drawn: (42.1) a larger train set can bring about a better recognition accuracy; (42.2) at the same condition (dimensionality, neighborhood size), improved LE always can get higher accuracy than LE.

42.4.3 USPS Dataset

USPS dataset consists of 10 different handwritten digits from 0 to 9, each digit is a 16×16 grayscale image. 1200 raw images are randomly selected from the dataset. The train set has 1000 samples, and the test set has 200 samples.

To verify the superiority of improved LE, we make several further experiments. In each experimental group, a comparison between LE and improved LE is conducted at the same condition (neighborhood size parameter k). In each experimental group, we only record the best recognition accuracy of each algorithm within the dimensionality ranges from 4 to 20. We will get the best recognition accuracy at each value of neighbor size parameter k eventually.

As demonstrated in Fig. 42.9, improved LE algorithm can get a better effect at the same value of neighborhood size parameter. With the increase of the value of k, the best accuracy start to decline. While the accuracy of improved LE descend slower than LE's. That is to say, the variation of value of k make less influence to improved LE algorithm. By contrast, LE algorithm is more sensitive to the parameter changes.

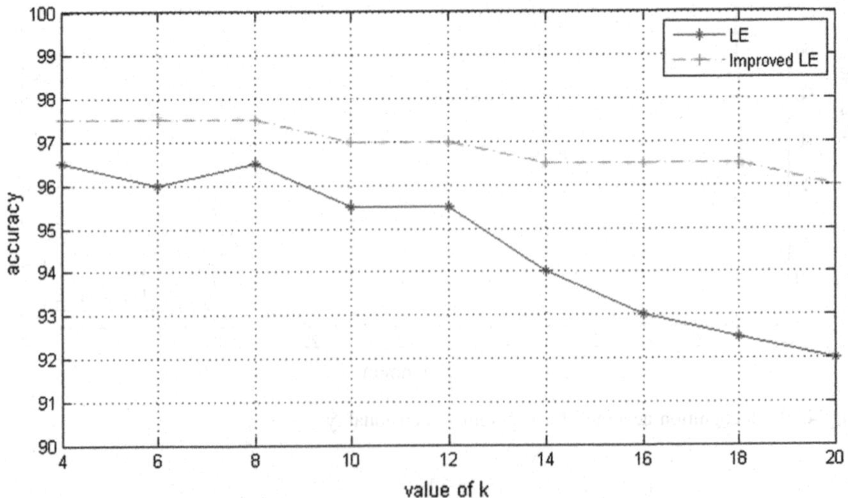

Fig. 42.9 The best recognition accuracy at different value of neighborhood size parameter

42.5 Conclusion

This paper presents a novel manifold learning methods for nonlinear dimensionality reduction. The proposed algorithm is based on improving LE by introducing a new neighbors selection method. Some compared experiments between our method and several popular manifold learning methods are conducted on synthetic and real-world datasets. The experimental results demonstrate the effectiveness and robustness of our algorithm. The proposed algorithm can deal with non-uniformly distributed data, and has robustness to the changes of neighborhood size parameter. One limitation of our algorithm, which exists in most popular manifold learning methods as well, is that it cannot embed new added high-dimensional samples into an existing low-dimensional manifold. Our future work will focus on resolving this problem by combining our algorithm with incremental learning.

Acknowledgments We would like to thank the supports by National Natural Science Foundation of China (61374135, 61203321), China Postdoctoral Science Foundation (2012M521676), China Central Universities Foundation (106112013CDJZR170005) and Chongqing Special Funding in Postdoctoral Scientific Research Project (XM2013007).

References

1. Hastie T, Tibshirani R, Friedman J et al (2005) The elements of statistical learning: data mining, inference and prediction. Math Intell 27(2):83–85
2. van der Maaten LJP, Postma EO, van den Herik HJ (2009) Dimensionality reduction: a comparative review. J Mach Learn Res 10(1–41):66–71
3. Tenenbaum JB, De Silva V, Langford JC (2000) A global geometric framework for nonlinear dimensionality reduction. Science 290(5500):2319–2323
4. Weinberger KQ, Saul LK (2006) Unsupervised learning of image manifolds by semidefinite programming. Int J Comput Vision 70(1):77–90
5. Coifman RR, Lafon S (2006) Diffusion maps. Appl Comput Harmon Anal 21(1):5–30
6. Roweis ST, Saul LK (2000) Nonlinear dimensionality reduction by locally linear embedding. Science 290(5500):2323–2326
7. Donoho DL, Grimes C (2003) Hessian eigenmaps: locally linear embedding techniques for high-dimensional data. Proc Natl Acad Sci 100(10):5591–5596
8. Belkin M, Niyogi P (2002) Laplacian eigenmaps and spectral techniques for embedding and clustering. In: Advances in neural information processing systems, pp 585–591
9. Zhang Z, Wang J, Zha H (2012) Adaptive manifold learning. IEEE Trans Pattern Anal Mach Intell 34(2):253–265
10. Wen GH, Jiang LJ, Wen J (2008) Dynamically determining neighborhood parameter for locally linear embedding. J Softw 19(7):1666–1673
11. Pan Y, Ge SS, Al Mamun A (2009) Weighted locally linear embedding for dimension reduction. Pattern Recogn 42(5):798–811

Chapter 43
A Distributed Charging and Discharging Coordination for Large-Population Plug-In Electric Vehicles

Huiling Li and Zhongjing Ma

Abstract Motivated by the economic and environmental benefits of plug-in electric vehicles (PEVs), the PEVs have been regarded as adjustable bilateral auxiliaries for the power grid. They can not only charge from the grid but also discharge back to the grid. This work studies the problems of PEV coordinations, considering the battery degradation cost. We analyze the characteristics of distributed strategy curve and propose an updated algorithm to implement the optimal distributed coordination in the case of the infeasibility of fully centralized formulation. As key contribution of this paper, we show that under certain mild conditions, the system converges to an optimal charging and discharging strategy. Simulation examples illustrate the results developed in the paper.

Keywords Plug-in electric vehicles · Charge and discharge · Distributed control · Dynamic algorithm · Peak-shift and valley-fill · Convergency

43.1 Introduction

Nowadays, PEVs have achieved attractive developments since PEVs can promisingly reduce the reliance on the exhaustible non-renewable energy sources. However, uncoordinated charging and discharging behaviors of PEVs may have negative impacts on the power grid [1, 2].

Recently lots of research has been dedicated to studying how to coordinate the PEVs' charging behaviors to mitigate the involved negative impacts properly. Clement-Nyns et al. [3] develops a dynamic programming method to optimize the PEVs' charging strategy and evaluates the impacts it has on the system. Sundstrom and Binding [4] formulates the charging coordinations as optimization problems with distribution and transmission capacity constraints, and uses linear programming methods to solve the problem. Galus and Andersson [5] introduces an optimal

H. Li (✉) · Z. Ma
School of Automation, Beijing Institute of Technology, Beijing 100081, China
e-mail: wendybit@126.com

© Springer-Verlag Berlin Heidelberg 2016

Y. Jia et al. (eds.), *Proceedings of the 2015 Chinese Intelligent Systems Conference*, Lecture Notes in Electrical Engineering 359,
DOI 10.1007/978-3-662-48386-2_43

model, which would benefit PEV users under the current relationship between the power plants and the PEVs.

Nevertheless, all these algorithms are centralized coordination methods which may be impractical to implement because of the autonomy of individual PEVs and the computational complexity of large-population PEVs. A novel distributed charging control strategy for the large-population PEVs problem is developed in [6] and the solution achieves a Nash Equilibrium. Under this strategy, the optimal solution for each PEVs is respect to the aggregated charging strategy of all the PEVs. Reference [7] presentes a decentralized water-filling-based algorithm to flatten the load curve of low-voltage transformers, which also achieved optimal charging timing. In paper [8], charging control of the PEVs in power grid is formulated as a class of Stackelberg Games which is normally that users set their strategy according to the electric price which is determined by the power plants.

However, all the research mentioned above only consider the process of charging, while PEVs on the other hand can operate as energy storage device [9], which means they can discharge and release power back to the grid. In [10] a real-time model of plug-in vehicles performing vehicles-to-grid (V2G) power transaction is presented. Under this method, power levels and charge or discharge times are scheduled smartly in order to maximize profits based on one-day ahead electricity price.

In this paper we further design an updated mechanism. Comparing with previous designs, we propose a decentralized coordination algorithm concerning both charging and discharging. The coexistence of charging and discharging may bring issues of the constraint on battery's State Of Charge(SOC). It is difficult to analyze the problem directly, therefor, we first introduce the best response of PEVs without constraints on SOC. Furthermore, we analyze the characteristics of the optimal strategy, and show that by adapting the proposed distributed algorithm, the system converges to an optimal result-a mutually beneficial solution under certain mild conditions.

The paper is organized as follows: in Sect. 43.2, we formulate a class of multi-objective coordination control issues concerning the tradeoff between battery degradation cost and charging cost. In Sect. 43.3, we design a distributed charging and discharging coordination for PEVs and analyse the characters of this trajectory. Then we formulate an distributed update algorithm to realize this trajectory. In Sect. 43.4, we prove the convergence of the distributed algorithm. Numerical examples are presented in Sect. 43.5. In Sect. 43.6, we draw conclusions of the paper.

43.2 Formulation of PEV Coordination Problems

We consider a PEV population with the size of N, with each PEV transmitting between charging and discharging. Let $\mathcal{T} \triangleq \{0, 1, \ldots, T\}$ denote the time interval. The state of charge of PEV n at time t, denoted by SOC_{nt}, is specified below:

$$SOC_{min} \leq SOC_{nt} \leq SOC_{max}, \tag{43.1}$$

where SOC_{min}, SOC_{max} represent the minimum and maximum of SOC for PEVs. For each PEV, we call $\boldsymbol{u}_n \equiv (u_{nt}; t \in \mathcal{T})$ an admissible charging and discharging control of PEV n, if

$$u_{nt} \in [-\alpha_n, \beta_n], \text{ and } \sum_{t \in \mathcal{T}} u_{nt} = \Gamma_n, \tag{43.2}$$

where α_n and β_n represent a uniform maximum discharge rate and maximum charge rate at any instant; and Γ_n represents the total fixed charging energy of individual PEV n. The set of admissible control for PEV n is denoted by $\mathcal{U}_n$.

Subject to a collection of admissible charging and discharging control $\boldsymbol{u} = (\boldsymbol{u}_n; n \in \mathcal{N})$, the system cost, denoted by J, composed of battery degradation cost and charging cost (and discharging benefit), is specified below:

$$J(\boldsymbol{u}) \triangleq \sum_{t \in \mathcal{T}} \Big\{ \sum_{n \in \mathcal{N}} f_n(u_{nt}) + p_t(d_t + \sum_{n \in \mathcal{N}} u_{nt}) \Big\}, \tag{43.3}$$

where p_t denotes the electricity price at instant t, d_t represents the aggregated inelastic based demand at t, and $f_n(u_{nt})$, represents the battery degradation cost of PEV n subject to u_{nt}.

In the rest of the paper we consider the following assumption:

- (A1) $f_n \in C^1$, for all $n \in \mathcal{N}$, is increasing and strictly convex on charge and discharge rate.

Concerning the electricity price, in this paper, we focus on methods that utilize real-time marginal electricity price information. Real time price models have been widely adapted for demand response management [6, 11, 12].

More specifically, the electricity price at an instant t, p_t, is determined by the total demand at that instant, i.e.

$$p_t = p(D_t^N(\boldsymbol{u}_t)), D_t^N(\boldsymbol{u}_t) \triangleq (d_t + z_t)/N, \text{ with } z_t = \sum_{n \in N} u_{nt}. \tag{43.4}$$

We assume that PEVs are plugged in the power grid at 20:00 PM and unplugged at 8:00 AM the next day. This is motivated by PEVs' behavior that they come back home at night and leave in the morning.

Comparing with the socially optimal strategy, peak-shift and valley-fill(PSVF) strategy only consider one objective: to fill the load valley and shift the load peak.

Figure 43.1 illustrates a specific numerical example of socially optimal and PSVF strategies based on the given inelastic basic demand curve which demonstrates the electric energy demand in the service area of Southern California Edison from 20:00 on December 29, 2014 to 8:00 on December 30, 2014.

Centralized charging and discharging coordination for large-population PEVs usually requires significant communication networking and centralized computing

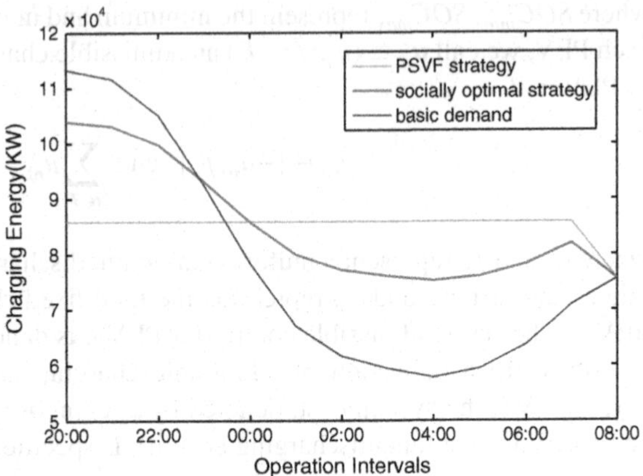

Fig. 43.1 Socially optimal and PSVF strategies for charging and discharging coordination with battery size of PEVs is 15 and an identical degradation cost for each PEV $f_n = 0.007 * u_{nt}^2$ and electricity price of $p = 0.012 * D^{0.8}$

resources, and may have difficulty in gaining public acceptance due to their autonomy property. Therefore, we will design an alternative distributed charging and discharging coordination method in the following section.

43.3 Distributed Coordination of PEVs

Suppose that the best behavior of each PEV deals with the tradeoff between the charging cost and the battery degradation cost, that is to say the best response of individual PEVn is to minimize the cost function, denoted by J_n, such that

$$J_n(\boldsymbol{u}) \triangleq \sum_{t \in \mathcal{T}} \left\{ f_n(u_{nt}) + p(\boldsymbol{u}_t) u_{nt} \right\}, \tag{43.5}$$

where $p(\boldsymbol{u}_t)$ denotes the electricity price at instant t with respect to charging and discharging behavior $\boldsymbol{u}_t$ at that instant.

We coordinate PEVs' charging and discharging timing such that the PEVs release power to the grid when the basic load is at its peak, and charge when the basic demand is in the valley. Take a look back at Fig. 43.1, since it is often the case that peak load occurs before the valley load, we give a rule about the optimal charging and discharging control u_{nt}^*, such that

$$u_{nt}^* \begin{cases} \leq 0, \text{ in case } t \in [0, t_1] \\ \geq 0, \text{ in case } t \in [t_1 + 1, T] \end{cases} . \tag{43.6}$$

43.3.1 Coordination of PEVs Without Constraint on SOC

We specify a charging and discharging coordination curve $\boldsymbol{u}_n^{\dagger}(z)$ of individual PEVs with respect to average curve z when we release the constraint of SOC in Lemma 43.3.1 below. There is no guarantee that $\boldsymbol{u}_n^{\dagger}(z)$ is an admissible charging and discharging strategy because it may cause over-charge and over-discharge.

Lemma 43.3.1 *Under Assumption (A1), if there is no constrain on SOCs, the following holds:*

(i) best strategy with respect to a fixed curve z is specified as $u_n^{\dagger}(z, A)$*, such that*

$$u_{nt}^{\dagger}(z, A) = [f_n']^{-1}(A^{\dagger} - p(z_t)), \tag{43.7}$$

where $[f_n']^{-1}$ *represents the inverse operator of the derivative of the function* f_n.

(ii) $\mathcal{U}_n \in \widetilde{\mathcal{U}}_n$ *and* $J_n(\boldsymbol{u}_n^{\dagger}) < J_n(\boldsymbol{u}_n^{*})$*, with the set of* $u_{nt}^{\dagger}$ *is denoted by* $\widetilde{\mathcal{U}}_n$.

It is straightforward to verify (i) by following the same technique applied in the proof of Lemma 43.3.2 in [12]. With $\mathcal{U}_n \in \widetilde{\mathcal{U}}_n$*, and* $J_n(\boldsymbol{u}_n^{\dagger})$ *and* $J_n(\boldsymbol{u}_n^{*})$ *share the same function, we can verify (ii).*

The analysis of coordination without constraint of SOC is the foundation of the following analysis.

43.3.2 Best Response of PEVs with Respect to z

Lemma 43.3.2 demonstrates the optimal coordination of PEVs with respect to curve z. We can obtain the characters of optimal coordinate curve of PEVs by this lemma.

Lemma 43.3.2 *The charging and discharging strategy that minimizes the local cost function (43.5) with respect to a fixed z can be derived as:*

$$u_{nt}^{*}(z, A(t)) = \begin{cases} min\{0, [f_n']^{-1}(A_1 - p(z_t))\}, & t \in [0, t_1] \\ max\{0, [f_n']^{-1}(A_2 - p(z_t))\}, & t \in [t_1 + 1, T] \end{cases}, \tag{43.8}$$

and we have $A_1 \geq A^{\dagger}$, $A_2 \leq A^{\dagger}$*, and* $|u_{nt}^{*}| \leq |u_{nt}^{\dagger}|$.

Proof We know that $u_{nt}^{*} \leq 0$, in case $t \in [0, t_1]$ from (43.6), following the same technique applied in the proof of Lemma 43.3.1 in [12], u_{nt}^{*} can be denoted by: $u_{nt}^{*}(z, A) = \min\{0, [f_n']^{-1}(A_1 - p(z_t))\}$, and when $t \in [t_1 + 1, t]$, u_{nt}^{*} can be denoted by: $u_{nt}^{*}(z, A) = \max\{0, [f_n']^{-1}(A_2 - p(z_t))\}$.

Suppose there exist one time instant t, such that $|u_{nt}^{*}| > |u_{nt}^{\dagger}|$. We assume that $t \in [0, t_1]$, then we have $u_{nt}^{*} < 0$, $u_{nt}^{*} < u_{nt}^{\dagger}$ and $u_{nt}^{*} = [f_n']^{-1}(A_1 - p(z_t))$.

Now we know that $u^*_{nt} = [f'_n]^{-1}(A_1 - p(z_t))$, $u^\dagger_{nt} = [f'_n]^{-1}(A^\dagger - p(z_t))$, $u^*_{nt} < u^\dagger_{nt}$. Because $[f'_n]^{-1}(x)$ is increasing on x, so $A_1 < A^\dagger$, and $\sum_{t=0}^{t_1} u^*_{nt} < \sum_{t=0}^{t_1} u^\dagger_{nt}$.

That means $\boldsymbol{u}^\dagger$ discharge less than $\boldsymbol{u}^*$ during $[0, t_1]$, and of course charge less during $[t_1+1, T]$. That means if $\boldsymbol{u}^*$ satisfy constraint (43.1), $\boldsymbol{u}^\dagger$ must be an admissible charging and discharging curve, which is better than $\boldsymbol{u}^*$ since we know that $J_n(\boldsymbol{u}^\dagger) < J_n(\boldsymbol{u}^*)$. Then there must exit $\boldsymbol{u}^* = \boldsymbol{u}^\dagger$, which is contradict with the hypothesis that $|u^*_{nt}| > |u^\dagger_{nt}|$. Such that$|u^*_{nt}| \le |u^\dagger_{nt}|$.

So $|u^*_{nt}| \le |u^\dagger_{nt}|$, and $A_1 \ge A^\dagger, A_2 \le A^\dagger$. □

43.3.3 Distributed Charging and Discharging Strategy Algorithm

We specify a distributed iterative update procedure to implement the optimal strategy in Algorithm 1 below.

Algorithm 1 (*Implementation of distributed coordination strategy*)

- Provide an initial aggregated charging and discharging strategy $\boldsymbol{z}^{(0)}$, such that $z_t^{(0)} = \sum_{n\in\mathcal{N}} u^{*,(0)}_{nt}$, for all $t \in \mathcal{T}$, where $\boldsymbol{u}^{*,(0)}_n = (\boldsymbol{u}^{*,(0)}_n; n \in \mathcal{N})$ is a collection of initial charging and discharging strategies;
- Set $k = 1$ and $\epsilon = \epsilon_0$ for some $\epsilon_0 > 0$;
- While $\epsilon > 0$. Obtain individual best response $\boldsymbol{u}^{*,(k)}_n$ w.r.z. $\boldsymbol{z}^{(k-1)}$, for all n, such that $\boldsymbol{u}^{*,(k)}_n = \text{argmin}_{u_n\in\mathcal{U}_n} \sum_{t\in\mathcal{T}} \{f_n(u_{nt}) + p(D^N_t)u_{nt}\}$, where $D^N_t = (d^N_t + z^{k-1}_t)/N$;
  Set $z_t(k) = \sum_{n\in\mathcal{N}} u^{*,k}_{nt}$, for all $t \in \mathcal{T}$;
  Update $\epsilon = \parallel z^k - z^{k-1} \parallel_1$;
  Update $k = k + 1$;

We will focus on identifying condition that guarantees the convergence of the distributed algorithm in the next section.

43.4 Convergence of Algorithm

We additionally define the charging and discharging strategy that minimizes the local cost function (43.5) with respect to a fixed curve z, such that $\boldsymbol{u}^*_n(z) \triangleq argmin_{u_n\in\mathcal{U}_n} J_n(\boldsymbol{u}_n; z)$. We define another local control strategy $\boldsymbol{v}_n(z, \hat{z})$ such that $v_{nt}(z, \hat{z}) = u_{nt}(\hat{z}, A^*(z)), t \in \mathcal{T}$. There is no guarantee that $\boldsymbol{v}_n(z, \hat{z})$ is an admissible charging and discharging strategy or $\sum(\boldsymbol{v}_n(z, \hat{z})) = \Gamma_n$.

Lemma 43.4.1 *Coordination strategy* $\boldsymbol{u}_n^*(z)$ *and* $\boldsymbol{u}_n^*(\hat{z})$ *satisfy the following inequality:*

$$||\boldsymbol{u}_n^*(z) - \boldsymbol{u}_n^*(\hat{z})||_1 \leq 4\sum_{t=0}^{T} |[f_n']^{-1}(A^*(z) - p(z_t)) - [f_n']^{-1}(A^*(z) - p(\hat{z}_t))|. \quad (43.9)$$

Proof There are the following three cases to consider. We set $A^* = A^*(z)$ and $\hat{A}^* = \hat{A}^*(z)$.

(I) With two fixed curves z and $\hat{z}$, $SOC_{nt_1} > SOC_{min}$ and $SOC_{n\hat{t}_1} > SOC_{min}$. Following the same technique applied in the proof of Lemma 43.3.2 in [12], we get: $||\boldsymbol{u}_n^*(z) - \boldsymbol{u}_n^*(\hat{z})||_1 \leq 2\sum_{t=0}^{T} |[f_n']^{-1}(A^*(z) - p(z_t)) - [f_n']^{-1}(A^*(z) - p(\hat{z}_t))|$.

(II) With two fixed curves z and $\hat{z}$, $SOC_{nt_1} > SOC_{min}$, $SOC_{n\hat{t}'} = SOC_{min}$.

$$u_{nt}^*(z, A^*) = [f_n']^{-1}(A^* - p(z_t));$$

$$\hat{u}_{nt}^*(\hat{z}, \hat{A}^*) = \begin{cases} \min\{0, [f_n']^{-1}(\hat{A}^*{}_1 - p(\hat{z}_t))\}, \text{ in case } t \in [0, \hat{t}_1] \\ \max\{0, [f_n']^{-1}(\hat{A}^*{}_2 - p(\hat{z}_t))\}, \text{ in case } t \in [\hat{t}_1, T] \end{cases}. \quad (43.10)$$

Since $||\boldsymbol{u}_n^*(z) - \boldsymbol{v}_n(z, \hat{z})||_1 = \sum_{t=0}^{T} |[f_n']^{-1}(A^*(z) - p(z_t)) - [f_n']^{-1}(A^*(z) - p(\hat{z}_t))|$, we need to prove:

$$||\boldsymbol{u}_n^*(z) - \boldsymbol{u}_n^*(\hat{z})||_1 \leq 4||\boldsymbol{u}_n^*(z) - \boldsymbol{v}_n(z, \hat{z})||_1. \quad (43.11)$$

We know that $\hat{A}_1^* \geq \hat{A}_2^*$ from Lemma 43.3.2.

(a) In case $A^* \geq \hat{A}_1^*$. By (43.10), and that $[f_n']^{-1}(x)$ is increasing on x, we have$v_{nt}(z, \hat{z}) = u_{nt}(\hat{z}, A^*) \geq \hat{u}_{nt}^*(\hat{z})$. Then we have,

$$\begin{aligned} ||\boldsymbol{v}_n(z, \hat{z}) - \boldsymbol{u}_n^*(\hat{z})||_1 &= |\sum v_{nt}(z, \hat{z}) - \sum u_{nt}^*(\hat{z})| \\ &= |\sum v_{nt}(z, \hat{z}) - \sum u_{nt}^*(z)| \leq ||\boldsymbol{v}_n(z, \hat{z}) - \boldsymbol{u}_n^*(\hat{z})||_1. \end{aligned} \quad (43.12)$$

(b) In case $A^* \geq \hat{A}_2^*$. Similar to (a), we can derive,

$$||\boldsymbol{v}_n(z, \hat{z}) - \boldsymbol{u}_n^*(\hat{z})||_1 \leq ||\boldsymbol{v}_n(z, \hat{z}) - \boldsymbol{u}_n^*(\hat{z})||_1. \quad (43.13)$$

(c) In case $\hat{A}_2^* < A^* < \hat{A}_1^*$. We can derive,

$$v_{nt}(z, \hat{z}) = u_{nt}(\hat{z}, A^*) \begin{cases} \leq u_{nt}^*(\hat{z}), t \in [0, \hat{t}_1] \\ \geq u_{nt}^*(\hat{z}), t \in [\hat{t}_1 + 1, T] \end{cases}. \quad (43.14)$$

That means $\boldsymbol{v}_n(z, \hat{z})$ discharge more energy than $\boldsymbol{u}_n^*(\hat{z})$ in the discharging process and charge more energy in the charging process. Then,$|v_{nt}(z, \hat{z})| \geq |u_{nt}^*(\hat{z})|$.

Now we have,

$$||\boldsymbol{v}_n(z,\hat{z}) - \boldsymbol{u}_n^*(\hat{z})||_1 = \sum |v_{nt}(z,\hat{z})| - \sum |u_{nt}^*(\hat{z})|$$
$$\leq \sum |v_{nt}(z,\hat{z})| - \sum |u_{nt}^*(z)| \leq ||\boldsymbol{v}_n(z,\hat{z}) - \boldsymbol{u}_n^*(\hat{z})||_1, \tag{43.15}$$

By (43.12) (43.13) (43.15), we have: $||\boldsymbol{v}_n(z,\hat{z}) - \boldsymbol{u}_n^*(\hat{z})||_1 \leq ||\boldsymbol{v}_n(z,\hat{z}) - \boldsymbol{u}_n^*(\hat{z})||_1$. Then $||\boldsymbol{u}_n^*(z) - \boldsymbol{u}_n^*(\hat{z})||_1 \leq ||\boldsymbol{v}_n(z,\hat{z}) - \boldsymbol{u}_n^*(\hat{z})||_1 + ||\boldsymbol{v}_n(z,\hat{z}) - \boldsymbol{u}_n^*(z)||_1 \leq 2||\boldsymbol{v}_n(z,\hat{z}) - \boldsymbol{u}_n^*(\hat{z})||_1 \leq 4||\boldsymbol{v}_n(z,\hat{z}) - \boldsymbol{u}_n^*(\hat{z})||_1$.

(III) With two fixed curves z and $\hat{z}$, $SOC_{nt_1} = SOC_{min}$, $SOC_{n\hat{t}} = SOC_{min}$.
In this case, $\boldsymbol{u}_n^*(z)$ and $\boldsymbol{u}_n^*(\hat{z})$ share the same discharging and charging energy because they have the same SOC_{n0}. Following the same technique applied in the proof of Lemma 43.3.2 in [12], we can verify the following results: $||\boldsymbol{u}_n^*(z) - \boldsymbol{u}_n^*(\hat{z})||_1 \leq 2\sum_{t=0}^{\max\{t_1,\hat{t}_1\}} |[f_n']^{-1}(A - p(z_t)) - [f_n']^{-1}(A - p(\hat{z}_t))|$
$+2\sum_{\min\{t_1,\hat{t}_1\}}^{T} |[f_n']^{-1}(A - p(z_t)) - [f_n']^{-1}(A - p(\hat{z}_t))|$
$\leq 4\sum_{t=0}^{T} |[f_n']^{-1}(A - p(z_t)) - [f_n']^{-1}(A - p(\hat{z}_t))|$.
In conclusion, (43.9) can be obtained following (I), (II) and (III).

□

Theorem 43.4.1 (Convergence of Algorithm)
Based on the statement in Lemma 43.4.1, we can give the convergence property. Under Assumption (A1), suppose p is increasing, and

$$|f_n'(x_1) - f_n'(x_2)| \geq a|x_1 - x_2|, |p(y_1) - p(y_2)| \leq b|y_1 - y_2|, \text{with } a > 4b > 0; \tag{43.16}$$

then the system converges to a unique collection of charging and discharging strategy.

With $|f_n'(x) - f_n'(y)| \geq a|x - y|$, *we have:* $|[f_n']^{-1}(\eta_1) - [f_n']^{-1}| \leq \frac{1}{a}$. *Then we can obtain that:*

$$|[f_n']^{-1}(A^*(z) - p(z_t)) - [f_n']^{-1}(A^*(z) - p(\hat{z}_t))|$$
$$\leq \frac{1}{a}|(A^*(z) - p(z_t)) - (A^*(z) - p(\hat{z}_t))| = \frac{1}{a}|p(z_t) - p(\hat{z}_t)|$$
$$\leq \frac{b}{a}|z_t - \hat{z}_t| < \frac{1}{4}|z_t - \hat{z}_t|, \text{ with } |p(y_1) - p(y_2)| \leq b|y_1 - y_2|$$

for all t, where the last inequality holds by $a > 4b > 0$.

By (43.16), together with (43.9), we have: $||\boldsymbol{u}_n^*(z) - \boldsymbol{u}_n^*(\hat{z})||_1 < ||z - \hat{z}||_1$. *Henceforth by implementing the distributed algorithm, the system converges to a strategy by applying the contraction mapping theorem.*

43.5 Numerical Examples

We specify several numerical examples over a time interval $T = 12h$ and the time interval $\Delta T = 1h$ with the basic demand trajectory illustrated in Fig. 43.1. The electricity price, p_t, by (43.4) and the increasing property of p_t, is specified as $p_t = 0.012D_t^{N0.8}$, and satisfies: $|p(y_1) - p(y_2)| = 0.012|y_1^{0.8} - y_2^{0.8}| \leq 0.012p'_{max}|y_1 - y_2| = 0.0067|y_1 - y_2|$.

We suppose that there are 10,000 PEVs and all of the PEVs share identical battery size, maximum SOC value, minimum SOC value and initial SOC value as 15 KWh, 90, 10 and 30 % respectively.

When the PEVs have an identical degradation cost function $f_n = 0.0135u_{nt}^2$, and $|f_n'(x_1) - f_n'(x_2)| = 0.027|x_1 - x_2|$. It is direct to verify the sufficient condition of (43.16) for the convergence of algorithm. If we decrease the coefficient of PEV degradation cost and condition (43.16) is not satisfied, such that $f_n = 0.004u_{nt}^2$, the system does not converge to any equilibrium under Algorithm 1. We display the updates of the PEV decentralized strategies in Fig. 43.2.

The total costs of the three charging strategies are shown in Table 43.1, by which we know that the decentralized control strategy by applying Algorithm 1 is near the socially optimal one and is better than the PSVF strategy.

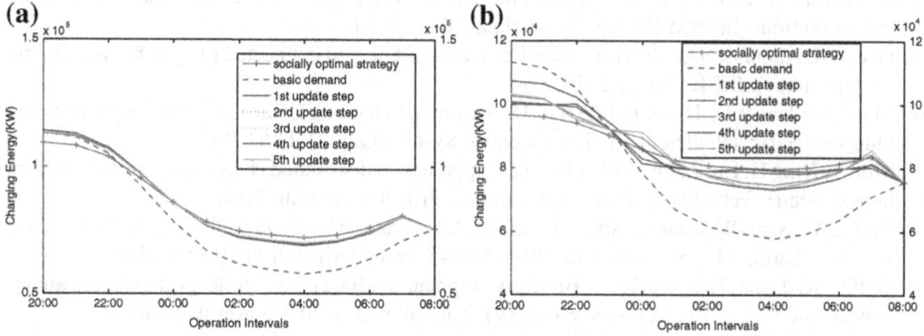

Fig. 43.2 Update procedure of PEVs' coordination strategy with: **a** $f_n = 0.0135u_{nt}^2$, **b** $f_n = 0.004u_{nt}^2$

Table 43.1 The total cost of the three charging and discharging strategies

Strategy	Total system cost
Socially optimal strategy	7.10×10^4
PSVF strategy	7.25×10^4
Distributed strategy	7.11×10^4

43.6 Conclusions and Future Research

In this paper we formulate a class of coordination issues for PEVs to shift peak load and fill valley load. We develop a distributed coordination algorithm to solve this problem based on that PEVs can charge and discharge. By applying this algorithm, the system can converge to an optimal strategy under certain mild conditions.

As ongoing research works, we are interested in: (i) extending the proof of convergency in the paper to an flexible load curve which covers 24 hours or even longer; (ii) further designing a novel distributed algorithm under which PEVs can join and leave from time to time.

References

1. Denholm P, Short W (2006) An evalution of utility system impacts and benefit of optimally dispatched plug-in hybrid electic vehicles. National Renewable Energy Labortary, Technical Report NREL 620:40293
2. Fernandez L, Roman T, Cossent R, Domingo C (2011) Assessment of the impact of plug-in electric vehicles on distribution networks. IEEE Trans Power Syst 26(1):206–213
3. Clement-Nyns K, Haesen E, Driesen J (2010) The impact of charging plug-in hybrid electric vehicles on a residential distribution grid. IEEE Trans Power Syst 25:371–380
4. Sundstrom O, Binding C (2010) Planning electric-drive vehicle charging under constrained grid conditions. In IBM-Zurich, Switzerland, Technical Report
5. Galus M, Andersson G (2008) Demand management of grid connected plug-in hybrid electric vehicles(phev). IEEE Energy 2030, 17–18
6. Ma Z, Callaway D, Hiskens I (2011) Decentralized charging control of large populations of plug-in electric vehicles. IEEE Trans Control Syst Technol 21(1):67–78
7. Mou Y, Xing H, Lin Z, Fu M (2014) A new approach to distributed charging control for plug-in hybrid electric vehicles. In Proceedings of the 33rd chinese control conference
8. Tushar W, Saad W, Poor H, Smith D (2012) A generationalized stackelberg game for electric vehicles charging in the smart grid. IEEE Trans Parallel Distrib Syst 196:541–549
9. Willett K, Jasna T, Letendre S, Brook A, Lipman T (2001) Vehicle-to-grid power: battery, hybrid, and fuel cell vehicles as resources for distributed electric power in california
10. Wenbo S, WS WV (2011) Real-time vehicle-to-grid control algorithm under price uncertainty. In IEEE International Conference on Smart Grid Communications, pp. 261–266
11. Gan L, Topcu U, Low S (2013) Optimal decentralized protocol for electric vehicle charging. IEEE Trans Power Syst 2011 28:940–951
12. Ma Z, Zou S, Liu X (2015) A distributed charging coordination for large-scale plug-in electric vehicles considering battery degradation costs. IEEE Trans Control Syst Technol, 23, pp. pre-print

Chapter 44
3D Model Classification Based on Transductive Support Vector Machines

Qiang Cai, Zhongping Si and Haisheng Li

Abstract With the development of graphical modeling methods and acquirement technology of 3D models, the numbers of 3D models increased exponentially. 3D model classification is crucial to effective management and precise retrieval. Due to the shortage of annotations, we adopted semi-supervised learning methods to solve this problem. This paper introduces Transductive Support Vector Machines (TSVMs) for 3D model classification. Our proposed method is verified efficient.

Keywords 3D model · Classification · Semi-supervised learning · Transductive support vector machines

44.1 Introduction

3D models have been widely applied in industrial design, virtual reality, television animation, molecular biology and other fields. 3D model retrieval has received extensive attention and development as an important part in the field of multimedia information retrieval. In the aspect of semantics, abundant existing 3D models have no semantic information, so they can't be managed and retrieved efficiently. Semantic learning mechanism combines computer vision and statistical learning theory to solve the problem of gap between underlying vision and high-level semantic information. Previous researches employed manual work, but it is time-consuming and inefficient. Therefore, for the purpose of better organization and retrieval, how to classify 3D models automatically has become an important research topic.

Q. Cai (✉) · Z. Si · H. Li
School of Computer and Information Engineering, Beijing Technology and Business University, Beijing 100048, China
e-mail: caiq@th.btbu.edu.cn

© Springer-Verlag Berlin Heidelberg 2016

Y. Jia et al. (eds.), *Proceedings of the 2015 Chinese Intelligent Systems Conference*, Lecture Notes in Electrical Engineering 359,
DOI 10.1007/978-3-662-48386-2_44

In order to realize automatic classification of 3D models, different annotation situations need to use different classification algorithms. When there is no labeled sample, unsupervised methods are applicable. These methods classify sample without any prior knowledge and cluster sample with features of sample. When there is large number of high quality annotations, first define semantic classification dictionary, then the models can be attributed to different semantic categories according to existing labels and supervised learning theory. There are some cases, in which only a small part of sample have category labels. How to make the remaining sample get category labels is the problem that the paper dealt.

In the statistics, induction inference firstly gets the rules from training sample and then utilizes the rules to determine test sample. Obviously, only using a small number of training sample leads to weak generalization ability of learning systems. Transductive inference is a kind of method that predicts test sample by observing specific training sample. Transductive inference utilizes information of unlabeled sample to find clusters and then to classify sample. Some scholars call these transductive inference methods as semi-supervised methods. The classic semi-supervised algorithms include Transductive Support Vector Machines algorithm, co-training algorithm and so on.

The structure of the paper is as follows, Sect. 44.2 provides a brief review of related works; Sect. 44.3 introduces Transductive Support Vector Machines and our experimental procedure; Sect. 44.4 shows our experiment results; Sect. 44.5 introduces conclusions and our future work; and in the next section, a review of the references is given.

44.2 Related Works

At present in the field of 3D model semantic analysis, many institutions at home and abroad are devoting themselves to the study of 3D model semantic annotation. Foreign institutions mainly include Princeton Shape Retrieval, Shape Retrieval Group, Columbia Robotics Lab and so on. Existing researches focus on automatic classification, relevant feedback and object ontology. Classification is the first step for 3D model retrieval. In addition to organize database, category information can help precise search in a class, which greatly narrows the search scope. We can use text to retrieve models if there is no appropriate 3D model as an input model.

Relative early researches on 3D model automatic classification employed support vector machines (SVMs), Bayes classification and K-neighbor methods, etc. In the recent literature, many new methods were proposed. For example, Gao et al. proposed a method based on Gaussian processes to improve the performance of 3D model retrieval systems. A new type of feature combined AC2 and D2 was proposed [1]. Gao et al. proposed another new type of feature based on 2D views (called ZA) in the next year [2]. Tian et al. proposed a graph-based semi-supervised learning algorithm. Each sample is a node, and the distance between two nodes determines the weight of an edge. Semantic labels spread and unlabeled sample acquire labels [3]. Li et al. realized 3D model classification based on nonparametric discriminant analysis with kernels [4]. Guo et al. developed a novel method of 3D object classification based on a two-dimensional hidden Markov model (2D HMM). Each object is decomposed by a spiderweb model and a shape function D2 is computed for each bin [5].

The setting of transductive inference was introduced by Vapnik [6]. And TSVMs was originally used by Joachims for text classification [7]. Joachims proved that TSVMs substantially improve the already excellent performance of SVMs for text classification.

44.3 The Application of Transductive Support Vector Machines in 3D Model Classification

44.3.1 Transductive Support Vector Machines

The dashed line is the solution of SVMs and the solid line is the solution of TSVMs. The examples surrounded by the red boxes need to exchange their labels. Figure 44.1 indicates that SVMs make the interval between positive sample and negative sample maximum. While TSVMs make the separating hyper plane gets through the low density area of all sample.

The principles of TSVMs are defined as follows: give a set of labeled sample S_{train} of n training examples:

$$(\vec{x}_1, y_1), (\vec{x}_2, y_2), \ldots, (\vec{x}_n, y_n), x_i \in R^m, y_i \in \{-1, +1\} \tag{44.1}$$

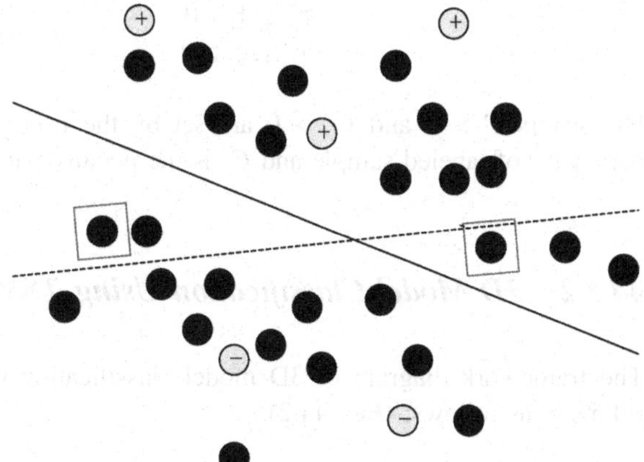

Fig. 44.1 The graphical representation of TSVMs

Each $\vec{x}_i \in X$ represents a feature vector, and y_i is the label of $\vec{x}_i$. In addition, give another set of unlabeled sample S_{test} of k testing examples from the same distribution with S_{train}:

$$\vec{x}_1^*, \vec{x}_2^*, \ldots, \vec{x}_k^* \tag{44.2}$$

Different from SVMs, TSVMs add constraints to unlabeled sample. Under the linearly separable condition, the training problem can be described as an optimization problem like Formula (44.3):

Minimize over $(y_1^*, \ldots, y_k^*, \vec{w}, b)$:

$$\frac{1}{2}||\vec{w}||^2$$

Subject to:

$$\begin{aligned} &\forall_{i=1}^n : y_i[\vec{w} \cdot \vec{x}_i + b] \geq 1 \\ &\forall_{j=1}^k : y_j^*[\vec{w} \cdot \vec{x}_j^* + b] \geq 1 \end{aligned} \tag{44.3}$$

To be able to deal with non-separable data, slack variables ξ_i and ξ_j^* are introduced. And the training problem can be described as an optimization problem like Formula (44.4):

Minimize over $(y_1^*, \ldots, y_k^*, \vec{w}, b, \xi_1, \ldots, \xi_n, \xi_1^*, \ldots, \xi_k^*)$:

$$\frac{1}{2}||\vec{w}||^2 + C\sum_{i=0}^{n}\xi_i + C^*\sum_{j=0}^{k}\xi_j^*$$

Subject to:

$$\begin{aligned} &\forall_{i=1}^n : y_i[\vec{w} \cdot \vec{x}_i + b] \geq 1 - \xi_i \\ &\forall_{j=1}^k : y_j^*[\vec{w} \cdot \vec{x}_j^* + b] \geq 1 - \xi_j^* \\ &\forall_{i=1}^n : \xi_i > 0 \\ &\forall_{j=1}^k : \xi_j^* > 0 \end{aligned} \tag{44.4}$$

Parameters $C > 0$ and $C^* > 0$ are set by the user. Parameter C is the penalty parameter of labeled sample and C^* is the penalty parameter of unlabeled sample.

44.3.2 3D Model Classification Using TSVMs

The framework diagram of 3D model classification using semi-supervised algorithms is as follows: (Fig. 44.2).

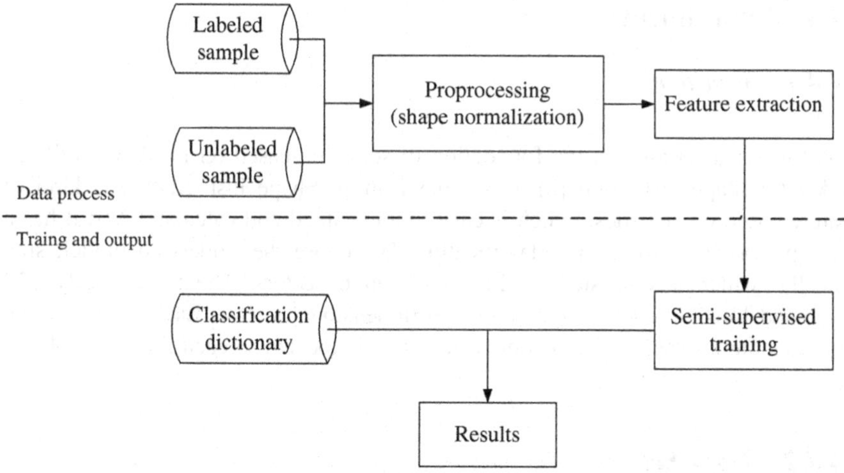

Fig. 44.2 Framework diagram of 3D model classification using semi-supervised algorithms

The implementing procedure of algorithms using TSVMs for 3D model classification is as follows:

(1) Sort data set and define semantic classification dictionary.
(2) 3D model preprocessing (translation, rotation and scaling transformation) and feature extraction.
(3) Assign values of C and C^{*}, and use SVM to train an initial classifier. According to the condition of our test sample, we used the ratio of positive and negative examples in the testing sample to calculate num_{+}.
(4) Classify the test examples using $<\vec{w}, b>$. The num_{+} test examples with the highest value of $\vec{w} \cdot \vec{x}_j^{*} + b$ are assigned to the class +; the remaining test examples are assigned to class—Assign a temp C_{temp}^{*}.
(5) Retrain all the examples. Exchange a pair of examples which satisfy certain rules that $y_m^{*} * y_l^{*} < 0$, $\xi_m^{*} > 0$, $\xi_l^{*} > 0$ and $\xi_m^{*} + \xi_l^{*} > 2$. Objective function value in optimization problems (4) decreases farthest. Repeat this step until there are no examples that satisfy the exchange conditions.
(6) Increase the value of C_{temp}^{*} evenly and return to step (5). Algorithm terminates and we can get results when $C_{temp}^{*} \geq C^{*}$.
(7) Realize multiple classification by "one versus rest" classification strategy. Classify each unlabeled example to the class with the maximum value of function.

44.4 Experiments

44.4.1 Feature

Content-based retrieval is the foundation of semantic-based retrieval. We utilized Ankerst's shape histogram [8] as sample feature. Shape histogram is a kind of feature based on statistics, which is easy to understand and calculate. And it has a good performance for rough classification. We chose the spiderweb model, specifically, which is consisted of 20 shells and 6 sectors. There are totally 120 combined bins, so each feature is a 120 dimensional vector. Each feature value indicates the number of 3D model surface points located in each bin (Fig. 44.3).

44.4.2 Data Set

Data set is Princeton Shape Benchmark (PSB) [9]. Classification information in PSB is in different levels. Due to the scarceness of 3D models in some categories, we select five classes contained 528 models from base, course1 and course2. Number of 3D models in each category is as follows: (Table 44.1).

44.4.3 Results

Our experiment tools are matlab, visual studio 2010 and SVM^{light} [10]. The multiple classification strategy is "one versus rest". The values of parameter p that represent fraction of unlabeled sample to be classified into the positive class are set to the ratio of positive and all test examples in the testing data. Other parameters are set to the default values of SVM^{light}.

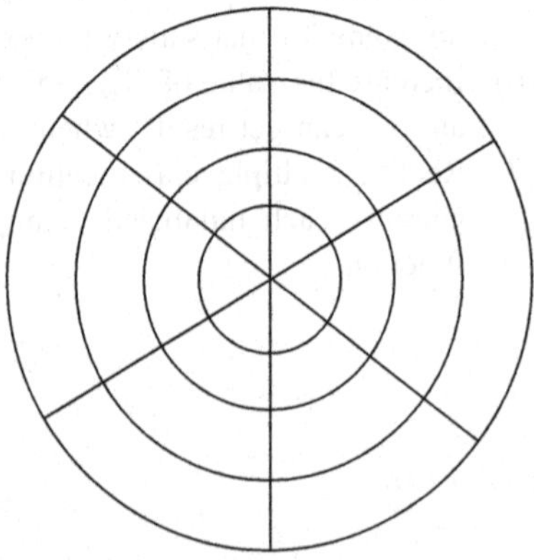

Fig. 44.3 Spiderweb model of shape histogram

Table 44.1 Number of 3D models in each category in PSB

Number	Plant	Car	Fighter_jet airplane	Building	Seat	Total
Total	138	113	100	100	77	528

Table 44.2 Precision and recall values

		SVM	TSVM
Plant	Prec	42.64	56.75
	Rec	78.39	46.62
Car	Prec	44.96	66.29
	Rec	79.57	59.14
Fighter_jet airplane	Prec	29.47	30.01
	Rec	3.75	36.25
Building	Prec	33.57	29.31
	Rec	10.63	29.38
Seat	Prec	48.34	35.15
	Rec	7.89	44.74

There are 100 examples (20 examples in each category) are labeled. In order to avoid accident results, we experimented for two times with different sample. We calculated the average values of two experiments. The results of our experiments are as follows: (Table 44.2).

From the experiment results, SVMs classified most examples into certain classes, which caused that recall values of most classes were extremely low. But TSVMs didn't lead to this situation because of the participation of unlabeled sample in training process.

Experiment results are showed in the form of histogram. From the Fig. 44.4, the values of average precision, recall and overall accuracy using TSVM are higher than SVM.

44.4.4 *Annotation*

According to our experiment results, each 3D model acquired a label. We realized 3D model annotation by mapping classification dictionary (Table 44.3).

44.5 Conclusions and Future Work

The paper proposed a semi-supervised algorithm to solve 3D model classification problem. TSVMs add unlabeled sample to training process and increase influence of test sample gradually. After limited times of iteration, optimal results are output.

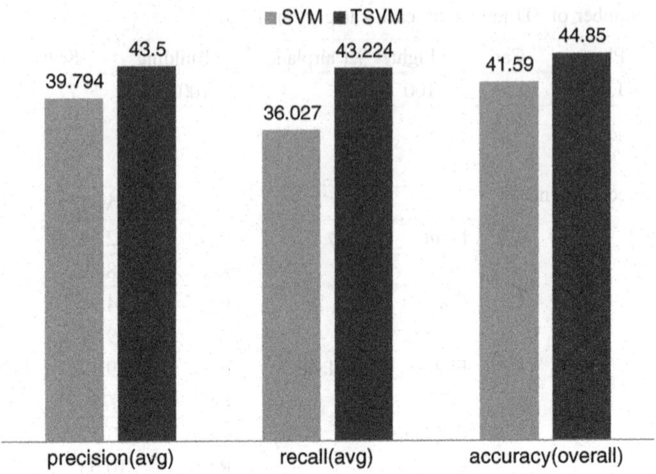

Fig. 44.4 Average precision, recall and overall accuracy using SVM and TSVM

Table 44.3 Annotation of 3D models

3D models						
Results	1	**3.164673**	−22.1314	−3.09462	−60.4279	−4.90657
	2	−24.4746	**14.20995**	−8.37084	87.35443	−3.53622
	3	−0.99976	−4.13294	**−1.66798**	−96.6018	−1.00012
	4	−16.3999	−28.4187	−6.96757	**166.436**	1.159751
	5	−77.7952	3.049675	−25.861	−723.95	**3.140866**
Annotation		plant	car	fighter_jet airplane	building	seat

Classification dictionary
1-plant, 2-car, 3-fighter_jet airplane, 4-building, 5-seat

TSVMs especially perform well compared with SVMs in dealing with small part of labeled sample and large part of unlabeled sample.

But TSVMs perform local combinatorial search with many iterations. In our experiment, running time using SVMs is about 0.04 s, while using TSVMs is about 68.38 s. So we next mainly concentrate on the optimization of TSVMs in the application of 3D model retrieval. We will try to decrease the complexity of TSVMs and improve the training ability of TSVMs. We will also study how to determine the fraction of unlabeled sample to be classified into the positive class.

References

1. Gao B, Yao F, Zhang S (2010) 3D model semantic classification and retrieval with Gaussian processes [J]. J Zhejiang Univ (Engineering Science) 12:006
2. Gao B, Zhang S, Pan X (2011) Semantic-Oriented 3D model classification and retrieval using Gaussian processes [J]. J Comput Inf Syst 7(4):1029–1037
3. Tian F, Xu KS, Xian ML et al (2011) 3D model multiple semantic automatic annotation for small scale labeled data set[C]. In: 2011 IEEE International conference on virtual reality and visualization (ICVRV), pp 193–198
4. Li JB, Sun WH, Wang YH et al (2013) 3D model classification based on nonparametric discriminant analysis with kernels [J]. Neural Comput Appl 22(3–4):771–781
5. Guo J, Zhou MQ, Li C et al (2013) 3D Object classification using a two-dimensional hidden Markov model [J]. Appl Mech Mater 411:2041–2046
6. Vapnik VN, Vapnik V (1998) Statistical learning theory [M]. Wiley, New York
7. Joachims T (1999) Transductive inference for text classification using support vector machines [C]. In: ICML, vol 99, pp 200–209
8. Ankerst M, Kastenmüller G, Kriegel HP et al (1999) 3D shape histograms for similarity search and classification in spatial databases[C].//Adv Spat Databases Springer, Berlin 1651:207–226
9. Princeton Shape Benchmark [EB/OL] (2015) http://shape.cs.princeton.edu/benchmark. Accessed 30 May 2015
10. Joachims T (2011) SVM^{light} Support vector machine [EB/OL]. http://svmlight.Joachims.org. Accessed 30 May 2015

Chapter 45
Robot Localization Based on Optical-Flow Sensor Array

Shaoxian Wang, Mingxiao He, Guoli Wang and Xuemei Guo

Abstract Recently, Optical-flow sensors commonly used as a PC input device have been explored in robot localization. One common difficulty is that ground-height variations in mobile robot moving can inevitably deteriorate the sensing performance of optical-flow sensors when used in a trivial fashion. In this paper, a novel deployment configuration is presented to build an optical-flow sensor array for robot localization, which is advantageous in the robustness to ground-height variation. In addition, the imaging capability of optical mouse sensors is exploited in developing a beacon based calibration approach for eliminating accumulative errors. The experimental results are reported to validate the proposed method.

Keywords Optical-flow mouse sensors · Mobile robot localization · Accumulative-error calibration

45.1 Introduction

Encoder, ultrasonic and GPS are commonly used on robotic positioning. An encoder is usually installed on a motor or a wheel to measure the rotation rate, which is converted into drift [1]. Once slip occurs, an encoder can no longer be used for localization. Ultrasonic positioning using beacons to response the ultrasonic signal and send ultrasonic signal back to the robot. Using the speed of sound and

This work was supported by Science and Technology Program of Guangzhou under Grant No. 201510010017, and by the SYSU-CMU Shunde International Joint Research Institute Free Application Project under Grant No. 20130201.

S. Wang · M. He · G. Wang · X. Guo (✉)
School of Information Science and Technology, Sun Yat-Sen University, Guangzhou 510006, China
e-mail: guoxuem@mail.sysu.edu.cn

© Springer-Verlag Berlin Heidelberg 2016

Y. Jia et al. (eds.), *Proceedings of the 2015 Chinese Intelligent Systems Conference*, Lecture Notes in Electrical Engineering 359,
DOI 10.1007/978-3-662-48386-2_45

the duration of transmission, the distance between robot and the beacon is easy to calculate [2]. Due to the low speed of sound, which means long delay, and the energy decay, ultrasonic cannot be used in large-space situation. And there is interference between robots using ultrasonic sensors. GPS is commonly used on car navigation. Launching satellites and construction projects are too expensive. And the error of GPS is larger than a few meters, which means GPS is not suitable for accurate localization.

The optical-flow sensor is widely used in the PC mouse. Via calculating the shifting of an image by comparing between two frames, the optical flow sensor is able to measure the displacement. That's why the researchers attempt to use it on robotic positioning and other fields. In [3] it uses optical-flow sensors in a catheter operating system to measure how deep the catheter has been inserted. In [4] it measures the rotation of a ball using optical-flow sensors. And [5] it uses optical-flow sensors to analyze the surface shape. The optical-flow sensors are forced to cling onto the floor by springs, which are used for the positioning of indoor robots [6]. In [7] Optical-flow sensor is modified by lens in order to increase the working height of the sensor. The experiment result of [8] shows that accuracy of optical-flow sensor array differs in different ways of arrangement. In [9, 10] the mathematical derivation shows that we can have the optimal accuracy when the center of optical-flow sensor array and the center of robot overlap.

In this paper, a design of sensor array is proposed, which is able to adjust to the variation of the working height. And to calibrate the accumulative error, a method of image beacons and the algorithm for recognizing the image of the beacon is proposed in this paper as well.

The content of this paper is arranged as follows. Section 45.2 introduces the design and arrangement of the sensor array. Section 45.3 introduces a beacon based calibration approach for eliminating accumulative errors. The experiment results are showed in sect. 45.4. And the last section summarizes this paper.

45.2 Arrangement of Sensor Array

Optical-flow sensors provide x and y displacement output but no rotation information. It is necessary to design a sensor array in order to get the rotation θ. And the velocity estimation algorithm will be proposed here as well.

45.2.1 Design of Sensor Array

Since the optical-flow is designed for PC mouse, it works only when it is clung to the working surface. Using fixed focus lens can increase its working height. However, the accuracy declines when the height changes. That why we use camera lens instead, which ensures the image will not become fuzzy when the height changes in a large range (Fig. 45.1).

Two sensors work as a pair. The heights of installation of two sensors are different from each other. One sensor is installed higher than the other (see Fig. 45.2). By this way, the heights from the working surface to the sensors can be figured out. The derivation is as follows.

As showed in Fig. 45.3, when CD is the working surface, and the robot's displacement is x_0. The outputs of two sensors are the same, marked as c_0.

$$x_0 = c_0 \cdot s_0 \tag{45.1}$$

s_0 is the scale factor when the working surface is CD.

When GI is the working surface, and the robot's displacement is Δx. The output of sensor A is c_1 and the scale factor is s_1, while the output of B is c_2 and the scale factor is s_2.

$$\Delta x = c_1 \cdot s_1 \tag{45.2}$$

$$\Delta x = c_2 \cdot s_2 \tag{45.3}$$

Fig. 45.1 The modified optical-flow sensor

Fig. 45.2 The installation height of a pair of optical-flow sensors

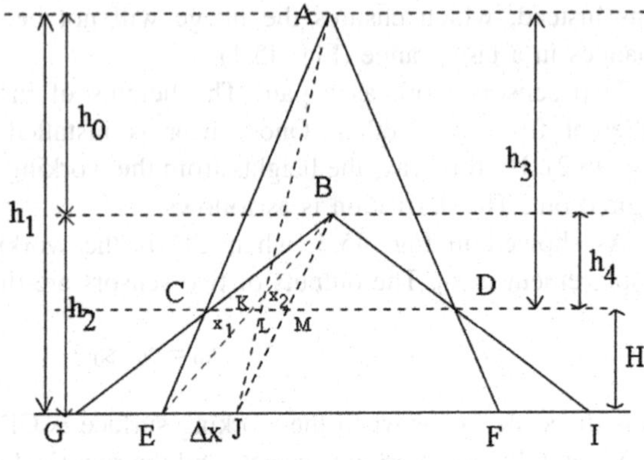

Fig. 45.3 Visual field of two sensor

If we mark the scale factor as s_0 when the working surface is CD, we get

$$x_1 = c_1 \cdot s_0 \tag{45.4}$$

$$x_2 = c_2 \cdot s_0 \tag{45.5}$$

where x_1 is the length of CL and x_2 is the length of KM. According to the property of similar triangles,

$$\frac{\Delta x}{x_1} = \frac{h_1}{h_3} \tag{45.6}$$

$$\frac{\Delta x}{x_2} = \frac{h_2}{h_4} \tag{45.7}$$

$$\Delta x = \frac{x_1 h_1}{h_3} \tag{45.8}$$

$$\frac{x_1}{x_2} = \frac{h_2 h_3}{h_1 h_4} = \frac{(H + h_4) h_3}{(H + h_3) h_4} \tag{45.9}$$

$$\frac{x_1}{x_2} = \frac{c_1 s_0}{c_2 s_0} = \frac{c_1}{c_2} \tag{45.10}$$

So, we can easily get Δx and H,

$$H = \frac{h_3 h_4 (c_1 - c_2)}{c_2 h_3 - c_1 h_4} \tag{45.11}$$

$$\Delta x = c_1 s_1 = \frac{x_1 h_1}{h_3} = \frac{c_1 s_0 (h_3 + H)}{h_3} = c_1 s_0 \left(1 + \frac{H}{h_3}\right) \tag{45.12}$$

The factor h_3, h_4 and s_0 can be measured by experiment, hence we are able to calculate the height of H and Δx according to the output of two sensors, even though the H changes. That's the reason why a pair of optical-flow sensors can adjust to the variation of height.

The sensor array consist of two pairs of sensors, in other word, there are four sensors installed on the robot. Assume that A and B, C and D are two pairs of sensors installed on the robot (Fig. 45.4). And their positions are marked as (x_i, y_i), $i = 1, 2, 3, 4$. And the center of rotation is marked as C. O is the center of the robot, and r is the length of CO. The arcs AA′, BB′, CC′, DD′ are marked as s_i, $i = 1, 2, 3, 4$. And their corresponding radii are R_i, $i = 1, 2, 3, 4$.

$$s_i = \theta R_i = \theta\sqrt{(r + x_i)^2 + y_i^2} \tag{45.13}$$

The distances that each pair of sensors traveled when the robot made a turn have to be the same, which is the precondition of the height adjustable algorithm we proposed before.

$$s_1 - s_2 = \theta R_1 - \theta R_2 = \theta\left(\sqrt{(r+x_1)^2 + y_1^2} - \sqrt{(r+x_2)^2 + y_2^2}\right) = 0 \tag{45.14}$$

$$r^2 + 2x_1 r + x_1^2 + y_1^2 = r^2 + 2x_2 r + x_2^2 + y_2^2 \tag{45.15}$$

$$\begin{cases} x_1 = x_2 \\ y_1 = \pm y_2 \end{cases} \tag{45.16}$$

Similarly,

$$\begin{cases} x_3 = x_4 \\ y_3 = \pm y_4 \end{cases} \tag{45.17}$$

It means that sensor A and B, C and D have to be symmetrical on the x axis. That why the sensors are arranged as Fig. 45.4a.

45.2.2 Velocity Estimation Algorithm

If the movement of the robot only consists of translation, the displacement is simply the average of the outputs of four sensors. When the robot rotates, we assume that the initial position of four sensors are (x_{pi}, y_{pi}), $i = 1, 2, 3, 4$ (Fig. 45.5).

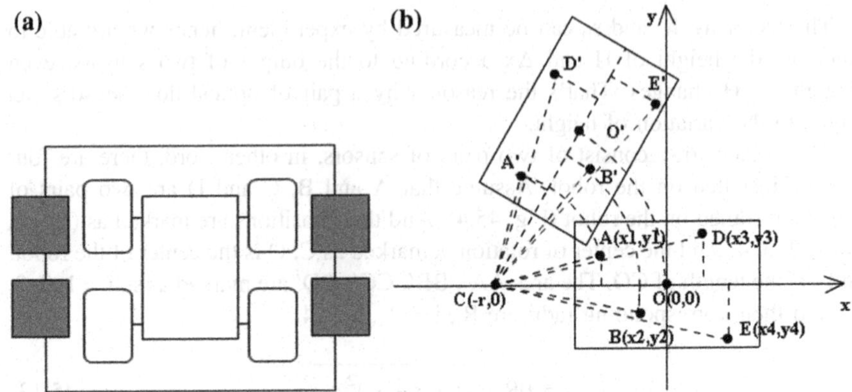

Fig. 45.4 The arrangement of sensor array. **a** Sensor array installed on robot **b** The changes of sensors' position

Fig. 45.5 Rotation of the robot

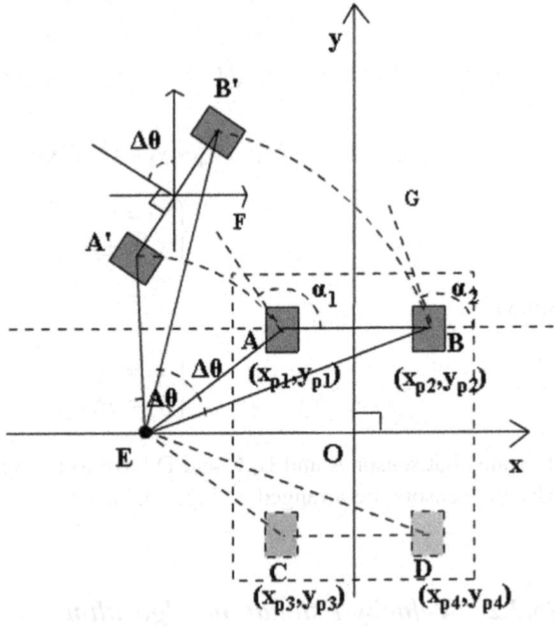

We can get α_i (i = 1, 2, 3, 4) from the x and y output of the optical-flow sensor.

$$\alpha_i = \begin{cases} \tan^{-1}\left(\frac{y_i}{x_i}\right) - \frac{\operatorname{sgn}(x_i) - 1}{2} \cdot \pi, & x_i \neq 0 \\ \operatorname{sgn}(y_i) \cdot \frac{\pi}{2}, & x_i = 0 \end{cases} \tag{45.18}$$

And the ∠AEB, marked as γ, is

$$\gamma = \alpha_1 - \alpha_2 \tag{45.19}$$

$$L_i = \sqrt{x_i^2 + y_i^2} \tag{45.20}$$

$$r_i = \frac{L_i}{|\Delta\theta|} \tag{45.21}$$

where L_i is the length of arcs AA′, BB′, CC′, DD′, and the corresponding radii are r_i, i = 1, 2, 3, 4. Thus, we have

$$\Delta\theta = \frac{\sqrt{L_1^2 + L_2^2 - 2L_1L_2\cos\gamma}}{d} \cdot \mathrm{sgn}(y_2 - y_1) \tag{45.22}$$

$$\mathbf{x} = \begin{bmatrix} x_1 \\ x_2 \\ x_3 \\ x_4 \end{bmatrix} = \begin{bmatrix} w_{x1} & & & 0 \\ & w_{x2} & & \\ & & w_{x3} & \\ 0 & & & w_{x4} \end{bmatrix} \begin{bmatrix} r_1 \\ r_2 \\ r_3 \\ r_4 \end{bmatrix} + \begin{bmatrix} x_{p1} \\ x_{p2} \\ x_{p3} \\ x_{p4} \end{bmatrix} \tag{45.23}$$

$$\mathbf{y} = \begin{bmatrix} y_1 \\ y_2 \\ y_3 \\ y_4 \end{bmatrix} = \begin{bmatrix} w_{y1} & & & 0 \\ & w_{y2} & & \\ & & w_{y3} & \\ 0 & & & w_{y4} \end{bmatrix} \begin{bmatrix} r_1 \\ r_2 \\ r_3 \\ r_4 \end{bmatrix} + \begin{bmatrix} y_{p1} \\ y_{p2} \\ y_{p3} \\ y_{p4} \end{bmatrix} \tag{45.24}$$

where

$$w_{xi} = [\sin(\alpha_i + \Delta\theta) - \sin\alpha_i] \cdot \mathrm{sgn}(\Delta\theta),\ i = 1,\ 2,\ 3,\ 4 \tag{45.25}$$

$$w_{yi} = [\cos\alpha_i - \cos(\alpha_i + \Delta\theta)] \cdot \mathrm{sgn}(\Delta\theta),\ i = 1,\ 2,\ 3,\ 4 \tag{45.26}$$

Then, the displacement of the center of the robot is (Δx, Δy)

$$\Delta x = \frac{1}{4}\sum_{i=1}^{4} x_i \tag{45.27}$$

$$\Delta y = \frac{1}{4}\sum_{i=1}^{4} y_i \tag{45.28}$$

We are able to get the position at t + 1 according to the position at t.

$$x_{t+1} = x_t + \sqrt{(\Delta x)^2 + (\Delta y)^2} \cdot \cos(\theta_t + \beta_{t+!}) \tag{45.29}$$

$$y_{t+1} = y_t + \sqrt{(\Delta x)^2 + (\Delta y)^2} \cdot \sin(\theta_t + \beta_{t+!}) \tag{45.30}$$

$$\theta_{t+1} = \theta_t + \Delta\theta \tag{45.31}$$

where

$$\beta_{t+!} = \begin{cases} \tan^{-1}\left(\frac{\Delta y}{\Delta x}\right) - \frac{\operatorname{sgn}(\Delta x) - 1}{2} \cdot \pi, & \Delta x \neq 0 \\ \operatorname{sgn}(\Delta y) \cdot \frac{\pi}{2}, & \Delta x = 0 \end{cases} \tag{45.32}$$

In this way, the global position of the robot can be calculated using the x and y output of four optical-flow sensors.

45.3 Accumulative Error Calibration

Optical-flow sensor can not only output the displacement, but also the image it shoots. Although the image is of low resolution, 18 × 18 in the sensor ADNS-2610 we used here, it's enough to provide the information of the working surface.

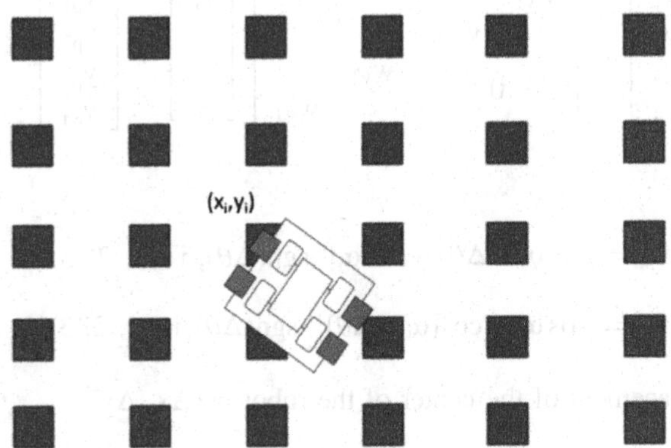

Fig. 45.6 Beacons on the floor

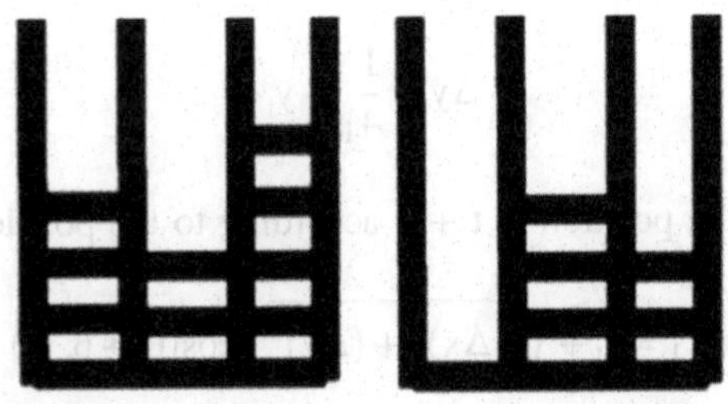

Fig. 45.7 The images of the beacons

We designed specific images as beacons, and pasted them on the floor (Fig. 45.6). Every time the sensors encounter the beacons, the global positions of the beacons are used to calibrate the accumulative error.

The beacons are showed in Fig. 45.7. It consists of four vertical lines and several short horizontal lines between every two vertical lines. The number of horizontal lines and their locations of one beacon are different from other beacons. That means

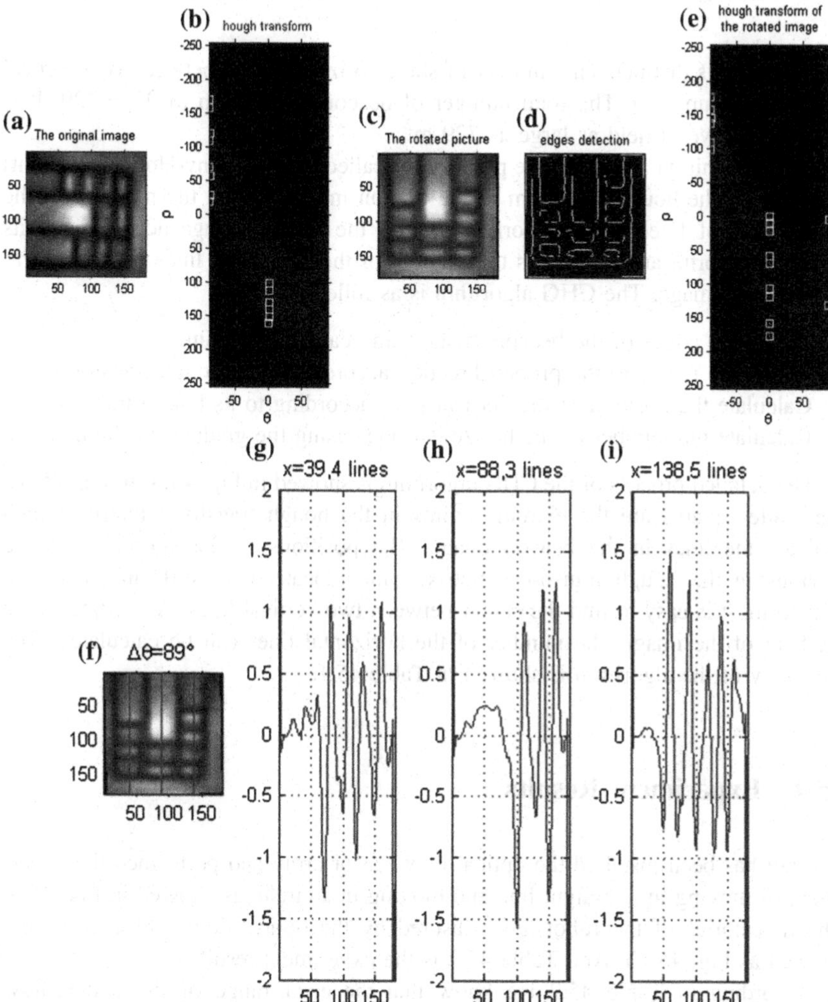

Fig. 45.8 **a** The original image. **b** The hough transform of the original image. **c** The rotated image. **d** The edges detection of the rotated image. **e** The hough transform of the rotated image, the positions of the *vertical lines* are obvious in the hough transform. **f** The rotated angle is $\Delta\theta = 89°$ and the center between two *vertical lines* are chosen. **g** The gradient of the image shows that there are 4 lines on x = 39. **h** The gradient of the image shows that there are 3 lines on x = 88. **i** The gradient of the image shows that there are 5 lines on x = 138

Table 45.1 The accuracy of the CHG algorithm

Angle	−180°	−90°	−45°	0°	90°	Accuracy (%)
The 143rd beacon	Correct	Correct	Correct	Correct	Correct	100
The 236th beacon	Correct	Correct	Incorrect	Correct	Correct	80
The 322nd beacon	Correct	Correct	Correct	Correct	Correct	100
The 435th beacon	Correct	Correct	Correct	Correct	Correct	100
The 531st beacon	Correct	Correct	Correct	Correct	Correct	100

every image is distinct. The number of short horizontal lines between two vertical lines can be up to 9. The total number of beacons can be up to $9^3 = 729$. It is enough to cover a field as large as 729 m^2.

The recognition algorithm we proposed is called CHG (Canny-Hough-Gradient) Algorithm. The hough transform of the beacon image contain the message of the distribution of lines. CHG algorithm rotates the beacon image according to its hough transform, and calculates the number of the horizontal lines using the gradient of the image. The CHG algorithm is as follows:

1. Detect the edges of the beacon image using canny algorithm.
2. Rotate the image to the proper direction according to its hough transform.
3. Calculate the position of the vertical lines according to its hough transform.
4. Calculate the number of the horizontal lines using the gradient of the image.

The detailed process of the CHG algorithm is showed in Fig. 45.8. In Fig. 45.8e, the white squares are the extreme points in the hough transform matrix, which indicate the lines in the beacon image. The positions of the vertical lines are obvious in the hough transform matrix, which locate at $\theta = 0°$ in the hough transform. It is easy to find the center between two vertical lines. According to the gradient of the image, the number of the horizontal lines can be calculated. The Accuracy of the algorithm is showed in Table 45.1.

45.4 Experiment Results

A robot has been installed the optical-flow sensor array and performed the experiment of moving in a straight line and moving in a circle, as showed in Fig. 45.9. The trajectories of the robot are measured by the optical-flow sensor array and showed as Fig. 45.10. And Table 45.2 is the experiment result.

According to Table 45.2 we know that the error range of the optical-flow positioning system is within ±3 cm, which is more accurate than GPS and ultrasonic positioning. The accuracy of optical-flow sensor array means it is able to serve the robot for indoor and outdoor localization.

Fig. 45.9 Motion experiment of robot with optical-flow sensor array

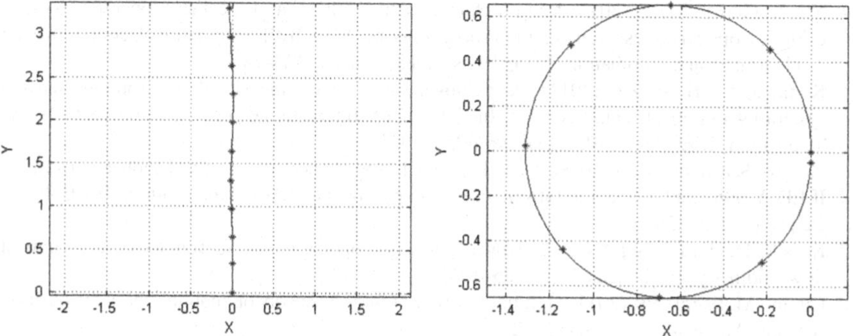

Fig. 45.10 The measured trajectory

Table 45.2 Experiment results

Number	1	2	3	4	5
X error range (m)	±0.0183	±0.0178	±0.0245	±0.0232	±0.0239
Y error range (m)	±0.0264	±0.0272	±0.0101	±0.0214	±0.0297

45.5 Conclusion

In this chapter, we modified the optical-flow sensor using camera lens and designed a special arrangement of the sensor array, in which every two optical-flow sensors worked as a pair. The imaging capability of optical mouse sensors is exploited and we proposed a beacon based calibration approach for eliminating accumulative errors. The beacons carry the message of their global position, which is used to calibrate the errors of the optical-flow positioning. And the CHG algorithm is proposed to identify the beacon image. Both the results of the experiment of the CHG algorithm and the positioning of the robot using optical-flow system are showed in this paper as well. Both of the CHG algorithm and the positioning system are reliable according to the experiment results.

References

1. Kim A-H, Seo B-S, Jeong K-W, Lee J-M (2012) Improvement of travel distance at the outdoor environment by using IMU and encoder. In: 12th International conference on control, automation and systems, IEEE Press, Jeju Island pp 1669–1674
2. Sekimori D, Miyazaki F (2005) Self-localization for indoor mobile robots based on optical mouse sensor values and simple global camera information. In: IEEE international conference on robotics and biomimetics, IEEE Press, Shatin, pp 605–610
3. Park J, Lee J (2011) A beacon color code scheduling for the localization of multiple robots. IEEE Trans Industr Inf 7(3):467–475
4. Yin X, Guo S, Ma X (2013) Motion characteristic evaluation of a catheter operating system using an optical mouse sensor. In: Proceedings of 2013 IEEE international conference on mechatronics and automation, IEEE Press, Takamatsu, pp 979–984
5. Kumagai M, Hollis RL (2011) Development of a three-dimensional ball rotation sensing system using optical mouse sensors. In: 2011 IEEE International conference on robotics and automation, IEEE Press, Shanghai, pp 5038–5043
6. Xin W, Katsunori S (2009) Surface shape analyzing device using optical mouse sensor, In: IEEE Youth conference on information, computing and telecommunication, IEEE Press, Beijin, pp 255–258
7. Ross R, Devlin J, Wang S (2012) Toward refocused optical mouse sensors for outdoor optical flow odometry. IEEE Sens J 12:1925–1932
8. Cimino M, Pagilla PR (2011) Optimal location of mouse sensors on mobile robots for position sensing. Automatica 47:2267–2272
9. Kim S, Kim H (2013) Optimal optical mouse placement for mobile robot velocity estimation. In: 2013 IEEE International conference on mechatronics (ICM), IEEE Press, Vicenza, pp 310–315
10. Kim S, Lim G, Kim H (2013) Scale-Invariant isotropy of an optical mouse array for mobile robot velocity estimation. In: 13th International conference on control, automation and systems, IEEE Press, Gwangju, pp 1229–1233

Chapter 46
Composite Continuous Anti-disturbance Autopilot Design for Missile System with Mismatched Disturbances

Xinge Liu, Haibin Sun and Jiayuan Shan

Abstract In this paper, the problem of autopilot design for missile system with mismatched disturbances is considered via a sliding mode control method and finite time disturbance observer. Firstly, a finite time disturbance observer is designed to estimate the mismatched disturbances. Secondly, based on disturbance estimation values and traditional sliding mode surface, a novel sliding mode surface is constructed. Thirdly, a composite continuous anti-disturbance autopilot is developed, which can guarantee system output converge to reference signal. Finally, a simulation result is presented to demonstrate the effectiveness of the proposed scheme.

Keywords Missile system · Anti-disturbance autopilot · Mismatched disturbances · Sliding mode control · Finite time disturbance observer

46.1 Introduction

In the past decades, the autopilot design problem for missile system has been attracting more and more attention [1]. Because of existing many kinds of disturbances in missile system, it needs to improve the anti-disturbance performance of autopilot. Disturbance observer based control method is a feasible way to improve the anti-disturbance performance of the missile system. Disturbance observer is presented

X. Liu
School of Automation, Beijing Institute of Technology, Beijing 100081, China
e-mail: lxg@bit.edu.cn

H. Sun (✉)
School of Engineering, Qufu Normal University, Shandong 276826, People's Republic of China
e-mail: seusunhaibin@gmail.com

J. Shan (✉)
School of Aerospace Engineering, Beijing Institute of Technology, Beijing 100081, China
e-mail: sjy1919@126.com

© Springer-Verlag Berlin Heidelberg 2016

Y. Jia et al. (eds.), *Proceedings of the 2015 Chinese Intelligent Systems Conference*, Lecture Notes in Electrical Engineering 359,
DOI 10.1007/978-3-662-48386-2_46

in 1980s, and it has been used to many kinds of fields, for example robotic systems [2], hard disk drive systems [3–5], grinding systems [6, 7], hypersonic vehicles [8, 9], spacecraft systems [10, 11], and the references therein. Based on nonlinear disturbance observer technique and nonlinear dynamic inversion method, a composite anti-disturbance autopilot scheme is proposed in [12]. In [13], a composite hierarchical anti-disturbance autopilot design scheme is proposed for missile system with multiple disturbances, where the feedback control law is designed based on nonlinear dynamic inversion.

In the above literature, matched disturbances are only considered for missile systems. In [14], a composite anti-disturbance autopilot design is investigated for missile system with mismatched disturbances, in which an appropriate disturbance compensation gain is designed to remove the influence of mismatched disturbances for missile system from system output channels. And the composite controller only guarantees the closed-loop system input-to-state stability.

In this paper, a novel composite continuous anti-disturbance autopilot is developed for missile system with mismatched disturbances. The composite continuous anti-disturbance autopilot is proposed via sliding mode control method and finite time disturbance observer, which can guarantee system output converge to zero in the presence of mismatched disturbances. By designing a novel sliding mode surface based on disturbance estimation values, the composite anti-disturbance autopilot is obtained. Using the Lyapunov function theory, the stability analysis of the closed-loop system is established. Finally, a simulation result is presented to demonstrate the effectiveness of the proposed scheme.

46.2 Model and Problem Formulation

According to [14], the longitudinal model of missile system with mismatched disturbances can be presented as

$$\begin{aligned}\dot{\alpha} &= f_1(\alpha) + b_1(\alpha)\delta + q + d_1,\\ \dot{q} &= f_2(\alpha) + b_2\delta + d_2,\end{aligned} \tag{46.1}$$

where α, δ, q denote attack angle, tail angle, and pitch rate. Nonlinear functions $f_1(\alpha)$, $f_2(\alpha)$, $b_1(\alpha)$ and b_2 are determined by aerodynamic parameter. d_1 and d_2 are the lumped disturbances, including unmodelled dynamics, external disturbances, and aerodynamic parameter uncertainties. For instance, when the velocity and the altitude of missile are 3 Mach and 6095 m (20,000 ft), respectively, and attack angle satisfies $|\alpha| \leq 20deg$, then $f_1(\alpha)$, $f_2(\alpha)$, $b_1(\alpha)$ and b_2 are given as follows

$$f_1(\alpha) = \frac{180QS}{\pi WV}\cos(\frac{\pi\alpha}{180})(1.03\times10^{-4}\alpha^3 - 9.45\times10^{-3}\alpha|\alpha| - 1.7\times10^{-1}\alpha),$$
$$f_2(\alpha) = \frac{180QSd}{\pi I_{yy}}(2.15\times10^{-4}\alpha^3 - 1.95\times10^{-2}\alpha|\alpha| + 5.1\times10^{-2}\alpha),$$
$$b_1(\alpha) = -3.4\times10^{-2}\frac{180gQS}{\pi WV}\cos(\frac{\pi\alpha}{180}),\ \ b_2 = -0.206\frac{180QSd}{\pi I_{yy}},$$

where W, I_{yy}, V, S, Q, d, g mean weight, pitch moment of inertia, velocity, reference area, reference diameter, dynamic pressure, and gravitational acceleration, respectively. The tail fin actuator dynamics are modelled by a first-order lag

$$\dot{\delta} = \frac{1}{t_1}(-\delta + v) + d_3, \tag{46.2}$$

where v, t_1 imply the commanded fin defection and time constant, respectively. d_3 is an external disturbance of system caused by the vibration of the airstream.

Assumption 1 ([17, 18]) The lumped disturbances satisfy the following condition $|\ddot{d}_i| \le L_i$, where L_i are known constants, $i = 1, 2, 3$.

By [12], the system output is defined as

$$y = \alpha + k_q q, \tag{46.3}$$

where k_q is a constant. Based on system equation and system output, the dynamic equation of system output can be presented as

$$\begin{aligned}\dot{y} &= y_1 + d_1 + k_q d_2,\\ \dot{y}_1 &= m + (b_1(\alpha) + k_q b_2)(\frac{1}{t_1}(-\delta + v)) + (b_1(\alpha) + k_q b_2)d_3 + m_1 d_1 + d_2,\end{aligned} \tag{46.4}$$

where

$$\begin{aligned}y_1 &= f_1(\alpha) + q + b_1(\alpha)\delta + k_q(f_2 + b_2\delta),\\ m &= m_1(f_1(\alpha) + q + b_1(\alpha)\delta) + f_2(\alpha) + b_2\delta,\ \ m_1 = (\frac{\partial f_1}{\partial\alpha} + k_q\frac{\partial f_2}{\partial\alpha} + \frac{\partial b_1}{\partial\alpha}\delta).\end{aligned}$$

Choose the tracking trajectory of system output as $y_r(t)$, and define the tracking error as

$$e_1 = y - y_r,\ \ e_2 = y_1 - \dot{y}_r. \tag{46.5}$$

The error system can be developed as

$$\dot{e}_1 = e_2 + d_1 + k_q d_2,$$
$$\dot{e}_2 = m - \ddot{y}_r + (b_1(\alpha) + k_q b_2)(\frac{1}{t_1}(-\delta + v)) + (b_1(\alpha) + k_q b_2)d_3 + m_1 d_1 + d_2. \tag{46.6}$$

46.3 Composite Anti-disturbance Autopilot Design

46.3.1 Finite Time Disturbance Observer Design

Borrowed from [17, 18], a finite time disturbance observer is designed as

$$\begin{aligned}
&\dot{z}_{0j} = v_{0j} + \bar{x}_j,\ \ \dot{z}_{1j} = v_{1j},\ \ \cdots,\ \ \dot{z}_{2j} = -\lambda_2 L_j \text{sign}(z_{2j} - v_{1j}),\\
&v_{0j} = -\lambda_0 L_j^{\frac{1}{3}} |z_{0j} - x_j|^{\frac{2}{3}} \text{sign}(z_{0j} - x_j) + z_{1j},\\
&v_{1j} = -\lambda_1 L_j^{1/2} |z_{ij} - v_{0j}|^{\frac{1}{2}} \text{sign}(z_{ij} - v_{0j}) + z_{2j},
\end{aligned} \tag{46.7}$$

where $j = 1, 2, 3$, $\bar{x}_1 = f_1(\alpha) + b_1(\alpha)\delta + q$, $\bar{x}_2 = f_2(\alpha) + b_2\delta$, $\bar{x}_3 = \frac{1}{t_1}(-\delta + v)$, $x_1 = \alpha$, $x_2 = q$, $x_3 = \delta$, $\lambda_0, \lambda_1, \lambda_2$ are the observer coefficients to be designed, $z_{01}, z_{02}, z_{03}, z_{1j}, z_{2j}$ are the estimates of α, q, δ, d_j, $\dot{d}_j$, respectively.

The observer error dynamics are obtained

$$\begin{aligned}
&\dot{e}_{0j} = -\lambda_0 L_j^{\frac{1}{3}} |e_{0j}|^{\frac{2}{3}} \text{sign}(e_{0j}) + e_{1j},\\
&\dot{e}_{1j} = -\lambda_1 L_j^{\frac{1}{2}} |e_{ij} - \dot{e}_{(0j}|^{\frac{1}{2}} \text{sign}(e_{ij} - \dot{e}_{(i-1)j}) + e_{(i+1)j},\\
&\dot{e}_{2j} = -\lambda_2 L_j \text{sign}(e_{2j} - \dot{e}_{1j}) - \ddot{d}_j,
\end{aligned} \tag{46.8}$$

where the observer errors are defined as $e_{01} = z_{01} - \alpha$, $e_{02} = z_{02} - q$, $e_{03} = z_{03} - \delta$, $e_{1j} = z_{1j} - d_j$, and $e_{2j} = z_{2j} - \dot{d}_j$. According to [17, 18] and *Assumption 1*, the observer error system is finite time stable, i.e., there exists a finite time instant t_1 such that $e_{1j} \equiv 0, e_{2j} \equiv 0, \cdots, e_{3j} \equiv 0, e_{0j} \equiv 0$ for $t > t_1$.

When $t > t_1$, the system (46.7) is changed to

$$\dot{z}_{0j} = z_{1j} + \bar{x}_j, \dot{z}_{1j} = z_{2j}. \tag{46.9}$$

46.3.2 Composite Continuous Anti-disturbance Autopilot Design

Introducing the estimation values of disturbances into system (46.6), yields

$$\begin{aligned}\dot{e}_1 &= e_2 + z_{11} + k_q z_{12} - e_{11} - k_q e_{12},\\ \dot{e}_2 &= m - \ddot{y}_r + (b_1(\alpha) + k_q b_2)(\frac{1}{t_1}(-\delta + v)) + (b_1(\alpha) + k_q b_2)z_{13} + m_1 z_{11} + z_{12}\\ &\quad -(b_1(\alpha) + k_q b_2)e_{13} - m_1 e_{11} - e_{12}.\end{aligned} \tag{46.10}$$

When $t > t_1$, $e_{11} = e_{12} = e_{13} = 0$, the system (46.10) reduces to

$$\begin{aligned}\dot{e}_1 &= e_2 + z_{11} + k_q z_{12}, \qquad (46.11)\\ \dot{e}_2 &= m - \ddot{y}_r + (b_1(\alpha) + k_q b_2)(\frac{1}{t_1}(-\delta + v)) + (b_1(\alpha) + k_q b_2)z_{13} + m_1 z_{11} + z_{12}.\end{aligned}$$

In the next, the composite continuous anti-disturbance autopilot is designed based on system (46.11).

Inspired by [15], a sliding mode surface with disturbance estimation value is designed as

$$s = c_1 e_1 + e_2 + z_{11} + k_q z_{12}. \tag{46.12}$$

where $c_1 > 0$.

The proposed composite continuous anti-disturbance autopilot is developed as

$$\begin{aligned}v = \delta - &\frac{t_1}{b_1(\alpha) + k_q b_2}(c_1 e_2 + m - \ddot{y}_r + z_{21} + k_q z_{22} + z_{12}\\ &+(b_1(\alpha) + k_q b_2)z_{13} + m_1 z_{11} + k_1 s + k_2 |s|^{\theta} \mathrm{sign}(s)),\end{aligned} \tag{46.13}$$

where $k_1 > 0$, $k_2 > 0$, $0 < \theta < 1$.

Theorem 1 *Consider the error system (46.6) with mismatched disturbances. If assumption 1 holds and a composite continuous anti-disturbance autopilot is designed as (46.13), then the system output asymptotically converges to zero.*

Proof First, we proof the system output can converge to zero when $t > t_1$. Choose the Lyapunov function as

$$V_s = \frac{1}{2}s^2. \tag{46.14}$$

Computing the first derivative of (46.14) along system trajectory (46.11), yields

$$\dot{V}_s = s(c_1 e_2 + m - \ddot{y}_r + (b_1(\alpha) + k_q b_2)(\frac{1}{t_1}(-\delta + v)) \\ +(b_1(\alpha) + k_q b_2) z_{13} + m_1 z_{11} + z_{12} + z_{21} + k_q z_{22}). \tag{46.15}$$

Substituting (46.13) into (46.15), we have

$$\dot{V}_s = s(-k_1 s - k_2 |s|^\theta \mathrm{sign}(s)) = -k_1 s^2 - k_2 |s|^{\theta+1} \\ = -2k_1 V_s - 2^{\frac{\theta+1}{2}} k_2 V_s^{\frac{\theta+1}{2}}. \tag{46.16}$$

According to [19], system trajectory can reach sliding mode surface in finite time. When $s = 0$, we obtain

$$\dot{e}_1 = -c_1 e_1, \tag{46.17}$$

which implies e_1 can converge to zero when $t > t_1$.

Secondly, we present the system trajectories (46.6) and (46.13) will not escape to the infinity in any time interval $[0, t_1]$. Define a finite time bounded function $B(e_1, \tilde{e}_2, s) = \frac{1}{2}(e_1^2 + \tilde{e}_2^2 + s^2)$, where $\tilde{e}_2 = e_2 + z_{11} + k_q z_{12}$. Note that $|s|^\theta \le 1 + |s|$. Taking the derivative of $B(e_1, \tilde{e}_2, s)$ along system trajectory and after some simple calculation, we obtain

$$\dot{B}(e_1, \tilde{e}_2, s) \le K_{v1} B(e_1, \tilde{e}_2, s) + L_{v1}, \tag{46.18}$$

where $K_{v1} = \max\{1 + \frac{k_q}{2}, \frac{k_1}{2} + 2 + \frac{\bar{b}}{2} + \frac{|m|_1}{2}, 1 + k_1 + \frac{3k_2}{2} + \frac{\bar{b}}{2} + \frac{|m_1|}{2}\}$, $L_{v1} = \max\{\frac{1}{2} e_{11}^2 + \frac{k_q}{2} e_{12}^2 + \frac{1}{2} + \frac{\bar{b} e_{13}}{2} + \frac{|m|_1}{2} e_{11}^2 + \frac{e_{12}^2}{2} + \frac{\bar{b}}{2} e_{13}^2 + \frac{|m_1|}{2} e_{11}^2 + \frac{1}{2} e_{12}^2\}$, $\bar{b} = 3.4 \times 10^{-2} \frac{180 g Q S}{\pi W V} + 0.206 \frac{180 Q S d}{\pi I_{yy}}$. K_{v1} and L_{v1} are bounded constants due to the boundness of e_{11}, e_{12}, e_{13}, and angle α. It can be concluded from (46.18) that $B(e_1, \tilde{e}_2), s$ and so e_1, $\tilde{e}_2$, s will not escape in finite time [20]. From the above analysis, we obtain system output y can track the reference signal y_r.

46.4 Simulation

In this section, the simulation result is presented to illustrate the effectiveness of the proposed method.

The parameters of missile system is listed in Table 46.1.

The initial state of missile system is chosen as $[\alpha, q, \delta] = [0, 0, 0]$deg. Select the reference output as y_r

Table 46.1 The physical meaning and values of missile system

Physical meaning	Symbol	Values
Mass	W	4,410 kg
Velocity	V	947.6 m/s
Moment of inertia	I_{yy}	$247.44\,\text{kg}\cdot\text{m}^2$
Dynamic pressure	Q	$293{,}638\,\text{N/m}^2$
Reference area	S	$0.04087\,\text{m}^2$
Reference diameter	d	0.229 m
Gravitational constant	g	$9.8\,\text{m/s}^2$
Time constant	t_1	0.1 s

$$y_r = \begin{cases} 10\text{ deg}, & 0 \le t < 4; \\ 0, & \text{else } t. \end{cases}$$

The controller parameters are listed as follows.
$k_1 = 10, k_2 = 20, c = 6, L_1 = L_2 = L_3 = 200, \lambda_0 = 2, \lambda_1 = 1.5, \lambda_2 = 1.1.$

Case I: External disturbance rejection ability
In this subsection, the tracking performance of missile system under external disturbances is tested. The external disturbances are taken as $d_1(t) = 0.2, d_2(t) = 0.3$, $d_3(t) = 0.2$ at $10 < t \le 15s$, $d_1(t) = 0.1 + 0.2\cos(t)$, $d_2(t) = 0.15 + 0.1\sin(t)$, $d_3(t) = 0.2 + 0.1\cos(t)$ at $15 < t \le 25s$. Curves of both the output and input under control method are given in Figs. 46.1 and 46.2.

It can be observed from Fig. 46.1 that the proposed method can guarantee system output track the reference signal and has a good anti-disturbance performance.

Case II: Robustness against model uncertainties
The robustness against model uncertainties of the proposed method is demonstrated in this part. To investigate the performance of robustness, we choose the model uncertainties as follows. Both $f_1(\alpha)$ and $f_2(\alpha)$ have variations of +20 %.

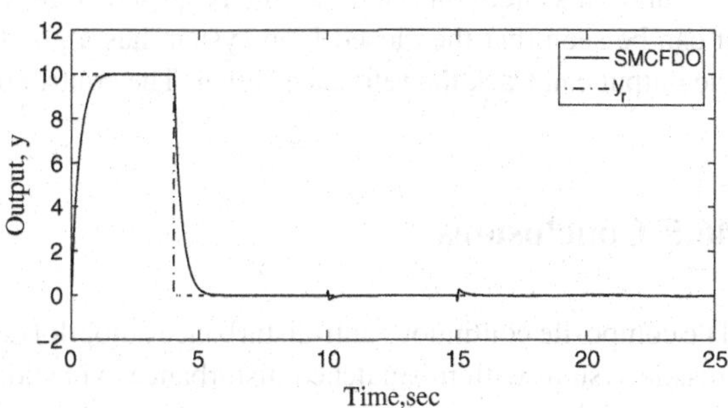

Fig. 46.1 Curves of system output y and reference output y_r under external disturbances

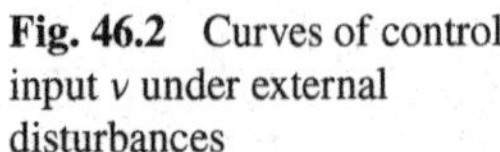

Fig. 46.2 Curves of control input v under external disturbances

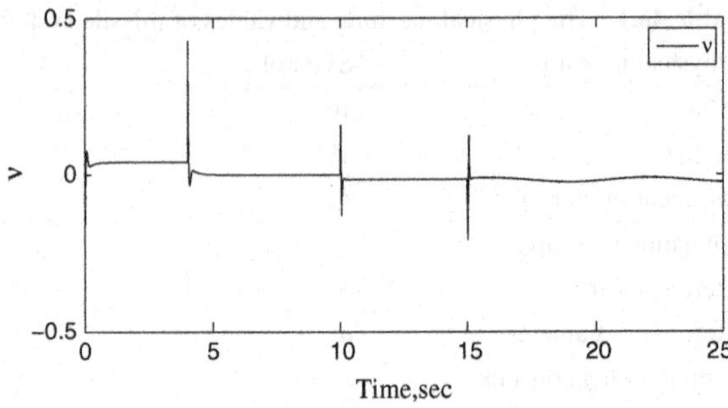

Fig. 46.3 Curves of system output y and reference output y_r under model uncertainties

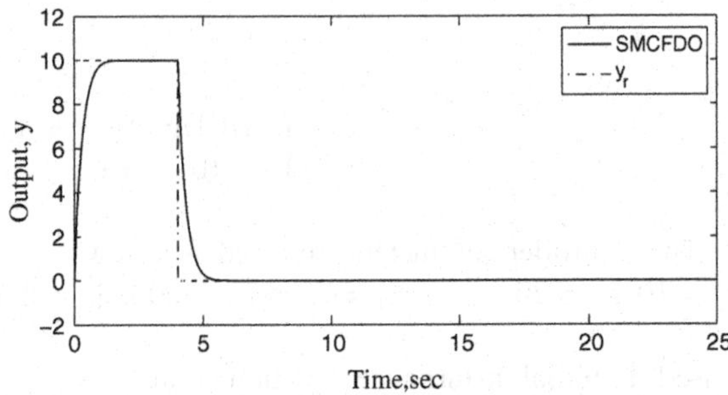

Fig. 46.4 Curves of control input v under model uncertainties

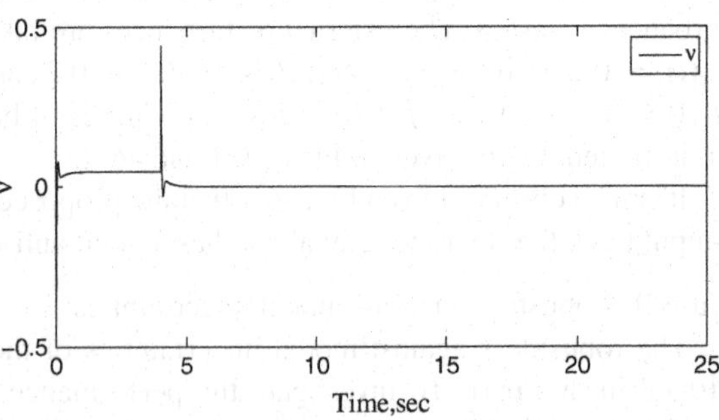

Curve of system output under the proposed control method is given in Fig. 46.3. It can be seen that the closed-loop system has a good robustness performance and the output can track the reference signal. The control input is presented in Fig. 46.4.

46.5 Conclusions

The composite continuous anti-disturbance autopilot design has been investigated for missile system with mismatched disturbances via sliding mode control method and finite time disturbance observer technique in this paper. Using the Lyapunov function theory, the stability of the closed-loop system has been analyzed. And the obtained autopilot can guarantee system output converge to zero. Finally, the simulation result has been given to demonstrate the effectiveness of the proposed method.

Acknowledgments This work was supported in part by the National Natural Science Foundation of China(Nos. 61203064,61403227).

References

1. Reichert R (1990) Dynamic scheduling of modern robust control for missile autopilot design. In: Proceedings of automatic control conference, pp 2368–2373
2. Chen WH, Ballance DJ, Gawthrop PJ, O'Reilly J (2000) A nonlinear disturbance observer for robotic manipulators. IEEE Trans Ind Electron 47(4):932–938
3. Guo L, Tomizuka M (1997) High-speed and high-precision motion control with an optimal hybrid feedforward controller. IEEE/ASME Trans Mechatron 2(2):110–122
4. Ishikawa J, Tomizuka M (1998) A novel add-on compensator for cancellation of pivot nonlinearities in hard disk drives. IEEE Trans Magn 34(4):1895–1897
5. Huang YH, Messner W (1998) A novel disturbance observer design for magnetic hard drive servo system with rotary actuator. IEEE Trans Magn 34(4):1892–1894
6. Chen XS, Yang J, Li SH, Li Q (2009) Disturbance observer based multi-variable control of ball mill grinding circuits. J Process Control 19(7):1205–1213
7. Yang J, Li SH, Chen XS, Li Q (2010) Disturbance rejection of ball mill grinding circuits using DOB and MPC. Powder Technol 198(2):219–228
8. Li SH, Sun HB, Sun CY (2012) Composite controller design for an airbreathing hypersonic vehicle. Proc Inst Mech Eng Part I: J Syst Control Eng 226(5):651–664
9. Sun HB, Li SH, Sun CY (2012) Finite time integral sliding mode control of hypersonic vehicles. Nonlinear Dyn 73:229–244
10. Sun HB, Li SH, Fei SM (2011) A composite control scheme for 6DOF spacecraft formation control. Acta Astronaut 69(7–8):595–611
11. Sun HB, Li SH (2013) Composite control method for stabilizing spacecraft attitude in terms of rodrigues parameters. Chin J Aeronaut 26(3):687–696
12. Chen WH (2003) Nonlinear disturbance observer-enhanced dynamic inversion control of missiles. J Guid Control Dyn 26(1):161–166
13. Guo L, Wen XY (2011) Hierarchical anti-disturbance adaptive control for non-linear systems with composite disturbances and applications to missile systems. Trans Inst Meas Control 33(8):942–956
14. Yang J, Chen WH, Li SH (2011) Non-linear disturbance observer-based robust control for systems with mismatched disturbances/uncertainties. IET Control Theory Appl 5(18): 2053–2062
15. Yang J, Li SH, Yu XH (2013) Sliding-mode control for systems with mismatched uncertainties via a disturbance observer. IEEE Trans Ind Electron 60(1):160–169
16. Yang J, Li SH, Su JY, Yu XH (2013) Continuous nonsingular terminal sliding mode control for systems with mismatched disturbances. Automatica 49(7):2287–2291
17. Levant A (2003) Higher-order sliding modes, differentiation and output-feedback control. Int J Control 76(9–10):924–941
18. Shtessel YB, Shkolnikov IA, Levant A (2007) Smooth second-order sliding modes: missile guidance application. Automatica 43(8):1470–1476
19. Bhat SP, Bernstein DS (2000) Finite-time stability of continuous autonomous systems. SIAM J Control Optim 38(3):751–766
20. Li SH, Tian YP (2007) Finite time stability of cascaded time-varying systems. Int J Control 80(4):646–657

Chapter 47
On Three-Player Potential Games

Xinyun Liu and Jiandong Zhu

Abstract In this paper, some new criteria for detecting whether a 3-player game is potential are proposed by solving potential equations. When a 3-player game is potential, the potential function is constructively expressed.

Keywords Three-player game · Potential game · Potential equation · Semi-tensor product

47.1 Introduction

Rosenthal first proposed the concept of potential game [1]. A game is said to be potential if it admits a potential function. For a potential game, the Nash equilibrium problem can be transformed into the maximum problem of the potential function. An natural problem is how to check whether a game is potential. Monderer and Shapley proposed necessary and sufficient conditions for potential games [2]. But it is required to verify all the simple closed paths with length 4 for any pair of players. Hino gave an improved condition for detecting potential games [3], which has a lower complexity than that of [2] due to that only the adjacent pairs of strategies of two players need to check. Game decomposition is also an important method for potential games [4, 5] and some new necessary and sufficient conditions are obtained for detecting whether a finite game is potential. In [6], some testing equations are proposed to verify potential games, but the number of the obtained verification equations is not the minimum.

Recently, Cheng developed a semi-tensor product method to deal with games including potential games, networked games and evolutionary games [7–10]. In [7], a linear system, called potential equation, is proposed, and then it is proved that the game is potential if and only if the potential equation is solvable. With a solution of the potential equation, the potential function can be directly calculated.

X. Liu · J. Zhu (✉)
Institute of Mathematics, School of Mathematical Sciences,
Nanjing Normal University, Nanjing 210023, People's Republic of China
e-mail: zhujiandong@njnu.edu.cn

© Springer-Verlag Berlin Heidelberg 2016

Y. Jia et al. (eds.), *Proceedings of the 2015 Chinese Intelligent Systems Conference*, Lecture Notes in Electrical Engineering 359,
DOI 10.1007/978-3-662-48386-2_47

A natural question is how to establish the connection between the potential equation proposed by Cheng and the other criteria of potential games. Moreover, an interesting problem is how to get the verification equations with the minimum number. In this paper, we investigate the solvability of the potential equation for three-player games. An equivalence transformation is constructed to convert the augmented matrix of the potential equation into a row echelon form. Based on this technique, some new necessary and sufficient conditions for potential games are obtained. For potential games, a new formula to calculate potential functions is proposed. Based on the obtained results, it is revealed the connection between the potential equation and the other results on potential games.

Throughout the paper, denote the $k \times k$ identity matrix by I_k, the ith column of I_k by δ_k^i, the n-dimensional column vector whose entries are all equal to 1 by $\mathbf{1}_k$, Kronecker product by $\otimes$ and the real number field by $\mathbb{R}$.

47.2 Preliminaries

Definition 1 (see [2]) A 3-payer finite game is a triple $\mathcal{G} = (\mathcal{N}, S, C)$, where

(i) $\mathcal{N} = \{1, 2, 3\}$ is the set of 3 players;
(ii) $S = S_1 \times S_2 \times S_3$ is the strategy set, where $S_1 = \{s_1^1, s_2^1, \dots, s_{k_1}^1\}$, $S_2 = \{s_1^2, s_2^2, \dots, s_{k_2}^2\}$ and $S_3 = \{s_1^3, s_2^3, \dots, s_{k_3}^3\}$ are the strategy sets of players 1, 2 and 3 respectively;
(iii) $C = \{c_1, c_2, c_3\}$ is the set of payoff functions, where every $c_i : S \to \mathbb{R}$ is the payoff function of players 1, 2 and 3 respectively.

Definition 2 (see [2]) A 3- player finite game $\mathcal{G} = (\mathcal{N}, S, C)$ is said to be potential if there exists a function $p : S \to \mathbb{R}$, called the potential function, such that

$$c_1(x_1, s_2, s_3) - c_1(y_1, s_2, s_3) = p(x_1, s_2, s_3) - p(y_1, s_2, s_3) \tag{47.1}$$
$$c_2(s_1, x_2, s_3) - c_2(s_1, y_2, s_3) = p(s_1, x_2, s_3) - p(s_1, y_2, s_3) \tag{47.2}$$
$$c_3(s_1, s_2, x_3) - c_3(s_1, s_2, y_3) = p(s_1, s_2, x_3) - p(s_1, s_2, y_3) \tag{47.3}$$

hold for all $x_1, y_1, s_1 \in S_1$, $x_2, y_2, s_2 \in S_2$ and $x_3, y_3, s_3 \in S_3$.

Lemma 1 *A 3-player finite game $\mathcal{G} = (\mathcal{N}, S, C)$ is potential if and only if there exist functions $d_1 : S_2 \times S_3 \to \mathbb{R}$, $d_2 : S_1 \times S_3 \to \mathbb{R}$, $d_3 : S_2 \times S_3 \to \mathbb{R}$, and $p : S \to \mathbb{R}$ such that*

$$c_1(x_1, x_2, x_3) = p(x_1, x_2, x_3) + d_1(x_2, x_3), \tag{47.4}$$
$$c_2(x_1, x_2, x_3) = p(x_1, x_2, x_3) + d_2(x_1, x_3), \tag{47.5}$$
$$c_3(x_1, x_2, x_3) = p(x_1, x_2, x_3) + d_3(x_1, x_2) \tag{47.6}$$

hold for all $x_1 \in S_1$, $x_2 \in S_2$, and $x_3 \in S_3$, where $p(x_1, x_2, x_3)$ is the potential function.

Definition 3 (see [7]) Assume $A \in \mathbb{R}^{m\times n}$, $B \in \mathbb{R}^{p\times q}$. Let $\alpha = \operatorname{lcm}(n, p)$ be the least common multiple of n and p. The left semi-tensor product of A and B is defined as $A \ltimes B = (A \otimes I_{\frac{\alpha}{n}})(B \otimes I_{\frac{\alpha}{p}})$.

Since the left semi-tensor product is a generalization of the traditional matrix product, the left semi-tensor product $A \ltimes B$ can be directly written as AB. Identifying each strategy s^i_j with the logical vector $\delta^j_{k_i}$ for $i = 1, 2, \ldots, n$ and $j = 1, 2, \ldots, k_i$, Cheng gave a new expression of the payoff functions shown in the following lemma using the left semi-tensor product [8].

Lemma 2 *(see [8]) Let $x_i \in S_i$ be any strategy expressed in the form of logical vectors. Then, for any payoff function c_i of a finite game $\mathcal{G}$ shown in Definition 1, there exists a unique row vector $V^c_i \in \mathbb{R}^{k_1k_2k_3}$ such that*

$$c_i(x_1, x_2, x_3) = V^c_i x_1 x_2 x_3, \tag{47.7}$$

where V^c_i is called the structure vector of c_i and $i = 1, 2, \ldots, n$.

In [8], the potential equation is proposed as follows:

$$\begin{bmatrix} -\Psi_1 & \Psi_2 & 0 \\ -\Psi_1 & 0 & \Psi_3 \end{bmatrix} \begin{bmatrix} \xi_1 \\ \xi_2 \\ \xi_3 \end{bmatrix} = \begin{bmatrix} (V^c_2 - V^c_1)^{\mathrm{T}} \\ (V^c_3 - V^c_1)^{\mathrm{T}} \end{bmatrix} \tag{47.8}$$

where $\Psi_1 = \mathbf{1}_{k_1} \otimes I_{k_2k_3}, \Psi_2 = I_{k_1} \otimes \mathbf{1}_{k_2} \otimes I_{k_3}, \Psi_3 = I_{k_1k_2} \otimes \mathbf{1}_{k_3}$

Theorem 1 *(see [10]) A finite game $\mathcal{G}$ shown in Definition 1 is a potential game if and only if the potential Eq. (47.6) has a solution ξ. Moreover, as (47.6) holds, the potential function p can be calculated by*

$$(V^p)^{\mathrm{T}} = (V^c_1)^{\mathrm{T}} - (\mathbf{1}_{k_1} \otimes I_{k_2k_3})\xi_1, \tag{47.9}$$

where V^p is the structure vector of p.

47.3 Main Results

Our main method is transforming the potential equation into a row echelon form. For constructing the transformation matrix, we first give a useful lemma as follows:

Lemma 3 *Let $B_k = [I_{k-1} \ \ -\mathbf{1}_{k-1}] \in \mathbb{R}^{(k-1)\times k}$ and $D_k = [I_{k-1} \ \ 0] \in \mathbb{R}^{(k-1)\times k}$. Then $B_k\mathbf{1}_k = D_k\delta^k_k = 0$, $B_kD^{\mathrm{T}}_k = D_kD^{\mathrm{T}}_k = I_{k-1}$, $I_k = D^{\mathrm{T}}_kB_k + \mathbf{1}_k(\delta^k_k)^{\mathrm{T}}$ and*

$$D_{k_1}^{\mathrm{T}} B_{k_1} \otimes \mathbf{1}_{k_2} (\delta_{k_2}^{k_2})^{\mathrm{T}} + I_{k_1} \otimes D_{k_2}^{\mathrm{T}} B_{k_2} = D_{k_1 k_2}^{\mathrm{T}} B_{k_1 k_2}. \tag{47.10}$$

Proof The equalities expect (47.10) are obvious. Moreover,

$$\begin{aligned} & D_{k_1}^{\mathrm{T}} B_{k_1} \otimes \mathbf{1}_{k_2} (\delta_{k_2}^{k_2})^{\mathrm{T}} + I_{k_1} \otimes D_{k_2}^{\mathrm{T}} B_{k_2} \\ &= (I_{k_1} - \mathbf{1}_{k_1} (\delta_{k_1}^{k_1})^{\mathrm{T}}) \otimes \mathbf{1}_{k_2} (\delta_{k_2}^{k_2})^{\mathrm{T}} + I_{k_1} \otimes D_{k_2}^{\mathrm{T}} B_{k_2} \\ &= I_{k_1} \otimes \mathbf{1}_{k_2} (\delta_{k_2}^{k_2})^{\mathrm{T}} - \mathbf{1}_{k_1 k_2} (\delta_{k_1 k_2}^{k_1 k_2})^{\mathrm{T}} + I_{k_1} \otimes D_{k_2}^{\mathrm{T}} B_{k_2} \\ &= I_{k_1 k_2} - \mathbf{1}_{k_1 k_2} (\delta_{k_1 k_2}^{k_1 k_2})^{\mathrm{T}} \\ &= D_{k_1 k_2}^{\mathrm{T}} B_{k_1 k_2}. \end{aligned} \tag{47.11}$$

Theorem 2 *Consider the 3-player finite game $\mathcal{G} = (\mathcal{N}, \mathcal{S}, \mathcal{C})$ described in Definition 1. Let $B_k = [I_{k-1} \ \ \mathbf{1}_{k-1}]$. Then $\mathcal{G}$ is a potential game if and only if one of the following statements holds:*

(i)

$$\begin{bmatrix} B_{k_1} \otimes B_{k_2} \otimes (\delta_{k_3}^{k_3})^T & -B_{k_1} \otimes B_{k_2} \otimes (\delta_{k_3}^{k_3})^T \\ B_{k_1} \otimes I_{k_2} \otimes B_{k_3} & 0 \\ 0 & I_{k_1} \otimes B_{k_2} \otimes B_{k_3} \end{bmatrix} \begin{bmatrix} (V_3^c - V_1^c)^T \\ (V_3^c - V_2^c)^T \end{bmatrix} = 0; \tag{47.12}$$

(ii)

$$\begin{cases} (B_{k_1} \otimes B_{k_2} \otimes (\delta_{k_3}^{k_3})^T)(V_2^c - V_1^c)^T = 0, \\ (I_{k_1} \otimes B_{k_2} \otimes B_{k_3})(V_3^c - V_2^c)^T = 0, \\ (B_{k_1} \otimes I_{k_2} \otimes B_{k_3})(V_3^c - V_1^c)^T = 0; \end{cases} \tag{47.13}$$

(iii)

$$\begin{cases} (B_{k_1} \otimes B_{k_2} \otimes I_{k_3})(V_2^c - V_1^c)^T = 0, \\ (I_{k_1} \otimes B_{k_2} \otimes B_{k_3})(V_3^c - V_2^c)^T = 0, \\ (B_{k_1} \otimes I_{k_2} \otimes B_{k_3})(V_3^c - V_1^c)^T = 0. \end{cases} \tag{47.14}$$

Proof Multiplying (47.8) on the left by $\begin{bmatrix} 0 & 1 \\ -1 & 1 \end{bmatrix} \otimes I_{k_1 k_2 k_3}$, we equivalently transform the potential Eq. (47.8) into

$$\begin{bmatrix} -\Psi_1 & 0 & \Psi_3 \\ 0 & -\Psi_2 & \Psi_3 \end{bmatrix} \begin{bmatrix} \xi_1 \\ \xi_2 \\ \xi_3 \end{bmatrix} = \begin{bmatrix} (V_3^c - V_1^c)^{\mathrm{T}} \\ (V_3^c - V_2^c)^{\mathrm{T}} \end{bmatrix}. \tag{47.15}$$

By Theorem 1, we only need to consider the solvability of (47.15). Construct a transformation matrix as follows

$$T = \begin{bmatrix} -(\delta_{k_1}^{k_1})^{\mathrm{T}} \otimes I_{k_2k_3} & 0 \\ 0 & -I_{k_1} \otimes (\delta_{k_2}^{k_2})^{\mathrm{T}} \otimes I_{k_3} \\ D_{k_1k_2}(D_{k_1}^{\mathrm{T}} B_{k_1} \otimes \mathbf{1}_{k_2} (\delta_{k_2}^{k_2})^{\mathrm{T}} \otimes (\delta_{k_3}^{k_3})^{\mathrm{T}}) & D_{k_1k_2}(I_{k_1} \otimes D_{k_1}^{\mathrm{T}} B_{k_1} \otimes (\delta_{k_3}^{k_3})^{\mathrm{T}}) \\ B_{k_1} \otimes B_{k_2} \otimes (\delta_{k_3}^{k_3})^{\mathrm{T}} & -B_{k_1} \otimes B_{k_2} \otimes (\delta_{k_3}^{k_3})^{\mathrm{T}} \\ B_{k_1} \otimes I_{k_2} \otimes B_{k_3} & 0 \\ 0 & I_{k_1} \otimes B_{k_2} \otimes B_{k_3} \end{bmatrix}, \quad (47.16)$$

where B_k and D_k are defined in Lemma 3. It is easy to check that T is a $2k_1k_2k_3$-dimensional square matrix. We first prove that T is nonsingular. Let

$$S^{\mathrm{T}} = \begin{bmatrix} -\mathbf{1}_{k_1}^{\mathrm{T}} \otimes I_{k_2k_3} & 0 \\ 0 & -I_{k_1} \otimes \mathbf{1}_{k_2}^{\mathrm{T}} \otimes I_{k_3} \\ D_{k_1k_2}(B_{k_1}^{\mathrm{T}} D_{k_1} \otimes I_{k_2} \otimes \mathbf{1}_{k_3}^{\mathrm{T}}) & D_{k_1k_2}(I_{k_1} \otimes B_{k_2}^{\mathrm{T}} D_{k_2} \otimes \mathbf{1}_{k_3}^{\mathrm{T}}) \\ D_{k_1} \otimes D_{k_2} \otimes \mathbf{1}_{k_3}^{\mathrm{T}} & 0 \\ D_{k_1} \otimes I_{k_2} \otimes D_{k_3} & 0 \\ 0 & I_{k_1} \otimes D_{k_2} \otimes D_{k_3} \end{bmatrix}. \quad (47.17)$$

By Lemma 3, it is easy to check that $TS = I_{2k_1k_2k_3}$, which implies that T is nonsingular. Multiplying (47.15) on the left by T, we get the equivalent equation of (47.15) as follows:

$$\begin{bmatrix} I_{k_2k_3} & 0 & -(\delta_{k_1}^{k_1})^{\mathrm{T}} \otimes I_{k_2} \otimes \mathbf{1}_{k_3} \\ 0 & I_{k_1k_3} & -I_{k_1} \otimes (\delta_{k_2}^{k_2})^{\mathrm{T}} \otimes \mathbf{1}_{k_3} \\ 0 & 0 & B_{k_1k_2} \\ 0 & 0 & 0 \end{bmatrix} \begin{bmatrix} \xi_1 \\ \xi_2 \\ \xi_3 \end{bmatrix} = T \begin{bmatrix} (V_3^c - V_1^c)^{\mathrm{T}} \\ (V_3^c - V_2^c)^{\mathrm{T}} \end{bmatrix}. \quad (47.18)$$

Considering the coefficient matrix of (47.18) is in a row echelon form, we know that (47.18) is solvable with respect to ξ if and only if (47.12) holds. Moreover, the equivalence of (47.12) and (47.13) is evident. To show the equivalence of (47.13) and (47.14), we only need to prove the first equation of (47.14) by virtue of (47.13). Assume that the equations in (47.13) hold. Then

$$\begin{aligned}
&(B_{k_1} \otimes B_{k_2} \otimes I_{k_3})(V_2^c - V_1^c)^{\mathrm{T}} \\
&= (B_{k_1} \otimes B_{k_2} \otimes (D_{k_3}^{\mathrm{T}} B_{k_3} + \mathbf{1}_{k_3}(\delta_{k_3}^{k_3})^{\mathrm{T}}))(V_2^c - V_1^c)^{\mathrm{T}} \\
&= (B_{k_1} \otimes B_{k_2} \otimes D_{k_3}^{\mathrm{T}} B_{k_3})(V_3^c - V_1^c)^{\mathrm{T}} \\
&\quad - (B_{k_1} \otimes B_{k_2} \otimes D_{k_3}^{\mathrm{T}} B_{k_3})(V_3^c - V_2^c)^{\mathrm{T}} \\
&\quad + (B_{k_1} \otimes B_{k_2} \otimes \mathbf{1}_{k_3}(\delta_{k_3}^{k_3})^{\mathrm{T}}))(V_2^c - V_1^c)^{\mathrm{T}} \\
&= (I_{k_1} \otimes B_{k_2} \otimes D_{k_3}^{\mathrm{T}})(B_{k_1} \otimes I_{k_2} \otimes B_{k_3})(V_3^c - V_1^c)^{\mathrm{T}} \\
&\quad - (B_{k_1} \otimes I_{k_2} \otimes D_{k_3}^{\mathrm{T}})(I_{k_1} \otimes B_{k_2} \otimes B_{k_3})(V_3^c - V_2^c)^{\mathrm{T}} \\
&\quad + (I_{k_1 k_2} \otimes \mathbf{1}_{k_3})(B_{k_1} \otimes B_{k_2} \otimes (\delta_{k_3}^{k_3})^{\mathrm{T}})(V_2^c - V_1^c)^{\mathrm{T}} \\
&= 0.
\end{aligned} \tag{47.19}$$

So the proof is completed.

Theorem 3 *If the 3-player finite game $\mathcal{G} = (\mathcal{N},\ \mathcal{S},\ C)$ described in Definition 1 is potential, then the potential function is given by*

$$p(x_1, x_2, x_3) = V^p x_1 x_2 x_3, \tag{47.20}$$

where

$$\begin{aligned}
(V^p)^{\mathrm{T}} &= (V_1^c)^{\mathrm{T}} + (\mathbf{1}_{k_1}(\delta_{k_1}^{k_1})^{\mathrm{T}} \otimes I_{k_2 k_3})(V_3^c - V_1^c)^{\mathrm{T}} \\
&\quad - (\mathbf{1}_{k_1}(\delta_{k_1}^{k_1})^{\mathrm{T}} \otimes D_{k_2}^{\mathrm{T}} B_{k_2} \otimes \mathbf{1}_{k_3}(\delta_{k_3}^{k_3})^{\mathrm{T}})(V_3^c - V_2^c)^{\mathrm{T}} \\
&\quad + c\mathbf{1}_{k_1 k_2 k_3}.
\end{aligned} \tag{47.21}$$

Proof The third equation of (47.18) is just

$$\begin{aligned}
[I_{k_1 k_2 - 1} \quad -\mathbf{1}_{k_1 k_2 - 1}]\xi_3 &= D_{k_1 k_2}(D_{k_1}^{\mathrm{T}} B_{k_1} \otimes \mathbf{1}_{k_2}(\delta_{k_2}^{k_2})^{\mathrm{T}} \otimes (\delta_{k_3}^{k_3})^{\mathrm{T}})(V_3^c - V_1^c)^{\mathrm{T}} \\
&\quad + D_{k_1 k_2}(I_{k_1} \otimes D_{k_1}^{\mathrm{T}} B_{k_1} \otimes (\delta_{k_3}^{k_3})^{\mathrm{T}})(V_3^c - V_2^c)^{\mathrm{T}}.
\end{aligned} \tag{47.22}$$

Solving the above equation, we have

$$\begin{aligned}
\xi_3 &= (D_{k_1}^{\mathrm{T}} B_{k_1} \otimes \mathbf{1}_{k_2}(\delta_{k_2}^{k_2})^{\mathrm{T}} \otimes (\delta_{k_3}^{k_3})^{\mathrm{T}})(V_3^c - V_1^c)^{\mathrm{T}} \\
&\quad + (I_{k_1} \otimes D_{k_1}^{\mathrm{T}} B_{k_1} \otimes (\delta_{k_3}^{k_3})^{\mathrm{T}})(V_3^c - V_2^c)^{\mathrm{T}} - c\mathbf{1}_{k_1 k_2},
\end{aligned} \tag{47.23}$$

where c is an arbitrary constant. Then, from the first equation of (47.18), it follows that

$$
\begin{aligned}
\xi_1 &= -((\delta_{k_1}^{k_1})^{\mathrm{T}} \otimes I_{k_2k_3})(V_3^c - V_1^c)^{\mathrm{T}} + ((\delta_{k_1}^{k_1})^{\mathrm{T}} \otimes I_{k_2} \otimes \mathbf{1}_{k_3})\xi_3 \\
&= ((\delta_{k_1}^{k_1})^{\mathrm{T}} \otimes I_{k_2k_3})(V_3^c - V_1^c)^{\mathrm{T}} \\
&\quad + ((\delta_{k_1}^{k_1})^{\mathrm{T}} \otimes D_{k_1}^{\mathrm{T}} B_{k_1} \otimes \mathbf{1}_{k_3}(\delta_{k_3}^{k_3})^{\mathrm{T}})(V_3^c V_2^c)^{\mathrm{T}} - c\mathbf{1}_{k_2k_3}.
\end{aligned}
\tag{47.24}
$$

Substituting ξ_1 into (47.9) yields (47.21).

Remark 1 Theorem 2 gives new necessary and sufficient conditions for finite potential games. The number of the equations in (47.12) is minimal. The equations in (47.14) establish the connection between the potential equation and the four-cycle conditions proposed in [2] and [6].

47.4 An Example

Consider a three-player finite game $\mathcal{G}$ with $k_1 = k_3 = 2$, $k_2 = 3$ and payoff matrix $C = (c_{i_1i_2i_3}^{\mu})$. Let the relative payoff matrix be

$$
R = \begin{bmatrix}
r_{111}^1 & r_{112}^1 & r_{121}^1 & r_{122}^1 & r_{131}^1 & r_{132}^1 & r_{211}^1 & r_{212}^1 & r_{221}^1 & r_{222}^1 & r_{231}^1 & r_{232}^1 \\
r_{111}^2 & r_{112}^2 & r_{121}^2 & r_{122}^2 & r_{131}^2 & r_{132}^2 & r_{211}^2 & r_{212}^2 & r_{221}^2 & r_{222}^2 & r_{231}^2 & r_{232}^2
\end{bmatrix},
$$

where each $r_{i_1i_2i_3}^{\mu} = c_{i_1i_2i_3}^3 - c_{i_1i_2i_3}^{\mu}$ for $\mu = 1, 2$. A computation shows that the coefficient matrix of (47.12) is

$$
\begin{bmatrix}
0 & 1 & 0 & 0 & 0 & -1 & 0 & -1 & 0 & 0 & 0 & 1 & 0 & -1 & 0 & 0 & 0 & 1 & 0 & 1 & 0 & 0 & 0 & -1 \\
0 & 0 & 0 & 1 & 0 & -1 & 0 & 0 & 0 & -1 & 0 & 1 & 0 & 0 & 0 & -1 & 0 & 1 & 0 & 0 & 0 & 1 & 0 & -1 \\
1 & -1 & 0 & 0 & 0 & 0 & -1 & 1 & 0 & 0 & 0 & 0 & 0 & 0 & 0 & 0 & 0 & 0 & 0 & 0 & 0 & 0 & 0 & 0 \\
0 & 0 & 1 & -1 & 0 & 0 & 0 & 0 & -1 & 1 & 0 & 0 & 0 & 0 & 0 & 0 & 0 & 0 & 0 & 0 & 0 & 0 & 0 & 0 \\
0 & 0 & 0 & 0 & 1 & -1 & 0 & 0 & 0 & 0 & -1 & 1 & 0 & 0 & 0 & 0 & 0 & 0 & 0 & 0 & 0 & 0 & 0 & 0 \\
0 & 0 & 0 & 0 & 0 & 0 & 0 & 0 & 0 & 0 & 0 & 0 & 1 & -1 & 0 & 0 & -1 & 1 & 0 & 0 & 0 & 0 & 0 & 0 \\
0 & 0 & 0 & 0 & 0 & 0 & 0 & 0 & 0 & 0 & 0 & 0 & 0 & 0 & 1 & -1 & -1 & 1 & 0 & 0 & 0 & 0 & 0 & 0 \\
0 & 0 & 0 & 0 & 0 & 0 & 0 & 0 & 0 & 0 & 0 & 0 & 0 & 0 & 0 & 0 & 0 & 0 & 1 & -1 & 0 & 0 & -1 & 1 \\
0 & 0 & 0 & 0 & 0 & 0 & 0 & 0 & 0 & 0 & 0 & 0 & 0 & 0 & 0 & 0 & 0 & 0 & 0 & 0 & 1 & -1 & -1 & 1
\end{bmatrix}.
$$

Thus, by Theorem 2, the game is a potential game if and only if

$$\begin{cases} r^1_{112}-r^1_{132}-r^1_{212}+r^1_{232}-r^2_{112}+r^2_{132}+r^2_{212}-r^2_{232}=0,\\ r^1_{122}-r^1_{132}-r^1_{222}+r^1_{232}-r^2_{122}+r^2_{132}+r^2_{222}-r^2_{232}=0,\\ r^1_{111}-r^1_{112}-r^1_{211}+r^1_{212}=0,\\ r^1_{121}-r^1_{122}-r^1_{221}+r^1_{222}=0,\\ r^1_{131}-r^1_{132}-r^1_{231}+r^1_{232}=0,\\ r^2_{111}-r^2_{112}-r^2_{131}+r^2_{132}=0,\\ r^2_{121}-r^2_{122}-r^2_{131}+r^2_{132}=0,\\ r^2_{211}-r^2_{212}-r^2_{231}+r^2_{232}=0,\\ r^2_{221}-r^2_{222}-r^2_{231}+r^2_{232}=0. \end{cases}$$

The minimal number of equations for detecting whether $\mathcal{G}$ is potential is 9.

47.5 Conclusions

For detecting whether a three-player game is potential, new necessary and sufficient conditions have been obtained by investigating the potential equations. The number of the obtained verification equalities is minimal. The connections between the potential equations and the existing results on potential games have been revealed.

References

1. Rosenthal RW (1973) A class of games possessing pure-strategy Nash equilibria. Int J Game Theory 2:65–67
2. Monderer D, Shapley LS (1996) Potential games. Games Econ Behav 14:124–143
3. Hino Y (2011) An improved algorithm for detecting potential games. Int J Game Theory 40(1):199–205
4. Candogan O, Menache I, Ozdaglar A, Parrilo PA (2011) Flows and decompositions of games: harmonic and potential games. Math Oper Res 36(3):474–503
5. Hwang S-H, Rey-Bellet L (2011) Decompositions of two player games: potential, zero-sum, and stable games. arXiv:1106.3552v2
6. Sandholm WH (2010) Decompositions and potentials for normal form games. Games and Econ Behav 70:446–456
7. Cheng D (2014) On finite potential games. Automatica 50(7):1793–1801
8. Cheng D, Xu T, Qi H (2014) Evolutionarily stable strategy of networked evolutionary games. IEEE Trans Neural Networks Learn Syst 25(7):1335–1345
9. Cheng D, Xu T, He F, Qi H (2014) On dynamics and Nash equilibriums of networked games. IEEE/CAA J Autom Sinica 1(1):10–18
10. Cheng D, He F, Qi H, Xu T (2015) Modeling, analysis and control of networked evolutionary games. IEEE Trans Autom Control. doi:10.1109/TAC.2015.2404471

Chapter 48
Flocking of Multi-Agent Systems with Multiple Virtual Leaders Based on Connectivity Preservation Approach

Zonggang Li, Chen Na and Guangming Xie

Abstract In this paper, we investigate the problem of flocking with multiple virtual leader based on connectivity preserving. The basic idea is stated as follows: according to design stable control law with a navigation feedback term the follow agents equipped with virtual leaders. The velocity of the center of the mass of agents will exponentially converge to weighted average velocity of the virtual leaders by the information navigation of virtual leaders, finally approach to the same. The certain distance between every agent is kept and eventually form flocking due to the introduction of the artificial potential function. And on this basis we assume that the initial network is connected and the control law make the network of the multi-agent system preserving connectivity.

Keywords Multi-agent · Flocking · Multiple virtual leaders · Connectivity preservation approach

48.1 Introduction

Flocking widely exists in nature. Biologists pay attention to the coordination mechanism between individuals in the group of Flocking [1, 2]. At the same time, the flocking of multi-agent system is also important in engineering field. Such as application of sense networks, robot team control and Unmanned Air Vehicles [3], all of this reflect the important projects application value of flocking. The classical flocking model was proposed from the act of bird flight in 1987s by Reynolds [1] form of three rules: (1) Separation: avoid collision with nearby agents; (2) Alignment:

Z. Li (✉) · C. Na
School of Mechatronic Engineering, Lanzhou Jiaotong University,
Lanzhou 730070, China
e-mail: lizongg@126.com

Z. Li · G. Xie
Intelligent Control Laboratory, College of Engineering, Peking University,
Beijing 100871, China

© Springer-Verlag Berlin Heidelberg 2016

Y. Jia et al. (eds.), *Proceedings of the 2015 Chinese Intelligent Systems Conference*, Lecture Notes in Electrical Engineering 359,
DOI 10.1007/978-3-662-48386-2_48

attempt to match velocity with nearby agents; (3) Cohesion: attempt stay close to nearby agents. Gazi and Passion [4] proposed a kind of cluster model by using the method of artificial potential function and studied stability. In 2006, Olfati Saber [5] design a decentralized control strategy with three rules of Reynolds to make the agent group of agents flock and avoid collision with obstacles. At the same time a flocking algorithm with a single virtual leader was put forward. Su [6, 8] proposed flocking with multiple virtual leaders based on connectivity preserving.

In this paper, we are interesting in the problem of flocking system with multiple virtual leaders based on the topological graph whose connectivity is preserved. Assume that the initial network topology of flocking system is connected and followers movement under the leading of their respective corresponding virtual leaders. Control law of agents is modeled by artificial potential function. The algorithm proposed in this paper is illustrated by the theoretical tools and matrix analysis theory. Group of agents leaded by their corresponding leader agents moving in the same speed under the connected initial network.

The content of this paper is organized as follows. In Sect. 48.2, preliminaries knowledge are introduced. Section 48.3 describes the dynamics modeling and problem statement of flocking system. Main results is given in Sect. 48.4. Section 48.5 is conclusion.

48.2 Graph Theory

Consider a system which contains N individuals of multi-agents. The relationship between every agent is descripted by information exchange topology, marked as $G = (V, E, A)$, which consisting of a set of vertices $V = \{v_1, v_2, \ldots, v_N\}$, whose elements represent agents in the group, and a set of edges $E = \left\{ (v_i, v_j)| \quad i, j = 1, \ldots, N \right\}$ means that there is an edge which began with v_i ended in v_j. We define adjacent matric $A = [a_{ij}]$, if $a_{ij} > 0$, then $(v_i, v_j) \in E$, otherwise $a_{ij} = 0$. The neighboring set of agent v_i is denoted as $N_i = \{v_j|(v_j, v_i) \in E, j = 1, \ldots, N\}$. Define $d_i = \sum_{j=1}^{N} a_{ij}$ is out-degree of the vertices v_i, and $D = diag\{d_1, \ldots, d_N\}$ is out-degree matrix of graph G, Laplacian matrix $L = D - A$. The path in G is a series of end to end edge such as $(v_i, v_j), (v_j, v_l), \ldots, (v_m, v_n)$. Graph contains a spanning tree if there is at least one vertex which has at least one path between other vertexes, and called vertex is node. If this entire vertexes are nodes, denote G is strongly connected. G is undirected if without spanning tree. If there is a path between vertexes v_i to v_j in graph $G(V, E)$, called v_j is reachable from v_i, otherwise unreachable. Assume $x_i \in R^n$ is state variable and $y_i \in R^m$ is output variable. When $t \to \infty$, if $x_i \to x_j$, then called the system is solvable state consensus problem; if $y_i \to y_j$, then called the system is solvable output consensus problem. Give a fixed point y_0; when $t \to \infty$, if $y_i \to y_0$, then called the system is solvable output tracking problem and definition of state tracking is similar, $i, j = 1, 2, \ldots, N$.

48.3 Dynamics Model and Problem Statement of Flocking System

48.3.1 Dynamics Model of Flocking System

In this paper, N follow agents and M virtual leader agents are considered which moving in an n-dimensional Euclidean space. The dynamic function of follow agents v_i is described by a double integrator of the form

$$\begin{array}{l} \dot{q}_i = p_i \\ \dot{p}_i = u_i \end{array}, \ i = 1, \ldots, N \tag{48.1}$$

where $q_i \in R^n$ is the position vector of agent v_i, $P_i \in R^n$ is its velocity vector and $u_i \in R^n$ is control input. The dynamic function of virtual leaders γ_i is described by a double integrator of the form

$$\begin{array}{l} \dot{q}_{\gamma_i} = p_{\gamma_i} \\ \dot{p}_{\gamma_i} = f_{\gamma_i}(q_{\gamma_i}, P_{\gamma_i}) \end{array}, \ \gamma_j = 1, \ldots, M \tag{48.2}$$

where $q_{\gamma_j}, p_{\gamma_j}, f_{\gamma_j}(q_{\gamma_j}, P_{\gamma_j}) \in R^n$ is the position vector, velocity vector and accelerate vector of the virtual leader which is tracked by agent v_i.

In this paper, the problem of control input is considered which make the velocity of every agent reach to the velocity of tracked leader. Every agent can only detect the velocity of virtual leader which is tracked by agent v_i, and the neighborhood region for agent v_i is defined as:

$$N_i(t) = \{j | \sigma(i,j)[t] = 1, j \neq i, j = 1, \ldots, N\}$$

Suppose that all agents own the same sensing radius r and the network which combined with node and edge E can be described by an undirected graph $G(t)$. The communication topological edge between the agent v_i according to the following rules [7, 8]:

(i). The initial links are generated by

$$E(0) = \left\{(v_i, v_j) | \parallel q_i(0) - q_j(0) \parallel < r, \ i,j = 1, \ldots, N.\right\};$$

(ii). If $(v_i, v_j) \notin E$ at t^-, but $(v_i, v_j) \in E$ at t, then exist information interaction between agent v_i and v_j at time t;

(iii). If $\parallel q_i(t) - q_j(t) \parallel \geq r$ at time t, then $(v_i, v_j) \notin E$.

Define a symmetric function $\sigma(i,j)$ to describe exist or without a link between agent v_i and v_j at time t; which is denoted as

$$\sigma(i,j)[t] = \begin{cases} 0, \; if \; \sigma(i,j)[t^-] = 0 \bigcap r - \epsilon \leq \| q_i(t) \\ \quad -q_j(t) \| < r \bigcup \| q_i(t) - q_j(t) \| \geq r \\ 1, \; if \; \sigma(i,j)[t^-] = 1 \bigcap r - \epsilon \leq \| q_i(t) \\ \quad -q_j(t) \| < r \bigcup \| q_i(t) - q_j(t) \| < r - \epsilon \end{cases} \tag{48.3}$$

We can see that there is hysteresis in the process of adding new links when the distant between two agents less than field radius.

In this paper, the purpose of control law that we proposed is to make the velocity of all agents to gradually converge to the velocity of their corresponding virtual leaders, and maintain an anticipant formation and a connected network by the guidance of the virtual leaders.

48.3.2 Problem Statement

We define the controller of following agents as:

$$u_i = \alpha_i + \beta_i + \gamma_i \tag{48.4}$$

where α_i is the potential gradient and the function of it is to make all agent reach expected formation and to preserve connected network; the second component β_i is velocity consensus term, which drives agents to reach the same velocity with their corresponding virtual leaders; the last component γ_i is navigational feedback term, which make agent v_i to track their corresponding virtual leaders. Then the following terms hold:

$$\begin{aligned} \alpha_i &= -\sum_{j \in N_i(t)}^{N} \nabla_{q_i} \Psi\left(\left\| q_{ij} - q_{\gamma_{ij}} \right\|\right) \\ \beta_i &= -\sum_{j \in N_i(t)} w_{ij}(p_{ij} - p_{\gamma_{ij}}) \\ \gamma_i &= -c_1(p_i - p_{\gamma_i}) + f_{\gamma_i}(q_{\gamma_i}, p_{\gamma_i}) \end{aligned} \tag{48.5}$$

where $q_{ij} = q_i - q_j, q_{\gamma_{ij}} = q_{\gamma_i} - q_{\gamma_j}$, denote $c_1 > 0$ is weights of the navigational feedback, weight $w_{ij} = w_{ji} > 0$, and p_γ is the velocity of virtual leader. We define the total energy is consist of the sum of the total relative artificial potential energy between every follow agent and the total relative kinetic energy between all following agents and their corresponding virtual leaders as following:

$$Q(q,p) = \frac{1}{2} \sum_{i=1}^{N} \left(\sum_{j \in N_i} \Psi\left(\left\| q_{ij} - q_{\gamma_{ij}} \right\|\right) + (p_i - p_{\gamma_i})^T (p_i - p_{\gamma_i}) \right) \tag{48.6}$$

where $q = [q_1^T, \ldots, q_n^T]^T \in R^{nN}, p = [p_1^T, \ldots, p_n^T]^T \in R^{nN}$, $Q = Q(q, p, q_\gamma, p_\gamma)$. Obviously, Q is a positive semi-definite function. When $\left\| q_{ij} - q_{\gamma_{ij}} \right\| \in [0, r]$, artificial potential function $\Psi(\left\| q_{ij} - q_{\gamma_{ij}} \right\|)$ is derivative, and the definition follow as:

(i) If $\left\| q_{ij} - q_{\gamma_{ij}} \right\| \to r$, then the artificial potential function $\frac{\partial \Psi(\left\| q_{ij} - q_{\gamma_{ij}} \right\|)}{\partial \left\| q_{ij} - q_{\gamma_{ij}} \right\|} > 0$;

(ii) $\lim\limits_{\left\| q_{ij} - q_{\gamma_{ij}} \right\| \to 0} \left[\frac{\partial \Psi(\left\| q_{ij} - q_{\gamma_{ij}} \right\|)}{\partial \left\| q_{ij} - q_{\gamma_{ij}} \right\|} \cdot \frac{1}{\left\| q_{ij} - q_{\gamma_{ij}} \right\|} \right]$ is a nonnegative finite value;

(iii) $\Psi(r) = \dot{Q} \in [Q_{\max}, +\infty)$, and $Q_{\max} = Q_0 + M\Psi(\|r - \varepsilon\|), M = \frac{(N-2)(N-1)}{2}$.

The condition (i) indicates that the potential function is increased and it can drives agent v_i produce attraction between corresponding virtual leaders γ_i.

The condition (ii) indicates that the distance between agent v_i and its corresponding virtual leader γ_i approximate to zero, the gradient of potential function is equivalent order or higher order infinitesimal. This property describes that there is a same position between agent v_i and virtual leader γ_i.

The condition (iii) indicates that the potential function is sufficiently large if the distance between agent v_i and its corresponding virtual leader γ_i trend to neighbourhood radius r. It can guarantee the links in the network connectivity preserving. The potential function which satisfies these conditions is given by

$$\Psi(\left\| q_{ij} - q_{\gamma_{ij}} \right\|) = \frac{\left\| q_{ij} - q_{\gamma_{ij}} \right\|^2}{r - \left\| q_{ij} - q_{\gamma_{ij}} \right\| + \frac{r^2}{Q'}} \tag{48.7}$$

48.4 Main Results and Theoretical Analysis

48.4.1 Main Results

The position and velocity of the center of mass (COM) of all agents in the group are denoted as

$$\bar{q} = \frac{\sum\limits_{i=1}^{N} q_i}{N}, \bar{p} = \frac{\sum\limits_{i=1}^{N} p_i}{N} \tag{48.8}$$

and the weighted average position and velocity of virtual leaders as

$$\bar{q}_{\gamma} = \frac{\sum_{i=1}^{N} q_{\gamma_i}}{N}, \bar{p}_{\gamma} = \frac{\sum_{i=1}^{N} p_{\gamma_i}}{N}, \gamma_i \in \{1, 2, \ldots, M\} \tag{48.9}$$

Theorem 1 *Consider a system with N agents and which described by a dynamics function (1), and the control input of each agent is (4). Suppose that the initial network is connected and the initial energy* $Q_0 = Q(q(0), p(0), q_{\gamma}(0), p_{\gamma}(0))$ *is a finite. Then we can conclude the following statement:*

(i) *For anytime t, the network topology graphs G(t) preserve connectivity from beginning to end;*
(ii) *The velocity of each agent approaches the desired velocity* p_{γ} *asymptotically.*
(iii) *If the velocity of the center of mass of all agents is equal to the weighted average velocity of virtual leaders, then the velocity of the center of mass of all agents will equal to the velocity of virtual leader all the way. If the velocities of them are not equality, then the velocity of the center of mass of all agents will convergence to the weighted average velocity of virtual leaders by exponent* c_1.

48.4.2 Theoretical Analysis

Firstly prove part (1) of Theorem 1. The position difference vector and the velocity difference vector between agent v_i and their virtual leaders are respectively labeled as:

$$\begin{cases} \tilde{q}_i = q_i - q_{\gamma_i} \\ \tilde{p}_i = p_i - p_{\gamma_i} \end{cases}, i = 1, \ldots, N; \tag{48.10}$$

Then

$$\begin{cases} \dot{\tilde{q}}_i = \tilde{p}_i \\ \dot{\tilde{p}} = u_i - f_{\gamma_i}(q_{\gamma_i}, p_{\gamma_i}) \end{cases}, i = 1, \ldots, N. \tag{48.11}$$

Function (48.11) can be inferred from (48.1) and (48.2). Thus the control input (48.4) can be rewritten as

$$\begin{aligned} u_i = & - \sum_{j \in N_i(t)} \nabla_{\tilde{q}_i} \Psi(\|\tilde{q}_{ij}\|) - \sum_{j \in N_i(t)} w_{ij}(\tilde{p}_{ij}) \\ & - c_1(p_i - p_{\gamma_i}) + f_{\gamma_i}(q_{\gamma_i}, p_{\gamma_i}) \end{aligned} \tag{48.12}$$

and the energy function (48.6) can be rewritten as

$$Q(\tilde{q}, \tilde{p}) = \frac{1}{2} \sum_{i=1}^{N} \left(\sum_{j \in N_i(t)} \Psi(\|\tilde{q}_{ij}\|) \right) + \sum_{i=1}^{N} \tilde{p}_i^T \tilde{p}_i \tag{48.13}$$

where $\tilde{q} = col(\tilde{q}_1, \ldots, \tilde{q}_N), \tilde{p} = col(\tilde{p}_1, \ldots, \tilde{p}_N)$, the derivative of energy function in $[t_{k+1}, t_k)$ is

$$\begin{aligned} \dot{Q}(\tilde{q}, \tilde{p}) &= \frac{1}{2} \sum_{i=1}^{N} \sum_{j \in N_i(t)} \dot{\Psi}(\left\| \tilde{q}_{ij} \right\|) + \sum_{i=1}^{N} \tilde{p}_i^T \dot{\tilde{p}}_i \\ &= -\tilde{p}^T [L(t) + c_1 I_N) \otimes I_n] \tilde{p} \end{aligned} \tag{48.14}$$

Since $L(t)$ is the Laplacian of the graph which is positive semi-definite [9], we have $\dot{Q}(t) \leq 0$ in $[t_0, t_1)$. So in $[t_0, t_1)$ we can get

$$Q(t) \leq Q_0 \leq Q_{\max}. \tag{48.15}$$

From the definition of artificial potential function, we have $\Psi(r) \leq Q_0 \leq Q_{\max}$. So the distance of all links is not greater than r in $[t_0, t_1)$. It indicate that all links are not increasing at time t_1. The hysteresis effect makes the potential energy which produced by adding new edge is finite. Furthermore, we assume that there are m_1 edges linked into the network, then

$$Q(t_1) = Q_0 + m_1 \Psi(\|r - \varepsilon\|) \leq Q_{\max} \tag{48.16}$$

On the basis of the same method we can conclude that the derivative of energy $Q(t)$ as follows.

$$\dot{Q}(t) = -p^T [L(t) + c_1 I_N) \otimes I_n] p \leq 0, \quad t \in \left[t_{k-1}, t_k\right), \; k = 1, 2, \ldots \tag{48.17}$$

Furthermore, we have

$$Q(t) \leq Q\left(t_{k-1}\right) \leq Q_{max}, \quad t \in \left[t_{k-1}, t_k\right), \; k = 1, 2, \ldots \tag{48.18}$$

Hence the distance of all links is not trend to r when $t \in \left[t_{k-1}, t_k\right)$ $(k = 1, 2, \ldots)$. It means that all links are not interrupt at t_k. So exist add edge delay effect and $Q\left(t_k\right) \leq Q_{max}$ at t_k. $G(0)$ is connected and all initial links $E(0)$ are not interrupt, so the network of system is connected for anytime $t \geq 0$.

Secondly prove part (2) of Theorem 1. We can gain from Eq. (48.12) that $\tilde{p}^T \tilde{p} \leq 2Q_0$ for any arbitrarily agent v_i. So the velocity difference between agent v_i and its corresponding virtual leader γ_i is not greater than $\sqrt{2Q_0}$. Then, the set

$$\Omega = \left\{ [\tilde{q}^T, \tilde{p}^T]^T \in R^{2nN} | Q \leq Q_{\max} \right\} \tag{48.19}$$

is invariant. According to LaSalle invariance principle [10], every solution which date from inner of Ω can be verged to maximum invariant set

$$S = \left\{ [\tilde{q}^T, \tilde{p}^T]^T \in R^{2nN} | \dot{Q} = 0 \right\}. \tag{48.20}$$

From the function (48.13), it is clear that

$$\begin{aligned}\dot{Q} &= \frac{1}{2}\sum_{i=1}^{N}\sum_{j\in N_i(t)}\dot{\Psi}(\left\|\tilde{q}_{ij}\right\|)+\sum_{i=1}^{N}\tilde{p}_i^T\dot{\tilde{p}}_i \\ &= -\tilde{p}^T[L(t)+c_1 I_N)\otimes I_n]\tilde{p} \\ &= -\tilde{p}^T[L(t)\otimes I_n]\tilde{p}-c_1\tilde{p}^T\tilde{p} \\ &= 0\end{aligned} \tag{48.21}$$

and then $\tilde{p}_1 = \cdots = \tilde{p}_N = 0$, furthermore, $p_i = p_{\gamma_i}, i = 1, \ldots, N$.

Lastly prove part (3) of Theorem 1. It is a constringent conclusion about the velocity of the center of mass in the group. According to the control protocol (48.4), function (48.8) and (48.11), we can obtain that

$$\begin{aligned}\dot{\bar{\tilde{p}}} &= \frac{\sum_{i=1}^{N}(u_i - f_{\gamma_i}\left(q_{\gamma_i}, p_{\gamma_i}\right))}{N} \\ &= -\frac{1}{N}\sum_{i=1}^{N}\left(\sum_{i=1}^{N}\nabla_{\tilde{q}_i}(\tilde{q}_{ij})+\sum_{j\in N_i(t)}w_{ij}(\tilde{p}_{ij})+c_1\tilde{p}_i\right) \\ &= 0\end{aligned} \tag{48.22}$$

So,

$$\dot{\bar{\tilde{p}}} = -c_1\bar{\tilde{p}} \tag{48.23}$$

The solution of (48.23) can be obtained in this

$$\bar{\tilde{p}} = \bar{\tilde{p}}(0)\cdot e^{-c_1 t} \tag{48.24}$$

From (48.24), it is clear that if the initial velocity of the center of mass of all agents is equal to the weighted average initial velocity of virtual leaders, then the velocity of the center of mass of all agents will equal to the velocity of virtual leader all the way. If the velocities of them are not equal, then the velocity of the center of mass of all agents will convergence to the weighted average velocity of virtual leaders by exponent c_1.

48.5 Conclusions

In this paper, a decentralized control strategy of flocking with multiple virtual leader based on connectivity preserving was proposed. This stable control strategy drives the velocity of the center of the mass of follow agents to exponentially converges to weighted average velocity of the virtual leader by the navigation of virtual leader, finally approach to the same. The certain distance between every agent is kept and eventually form flocking due to the introduction of the artificial potential function.

It was show that the system network can be preserve connected and the velocity of the center of the mass of all agents will exponentially converge to weighed average velocity of the virtual leaders.

References

1. Reynolds C (1987) Folcks, birds and schools: a distributed behavioral model. Comput Gr 21(1):25–34
2. Vicsek T, Czirok A, Jacob EB, Cohen I, Schochet O (1995) Novel type of phase transitions in a system of self-driven particles. Phys Rev Lett 75(6):1226–1229
3. Shi DH, Wang L, Chu T (2006) Virtue leader approach to coordinated control of multiple mobile agents with asymmetric interactions [J]. Physica 213(1):51–65
4. Gazi V, Passino K (2004) Stability analysis of foraging swarms. IEEE Trans Syst Man Cybern-Part B Cybern 43(1):539–557
5. Olfati SR (2006) Flocking for multi-agent dynamic systems:algorithms and theory[J]. IEEE Trans Autom Control 51(3):401–420
6. Housheng S, Xiaofan W, Wen Y (2008) Flocking in multi-agent systems with multiple virtual leader. Asian J. Control [J] 2, 238–245
7. Ji M, Egerstedt M (2007) Distributed coordination control of multiagent systems while preserving connectedness. IEEE Trans Robotics 23:693–703
8. Housheng S, Xiaofan W, Guanrong C (2009) A connectivity-preserving flocking algorithm for multiagent systems based only on position measurement. Int J Control [J] 82(7), 1334–1343, July 2009
9. Horn RA, Johson CR (1987) Matrix Analy. Cambridge University Press, Cambridge
10. Khalil HK (2002) Nonlinear systems, 3rd edn. Prentice Hall, Upper Saddle River

Chapter 49
An Adaptive Weighted One-Class SVM for Robust Outlier Detection

Jinhong Yang, Tingquan Deng and Ran Sui

Abstract This paper focuses on outlier detection from the perspective of classification. One-class support vector machine (OCSVM) is a widely applied and effective method of outlier detection. Unfortunately experiments show that the standard one-class SVM is easy to be influenced by the outliers contained in the training dataset. To cope with this problem, a robust OCSVM is presented in the paper. In consideration that the contribution yielded by the outlying instances and the normal data is different, a robust one-class SVM which assigns an adapting weight for every object in the training dataset was proposed in this paper. Experimental analysis shows the better performances of the proposed weighted method compared to the conventional one-class SVM on robustness.

Keywords One-class support vector machine · Outlier detection · Robustness · Adaptive weighting

49.1 Introduction

Outliers are instances in a dataset, which deviate greatly from the majority of the data [1]. Outlier detection is an important problem in community of data mining and artificial intelligence. It is successfully applied in fields including fraud detection [2], intrusion detection [3], medical diagnosis [4], faulty detection [5].

There are a lot of prevailing works on outlier detection which focus on describing the intrinsic characteristics of outliers, such as statistical methods [6], distance-based methods [7], clustering-based methods [8], density-based methods [9], depth-based

---

J. Yang (✉)
College of Computer Science and Technology, Harbin Engineering University, Harbin 150001, China
e-mail: yangjinhong.66@163.com

T. Deng · R. Sui
College of Science, Harbin Engineering University, Harbin 150001, China

© Springer-Verlag Berlin Heidelberg 2016
Y. Jia et al. (eds.), *Proceedings of the 2015 Chinese Intelligent Systems Conference*, Lecture Notes in Electrical Engineering 359,
DOI 10.1007/978-3-662-48386-2_49

methods [10], rough set based methods [11]. They are successfully unsupervised approaches though they suffer from various weaknesses. Here, a different perspective called one-class classification learning was investigated in outlier detection. Inspired by the unusual definition about outlier: "An outlier is an observation that lies outside the overall pattern of distribution", we can use one-class SVM(OCSVM) to achieve outlier detection [12]. In one-class classification, usually only one target class of data is available, and others are too difficult to characterize or expensive to acquire. One-class SVM learns an optimal decision boundary enclosing the normal instances, and the rest data outside the boundary are naturally identified as outliers. This idea makes OCSVM suitable for outlier detection. Moreover, one-class SVM theory is heavily investigated and it comes with a convex optimization objective ensuring that the global optimum will be reached. Finally, some kernels even allow OCSVM to be considered as a dimensionality reduction technique [5]. Thus it is argued that it can be used to overcome the curse of dimensionality, which makes OCSVM theoretically very attractive for the outlier detection problem.

Many experimental results show that the decision of the traditional OCSVM is sensitive to the outliers lying in the training dataset [5]. In order to weaken the effect of outliers on the model performance and get a better classification performance, the term of slack variable ξ_i is weighted. The proposed method thinks that different data points have different degrees of influence on the decision boundary. So a weight is assigned to every instance in the training dataset to measure the contribution of various points in decision formation. Especially, the outlying samples should contribute less to the decision compared to the normal points. The weights of all samples are adaptively computed according to the distance to the center.

49.2 One-Class SVM Model

Different from the traditional SVMs, one-class SVM proposed by Schölkopf attempts to learn a decision boundary (a hyperplane) that achieves the maximum margin between the points and the origin in the high dimensional feature space [12]. Considering a dataset $X = [x_1, x_2, \ldots, x_N]^T \in \mathfrak{R}^{N \times M}$, in order to obtain the decision boundary of the target class, the following optimization problem is formulated:

$$\begin{aligned} &\min_{W \in \mathfrak{I}, \xi \in \mathfrak{R}^N, \rho \in \mathfrak{R}} \quad \frac{1}{2}\|W\|^2 + \frac{1}{N\nu}\sum_{i=1}^{N} \xi_i - \rho \\ &\text{Subject to} \quad W \cdot \Phi(x_i) \geq \rho - \xi_i, \quad \xi_i \geq 0, \quad i = 1, 2, \ldots, N \end{aligned} \tag{49.1}$$

where $\Phi: \mathfrak{R} \to \mathfrak{I}$ is a mapping which projects the primary data sets to a high dimensional feature space, transforming the non-linear separable data into linear separable data. $\boldsymbol{x}_\text{i}$ is the slack variable for point x_i that allows it to lie outside of the decision boundary and $\boldsymbol{\xi} = \{\xi_1, \xi_2, \ldots, \xi_N\}$. Moreover, N is the number of data points and $\nu \in (0, 1]$ is the regularization parameter that represents an upper bound on the fraction of outlying points and a lower bound on the number of support vectors. W and ρ are the parameters which decision boundary and they are the target variable of the optimization problem.

To avoid the direct computation of the nonlinear mapping Φ, kernel functions are introduced to map the data samples to a high dimensional feature space in support vector machine. The element of the kernel function is expressed as:

$$k(x_i, x_j) = \Phi(x_i) \cdot \Phi(x_j) \tag{49.2}$$

In practice, by introducing Lagrange multiplier α_i, the quadratic program is solved via its dual:

$$\begin{aligned} &\min_{\alpha} \frac{1}{2} \sum_{i=1}^{N} \sum_{j=1}^{N} \alpha_i \alpha_j k(x_i, x_j) \\ &\text{Subject to} \quad 0 \le \alpha_i \le 1/N\nu, \quad \sum_{i=1}^{N} \alpha_i = 1 \quad Z \end{aligned} \tag{49.3}$$

The data points are classified into three kinds: (49.1) the points with $\alpha_i = 0$ locate within the boundary; (49.2) the points with $0 < \alpha_i < 1/N\nu$ are one the boundary; (49.3) the points which satisfy $\alpha_i = 1/N\nu$ are outside of the boundary. The points with non-zero α_i are called support vectors. Through solving the optimization problem in (49.4), $\boldsymbol{\alpha} = [\alpha_1, \ldots, \alpha_N]^T$ can be obtained.

The optimal decision boundary is finally determined by the support vector expansion:

$$f(x) = sgn(\sum_{i=1}^{N} \alpha_i k(x_i, x) - \rho) \tag{49.4}$$

Figure 49.1 illustrates the classification results of a 2D training dataset by the one-class SVM respectively in the original space (the left) and in the kernel feature

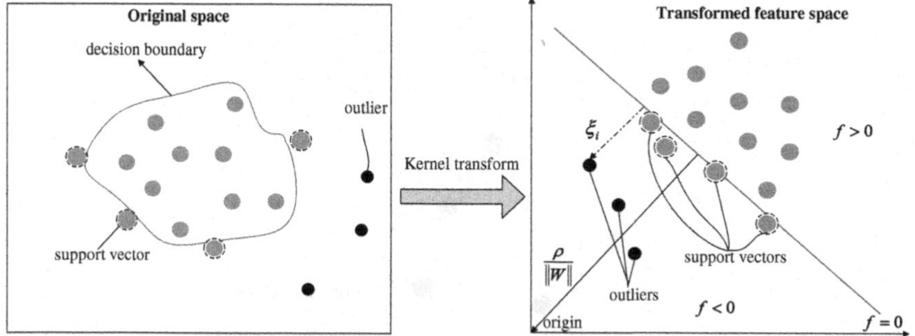

Fig. 49.1 One-class SVM classification to a 2D training dataset

space (the right). The OC-SVM classifier learns a boundary that encloses the blue normal data class. The black points lying outside the boundary are considered as outliers. The blue points with black edges are support vectors which fall on the boundary.

49.3 Influences of Outliers

As the aforementioned, non-zero slack variables ξ_i make it possible for outlying points to fall outside of the decision boundary. The penalty factor $1/N\nu$ in optimization model (49.1) represents the degree of the trade-off between maximum margin and minimum classification error. Unfortunately the normal data and the outlying data share the same penalty factor in general one-class SVM. It is obviously that the number of outliers is far less than that of normal points in practical datasets. Thus, the total error generated by the normal data is far greater than that yielded by the outliers. That means a larger error penalty is imposed to the normal data points, making the decision boundary shift to the outliers. The diagram of the influences of outliers on one-class SVM is illustrated in Fig. 49.2. Two outlying points are added onto the same training dataset in Fig. 49.1. The decision boundary shifts towards the outliers and the primary outliers become the support vectors. The outliers make a large contribution to the location of the decision boundary and the classification accuracy. In other words, the traditional OC-SVM is not robust to the outliers contained in the training dataset. To overcome this problem, a robust one class SVM which takes more adaptive information into account will be discussed in this paper.

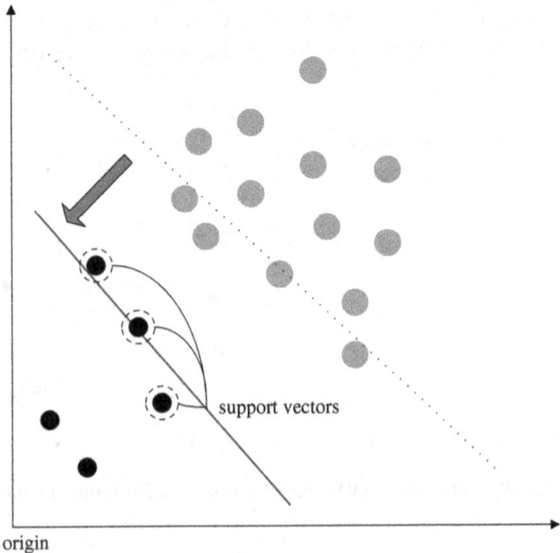

Fig. 49.2 Influences of outliers to the decision boundary

49.4 Robust One-Class SVM

In this section we consider a robust one-class SVM that is suitable to outlier detection. The basic idea of the robust one-class SVM is that the outlying data should contribute less to the decision boundary. Actually different points present various degrees of significance to the classification decision. This paper adaptively assigns a weight for the N training samples in consideration that bigger weights are to points close to the center. Then robustness can be achieved with properly chosen weights.

By assigning a weight ω_i to the error term of each input data, the robust one-class SVM model is formulated as:

$$\min_{W \in \Im, \xi \in \Re^N, \rho \in \Re} \frac{1}{2}\|W\|^2 + \frac{1}{N\nu}\sum_{i=1}^{N} \omega_i \xi_i - \rho \tag{49.5}$$

$$\text{Subject to} \quad W \cdot \Phi(x_i) \geq \rho - \xi_i, \quad \xi_i \geq 0, \quad i = 1, 2, \ldots, N$$

The adaptive weights can measure the importance of each data sample to the optimization problem. This provides a more flexible scheme compared to the traditional OCSVM where the overall penalization for all samples is only a constant. For a data sample, the larger the distance to the center is, the more possible is an outlier. Here, an outlier should have smaller influence to the decision than a normal point. By this virtue, the weight ω_i for object x_i is data-dependent and it is adaptively determined by

$$\omega_i = e^{-\frac{\hat{d}^2(x_i, C)}{2}} \tag{49.6}$$

in which $\hat{d}(x_i, C)$ is the normalized distance between x_i and the center $\boldsymbol{C}$ of the dataset which is defined as

$$\hat{d} = \frac{d}{d_{\max}} \tag{49.7}$$

And the distance from x_i to the center $\boldsymbol{C}$ is

$$\begin{aligned} d(x_i, C) &= \left\| \Phi(x_i) - 1/{-}0ptN \sum_{i=1}^{N} \Phi(x_i) \right\|^2 \\ &= \Phi(x_i) \cdot \Phi(x_i) - 2/{-}0ptN \sum_{j=1}^{N} \Phi(x_i) \cdot \Phi(x_j) + 1/{-}0ptN \sum_{j=1}^{N} \Phi(x_j) \cdot 1/{-}0ptN \sum_{j=1}^{N} \Phi(x_j) \\ &\approx k(x_i, x_i) - 2/{-}0ptN \sum_{j=1}^{N} k(x_i, x_j) \end{aligned} \tag{49.8}$$

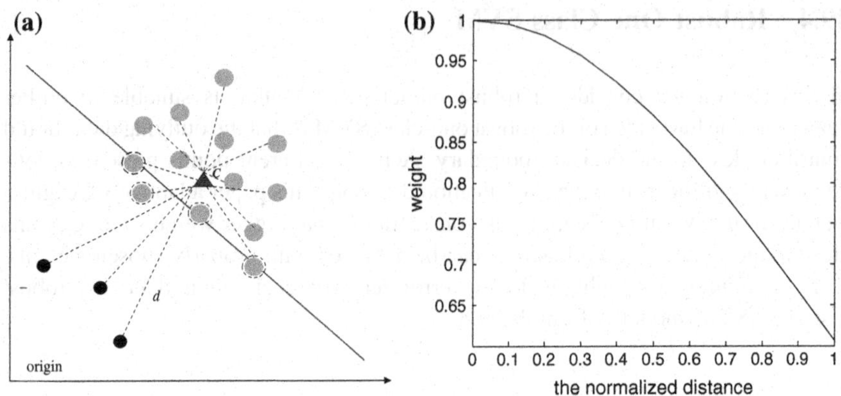

Fig. 49.3 The schematic of weighted one-class SVM. **a** The distances to the center. **b** The relationship between weight and distance

where the term $1/-0ptN\sum_{j=1}^{N}\Phi(x_j)\cdot 1/-0ptN\sum_{j=1}^{N}\Phi(x_j)$ is a constant and an approximation form that was introduced by [13] is computed in stead of the actual distance. It is easy to see that outlying instances are far away from the center in Fig. 49.3a. According to Fig. 49.3b, for any data x_i, we have $0\le\omega_i\le 1$, and the weight is monotonically decreasing with the distance. So it is reasonable that the instances far from the center receive smaller weights.

To solve the dual problem of the quadratic programming in (49.5), a Lagrange function is constructed as follows:

$$L(W,\rho,s,\alpha,\beta)=\frac{1}{2}\|W\|^2+\frac{1}{N\nu}\sum_{i=1}^{N}\omega_i\xi_i-\rho-\sum_{i=1}^{N}\alpha_i(W\cdot\Phi(x_i)-\rho+\xi_i)-\sum_{i=1}^{N}\beta_i\xi_i \tag{49.9}$$

Based on the Karush–Khun–Tucker (KKT) conditions, the dual form of the Lagrangian is written as follows:

$$\begin{aligned}&\min_{\alpha}\tfrac{1}{2}\sum_{i=1}^{N}\sum_{j=1}^{N}\alpha_i\alpha_j k(x_i,x_j)\\&\text{Subject to}\quad 0<\alpha_i<\tfrac{1}{N\nu}\cdot\omega_i,\ \sum_{i=1}^{N}\alpha_i=1,\quad i=1,2,\ldots,N\end{aligned} \tag{49.10}$$

After getting the optimal solution $\boldsymbol{\alpha}^*=[\alpha_1^*,\alpha_2^*,\ldots,\alpha_N^*]^T$ and ρ^* in quadratic programming (49.10), the corresponding decision function for a certain input data x is

$$f(x) = sgn(\sum_{i=1}^{N} \alpha_i^* k(x_i, x) - \rho^*) \tag{49.11}$$

The criterion of outlier identification is:
if $f(x) = 1$, x is a normal point;
if $f(x) = -1$, x is an outlier.

With this weighted method which makes a difference of the normal and the outlying samples, we can reduce the effect of the outliers on the construction of the decision function to get a robust solution.

49.5 Experiments

In the experiment, several publicly available datasets from UCI machine learning repository are used to evaluate the robustness of the weighted outlier detection method proposed in this paper, and then the performance of the robust one-class SVM with the standard one-class SVM [12] was compared.

As known, most of the datasets of UCI repository are traditionally dedicated for classification tasks. Hence they have to be pre-processed in order to serve for the evaluation of unsupervised outlier detection algorithms. This is typically performed by picking a meaningful outlying class and randomly sampling the outliers to a small fraction. The preprocessing selects particular classes as outliers and samples it down to a small fraction in order to meet the requirements for unsupervised anomaly detection. And the preprocessing details to each dataset are presented in Table 49.1.

The experiment is executed on platform of Matlab 2014b. Guassian kernel function is utilized in this experiment. And the comparison experiment is respectively executed on Iris, Ionosphere, Shuttle and Breast cancer wisconsin datasets under different parameters. Three indexes of false positive rate (FPR) and true positive rate (TPR) as well as mean average precision (MAP) are utilized to evaluate the performance of outlier detection algorithms.

The comparison experiment is conducted under various parameter ν and the results are given in Table 49.2. The experimental value of ν is respectively set to 0.01, 0.015, 0.02, 0.025, 0.03, 0.035, 0.04, 0.045, 0.05, 0.055, 0.06, 0.065, 0.07 and the width $\sigma^2 = 0.8995$. The indexes in Table 49.2 is an average statistical results

Table 49.1 Description of datasets

Dataset	Original size	Attributes	Outlier class(es)	Resulting dataset size	Sampled outliers percentage (%)
Iris	150	4	Virginica	105	4.8
Ionosphere	351	34	b	233	3.4
Shuttle	58000	9	2, 4, 5	46987	2.3
Breast cancer wisconsin	569	32	M	299	1.8

Table 49.2 Comparisons of Experimental results between weighted OCSVM and traditional OCSVM

Dataset	Weighted OCSVM			Traditional OCSVM		
	MAP (%)	TPR (%)	FPR (%)	MAP (%)	TPR (%)	FPR (%)
Iris	91	93.1	5	86.4	89.5	10.2
Ionosphere	98.7	100	2	90.4	92	6
Shuttle	89.3	92.9	7.3	88.1	90.4	12
Breast cancer wisconsin	86.2	87.65	4.3	80.88	81.2	9

on different parameter settings. It is clear that the mean average precisions (MAPs) of weighted one-class SVM on four datasets are universally larger than those of traditional OCSVM. The true positive rate (TPR) of weighted OCSVM even reaches up to 100 % on Ionosphere dataset which guarantees all actual outliers are detected. Otherwise, the FPR of the proposed algorithm is significantly smaller than that of the traditional OCSVM.

The ROC curves of weighted OCSVM and the traditional OCSVM on four datasets are respectively shown in Fig. 49.4. The area below the ROC curves means

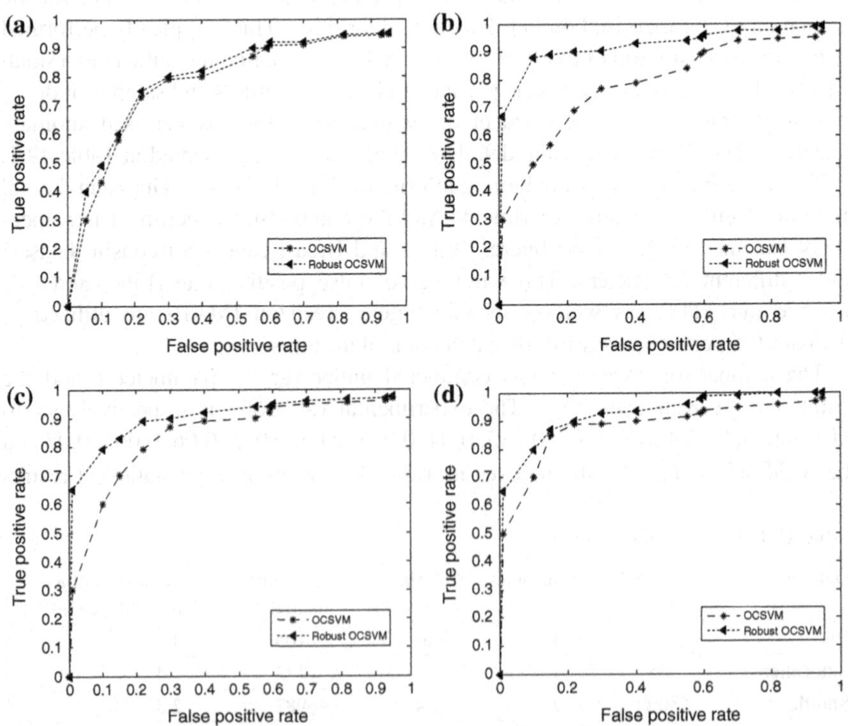

Fig. 49.4 The ROC curve comparisons of the robust OCSVM and the traditional OCSVM on four datasets. **a** Iris. **b** Ionosphere. **c** Shuttle. **d** Breast cancer wisconsin

the performance of algorithms. The larger the area is, the better the algorithm is. In general, the ROC area of the proposed weighted OCSVM is bigger than that of the traditional OCSVM. Especially for Ionosphere dataset, the ROC value of weighted one-class SVM reaches 0.9895, which is significantly larger than that of OCSVM. Thus, it is evident that the proposed method performs better than the traditional one-class SVM on classification accuracy and robustness.

49.6 Conclusions

A weighted robust one-class SVM is presented in this paper. The proposed OCSVM assigns a weight to each sample in the training dataset. Thus, different instances have various contributions to the decision. Outlying objects are generally far away from the center. The weights given in this paper is proved to be monotonically decreasing with the distance to the center. Comparison experiments show that weighted OC-SVM has a better performance than the traditional OCSVM on classification flexibility as well as robustness.

Acknowledgements This paper is supported by the National Natural Science Foundation of China under grant 11471001.

References

1. Hawkins D (1980) Identifications of outliers. Chapman and Hall, London
2. Christou IT, Bakopoulos M, Dimitriou T et al (2011) Detecting fraud in online games of chance and lotteries. Expert Syst Appl 38(10):13158–13169
3. Casas Pedro, Mazel Johan, Owezarski Philippe (2012) Unsupervised network intrusion detection systems: detecting the unknown without knowledge. Comput Commun 35 (7):772–783
4. Hauskrecht Milos, Batal Iyad, Valko Michal et al (2013) Outlier detection for patient monitoring and alerting. J Biomed Inform 46(1):47–55
5. Yin S, Zhu X, Jing C (2014) Fault detection based on a robust one class support vector machine. Neurocomputing 145:263–268
6. Barnett V, Lewis T (1994) Outliers in statistical data. Wiley, New York
7. Knorr E, Ng R, Tucakov V (2000) Distance-based outliers: algorithms and applications. VLDB J Very Large Databases 8(3–4):237–253
8. Xue Z, Shang Y, Feng A (2010) Semi-supervised outlier detection based on fuzzy rough C-means clustering. Math Comput Simul 80:1911–1921
9. Breunig MM, Kriegel H-P, Ng RT, Sander J (2000) LOF: identifying density-based local outliers. In: Proceedings of the 2000 ACM SIGMOD international conference on management of data, USA
10. Johnson T, Kwok I, Ng RT (1998) Fast computation of 2-dimensional depth contours. In: Proceedings of the 4th international conference on knowledge discovery and data mining, New York, pp 224–228
11. Albanese A, Pal SK, Petrosino A (2014) Rough sets, kernel set, and spatiotemporal outlier detection. IEEE Trans Knowl Data Eng 26(1):194–207

12. Scholkopf B, Platt JC, Shawe-Taylor JC et al (2001) Estimating the support of a high-dimensional distribution. Neural Comput 13(7):1443–1471
13. Amer M, Goldstein M, Abdennadher S Enhancing one-class support vector machines for unsupervised anomaly detection. In: ODD 13: Proceedings of the ACM SIGKDD workshop on outlier detection and description

Chapter 50
An Improved Kalman Filter for Fractional Order System with Measurement Lévy noise

Yi Wang, Yonghui Sun, Zhenyang Gao, Xiaopeng Wu and Chao Yuan

Abstract In this paper, taking the measurement noise as the non-Gaussian Lévy noise, an improved Kalman filter for discrete linear stochastic fractional order system is proposed. By eliminating the maximum of the noise, the Lévy noise can be approximated by a series of Gaussian white noises. Then, based on the principle of least square, an improved Kalman filter is developed for discrete linear stochastic fractional order system with measurement Lévy noise. Finally, simulation results are provided to illustrate the effectiveness and usefulness of the proposed filter designing algorithm, where a better filtering performance could be found.

Keywords Fractional order system · Kalman filter · Lévy noises · State estimation

50.1 Introduction

The history of fractional order calculus can date back to eighteenth century, when two mathematicians, Euler and Lagrange developed the concept of fractional calculus. Then, in the nineteenth century, Liouville, Riemann and Holmgren established the theory of fractional order system [1].

In recent decades, the fractional order system has attracted an increasing attention [2], a large number of results about the stability and control of fractional order system have been developed. In [3], the stability of fractional order system with time delay and the designing of the fractional order PID controller were discussed. The authors in [4] discussed the problem of robust stability and stabilization problems of fractional order chaotic system. In [5], the authors investigated the robust stability of fractional order system with parameter uncertainties.

On the other hand, the study on state estimation and parameter identification of fractional order system also has gained some special interest. In [6], the authors mainly discussed parameter identification of the continuous fractional system with

Y. Wang · Y. Sun (✉) · Z. Gao · X. Wu · C. Yuan
College of Energy and Electrical Engineering, Hohai University, Nanjing 210098, China
e-mail: sunyonghui168@gmail.com

© Springer-Verlag Berlin Heidelberg 2016

Y. Jia et al. (eds.), *Proceedings of the 2015 Chinese Intelligent Systems Conference*, Lecture Notes in Electrical Engineering 359,
DOI 10.1007/978-3-662-48386-2_50

time delay. In [7], the authors considered the problem of parameter identification using adjustable fractional order differentiator in fractional order system. In [8], a Kalman filtering algorithm was proposed for the fractional order singular system. In [9], the state estimation for the noisy chaotic secure system was realized based on the extended fractional Kalman algorithm. Compared with integer-order Kalman filters [10, 11], the fractional ones show better performances to satisfy the requirements of the designers, a detailed introduction was provided in [12]. In [13], the authors proposed the fractional order Kalman filter (FKF) for linear fractional order system, then the fractional order Kalman filter was used in singular systems [8]. Furthermore, the extended fractional order Kalman filter (EFKF) and the unscented fractional order Kalman filter (UFKF) were proposed for nonlinear fractional order systems [14], then the EFKF was used to chaotic cryptography in noisy environment [15].

It needs to be pointed out that most of the existing results mainly assumed that the stochastic fractional order system perturbed by Gaussian white noise. However, in most cases, non-Gaussian noises are widely existing in some practical systems, especially non-Gaussian Lévy noises. The statistical properties of non-Gaussian Lévy noises will greatly influence the performance of fractional order Kalman filter. Theoretically, Lévy noises can be approximated by the increments of the corresponding Lévy process per time step [16], however, due to the infinite variance, it is difficult to estimate the states by using the traditional algorithms when considering Lévy noises. In [17], the authors considered the optimal filtering problem with Lévy noise in the Gaussian signal. The authors in [18] discussed the approximating method for small jump process in Lévy sequences. In [19], the authors investigated the KF algorithm with Lévy noises instead of Gaussian white noises. Generally, for a practical system, the system state is difficult to measure in practice with Gaussian noises, although a state observer could be designed, not to mention the non-Gaussian Lévy noises in the measuring process. However, to the best of the authors' knowledge, there are few results considering filter designing for fractional order systems with Lévy noises.

Based on the above discussions, in this paper, taking the measurement noise as the Lévy noise, for discrete linear fractional order system, an improved fractional order Kalman filter algorithm is proposed. Simulation results are provided to demonstrate the usefulness and effectiveness of proposed filter designing strategy.

50.2 Fractional Order System and Kalman Filter

In this section, the conventional fractional order Kalman filter algorithm will be reviewed, and the related definitions and Lemmas will be introduced.

Definition 1 ([2]) The fractional order Grünwald-Letnikov difference is given by the following equation

$$\Delta^{\alpha} x_k = \frac{1}{h^{\alpha}} \sum_{j=0}^{k} (-1)^j \binom{\alpha}{j} x_{k-j}, \tag{50.1}$$

where α is the fractional order, h is the sampling time, and k is the number of samples. The factor $\binom{\alpha}{j}$ represents

$$\binom{\alpha}{j} = \begin{cases} 1, & \text{if } j = 0, \\ \frac{\alpha(\alpha-1)\cdots(\alpha-j+1)}{j!}, & \text{if } j > 0. \end{cases}$$

Based on Definition 1, it's possible to obtain the discrete equivalent of derivative (when α is positive), the discrete equivalent of integration (when α is negative) or the original function (when α equals to 0).

Definition 2 ([12]) The discrete linear stochastic fractional order system can be formulated as:

$$\Delta^{\alpha} x_{k+1} = A x_k + B u_k + w_k, \tag{50.2}$$

$$x_{k+1} = \Delta^{\alpha} x_{k+1} - \sum_{j=1}^{k+1} (-1)^j \gamma_j x_{k+1-j}, \tag{50.3}$$

$$y_k = C x_k + v_k, \tag{50.4}$$

where x_k is the state vector, u_k is the system input, y_k is the measurement output, A, B and C are the known constant matrices with appropriate dimensions, w_k and v_k represent system noise and measurement noise at the time instant k, respectively, which are always assumed to be the Gaussian white noises with zero means characterized by the following covariance matrices

$$E[w_i w_j^T] = \begin{cases} Q_i, & \text{if } i = j, \\ 0, & \text{if } i \neq j, \end{cases}$$

$$E[v_i v_j^T] = \begin{cases} R_i, & \text{if } i = j, \\ 0, & \text{if } i \neq j, \end{cases}$$

and

$$\gamma_j = \text{diag}\left([\binom{\alpha_1}{j} \binom{\alpha_2}{j} \ldots \binom{\alpha_N}{j}]\right),$$

where Q_i is the system noise covariance matrix at time instant i, R_i is the measurement noise covariance matrix at time instant i, $\alpha_1, \alpha_2, \ldots, \alpha_N$ are the orders of the fractional order system, $j = 1, 2, \ldots, k+1$.

Lemma 1 *([20]) If A, C, BCD are nonsingular square matrix (the inverse exists), then*

$$(A + BCD)^{-1} = A^{-1} - A^{-1}B(C^{-1} + DA^{-1}B)^{-1}DA^{-1}.$$

Theorem 1 *([12]) For discrete linear stochastic fractional order systems (50.2)–(50.4), the following fractional order Kalman filter is given by the following sets of formulas:*

$$\Delta^{\alpha}\tilde{x}_{k+1} = A\hat{x}_k + Bu_k, \tag{50.5}$$

$$\tilde{x}_{k+1} = \Delta^{\alpha}\tilde{x}_{k+1} - \sum_{j=1}^{k+1}(-1)^j\gamma_j\hat{x}_{k+1-j}, \tag{50.6}$$

$$\tilde{p}_{k+1} = A\hat{p}_kA^T + Q_k + \sum_{j=1}^{k+1}\gamma_j\hat{p}_{k+1-j}\gamma_j^T, \tag{50.7}$$

$$\hat{x}_k = \tilde{x}_k + K_k(y_k - C\tilde{x}_k), \tag{50.8}$$

$$\hat{p}_k = (I - K_kC)\tilde{p}_k, \tag{50.9}$$

where

$$K_k = \tilde{p}_kC^T(C\tilde{p}_kC^T + R_k)^{-1}, \tag{50.10}$$

with initial conditions $\tilde{x}_0$ and $\tilde{p}_0$, v_k and w_k are assumed to be independent and with zero expected value.

Remark 1 Based on the conventional KF for fractional order system, the system state could be estimated under Gaussian white noises. However, in most of the actual systems, the non-Gaussian noises cannot be ignored, the infinite variance of the non-Gaussian noises will greatly affect the accuracy of fractional order Kalman filter. Thus it is necessary to develop an improved fractional order Kalman filter with measurement Lévy noises.

50.3 An Improved Fractional Kalman Filter

In this section, an improved fractional Kalman filter algorithm is proposed. As we know, the non-Gaussian Lévy noise is generally existing in actual systems, which has infinite variance, so it is necessary to develop fractional order Kalman filter algorithm with measurement Lévy noises. By the Lévy-Ito theorem [17], a Lévy process consists of the sum of a Brownian motion process and a pure jump process. It is known that the Gaussian process can be approximated by the increments of Brownian motion per time step. Therefore, in this paper, a Gaussian process and a jump process are employed to approximate a Lévy process, which also enables us to decompose a Lévy noise into a Gaussian white noise with some extremely large values.

Let system noise w_k be Gaussian white noise and measurement noise $\bar{v}_k$ be Lévy noise. The fractional order system obtained can be rewritten by

$$\Delta^{\alpha} x_{k+1} = Ax_k + Bu_k + w_k, \tag{50.11}$$

$$x_{k+1} = \Delta^{\alpha} x_{k+1} - \sum_{j=1}^{k+1} (-1)^j \gamma_j x_{k+1-j}, \tag{50.12}$$

$$\bar{y}_k = Cx_k + \bar{v}_k. \tag{50.13}$$

Due to the noises $\bar{v}_k$ is Lévy noise that is unknown, it is difficult to obtain $\bar{y}_k$ directly. As discussed before, the Lévy noise can be decomposed into a series of Gaussian white noise with some extremely large values, similarly, $\bar{y}_k$ could be approximated by the following equations

$$\bar{y}_k = \begin{cases} C\tilde{x}_k + \delta \cdot \mathrm{sign}(y_k - C\tilde{x}_k), & \text{if } |(y_k - C\tilde{x}_k)| \geq \delta, \\ y_k, & \text{if } |(y_k - C\tilde{x}_k)| < \delta, \end{cases} \tag{50.14}$$

δ are positive threshold values.

It is worth pointing out that piecewise equation of $\bar{y}_k$ in equations (50.14) enables us to obtain the corresponding value based on previous predicted value, which helps us to deduce the following modified fractional order Kalman filter algorithm.

Theorem 2 *For the discrete linear fractional order system with measurement Lévy noise (50.11)–(50.13), the improved fractional order Kalman filter is given by*

$$\Delta^{\alpha} \tilde{x}_{k+1} = A\hat{x}_k + Bu_k, \tag{50.15}$$

$$\tilde{x}_{k+1} = \Delta^{\alpha} \tilde{x}_{k+1} - \sum_{j=1}^{k+1} (-1)^j \gamma_j \hat{x}_{k+1-j}, \tag{50.16}$$

$$\tilde{p}_{k+1} = A\hat{p}_k A^T + Q_k + \sum_{j=1}^{k+1} \gamma_j \hat{p}_{k+1-j} \gamma_j^T, \tag{50.17}$$

$$\hat{x}_k = \tilde{x}_k + K_k(\bar{y}_k - C\tilde{x}_k), \tag{50.18}$$

$$\hat{p}_k = (I - K_k C)\tilde{p}_k, \tag{50.19}$$

where

$$K_k = \tilde{p}_k C^T (C\tilde{p}_k C^T + \bar{R}_k)^{-1},$$
$$\bar{R}_k = (\bar{y}_k - C\tilde{x}_k)(\bar{y}_k - C\tilde{x}_k)^T + C\tilde{p}_k C^T.$$

Remark 2 In [19], the authors developed the Kalman filter for discrete linear system under Lévy noises, where the Lévy noises were considered. However, the developed results did not consider the fractional order system with Lévy noises.

50.4 Illustrative Examples

In this section, a numerical example is provided to illustrate the usefulness and effectiveness of the developed results.

Example 1 Consider the following discrete fractional order system with system Gaussian white noise and measurement Lévy noise

$$
\begin{aligned}
\Delta^{\alpha} x_{k+1} &= \begin{bmatrix} 0 & 1 \\ -0.1 & -0.2 \end{bmatrix} x_k + \begin{bmatrix} 0 \\ 1 \end{bmatrix} u_k + w_k, \\
x_{k+1} &= \Delta^{\alpha} x_{k+1} - \sum_{j=1}^{k+1} (-1)^j \gamma_j x_{k+1-j}, \\
\bar{y}_k &= \begin{bmatrix} 0.1 & 0.3 \\ 1 & 0 \end{bmatrix} x_k + \begin{bmatrix} \bar{v}_k^1 \\ \bar{v}_k^2 \end{bmatrix},
\end{aligned}
$$

where the fractional order and the threshold value are given by

$$
\alpha = \begin{bmatrix} 0.7 \\ 1.2 \end{bmatrix}, \delta = 20.
$$

In this example, without loss of generality, $\bar{v}_k^1$ is symmetric α-stable Lévy noise with the index of stability μ=1.5 and the scale parameter $\sigma = 5$, $\bar{v}_k^2$ is a symmetric α-stable Lévy noise with the index of stability $\mu = 1.3$ and the scale parameter σ=5. Since the measurement noise has infinite variances, the conventional fractional order Kalman filter algorithm proposed in [12] failed to estimate the system states. By using the improved FKF developed in this paper, the states of the fractional order system under Lévy noises can be estimated with a high accuracy. In the improved fractional order Kalman filter algorithm, the real value and the estimated value of the system are shown in Figs. 50.1 and 50.2.

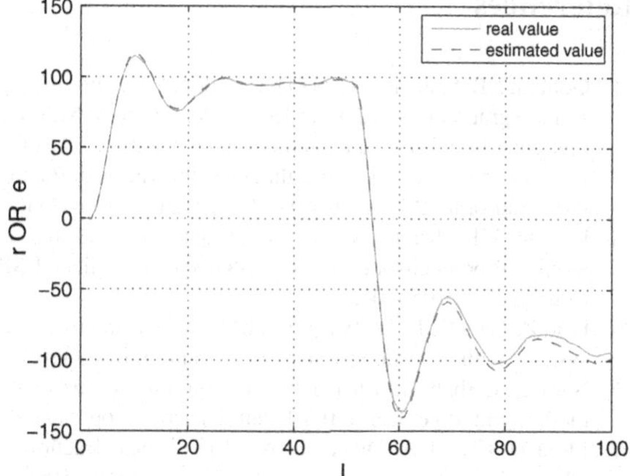

Fig. 50.1 Real value and estimated value of $x_1(\mu = 1.5, \sigma = 5)$

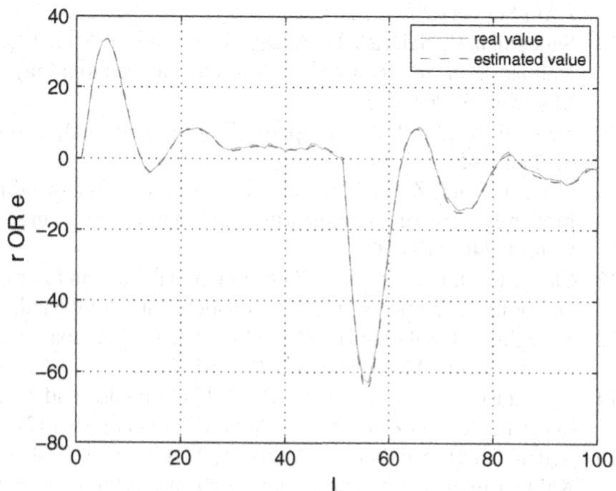

Fig. 50.2 Real value and estimated value of $x_2(\mu = 1.3, \sigma = 5)$

50.5 Conclusions

In this paper, taking the measurement noise as the non-Gaussian Lévy noise in the discrete linear stochastic fractional order system, the Kalman filter designing problem with measurement Lévy noise has been discussed. By eliminating maximum value of the Lévy noise, based on the least square principle, an improved fractional Kalman filter algorithm with non-Gaussian Lévy noise was developed. Finally, a numerical example was provided to show the effectiveness of the proposed filter designing strategy under non-Gaussian measurement Lévy noises.

Acknowledgments The work was supported in part by the National Natural Science Foundation of China under Grant 61104045, in part by the 111 Project (B14022), and in part by the Fundamental Research Funds for the Central Universities of China under Grant 2014B08014.

References

1. Odlham KB, Spaniar J (1974) The fractional calculus: theory and applications of differentiation and integration to arbitrary order. Academic Press, New York
2. Podlubny I (1999) Fractional differential equations. Academic Press, New York
3. Hamamci SE (2007) An algorithm for stabilization of fractional-order time delay systems using fractional-order PID controllers. IEEE Trans Autom Control 52(10):1964–1969
4. Faieghi MR, Kuntanapreeda S, Delavari H, Baleanu D (2014) Robust stabilization of fractional-order chaotic systems with linear controllers: LMI-based sufficient conditions. J Vib Control 20(7):105–1042
5. Liao Z, Peng C, Li W, Wang Y (2011) Robust stability analysis for a class of fractional order systems with uncertain parameters. J Franklin Inst 348(6):1101–1113
6. Narang A, Shah SL, Chen T (2011) Continuous-time model identification of fractional-order models with time delays. IET Control Theory Appl 5(7):900–912
7. Idiou D, Charef A, Djouambi A (2013) Linear fractional order system identification using adjustable fractional order differentiator. IET Signal Proc 8(4):398–409
8. Ashayeri L, Shafiee M, Menhaj M (2013) Kalman filter for fractional order singular systems. J Am Sci 9(1):209–216
9. Sadeghian H, Salarieh H, Alasty A, Meghdari A (2013) On the fractional-order extended Kalman filter and its application to chaotic cryptography in noisy environment. Appl Math Model 38(3):961–973
10. Kalman RE (1960) A new approach to linear filtering and prediction problems. J Fluids Eng 82(1):35–45
11. Liang J, Wang Z, Liu Y (2014) State estimation for two-dimensional complex networks with randomly occurring nonlinearities and randomly varying sensor delays. Int J Rob Nonlinear Control 24(1):18–38
12. Sierociuk D, Dzieliński A (2006) Fractional Kalman filter algorithm for the states, parameters and order of fractional system estimation. Int J Appl Math Comput Sci 16(1):129–140
13. Sadeghian H, Salarieh H (2013) On the general Kalman filter for discrete time stochastic fractional systems. Mechatronics 23(7):764–771
14. Águila RC, Carazo AH, Pérez JL (2012) Extended and unscented filtering algorithms in nonlinear fractional order systems. Appl Math Sci 6(30):1471–1486
15. Sadeghian H, Salarieh H, Alasty A, Meghdari A (2014) On the fractional-order extended Kalman filter and its application to chaotic cryptography in noisy environment. Appl Math Model 38(3):961–973
16. Applebaum D (2009) Lévy processes and stochastic calculus. Cambridge University Press, Cambridge
17. Ahn A, Feldman R (1999) Optimal filtering of a gaussian signal in the presence of Lévy noise. SIAM J Appl Math 60(5):359–369
18. Asmussen S, Rosinski J (2001) Approximation of small jumps of Lévy processes with a view towards simulation. J Appl Probab 38:482–493
19. Sun X, Duan J, Li X, Wang X (2013) State estimation under non-Gaussian Levy noise: a modified Kalman filtering method. In: Proceedings of the Banach Center, in press. arXiv:1303.2395
20. Nering ED (1963) Linear algebra and matrix theory. Wiley, New York

Chapter 51
Nonlinear Control of Inertial Wheel Pendulum and Its Implementation Based on STM32

Yan Lixia and Ma Baoli

Abstract The swing-up and balance control problems of inertial wheel pendulum are investigated in this work. We firstly introduce the dynamic model of inertial wheel pendulum and then propose a swing-up control law by using an energy function. Theoretical analysis shows that the system gradually moves into an invariant set during swing-up process. Later, we propose the balance control law with approximate linearization and introduce a self-built experimental platform based on STM32 microprocessor. The final combination of swing-up control law and balance control law leads to a hybrid control law. We verify the efficiency of the proposed control law on the apparatus.

Keywords Inertial wheel pendulum · Nonlinear control · STM32

51.1 Introduction

The inertial wheel pendulum (IWP) is a type of pendulum with a stable pivot and a rotary wheel mounted on the end, as shown in Fig. 51.1. Swinging the rod up from bottom and balancing it at the upright vertical position are the main problems researched on IWP.

The model of IWP is initially proposed in [1]. In [2], a collocated state transformation and a strict feedback form are introduced, which make the IWP be globally stabilized with aggressive control input. However, the limited torque of motor means that the IWP can not swing up in a single swing and whether the global stabilized control law works is uncertain. Passivity-based swing-up control laws and switching schemes are introduced in [3] and [4], both of which launch experiments

Y. Lixia · M. Baoli (✉)
Seventh Research Division, School of Automation Science and Electrical Engineering,
Beihang University, Beihang University, Beijing, China
e-mail: mabaoli@buaa.edu.cn

Y. Lixia
e-mail: robotyanlx@yahoo.com

© Springer-Verlag Berlin Heidelberg 2016

Y. Jia et al. (eds.), *Proceedings of the 2015 Chinese Intelligent Systems Conference*, Lecture Notes in Electrical Engineering 359,
DOI 10.1007/978-3-662-48386-2_51

on well-calibrated platforms. Under the assumption that the rod's angle is the only measurable state, a sliding-mode swing-up control law is proposed in [5], because the wheel's angular speed is absent, such controller often leads to a rotating wheel in steady state. In [6], a generalized PI control law is proposed to design a swing-up reference trajectory for IWP and turns the swing-up process into the trajectory tracking control. A two relay controller and a oscillating reference trajectory are introduced in [7]. May the trajectory tracking of swing-up be a good way, but the reference trajectory design usually costs many efforts. Some classical swing-up strategies of IWP can be seen in [8]. Furthermore, applications of IWP are also researched, such as a IWP mounted on a movable LEGO cart proposed in [9], a 3D Cubli introduced in [10].

Considering more about the reality, we design a hybrid control scheme for IWP. We propose a swing-up control law that makes the rod run a homoclinic orbit with the angular speed of wheel descending, which implies that the IWP enters into an invariant set asymptotically. Then, we linearize the IWP at the upright vertical position and show that a PD-like state feedback control law can be taken to balance the rod. The self-built experimental platform based on a cheap STM32 processor and its technical details are also presented. Through combination of the swing-up and the balance control law, we conduct the experiment with this hybrid control law on our apparatus.

The rest of the paper is organized as follow. In Sect. 51.2, problem formation is given. The swing-up control and balance control are described in Sect. 51.3. The self-built experimental platform is introduced in Sect. 51.4 along with experimental results. Conclusion and future work are presented in Sect. 51.5.

51.2 Problem Formation

The simplified model of inertial wheel pendulum is shown in Fig. 51.1, where

m_1 – *the total mass of rod and motor*
l_1 – *the length between pivot and the mass center of* m_1
I_1 – *the rotational inertia of* m_1

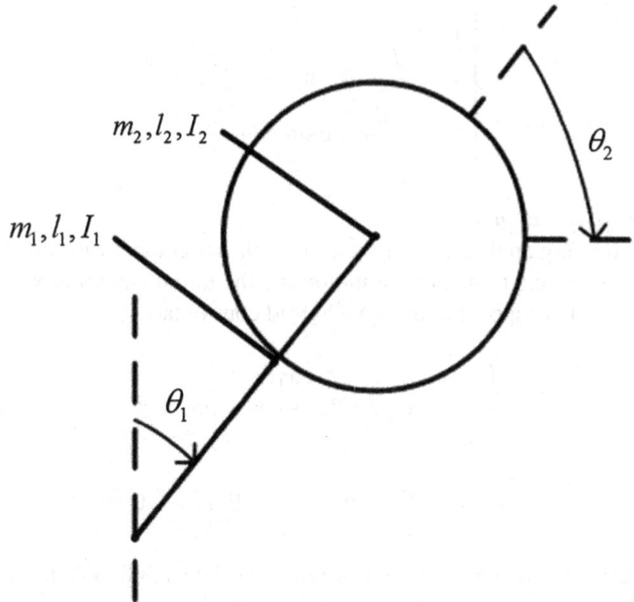

Fig. 51.1 Model of inertial wheel pendulum

m_2 – *the mass of the wheel*
l_2 – *the length of the rod*
I_2 – *the rotational inertia of the wheel*
θ_1 – *the angle of the rod*
θ_2 – *the angle of the wheel*

Based on Lagrange method, the dynamics model of IWP can be given as [1]

$$\begin{cases} d_{11}\ddot{\theta}_1 + d_{12}\ddot{\theta}_2 - \bar{m}g\sin\theta_1 = 0 \\ d_{21}\ddot{\theta}_1 + d_{22}\ddot{\theta}_2 = \tau \end{cases} \tag{51.1}$$

where control input τ is the motor torque and

$$\begin{cases} d_{11} = m_1 l_1^2 + m_2 l_2^2 + I_1 + I_2 \\ d_{12} = d_{21} = d_{22} = I_2 \\ \bar{m} = (m_1 l_1 + m_2 l_2) \end{cases} \tag{51.2}$$

Leave out the angle of the wheel because of its symmetry and define $x = [x_1, x_2, x_3]^T = [\theta_1, \dot{\theta}_1, \dot{\theta}_2]^T$. Derive from the linear equation set (51.1), we get the reduced order model

$$\begin{cases} \dot{x}_1 = x_2 \\ \dot{x}_2 = \dfrac{d_{22}}{D}\bar{m}g\sin x_1 - \dfrac{d_{12}}{D}\tau \\ \dot{x}_3 = -\dfrac{d_{12}}{D}\bar{m}g\sin x_1 + \dfrac{d_{11}}{D}\tau \end{cases} \tag{51.3}$$

where $D = d_{11}d_{22} - d_{12}d_{21} > 0$.

Limit the rotating angle range of the rod in $[0, 2\pi)$ in clockwise direction. With considering the limited torque of the motor and the initial condition $x = [\pi, 0, 0]^T$, the basic task in this paper is to design a hybrid control law

$$\tau = \begin{cases} f_1(x_1, x_2, x_3), \epsilon < x_1 < 2\pi - \epsilon \\ f_2(x_1, x_2, x_3), 2\pi - \epsilon \le x_1 \ or \ x_1 \le \epsilon \end{cases} \tag{51.4}$$

such that

$$\lim_{t\to\infty} x_1 = 0, \ \lim_{t\to\infty} x_2 = 0, \ \lim_{t\to\infty} x_3 = 0 \tag{51.5}$$

where f_1 is the smooth swing-up control law, f_2 is the smooth balance control law, and ϵ is a small positive constant designed for switching.

51.3 Control of Swing-Up and Balance

We investigate the swing-up and balance control of the IWP in this section. To construct a hybrid control law, we decide to design the swing-up and balance control law separately. The swing-up control law will be switched to balance control law once the IWP enters a certain vicinity of the upright vertical position.

51.3.1 Swing-Up Control

Inspired by the energy control strategy of swinging up a linear inverted pendulum proposed in [11], we consider the energy function

$$E = \frac{1}{2}d_{11}x_2^2 + \bar{m}g\left(\cos x_1 - 1\right) \tag{51.6}$$

Unlike the positive definite function in passivity-based control of IWP, (51.6) stays negative before its actual energy reaches $2\bar{m}g$. Moreover, on the manifold $E = 0$, we obtain

$$\frac{1}{2}d_{11}x_2^2 = \bar{m}g\left(1 - \cos x_1\right) \tag{51.7}$$

which shows that the kinetics energy of IWP equals the energy needed to swing to the upright position from its present position. Concerning the angular speed of the wheel, we choose a positive definite function

$$V = \frac{1}{2}kE^2 + \frac{1}{2}k_v x_3^2 \tag{51.8}$$

where $0 < k_v < \dfrac{2k\bar{m}gd_{12}^2}{d_{11}}$. Then we compute the derivative of V as

$$\dot{V} = -(d_{12}kEx_2 - k_v x_3)\dot{x}_3 \tag{51.9}$$

Choose a control input

$$\tau = \left((d_{12}kEx_2 - k_v x_3) + \frac{d_{12}}{D}\bar{m}g \sin\theta_1 \right) \frac{D}{d_{11}} \tag{51.10}$$

we transform (51.9) into

$$\dot{V} = -(d_{12}kEx_2 - k_v x_3)^2 \leq 0 \tag{51.11}$$

Generally, a negative semi-definite $\dot{V}$ hints us to consider the LaSalle invariant principle to show that $E \to 0$ and $x_3 \to 0$, but it is unfortunate that the IWP will never start to swing once initialized right at $x = [\pi, 0, 0]^T$, since $x_2 = 0$ causes $\dot{x}_3 = 0$ and results with $x_3 \equiv 0$. The gravity, whose tangential component remains zero due to $x_1 = \pi$, can not help either. Therefore, we need to dig out more properties of the closed-loop system. Take (51.10) into (51.3), we get

$$\begin{cases} \dot{x}_1 = x_2 \\ \dot{x}_2 = \dfrac{1}{d_{11}}\bar{m}g \sin x_1 - \dfrac{d_{12}}{d_{11}} \left(d_{12}kEx_2 - k_v x_3\right) \\ \dot{x}_3 = d_{12}kEx_2 - k_v x_3 \end{cases} \tag{51.12}$$

Lemma 1 *Under the control law (51.10),* $(x_1, x_2, x_3) = (\pi, 0, 0)$ *is an unstable equilibrium point for the closed-loop system (51.12).*

Proof To linearize (51.3) near $(x_1, x_2, x_3) = (\pi, 0, 0)$, we firstly take zero-shifting state $z = [z_1, z_2, z_3]^T = [x_1 - \pi, x_2, x_3]^T$. Linearize z at origin as

$$\begin{bmatrix} \dot{z}_1 \\ \dot{z}_2 \\ \dot{z}_3 \end{bmatrix} = \begin{bmatrix} 0 & 1 & 0 \\ -\dfrac{\bar{m}g}{d_{11}} & \dfrac{2\bar{m}gkd_{12}^2}{d_{11}} & \dfrac{d_{12}k_v}{d_{11}} \\ 0 & -2d_{12}k\bar{m}g & -k_v \end{bmatrix} z^T = Az^T \tag{51.13}$$

Compute the characteristic polynomial of A as

$$|sI - A| = s^3 + \left(k_v - \frac{2k\bar{m}gd_{12}^2}{d_{11}}\right)s^2 + \frac{\bar{m}g}{d_{11}}s + \frac{\bar{m}gk_v}{d_{11}} \tag{51.14}$$

and we get

$$\begin{vmatrix} 1 & \frac{\bar{m}g}{d_{11}} \\ k_v - \frac{2k\bar{m}gd_{12}^2}{d_{11}} & \frac{\bar{m}gk_v}{d_{11}} \end{vmatrix} \frac{1}{k_v - \frac{2k\bar{m}gd_{12}^2}{d_{11}}} < 0 \tag{51.15}$$

By using Routh Criteria, we know that A is not Hurwitz and verify $z = 0$ an unstable point for system (51.13). □

To utilize the instability of the IWP at $z = 0$, enlightened by Lemma 1, we initialize a minor motor torque for some time and then apply the real swing-up control law (51.10), within that the IWP is forced to depart from $z = 0$ and start to swing. Assume that the IWP is successfully triggered to swing from the bottom, consider the set

$$S = \{x \mid \dot{V} = 0\} \tag{51.16}$$

we obtain

$$\begin{cases} \dot{x}_3 = 0, \dot{E} = 0 \\ d_{12}Ex_2 - k_v x_3 = 0 \end{cases} \tag{51.17}$$

and $(E, x_3) \equiv (E_c, x_{3c})$ in the set S. There are four cases left to be discussed:

(1) $E_c \neq 0$ and $x_{3c} \neq 0$, we get none-zero constant $x_2 \equiv \frac{k_v x_{3c}}{d_{12}E_c}$ and a changing E, which contradicts $E \equiv E_c$.
(2) $E_c \neq 0$ and $x_{3c} = 0$, we get $x_2 = 0$, which only happens before triggering the swing-up process.
(3) $E_c = 0$ and $x_{3c} \neq 0$, which contradict (51.17) clearly.
(4) $E_c = 0$ and $x_{3c} = 0$, which establishes from every aspect.

Thus $S = \{x \mid E = 0 \textit{ and } x_3 = 0\}$ for a triggered IWP. With LaSalle invariant principle [12], we know that x would asymptotically move into the set S and

$$\lim_{t\to\infty} E = 0, \ \lim_{t\to\infty} x_3 = 0 \tag{51.18}$$

51.3.2 Balance Control

Linearize the system (51.3) at $x = [0, 0, 0]^T$ and get

$$\dot{x} = \begin{bmatrix} 0 & 1 & 0 \\ \frac{\bar{m}gd_{22}}{D} & 0 & 0 \\ \frac{-\bar{m}gd_{21}}{D} & 0 & 0 \end{bmatrix} x + \begin{bmatrix} 0 \\ \frac{-d_{12}}{D} \\ \frac{d_{11}}{D} \end{bmatrix} \tau = A_0 x + B_0 \tau \tag{51.19}$$

Through simple computation, we obtain $rank[B_0, A_0B_0, A_0^2B_0] = 3$, which verifies that the linearized system (51.19) is controllable. Thus we choose the state feedback control law below to balance the rod.

$$\tau = k_1x_1 + k_2x_2 + k_3x_3 \tag{51.20}$$

where k_1, k_2, k_3 are constant gains and $K = [k_1, k_2, k_3]$, which should be chosen such that $A_0 + B_0K$ is hurwitz.

Furthermore, as x gets closer to the set S during swing-up process, the maximum height of the swinging rod would approach top position while the rotating speed of the wheel descending to zero. Thus, adopting a crude switch from the swing-up control law to the balance control law near the upright vertical position is acceptable.

51.4 Experiment

In this section, we introduce the experiment based on a self-built hardware platform in the dorm room of the author. The parameters used in the experiment are shown in Table 51.1.

A popular STM32F407 microprocessor, a 90-watt DC motor and an amplifier with current-loop driven mode are adopted in our scheme. The angle of the rod is measured by a 1000-resolution absolute encoder, while a 512-resolution increasing quadrature encoder is applied to get the angle of motor. The signal of absolute encoder is detected by 10 parallel general IO pins of microprocessor. The increasing encoder sends out A/B quadrature pulses that automatically counted by the 16-bit counter of microprocessor with a overflow handle algorithm. In every 20-ms control interval, we sample the angle of the wheel 4 times while twice for the rod, and up-

Table 51.1 Parameters of the inertial wheel pendulum platform

Parameter	Value	Units	Parameter	Value	Units
l_1	0.137	m	I_2	0.00153	$kg \times m^2$
l_2	0.145	m	d_{11}	0.0155	$kg \times m^2$
m_1	0.383	kg	d_{12}	0.00153	$kg \times m^2$
m_2	0.308	kg	d_{21}	0.00153	$kg \times m^2$
I_1	0.000276	$kg \times m^2$	d_{22}	0.00153	$kg \times m^2$

date control input once. Both angular speed of the rod and the wheel are computed by difference within their sample interval. The arithmetic control input is computed out in the microprocessor and then transformed into two-way differential analog voltage signals to drive the amplifier and motor connected with it. To depict the dynamic process of IWP, the author use the serial port of microprocessor to send back the state information to his computer. Let single arrows represent the flow direction of energy or signal, the hardware structure is shown in Fig. 51.2.

The author implemented the control algorithm by using C-language, with Keil-μvision, and then download it into the microprocessor with a ST-Link debugger. For the step of adjusting the control parameters, we firstly trigger the swing-up process by applying a small torque within 0.1 s and then switch to the swing-up algorithm. Once the IWP enters the region of $\pm 10°$ ($\epsilon = 0.17$) relative to upright vertical position, we crudely switch from the swing-up control law to the balance control law. Through a short-time trial and error, we get the gain constants as

$$\begin{cases} k = 273.8, k_v = 0.062 \\ k_1 = 8, k_2 = 0.08, k_3 = 50 \end{cases} \tag{51.21}$$

which successfully swing up the rod and balance it at the upright vertical position. The experimental results are shown in Fig. 51.3, the snapshots of experimental video are shown in Fig. 51.4.

The experimental results show that the rod swings up to the upright region in only four swings, the position and speed of rod approach zero asymptotically after switching to the balance control law, while the wheel results with a none-zero rotating speed due to the unbalance offset of amplifier, friction of pivot and the affect of connected wires.

51.5 Conclusion

The control of inertial wheel pendulum is investigated in this work. Under the proposed swing-up control law, the IWP will run into a homoclinic orbit with a de-

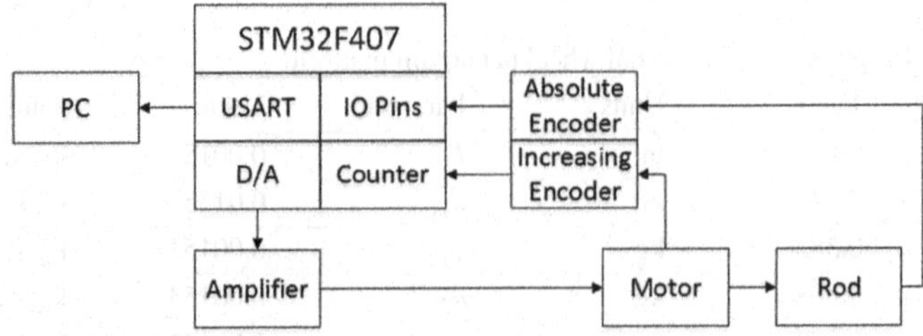

Fig. 51.2 Hardware structure of self-built IWP platform

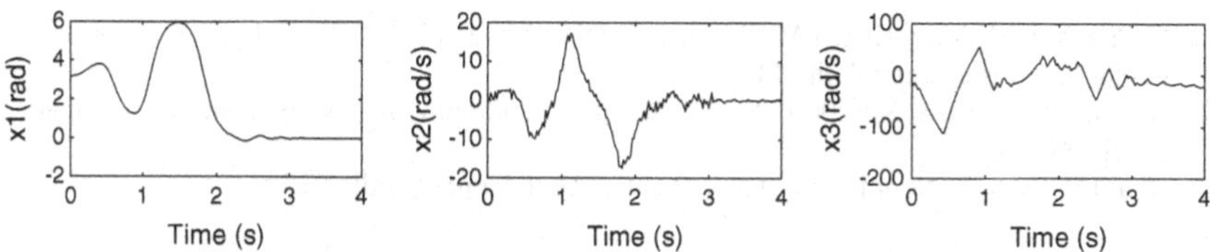

Fig. 51.3 Experimental results of inertial wheel pendulum

Fig. 51.4 Snapshots of experimental video

celerating wheel. The balance control law is quite similar with PD control which is popular among engineers. With a crude switching scheme, the validity of the proposed hybrid control law is verified by our apparatus. Moreover, the STM32-based hardware scheme shows a new implementation of IWP and a cost-saving way for DIY fans. To move on, the author would like to consider the inertial wheel pendulum with a movable pivot and update the apparatus in future work.

References

1. Spong MW, Vidyasagar M (1989) Robot dynamic and control. Wiley, NewYork
2. Olfati-Saber R (2001) Global stabilization of a flat underactuated system: the inertia wheel pendulum. IEEE Conf Decis Control 4:3764–3765
3. Spong MW, Corke P, Lozano R (2001) Nonlinear control of the reaction wheel pendulum. Automatica 37(11):1845–1851
4. Block DJ, Astrom KJ, Spong MW (2007) The reaction wheel pendulum. Synth Lect Control Mechatron 1(1):1–105
5. Hernandez VM (2003) A combined sliding mode-generalized pi control scheme for swinging up and balancing the inertia wheel pendulum. Asian J Control 5(4):620–625
6. Hernandez VM, Sira-Ramirez H (2003) Generalized PI control for swinging up and balancing the inertia wheel pendulum. Proc Am Control Conf 4:2809–2814
7. Iriarte Rafael, Aguilar Luis T, Fridman L (2013) Second order sliding mode tracking controller for inertia wheel pendulum. J Frankl Inst 350(1):92–106
8. Srinivas KN, Behera L (2008) Swing-up control strategies for a reaction wheel pendulum. Int J Syst Sci 39(12):1165–1177
9. Bobstov AA, Pyrkin AA, Kolyubin SA (2009) Adaptive stabilization of a reaction wheel pendulum on moving LEGO platform. ISIC, pp 1218–1223

10. Gajamohan M, Muehlebach M, Widmer T, D'Andrea R (2013) The cubli: a reaction wheel based 3D inverted pendulum. IMU 2(2)
11. Astrom KJ, Furuta K (2000) Swinging up a pendulum by energy control. Automatica 36(2):287–295
12. Khalil HK (2011) Nonlinear systems. Prentice Hall, Upper Saddle River

Chapter 52
Intermittent Synchronization of Cascaded Boolean Networks with Time Delays

Zhengdong Yang and Tianguang Chu

Abstract In this paper, we considered the intermittent synchronization of cascaded Boolean networks (CBNs) with time delays and get its criteria based on the results of synchronization of delay-free CBNs. We further discuss complete synchronization of CBNs with time delays from intermittent synchronization.

Keywords Cascade Boolean networks · Cyclic pattern · Time delay · General synchronization · Semi-tensor product

52.1 Introduction

Boolean networks (BNs) is an important model to describe the operation of gene regulatory networks [1]. A BN consists of a directed graph with nodes that represent genes or other elements. Each node assumes only two states: "on" or "off", referring occurrence of a gene transcription or not. In a BN, every node gets input from its neighbouring nodes and updates its state simultaneously according to their interactions described by Boolean functions.

An interesting issue in study of BNs addresses the synchronization of coupled BNs due to its potential applications [2, 3]. Recently, some analytical results for synchronization of deterministic BNs connected in "drive-response" configuration have been obtained, [4–6], by using the theory of semi-tensor product (STP) of matrices [7, 8]. These results mainly focus on the delay-free case. It is well known that the information flow in complex systems is not instantaneous in general but often involves time delays. This has raised increasing interest in the study of issues concerning synchronization of delayed BNs [9, 10].

Z. Yang (✉) · T. Chu
College of Engineering, Peking University, Beijing 100871,
People's Republic of China
e-mail: yangzd1980@sina.com

© Springer-Verlag Berlin Heidelberg 2016

Y. Jia et al. (eds.), *Proceedings of the 2015 Chinese Intelligent Systems Conference*, Lecture Notes in Electrical Engineering 359,
DOI 10.1007/978-3-662-48386-2_52

In this paper, we extend the study of general synchronization of CBNs with respect to (w.r.t.) the domain of attractions (DAs) presented in [11] to delayed CBNs. Particularly, we will reveal intermittent synchronization occurring in CBNs with finite delays and discuss its relation with complete synchronization.

52.2 Problem Statement

Our study is based on the algebraic form of BNs in terms of STP theory. We first recall some preliminaries of STP [7, 8].

Definition 1 Let A an $n \times m$ matrix and B a $p \times q$ matrix, the semi-product of A and B is defined as

$$A \ltimes B = (A \otimes I_{l/m})(B \otimes I_{l/p}),$$

where $\otimes$ is the Kronecker product and $l = lcm(m, p)$ is the least common multiple of m and p.

Clearly, STP reduces to the conventional matrix product when $m = p$. We can have a vector expression for a logic variable by using STP technique. To see this, we assign logical value with vectors by letting $T = 1 \sim \delta_2^1$ and $F = 0 \sim \delta_2^2$, where $\delta_2^1 = [1\ \ 0]^T$ and $\delta_2^2 = [0\ \ 1]^T$. Then a logical variable $X(t)$ can assume vector values as below:

$$X(t) \in D := \Delta_2 = \{\delta_2^1, \delta_2^2\}.$$

Any logical function $L(X_1, \ldots, X_n)$ with logical arguments $X_1, \ldots, X_n$ can be expressed in a multi-linear form as

$$L(X_1, \ldots, X_n) = M_L X_1 \cdots X_n,$$

where $M_L \in L_{2\times 2^n}$ is uniquely determined by $L(X_1, \ldots, X_n)$ and referred to as the structure matrix of L, see [8].

In this manner, a cascaded system of two CBNs with time delays can be expressed as following,

$$u_i(t+1) = g_i(u_1(t-\tau), \ldots, u_n(t-\tau)), \tag{52.1}$$

$$x_j(t+1) = f_j(u_1(t-\tau), \ldots, u_n(t-\tau), x_1(t-\tau'), \ldots, x_n(t-\tau')), \tag{52.2}$$

where $u_i \in D, i = 1, 2, \ldots, m$ and $x_j \in D, j = 1, 2, \ldots, n$ represent node state variables of the drive BNs (52.1) and the response BNs (52.2), respectively; g_i, f_j are Boolean functions, $g_i : D^m \longrightarrow D, i = 1, 2, \ldots, m; f_j : D^{n+m} \longrightarrow D, j = 1, 2, \ldots, n$; and τ and τ' are nonnegative integers signifying time delays; $t \in N = \{1, 2, \ldots\}$.

By invoking the STP technique, we can convert CBNs (52.1) and (52.2) into the following algebraic form

$$\begin{cases} u(t+1) = \quad Gu(t-\tau), & u \in D^m \\ x(t+1) = Lu(t-\tau)x(t-\tau'), & x \in D^n \end{cases} \tag{52.3}$$

where, $u(t) = u_1(t) \cdots u_m(t) \in \Delta_{2^m}$, $x(t) = x_1(t) \cdots x_n(t) \in \Delta_{2^n}$. $G \in \mathcal{L}_{2^m \times 2^m}$, $L \in \mathcal{L}_{2^m \times 2^{n+m}}$. $Lu(t-\tau)$ is the input-determined transition matrix [7]. For simplicity, we assume that the BN (52.1) has the same number of nodes as BN (52.2) does, i.e., $m = n$. We will show that the initial value sequence$\{u(0)\}_\tau = \{u(-\tau), u(1-\tau), \ldots, u(0)\}$, $\{x(0)\}_{\tau'} = \{x(-\tau'), x(1-\tau'), \ldots, x(0)\}$ and the relationship between τ and τ' will influence the patterns of synchronization phenomenon of CBNs with delays. The primary objective of this paper is to give the criteria of intermittent synchronization of the coupled systems.

To give the precise definition of intermittent sychronization, let us follow [11], assuming that for the delay-free case; i.e., $\tau = \tau' = 0$, BN (52.1) has p distinct DAs, $S_1, \ldots, S_p$. We set the coupled system without delays, $y(t) = u(t) \ltimes x(t)$, as BN (3'). For each DA, S_i, $1 \le i \le p$, BN (3') has q_i corresponding input-determined DAs, $S_{i1}, \ldots, S_{iq_i}$.

Definition 2 We call the intermittent synchronisation of CBNs (52.1) and (52.2) occurs w.r.t. the given DAs, S_i and S_{ij}, if there exist subsets $\{u(m-\tau), 0 \le m \le \tau\} \subset \{u(0)\}_\tau$, $\{x(n-\tau'), 0 \le n \le \tau'\} \subset \{x(0)\}_{\tau'}$, and a $k \in N$, such that

$$\begin{aligned} &x(T_{in}; x(-\tau'), x(1-\tau'), \ldots, x(0); u(-\tau), u(1-\tau), \ldots, u(0)) \\ &= u(T_{in}; u(-\tau), u(1-\tau), \ldots, u(0)), \quad t \ge k, \end{aligned}$$

whenever $\{u(m-\tau)\} \in S_i$ and $\{x(n-\tau')\} \in \{D\}_{S_{ij}}$, where $\{D\}_{S_{ij}}$ is an initial value set specified by S_{ij}, $T_{in} = T_1 + aT_a$, and $T_1, T_a \in N, a = 1, 2, \cdots$.

In Definition 2, T_1 is the transition time of the coupled system entering intermittent synchronisation, T_a is the fixed time interval for the occurrence of intermittent synchronisation.

For complete synchronisation of CBNs (52.1) and (52.2) w.r.t. given DAs, we have following definition.

Definition 3 We call the complete synchronisation of CBNs (52.1) and (52.2) w.r.t. the given DAs, S_i and S_{ij}, occurs, if there exists a $k \in N$, such that

$$\begin{aligned} &x(t; x(-\tau'), x(1-\tau'), \ldots, x(0); u(-\tau), u(1-\tau), \ldots, u(0)) \\ &= u(t; u(-\tau), u(1-\tau), \ldots, u(0)), \quad t \ge k, \end{aligned}$$

for any $\{u(0)\}_\tau \in S_i$ and any $\{x(0)\}_{\tau'} \in \{D\}_{S_{ij}}$.

52.3 Main Results

52.3.1 Delay-Free Formulation of Drive BN

Consider the delay-free case of drive BN (52.1′)

$$u(t+1) = Gu(t), \quad u \in D^n \tag{52.1′}$$

From [8], we know that BN (52.1′) has distinct DAs and each DA has only a cycle or a fixed point in it. The state of BN (52.1′) depends on the initial state $u(0)$. We divide BN (52.1) to $\tau + 1$ systems $U_m(t'_m)$ in time sequence.

$$U_m(t'_m + 1) = GU_m(t'_m), \; m = 1, 2, \ldots, \tau + 1 \tag{52.4}$$

where $U_m(t'_m) = u_1(t'_m) \ltimes \cdots \ltimes u_m(t'_m) \in D^m$. Therefore, $u(t)$ can be expressed as repeat of system sequence $U_m(t'_m)$. The step time t'_m of the m_{th} system, $U_m(t'_m)$, is $t = t'_m(0) + (\tau + 1)t'_m$, where the $t'_m(0)$ is the starting time the system $U_m(t'_m)$ begins, $t'_m(0) = -\tau - 1 + m$. The initial value of each $U_m(t'_m)$ is $U_m(0) = u(-\tau - 1 + m)$.

Obviously, each $U_m(t'_m)$ in (52.4) is delay-free as BN (52.1′). We can get the solution of $U_m(t'_m)$ by solving BN (52.1′) [8]. Then, we get the solution of delayed BN (52.1) by putting $U_m(t'_m)$ together repeating in time sequence.

52.3.2 Intermittent Synchronization

We have following results of intermittent synchronization for CBNs (52.1) and (52.2).

I. Case $\tau = \tau'$
Consider the delay-free case of the BN (52.2)

$$x(t+1) = Lu(t)x(t), \quad x \in D^n. \tag{52.2′}$$

We firstly divide CBNs (52.1) and (52.2) to systems $U_m(t'_m), X_n(t''_n), m = 1, \ldots, \tau+1$; $n = 1, \ldots, \tau' + 1$, in time sequence, and get

$$\begin{cases} U_m(t'_m + 1) = GU_m(t'_m), & U_m \in D^n \\ X_n(t''_n + 1) = LU_m(t'_m)X_i(t''_n), & X_n \in D^n \end{cases} \tag{52.5}$$

where $U_m = u_1(t'_m) \ltimes \cdots \ltimes u_n(t'_m)$, $X_n = x_1(t''_n) \ltimes \cdots \ltimes x_n(t''_n)$. We denote t'_m, t''_n as step time for each $U_m(t'_m)$ and $X_n(t''_n)$, we have

$$t = t'_m(0) + (\tau' + 1)t'_m, \qquad t = t''_n(0) + (\tau + 1)t''_n, \tag{52.6}$$

where $t'_m(0)$ and $t''_n(0)$ means the starting time of $U_m(t'_m)$ and $X_n(t''_n)$, respectively. Because $\tau = \tau'$, $U_m(t'_m)$ and $X_n(t''_n)$ have one-to-one correspondence, we get $t'_m = t''_n$, if $m = n$. By merging subscript m and n as m,

$$\begin{cases} U_m(t'_m+1) = \quad GU_m(t'_m) \\ X_m(t'_m+1) = LU_m(t'_m)X_m(t'_m), \end{cases} \tag{52.7}$$

with the initial value $U_m(0) = u((m-1)-\tau)$ and $X_m(0) = x((m-1)-\tau)$ we get a new coupled system without delays, with $t'_m = (t - t_m(0)')/(\tau+1)$, where $t'_m(0) = -\tau - 1 + m$. Obviously, (52.7) is identical to delay-free system CBNs (52.1′) and (52.2′). From [11], if complete synchronization in (52.7) occurs within specified DAs, S_i and S_{ij}, then CBNs (52.1) and (52.2) will achieve intermittent synchronization, the fixed time interval is $\tau + 1$.

Theorem 1 *Let $\tau = \tau'$. Assume that complete synchronisation of CBNs (52.1′) and (52.2′) occurs w.r.t. S_i and S_{ij}. If the initial values of $u(-(\tau+1)+m)$ and $x(-(\tau+1)+n)$ satisfy*

a. $m = n$,

b. $u(-(\tau+1)+m) \in S_i$,

c. $u(-(\tau+1)+m) \ltimes x(-(\tau+1)+n)) \in S_{ij}, \qquad 1 \le m, n \le \tau+1$

Then intermittent synchronization between CBNs (52.1) and (52.2) occurs w.r.t. S_i and S_{ij}, and the fixed time interval is $(\tau+1)$.

The proof can be yield from the observation that $\{D\}_{S_{ij}} = \{\bigcup_{u\in S_i} \{x_u : u \ltimes x \in S_{ij}\}\}$, x_u means each x chosen must subject to u at each time. We have complete synchronisation if Theorem 1 holds for all $m = 1, 2, \ldots, \tau+1$.

II. Case $\tau < \tau'$ & $\tau = \tau' mod(\tau+1)$

Similarly, we divide BN (52.1) and (52.2) into systems, $U_m(t'_m)$ and $X_n(t''_n)$, in time sequence respectively as (52.5). We assume $\tau' = ak_{ui}(\tau+1)+\tau$, $a \in Z_0^+$, k_{ui} is the cycle length of BN (52.1′) with specific DA, S_i. Considering the correspondence between $U_m(t'_m)$ and $X_n(t''_n)$, we knows that a subsystem U_m will drive ak_{ui}+1 systems X_n in the dynamical process. We denote the systems X_n driven by U_m as a group,

$$U_m(t'_m) \to \{X_n(t''_n) : n = m, (\tau+1)+m, 2(\tau+1)+m, \ldots, ak_{ui}(\tau+1)+m\}_{U_m}.$$

and we get $n = a_0 k_{cui}\tau + m$, $a_0 \in Z_0^+$, $a_0 \le a$. For each $U_m(0)$, $X_n(0)$, we have $U_m(0) = u(-\tau+(m-1)), X_n(0) = x(-\tau'+(n-1))$, with their starting time $t'_m(0) = -\tau+(m-1), t''_n(0) = -\tau'+(n-1)$. Then we show the system (52.5) in this case is equivalent to delay-free BN (3') under following condition.

Assume $U_m(0)$ is the point on its cycle $C_{ui} = \{u_{i1}, u_{i2}, \ldots, u_{ik_{ui}}\}$, where k_{ui} is the cycle length. The initial state of $U_m(0)$ which drives $X_n(0)$ is

$$U_m(0) = GU_m(a_0) = G^{a_0}U_m(0).$$

From (52.6), we have $t = a_0\tau + m, t'_m(0) = a_0, t''_n(0) = 0$. Then we obtain $t = a_0\tau + m + (\tau' + 1) = a_0\tau + m + (ak_{i0}(\tau + 1) + \tau + 1)$, $t'_m = a_0 + (ak_{ui} + 1)$. For $U_m(0)$ is on the cycle of C_{ui}, we get

$$U_m(t'' + 1) = U_m(t' + 1).$$

Simultaneously, we get $U_m(t'' + 1) = GU_m(t'') = U_m(t' + 1) = GU_m(t')$. Therefore, (52.5) can be rewritten as following

$$\begin{cases} U_m(t'' + 1) = \quad GU_m(t'') \\ X_m(t'' + 1) = LU_m(t'')X_m(t''). \end{cases} \tag{52.8}$$

It is just as delay-free coupled system BN (52.1′) and (52.2′) and complete synchronization within specified DAs occurs, if $U_m(0) = u((-\tau + 1) + m) \in S_i, X_n(0) = x((-\tau + 1) + n) \in S_{ij}$ [11]. Therefore, the intermittent synchronization of CBNs (52.1) and (52.2) occurs and the fixed time interval is $(\tau' + 1)$.

When $U_m(0)$ is not on the cycle, there must exist a transition time T_p, $U_m(T_p)$ just enters its cycle. If $U_m(T_p) \ltimes X_n(T_p) \in S_{ij}$, (52.5) can also be written as (52.8).

Theorem 2 *Let $\tau < \tau'$ and $\tau = \tau' mod(\tau+1)$. Assume that complete synchronisation of CBNs (52.1′) and (52.2′) occurs w.r.t. S_i and S_{ij}, and there is a transition time T_p, $U_m(T_p)$ is on its cycle. If the initial values of $u(-(\tau + 1) + m)$ and $x(-(\tau + 1) + n)$ satisfy*

a. $n = n_0, (\tau + 1) + n_0, \ldots, ak_{ui}(\tau + 1) + n_0,$ $\quad 1 \le n_0 < \tau' + 1$
b. $U_m(T_p) \in S_i,$ $\quad 1 \le m \le \tau + 1$
c. $U_m(T_p) \ltimes X_n(T_p) \in S_{ij},$ $\quad 1 \le n \le \tau' + 1$
d. $\tau' = ak_{ui}(\tau + 1) + \tau,$ $\quad a \in Z_0^+$
e. $n = a_0k_{ui}\tau + m,$ $\quad a_0 \in Z_0^+, a_0 \le a$

Then intermittent synchronization between CBNs (52.1) and (52.2) occurs w.r.t. S_i and S_{ij},

The proof can be yield from the observation that $\{D\}_{S_{ij}} = \{\bigcup_{u \in S_i} \{x_u : u_m(T_p) \ltimes x_n(T_p) \in S_{ij}\}\}$, where the initial value of $u_m(T_p), x_n(T_p)$ are $u(-\tau + m + 1), x(-\tau' + n + 1)$, respectively. We have complete synchronisation if Theorem 2 holds for all $m = 1, 2, \ldots, \tau + 1$.

III. Case $\tau > \tau'$ & $\tau = \tau' mod(\tau' + 1)$
Similarly, we divide BN (52.1), (52.2) into systems, $U_m(t'_m)$ and $X_n(t''_n)$, in time sequence respectively as (52.5). We set $\tau = ak_{ui}(\tau' + 1) + \tau$, $a \in Z_0^+$, k_{ui} is the cycle length of BN (52.1′) with specific DA, S_i. Considering the correspondence

between $U_m(t'_m)$ and $X_n(t''_n)$, we knows that $ak_{ui}+1$ systems U_m will drive a systems X_n in the dynamical process. We denote the systems U_m driving X_n as a group,

$$\{U_m(t'_m) : m,\ (\tau'+1)+m,\ 2(\tau'+1)+m, \ldots, ak_{ui}(\tau'+1)+m\}_{X_n} \to X_n(t''_n),$$

we get $m = a_0 k_{cui}\tau + n,\ a_0 \in Z_0^+,\ a_0 \le a$. Then we present analysis that the system (52.5) is equivalent to delay-free coupling BN (3') with complete synchronization within specific DAs under following condition.

Assume there exists finite initial state subsequence$\{U_m(0), m = m_0, m_0+(\tau'+1), \ldots, m_0+(k_{ui}+1)(\tau'+1)\} \in \{u(0)\}_\tau$ are all on cycle $C_{ui} : \{u_{i1}, u_{i2}, \ldots, u_{ik_{ui}}\}$, k_{ui} is the cycle length, and they are arranged in cyclic succession. We easily have, $m_0 = a_0\tau' + n,\ a_0 \in Z_0^+,\ a_0 \le a$. We get $t = a_0\tau + m,\ t'_m = a_0,\ t''_n = 0$, and $t = a_0\tau + m + (\tau'+1) = a_0\tau + m + (ak_{i0}(\tau+1)+\tau+1), t'_m = a_0 + (ak_{ui}+1)$.

From above, we have $U_m(t'+1) = U_m(t''+1)$ and we get

$$U_m(t''+1) = GU_m(t'') = U_m(t'+1) = GU_m(t').$$

System (52.5) is equivalent to (52.8).

If $\{U_m(0)\} \in S_i$ and $U_m(0) \ltimes X_n(0) \in S_{ij}$, the complete synchronization occurs within specified DAs in system (52.5). Therefore, the intermittent synchronization of CBNs (52.1) and (52.2) occurs and the fixed time interval is $(\tau'+1)$.

When $U_m(0),\ m = m_0,\ m_0+(\tau'+1),\ m_0+2(\tau'+1), \ldots,\ m_0+(k_{ui}+1)(\tau'+1)$ is not or not all on the cycle, there must exists a transition time T_p, all $U_m(T_p)$ just enters their cycles and if their states are arranged in cyclic succession, (52.5) can also be written as (52.8).

Theorem 3 *Let $\tau > \tau'$ and $\tau = \tau' mod(\tau'+1)$. Assume that complete synchronisation of CBNs (52.1′) and (52.2′) occurs w.r.t. S_i and S_{ij}, and there is a transition time T_p, all $U_m(T_p)$s are on their cycles and arranged in cyclic succession. If the initial values of $u(-(\tau+1)+m)$ and $x(-(\tau+1)+n)$ satisfy*

$$\begin{aligned}
&a.\ \ m = m_0, m_0+(\tau'+1), \ldots, m_0+(k_{ui}+1)(\tau'+1), && 1 \le m_0 < \tau+1\\
&b.\ \ U_m(T_p) \in S_i, && 1 \le m \le \tau+1\\
&c.\ \ U_m(T_p) \ltimes X_n(T_p) \in S_{ij}, && 1 \le n \le \tau'+1\\
&d.\ \ \tau = ak_{ui}(\tau'+1)+\tau', && a \in Z_0^+\\
&e.\ \ m_0 = a_0\tau' + n, && a_0 \in Z_0^+,\ a_0 \le a
\end{aligned}$$

Then intermittent synchronization between CBNs (52.1) and (52.2) occurs w.r.t. S_i and S_{ij}, and the fixed time interval is $(\tau'+1)$.

The proof can be yield from the observation that $\{D\}_{S_{ij}} = \{\bigcup_{u \in S_i} \{x_u : u_m(T_p) \ltimes x_n(T_p) \in S_{ij}\}\}$, where the initial value of $U_m(T_p), X_n(T_p)$ are $u(-\tau+m+1), x(-\tau'+1+n)$ respectively. Furthermore, $u(-\tau+1+m)$ chosen must be made to satisfy $U_m(T_p)$s arranged in cyclic succession. We have complete synchronisation if Theorem 3 holds for all $n = 1, 2, \ldots, \tau' + 1$.

IV. Case $\tau \neq \tau' \ mod(\tau' + 1)$
Under this circumstance, CBNs are hard to achieve intermittent or complete synchronization unless the corresponding CBNs without delays have some special dynamical property, we have the following theorem and its proof is obvious.

Theorem 4 *Let* $\tau \neq \tau' \ mod(\tau' + 1)$. *Assume that BN (52.1′) has a fixed point* C_{up} *as global attractor, and BN (52.2′) has a global input-determined a fixed point* C_{xp}. *If* $C_{up} = C_{xp}$, *the delayed systems, CBNs (52.1) and (52.2), will achieve complete synchronisation.*

References

1. Kauffman SA (1969) Metabolic stability and epigenesis in randomly constructed genetic nets[J]. J Theor Biol 22(3):437–467
2. Zhou C, Zemanova L, Zamora-Lopez G, Hilgetag CC, Kurths J (2007) Structure-function relationship in complex brain networks expressed by hierarchical synchronization[J]. New J Phys 9:178
3. Garcia-Ojalvo J, Elowitz MB, Strogatz SH (2004) Modeling a synthetic multicellular clock: repressilators coupled by quorum sensing[P]. Proc Natl Acad Sci USA 101(30):10955–10960
4. Li R, Chu TG (2012) Complete synchronization of Boolean networks[J]. IEEE Trans Neural Networks Learn Syst 23(5):840–846
5. Li R, Chu TG (2012) Synchronization in an array of coupled Boolean networks[J]. Phys Lett A 376(45):1486–1493
6. Li R, Yang M, Chu TG (2012) Synchronization of Boolean networks with time delays[J]. Appl Math Comput 219(3):917–927
7. Cheng D, Qi H (2010) A linear representation of dynamics of Boolean networks[J]. IEEE Trans Autom Control 55(10):2251–2258
8. Cheng D, Qi H, Li Z (2011) Analysis and control of Boolean networks: a semi-tensor product approach[M]. Springer, London
9. Li Rui, Yang Meng, Chu Tianguang (2012) Synchronization of Boolean networks with time delays[J]. Appl Math Comput 219(3):917–927
10. Li Fangfei, Xiwen Lu (2013) Complete synchronization of temporal Boolean networks[J]. Neural Networks 44:72–77
11. Yang Z, Chu T (2015) General synchronization of cascaded boolean networks within different domains of attraction. In: Proceedings of the International Conference of Control, Dynamic Systems, and Robotics. Ottawa, Ontario, Canada, p 173

Chapter 53
Overhead Transmission Line Condition Evaluation Based on Improved Scatter Degree Method

Guoqiang Sun, Chao Zhu, Ming Ni and Zhinong Wei

Abstract As an important part of the transmission network, the running state of overhead transmission line will directly affect the reliability of the whole power system. With a single element reliability model, it is hard to exactly reflect the true condition of the complicated system with multi hierarchies and multi indicators, which means it is difficult to conduct an accurate comprehensive evaluation on overhead transmission line. This paper proposes an overhead transmission line evaluation model based on improved scatter degree method. The proposed model combines the subjective and objective combination weighting method, which can not only calculate the line units, but also evaluate the running state of the whole overhead transmission line. Finally, this model is used to assess overhead transmission lines in a certain area and results show that the proposed method is effective and feasible.

Keywords Improved scatter degree method · Condition evaluation · The subjective weight · The objective weight

53.1 Introduction

Overhead transmission line is an important part of the power system. The running state will directly affect the reliability of the operation of the whole power system. At present, many scholars have studied the condition assessment of overhead transmission line. A method based on circuit element reliability model was proposed to study the overhead transmission line [1–3], and the analysis process of the method includes state selection, state estimation and indicator calculation. Compared with the circuit element reliability model, the transmission line risk

G. Sun · C. Zhu (✉) · Z. Wei
College of Energy and Electrical Engineering, Hohai University, Nanjing 210098, China
e-mail: zhuchao198811@163.com

M. Ni
State Grid Electric Power Research Institute, Nanjing 210003, China

© Springer-Verlag Berlin Heidelberg 2016
Y. Jia et al. (eds.), *Proceedings of the 2015 Chinese Intelligent Systems Conference*, Lecture Notes in Electrical Engineering 359,
DOI 10.1007/978-3-662-48386-2_53

assessment model is a short time evaluation model [4–7], and the line risk assessment indicators are obtained by putting the time-varying circuit element failure rate model into the line risk assessment model. However, the reliability model and the risk assessment model both use a single circuit element model to replace the whole evaluation system, which is hard to exactly reflect the true state of the complicated system with multi hierarchies and multi indicators. As a result, the evaluation results of these methods have some limitations.

This paper proposes an overhead transmission line condition assessment model based on improved scatter degree method. First, the order relation method is used to calculate the subjective weights of each evaluation indicator and line unit. Second, the scatter degree method is adopted to calculate the objective weights. At last, the function value is calculated according to the linear evaluation function and the line running state is evaluated based on the comprehensive evaluation function value.

53.2 The Establishment of the Evaluation Indicator System of Overhead Transmission Line

53.2.1 The Establishment of the Line Evaluation Indicator System

The establishment of the condition evaluation indicator system is the premise and foundation of comprehensive evaluation. At present, the main monitoring items of online monitoring system are conductor sag, conductor temperature, insulators for wind, micro meteorological and so on. According to “Guide for Condition Evaluation of Overhead Transmission Line” and “Patrol system for overhead lines”, which are promulgated in 2007, this paper brings the foundation, tower, conductor, insulator, ancillary facilities of the five main line units into the overhead transmission line condition evaluation index system. The indicator system is shown in Table 53.1.

53.2.2 Data Preprocessing

The condition evaluation indicator system of overhead transmission line is complex and the monitoring of the indicators mainly relies on line inspection. The collected data from online monitoring system is quantitative data, such as the temperature of conductor and the insulator wind angle. While the line inspection data is qualitatively used to describe some mechanical components, such as the tower corrosion. In order to make the integration of inspection data and online monitoring data, the qualitative data needs to be quantified.

Table 53.1 Condition evaluation index system of overhead transmission line

Line unit	Line unit index
Foundation $s_1^{(1)}$	The surface damage of tower foundation x_1
	The damage of foundation slope and flood control facilities x_2
	The basic situation of metal corrosion x_3
	Anti-collision facilities x_4
Tower $s_2^{(1)}$	The inclined situation of tower x_5
	The construction deletion of tower and steel tube tower x_6
	The corrosion situation of tower and steel bar x_7
	The damage of cable corrosion x_8
	Concrete crack x_9
Conductor $s_3^{(1)}$	Corrosion, broken and flashover x_{10}
	Foreign bodies in suspension x_{11}
	Abnormal vibration, galloping and icing x_{12}
	Conductor sag x_{13}
Insulator $s_4^{(1)}$	Insulator pollution situation x_{14}
	Inclination of the insulator string x_{15}
	Porcelain insulator glaze broken situation x_{16}
Ancillary Facilities $s_5^{(1)}$	The defect of tower marking x_{17}
	The damage of lightning protection facilities x_{18}
	The damage of anti-bird facilities x_{19}
	The defect of guardrail and ladder stand x_{20}

According to "Guide for Condition Evaluation of Overhead Transmission Line", the line state deterioration degree is divided between 0–10 points. The more points, the more serious the state deterioration degree will be, and the indicator state will be worse. Then, the collected inspection data is compared with the evaluation criterion, and quantitated into 4 sections: 0–2 shows that the state is normal, 2–4 shows that the state needs attention, 4–8 shows that the state is abnormal, and 8–10 shows that the state is critical.

Most of the indicators are very-small indicators, which means that the smaller the index is, the better the line state will be. In order to make the calculation and analysis easy, this paper unifies all indicators to very-small index. For a spot of very-large and intermediate indicators, the following formulas are presented to unify them.

For very-large index:

$$x' = M - x \tag{53.1}$$

where M is the allowable upper bound of indicator x. For convenience, the x is usually transformed to x.

For intermediate index:

$$y^* = \begin{cases} \frac{2[y-m]}{M-m} \left(m \le y \le \frac{M+m}{2}\right) \\ \frac{2[M-y]}{M-m} \left(\frac{M+m}{2} \le y \le M\right) \end{cases} \tag{53.2}$$

where M is the allowable upper bound of indicator y, m is the allowable lower bound of indicator y. For convenience, the y^* is usually transformed to y.

At last, to dimensionless all indicators, the linear scaling method is adopted as follows:

$$x^* = \frac{x}{x^s} \tag{53.3}$$

where, x^s can be the minimum, maximum or average value. In this paper, for the very-small and transformed very-large index, x^s is 10. For intermediate index, considering only the conductor sag is belonged to the middle type indicator, x^s can be taken as the designed sag of conductor.

53.3 Determine the Weights of Evaluation Indicators

Determining the weight coefficients is the core problem in comprehensive evaluation. In general, there are three methods to determine weight coefficients. The first is subjective weighting method based on the principle of "function driven", such as analytic hierarchy process, the least square method, and order relation method. Although the method directly reflects the evaluator's subjective judgment and intuition, it may have certain subjectivity in comprehensive evaluation results or sorting. The second is the objective weighting method based on the principle of "difference driven", such as the variation coefficient method, entropy method, and the scatter degree method. Although it usually uses the complete mathematical theory and method, it ignores the subjective information of decision makers. The third is the integrated weighting method which can combine the objective and subjective weighting methods from logic. The method can reflect the principle of "function driven" and "difference driven", and display the subjective preferences of decision makers based on data mining.

Three commonly used integrated weighting methods are presented as follows: "addition" integrated method, "multiplication" integrated method, and improved scatter degree method. The first two methods are both based on Lagrange extreme value theory to weight the subjective and objective weights. Compared to the first two methods, the improved scatter degree method is a kind of dynamic weighting method. The observed data are weighted twice: the first weighting is based on the important degree of every evaluation indicators relative to the evaluation target, and the second weighting is to reflect the difference between each evaluation objects from the whole, which makes a certain distinction between objects. As a result, this paper uses the improved scatter degree method.

53.3.1 Determine the Subjective Weights by Order Relation Method

Order relation method is a new method that does not need consistency check. It is a subjective weighting method which first sorts the evaluation index qualitatively, then judges the importance of adjacent indicator, and finally calculates the indicator quantitatively [8].

The method has the advantages of no need of judgment matrix or consistency checking. Compared with the judgment matrix structured by analytic hierarchy process (AHP), the computation reduced rapidly. The calculation process is described as follows:

1. Determine the order relation. Through the analysis of the order relation and the adjustment of the position number, the order relation of the indicators can be determined as $x_1 \succ x_2 \succ \cdots x_m$, where m is the number of indicators.
2. Determine the relative importance between adjacent indicator. The importance ratio of x_{j-1} and x_j is judged by experts as follows:

$$w_{j-1}/w_j = r_j, j = m, m-1, \ldots, 2 \tag{53.4}$$

   The assignment of r_j can refer to Table 53.2.
3. Calculate the weight coefficient w_j. If the experts have given the assignment of r_j after rational judgment

$$w_j = (1 + \sum_{j=2}^{m} \prod_{k=j}^{m} r_i)^{-1} \tag{53.5}$$

   Then the other subjective weights can be calculated by w_j

$$w_{j-1} = r_j w_j \tag{53.6}$$

Table 53.2 Assignment reference table of r_j [9]

R_j	Illustration
1.0	Index x_{j-1} and index x_j has the same importance
1.2	Index x_{j-1} is a little more important than index x_j
1.4	Index x_{j-1} is obviously important than index x_j
1.6	Index x_{j-1} is strong important than index x_j
1.8	Index x_{j-1} is extreme important than index x_j

53.3.2 Determine the Objective Weights by Scatter Degree Method

The line evaluation indicator system shown in Table 53.1 indicates that each overhead transmission line can be regarded as a large system. The system is composed of five subsystems, and each subsystem corresponds to different amounts of state index.

1. For large systems: S_1, S_2, …, S_n, each system includes foundation, tower, conductor, insulator and ancillary facilities.
2. The observation data of these n systems: $\{x_{ij}^{(1,t)}\}$ is the observation value of indicator j from subsystem t of the first level in system i. These observations are assumed to be standard observation value which has been minimized and dimensionless.

Regard the linear weighted function which consists of the indicator observation value $x_{i1}^{(1,t)}$, $x_{i2}^{(1,t)}$, …, $x_{imt}^{(1,t)}$ from subsystem t in system i as the comprehensive evaluation function:

$$y_i^{(1,t)} = w_1 x_{i1}^{(1,t)} + \cdots + w_{mt} x_{imt}^{(1,t)} = w_{mt}^T x_i^{(1,t)} \tag{53.7}$$

where mt means the indicator number of subsystem t, $\mathbf{w}_{mt} = (w_1, w_2, \ldots, w_{mt})^T$ is the subjective weights calculated by the order relation method.

Assuming

$$Y^{(1,t)} = (y_1^{(1,t)}, y_2^{(1,t)}, \ldots y_n^{(1,t)})^T \tag{53.8}$$

$$X^{(1,t)} = (x_1^{(1,t)}, x_2^{(1,t)}, \ldots x_n^{(1,t)})^T \tag{53.9}$$

Then

$$Y^{(1,t)} = \mathbf{w}_{mt}^{\mathrm{T}} X^{(1,t)} \tag{53.10}$$

The difference from $\boldsymbol{x}_1^{(1,t)}$, $\boldsymbol{x}_2^{(1,t)}$, …, $\boldsymbol{x}_n^{(1,t)}$ to $\boldsymbol{x}^{(1,t)}$ can be expressed by total deviation sum of squares:

$$\max \sigma^2 = \frac{1}{n} \sum_{i=1}^{n} (y_i^{(1,t)} - \bar{y}^{(1,t)})^2 \tag{53.11}$$

Due to the original data has been standardized, the value of $\bar{y}^{(1,t)}$ is zero.

Then

$$\max \sigma^2 = \frac{1}{n} \sum_{i=1}^{n} (y_i^{(1,t)})^2 = w_{mt}^T H^{(1,t)} w_{mt} \tag{53.12}$$

where $\boldsymbol{H}^{(1,t)} = \boldsymbol{X}^{(1,t)}(\boldsymbol{X}^{(1,t)})^{\mathrm{T}}$ is real symmetric matrix. If $\boldsymbol{X}^{(1,t)}(\boldsymbol{X}^{(1,t)})^{\mathrm{T}} = 1$, then the maximum value of (53.12) can be calculated by choosing $\boldsymbol{w}_{mt}$ to match the formula:

$$\begin{aligned} &\max w_{mt}^T H^{(1,t)} w_{mt} \\ &s.t.\, w_{mt}^T w_{mt} = 1 \\ &w_{mt} > 0 \end{aligned} \tag{53.13}$$

When $\boldsymbol{w}_{mt}$ is the standard feature vector of the largest eigenvalue of $\boldsymbol{H}^{(1,t)}$, formula (53.13) will get the maximum value. And the new weight $\boldsymbol{w}_{mt}'$ is the objective weight of subsystem t got by the scatter degree method.

Note that the objective weight $\boldsymbol{w}_{mt}'$ does not has the "inheritance", and it will change when indicator value $\boldsymbol{X}^{(1,t)}$ changes.

53.4 Condition Assessment of Overhead Transmission Line

According to Table 53.1, take a certain area of n overhead transmission lines as the objects of assessment and record them as $S_1, S_2, \ldots, S_n$. Evaluate these n overhead transmission lines and their subsystems.

53.4.1 Comprehensive Evaluation Function of the First Level Subsystem

Before establishing the comprehensive evaluation function, the indicators need to be weighted:

$$x_{ij}^{*(1,t)} = w_j^{(1,t)} x_{ij}^{(1,t)} \tag{53.14}$$

where $w_j^{(1,t)}$ is the subjective weight calculated by formula (53.5).

Regard the linear weighted function which consists of $x_{i1}^{*(1,t)}, x_{i2}^{*(1,t)}, \ldots, x_{imt}^{*(1,t)}$ as the comprehensive evaluation function:

$$y_i^{(1,t)} = \sum_{j=1}^{mt} b_j^{(1,t)} x_{ij}^{*(1,t)} \tag{53.15}$$

where $b_j^{(1,t)}$ is the constant and used as the objective weight calculated by scatter degree method.

Assuming

$$b^{(1,t)} = (b_1^{(1,t)}, \ldots, b_{mt}^{(1,t)})^T \tag{53.16}$$

$$y^{(1,t)} = (y_1^{(1,t)}, \ldots, y_n^{(1,t)})^T \tag{53.17}$$

$$A^{(1,t)} = \begin{bmatrix} x_{11}^{(1,t)} & \cdots & x_{1mt}^{(1,t)} \\ \cdots & \cdots & \cdots \\ x_n^{(1,t)} & \cdots & x_{nmt}^{(1,t)} \end{bmatrix} \tag{53.18}$$

Then formula (53.15) will be abbreviated as

$$\mathbf{y}^{(1,t)} = A^{(1,t)} b^{(1,t)} \tag{53.19}$$

According to scatter degree method, $\boldsymbol{b}^{(1,t)}(\boldsymbol{b}^{(1,t)}$ can be obtained as the standard feature vector of the largest eigenvalue of $\boldsymbol{H}^{(1,t)} = (\boldsymbol{A}^{(1,t)})^T \boldsymbol{A}^{(1,t)})$. Put $\boldsymbol{b}^{(1,t)}$ into formula (53.15), then the comprehensive evaluation function of subsystem $s_t^{(1)}$ will be obtained.

53.4.2 *Comprehensive Evaluation Function of the Large System S_i*

Before establishing the comprehensive evaluation function of the large system through functions of the subsystems, the functions of subsystems need to be weighted:

$$y_i^{*(1,t)} = w^{(1,t)} y_i^{(1,t)} \tag{53.20}$$

where $w^{(1,t)}$ is the weight corresponding $y_i^{(1,t)}$.

Assuming

$$\mathrm{Y}_i = \sum_{t=1}^{5} b_t y_i^{*(1,t)} \tag{53.21}$$

where $\mathrm{Y_i}$ is the comprehensive evaluation function of the large system S_i, and b_t is the constant.

53.4.3 Line Condition Assessment Based on Comprehensive Evaluation Function Value

Due to each indicator of overhead transmission lines is dimensionless value, and the weight of each indicator is the standard weight (sum of the indicator weights is 1), the comprehensive evaluation function value should be a number between 0 and 1. The indicators are very-small index, which means that the smaller the indicator is, the better the line state will be.

According to "Guide for Condition Evaluation of Overhead Transmission Line", the running state of the line is divided into four states: normal, attention, abnormal and serious. This paper attempts to link comprehensive evaluation function value and these four states. According to the method proposed in Sect. 1.2, the comprehensive evaluation function value may be divided into four parts: 0–0.2, 0.2–0.4, 0.4–0.8, 0.8–1.0, which respectively corresponds to the normal, attention, abnormal and serious state. A real example is presented to demonstrate the reasonability of the proposed method.

53.5 The Example Analysis

The improved scatter degree method is used to evaluate 3 overhead transmission lines of 110 kV in Jiangsu province. Each line consists of 5 subsystems: foundation, tower, conductor, insulator, and ancillary facilities. The large system totally has 20 evaluation indicators and the indicator system is shown in Table 53.1. The data is collected and pretreated according to Sect. 1.2. Type (53.5) and (53.6) are applied to establish the line evaluation indicator subjective weights based on the order relation method. The result is shown in Table 53.3.

The comprehensive evaluation value of the first level subsystems is calculated according to Sect. 1.3, as shown in Table 53.4.

Type (53.5) and (53.6) are applied again to establish the subsystem subjective weights: 0.1882, 0.2258, 0.2258, 0.2258 and 0.1344. At last, the comprehensive evaluation value of the whole system is calculated according to Sect. 53.2, as shown in Figs. 53.1 and 53.2.

It can be seen from Table 53.4 and Fig. 53.1, the comprehensive evaluation value of subsystems of line 2 is bigger than that of the other two lines, and the final

Table 53.3 Evaluation index of subjective weights based on the order relation method

The first subsystem	Weights
Foundation $S_1^{(1)}$	0.2609 0.2609 0.26029 0.2173
Tower $S_2^{(1)}$	0.2 0.2 0.2 0.2 0.2
Conductor $S_3^{(1)}$	0.25 0.25 0.25 0.25
Insulator $S_4^{(1)}$	0.4118 0.2941 0.2941
Ancillary facilities $S_5^{(1)}$	0.2728 0.2728 0.2272 0.2272

Table 53.4 The comprehensive evaluation value of the sub system

Subsystem	Line 1	Line 2	Line 3
Foundation	0.09	0.11	0.06
Tower	0.08	0.12	0.05
Conductor	0.11	0.17	0.09
Insulator	0.08	0.13	0.06
Ancillary facilities	0.06	0.11	0.05

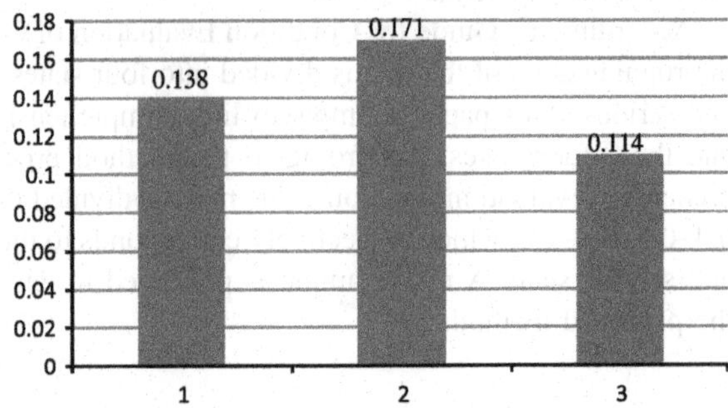

Fig. 53.1 The evaluation result based on this method

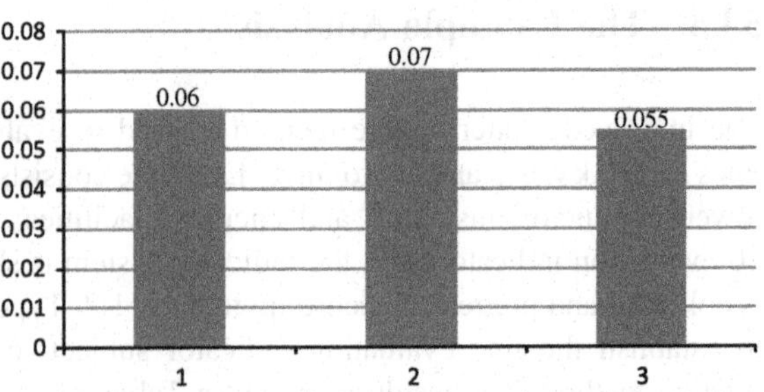

Fig. 53.2 The evaluation result based on AHP

system function value of line 2 is the biggest too. Considering the comprehensive evaluation function value is a very-small number, the state of line 3 is best due to its smallest function value while the state of line 2 is the worst.

As shown in Fig. 53.2, the condition evaluation result based on AHP is the same as the method proposed in this paper, which can both conclude that the state of line 3 is the best and the line 2 is the worst. However, the comprehensive evaluation value obtained by those two methods is different. Although AHP is a kind of comprehensive evaluation method which can analyze the multi-level and multi index system, it only weights the indicators once and can't make a difference in the evaluation objects, and its assessment result is too conservative. On the contrary, the improved scatter degree method can not only weight the indicators according to the importance of the indicators, and avoid the influence of some unimportant index on the evaluation result, but also make a difference in the evaluation objects and make them easy to compare.

53.6 Conclusion

This paper proposes an overhead transmission line condition assessment model based on an improved scatter degree method. First, the order relation method is used to calculate the subjective weights of each evaluation indicator and line unit. Secondly, the scatter degree method is used to calculate the objective weights. At last, the function value is calculated according to the linear evaluation function, and the line running state is evaluated based on the comprehensive evaluation function value. Through case analysis, results show that this method can synthesize the indicators effectively. The proposed method can not only evaluate each unit of the overhead transmission line, but also evaluate the whole line comprehensively. Besides, it is sensitive to the data.

Acknowledgment The work was supported in part by the National Natural Science Foundation of China under Grants 51277052, 61104045, and 51107032, and in part by the State Grid Corporation of China project: Key Technologies for Power System Security and Stability Defense Considering the Risk of Communication and Information Systems.

References

1. Yuan Z, Kaigui X (2011) Uncertainty of reliability parameters in probability risk assessment of bulk power systems. Autom Electric Power Syst 35(4):6–11
2. Yuanzhang S, Haitao L, Lin C et al (2008) A scheme for online Short-term operational reliability evaluation. Autom Electric Power Syst 32(3):4–8
3. Ji G, Zhang B, Wu W et al (2013) A new time-varying component outage model for power system reliability analysis. Proc CSEE 33(1):56–62
4. Wu W, Ning L, Zhang B et al (2008) Online operation Risk assessment for power system static security considering secondary devices models. Autom Electric Power Syst 32(7):10–15
5. Ning L, Wu W, Zhang B (2009) Analysis of a time-varying power component outage model for operation risk assessment. Autom Electric Power Syst 33(16):7–12
6. Feng Y, Wu W, Sun H et al (2005) A preliminary investigation on power system operation risk evaluation in the modern energy control center. Proc CSEE 25(13):73–79
7. Ji G, Wu W, Zhang B et al (2013) A time-varying component outage model used in power system condition-based maintenance. Proc CSEE 33(0):1–8
8. Sen O, Ruiyi H, X Cheng (2013) Comprehensive evaluation for power quality based on stepwise scatter degree method. J South China Univ Technol: Nat Sci Ed 41(5):93–98
9. Chen K, Huang J et al (2014) Economic operation evaluation for distribution network based on improved scatter degree method. East China Electric Power 42(6):1075–1078

Chapter 54
A Novel Driving Method for Switching Control Pulse Signal

Jin-Yan Zheng, Yu Fang, Pan Xu and Qi-Qi Zhao

Abstract A novel driving method implemented with high frequency electromagnetic isolation is proposed in this paper. And the hardware designing schematic diagram and the software configuration method are presented. Then the corresponding operating principle is analyzed in detail based on digital control chip and analog circuits. The proposed driving method can be applied to real-time varying pulse width of the control signal and the duty-cycle ratio of this control pulse can be regulated in the range of 0–100 %. This novel driving method can achieve not only the advantages of high insulation and high common-mode suppression, but also the merits of low cost and fast dynamic response speed, and it is suitable for the isolation and amplification of high frequency switching control pulse signal. Finally, the experimental results show that the proposed driving method in this paper is effective.

Keywords Electromagnetic isolation · Driving circuits · Duty cycle ratio · Digital control

J.-Y. Zheng (✉) · Y. Fang · P. Xu · Q.-Q. Zhao
College of Information Engineering, Yangzhou University, Yangzhou 225127, China
e-mail: zhengjinyannt@163.com; 1964865934@qq.com

Y. Fang
e-mail: yzfangyu@126.com

P. Xu
e-mail: polar_xupan@163.com

Q.-Q. Zhao
e-mail: zhao_qiqi@126.com

© Springer-Verlag Berlin Heidelberg 2016

Y. Jia et al. (eds.), *Proceedings of the 2015 Chinese Intelligent Systems Conference*, Lecture Notes in Electrical Engineering 359,
DOI 10.1007/978-3-662-48386-2_54

54.1 Introduction

In recent years, the power electronic technology has developed rapidly in the areas of new energy power generation and the electric vehicles charger, thus increasing the use of power electronic devices. Higher requirements of the high power density, high reliability and high cost performance of the power electronic devices have been put forward, and the power is developing towards the direction of high frequency. The function of the digital control chip with high speed is becoming more and more powerful, and the digitalization of power supply has become the trend of development. Digital control has brought new research ideas for the design of power electronic equipment [1, 2].

In the technology of power conversion in power electronics, the research on the driving circuit, which functioned as a connecting hub of the power circuit and control circuit, has a great significance. In a converter, according to the circuit topology, the driving mode of the power switch device has two kinds of method including direct driving and isolated driving. In many applications, especially in the high power converter, we generally have to realize the electrical isolation between the power circuit and control circuit, so isolated driving is often needed [3, 4].

Isolated driven is also divided into two ways of electromagnetic isolation and photoelectric isolation. Among them, photoelectric isolation, because of its advantages of small volume, simple structure and so on, has been widely used. However, it also has shortcomings of poor common-mode rejection ability, slow transmission speed and high cost, which is not conducive to the high power density and high performance of the converter. Electromagnetic isolation driving, with pulse transformer as its isolation element, has fast response speed, high dielectric strength between primary and secondary windings, high common-mode suppression and low cost and is usually used in switching power supply with high power density. However, due to the magnetic saturation characteristics of pulse transformer, on the occasion when the duty ratio is less than 50 %, the traditional electromagnetic isolation driving mode is usually limited in application, while for the inverter and the rectifier, pulse control signal generated by controller continuously changes in a low frequency period, which will lead to the saturation of the pulse transformer easily [5, 6]. In order to solve this problem, we usually add a high frequency modulation circuit to the electromagnetic isolation driving circuit, but this will introduce high-frequency noise, increase the cost of hardware and make the driving circuit complicated [7].

Aiming at this situation, combined with digital control technology, a novel electromagnetic isolation driving method is proposed in this paper, which can realize the adjustment of duty cycle ratio in the range of 0–100 %, and can be continuously changed in a low frequency period.

54.2 The Principle of the Novel Electromagnetic Isolation Driving Circuit

54.2.1 Hardware Configuration of the Driving Circuit

As shown in Fig. 54.1, PWM1A and PWM2A are pulse signals from the digital control chip; Q_{1A} and Q_{2A} are P-channel MOS transistors, Q_{1B} and Q_{2B} are N-channel MOS transistors. P-channel MOS transistor and N-channel MOS tube, two as a group, are combined into a totem pole to achieve the enhanced input signal. Therefore, when the P-channel MOS transistor is turned on, the amplitude of the pulse signal sent to the transformer will not be reduced a lot. C_1 is the driving capacitor, R_3 is the discharge resistor of C_1; T_1 is pulse isolation transformer; U_1 is an OR logic gate; U_2 is a driving chip; V_3 is a regulator whose voltage value is 2.7 V.

54.2.2 The Working Principle of Driving Circuit

Target pulse control signal is divided into two equal-width pulse control signal PWM1A and PWM2A in the digital control chip, as shown in Fig. 54.2. Supposing

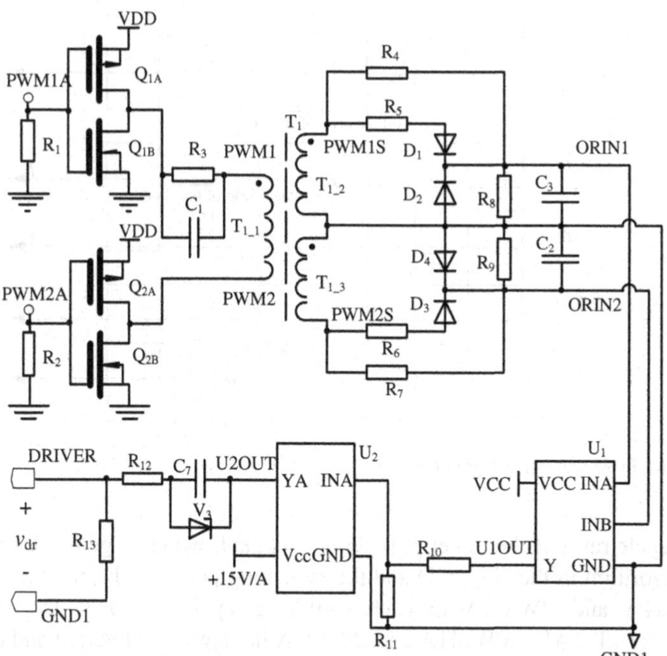

Fig. 54.1 The schematic diagram of novel driving circuit

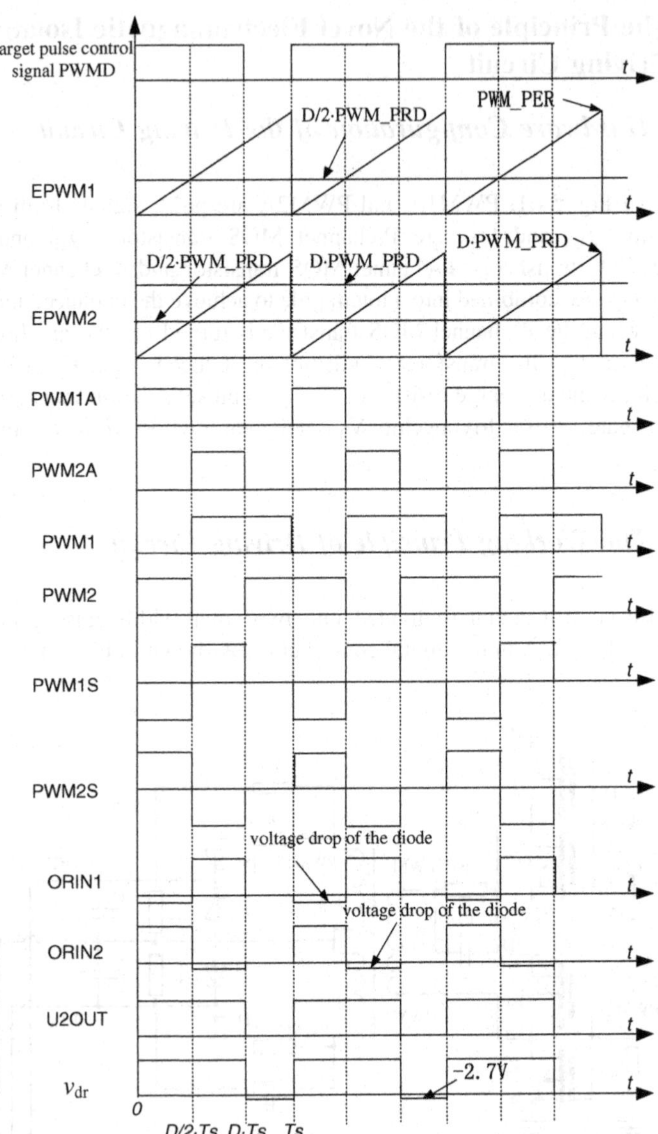

Fig. 54.2 Main waveforms of principles

the duty cycle ratio of the target pulse control signal, which is generated by digital control algorithm in the chip is D and the switching period is Ts, the time zones of the PWM1A and PWM2A in each switching cycle are $0 \sim D \cdot Ts/2$ and $D \cdot Ts/2 \sim D \cdot Ts$; After PWM1A and PWM2A having been inverted and enhanced the current driving capability by totem pole shown in Fig. 54.1, PWM1 and PWM2 are formed. Then PWM1 and PWM2 separately act on the two inputs in the primary side of the pulse transformer. As we can see from Fig. 54.1, for the primary side of the pulse transformer, PWM1 and PWM2 are common-mode inputs, so only those parts whose levels are different in PWM1 and PWM2 signals in Fig. 54.2 excite the primary winding T1_1 of the transformer. It is easy to know that the direction of excitation current alternately turns positive and negative, and the positive and negative excitation time is the same with the time of pulse width of PWM1A and PWM2A. Therefore, the width of the pulse signals PWM1S and PWM2S generated by secondary windings is the same with the corresponding PWM2A and PWM1A, as shown in Fig. 54.2. Obviously, voltages of PWM1 and PWM2 are equal in magnitude and pulse width, so primary side of the pulse transformer can achieve a volt-second balance and the pulse transformer is not saturated.

According to the analysis above, PWM1A and PWM2A are got when target pulse control signal in a switching period is divided into two parts. So, even if the width of the pulse control signal changes continuously, driven by the method of Fig. 54.1 can also ensure that the pulse transformer is not saturated. Therefore, the circuit in Fig. 54.1 can be used as the driving circuit of the pulse control signal generated by SPWM modulation or SVPWM modulation and achieve the duty ratio of 0 to 100 %. That is to say, the duty ratio is not limited by the traditional electromagnetic isolation driving to a maximum of 50 % or by the changes in a large and continuous range.

Between the isolation transformer T_1 and OR logic gate U_1, R_5, D_1, R_8 and secondary windings T_{1_2} constitutes the signal channel when the input signal of OR gate ORIN1 is at a high level; while R_6, D_3, R_9 and secondary windings T_{1_3} constitutes the signal channel when the input signal of OR gate ORIN2 is at a high level. When the input signal of the OR gate is high, R_5 and R_6 respectively limit the current for diode D_1 and D_3. In Fig. 54.1, D_2, R_4 and secondary windings T_{1_2} constitutes the signal channel when the input signal of OR gate ORIN1 is at a low level; while D_4, R_7 and secondary windings T_{1_3} constitutes the signal channel when the input signal of OR gate ORIN2 is at a low level. It can be seen from Fig. 54.1 that diodes D_1 and D_3 can prevent the intrusion of large current when the corresponding input signal of OR gate is at a low level; The conduction of D_2 and D_4 respectively play the role of bypass for resistance R_8 and R_9, which can ensure that the corresponding low level is sent to the OR gate in the subsequent stage in a stable and reliable way. As we can see, D_2 and D_4 also play the role of clamping for low level, preventing the input of the OR logic gate appear high negative pressure and be damaged. C_2 and C_3 are used as filter capacitors; resistors R_8 and R_9 are used for absorbing the charge on capacitors C_2 and C_3; resistors R_4 and R_7 respectively limit the current for diode D_2 and D_4.

As shown in Fig. 54.2, the ORIN1 and ORIN2 signals will go through OR logic gate U_1, thus the signal U2OUT can be obtained by adding the equally divided pulse control signals. The pulse width of the signal U2OUT is equal to that of the target pulse control signal PWMD in a digital control chip. U2OUT generates a driving signal v_{dr} of the switch tube through driving chip U_2. V_3 in Fig. 54.1 is the regulator. As the capacitor C_7 will be charged and then be clamped in the regulation value of V_3 when signal U2OUT is high, the regulation value of V_3 will provide switch with negative voltage when signal U2OUT is low, which is helpful to the reliable and fast turn-off of the switching tube.

In Fig. 54.2, PWMD is the target pulse control signal, EPWM1 and EPWM2 are the EPWM modules of digital control chip, PWM1A and PWM2A are digital output signals of digital control chip, PWM1 and PWM2 are input signals in the primary side of the pulse transformer, PWM1S and PWM2S are output signals in the secondary side of the pulse transformer, ORIN1 and ORIN2 are input signals of OR logic gate, U2OUT is output signal of OR logic gate, v_{dr} is a driving signal which is sent to the switch, PWM_PRD is the counting peak of period register in EPWM module and D is duty cycle ratio of the target pulse control signal.

54.2.3 The Design of the Pulse Transformer

In this paper, the pulse transformer in the driving circuit has two outputs, and they are connected with the OR logic gate, thus achieving the sum of the two output signals. The total output power of pulse transformer P_{TO} is 2 W. We select MnZn power ferrite of Dongci, which is made in DMR40 and whose saturation power density is 0.051 T and model is EP7DMR40, as the pulse transformer in this paper. In order to reduce the loss of driving that excessive magnetic swing will bring, the maximum working magnetic induction Bm is selected as 0.075 T. According to the working characteristics of the pulse transformer in this paper, magnetic amplitude of the pulse transformer ΔB = 2B m = 0.15 T [8, 9].

Determine the core size, as represented by the Eq. (54.1):

$$AP = AeAw = \left(\frac{2P_{To}}{K\Delta Bf_T}\right)^{4/3} \tag{54.1}$$

where P_{TO} is output power; ΔB is magnetic swing; K is 0.017;

Supposing the switching frequency is 45 kHz.

Then: $AP = 0.452 \times 10^{-10}\ \mathrm{m}^4$, the cross-sectional area of window is $1.098 \times 10^{-6}\ \mathrm{m}^4$ after calculating according to the core of the selected dongci EP7. Therefore, it has a large margin.

Turns of the primary side is calculated as shown in the Eq. (54.2):

$$N_1 = \frac{U_i T_{on}}{\Delta BS} \tag{54.2}$$

where, duty cycle ratio of the pulse is 0.5; the switching frequency is 45 kHz, namely $T_{on} = 0.5/(45 \times 10^3)$; U_i is the input voltage of the primary side and it is 3.3 V in this paper; core cross-sectional area S of EP7DMR40 is $10.7 \times 10^{-6}\ \mathrm{m}^2$;

Turns of the primary side N_1 is 24 through calculation. As the turns ratio of pulse transformer is 1:1:1, turns of the two secondary windings is also 24.

Wire diameter is calculated as shown in the Eq. (54.3):

$$S_{line} = \frac{1}{4}\pi d^2 = \frac{I_{TRMS}}{j} \tag{54.3}$$

Therefore, the wire diameter of primary and secondary side can be selected as 0.2 mm.

Verifying the window utilization factor:

$$K_u = \frac{\sum S_{\text{line}} \times N}{Q} = 0.21 < 0.3 \tag{54.4}$$

All these can illustrate that the design of pulse transformer is reasonable.

54.3 Digital Configuration of Pulse Control Signal

In order to obtain PWM1A and PWM2A in Fig. 54.1, we need to configure the PWM module in the digital control chip. In this paper, TI company's DSP chip TMS320F28035 is used to realize the control.

PWM1A and PWM2A are generated by two modules of EPWM1 and EPWM2 in TMS320F28035. With the same time base, carrier mode of the two EPWM modules is configured with counting mode, so counting peak of period register in these two EPWM modules is equal to the value of switching cycle PWM_PRD. PWM1A is obtained by the comparison between the value of counter in EPWM1 and comparison register A in this module. The counter in EPWM2 is compared respectively with comparison register A and comparison register B to get the corresponding up and down edge of the PWM2A. Comparison register A in EPWM1 module is assigned to (D/2) · PWM_PRD, comparison register A in EPWM2 module is assigned to (D/2) · PWM_PRD and comparison register B is assigned to D · PWM_PRD. As shown in Fig. 54.2, when the value of counter in EPWM1 module is smaller than the value in comparison register A, the output of EPWM1 module is at a high level, or PWM1A is high. And when the value of counter in EPWM1 module is larger than the value in comparison register A, the output of EPWM1 module is at a low level, or PWM1A is low. In Fig. 54.2, when the value of counter in EPWM2 module is larger than the value in comparison register A and is smaller than the value in comparison register B, the output of EPWM2 module is at a high level, or PWM2A is high. When the value of counter in EPWM2 module is smaller than the value in comparison register A or is greater than the value in comparison register B, the output of EPWM2 module is at a low level, or PWM2A is low. From the above configuration, we can devide target pulse control signal PWMD equally and get the corresponding PWM1A and PWM2A [10].

54.4 The Experimental Results

The driving circuit is used in pre-VIENNA rectifier of the electric vehicle charger and the circuit is shown in Fig. 54.3. Prototype parameters: $L_1 = L_2 = L_3 = 0.8$ mH, $C_1 = C_2 = 1080$ uF. The effective value of phase voltage in input grid is 220 V. The DC output voltage U_{dc} is 800 V. The output power is 5 kW, the switching frequency fs is 45 kHz. According to the driving requirements of switch in VIENNA rectifier, two switches in bi-directional switch use the same driving signal to achieve control, so the circuit shown in Fig. 54.3 requires the use of three groups of the driving circuit shown in Fig. 54.1.

The OR gate in Fig. 54.1 uses NXP company's chip 74HC1G32; the driving chip uses MIC- ROCHIP company's chip TC4424A; resistances $R_1 = R_2 = 2$ KΩ, $R_3 = 51$ Ω, $R_5 = R_6 = 10$ Ω, $R_4 = R_7 = R_8 = R_9 = 1$ KΩ, $R_{10} = 200$ Ω, $R_{11} = R_{13} = 5.1$ KΩ, $R_{12} = 5$ Ω; capacitors $C_1 = 1$ uF, $C_2 = C_3 = 100$ pF; we adopt dual closed-loop control and SPWM modulation in digital controller and get the corresponding experimental results as follows.

Figure 54.4 gives respectively the waveforms of the pulse control signals PWM1A and PWM2A, which are outputs of DSP, and driving signal v_{drA} on the switch of phase A. Figure 54.5 shows the waveforms of the pulse control signals PWM1A and PWM2A, and the input signals ORIN1 and ORIN2 of OR gate.

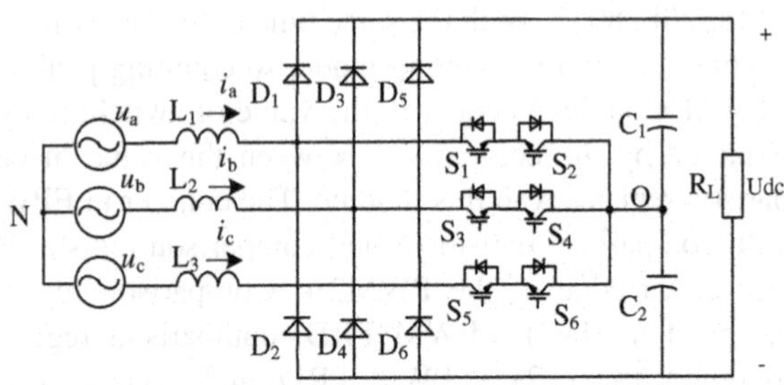

Fig. 54.3 The main circuit of VIENNA rectifier

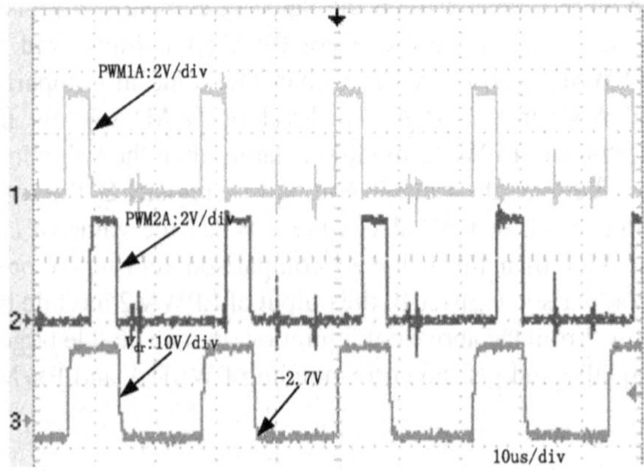

Fig. 54.4 The waveforms of switching control pulse signals and the corresponding driving signals

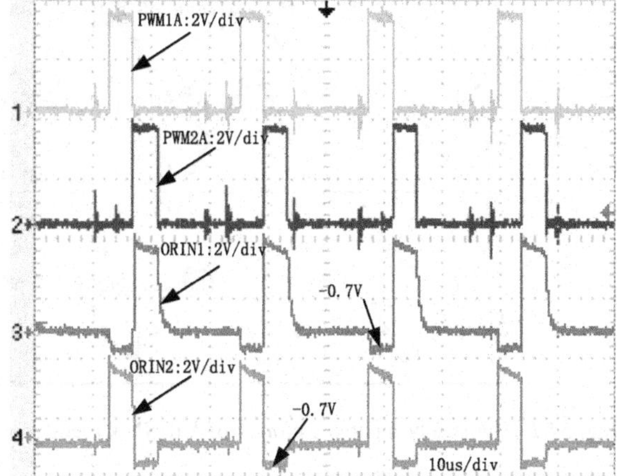

Fig. 54.5 The waveforms of switching control pulse signals and the input signals of OR-gate

Figure 54.6 gives waveforms of the output signal U2OUT of the OR gate and driving signal v_{drA} of the switch. Figure 54.7 shows the steady-state waveforms of the main power when power output is 5 kW, respectively are the input voltage waveform of B phase, the input current waveform of B phase, the input voltage waveform of C phase, the input current waveform of C phase. It can be seen from the waveforms that VIENNA converter can realize the sinusoidal current and unity power factor in the grid side. The experiments above illustrate that the proposed driving method in this paper can achieve high-frequency isolation driving, and provide reliable and efficient driving for the converter whose pulse control signals continuously vary.

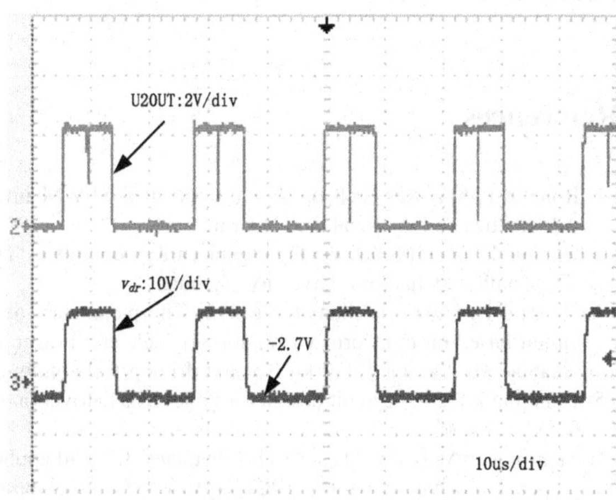

Fig. 54.6 The output waveform of OR gate and the driving waveform

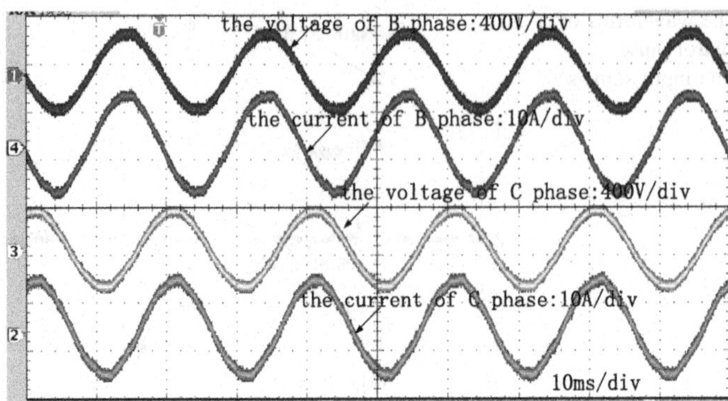

Fig. 54.7 The steady input waveforms of VIENNA rectifier

54.5 Conclusion

A novel high frequency electromagnetic isolation driving method is presented in this paper. And the driving circuit of hardware and digital configuration methods of pulse control signal are given. Moreover, the high-frequency electro-magnetic isolation driving circuit studied in this paper is applied in the three-phase three-level Vienna rectifier, thus realizing the high power factor correction. It is obvious that the driving method studied in this paper, which has the characteristics of fast performance and low cost, can regulate the duty cycle ratio in the range from 0 to 100 % and can be used as driving circuit for the width-varying pulse control signal. The proposed driving circuit and method in this paper can be extended to high-frequency converters to achieve high power density and high cost performance.

References

1. Rong Z (2006) Research on digital controlled SPWM inverters [D]. Nanjing University of Aeronautics and Astronautics, Nanjing
2. Guohua Z (2007) Study on the digital current control technology of switching power supply [D]. Southwest Jiaotong Uiversity, Sichuan
3. Hongfei L, Qifeng L, Jianyu P (2012) The parameters design for power MOSFET driving circuit based on transformer isolation [J]. Telecom Power Techonol 03:31–32 + 51
4. Huabiao W, Yaning C (2006) Isolated driving technology of IGBT and MOSFET. 05:43–45
5. Dianqing Z (2013) Circuit design for novel IGBT driver on pulse transformer [J]. World Invert 03:62–64 + 61
6. Lianrong Z, Meizhen P (2009) The characteristics and application technology on high-voltage suspension driving circuit of IR2110 [J]. Electron Compon Device Appl 11(4):30–33
7. Zhaoan W, Jinjun L (2010) Power electronic technology [M], 5th edn. China Machine Press, Beijing

8. Zhiyu H, Zhanghao Q, Zhenglong X (2014) Optimization design of isolated driver for pulse transformer based on bridge topology. Electr Driv 44(2): 75–79
9. Jiemin Z, Xiuke Z, Siyu T (2004) Theory and design of magnetic components on switch power supply [M]. Beijing University of Aeronautics and Astronautics Press, Beijing
10. Heping L, Ping L, Huabing W, Liping Y (2011) The principle and application of digital signal controller-based on TMS320F2808 [M]. Beijing University of Aeronautics and Astronautics Press, Beijing

Chapter 55
Study of Transformer Fault Diagnosis Based on DGA and Coupled HMM

Hao Guo, Zhinong Wei, Ming Ni, Guoqiang Sun and Yonghui Sun

Abstract In this paper, a coupled hidden Markov model (CHMM) based on dissolved gas analysis (DGA) is proposed for transformer fault diagnosis in power systems. By using the theory of scalar quantization, the collected date of dissolved gas is preprocessed and put into two observation channels in the coupled two-chain hidden Markov model (HMM) with the form of gas content and ratio. Transformed to the form of HMM equivalently, CHMM can be easily trained and work as the condition classifier for power transformer. Finally, it follows from the experimental results and comparison with HMMs that the proposed model is successful and effective.

Keywords Power transformer · Fault diagnosis · HMM · Coupled HMM · DGA

55.1 Introduction

Power transformer is one of the most critical equipment in power system. It is also the important insurance for the safe, economic, reliable operation and high quality of the power system. Therefore, preventing and reducing the incidence of accidents is of importance [1]. Power transformer has always been the popular subject of electrical equipment in the field of condition monitoring and fault diagnosis. Accurate diagnosis of the internal latent faults is significant for improving the level of stable operation of power system [2].

H. Guo (✉) · Z. Wei · G. Sun · Y. Sun
College of Energy and Electrical Engineering, Hohai University,
Nanjing 211100, China
e-mail: guohaoking@163.com

M. Ni
State Grid Electric Power Research Institute, Nanjing 210003, China

© Springer-Verlag Berlin Heidelberg 2016

Y. Jia et al. (eds.), *Proceedings of the 2015 Chinese Intelligent Systems Conference*, Lecture Notes in Electrical Engineering 359,
DOI 10.1007/978-3-662-48386-2_55

Owing to the influence of electricity or heat, transformer insulating oil will cleave and produce gas, thus technology of dissolved gas analysis (DGA) has become an effective method to detect latent fault. Traditional diagnostic methods based on DGA can be divided into three categories, including the warning value, the ratio method and the graphical methods [3]. Artificial intelligent methods and statistical learning methods have achieved satisfactory results in the fault diagnosis of transformers for the past few years. Although artificial neural network has a great learning ability, the question of excessive fitting is still existing [4]. Faced with the dramatic increasing date, the effect of support vector machine will be cut down [5]. Expert system can conduct a comprehensive analysis of large amounts of data, while the reasoning contains some uncertainties [6]. Transformer fault diagnosis based on hidden Markov model (HMM) is presented, which has been proved that it does well in fault classification [7]. In the field of rolling element bearing, coupled hidden Markov model (CHMM) is used to assess the condition and performs better than HMM [8].

In this paper, a transformer fault diagnosis model based on DGA and CHMM is proposed. To prove the better feasibility and effectiveness, it is compared with traditional HMM in the case study.

55.2 Preliminaries

CHMM is a probabilistic model to describe the statistical properties of two or more interrelated stochastic processes [9]. It can be seen as a multi-chain HMM which can introduce the conditional probabilities into state sequences of HMM.

The HMM can perform as a probabilistic finite state system where the actual states are not observable directly. They can only be estimated by using observables associated with the hidden states. An HMM is characterized as follows [10]:

(1) N, the number of hidden states in the model. $S=\{S_1, S_2, \ldots S_N\}$ is a set of N possible states, and the state at time t is q_t.
(2) M, the number of distinct observation symbols. $V=\{V_1, V_2, \ldots, V_M\}$ is a set of M possible observation symbols, and the observation symbol at time t is o_t.
(3) The initial state distribution $\pi = \{\pi_i\}$, where
$$\pi_i = \mathrm{P}(q_1 = S_i),\ 1 \le i \le N \tag{55.1}$$
(4) The state transition probability distribution $\mathrm{A}=\{a_{i,j}\}$, where
$$a_{i,j} = \mathrm{P}(q_{t+1} = S_j | q_t = S_i),\ 1 \le i,j \le N \tag{55.2}$$
(5) The observation symbol probability distribution in state j, $\mathrm{B}=\{b_j(k)\}$, where
$$b_j(k) = \mathrm{P}(o_t = v_k | q_t = S_j),\ 1 \le j \le N,\ 1 \le k \le M \tag{55.3}$$

The HMM implementation usually involves the following three fundamental problems [10]:

(1) Evaluation Problem: Given an HMM and a sequence of observations, what is the probability that the observed sequence was generated by the model? The Forward-Backward Algorithm can solve this problem.
(2) Decoding Problem: Given an HMM and a sequence of observations, what is the most likely hidden state sequence that could produce the observations? It is usually implemented using the Viterbi Algorithm.
(3) Learning Problem: Given a set of observation sequences, find the HMM that best explains the observation sequences. It is usually solved by using the Baum-Welch Algorithm.

Here a coupled two-chain HMM is taken for example to expound the model. Each HMM has a sequence of hidden states and a sequence of observation symbols. The number of hidden states in each HMM may be the same or different, it all depends on the actual demand. The observation symbol at time t in each HMM is associated with the state at time t, however, symbol at t ($t > 1$) not only depends on the state of the HMM at time $t - 1$,but also depends on the state of the other HMM. Like the HMM, CHMM can be characterized with the notation.

Each HMM has a sequence of hidden states and a sequence of observation symbols. The number of hidden states in each HMM may be the same or different, it all depends on the actual demand. The observation symbol at time t in each HMM is associated with the state at time t, however, symbol at t($t > 1$) not only depends on the state of the HMM at time $t - 1$, but also depends on the state of the other HMM. Like the HMM, CHMM can be characterized with the notation

$$\lambda_{CHMM} = (\pi, \mathrm{A}, \mathrm{B}) \tag{55.4}$$

55.3 Solution Strategy

55.3.1 Diagnosis Model

IEC Standard 599 divides the running condition of transformer into five categories. They are the condition of "Normal", "Low Overheating"(LO), "High Overheating"(HO), "Low Energy Discharge"(LED) and "High Energy Discharge or Arcing" (HEDA).

The method of transformer fault diagnosis based on CHMM deals with pattern classification of different fault types. Figure 55.1 shows the diagnosis model. The method involves the part of model training and the part of model testing. During the section of training, date which is known for the transformer's condition is used to train CHMM. Different kinds of date gets its parameters for one of CHMMs, λ_i, where $\{\lambda_i, i \in \{\text{Normel, OH, HO, LED, HEDA}\}\}$. As soon as the training is

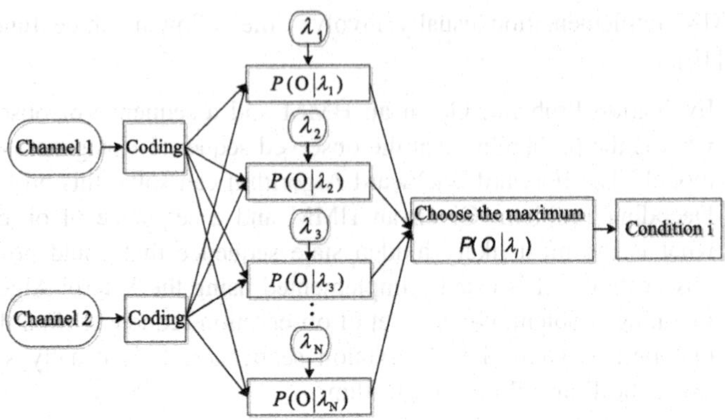

Fig. 55.1 Diagnosis Model with CHMM

finished, data for test can be entered into the model. Each CHMM on behalf of different condition then outputs the likelihood probability. The maximum one of the likelihood probability reveals the most probable state of the transformer.

55.3.2 Date Pregrocessing

Commonly used gases include hydrogen (H_2), methane (CH_4), acetylene (C_2H_2), ethylene (C_2H_4), ethane (C_2H_6) in DGA. For the CHMM with two chains, the value of gas content is inputted into channel 1. The observation symbol vector is noted as $X_1 = [H_2, CH_4, C_2H_6, C_2H_4, C_2H_2]$. In reference to Doernenberg Ratio Method, ratio of gases is inputted into channel 2. The observation symbol vector is noted as $X_2[CH_4/H_2, C_2H_2/C_2H_4, C_2H_2/CH_4, C_2H_6/C_2H_2, C_2H_6/C_2H_4]$.

The number of observation symbol in discrete HMM needs to be finite. As the value of gas content is generally not uniform, it is essential to preprocess the date. The first step is normalization. There are five figures of gases in a observation symbol vector. The normalization is a calculation of each value of the five by their sum. Channel 1 and Channel 2 do the same work. As a result, the observation symbols of the two channels are limited from 0 to 1. In the second step, scalar quantization is adopted with Lloyds Algorithm in order to code the observation symbol after normalization [11]. The amplitude of observation symbol is divided into M parts on average, where M matches the number of distinct observation symbols per. Then the value gets its code from 1 to M.

55.3.3 CHMM Implement

The hidden states are converted to composite states. Conversion process is shown in Fig. 55.2. Figure 55.2a is the CHMM structure adopted in this paper. Figure 55.2b is the equivalent HMM. Composite state $\{A3, B1\}$ in Fig. 55.2b is removed artificially to simplify the structure. It conforms to the related literature [12].

Observation symbols are also converted to composite ones. Scalar quantization is carried out to observation symbols of each channel individually. Then the code of each channel is combined according to the composite code table in Fig. 55.3. The accuracy of scalar quantization will cut down if the code number, M is selected unreasonably. In this paper, number of observation symbols in channel 1 is set as 5. It's the same to channel 2. Finally, the number of observation symbols for CHMM is 25.

The new HMM can be trained by the Baum-Welch Algorithm. At the first stage of CHMM's training, the initial state distribution, π and state transition probability distribution, A can be chosen randomly or evenly. In this paper, A can be set from the model in Fig. 55.2b. The observation symbol probability distribution, B has great impact on the effect of training. Figure 55.4 shows the improved method of choosing B. During the period of testing, the model will output likelihood probability using Forward-Backward Algorithm like HMM.

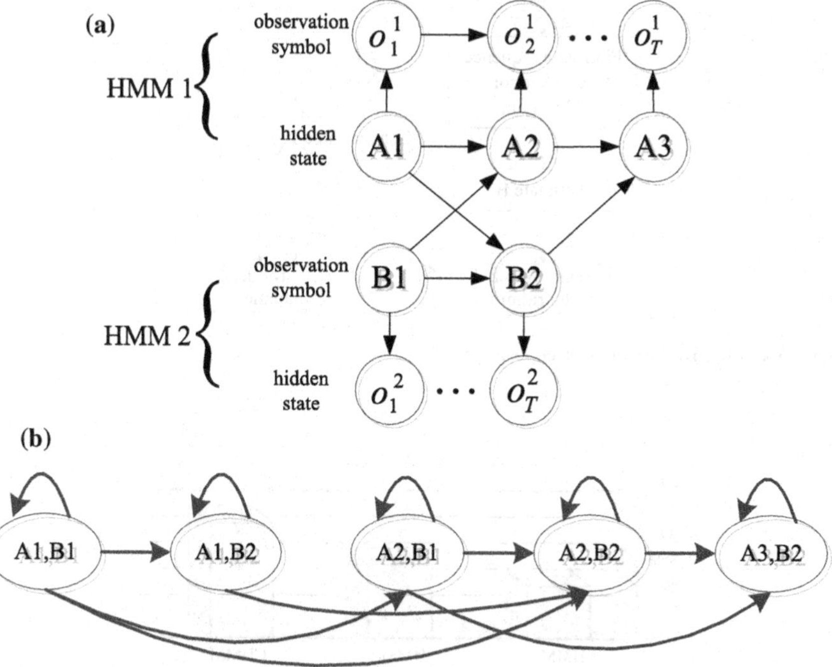

Fig. 55.2 Conversion Process of hidden states. **a** CHMM in this paper. **b** Equivalent HMM

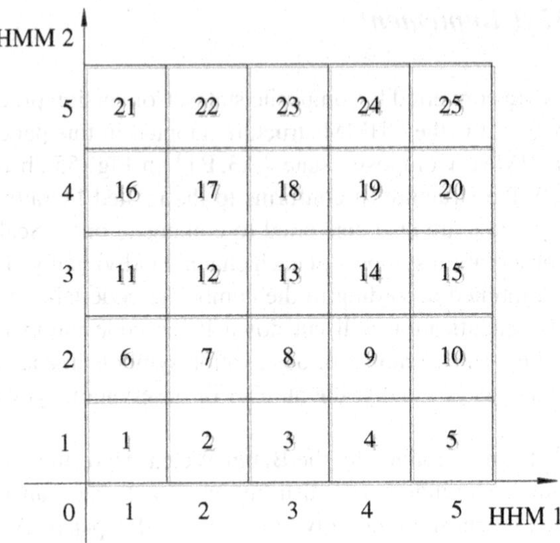

Fig. 55.3 Code table of CHMM

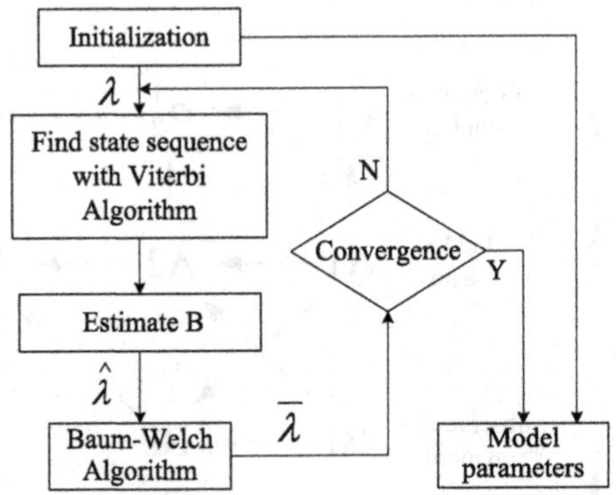

Fig. 55.4 Improved method of choosing B

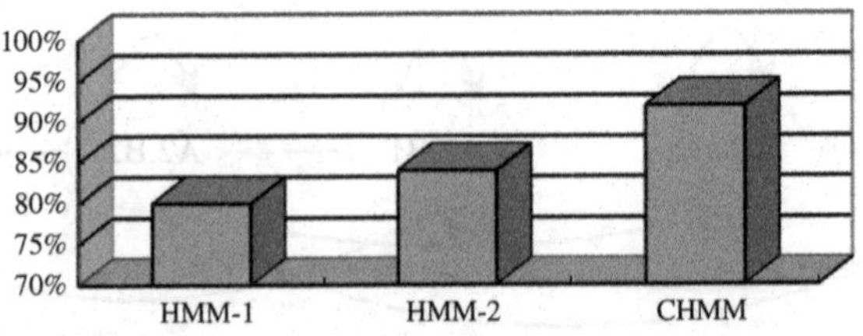

Fig. 55.5 Average recognition rate comparison

55.4 Simulation Results

In this paper, a large number of transformer oil chromatographic detection recording is collected, and corresponding running condition is also determined. 300 groups of representative samples are chosen for the diagnosis model training and testing. Specifically, for each type of transformer condition 50 groups of corresponding date are used for training, and 10 groups of corresponding date are selected for testing. Output of diagnosis model is given in Table 55.1. Results of model testing are shown in Table 55.2.

As shown in Table 55.1, one can judge the condition of a transformer by comparing the output value in the form of probability. Table 55.2 reveals that most condition of transformer can be classified correctly, apart from a few ones. Especially, recognition rate of condition on "Normal" and "High Energy Discharge or Arcing" can reach up to 100 %, and the lowest recognition rate stays at 80 %. The average recognition rate accounts for 92 %. It follows from these results that fault diagnosis based on HMM is effective.

To further verify the feasibility and effectiveness of the proposed method, more tests are carried out. We compare the recognition rate of CHMM with the traditional HMM. Relevant literature indicates that the recognition rate is insensitive to the number of hidden states [10]. Considering the kind of gases used in this paper is 5 and the number of observation symbols is 25, the number of hidden states for HMM is defined as 4. The number of observation symbols is the same with CHMM. The structure of HMM is the left-right model. HMM-1 takes the value of gas content as

Table 55.1 Output of diagnosis model

Number	Normal	LO	HO	LED	HEDA	Recognition result	Actual result
1	−8.48	−16.14	−16.92	−Inf	−18.61	Normal	Normal
2	−12.37	−11.76	−13.25	−15.11	−12.17	LO	LO
3	−Inf	−12.73	−10.17	−13.44	−14.36	HO	HO
4	−13.41	−15.46	−10.45	−6.43	−8.46	LED	LED
5	−14.35	−15.32	−15.66	−18.54	−11.39	HEDA	HEDA

Table 55.2 Results of model testing

Status	Number of samples		Recognition result		Recognition rate (%)
	Training	Testing	Correct	Wrong	
Normal	50	10	10	0	100
LO	50	10	8	2	80
HO	50	10	9	1	90
LED	50	10	9	1	90
HEDA	50	10	10	0	100
Total	250	50	46	4	92

Table 55.3 Results Comparison

Status	Number		Recognition rate		
	Training samples	Testing samples	HMM-1 (%)	HMM-2 (%)	CHMM (%)
Normal	50	10	90	90	100
LO	50	10	70	80	80
HO	50	10	80	80	90
LED	50	10	80	80	90
HEDA	50	10	80	90	100
Total	250	50	80	84	92

observation symbols, while HMM-2's is composite observation symbols used in CHMM. Trained and tested by the same date samples, different models make great difference, which can be seen in Table 55.3 and Fig. 55.5.

All the three models have the acceptable effect in the recognition rate. The average recognition rate of HMM-1 is 80 %, while is 84 and 92 % in HMM-2 and CHMM, respectively. Experiments verify the feasibility and effectiveness of HMM and CHMM in the field of transformer fault diagnosis.

Recognition rate of HMM-2 is 4 % higher than the one of HMM-1, though the gap is small. The reason is that HMM-1 only uses the observation symbols from single channel, while the composite symbols used in HMM-2 contain more feature and information. In other words, diagnosis model performs better using date from multi-channel.

As the effect of diagnosis is concerned, CHMM is significantly higher than the others. CHMM not only makes better use of the two-channel date, but also uses the structure of two HMMs, so more detailed characteristics of dissolved gas is reflected.

55.5 Conclusions

Power transformer is the critical equipment in power systems, which needs real-time and accurate diagnosis when fault occurs. In this paper, a coupled hidden Markov model based on DGA has been proposed for transformer fault diagnosis. The model can be divided into three parts, which are data preprocessing, model training and model testing. In the part of data preprocessing, date about dissolved gas is converted to the form of gas content and ratio after normalization and scalar quantization. During the period of model training, CHMM is transformed to the form of HMM equivalently and trained with the algorithm of HMM. In simulation results, transformer fault can be determined by comparing the likelihood probability of the model. Experiment results and comparison with HMMs proves that the proposed model is reliable and effective.

Acknowledgments The work was supported in part by the National Natural Science Foundation of China under Grants 51277052, 61104045, and 51107032, and in part by the State Grid Corporation of China project: Key Technologies for Power System Security and Stability Defense Considering the Risk of Communication and Information Systems.

References

1. Richardson ZJ, Fitch J, Tang WH et al (2008) A probabilistic classifier for transformer dissolved gas analysis with a particle swarm optimizer. IEEE Trans Power Deliv 23 (2):751–759
2. Yongliang L, Kejun L et al (2013) A transformer diagnosis method based on optimized feature selection methods and fast relevance vector machine. Power Syst Technol 37(11):3262–3267
3. Duval M (2002) A review of faults detectable by gas-in-oil analysis in transformers. IEEE Electr Insul Mag 18(3):8–17
4. Weigen C, Yun L, Degang Gan et al (2012) Method to identify developing stages of air-gap discharge in oil-paper insulation based on cluster-wavelet neural network. Power Syst Technol 36(7):126–132
5. Yujuan Y, M W, Jinjiang Z (2012) An autonomic kernel optimization method to diagnose transformer faults by multi-kernel learning support vector classifier based on binary particle swarm optimization. Power Syst Technol 36(7):249–254
6. Ahmed MR, Geliel MA, Khalil A (2013) Power transformer fault diagnosis using fuzzy logic technique based on dissolved gas analysis. In: 21st Mediterranean conference on control and automation (MED) 2013, IEEE, pp 584–589
7. Xiliang L, Shiqing Y, Qinghua C (2006) Design and realization of HMM-based power transformer fault diagnosis system software. Ind Control Comput 19(4):36–37
8. Wen-bin X, Jin C, Yu Z (2012) Application of coupled hidden markov models in bearing fault diagnosis. Noise Vib Control 31(6):161–164
9. Brand M (1997) Coupled hidden Markov models for modeling interacting processes,
10. Rabiner L (1989) A tutorial on hidden Markov models and selected applications in speech recognition. Proc IEEE 77(2):257–286
11. Winger LLR (2001) Linearly constrained generalized Lloyd algorithm for reduced codebook vector quantization. IEEE Trans Signal Process 49(7):1501–1509
12. Jun H, Hua Z, Ji-zhong L (2009) Multi-sensor asynchronous information fusion classification strategy based on coupled HMM. Appl Res Comput 26(8)

Chapter 56
On Regular Subspaces of Boolean Control Networks

Jiandong Zhu and Pengjing Jü

Abstract This paper investigates some fundamental problems on regular subspaces of Boolean control networks (BCNs). A new necessary and sufficient condition for regular subspaces is obtained. A new method to compute complementary subspaces is proposed. An example is given to illustrate the obtained theoretical results.

Keywords Boolean control network · Semi-tensor product · Regular subspace

56.1 Introduction

Boolean networks (BNs) are composed of a group of dynamical equations described by logical functions, which are first proposed by Kauffman to model cell regulation networks [1–3]. BNs with inputs and outputs are called Boolean control networks (BCNs), and they have attracted much attention from researchers in the files of biology, physics and control theory [4–7]. In recent years, Cheng and his colleagues developed a general theoretic framework for BNs and BCNs using the semi-tensor product of matrices [7–14]. With the linear algebraic expressions of BCNs, many classical control problems have be generalized to BCNs e.g. controllability [7], [15], observability [7, 16], observers [17], stabilization [12, 18], disturbance decoupling [13], optimal control [19, 20], system decomposition [21, 22] and so on.

In the traditional control theory, coordinate transformation is an important tool for system analysis and control design. For example, the pole assignment problem can be solved under some canonical forms obtained by coordinate transformations. Similarly, the concept of logical coordinate transformation is proposed for BCNs [10]. Two interesting problems are whether one can construct a logical coordinate transformation from some given logical functions and how to construct it. This is just

J. Zhu (✉) · P Jü
Institute of Mathematics, School of Mathematical Sciences,
Nanjing Normal University,
Nanjing 210023, People's Republic of China
e-mail: zhujiandong@njnu.edu.cn

© Springer-Verlag Berlin Heidelberg 2016
Y. Jia et al. (eds.), *Proceedings of the 2015 Chinese Intelligent Systems Conference*, Lecture Notes in Electrical Engineering 359,
DOI 10.1007/978-3-662-48386-2_56

the motivation of introducing the concept of regular subspace [10, 11], which is very useful to topology structure analysis [9], system decomposition [10] and disturbance decoupling problem [13]. So it is meaningful to investigate some new properties of regular subspaces.

In [10], a necessary and sufficient condition of regular subspaces has been obtained and a procedure to compute complementary subspaces is given. In this paper, we obtain a new necessary and sufficient condition and subsequently give a simple and straightforward method to construct a complementary subspace.

Notations: Denote the real number field by $\mathbb{R}$ and the set of all the $m \times n$ real matrices by $\mathbb{R}_{m\times n}$. Let Col(A) be the set of all the columns of matrix A and denote the ith column of A by $\mathrm{Col}_i(A)$. Set $\Delta_k = \{\delta_k^i | i = 1, 2, \ldots, k\}$, where $\delta_k^i = \mathrm{Col}_i(I_k)$ with I_k the $k \times k$ identity matrix. For simplicity, we denote $\Delta := \Delta_2 = \{\delta_2^1, \delta_2^2\}$. A matrix $L \in \mathbb{R}_{m\times n}$ is called a logical matrix if $\mathrm{Col}(L) \subset \Delta_m$. Obviously, the logical matrix L satisfies $\mathbf{1}_m^{\mathrm{T}} L = \mathbf{1}_n^{\mathrm{T}}$, where $\mathbf{1}_n$ denote the n–dimensional column vector of ones. The set of all the $m \times r$ logical matrices is denoted by $\mathcal{L}_{m\times r}$. For simplicity, we denote the logical matrix $L = [\delta_m^{i_1}\ \delta_m^{i_2}\ \ldots\ \delta_m^{i_r}]$ by $\delta_m[i_1\ i_2\ \ldots\ i_r]$.

56.2 Preliminaries

A BCN is described by a group of dynamical logical equations with inputs and outputs formulated as follows:

$$\begin{cases} X(t+1) = f(X(t), U(t)), \\ Y(t) = h(X(t)), \end{cases} \tag{56.1}$$

where $X = [x_1, x_2, \ldots, x_n]^{\mathrm{T}}$ is the logical state vector, $U(t) = [u_1, u_2, \ldots, u_p]^{\mathrm{T}}$ the logical control vector, $Y(t) = [y_1, y_2, \ldots, y_q]^{\mathrm{T}}$ the logical output vector, x_i, u_j, $y_k \in \mathscr{D} = \{\text{True, False}\}$. In (56.1), $f : \mathscr{D}^n \times \mathscr{D}^p \to \mathscr{D}^n$ and $h : \mathscr{D}^n \to \mathscr{D}^q$ are logical mappings.

Semi-tensor product of matrices is an important tool to investigate BCNs. In [7], the semi-tensor product of matrices $A \in \mathbb{R}^{m\times n}$ and $B \in \mathbb{R}^{p\times q}$ is defined by

$$A \ltimes B = (A \otimes I_{\alpha/n})(B \otimes I_{\alpha/p}), \tag{56.2}$$

where α is the least common multiple of n and p. Since the semi-tensor product is a generalization of the conventional matrix product, we can directly write $A \ltimes B$ as AB. The semi-tensor product satisfies distributive law and associative law as the conventional matrix product. Moreover, if $z \in \mathbb{R}^t$ is a column vector, the following pseudo-commutative property (Proposition 2.18 of [11]) holds:

$$zA = (I_t \otimes A)z. \tag{56.3}$$

If $z \in \Delta_k$, then

$$z^2 = \Phi_k z, \tag{56.4}$$

where $\Phi_k = \text{block-diag}\{\delta_k^1, \delta_k^2, \ldots, \delta_k^k\}$ is called the power-reducing matrix (Proposition 3.2 of [11]).

Representing the logical values True and False by δ_2^1 and δ_2^2 respectively, Cheng and Qi [8] obtained an algebraic form of the BCN (56.1) as follows:

$$\begin{cases} x(t+1) = Fu(t)x(t), \\ y(t) = Hx(t), \end{cases} \tag{56.5}$$

where $x = \ltimes_{i=1}^n x_i$, $u = \ltimes_{j=1}^p u_j$, $y = \ltimes_{k=1}^q y_k$, x_i, u_j, $y_k \in \Delta$, $F \in \mathcal{L}_{2^n \times 2^{n+p}}$ and $H \in \mathcal{L}_{2^q \times 2^n}$.

A transformation $Z = g(X)$ is called a *logical coordinate transformation* if $g : \mathcal{D}^n \to \mathcal{D}^n$ is a bijection [10]. Let $z = Gx$ be the algebraic form of the logical coordinate transformation $Z = g(X)$, where G is a permutation matrix (invertible logical matrix). Then BCN (56.5) becomes

$$\begin{cases} z(t+1) = GF(I_p \otimes G^{\mathrm{T}})u(t)z(t), \\ y(t) = HG^{\mathrm{T}}z(t). \end{cases} \tag{56.6}$$

Lemma 1 *Let R be a $n \times n$ logical matrix. Then R is a permutation matrix if and only if $R\mathbf{1}_n = \mathbf{1}_n$.*

Lemma 2 (Theorem 2 of Sect. 2 in [23]) *Let A, B and C be matrices with appropriate sizes. Then $V_c(ABC) = (C^{\mathrm{T}} \otimes A)V_c(B)$, where V_c is the column-stacking operator.*

Lemma 3 *Let S be a $m \times n$ matrix with non-negative integral entries. Then S is a logical matrix if and only if $\mathbf{1}_m^{\mathrm{T}} S = \mathbf{1}_n$.*

Lemma 4 *[23] Set A, B, C, D have proper dimensions, then*

$$AC \otimes BD = (A \otimes B)(C \otimes D).$$

Set $X \in \triangle_m$, $Y \in \triangle_n$. Then $\mathbf{1}_m^{\mathrm{T}} X = 1$ and $\mathbf{1}_n^{\mathrm{T}} Y = 1$. So we can easily get the results as follows:

Lemma 5 *If $X \in \triangle_m$, $Y \in \triangle_n$, then*

$$X = (I_m X) \otimes (\mathbf{1}_n^{\mathrm{T}} Y) = (I_m \otimes \mathbf{1}_n^{\mathrm{T}})(XY), \tag{56.7}$$

$$Y = (\mathbf{1}_m^{\mathrm{T}} X) \otimes (I_n Y) = (\mathbf{1}_m^{\mathrm{T}} \otimes I_n)(XY). \tag{56.8}$$

Lemma 6 *[21] Assume that* $M_1, M_2, \dots, M_l \in \mathcal{M}_{m\times n}$ *are non-negative matrices. If*

$$\mathbf{1}_m^{\mathrm{T}} M_k = \mathbf{1}_n^{\mathrm{T}}, \ \forall\, k = 1, 2, \dots, l \tag{56.9}$$

and there exists a logical matrix $G \in \mathcal{L}_{m\times n}$ *satisfying*

$$M_1 + M_2 + \cdots + M_l = lG,$$

Then

$$M_1 = M_2 = \cdots = M_l = G.$$

56.3 Regular Subspaces

Before the main result of this section, we first give some basic concepts of state space of BCNs.

Definition 1 ([10]) Consider BCN (56.1). The *state space* $\mathcal{X}$ is defined as the set of all logical functions of $\{x_1, x_2, \dots, x_n\}$ denoted by $\mathcal{F}_l\{x_1, x_2, \dots, x_n\}$. Let $z_1, z_2, \dots, z_k \in \mathcal{X}$. The *subspace generated by* $\{z_1, z_2, \dots, z_k\}$ is defined as the set of logical functions of $\{z_1, z_2, \dots, z_k\}$, denoted by $\mathcal{Z} = \mathcal{F}_l\{z_1, z_2, \dots, z_k\}$.

Definition 2 ([10]) A subspace $\mathcal{Z} = \mathcal{F}_l\{z_1, z_2, \dots, z_k\}$ is called a *regular subspace of dimension* k if there are $z_{k+1}, z_{k+2}, \dots, z_n \in \mathcal{X}$ such that $Z = (z_1, z_2, \dots, z_n)^{\mathrm{T}}$ is a logical coordinate transformation. $\{z_1, z_2, \dots, z_k\}$ is called a *sub-basis* of $\mathcal{Z}$. $\mathcal{F}_l\{z_{k+1}, z_{k+2}, \dots, z_n\}$ is called the *complementary space* of $\mathcal{Z}$.

Theorem 1 *Let* $\mathscr{X} = \mathscr{F}_l(x_1, x_2, \dots, x_n)$ *be the state space of the BCN (56.1). Consider a subspace* $\overline{\mathscr{Z}} = \mathscr{F}_l(z_1, z_2, \dots, z_s) \subset \mathscr{X}$. *Let*

$$x = x_1 x_2 \dots x_n, \ \ \bar{z} = z_1 z_2 \dots z_s = Mx,$$

where M is a logical matrix. Then $\overline{\mathscr{Z}}$ *is a regular subspace with sub-basis* $z_1, z_2, \dots, z_s$ *if and only if there exists a logical matrix N such that*

$$MN^{\mathrm{T}} = J_{2^s, 2^{n-s}}, \tag{56.10}$$

where $J_{2^s, 2^{n-s}}$ *denotes the* $2^s \times 2^{n-s}$ *matrix of ones.*

Proof Let $z_{s+1}, z_{s+2}, \dots, z_n \in \mathscr{X}$ be $n - s$ logical functions with the algebraic form $\tilde{z} = z_{s+1} z_{s+2} \dots z_n = Nx$. Then $\overline{\mathscr{Z}}$ is a regular subspace with sub-basis $z_1, z_2, \dots, z_s$ if and only if there exists a logical matrix N such that the transformation $z = \bar{z}\tilde{z} = MxNx$ is a logical coordinate transformation. By (56.2)–(56.4), we get

$$\begin{aligned} MxNx &= M(I_n \otimes N)\Phi_{2^n}x \\ &= (M \otimes I_{2^{n-s}})(I_n \otimes N)\Phi_{2^n}x \\ &= (M \otimes N)\Phi_{2^n}x. \end{aligned} \tag{56.11}$$

Thus, $z = MxNx$ is a logical coordinate transformation if and only if the transformation matrix $(M \otimes N)\Phi_{2^n}$ is a permutation matrix, that is,

$$(M \otimes N)\Phi_{2^n}\mathbf{1}_{2^n} = \mathbf{1}_{2^n}, \tag{56.12}$$

which is due to Lemma 1. A straightforward computation shows that

$$\Phi_{2^n}\,\mathbf{1}_{2^n} = \begin{bmatrix} \delta_{2^n}^1 & & & \\ & \delta_{2^n}^2 & & \\ & & \ddots & \\ & & & \delta_{2^n}^{2^n} \end{bmatrix} \begin{bmatrix} 1 \\ 1 \\ \vdots \\ 1 \end{bmatrix} = \begin{bmatrix} \delta_{2^n}^1 \\ \delta_{2^n}^2 \\ \vdots \\ \delta_{2^n}^{2^n} \end{bmatrix} = V_c(I_{2^n}). \tag{56.13}$$

Substituting (56.13) into (56.12) yields

$$(M \otimes N)V_c(I_{2^n}) = \mathbf{1}_{2^n}. \tag{56.14}$$

By Lemma 2, (56.14) can be rewritten as $NM^{\mathrm{T}} = J_{2^{n-s},2^s}$, which is just (56.10). Thus Theorem 1 is proved.

Theorem 2 *With the conditions and notions of Theorem 1,* $\overline{\mathscr{X}}$ *is a regular subspace with sub-basis* $z_1, z_2, \ldots, z_s$ *if and only if*

$$M\mathbf{1}_{2^n} = 2^{n-s}\mathbf{1}_{2^s} \tag{56.15}$$

or equivalently

$$MM^{\mathrm{T}} = 2^{n-s}I_{2^s} \tag{56.16}$$

Proof Since M is a logical matrix, it is easy to check that (56.15) is equivalent to (56.16).

(Necessity) Multiplying (56.10) on the right by $\mathbf{1}_{2^{n-s}}$ yields

$$MN^{\mathrm{T}}\mathbf{1}_{2^{n-s}} = 2^{n-s}\mathbf{1}_{2^s}. \tag{56.17}$$

Considering N is a logical matrix, by Lemma 3, we have that $N^{\mathrm{T}}\mathbf{1}_{2^{n-s}} = \mathbf{1}_{2^n}$. Thus (56.15) follows from (56.17).

(Sufficiency) Since (56.15) holds, each row of M has exact 2^{n-s} nonzero elements equal to 1. Thus there exists a permutation matrix T such that

$$MT = \mathbf{1}^{\mathrm{T}}_{2^{n-s}} \otimes I_{2^s}. \tag{56.18}$$

Choose

$$N = (I_{2^{n-s}} \otimes \mathbf{1}^{\mathrm{T}}_{2^s})T^{\mathrm{T}}. \tag{56.19}$$

A straightforward computation shows that N is a logical matrix and

$$MN^{\mathrm{T}} = (\mathbf{1}^{\mathrm{T}}_{2^{n-s}} \otimes I_{2^s})(I_{2^{n-s}} \otimes \mathbf{1}_{2^s}) = \mathbf{1}^{\mathrm{T}}_{2^{n-s}} \otimes \mathbf{1}_{2^s} = J_{2^s, 2^{n-s}}. \tag{56.20}$$

Thus, by Theorem 1, $\overline{\mathscr{Z}}$ is a regular subspace with sub-basis $z_1, z_2, \dots, z_s$.

Remark 1 In this section, Theorem 1 is a new result on regular subspaces, which exactly reveals the relationship between a regular subspace and its complementary subspace. But Theorem 2 is not new, which is essentially Theorem 11 of [10]. Actually, here we get a new proof and a new method to compute the complementary subspace of a regular subspace.

56.4 An Example

Consider the third-order BCN

$$\begin{cases} x_1(t+1) = \neg(x_1(t) \leftrightarrow x_2(t)), \\ x_2(t+1) = \neg(x_2(t) \leftrightarrow x_3(t)), \\ x_3(t+1) = x_1(t) \wedge u(t), \\ y(t) = x_1(t) \leftrightarrow x_2(t), \end{cases}$$

where x_1, x_2 and x_3 are the states, u the control and y the output. Consider the subspace $\overline{\mathscr{Z}} = \mathscr{F}_l(z_1, z_2)$ of the state space $\mathscr{X} = \mathscr{F}_l(x_1, x_2, x_3)$, where

$$\begin{cases} z_1(t) = x_1(t) \leftrightarrow x_2(t), \\ z_2(t) = x_2(t)\bar{\vee}x_3(t)), \end{cases}$$

or the algebraic form $\bar{z}(t) = z_1(t)z_2(t) = Mx$, with $M = \delta_4[2\ 1\ 3\ 4\ 4\ 3\ 1\ 2]$. It is easy to check that $M\mathbf{1}_8 = 2\mathbf{1}_4$. Thus $\overline{\mathscr{Z}}$ is a regular subspace. By (56.18), we have

$$MT = \delta_4[1\ 2\ 3\ 4\ 1\ 2\ 3\ 4]. \tag{56.21}$$

Thus a permutation matrix T is solved from (56.21) as $T = \delta_8[2\ 1\ 3\ 4\ 7\ 8\ 6\ 5]$. By (56.19), we get $N = (I_2 \otimes \mathbf{1}^{\mathrm{T}}_4)T^{\mathrm{T}} = \delta_2[1\ 1\ 1\ 1\ 2\ 2\ 2\ 2]$, which implies that $z_3 = Nx = x_1$. Thus a logical coordinate transformation is obtained as

$$
\begin{aligned}
z_1(t) &= x_1(t) \leftrightarrow x_2(t),\\
z_2(t) &= x_2(t)\bar{\vee}x_3(t),\\
z_3(t) &= x_1(t).
\end{aligned}
$$

It should be noted that the complementary subspace of $\overline{\mathscr{X}}$ is not unique. From (56.21), we can get another permutation matrix solution $T = \delta_8[7\ 1\ 6\ 4\ 2\ 8\ 3\ 5]$. Then $N = (I_2 \otimes \mathbf{1}_4^{\mathrm{T}})T^{\mathrm{T}} = \delta_2[1\ 2\ 2\ 1\ 2\ 1\ 1\ 2]$, which yields $z_3 = (x_1 \wedge (x_2 \leftrightarrow x_3)) \vee (\neg x_1 \wedge (x_2 \bar{\vee} x_3))$.

56.5 Conclusions

This paper has investigated regular subspaces for Boolean control networks. It has been exactly revealed the relationship between a regular subspace and its complementary subspaces. A new computation method to get complementary subspaces has been obtained. Our future work will focus on the applications of the obtained theoretical results to some models of genetic regulation networks.

Acknowledgments This work is supported in part by National Natural Science Foundation (NNSF) of China under Grant 11271194.

References

1. Kauffman SA (1969) Metabolic stability and epigenesis in randomly constructed genetic nets. J Theoret Biol 22(3):437–467
2. Faure A, Naldi A, Chaouiya C, Thieffry D (2006) Dynamical analysis of a generic Boolean model for the control of the mammalian cell cycle. Bioinformatics 22(14):e124–e131
3. Shmulevich I, Kauffman SA (2004) Activities and sensitivities in Boolean network models. Phys Rev Lett 93(4):048701
4. Akutsu T, Hayashida M, Ching W, Ng MK (2007) Control of Boolean networks: hardness results and algorithms for tree structured networks. J Theoret Biol 244(4):670–679
5. Datta A, Choudhary A, Bittner ML, Dougherty ER (2004) External control in Markovian genetic regulatory networks: the imperfect information case. Bioinformatics 20(6):924–930
6. Rosin DP, Rontani D, Gauthier DJ, Schöll E (2013) Control of synchronization patterns in neural-like Boolean networks. Phys Rev Lett 110:104102
7. Cheng D, Qi H (2009) Controllability and observability of Boolean control networks. Automatica 45(7):1659–1667
8. Cheng D, Qi H (2010) A linear representation of dynamics of Boolean networks. IEEE Trans Autom Control 55(10):2251–2258
9. Cheng D, Qi H (2010) State-space analysis of Boolean networks. IEEE Trans Neural Netw 21(4):584–594
10. Cheng D, Li Z, Qi H (2010) Realization of Boolean control networks. Automatica 46(1):62–69
11. Cheng D, Qi H, Li Z (2011) Analysis and control of Boolean networks: a semi-tensor product approach. Springer, London
12. Cheng D, Qi H, Li Z, Liu JB (2011) Stability and stabilization of Boolean networks. Int J Robust Nonlinear Control 21(2):134–156

13. Cheng D (2011) Disturbance decoupling of Boolean control networks. IEEE Trans Autom Control 56(1):2–10
14. Cheng D, Zhao Y (2011) Identification of Boolean control networks. Automatica 47(4): 702–710
15. Laschov D, Margaliot M (2012) Controllability of Boolean control networks via the perron-frobenius theory. Automatica 48(6):1218–1223
16. Laschov D, Margaliot M, Even G (2013) Observability of Boolean networks: a graph-theoretic approach. Automatica 48(8):2351–2362
17. Fornasini E, Valcher ME (2013) Observability, reconstructibility and state observers of Boolean control networks. IEEE Trans Autom Control 58(6):1390–1401
18. Fornasini E, Valcher ME (2013) On the periodic trajectories of Boolean control networks. Automatica 49(5):1506–1509
19. Laschov D, Margaliot M (2011) A maximum principle for single-input Boolean control networks. IEEE Trans Autom Control 56(4):913–917
20. Fornasini E, Valcher ME (2014) Optimal control of Boolean control networks. IEEE Trans Autom Control 59(5):1258–1270
21. Zou Y, Zhu J (2014) System decomposition with respect to inputs for Boolean control networks. Automatica 50(4):1304–1309
22. Zou Y, Zhu J (2015) Kalman decomposition for Boolean control networks. Automatica 54: 65–71
23. Magnus JR, Neudecker H (2007) Matrix differential calculus with applications in statistics and econometrics. Wiley, New York

Chapter 57
Research on Evolution and Simulation of Transaction Process in Cloud Manufacturing

Chun Zhao, Lin Zhang, Bowen Li, Jin Cui, Lei Ren and Fei Tao

Abstract Cloud manufacturing is a new service-oriented networked manufacturing mode. The participants of cloud manufacturing platform conduct transactions in an open environment. This paper focuses on the cooperation of supply-sides and demand-sides in cloud manufacturing platform from the perspective of evolutionary simulation. By combining the agents and evolutionary simulation, we design and implement cloud manufacturing platform simulation model based on service agents. Meanwhile, transactions and resource utilization based on cloud manufacturing and traditional manufacturing modes have been compared through evolutionary simulation. The experimental results demonstrate that cloud manufacturing simulation platform based on service agents could represent the main characteristics of cloud manufacturing transactions. And in this way, the advantages of cloud manufacturing cloud also be illustrated.

Keywords Cloud manufacturing · Simulation · Agent, Multi-thread

57.1 Introduction

Compared with the traditional manufacturing modes, cloud manufacturing platform provides a new mode which is dynamic, intelligent, distributed and applies networked manufacturing [1–3]. In cloud manufacturing platform, resources are shared and provided to the platform users as services. Moreover, requirement publishing is also available in this platform. The demand-side could release requirements without constraints and supply-side all over the world could get the information of these requirements. Consequently, a convenient trading environment is built in this platform.

C. Zhao · L. Zhang (✉) · B. Li · J. Cui · L. Ren · F. Tao
School of Automation Science and Electrical Engineering, Beihang University, Beijing 100191, China
e-mail: johnlin9999@163.com

© Springer-Verlag Berlin Heidelberg 2016

Y. Jia et al. (eds.), *Proceedings of the 2015 Chinese Intelligent Systems Conference*, Lecture Notes in Electrical Engineering 359,
DOI 10.1007/978-3-662-48386-2_57

Within this platform, cooperation and evolution among enterprise users is a significant research point. The enterprise users in cloud manufacturing are autonomous and dynamic, and rules are required for trading establishment.

There exist some research results in the corresponding field. A new idea about enterprise modeling and evolution in cloud manufacturing has been presented [14]. Some factors have also been studied which could affect the adoption of cloud computing by enterprises, such as the relative advantage, complexity, compatibility, upper manager support, enterprise scale, technology readiness, competition of the trading partners [12]. The decision model of entering or exiting the collaborative network has also been studied [11]. Cloud services trading in manufacturing process has also been analyzed [7, 13].

However, the evolution of the service agent simulation is a complicated system engineering. To conduct this simulation process, we need to employ a lot of computer techniques, such as multi-thread, database, distribution, etc. The usage of agent based on multi-thread could reach a better performance to stimulate the collaborative process. The employment of database could achieve complex data structures and help to describe the complicated relationship among services, enterprises, requirements and transactions. Additionally, statistics and search could also be implemented quickly. The simulation platform based on agents is able to simulate the parallel events and better performance could be attained by adjusting the parameters of random factors.

In this paper we introduce a kind of simulation platform based on agents. In this platform, we first propose a method to evaluate the provider, and then compare the results corresponding to several evolution rules. By analyzing the results, some dominant factors have been discovered.

57.2 Simulation Method

Simulation platform established base on the structure of Cloud Manufacturing System. Each user abstracted into an independent agent, in order to realize of service publish, provider selection, self-trading etc. Each agent worked as a thread in the simulation platform, independently implement the function of communication and collaboration each other. And through the different rule to form individual characteristics among the agents, different characteristics of agent in the provider selection will get different results. In addition, simulation platform inner join of real-time sampler, the real-time status of different time t can be collected.

57.2.1 Simulation Platform Model

Simulation platform is composed of platform manager and agent network, agent will send the message to platform manager, such as demand information and searching

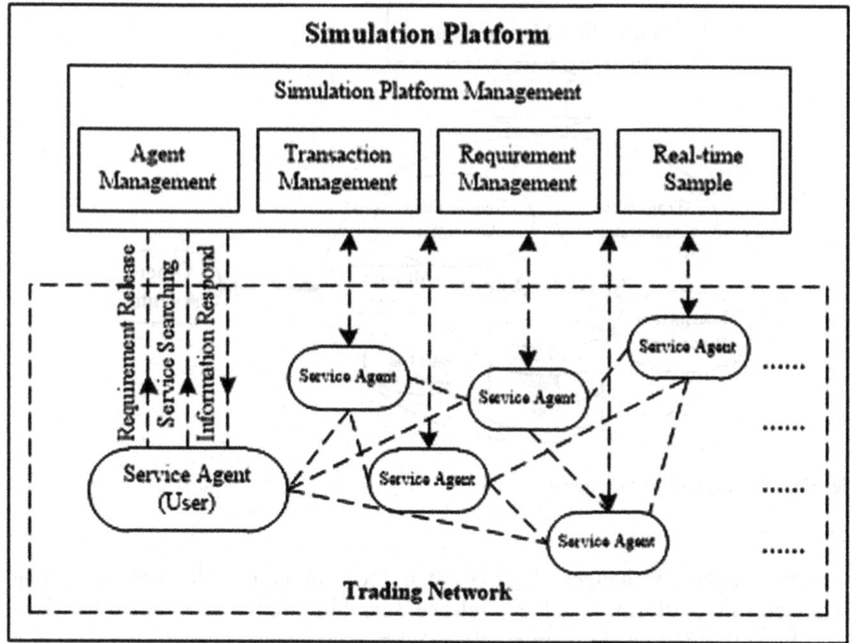

Fig. 57.1 Control allocation strategy

information. According to the information it will communicate with other agents, to bidding, selection and transaction, as shown in Fig. 57.1.

Cloud manufacturing platform to achieve the management of enterprises, transactions and services, where users can trade freely. Simulation platform based on Cloud manufacturing structure simulates the transactions of enterprise through the multi-agent system. As shown in Fig. 57.1, the simulation platform is built in a simulation environment above the cloud manufacturing platform. In cloud manufacturing platform to build enterprise relationship network as the foundation, formed an independent agent each enterprise within the network. Service agents communicate with cloud manufacturing platform independently, to realize demand release, service search. And according to the platform's information feedback get association information of enterprise, the realization of independent communication between enterprises.

57.2.2 Agent Package

Service agent as a enterprise users in cloud manufacturing environment, is the key part of the simulation platform. Trading agent realized by thread in computer system, can create a maximum of two thousand threads in a process.

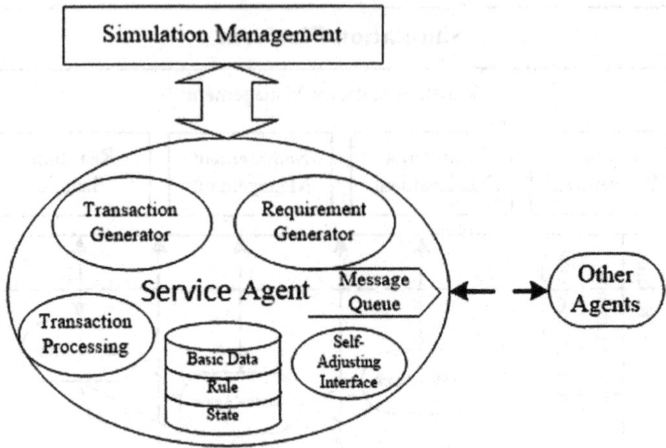

Fig. 57.2 Package of service agent

Service agent exchanges data between the simulation platform and cloud manufacturing platform, to achieve release, searching, registration and other functions. In addition, it need to communicate with other agents, to achieve transaction, query and other functions. Figure 57.2 shows, the Service agent has three basic functions: requirement release, transaction generation and transaction processing.

Requirement generator is the function of service agents release the demand, when a service agent is in idle state of demand, requirement generator will generate a random demand. Bulletin will record the basic information of demand and demander.

Transaction generator is function that service agents generate the transaction, when a service agent is in idle state of transaction, the transaction generator will randomly generates a transaction event.

Transaction processing module is the function that service agent carry out a transaction, the module supports the message transmit and the process driving in the trading. Demander and provider follow their own process respectively. In the implementation of system, Transaction process module take a corresponding mark for each agent, so as to guide the next step of work, but does not provide decision and tactics.

Agent includes a database, to record the basic information, rule, and state information. The basic information is the associated information of the enterprise, including the variation of service fee, variation of service time, variation of QoS and position. The rule records the weights of above four parameters. State information records the three class state: requirement status, working status, activation status.

In addition, The agents use the message to communication each other. Any request or reply will be achieved by message.

57.3 Simulation Models

In this simulation platform, we introduce two models support platform: agent transaction flow and provider selection model. Agent transaction flow is the flow that both sides in the transaction process transactions. And the provider selection model is the rule how to select the service provider from demand requesters.

57.3.1 Flow of Transaction

Flow of transaction is divided into two parts as shown in Fig. 57.3. The first part, the service agent as a service demander, in a wait state, receives the transaction request, and records the service agent, until the deadline of requirement. When the deadline of requirement is reached, transaction processing module will aggregate all of the transaction request agent, sort the list of agent according to the rule of provider

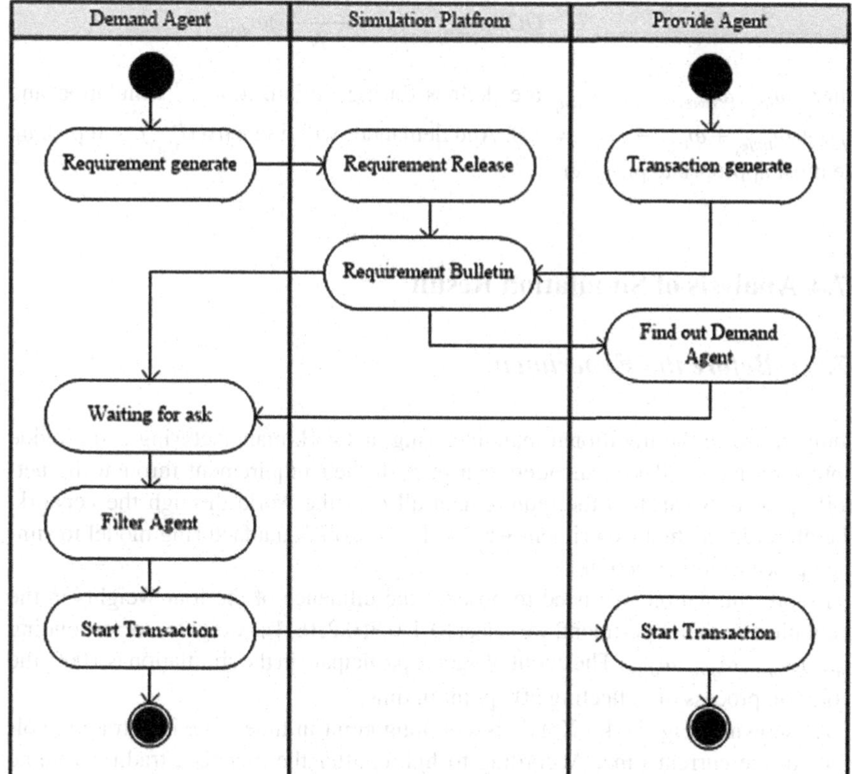

Fig. 57.3 Agent transaction flow

selection, and send message to the best agent, at the same time, turning the current demand on bulletin into transaction status. The second part, the service agent as a service provider, in a transaction wait state. When it received a transaction respond, the success of the transaction. The transaction processing module will turn the state into busy state, and begin to transaction count, until the end of the transaction.

57.3.2 Provider Selection Model

The provider selection model is the comprehensive utility demander assessment provider. While the time of requirement from the demander to limit, demander will evaluate each bid provider. The comprehensive utility evaluation model are as follows:

$$U_{(i,j)} = (1 - Fee_i)\omega_{fee_j} + (1 - \frac{Time_i}{N})\omega_{time_j} + QOS_i\omega_{qos_j} + (1 - \frac{Dist_{(i,j)}}{N})\omega_{dist_j} \tag{57.1}$$

where $\omega_{fee_j}, \omega_{time_j}, \omega_{qos_j}, \omega_{dist_j}$ are defines during initialization of simulation and $\omega_{fee_j} + \omega_{time_j} + \omega_{qos_j} + \omega_{dist_j} = 1.0$. And demander will find $max(U^k_{(i,j)})$, and pick up the most appropriate provider.

57.4 Analysis of Simulation Result

57.4.1 Before the Experiment

Compared with the traditional manufacturing, network manufacturing can provide more information. The demanders can publish their requirement through the network, providers can find the requirement all over the world through the network. Therefore, In the first experiment we use the network manufacturing model to simulate process of transaction.

Before simulation, We need to observe the influence of the four weights in the simulation. In this experiment, we select $[0.4, 0.3, 0.2, 0.1]$ as weights, corresponding $\omega_{fee_j}, \omega_{time_j}, \omega_{qos_j}, \omega_{dist_j}$. The count of agents participate in the simulation is 1000, the evolution process of collecting 200 point in time.

As shown in Fig. 57.4a, X axis is sampling point in time, Y axis is trading volumes in the current time. According to figure, after the first rise, trading volume is stable between 20 to 50. Then, we simulated 24 types of combination, include $[0.4, 0.3, 0.2, 0.1]$, $[0.4, 0.2, 0.3, 0.1]$, ..., $[0.1, 0.2, 0.3, 0.4]$ etc.

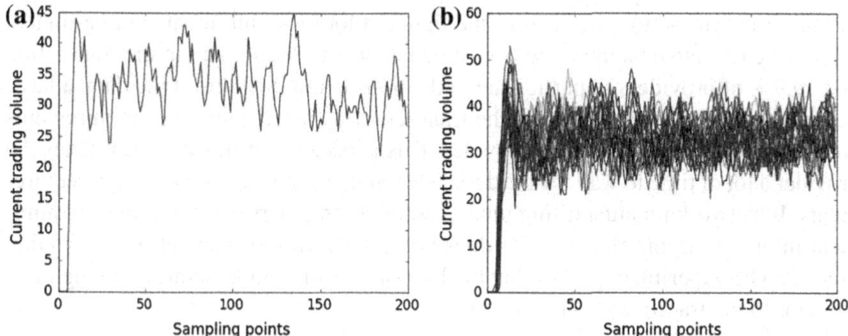

Fig. 57.4 **a** Real-time transaction volume statistics under a type of weights; **b** Real-time transaction volume statistics under 24 types of weights

As shown in Fig. 57.4b, we can gain a conclusion that the weights of selection model is not effect the simulation result. Therefore, the following experiment will simulate in mix with 24 types weight.

57.4.2 Analysis of the Bidding Mode

According to the conclusion above, we will simulation in mix with 24 types weight. There are 1000 agents, 50 types service, 20 types interface in this experiment. And service is single interface. We sampled 400 time points to get the result like Fig. 57.5. There are two lines in the figure. The starts refers to the number of real-time release requirements of demanders. The diamonds refers to the number

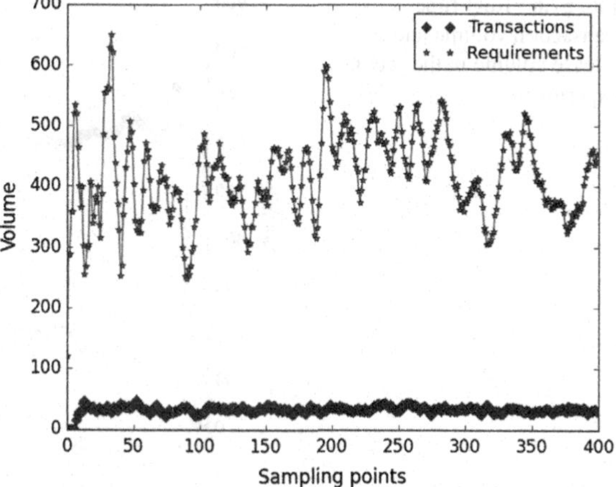

Fig. 57.5 Real-time transaction volume and release volume in the first experiment

of real-time transaction. Seen from the figure, a lot of requirements not be traded. According to statistics, there are 6 % of demanders to participate in the transaction and 19.9 % of providers to participate in the transaction. Seen from the data, a large number of agents not involved in the transaction. A large number of requirements are not find the suitable service provider. This is because in this model, It takes the provider a lot of time to wait for the demander choices, thus missed the other requirements. In network manufacturing mode, although providers can get a lot of requirement information, but the provider does not track each requirement of the bidding process. The experiment proves that the lack of network manufacturing, though it is superior to the traditional manufacturing.

57.4.3 Analysis of the Extended Bidding Mode

The second experiment changed some strategy base on the first one. In the first experiment, provider sends the application after he found the first appropriate requirement, then waits for demander response, and did not do other operations during this period. In the second experiment, provider constant search for new requirement and apply during he waiting for demander response, he will begin trading at once until receiving the response. So, the transaction process is dominated by the demander becomes the mutual choice of both parties.

In this experiment we will simulation in mix with 24 types weight. There are 1000 agents, 50 types service, 20 types interface in this experiment. And service is single interface. As shown in Fig. 57.6, There are two lines in the figure. The starts refers to the number of real-time release requirements of demanders. The diamonds refers to the number of real-time transaction. Seen from the figure, The number of deals increased immediately after beginning a period of time, and kept stable. The

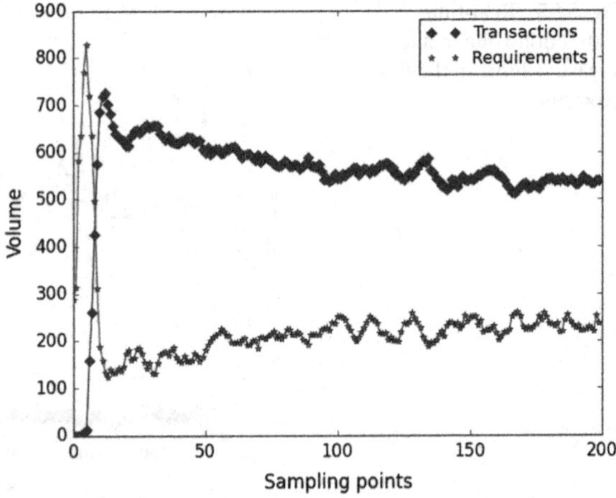

Fig. 57.6 Real-time transaction volume and release volume in the second experiment

number of requirement drops rapidly and kept lower than the number of deals. This shows that most of the requirement are successful trading, And the bargain rate was increased.According to statistics, there are 80.8 % of demanders to participate in the transaction and 92.3 % of providers to participate in the transaction. Seen from the figure, a large number of agents involved in the transaction. The participation rate of agent greatly improved than the first experiment. In cloud manufacturing mode, most of the requirement can be successfully traded, makes up the deficiency in the network manufacturing. Cloud manufacturing platform not only helps demander release requirement information, can also helps the provider to quickly connect with demander.

57.5 Conclusions

This paper concentrates on the cooperation and transaction of both supply-sides and demand-sides in cloud manufacturing platform through evolutionary simulation. We have designed an evolutionary simulation platform of cloud manufacturing transaction based on service agents and propose a new method to package service agents. The evolving process of enterprise transaction on the basis of demand-oriented trading mode has been researched. In addition, the effects on the enterprise cooperation of typical networked manufacturing modes and cloud manufacturing mode have been compared.

The evolutionary simulation platform of cloud manufacturing transaction based on service agents proposed in this paper embodies the typical features of cloud manufacturing such as intelligent and efficient. The experimental results demonstrate that cloud manufacturing mode could promote the transaction between supply-sides and demand-sides more effectively than typical traditional networked manufacturing modes.

Although some important mechanisms identified by this paper could be quite necessary to make the cloud manufacturing platform more efficient, further work still remains to be done. In terms of service agent interface, multi-input and multi-output will be taken into consideration. And the number of service agents in the simulation platform would be expanded. Moreover, more transaction modes would be analyzed to research the corresponding evolution process.

Acknowledgments Supported by the National Natural Science Foundation of China (Grant No. 61374199). The Beijing Municipal Natural Science Foundation (Grant No. 4142031). The National High-Tech Research and Development Plan of China under (Grant No. 2015AA042101)

References

1. Bohu L, Lin Z, Shilong W et al (2010) Cloud manufacturing: a new service-oriented manufacturing model[J]. Comput Integr Manuf Syst 16(1):1–7, 16(Chinese)
2. Lin Z, Yongliang L, Fei T et al (2010) Study on the key technologies for construction of manufacturing cloud[J]. Comput Integr Manuf Syst 15(11):2510–2520 (in Chinese)
3. Bohu L, Lin Z, Lei R et al (2011) Further discussion on cloud manufacturing[J]. Comput Integr Manuf Syst 17(3):449–457 (in Chinese)
4. Schriber TJ (2009) Simulation for the masses spreadsheet-based Monte Carlo simulation[C]. In: Proceedings of the 2009 Winter Simulation Conference, WSC
5. Stirling WCA (2013) Game-theoretic social model for multiagent systems. In: Proceedings of the 2013 IEEE international conference on systems, man, and cybernetics (SMC 2013)
6. Yongkui L, Lin Z, Fei T, Long W (2013) Development and Implementation of cloud manufacturing: an evolutionary perspective[C]. In: Proceedings of the 8th ASME 2013 manufacturing science and engineering conference, 10–14 June 2013, Madison, Wisconsin
7. Cheng Y, Tao F et al (2013) Energy-aware resource service scheduling in cloud manufacturing system based on utility evaluation [J]. J Eng Manuf (Proc ImechE Part B: J Eng Manuf),10. 1177/0954405413492966 August 2013 (Published online)
8. Huang B, Li C, Yin C, Zhao X (2013) Cloud manufacturing service platform for small- and medium-sized enterprises[J]. Int JAdy Manuf technol 65:1261–1272
9. Tao F, Zhao D, Zhang L (2010) Resource service optimal-selection based on intuitionistic fuzzy set and non-functionality QoS in manufacturing grid system[J]. Knowl Inf Syst 25(1):185–208 October
10. Valilai OF, Houshmand M (2013) A collaborative and integrated platform to support distributed manufacturing system using a service-oriented approach based on cloud computing paradigm[J]. Robot Comput-Integr Manuf 29:110–127
11. Renna P (2013) Decision model to support the SMEs decision to participate or leave a collaborative network[J]. Int J Prod Res 51(7):1973–1983
12. Low C, Chen Y, Wu M (2011) Understanding the determinants of cloud computing adoption[J]. Ind Manag Data Syst 111(7):1006–1023
13. Cheng Y, Zhang Y, Lv L et al (2012) Analysis of cloud service transaction in cloud manufacturing[C]. In: Proceedings of the 10th IEEE international conference on industrial informatics. IEEE, 2012, pp. 320–325
14. Yongkui L, Lin Z, Fei T, Long W (2009) Effects of global resource service sharing in cloud manufacturing under the Gale-Shapley allocation rule[J]. Int J Comput Integr Manuf 2009:1–22

Chapter 58
Robust Output Feedback Control of Civil Aircrafts with Unknown Disturbance

Mengjie Wei, Mou Chen and Qingxian Wu

Abstract In this paper, the disturbance attenuation problem is investigated for the flight control system of civil aircrafts. The unknown disturbance is generated by a linear exogenous system, and the civil aircraft dynamic model with unknown disturbance is established. The disturbance is augmented to the state vector and an augmented state observer is designed for estimating the unmeasured augment state vectors. Based on the designed augment state observer, the robust output feedback control scheme is developed for the civil aircrafts with unknown disturbance. Focusing on the B747-100/200, simulation results show that the developed controller could attenuate disturbance effectively.

Keywords Civil aircrafts · Augmented state observer · Unknown disturbance · Disturbance attenuation · Robust control

58.1 Introduction

Flight safety is an important issue in aviation industry which receives many attentions of scholars. Unknown disturbance commonly encountered in actual flight, and it produces a negative impact on structural strength, flight quality and flight path tracking of aircraft. In the process of flying, unknown disturbance often leads civil aircrafts to deviate from the scheduled path. It may also affect the normal operation of the civil aircrafts, and even endanger the flight safety. Hence, the unknown disturbance not only reduces the comfort of the passengers, but also threats the aircraft safety. Therefore, the research on unknown disturbance of civil aircrafts is significant for the reliable flight control system. Until now, some efforts have been devoted to solving the disturbance attenuation problem, such as the

M. Wei (✉) · M. Chen · Q. Wu
College of Automation Engineering, Nanjing University of Aeronautics and Astronautics, 210016 Nanjing, China
e-mail: chenmou@nuaa.edu.cn

© Springer-Verlag Berlin Heidelberg 2016

Y. Jia et al. (eds.), *Proceedings of the 2015 Chinese Intelligent Systems Conference*, Lecture Notes in Electrical Engineering 359,
DOI 10.1007/978-3-662-48386-2_58

optimal control-based designs [1], sliding mode control-based designs [2] and adaptive-based designs [3]. In [4], a disturbance-observer-based (DOB) adaptive sliding mode controller is proposed for a class of uncertain nonlinear system. The robust bounded flight control scheme is developed for the uncertain longitudinal flight dynamics to estimate the compounded disturbance including the unknown external disturbance and the effect of the control input saturation in [5]. It is of highly significance to develop an effective disturbance attenuation scheme to suppress the unknown disturbance of civil aircrafts.

With the development of disturbance attenuation technologies, the disturbance attenuation problem of civil aircrafts has been studied, and many relevant achievements have been obtained since the 1970s [6]. In the early 1970s, the open-loop control method is designed in Boeing-52 in the United States, suppressing the flutter caused by the unknown disturbance effectively. In the 1980s, the German Aerospace Research Institute designed the LARS (Load Alleviation and Ride Stabilization), which contains an open-loop system and a closed-loop system, improving the comfort of the passengers greatly. Since the 1990s, engineers in United States developed the gust load alleviation systems for Boeing-787 with the usage of laser radar. The systems combined with aileron, spoiler and elevons are adopted to offset the disturbance. Although a lot of achievements have been applied in the design of flight control system, open problems still exist, especially for the model uncertainties or unmeasured system states.

This paper focuses on the design of flight control system of civil aircrafts considering uncertainties and unknown disturbance. A robust control method based on augmented state observer is proposed. The unknown disturbance can be augmented to the state vector, and the disturbance value can be estimated through state observer. The controller is obtained by solving the linear matrix inequality (LMI). The robust control method based on augmented state observer could attenuate the unknown disturbance effectively.

58.2 Problem Descriptions

The body axis system is widely used in analyzing the external disturbance of civil aircrafts. When the civil aircraft is affected by the unknown disturbance, the external disturbance should be considered in the aircraft model. In this paper, the short periodic motion mode of civil aircrafts with unknown disturbance is analyzed. The equation for short periodic motion mode [7] can be written as

$$\dot{x}(t) = Ax(t) + Bu(t) + d \tag{58.1}$$

where $x = [q \quad \alpha]^T$ is the state vector, q is the pitching angle rate, α is the angle of attack. $u = \delta_e$ is the control input vector which is produced by the rudder. $d = [d_1\ d_2]^T$ is the unknown disturbance vector. $A = \begin{bmatrix} M_q & M_\alpha \\ 1 & Z_\alpha \end{bmatrix}$, $B = \begin{bmatrix} M_{\delta_e} \\ Z_{\delta_e} \end{bmatrix}$, M_q,

M_α, M_{δ_e} are the pitching-moment coefficient caused by pitching angle rate, angle of attack and elevator, respectively, and Z_α, Z_{δ_e} are the lift coefficient caused by angle of attack and elevator.

Suppose that the unknown disturbance d of system is generated by the following linear exogenous system

$$\begin{cases} \dot{\xi} = W\xi \\ d = V\xi \end{cases} \tag{58.2}$$

where $\xi \in R^2$ and $d \in R^2$. W and V are matrices with corresponding dimensions. As shown in [8–10], a wide class of real engineering disturbance can be represented by this disturbance model (58.2).

The control objective of the system is designing an augmented state observer and then the output feedback controller will be designed for civil aircrafts.

58.3 Control Design Based on Augmented State Observer

In this section, an augmented state observer and an output feedback controller will be developed for civil aircrafts. The following lemmas are needed for the robust output feedback design.

Lemma 1 ([11]) *H, M, N are the constant matrices of appropriate dimension, and* $M^TM \le I$, *I is the unit matrix of appropriate dimension. Then, there exists* $\bar{\alpha} > 0$ *which makes* (58.3) *hold.*

$$HMN + (HMN)^T \le \bar{\alpha}^{-1}HH^T + \bar{\alpha}N^TN \tag{58.3}$$

Lemma 2 ([12] Shur complement) *For a given symmetric matrix* $S = \begin{bmatrix} S_{11} & S_{12} \\ S_{21} & S_{22} \end{bmatrix}$, *where* S_{11} *is* $r \times r$ *dimension. The following three conditions are equivalent.*

$$(1)\, S < 0 \quad (2)\, S_{11} < 0, S_{22} - S_{12}^T S_{11}^{-1} S_{12} < 0 \quad (3)\, S_{22} < 0, S_{11} - S_{12} S_{22}^{-1} S_{12}^T < 0 \tag{58.4}$$

According to the above lemmas, the augmented state observer and the output feedback controller will be designed as the following.

Considering the uncertainties in the system, the model (58.1) can be described as

$$\begin{cases} \dot{x} = (A + \Delta A)x + Bu + d \\ y = Cx \end{cases} \tag{58.5}$$

where $x \in R^2$ is the state vector; $u \in R$ is the control input; $y \in R$ is the output vector. A, B, C, ΔA are the constant matrices of appropriate dimension, respectively. Suppose $\Delta A = E\Sigma F$, E and F are known matrices, Σ is an unknown matrix and $\Sigma^T\Sigma \leq I$.

The state vector ξ of the system expression (58.2) is treated as part of the augmented state vector. Letting $z = \begin{pmatrix} x \\ \xi \end{pmatrix}$ and considering (58.5), the augmented system is constructed as

$$\begin{cases} \dot{z} = (\bar{A} + \Delta\bar{A})z + \bar{B}u \\ y = \bar{C}z \end{cases} \tag{58.6}$$

where $\bar{A} = \begin{pmatrix} A & V \\ 0 & W \end{pmatrix}$, $\Delta\bar{A} = \begin{pmatrix} \Delta A & 0 \\ 0 & 0 \end{pmatrix}$, $\bar{B} = \begin{pmatrix} B \\ 0 \end{pmatrix}$, $\bar{C} = (C \quad 0)$ and $\Delta\bar{A} = \bar{E}\Sigma\bar{F}$.

The state observer and the output feedback controller are designed as follows,

$$\begin{cases} \dot{\hat{z}} = \bar{A}\hat{z} + \bar{B}u + L[y - \hat{y}] \\ u = -K\hat{z} \end{cases} \tag{58.7}$$

where $\hat{z}$ is the observation vector, L is the observer gain and K is a controller design parameter.

Assuming the estimated error of the state is $e = z - \hat{z}$, the closed-loop system equation and error equation can be written as follow

$$\begin{cases} \dot{z} = [\bar{A} + \Delta\bar{A} - \bar{B}K]z + \bar{B}Ke \\ \dot{e} = \dot{z} - \dot{\hat{z}} = \Delta\bar{A}z + (\bar{A} - L\bar{C})e \end{cases} \tag{58.8}$$

The analysis and design above can be summarized as the following theorem.

Theorem *For the linear uncertain system* (58.6), *there exists output feedback control law* $u = -K\hat{z}$ *based on the augmented state observer* (58.7). *The sufficient condition of closed-loop system stability is that there exist symmetric positive definite matrices* $X > 0$, $P_2 > 0$, *matrices* Y, Z, K, L *and positive constant* β, *which make the matrix inequality* (58.9) *hold.*

$$\begin{bmatrix} X\bar{A}^T + \bar{A}X - Y^T\bar{B}^T - \bar{B}Y & \bar{B}Y & \bar{E} & X\bar{F}^T & 0 \\ * & \bar{A}^T P_2 - \bar{C}^T Z^T + P_2\bar{A} - Z\bar{C} & 0 & 0 & P_2\bar{E} \\ * & * & -\beta I & 0 & 0 \\ * & * & * & -\frac{1}{2}\beta^{-1}I & 0 \\ * & * & * & * & -\beta I \end{bmatrix} < 0 \tag{58.9}$$

where, * represents the symmetric transposed matrix and $X = P_1^{-1}$, $Y = KX$, $Z = P_2L$.

Proof The Lyapunov function is constructed as follows

$$V(z(t),t) = z^T(t)P_1z(t) + e^T(t)P_2e(t) \tag{58.10}$$

□

Then, we have

$$\begin{aligned}\dot{V} &= \dot{z}^T(t)P_1z(t) + z^T(t)P_1\dot{z}(t) + \dot{e}^T(t)P_2e(t) + e^T(t)P_2\dot{e}(t) \\ &= z^T(\bar{A} - \bar{B}K)^TP_1z + z^TP_1(\bar{A} - \bar{B}K)z + z^TP_1\bar{B}Ke + e^T(\bar{B}K)^TP_1z \\ &+ z^T\Delta\bar{A}^TP_1z + z^TP_1\Delta\bar{A}z + e^T\bar{A}^TP_2e - e^T(L\bar{C})^TP_2e + z^T\Delta\bar{A}^TP_2e \\ &+ e^TP_2\bar{A}e - e^TP_2L\bar{C}e + e^TP_2\Delta\bar{A}z\end{aligned} \tag{58.11}$$

According to the Lemma 1, we obtain

$$\begin{aligned}&z^T(\Delta\bar{A})^TP_1z + z^TP_1\Delta\bar{A}z \le z^T(\beta^{-1}P_1\bar{E}\bar{E}^TP_1 + \beta\bar{F}^T\bar{F})z \\ &z^T(\Delta\bar{A})^TP_2e + e^TP_2\Delta\bar{A}z \le \beta^{-1}e^TP_2\bar{E}\bar{E}^TP_2e + \beta z^T\bar{F}^T\bar{F}z\end{aligned} \tag{58.12}$$

Hence, we have

$$\begin{aligned}\dot{V} &\le z^T\left[(\bar{A} - \bar{B}K)^TP_1 + P_1(\bar{A} - \bar{B}K) + \beta^{-1}P_1\bar{E}\bar{E}^TP_1 + 2\beta\bar{F}^T\bar{F}\right]z \\ &+ e^T\left[\bar{A}^TP_2 - (L\bar{C})^TP_2 + \beta^{-1}P_2\bar{E}\bar{E}^TP_2 + P_2\bar{A} - P_2L\bar{C}\right]e + z^TP_1\bar{B}Ke + e^T(\bar{B}K)^TP_1z \\ &= \begin{bmatrix} z \\ e \end{bmatrix}^T \bar{\bar{\Omega}} \begin{bmatrix} z \\ e \end{bmatrix}\end{aligned} \tag{58.13}$$

where $\bar{\bar{\Omega}} = \begin{bmatrix} (\bar{A} - \bar{B}K)^TP_1 + P_1(\bar{A} - \bar{B}K) + \beta^{-1}P_1\bar{E}\bar{E}^TP_1 + 2\beta\bar{F}^T\bar{F}P_1\bar{B}K \\ (\bar{B}K)^TP_1\bar{A}^TP_2 - (L\bar{C})^TP_2 + \beta^{-1}P_2\bar{E}\bar{E}^TP_2 + P_2\bar{A} - P_2L\bar{C} \end{bmatrix}$.

Left and right multiplications by *diag*(*X*, *I*) on both sides of Ω, respectively, and letting $X = P_1^{-1}$, $Y = KX$, $Z = P_2L$, it yields

$$\bar{\Omega} = \begin{bmatrix} X(\bar{A} - \bar{B}K)^T + (\bar{A} - \bar{B}K)X + \beta^{-1}\bar{E}\bar{E}^T + 2\beta X\bar{F}^T\bar{F}X\bar{B}Y \\ Y^T\bar{B}^T\bar{A}^TP_2 - \bar{C}^TZ^T + \beta^{-1}P_2\bar{E}\bar{E}^TP_2 + P_2\bar{A} - Z\bar{C} \end{bmatrix} \tag{58.14}$$

Considering Lemma 2, (58.9), (58.13) and (58.14), we have

$$\dot{V} \le \begin{bmatrix} z \\ e \end{bmatrix}^T \begin{bmatrix} z \\ e \end{bmatrix} < 0 \tag{58.15}$$

where $\Omega = \begin{bmatrix} X\bar{A}^T + \bar{A}X - Y^T\bar{B}^T - \bar{B}Y & \bar{B}Y & \bar{E} & X\bar{F}^T & 0 \\ * & \bar{A}^T P_2 - \bar{C}^T Z^T + P_2\bar{A} - Z\bar{C} & 0 & 0 & P_2\bar{E} \\ * & * & -\beta I & 0 & 0 \\ * & * & * & -\frac{1}{2}\beta^{-1} I & 0 \\ * & * & * & * & -\beta I \end{bmatrix}$.

Therefore, the closed-loop system is stable.

58.4 Simulation

Consider the augmented system of a civil aircraft as

$$\begin{cases} \dot{z} = \left[\begin{pmatrix} A & V \\ 0 & W \end{pmatrix} + \begin{pmatrix} \Delta A & 0 \\ 0 & 0 \end{pmatrix} \right] z + \begin{pmatrix} B \\ 0 \end{pmatrix} u \\ y = \begin{pmatrix} C & 0 \end{pmatrix} z \end{cases} \tag{58.16}$$

where $z = [q \quad \alpha \quad d_1 \quad d_2]^T$, $u = \delta_e$, $\Delta A = E\Sigma F$.

The model of B747-100/200 [13] is a common and open simulation model, when setting the balancing point $H = 7000$ m, $V = 240$ m/s, the system matrix A and input matrix B of the longitudinal dynamic model can be written as follows

$$A = \begin{bmatrix} -0.728 & -1.2025 \\ 1.0019 & -0.515 \end{bmatrix} \quad B = \begin{bmatrix} -4.6099 \\ -0.0944 \end{bmatrix}$$

The unknown disturbance of the civil aircrafts can be generated by a linear exogenous neutral stable system described by (58.2) with

$$W = \begin{bmatrix} 0 & 1 \\ -0.1 & -0.6 \end{bmatrix} \quad V = \begin{bmatrix} 1 & 0 \\ 0 & 1 \end{bmatrix}$$

Considering parameter perturbation of $\bar{A}$, $\bar{E}$, $\bar{F}$ can be selected as follows

$$\bar{E} = \begin{bmatrix} -0.0728 & 0 & 0 & 0 \\ 0 & -0.0515 & 0 & 0 \\ 0 & 0 & 0 & 0 \\ 0 & 0 & 0 & 0 \end{bmatrix} \quad \bar{F} = \begin{bmatrix} 1 & 0 & 0 & 0 \\ 0 & 1 & 0 & 0 \\ 0 & 0 & 0 & 0 \\ 0 & 0 & 0 & 0 \end{bmatrix}$$

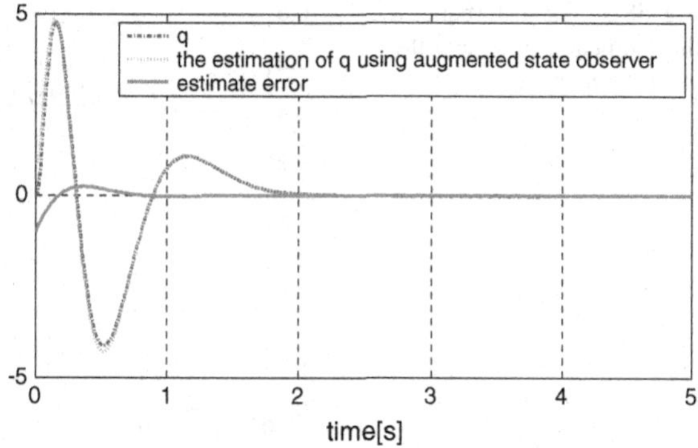

Fig. 58.1 Pitching angle rate q, the estimation of q and the estimate error

By solving the linear matrix inequality (58.9), the controller and observer gain can be obtained as follows

$$\beta = 0.9529 \quad K = [-0.539 \quad -0.0771 \quad -0.1414 \quad -0.3389]$$

$$L = \begin{bmatrix} 0.5270 \\ 0.2040 \\ 2.7223 \\ -0.1900 \end{bmatrix}$$

The robust controller is designed according to (58.7), and the simulation results are presented to demonstrate the effectiveness of the developed augmented state observer in Figs. 58.1, 58.2, 58.3 and 58.4. It can be observed that the state values of the proposed augmented state observer indeed converge uniformly to the original state values. It is shown in Fig. 58.1 that the observation error of the pitching angle rate q between the estimated state and the true state is convergent to zero as far as possible; and from Fig. 58.2, we note that the estimate error of α with a relatively high vibration curve in the first 0.6 s and quickly converge to zero. The disturbance estimate ability of the developed augmented state observer is shown in Figs. 58.3 and 58.4, we can see that the disturbance estimate errors converge to zero within

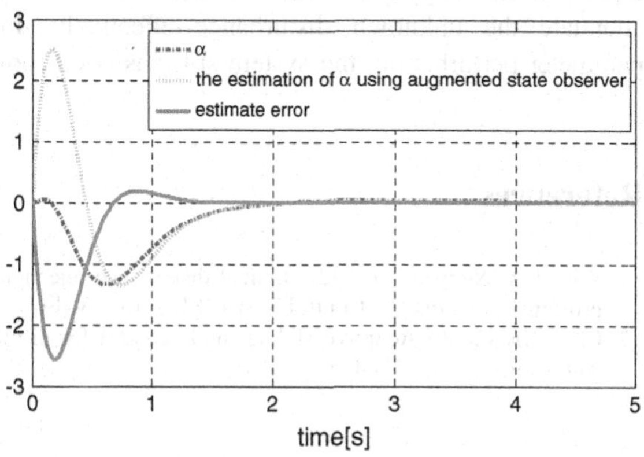

Fig. 58.2 Angle of attack α, the estimation of α and the estimate error

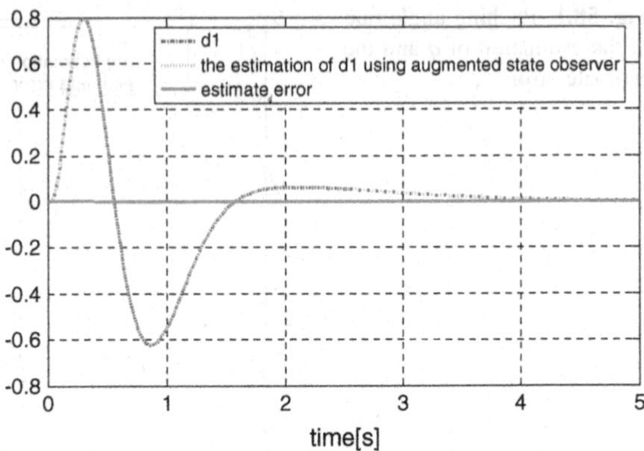

Fig. 58.3 The disturbance d_1, the estimation of d_1 and the estimate error

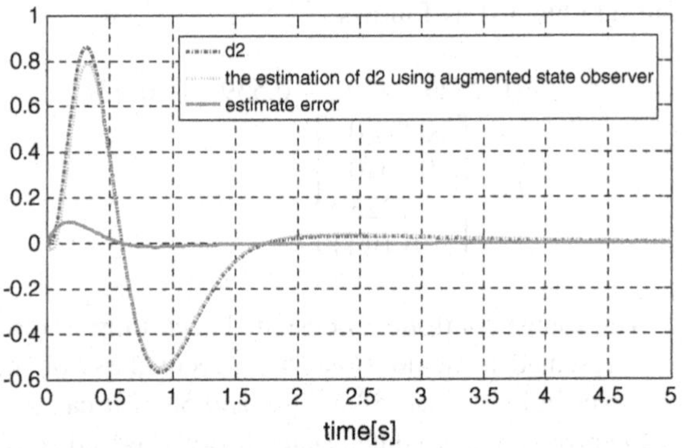

Fig. 58.4 The disturbance d_2, the estimation of d_2 and the estimate error

1 s. The simulation results show that the designed controller can improve the dynamic performance obviously with short settling time, smooth transitional process and robustness against parameter perturbation.

From the plots above, we can see that the augmented state observer can approximate the system state effectively, and the designed robust control scheme based on the augmented state observer is valid. The designed controller can attenuate the unknown disturbance effectively. In the case of aerodynamic parameter perturbation, the system still has good robustness.

References

1. Wise KA, Nguyen T (1992) Optimal disturbance rejection in missile autopilot design using projective controls [J]. Control Syst IEEE 12(5):43–49
2. Chen XK (2006) Adaptive sliding mode control for discrete-time multi-output systems [J]. Automatica 42(3):427–435

3. Yu L, Gang T (2007) An adaptive disturbance rejection algorithm for MIMO system with an aircraft flight control application [C]. In: Proceeding of AIAA guidance, navigation, and control conference, pp 1–17
4. Chen M, Chen WH (2010) Sliding mode control for a class of uncertain nonlinear system based on disturbance observer [J]. Int J Adapt Control Signal Process 24(1):51–64
5. Chen M, Jiang B (2014) Robust bounded control for uncertain flight dynamics using disturbance observer [J]. J Syst Eng Electron 25(4):640–647
6. Regan CD, Jutte CV (2012) Survey of applications of active control technology for gust alleviation and new challenges for lighter-weight aircraft [R]. Dryden Flight Research Center, Edwards
7. Xiao Y, Jin CJ (1993) Atmospheric disturbances in the flight principle [M]. National Defence Industry Press, Beijing
8. Chen WH (2004) Disturbance observer based control for nonlinear systems [J]. IEEE Trans Mechatron 9(4):706–710
9. Isidori A, Byrnes CI (1990) Output regulation of nonlinear systems [J]. IEEE Trans Autom Control 35(2):131–140
10. Huang J (1995) On the minimal robust servo-regulator for nonlinear systems [J]. Syst Control Lett 26(3):313–320
11. Lien CH (2004) Robust observer-based control of systems with state perturbations via LMI approach [J]. IEEE Trans Autom Control 49(8):1365–1370
12. Li Yu (2002) Robust control linear matrix inequality (LMI) approach [M]. Tsinghua University Press, Beijing
13. Hanke CR (1971) The simulation of a jumbo jet transport aircraft: mathematical model, vol 1 [R]. Boeing Company, Kansas

Chapter 59
Photovoltaic System Power Generation Forecasting Based on Spiking Neural Network

Tong Chen, Guoqiang Sun, Zhinong Wei, Huijie Li, Kwok W. Cheung and Yonghui Sun

Abstract A forecasting model based on Spiking neural network (SNN) was proposed to tackle with the problem of the forecasting of photovoltaic system (PVS) power generation. This neural network uses temporal encoding scheme with precise times of spikes, which is closer to the real biological neural system and has powerful computing ability. Considering the main influencing factors such as season types, weather types, sunshine intensity and temperature etc., this model use the method of grey correlation analysis to select similar days. The high accuracy and robust applicability of the proposed forecasting model are verified by the simulation using actual operating data of PVS.

Keywords Photovoltaic system · Spiking neural network · Similar day selection algorithm · Power generation forecasting

59.1 Introduction

The randomness and volatility of the PVS will have an impact on the system operation, and thus endangers the security and stability of the grid. Thus, the high forecast accuracy of power generation can not only reduce the negative impact of large-scale photovoltaic power systems, but also improve the security and stability of power system [1].

Although some extensive researches have been carried out, traditional models of power generation, for example, markov chain, auto regressive moving average

T. Chen (✉) · G. Sun · Z. Wei · Y. Sun
College of Energy and Electrical Engineering, Hohai University, Nanjing 211100, China
e-mail: ntuchentong@163.com

H. Li
ALSTOM GRID Technology Center Co., Ltd, Shanghai 201114, China

K.W. Cheung
ALSTOM Grid Inc., Redmond, WA 98052, USA

© Springer-Verlag Berlin Heidelberg 2016

Y. Jia et al. (eds.), *Proceedings of the 2015 Chinese Intelligent Systems Conference*, Lecture Notes in Electrical Engineering 359,
DOI 10.1007/978-3-662-48386-2_59

(ARMA), their forecast accuracy is not high because of ignoring the changing weather and the nonlinearity of system [2, 3]. Using the model of support vector machine (SVM), it is possible to overcome the drawbacks of small sampling, nonlinearity and high dimension with long computation time. Nevertheless, the SVM should be given an error parameter c and its kernel function must satisfy the Mercer condition [4]. Additionally, artificial neural networks (ANN) especially BP-ANN which has the ability of strong self-learning, adaptive and fault tolerance being widely used in PV power forecast could mimic the human brain training available information as well as execute complex nonlinear mapping [5]. The demerits of using BP-ANN are its low forecast accuracy and weak computation ability.

Some researchers, recently, have found that SNN, or Third-generation neural network [6], uses the temporal encoding scheme to transmit and calculate information, which enables SNN to reflect the actual biological neural systems more realistic. SNN has good performance in pattern recognition and classification [7, 8], and is especially suitable for solving the problem of high-dimensional clustering and nonlinear classification [9, 10]. Furthermore, it has been demonstrated that SNN can realize the function of any feedforward sigmoid neural network and can approximate any continuous function, which makes SNN stronger than other neural networks in computing ability and applicability. Using neural network to forecast the time series, in fact, is to approximate the nonlinear and high-dimensional function. Thus, this paper proposed a forecasting model based on SNN and tested on the actual PVS to verify its validity.

59.2 Spiking Neural Networks

Spiking neuron is the basic unit of SNN and the model of spiking neuron used in this paper is spike response model (SRM). The detailed structure and working principle of SRM can be found in [10]. This paper adopts 3 layers feedforward SNN. In the network, an individual connection consists of a fixed number of m synaptic terminals, where each terminal serves as a sub-connection that is associated with a different delay and weight (Fig. 59.1).

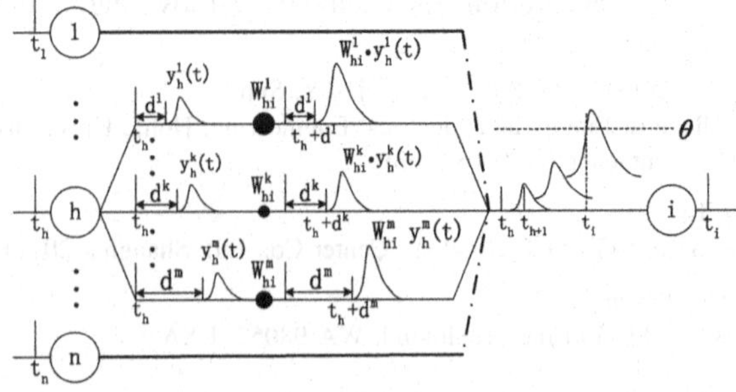

Fig. 59.1 Connectivity between hth and ith neuron with multiple delayed synaptic terminals

SNN adopts temporal encoding scheme which takes the firing time of spiking neuron as input and output signals directly. The neurons in the network generate action potentials, or spikes, when the internal neuron state variable, called "membrane potential", crosses the threshold θ.

The training algorithm in SNN is SpikeProp which is an error-backpropagation training algorithm. It is a supervised learning method and was proposed by Bohte [11].

59.3 Forecasting Model Based on SNN

59.3.1 Similar Day Selection of the Forecasting Model

The photovoltaic array output power [12] in per unit area is calculated as follows:

$$p_s = \eta SI[1 - 0.005(t_0 + 25)] \tag{59.1}$$

where η is the conversion efficiency of photovoltaic cells (%), S is the area of the PV array (m^2), I is the intensity of solar radiation (km/m^2), and t_0 is the ambient temperature (°C).

As shown in formula (59.1), the main influence factors affecting the PVS power generation are η, S I and t_0. For the same PVS, η and S have been included in the historical generation data, so they will no longer be considered. Otherwise, even in the same season types, I t_0 and other factors of different weather types are often different, and the outputs of the PVS power generation are also different. So the weather type is one of the major factors affecting the forecasting accuracy of power generation.

The similar day is determined by gray correlation analysis and the specific steps are as follows:

Step 1: According to the season and weather forecasting information of the forecasting day, the historical days which are similar to the forecasting day are chosen from the total sample to constitute the initial sample, which includes 4 weather types: sunny, overcast, cloudy and rainy days.

Step 2: Constitute the daily weather feature vector $X_i = \left[T_{ih}, T_{il}, \overline{T_i}\right]$, where T_{ih}, T_{il}, $\overline{T_i}$ represents the maximum, minimum and average temperature of the ith historical day respectively (°C).

Step 3: Calculating the correlation coefficient between the forecasting day and the kth meteorological characteristic component of the ith history day:

$$\varepsilon_i(k) = \frac{\min\limits_i \min\limits_k |x'(k) - x_i'(k)| + \rho \max\limits_i \min\limits_k |x'(k) - x_i'(k)|}{|x'(k) - x_i'(k)| + \rho \max\limits_i \max\limits_k |x'(k) - x_i'(k)|} \tag{59.2}$$

where $x'(k)$ and $x_i'(k)$ represents the kth meteorological characteristic component of the normalized forecasting day and the ith history day respectively, ρ is a constant, 0.5 is selected in the paper.

Step 4: Calculate the total correlation degree R_i, and select the historical generation day which satisfies the index of $R_i \geq 0.8$ to sort by the date order, and then select b days which is closest to the forecasting day. We set b = 6 in the paper.

$$R_i = \frac{1}{m} \sum_{k=1}^{m} \varepsilon_i(k) \tag{59.3}$$

where m is the number of the components of the meteorological feature vector.

59.3.1.1 Input and Output Encoding in Forecasting Model

The information in SNN is transmitted and calculated based on accurate spike time, so it is necessary to convert the analog data into spike time. In this paper, time-to-first-spike method [10] is used for input and output encoding in forecasting model. The conversion formula is as follows:

$$T = T_{\max}(1 - p^*) \tag{59.4}$$

where p^* is the normalized analog sample data, $T_{\max}$ is the maximum spike time, and T represents the spike time of neurons, the unit is ms.

In summary, the flow chart of PVS power generation forecasting based on SNN is presented in Fig. 59.2.

59.3.2 Determine the Forecasting Model Structure

(1) Determine the number of neurons in input and output layers
The number of neurons in the input layer is n, corresponding to the number of input variables, and q is the number of neurons in the output layer, corresponding to the number of output variables. Input and output variables of forecasting model are shown in Table 59.1.
(2) Determine the number of neurons in hidden layer.

The number of neurons in hidden layer is an important part of the network structure. In this paper, we use empirical formula (59.5) to determine the best number of neurons as 15 after several tests, where a is an integer between [1, 10].

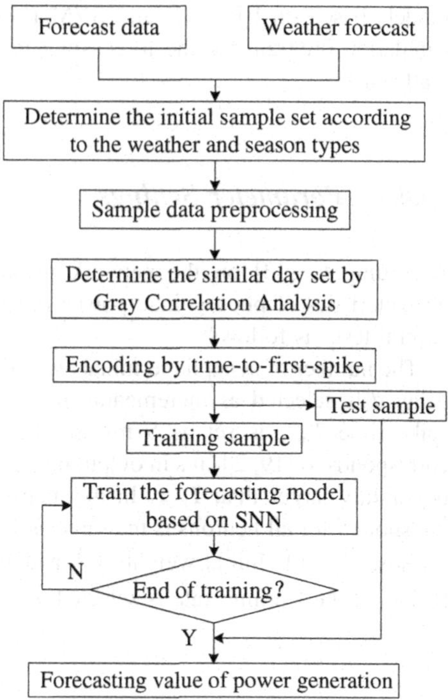

Fig. 59.2 Flow chart of forecasting method

Table 59.1 Input and output variables of forecasting model

Input/output variables	Variable names
x_1–x_{10}	Hourly power generation from 08:00 to 17:00 on similar days
x_{11}–x_{13}	The maximum temperature T_{ih}, the minimum temperature T_{il} and the average temperature $\overline{T}_i$ of similar days
x_{14}–x_{16}	The maximum temperature T_{oh}, the minimum temperature T_{ol} and the average temperature $\overline{T}_o$ of the forecasting day
y_1–y_{10}	Hourly power generation from 08:00 to 17:00 on the forecasting day

Note x represents input variables, y represents output variables

$$p=\sqrt{n+q}+a \tag{59.5}$$

59.4 Case Study

In order to verify the effectiveness of the proposed forecasting model, a real PVS in Jiangsu Province China is adopted as a research object. Historical generation data and its corresponding meteorological data from January 1 to December 31, 2011 are selected to forecast power generation under four different weather types. Forecasting models based on BP-ANN and SVM respectively are used for comparison. The simulation program for the forecasting models is performed on Matlab (R2014a) platform.

59.4.1 Parameter Settings

According to [13] on the network parameter settings, combined with its own characteristics of photovoltaic power generation, parameters are set for SNN after several tests as follows:

The number of synaptic terminals m is set as 16, and the corresponding synapses delay d is selected as incremental integer value between [1, 16]; The maximum spike time $T_{\max}$ is set as 5 ms, so the time interval [0, 5] ms in input layer corresponds to [19, 24] ms in output layer; The time decay constant τ of PSP should be slightly larger than $T_{\max}$. In this paper, it is selected as 6 ms. The excitation threshold θ for all neurons is the same as 1 mv; The iteration step size of SpikeProp is chosen as 0.1; Initial weights W_{hi}^{k} and W_{ij}^{k} can be arbitrary value in the interval [0 1] and the learning rate is selected as 0.05.

59.4.2 Simulation Results and Evaluation

Three forecasting models are used to forecast the power generation of every day in November 2011, six days in this month are sunny, two are overcast, seven are rainy and the other days are cloudy. One day of each four weather types is selected to be analyzed qualitatively. The forecasting results and the real value are compared in Fig. 59.3.

Figure 59.3a shows the power generation forecasting curve and the actual curve in November 15, which is a sunny day. It can be observed that the curve of the power is regular. Also, the three forecasting models show satisfying results in approximating the actual curve. (b) shows the forecasting results in November 18, an overcast day. Due to the reduced training samples on such weather, as well as variable weather conditions, the forecasting results deviate from the actual results in the period 10:00 to 13:00. However, SNN shows better performance than others and reduced the errors apparently in the period 11:00 to 13:00. (c) shows the forecasting results in November 6, a rainy day. There is more uncertainty and randomness on rainy days in PVS. The forecasting curve of SNN can approximate the actual curve with better accuracy from 10:00 to 13:00, in which period the actual curve fluctuates. This exemplifies the better learning and reflection ability of SNN. (d) shows the forecasting results in November 22, a cloudy day. Compared with sunny days, it is difficult to forecast the thickness of the cloud and its mobile trend. As a result, none of the three models can forecast accurately whole time.

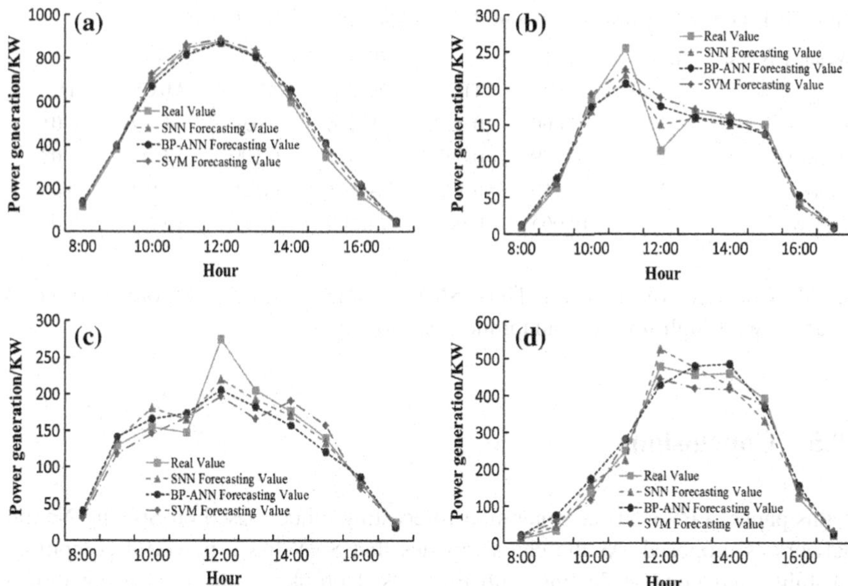

Fig. 59.3 The forecasting results. **a** The forecasting results of sunny days. **b** The forecasting results of overcast days. **c** The forecasting results of rainy days. **d** The forecasting results of cloudy days

Nevertheless, forecasting curve obtained from SNN can reflect the general variation tendency.

Mean absolute percent error (MAPE) and Theil inequality coefficient (TIC) [14] are adopted as two evaluation indexes to give a quantitative analysis on the forecasting errors. MAPE is expressed in the form of percentage. The forecasting errors corresponding to Fig. 59.3 are listed in Table 59.2 respectively. Since the dispersion of error in PVS is large, statistical errors in a month are given in Table 59.3.

Tables 59.2 and 59.3 show that MAPE and TIC are all relatively small under the three different forecasting models on sunny days, no matter a single day or multiple days are calculated. However, the MAPE and TIC on the other three weather types appear to be larger, compared with those on sunny days due to the fluctuation of power generation. On the whole, SNN shows better performance than SVM and

Table 59.2 Forecasting error statistics of three forecasting models for single day

Fig. no.	Weather types	SVM		BP-ANN		SNN	
		MAPE	TIC	MAPE	TIC	MAPE	TIC
3(a)	Sunny	8.575	0.029	9.127	0.030	5.527	0.016
3(b)	Overcast	15.660	0.090	16.896	0.095	12.25	0.066
3(c)	Rainy	14.671	0.101	16.046	0.090	10.208	0.068
3(d)	Cloudy	18.967	0.046	19.567	0.051	14.139	0.050

Table 59.3 Forecasting error statistics of three forecasting models for multi-day

Weather types	Days	SVM		BP-ANN		SNN	
		MAPE	TIC	MAPE	TIC	MAPE	TIC
Sunny	6	7.060	0.027	9.696	0.036	4.476	0.016
Overcast	2	16.449	0.097	17.651	0.109	13.963	0.083
Rainy	7	21.469	0.125	24.501	0.126	16.089	0.087
Cloudy	15	19.065	0.083	21.505	0.101	15.341	0.063

BP-ANN in terms of precision. Thus, SNN is more accurate and more adaptive to weathers with high level of randomness and fluctuation.

59.5 Conclusions

In this paper, a PVS power generation forecasting model based on Spiking Neural Network is proposed. As the characteristics that SNN has a powerful computing capability and good at dealing with time-based problems, the forecasting model based on it has better learning and mapping capability. In the process of modeling, the gray correlation analysis is adopted to determine the similar days, which makes the model more adaptive to different weather types. The results show that, the SNN forecasting model has higher forecasting accuracy and applicability compared with BP-ANN and SVM. Therefore, the proposed PVS power generation forecasting model based on SNN in this paper has a certain value for the photovoltaic generation system in making power generation schedules to meet the needs of the power system.

Acknowledgments The work was supported in part by the National Natural Science Foundation of China under Grants 51277052, 61104045, and 51107032, and in part by the State Grid Corporation of China project: Key Technologies for Power System Security and Stability Defense Considering the Risk of Communication and Information Systems.

References

1. Zhengming Z, Yi L, Fanbo H (2011) Overview of Large-scale grid-connected photovoltaic power plants [J]. Autom Electr Power Syst 35(12):101–107
2. Ming D, Ningzhou X (2011) A method to forecast short-term output power of photovoltaic generation system based on Markov chain [J]. Power Syst Technol 35(1):152–157 (In Chinese)
3. Xuyang G, Kaigui Xie, Bo H et al (2013) A time-interval based probabilistic production simulation of power system with grid-connected photovoitaic generation [J]. Power Syst Technol 37(6):1499–1505 (In Chinese)

4. Jie S, Weijei L, Yongqian L et al (2012) Forecasting power output of photovoltaic systems based on weather classification and support vector machines [J]. IEEE Trans Ind Appl 48 (3):1064–1069
5. Xiaoling Y, Junhua S, Jieyan Xu (2013) Short-term power forecasting for photovoltaic generation considering weather type index. Proc CSEE 33(34):57–64
6. Maass W (1997) Networks of spiking neurons: the third generation of neural network models [J]. Neural Netw 10(9):1659–1671
7. Rowcliffe P, Feng J (2008) Training spiking neuronal networks with applications in engineering tasks [J]. IEEE Trans Neural Netw 19(9):1626–1640
8. Natschläer T, Ruf B (1998) Spatial and temporal pattern analysis via spiking neurons. Netw Comput Neural Syst 9(3):319–332
9. Bohte SM, Poutre HL, Kok JN (2002) Unsupervised clustering with spiking neurons by sparse temporal coding and multi-layer spike neural network. IEEE Trans Neural Netw 13 (2):426–435
10. Gerstner W, Kistler WM (2002) Spiking neuron models. Single neurons, populations, plasticity. Cambridge University Press, New York
11. Bohte SM, Kok JN, La Poutre H (2002) Error-backpropagation in temporally encoded networks of spiking neurons [J]. Neurocomputing 48(1):17–37
12. Yona A, Senjyu T, Saber AY et al (2008) Application of neural network to 24-hour-ahead generating power forecasting for PV system [C]. In: Power energy society general meeting-conversion and delivery of electrical energy in the 21st century, 2008 IEEE, IEEE, pp 1–6
13. Ghosh-Dastidar S, Adeli H (2007) Improved spiking neural networks for EEG classification and epilepsy and seizure detection [J]. Integr Comput Aided Eng 14(3):187–212
14. Danhui Yi (2008) Data analysis and application of Eviews [M]. China Renmin University Press, Beijing

Chapter 60
Stochastic Stabilization of a Class of Nonlinear Systems with Sampled Data Control

Yueling Shen and Wenbing Zhang

Abstract In this paper, the stabilization problem is investigated for a class of nonlinear stochastic systems with sampled data control. By means of the Lyapunov stability theory, a sufficient condition is derived to ensure that the considered stochastic nonlinear system with sampled data is asymptotically stable in mean square. The allowable bound of the sampling interval is obtained and the control gain matrix is solved in terms of convex optimization methods. Finally, a numerical example is presented to further demonstrate the effectiveness of the proposed approach.

Keywords Stabilization · Stochastic system · Sampled data control

60.1 Introduction

Stochastic nonlinear systems have wide application in many fields, such as mechanical systems, economic, biology, control engineering [1–3]. Hence, it is of great importance to investigate the stochastic nonlinear system. When investigating stochastic nonlinear systems, an important issue is the stability of stochastic systems since stochastic stability are widely observed in analysis and synthesis of networked control systems and the consensus of multi-agent systems [4, 5].

On the other hand, one of the most important issues in the study of stochastic nonlinear systems is automatic control, and in the past decade, there are many control strategies have been proposed to control the nonlinear systems, such as: feedback control [6], sampled data control [7, 8] and impulsive control [9, 10]. With the development of computer technology, digital control has widely been used in

Y. Shen
Mechanical and Electrical Engineering College, Nantong Polytechnic College, Jiangsu, China
e-mail: 115195443@qq.com

W. Zhang (✉)
Department of Mathematics, YangZhou University, Jiangsu 225002, China
e-mail: zwb850506@126.com

© Springer-Verlag Berlin Heidelberg 2016

Y. Jia et al. (eds.), *Proceedings of the 2015 Chinese Intelligent Systems Conference*, Lecture Notes in Electrical Engineering 359,
DOI 10.1007/978-3-662-48386-2_60

control systems. Thus, stability/stabilization of networked control systems with sampled data control is a hot research topic now. For instances, in [11], an input delay approach was proposed to investigate the sampled data stabilization problem of linear systems. In [12], By using the impulsive system approach, the stability of uncertain sampled data system was investigated. Very recently, in [13], the stability problem was investigated for a class of linear systems with asynchronous samplings.

Although, there are many resulting concerning on the stability/stabilization of networked control systems with sampled data control, they mainly focus on the linear systems. It is well-known that, nonlinear phenomenon widely exist in practical systems, and the nonlinear effects may lead to network instability. Hence, it is important to investigate the sampled-data stabilization problem of nonlinear systems.

On the other hand, in practical systems, stochastic effects can not be avoided since the communication in dynamical networks is often subjected to an external noisy environment. However, to the best of our knowledge, there are few results concerning the stabilization problem for nonlinear stochastic systems with sampled data control primarily due to the mathematical difficult in dealing with the nonlinear effects.

Based on the above discussion, the main purpose of this paper is to investigate the stochastic stabilization problem for a class of nonlinear systems with sampled data control. By means of the Lyapunov function approach, a sufficient condition is derived to ensure that the considered nonlinear system with sampled data is asymptotically stable in mean square. Finally, an example is given to illustrate the main results obtained in this paper.

Notations: The notations used in this paper is standard.

60.2 Preliminaries

Before the main results, we give the model formulation, lemmas and definitions in this section.

Consider the following nonlinear systems

$$\dot{x}(t) = f(x(t)) + u(t) + g(x(t))d\omega(t), \tag{60.1}$$

where $x(t) = [x_1(t), x_2(t), \ldots, x_n(t)]^T \in \mathbb{R}^n$ is the state vector. $u(t)$ is the control input. $f(x(t)) = (f_1(x(t)), f_2(x(t)), \ldots, f_n(x(t)))^T$ is a nonlinear function with $f_i(0) = 0$. $\omega(t)$ is a one-dimensional Brownian motion. Assume that $g(t, x(t))$ satisfies locally Lipschitz continuous and linear growth conditions. Moreover, $g(x(t))$ satisfies

$$\mathrm{trace}[g^T(x(t))g(x(t))] \leq x^T(t)Lx(t), \tag{60.2}$$

where L is a constant matrix. The initial value of system (60.1) is given by $x(t) = \phi(t)$. The main purpose of this paper is to investigate the stability of the stochastic nonlinear system with sampled data control. Consider the following sampled data control strategy:

$$u(t) = Kx(t_k), t \in [t_k, t_{k+1}) \tag{60.3}$$

where $K \in \mathbb{R}^{n \times n}$ is the control gain matrix, $\{t_k\}_{k=1}^{+\infty}$ denotes the sampled instants satisfying $0 < t_1 < t_2 <, \ldots, < t_k, \lim_{k \to +\infty} t_k = \infty$. Without loss of generality, in this paper, we assume that $t_0 = 0$.

From (60.1) and (60.3), we can obtain

$$\begin{aligned} \dot{x}(t) =& f(x(t)) + Kx(t) + K(x(t_k) - x(t)) \\ &+ g(x(t))d\omega(t), t \in [t_k, t_{k+1}). \end{aligned} \tag{60.4}$$

The following basic definition, lemmas and assumptions are needed in deriving the main results of this paper.

Definition 1 The nonlinear system in (60.1) is said to be asymptotically stable in mean square, if the following inequality holds

$$\lim_{t \to +\infty} \mathbb{E}\|x(t)\| = 0, \forall t \geq t_0. \tag{60.5}$$

Lemma 1 (Jensens inequality). *For any positive definite matrix $\bar{M} > 0$, a scalar $\bar{\gamma} > 0$ and a vector function $\varpi : [0, \bar{\gamma}] \to \mathbb{R}^n$ such that the integrations concerned are well defined, then the following inequality holds:*

$$\left(\int_0^{\bar{\gamma}} \varpi(s)ds\right)^T \bar{M} \left(\int_0^{\bar{\gamma}} \varpi(s)ds\right) \leq \bar{\gamma} \int_0^{\bar{\gamma}} \varpi(s)^T \bar{M} \varpi(s)ds.$$

Lemma 2 ([14]) *The following linear matrix inequality*

$$\begin{bmatrix} Q(x) & S(x) \\ S^T(x) & R(x) \end{bmatrix} > 0,$$

where $Q(x) = Q^T(x)$ and $R(x) = R^T(x)$, is equivalent to either of the following conditions

(1) $Q(x) > 0, R(x) - S^T(x)Q(x)^{-1}S(x) > 0$;
(2) $R(x) > 0, Q(x) - S(x)R(x)^{-1}S^T(x) > 0$.

Assumption 1 The nonlinear function $f(\cdot)$ satisfies the following Lipschitz condition

$$\|f(x) - f(y)\| \leq l_1 \|x - y\|, \tag{60.6}$$

$\forall x, y \in \mathbb{R}^n$, and l_1 is a positive constant.

For simplicity, the periodic sampled data control is used in this paper, i.e., the sampling interval satisfies *Assumption*

$$t_k - t_{k-1} = \tau$$

where, τ is a positive constant.

Remark 1 In this paper, the periodic sampled data control strategy is used to control the stochastic nonlinear system in (60.1). Hence, in order to ensure that the stochastic nonlinear system in (60.1) is asymptotically stable in mean square. the sampling interval is needed to characterize the frequency of the sampled data control. In the following, we will derive a sufficient condition such that the consider system in (60.1) is asymptotically stable in mean square and the allowable upper bound of the sampling interval τ will be obtained.

60.3 Main Results

In this section, the stochastic stabilization problem of the nonlinear system in (60.1) with sampled data control is investigated.

Theorem 1 *Consider the stochastic nonlinear system in (60.1) with sampled data control. If for a positive constant μ, there exist positive definite matrices P, such that the following inequalities hold:*

$$\lambda_1 I \leq P\lambda_2 \leq I, \tag{60.7}$$

$$\begin{bmatrix} 2PY + P + \lambda_2 l_1^2 + \lambda_2 \|L\| + \mu P & 2Y \\ 2Y^T & -2I \end{bmatrix} < 0, \tag{60.8}$$

$$-\mu + (\beta + \gamma)\tau^2 < 0, \tag{60.9}$$

where $\beta = \frac{l_1^2}{\lambda_1}$, $\gamma = \frac{\|K\|^2}{\lambda_1}$. Then the stochastic system in (60.1) with sampled data control will be asymptotically stable in mean square. And the control gain matrix $K = P^{-1}Y$.

Proof Consider the following Lyapunov function:

$$V(t) = x^T(t)Px(t). \tag{60.10}$$

Then, for any $t \in [t_{k-1}, t_k)$, taking the derivative of $V(t)$, along the trajectories of (60.4), we have for $t \in [t_k, t_{k+1})$,

$$\begin{aligned}\mathcal{L}V(t) =&2x^T(t)P[f(x(t)) + Kx(t)]\\ &+ \mathrm{trace}[g^T(x(t))Pg(x(t))]\\ &+ 2x^T(t)PK(x(t_k) - x(t)).\end{aligned} \quad (60.11)$$

In view of Assumption 1 and $2ab \le a^2 + b^2$, we have

$$\begin{aligned}2x^T(t)Pf(x(t)) \le& x^T(t)Px(t)\\ &+ f^T(x(t))Pf(x(t))\\ \le& x^T(t)Px(t)\\ &+ \lambda_{\max}(P)l_1^2x^T(t)x(t).\end{aligned} \quad (60.12)$$

From (60.2), one has

$$\mathrm{trace}[g^T(x(t))Pg(x(t))] \le \lambda_{\max}(P)x^T(t)Lx(t). \quad (60.13)$$

In addition

$$\begin{aligned}&-2x^T(t)PK(x(t) - x(t_k))\\ =&-2x^T(t)PK\int_{t_k}^{t}[f(x(s)) + Kx(t_k)]ds\\ &+\int_{t_k}^{t} g(x(s))d\omega(s).\end{aligned} \quad (60.14)$$

From Lemma 1, we have

$$\begin{aligned}&-2x^T(t)PK\int_{t_k}^{t} f(x(s))ds\\ \le& x^T(t)PKK^TP + \int_{t_k}^{t} f^T(x(s))f(x(s))ds\\ \le& x^T(t)PKK^TP + \frac{l_1^2}{\lambda_1}(t - t_k)^2V_{\max}(s),\end{aligned} \quad (60.15)$$

$$\begin{aligned}&-2x^T(t)PK\int_{t_k}^{t} K(x(t_k))ds\\ \le& x^T(t)PKK^TP + \int_{t_k}^{t} x^T(t_k)K^TKx(t_k)ds\\ \le& x^T(t)PKK^TP + \frac{\|K\|^2}{\lambda_1}(t - t_k)^2V(t_k),\end{aligned} \quad (60.16)$$

where $V_{\max}(s) = \max_{s\in[t_k,t]} V(s)$. From (60.12) to (60.16), we have

$$\begin{aligned}\mathbb{E}\mathcal{L}V(t) \leq & x^T(t)[P + \lambda_2(l_1^2 I_n + \|L\| I_n) \\ & + 2PKP^{-1}K^T P]x(t) \\ & + \beta(t - t_k)^2 V_{\max}(s) \\ & + \gamma(t - t_k)^2 V(t_k). \end{aligned} \tag{60.17}$$

Let $\vartheta = (\beta + \gamma)(t - t_k)^2$. In the following, we will prove that

$$\max_{s\in[t_k,t_{k+1}]} V(s) = V(t_k). \tag{60.18}$$

If (60.18) is not true, then there exists some $t_* \in [t_k, t_{k+1}]$, such that

$$V(t_*) > V(t_k). \tag{60.19}$$

Note that for $t \in [t_k, t_{k+1})$, $t - t_k < \tau$. From (60.9), we know that $-\mu + \vartheta < 0$. In view of (60.17), we have

$$\mathbb{E}\mathcal{L}V(t_k) \leq (-\mu + \vartheta)V(t_k) < 0, \tag{60.20}$$

which means that $V(t)$ will decrease in a short time starting from t_k and therefore, there exist some $t_{**} \in [t_k, t_*]$ such that

$$\begin{aligned} & V(t_{**}) = V(t_k), \\ & \mathbb{E}\mathcal{L}V(t_{**}) > 0, \\ & V(t) \leq V(t_k), t \in [t_k, t_{**}]. \end{aligned} \tag{60.21}$$

It follows from (60.9) that

$$\begin{aligned}\mathbb{E}\mathcal{L}V(t_{**}) \leq & -\mu V(t_{**}) + \vartheta V(t_{**}) \\ < & (-\mu + \vartheta)V(t_{**}) \\ < & 0, \end{aligned} \tag{60.22}$$

which contradicts to $\mathbb{E}\mathcal{L}V(t_{**}) > 0$. Thus, we can conclude that (60.18) is true. In view of (60.17) and (60.18), we have, for $t \in [t_k, t_{k+1})$

$$\begin{aligned}\mathbb{E}\mathcal{L}e^{\mu t}\frac{V(t)}{V(t_k)} = & \mathbb{E}e^{\mu t}\frac{\mathcal{L}V(t)}{V(t_k)} + \mu\mathbb{E}e^{\mu t}\frac{V(t)}{V(t_k)} \\ \leq & -\mu e^{\mu t}\frac{V(t)}{V(t_k)} + \vartheta e^{\mu t} + \mu e^{\mu t}\frac{V(t)}{V(t_k)} \\ \leq & \vartheta e^{\mu t}. \end{aligned} \tag{60.23}$$

Integrating both side of (60.23) from t_k to t_{k+1}, we have:

$$\begin{aligned} & e^{\mu t_{k+1}} \frac{V(t_{k+1})}{V(t_k)} - e^{\mu t_k} \frac{V(t_k)}{V(t_k)} \\ & \leq \frac{\vartheta}{\mu}(e^{\mu t_{k+1}} - e^{\mu t_k}). \end{aligned} \tag{60.24}$$

From (60.24) and note that $t_k - t_{k-1} = \tau$, one has

$$\frac{V(t_{k+1})}{V(t_k)} \leq e^{-\mu\tau} + \frac{\vartheta}{\mu}(1 - e^{-\mu\tau}). \tag{60.25}$$

From (60.9), we have $-\mu + \vartheta < 0$ and therefore $\vartheta < \mu$. Hence

$$\frac{\vartheta}{\mu} < 1. \tag{60.26}$$

In view of (60.25) and (60.26), we have

$$\begin{aligned} \frac{V(t_{k+1})}{V(t_k)} & \leq e^{-\mu\tau} + \frac{\vartheta}{\mu}(1 - e^{-\mu\tau}) \\ & =: \varrho < e^{-\mu\tau} + 1 - e^{-\mu\tau} = 1. \end{aligned} \tag{60.27}$$

Hence, $\mathbb{E}V(t_k)$ converges to zero as k tends to infinity. This completes the proof. This completes the proof.

Remark 2 In Theorem 1, the stabilization problem is studied for a class of stochastic nonlinear systems with sampled data control. The design problem of the sampled data controller is solvable if (60.8), (60.9) and (60.10) are feasible. From (60.8), (60.9) and (60.10), it is clear that we need to obtain the control gain matrix K first and then find the suitable sampling interval τ.

60.4 Numerical Examples

In this section, an example is given to illustrate the results in the previous section.

Example 1 Consider the following stochastic system with sampled data control

$$\begin{aligned} \dot{x}(t) = & \begin{bmatrix} -0.5 & 0.1 \\ -0.3 & 0.3 \end{bmatrix} f(x(t)) \\ & + g(t, x(t))d\omega(t), t \neq t_k, \end{aligned} \tag{60.28}$$

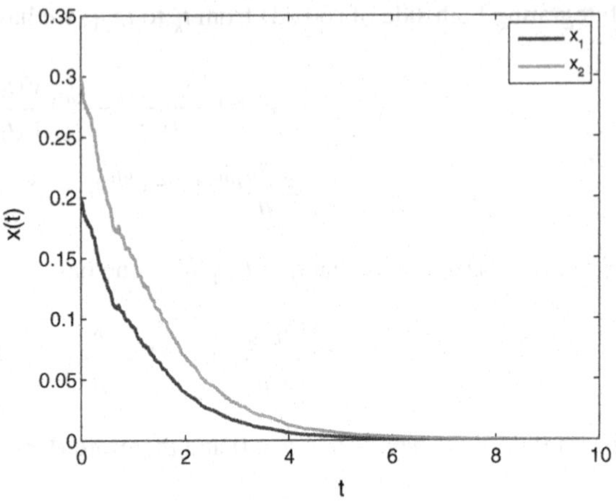

Fig. 60.1 State trajectory of (60.28) with sampled data control

$g(t, x(t)) = 0.2x(t), f(x(t)) = \tanh(0.8x(t))$. Then, we can get $L = \begin{bmatrix} 0.04 & 0 \\ 0 & 0.04 \end{bmatrix}$. Let $\mu = 0.5$, solving Theorem 1, we have $\tau \leq 0.0624$ and $K = \text{diag}\{-5.1824, -5.1824\}$. Let $\tau = 0.06$, the corresponding trajectories of the stochastic nonlinear systems with sampled data in (60.28) are simulated in Fig. 60.1.

From Fig. 60.1, we can see that the nonlinear system in (60.28) is asymptotically stable in mean square with the sampling interval $t_k - t_{k-1} = 0.06$. The simulations confirm the results of this paper well.

60.5 Conclusion

In this paper, the stabilization problem is investigated for a class of nonlinear system with sampled data control. By means of the Lyapunov function approach, a sufficient condition is derived to ensure that the stochastic nonlinear system with sampled data control is asymptotically stable in mean square. An example is given to illustrate the effectiveness of the main results derived in the paper. It is interesting to study stability of vehicle suspension systems with sampled data control [15, 16] in future.

Acknowledgments This work was supported in part by the National Natural Science Foundation of China under Grant Nos. (11426196), in part by the Natural Science Foundation of the Higher Education Institutions of Jiangsu Province, China (Grant No. 14KJB120014).

References

1. Arnold L (1974) Stochastic differential equations-theory and applications. Wiley, New York
2. Hespanha JP, Naghshtabrizi P, Xu Y (2007) A survey of recent results in net-worked control systems. Proc IEEE 95:138–162
3. Teel AR, Subbaraman A, Sferlazza A (2014) Stability analysis for stochastic hybrid systems: a survey. Automatica 50:2435–2456
4. Luo XMQ, Shen Y (2011) Generalised theory on asymptotic stability and boundedness of stochastic functional differential equations. Automatica 47:2075–2081
5. Hespanha JP (2014) Modeling and analysis of networked control systems using stochas-tic hybrid systems. Ann Rev Control Vol Accept
6. Peng C, Han Q, Yue D (2013) Communication-delay-distribution-dependent de- centralized control for large-scale systems with ip-based communication networks. IEEE Trans Control Syst Technol 21:820–830
7. Zhang W, Yu L (2010) Stabilization of sampled-data control systems with control inputs missing. IEEE Trans Autom Control 55:447–452
8. Chen W, Zheng W (2011) An improved stabilization method for sampled-data control systems with control packet loss. IEEE Trans Autom Control 57:2378–2384
9. Yang Z, Xu D (2007) Stability analysis and design of impulsive control systems with time delay. IEEE Trans Autom Control 52:1448–1454
10. Liu X, Wang Q (2008) Impulsive stabilization of high-order hopfieldtype neural networks tith time-varying delays. Neural Networks 19:71–79
11. Fridman E, Seuret A, Richard JP (2004) Robust sampled-data stabilization of linear systems: an input delay approach. Automatica 40:1441–1446
12. Naghshtabrizi P, Hespanha JP, Teel AR (2008) Exponential stability of impul-sive systems with application to uncertain sampled-data systems. Syst Control Lett 57:378–385
13. Seuret A (2012) A novel stability analysis of linear systems under asynchronous sam-plings. Automatica 48:177–182
14. Boyd S, Ghaoui LE, Balakrishnan V (1994) Linear matrix inequalities in system and control theory. SIAM, Philadelphia
15. Li H, Jin X, Lam HK, Shi P (2014) Fuzzy sampled-data control for uncertain vehicle suspension systems. IEEE Trans Cybern 44:1111–1126
16. Li H, Yu J, Hilton C, Liu H (2013) Adaptive sliding-mode control for nonlinear active suspension vehicle systems using t-s fuzzy approach. IEEE Trans Industr Electron 60:3328–3338

Chapter 61
A Reconfigurability Evaluation Method for Satellite Control System

Heyu Xu, Dayi Wang and Wenbo Li

Abstract The reconfigurability of satellite control systems is a fundamental index to the capability of spacecraft stability after the faults occur in the process of satellite operating in-orbit. This paper addresses the problem of reconfigurability evaluation for faulty control systems. Firstly, The reconfiguration indicator based on system stability is presented by regarding the faults that occur in the system as model uncertainties. And then the method for normalized coprime factorization is introduced to describe the maximal reconfigurability boundary, which causes the control system instability when the fault exceeds the allowable range of the given controller and model. Finally, the efficacy of the proposed method is tested through an numerical simulation on the Hubble telescope.

Keywords Reconfigurability evaluation · Control system · Coprime factorization

61.1 Introduction

The control system is one of the most important and the most complex systems in the spacecraft. Since the crucial importance of the undertaken tasks of control system, it is catastrophic if faults occur in the control system of spacecraft. In order to avoid the occurrence of the above phenomenon, the system has been required to

---

H. Xu · D. Wang (✉) · W. Li
Beijing Institute of Control Engineering, Beijing 100190, China
e-mail: dayiwang@163.com

H. Xu
e-mail: 694936963@qq.com

W. Li
e-mail: liwenbo_502@163.com

H. Xu · D. Wang · W. Li
Science and Technology on Space Intelligent Control Laboratory, Beijing 100190, China

© Springer-Verlag Berlin Heidelberg 2016

Y. Jia et al. (eds.), *Proceedings of the 2015 Chinese Intelligent Systems Conference*, Lecture Notes in Electrical Engineering 359,
DOI 10.1007/978-3-662-48386-2_61

be reconfigurable. In other words, the system can be reconfigured after the faults happen. To improve the satellite's capacity for failures, the system must be reconfigurable. The reconfigurability of spacecraft is defined as follows. The ability to recover functionality in whole or in part is by means of changing the configuration when fault occurs. Precisely because of this it is essential to research the reconfigurability of the system to improve the ability of handling failure.

In order to improve design insights into the synthesis of controller reconfiguration for reconfigurable systems, control reconfigurability of linearized systems is analyzed. The stability is the most important condition for systems. Therefore, whether the system is stable, what kinds of system stability, how much the stability margin is play an important role for reconstruction of system. At the same time, oriented to control systems, a reconfiguration goal is identified, i.e. stabilization margin [1]. In this way, when the controller is fixed, the parameters of controller needn't be adjusted if the fault system satisfied the stabilization goal. Lastly, the simulation experiments have demonstrated the validity and practicability of the conclusion of this paper.

The main content of this paper is to regard the fault of the satellite control system as the uncertainty of a system and to research the reconfigurability of linearized systems based on maximal fault stability margin.

The fault represented by additive/multiplicative fault and the coprime factorization respectively [3]. The relation between these two method is also investigated. The additive/multiplicative fault can be represented by transfer function forms, and it requires the fault plant and the nominal plant to have the same number of closed right plane poles. The coprime factorization doesn't require the fault plant and the nominal plant to have the same number of closed right plane poles. Therefore, this paper researches the maximal fault stability margin by means of coprime factorization which gives a specific index to the solution of controller.

At last, this method can be applied into Hubble telescope. When the controller is designed, namely when the controller is fixed, the parameters of controller needn't be adjust if the fault have not surpass the maximal fault stability margin. This simulation experiments have demonstrated the validity and practicability of the conclusion of this paper.

The contribution of this paper is to propose a method for quantitative reconfigurability evaluation of closed-loop control system. The principle of the method for normalized coprime factorization is presented, in order to describe the maximal reconfigurability boundary of the given controller and system model.

61.2 Additive/Multiplicative Fault

61.2.1 Additive Fault

A structure chart of additive fault of the closed-loop control system is shown in Fig. 61.1a, the transformed equivalent structure diagram as shown in Fig. 61.1b [7].

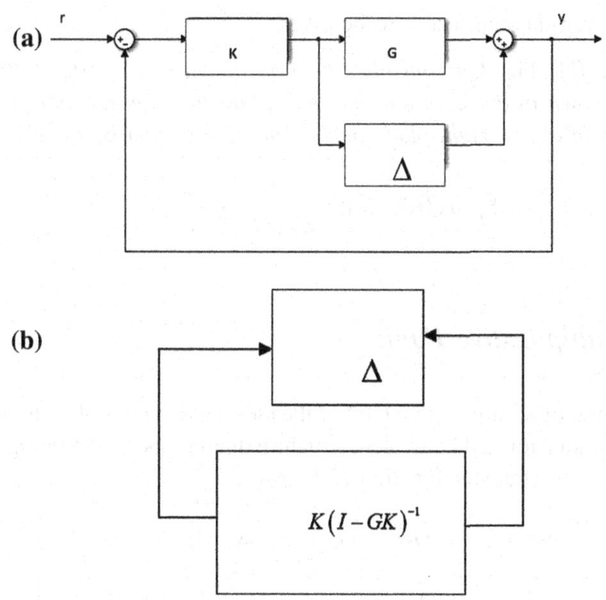

Fig. 61.1 **a** Additive fault. **b** Additive fault

Assume that the transfer function as follows:

$$G_{\Delta}(s) = G_0(s) + \Delta_a(s) \tag{61.2.1}$$

$G_0(s)$ is the nominal model, $\Delta_a(s)$ is the additive fault, $\Delta_a(s)$ satisfied:

$$\sigma_{\max}(\Delta_a(j\omega)) \le |r(j\omega)| \tag{61.2.2}$$

where $\sigma_{\max}(\Delta_a(j\omega))$ is the largest singular value of $\Delta_a(j\omega)$, i.e.

$$\sigma_{\max}(\Delta_a(j\omega)) \le \{\lambda_{\max}\Delta_a(j\omega)^*\Delta_a(j\omega)\}$$

$A(G_0, r)$ is given by:

$$A(G_0, r) \triangleq \{G(s) = G_0(s) + \Delta_a(s) | \sigma_{\max}(\Delta_a(j\omega)) \le |r(j\omega)|, \forall\omega\} \tag{61.2.3}$$

Theorem 2.1 ([1]) *Let controller K(s) as shown in Fig.* 61.1a*, the object set* $A(G_0, r)$ *as shown in* (61.2.3)*, and the fault plant and the nominal plant have the same number of closed right plane poles. Then the system is stable if and only if:*

$$① \Delta_a = 0,\ ② \left\| r(s)K(s)[I + G_0(s)K(s)]^{-1} \right\|_{\infty} < 1 \tag{61.2.4}$$

The following theorem can also be given:

Theorem 2.2 ([1, 11]) *Let controller K(s) as shown in Fig.* 61.1*a, the object set* $A(G_0, r)$ *as shown in* (61.2.3)*, and the fault plant and the nominal plant have the same number of closed right plane poles. Then the system is stable if and only if*:

$$\text{①}\ \Delta_a = 0\ \text{②}\ \bar{\sigma}[\Delta(s)] < \frac{1}{\bar{\sigma}[K(I+G_0K)^{-1}](s)}, \forall s \tag{61.2.5}$$

61.2.2 Multiplicative Fault

A structure chart of multiplicative fault of the closed-loop control system is shown in Fig. 61.2a, the transformed equivalent structure diagram is shown in Fig. 61.2b [12].

Assume that the transfer function as follows:

$$G(s) = G_0(s)[I + \Delta_a(s)] \tag{61.2.6}$$

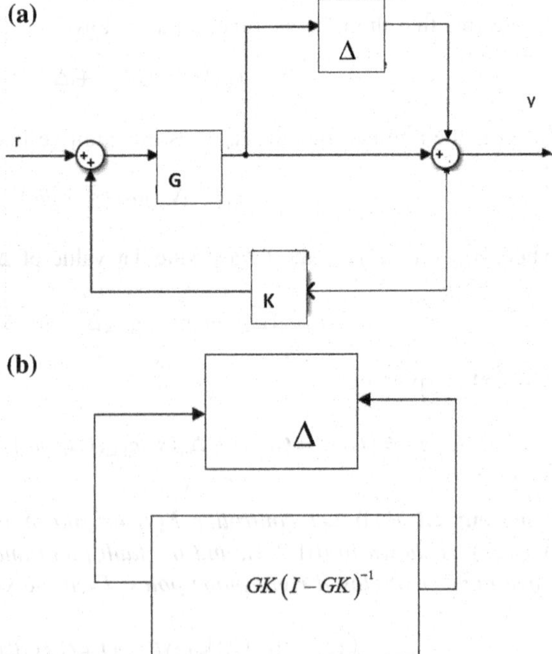

Fig. 61.2 **a** Multiplicative fault. **b** Multiplicative fault

$G_0(s)$ is the nominal model, $\Delta_a(s)$ is the additive fault, $\Delta_a(s)$ satisfied:

$$\sigma_{\max}(\Delta_a(j\omega)) \le |r(j\omega)| \tag{61.2.7}$$

where $\sigma_{\max}(\Delta_a(j\omega))$ is the largest singular value of $\Delta_a(j\omega)$, i.e.

$$\sigma_{\max}(\Delta_a(j\omega)) \le \{\lambda_{\max}\Delta_a(j\omega)^*\Delta_a(j\omega)\}$$

$A(G_0, r)$ is given by:

$$A(G_0, r) \triangleq \{G(s) = G_0(s)[I + \Delta_a(s)] | \sigma_{\max}(\Delta_a(j\omega)) \le |r(j\omega)|, \forall\omega\} \tag{61.2.8}$$

Theorem 2.3 ([1]) *Let controller K(s) as shown in Fig. 61.2a, the object set $A(G_0, r)$ as shown in* (61.2.8), *and the fault plant and the nominal plant have the same number of closed right plane poles. Then the system is stable if and only if:*

$$① \ \Delta_a = 0 \ ② \ \left\| r(s) = [G_0(s)K(s)[I - G_0(s)K(s)]^{-1} \right\|_\infty < 1 \tag{61.2.9}$$

The following theorem can also be given:

Theorem 2.4 ([1]) *Let K controller (s) as shown in Fig. 61.2a, the object set $A(G_0, r)$ as shown in* (61.2.9), *and the fault plant and the nominal plant have the same number of closed right plane poles. Then the system is stable if and only if:*

$$① \ \Delta_a = 0 \ ② \ \bar{\sigma}[\Delta(s)] < \frac{1}{\bar{\sigma}[G_0K(I - G_0K)^{-1}(s)]}, \forall s \tag{61.2.10}$$

So, when the controller is known, for additive or multiplicative fault, we can use the above theorem to calculate the control object of the fault area. The controller is designed by PID controller as an example. The calculation steps of fault coverage are as follows:

① Making Fig. 61.1a or 61.2a respectively into Figs. 61.1a and 61.2b;
② Then $G(s) = K(s)[I + G_0(s)K(s)]^{-1}$ or $G(s) = G_0(s)K(s)[I - G_0(s)K(s)]^{-1}$;
③ When the PID controller can make the stability of the nominal model, if $\Delta_a(s)$ has no closed right plane poles, then the margin is: $\|\Delta_a(s)\|_\infty < \frac{1}{\|G(s)\|_\infty}$

However, when the fault plant and the nominal plant doesn't have the same number of closed right plane poles, this method can't be used. Therefore, coprime factorization method will be discussed in the following sections.

61.3 The Method for Coprime Factorization Fault and Stability Margin

Considering the additive fault with maximal fault margin. If the nominal object described by left coprime factorization, fault object also use the left coprime factorization described, i.e. [10]

① $G=\tilde{M}^{-1}\tilde{N}$, $\tilde{M},\tilde{N}\in RH_{\infty}$ and they are coprime.
② $\Delta_{\tilde{M}},\Delta_{\tilde{N}}$ are described as coprime factorization.

Then the fault plant can be expressed as:

$$G_{\Delta}=(\tilde{M}+\Delta_{\tilde{M}})^{-1}(\tilde{N}+\Delta_{\tilde{N}}),\ \begin{bmatrix}\Delta_{\tilde{M}}\\ \Delta_{\tilde{N}}\end{bmatrix}\in RH_{\infty}$$

This kind of object can be described as follows:

$$\{G_{\Delta}=(\tilde{M}+\Delta_{\tilde{M}})^{-1}(\tilde{N}+\Delta_{\tilde{N}}):\left\|\begin{bmatrix}\Delta_{\tilde{M}}\\ \Delta_{\tilde{N}}\end{bmatrix}\right\|<\varepsilon,\ \begin{bmatrix}\Delta_{\tilde{M}}\\ \Delta_{\tilde{N}}\end{bmatrix}\in RH_{\infty}\} \tag{61.3.1}$$

Lemma 3.1 ([2, 8]) *The following conditions are equivalent:*

① $\begin{bmatrix}M & U\\ N & V\end{bmatrix}$ *is invertible in* RH_{∞}
② $\begin{bmatrix}\tilde{V} & -\tilde{U}\\ -\tilde{N} & \tilde{M}\end{bmatrix}$ *is invertible in* RH_{∞}
③ $\tilde{M}V-\tilde{N}U$ *is invertible in* RH_{∞}
④ $\tilde{V}M-\tilde{U}N$ *is invertible in* RH_{∞}.

Let the fault plant have a standard for coprime factorization:

$$\{G_{\Delta}=(\tilde{M}+\Delta_{\tilde{M}})^{-1}(\tilde{N}+\Delta_{\tilde{N}}):\left\|\begin{bmatrix}\Delta_{\tilde{M}}\\ \Delta_{\tilde{N}}\end{bmatrix}\right\|<\varepsilon,\ \begin{bmatrix}\Delta_{\tilde{M}}\\ \Delta_{\tilde{N}}\end{bmatrix}\in RH_{\infty}\} \tag{61.3.2}$$

Assume that the controller K can make the nominal system in stability, controller K can make it stable for all fault plant described as (61.3.2) if and only if:

$$\varepsilon_{\max}^{-1}=\inf_{K}\left\|\begin{bmatrix}I\\ K\end{bmatrix}(I+GK)^{-1}\tilde{M}\right\|_{\infty} \tag{61.3.3}$$

When the object is coprime factorization the same conclusion still established, i.e.

$$\{G_\Delta = (N+\Delta_N)(M+\Delta_M)^{-1} : \|[\Delta_N \quad \Delta_M]\| < \varepsilon, [\Delta_N \quad \Delta_M] \in RH_\infty\}.$$

Then, controller K is obtained, which makes all objects described in (61.3.2) stable if and only if: [3, 6]

$$\|M^{-1}(I+KG)^{-1}[K \quad I]\| \le \varepsilon^{-1} \tag{61.3.4}$$

Now, we extend the nominal system as the form of H_∞:

$$P = \begin{bmatrix} P_{11} & P_{12} \\ P_{21} & P_{22} \end{bmatrix} \triangleq \begin{bmatrix} \begin{bmatrix} 0 & M^{-1} \end{bmatrix} & M^{-1} \\ \begin{bmatrix} I & G \end{bmatrix} & G \end{bmatrix} \tag{61.3.5}$$

After that, the object can be described as the following form by linear fractional transformation:

$$\begin{gathered} \Omega_u\left(P, \begin{bmatrix} \Delta_n \\ -\Delta_m \end{bmatrix}\right) \triangleq P_{22} + P_{21}\begin{bmatrix} \Delta_N \\ -\Delta_M \end{bmatrix}\left(I - P_{11}\begin{bmatrix} \Delta_N \\ -\Delta_M \end{bmatrix}^{-1} P_{12}\right) = (N+\Delta_N)(M+\Delta_M)^{-1} \\ \Omega_l(P,K) \triangleq P_{11} + P_{12}K(I-P_{22}K)^{-1}P_{21} = M^{-1}(I-KG)^{-1}[K \quad I] \end{gathered} \tag{61.3.6}$$

As the description above, the theorem can be obtained as following:

Theorem 3.2 ([9, 10]) The maximum stable boundary to meet the above conditions can be expressed as:

$$\varepsilon_{\max}^{-1} = \inf_K \|\Omega_l(P,K)\|_\infty \tag{61.3.7}$$

Therefore, we can gain the stability margin of the fault plant controlled by the fixed controller. The details are as follows [4, 5]:

① Stable controller k is well designed. Make the nominal system G_0 to coprime factorization as $G_0 = NM^{-1} = \tilde{M}^{-1}\tilde{N}$.

② Find $\varepsilon^{-1} = \left\| \begin{bmatrix} I \\ K \end{bmatrix} (I+G_0K)^{-1}\tilde{M}^{-1} \right\|_\infty$

③ Then the stable range is $\left\| \begin{bmatrix} \Delta_{\tilde{M}} \\ \Delta_{\tilde{N}} \end{bmatrix} \right\|_\infty < \varepsilon$

This method for fixed controller fault margin is better than the additive/multiplicative method for finding maximal fault stability margin. Unlike additive/multiplicative fault, the coprime factorization fault part is in the internal fault object factorization. So the fault $G_\Delta - G$ does not have to be stable.

61.4 Illustrative Example

This section takes the Hubble telescope as the research object. Based on the attitude control system, the fault analysis on the control system stability and the impact of the changes are studied. Calculating the maximal fault stability margin and designing the PID controller. Finally the simulation results prove that when the fault have not surpass the maximal fault stability margin the system can stay stable.

The rigid model of satellite control system structure is shown in Fig. 61.3 [7]:

The nominal model is $G = \frac{12788}{s^2 + 159.8s + 12788}$. By the method provided by the third section we can obtain maximal fault stability margin $\varepsilon_{\max} = 0.8593$. The parameters of the PID controller is $K_D = 2.65, K_P = 0.5, K_I = 0.05$. And the PID controller can be described as:

$$K = \frac{250.5s^2 + 50.05s + 5}{s^2 + 100s}.$$

The stability margin of the fault plant controlled by the fixed controller is $\varepsilon = 0.6164 < \varepsilon_{\max}$, which means this controller can make it stable. Also, the unit step response curve without fault shown in Fig. 61.4 can prove it.

The nominal system of G's regular coprime factorization is $G = \tilde{M}^{-1}\tilde{N}$, where

$$\tilde{M} = \frac{-s^2 - 159.8s - 12780}{s^2 + 190.1s + 18070}, \quad \tilde{N} = \frac{-12780}{s^2 + 190.1s + 18070}$$

Assume that a fault model is follows: $G_{\Delta_1} = \frac{12778}{s^3 + 159.8s^2 + 3s + 12778}$, whose regular coprime factorization is $G_{\Delta_1} = \tilde{M}_1^{-1}\tilde{N}_1$, where

$$\tilde{M}_1 = \frac{-s^3 - 159.8s^2 - 3s - 12780}{s^3 + 168.4s^2 + 1415s + 18070}, \quad \tilde{N}_1 = \frac{-12780}{s^3 + 168.4s^2 + 1415s + 18070}$$

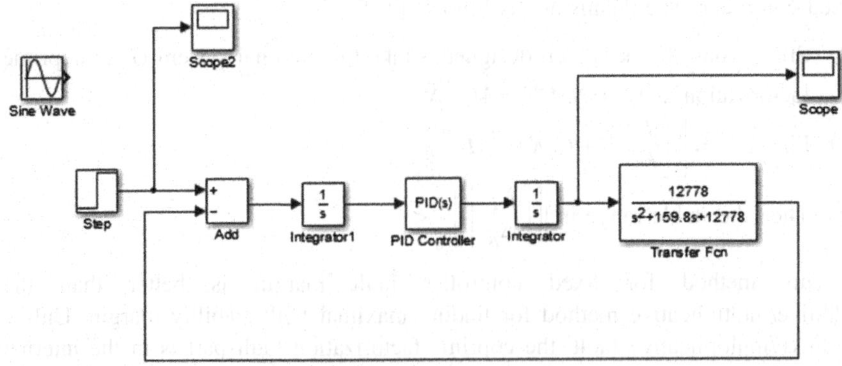

Fig. 61.3 Rigid model of satellite control system structure

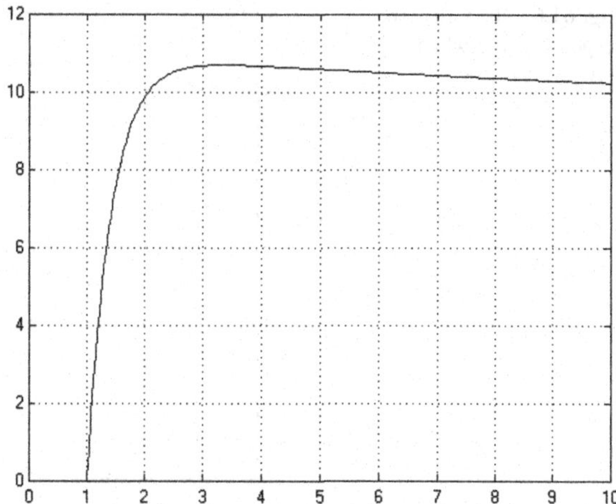

Fig. 61.4 The unit step response curve without fault

In this way, the fault $\Delta_{\tilde{M}_1} = \tilde{M}_1 - M, \Delta_{\tilde{N}_1} = \tilde{N}_1 - N$ can be calculated respectively. By calculation the $\left\| \Delta_{\tilde{M}_1} \quad \Delta_{\tilde{N}_1} \right\| = 1.4445 > 0.6164$. Based on the theorem mentioned in third chapter, we can judge the controller cannot control the fault object. Also, the simulation results prove that the controller K does not make the fault object stabilized, as shown in Fig. 61.5.

Change the fault system to $G_{\Delta_2} = \frac{20000}{s^3 + 159.8s^2 + 20000}$, whose normalized coprime factorization is $G_{\Delta_2} = \tilde{M}_2^{-1} \tilde{N}_2$, where

$$\tilde{M}_2 = \frac{-s^2 - 159.8s - 20000}{s^2 + 205.2s + 18280}, \tilde{N}_2 = \frac{-20000}{s^2 + 205.2s + 18280}$$

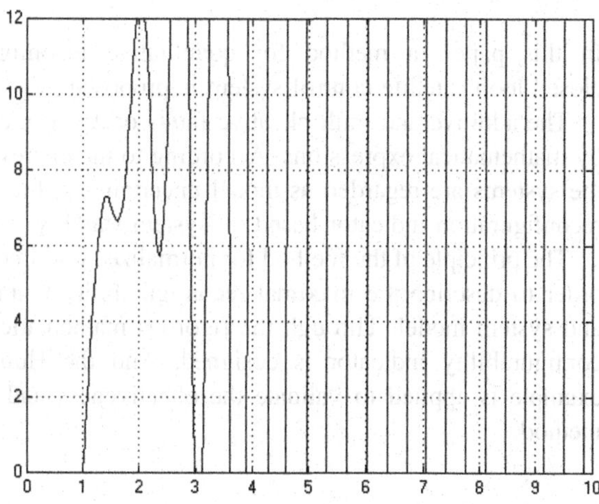

Fig. 61.5 The unit step response curve when $\left\| \Delta_{\tilde{M}_1} \quad \Delta_{\tilde{N}_1} \right\| > \varepsilon$

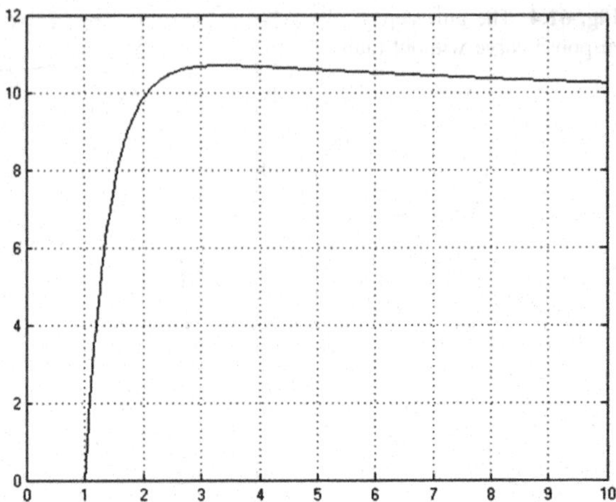

Fig. 61.6 The unit step response curve when $\| \Delta_{\tilde{M}_1} \quad \Delta_{\tilde{N}_1} \| < \varepsilon$

In this way, the fault $\Delta_{\tilde{M}_2} = \tilde{M}_2 - M, \Delta_{\tilde{N}_2} = \tilde{N}_2 - N$ can be calculated respectively. Then calculate the $\| \Delta_{\tilde{M}_2} \quad \Delta_{\tilde{N}_2} \| = 0.3043 < 0.6164$. Based on the theorem mentioned in third chapter, we can judge that the controller can control the fault object. Also, the simulation results prove that the controller K does make the fault object stabilized, is shown in Fig. 61.6.

In summary, this chapter gives a simulation of the maximal fault stability margin of the Hubble telescope. Simulation results show that when the fault have not surpass the maximal fault stability margin the system can stay stable. Otherwise, the parameters of controller should be adjusted. This simulation experiments have demonstrated the validity and practicability of the conclusion of this paper.

61.5 Conclusions

In this paper, a method for quantitative reconfigurability evaluation of the closed-loop satellite control system is proposed.

The additive and multiplicative faults occured in control systems are described by mathematical expressions. According to the expressions, the faults that occur in the systems are regarded as model uncertainties, for the purpose of providing the reconfiguration indicator based on system stability.

The principle of the method for normalized coprime factorization is presented, in order to describe the maximal reconfigurability boundary of the given controller and system model. Through the rigorous mathematical proof, the quantitative reconfigurability indicator is designed. And the Hubble telescope as a numeral example is applied to validate the effectiveness and correctness of the proposed method.

References

1. Doyle JC, Stein G (1981) Multivariable feedback design: Concepts for a classical/modern synthesis. IEEE Trans Auto Control
2. Kemin Z, Doyle JC, Jianqin M (2002) Robust and optimal control. National Defense Industry Press
3. Sefton JA, Ober R, Glover K (1990) Robust stabilization in the presence of coprime factor perturbations. In: Proceedings of the 29th IEEE conference on decision and control, IEEE, pp 1197–1198
4. Lixia X (2005) Uncertainty analysis of linear time-invariant systems. Tianjin University
5. Vidyasagar M, Kimura H (1986) Robust controllers for uncertain linear multivariable systems. Automatica 22(86):85–94
6. Sefton JA, Ober R, Glover K (1990). Robust stabilization in the presence of coprime factor perturbations. In: IEEE conference on decision and control, IEEE, pp 1197–1198
7. Xiang W (2012) The study of satellite attitude control system stability. Harbin Institute of Technology
8. Doyle JC, Francis BA, Tannenbaum A (1992) Feedback control theory. Macmillan Publishing Company, New York
9. EL-Sakkary AK (1985) The gap metric: robustness of stabilization of feedback systems. IEEE Trans Autom Control 30(3):240–247
10. Vidyasagar M (1985) Control system synthesis: a factorization approach. Synth Lect Controls Mechatron 22(1):500–501
11. Zhu SQ, Hautus MLJ, Praagman C (1988) Sufficient conditions for robust BIBO stabilization: given by the gap metric. Syst Control Lett 11(1):53–59
12. Gu G, Qiu L (1997) Connection of multiplicative/relative perturbation in coprime factors and gap metric uncertainty. Automatica 34(5):603–607

Chapter 62
LS-SVM Generalized Predictive Control Based on PSO and Its Application of Fermentation Control

Li Huang, Zhaohua Wang and Xiaofu Ji

Abstract Fermentation process is a complex time-varying, nonlinear and multi-variable biochemical process. The traditional fed-batch fermentation conditions are difficult to satisfy the control request. A control algorithm based on Generalized Predictive Control (GPC) is proposed. Firstly, the algorithm utilizes least square support vector machine (LS-SVM) and GPC to construct the prediction model and forecast the output value. And then, the particle swarm optimization (PSO) algorithm is applied to realize rolling optimization and obtain the control values. Finally, the control algorithm is applied to control the substrate concentration (S) of lysine fermentation. The simulation results show that the LS-SVM Generalized Predictive Control based on PSO has an excellent adaptive ability with rapid control response speed, high precision, and good performance.

Keywords Generalized predictive control · Least square support vector machine · Particle swarm optimization · Fermentation

62.1 Introduction

Biological fermentation process is a complex biochemical reaction process which is high nonlinear, time-varying, uncertain, multivariable and hysteretic. For involving the growth and reproduction of lives, the internal mechanism is very complicated.

L. Huang (✉) · X. Ji
School of Electrical and Information Engineering, Jiangsu University,
Jiangsu 212013, Zhenjiang, China
e-mail: lihuang@ujs.edu.cn

X. Ji
e-mail: xfji@msn.com

Z. Wang
School of Management, Jiangsu University, Jiangsu 212013, Zhenjiang, China
e-mail: wzhaohua@163.com

© Springer-Verlag Berlin Heidelberg 2016

Y. Jia et al. (eds.), *Proceedings of the 2015 Chinese Intelligent Systems Conference*, Lecture Notes in Electrical Engineering 359,
DOI 10.1007/978-3-662-48386-2_62

Traditional fermentation process is fed-batch fermentation. Some key parameters are very difficult to control. Related parameters are usually adjusted depending on expert experience. The traditional method is difficult to satisfy optimal control request of fermentation. It is imperative to seek new control method to solve the problem.

Predictive control based on predictive model is an advanced process control technology. So the control is known as model predictive control (MPC) and it is classified into many types such as model algorithmic control (MAC) [1], dynamic matrix control (DMC) [2, 3], generalized predictive control (GPC) [4, 5] and so on. GPC algorithm that was developed from adaptive control was proposed by Clarke and his co-workers in the 1980. By using of CARIMA model to describe controlled object, the algorithm has good robustness.

Now, the existing algorithms of predictive control are mostly for linear or weakly nonlinear system. However, some strong nonlinear system such as fermentation process, which is very difficult to find a perfect control strategy. Li and Lu [6] achieved good control effect by combining predictive control algorithm and inverse system. Because the method was based on system mechanism model, the application range was limited seriously by controlled object. To solve the nonlinear control problem, Lin and Lee [7] integrated predictive control with neural net-work. However, the neural network had some disadvantages such as over-fitting and easy falling into local extremum. In this paper, the closed loop control method based on GPC and PSO rolling optimization is put forward. Firstly, the nonlinear model of the controlled object is established by LS-SVM with the radial basis function (RBF) kernel. Secondly, the GPC algorithm is employed to implement the predictive control of the controlled object. And then, control variables of the nonlinear system were obtained by using PSO rolling optimization. Finally, the method was applied in the process of biological fermentation.

62.2 LS-SVM

The support vector machine (SVM) based on VC-Dimension and Structural Risk Minimization (SRM), is a kind of intelligent learning method, which is put forward by Vapnik [8] and his co-workers in 1999. Thereafter, Suykens [9] presented the least squares support vector machine (LS-SVM), which using equality constraints to replace the inequality constrains of the standard SVM. In LS-SVM, the least squares linear system is used as loss function, the precision is higher, convergent speed is faster. So LS-SVM [10, 11] is widely used in the process modeling and control. The model of lysine biological fermentation process could be described as $\{(\mathbf{x_i}, y_i)|i = 1, 2, \ldots, l, \mathbf{x_i} \in \mathbb{R}^{n_u + n_y}, y_i \in \mathbb{R}\}$, where l is training sample set, n_u and n_y are order of input and output respectively. $\varphi(\bullet)$: $\mathbb{R}^n \to \mathbb{R}^H$ is a nonlinear mapping which maps the input space into a higher dimension feature space. Data approximation model based on LS-SVM is expressed as

$$\begin{cases} \min\limits_{w,b,\xi} \ J(w,\xi) = \frac{1}{2} w^T w + \frac{1}{2}\gamma \sum\limits_{i=1}^{l} \xi_i^2 \\ \text{s.t.} y_i = w^T \varphi(\mathbf{x_i}) + b + \xi_i, \quad i = 1, 2, \ldots l \end{cases} \tag{62.1}$$

where w is weight vector, γ is regularization parameter, ξ_i is relaxation factor, b is deviation.

Here introduce Lagrange operator a_i, then

$$L(w, b, \xi, a) = \frac{1}{2} w^T w + \frac{1}{2}\gamma \sum_{i=1}^{l} \xi_i^2 - \sum_{i=1}^{l} a_i \left[w^T \varphi(\mathbf{x_i}) + b + \xi_i - y_i \right] \tag{62.2}$$

Based on KKT conditions, $\frac{\partial L}{\partial w} = \frac{\partial L}{\partial b} = \frac{\partial L}{\partial \xi} = \frac{\partial L}{\partial a} = 0$, so

$$\begin{bmatrix} 0 & \vec{\mathbf{1}}^{\mathrm{T}} \\ \vec{\mathbf{1}} & \mathbf{K} + \gamma^{-1}\mathbf{I} \end{bmatrix} \begin{bmatrix} b \\ \mathbf{a} \end{bmatrix} = \begin{bmatrix} 0 \\ \mathbf{y} \end{bmatrix} \tag{62.3}$$

where $\vec{\mathbf{1}} = [1, \cdots, 1]^T$, $\boldsymbol{I}$ is unit matrix of l order.

$\mathbf{y} = [y_1, \cdots, y_l]^T$; $i, j = 1, 2, \ldots, l$. RBF kernel function is chosen with kernel parameter σ^2.

$$K = \exp\left[- \|\mathbf{x} - \mathbf{x_i}\|^2 / (2\sigma^2) \right] \tag{62.4}$$

Then the model of fermentation controlled system is given by

$$y(\mathbf{x}) = \sum_{i=1}^{l} a_i \mathbf{K}(\mathbf{x}, \mathbf{x_i}) + b = \sum_{i=1}^{l} a_i \exp\left[- \|\mathbf{x} - \mathbf{x_i}\|^2 / (2\sigma^2) \right] + b \tag{62.5}$$

62.3 LS-SVM Predictive Control Based on PSO Rolling Optimization

(1) Modeling of LS-SVM generalized predictive control

From (62.6), the following input-output vector $\boldsymbol{x}$ is the regression vector of lysine fermentation process.

$$\mathbf{x} = \left[u(k-1), \ldots, u(k-n_u), y(k-1), \ldots, y(k-n_y) \right] \tag{62.6}$$

where k is sampling time. $u(k)$ and $y(k)$ are input and output of fermentation process respectively.

(62.5) is a nonlinear LS-SVM model of fermentation process. By doing Taylor expansion at the sampling point *x*(*k*), linear model could be obtained from

$$\begin{aligned} y(k) &= y(x)|_{x=x_k} + \left.\frac{\partial y}{\partial x(1)}\right|_{x=x_k} [x(1) - x_k(1)] + \cdots \\ &\qquad + \left.\frac{\partial y}{\partial x(n_u+n_y)}\right|_{x=x_k} [x(n_u+n_y) - x_k(n_u+n_y)] \\ &= y(x)|_{x=x_k} - \left.\frac{\partial y}{\partial x(1)}\right|_{x=x_k} x_k(1) - \cdots - \left.\frac{\partial y}{\partial x(n_u+n_y)}\right|_{x=x_k} x_k(n_u+n_y) \\ &\qquad + \left.\frac{\partial y}{\partial x(1)}\right|_{x=x_k} x(1) + \cdots + \left.\frac{\partial y}{\partial x(n_u+n_y)}\right|_{x=x_k} x(n_u+n_y) \\ &= q + b_0 u(k-1) + \cdots + b_{n_u-1} u(k-n_u) - a_1 y(k-1) - \cdots - a_{n_y} y(k-n_y) \end{aligned} \tag{62.7}$$

Namely

$$A(z^{-1})y(k) = B(z^{-1})u(k-1) + q \tag{62.8}$$

where

$$A(z^{-1}) = 1 + a_1 z^{-1} + \ldots + a_{n_y} z^{-n_y}, \quad B(z^{-1}) = b_0 + b_1 z^{-1} + \ldots + b_{n_u-1} z^{-(n_u-1)}.$$

Simultaneously, considering error and random disturbance of fermentation process, the above equation is modified by

$$A(z^{-1})y(k) = B(z^{-1})u(k-1) + \frac{C(z^{-1})\omega(k)}{\Delta} + q \tag{62.9}$$

Multiplying both sides of (62.9) by $E_j(z^{-1})\Delta$, we can get

$$\begin{aligned} &E_j(z^{-1})A(z^{-1})\Delta y(k+j) \\ &= E_j(z^{-1})B(z^{-1})\Delta u(k+j-1) \\ &+ E_j(z^{-1})C(z^{-1})\omega(k) + E_j(z^{-1})\Delta q \end{aligned} \tag{62.10}$$

where q is constant at current sample point, so $\Delta q = 0$. According to GPC algorithm, the Diophantine equation can be introduced. Then the above equation is simplified as

$$y = G\boldsymbol{\Delta u} + Fy(k) + H\Delta u(k-1) \tag{62.11}$$

where P is prediction horizon. M is control horizon.

$$y=[y(k+1),\quad \cdots,\quad y(k+P)]^T,$$

$$\boldsymbol{\Delta u}=[\Delta u(k),\quad \cdots,\quad \Delta u(k+M-1)]^T,\ G=\begin{bmatrix} g_0 & 0 & \cdots & 0 \\ g_1 & g_0 & \cdots & \vdots \\ \vdots & & \ddots & 0 \\ g_{P-1} & g_{P-2} & \cdots & g_{P-M} \end{bmatrix},$$

$$F=\left[F_1(z^{-1}),\quad \cdots,\quad F_P(z^{-1})\right]^T,\ H=\left[H_1(z^{-1}),\quad \cdots,\quad H_P\left(z^{-1}\right)\right]^T.$$

(2) **PSO rolling optimization**

In the process of biological fermentation, the LS-SVM generalized predictive control based on PSO rolling optimization could be realized as follows

Step 1: All the parameters of GPC, LS-SVM and PSO are initialized in this step.

Step 2: Using sample data, LS-SVM predictive model (62.7) is trained. Combined with GPC algorithm, the output of model is obtained according to (62.11).

Step 3: According to (62.12), reference trajectory is get

$$y_r(k+j)=\alpha y_r(k+j-1)+(1-\alpha)y_{sp} \tag{62.12}$$

where $\alpha \in [0,1)$ is softness factor. y_{sp} is set value.

Step 4: Objective function (62.13) is optimized using PSO algorithm. Positions x_i and velocities v_i of the population are updated using (62.14).

$$J=E\left\{\sum_{j=1}^{P}(y(k+j)-y_r(k+j))^2+\sum_{j=1}^{M}\lambda(j)(\Delta u(k+j-1))^2\right\} \tag{62.13}$$

$$\begin{aligned} v_{id}^{k+1} &= \omega v_{id}^{k}+c_1 r_1(p_{id}^{k}-x_{id}^{k})+c_2 r_2(p_{gd}^{k}-x_{id}^{k}) \\ x_{id}^{k+1} &= x_{id}^{k}+v_{id}^{k+1} \end{aligned} \tag{62.14}$$

where $\lambda(j)$ is control coefficient. E is mathematical expectation. The individual best value p_i is set to its current position value. And the global best value p_g is set to the best value of p_i. $x_i=[\Delta u^*(k),\ldots,\Delta u^*\ (k+M-1)]_i$. $i=1,2,\ldots,m$. $k=1,2,\ldots,iter_{\max}$. $d=1,2,\cdots,M$. c_1 and c_2 are constants; r_1 and r_2 are random numbers in the range [0, 1]; ω is the inertia weight.

Step 5: The fermentation control variable $u(k)$ is obtained by (62.15).

$$u(k)=u(k-1)+\Delta u^*(k) \tag{62.15}$$

Step 6: System state $y(k)$ is recorded. Update date and given $k=k+1$. Then return step 2 until the end of fermentation.

Step 7 Output $u(k)$ and exit.

62.4 Experiment and Simulation

(1) Experiment background

In order to verify the effectiveness of the above control method, the algorithm is applied to control lysine fermentation process. In the experiment, substrate concentration S of lysine fermentation is considered as controlled object. Control variables are glucose feeding rates f_{gl}, ammonia feeding rates f_a and speed of the stirring motor r. The capacity of fermentation device is 30 L. Fermentation process bed pressure is kept at 0.1 Mpa. The temperature is controlled in the range of (30 ± 0.5) °C ,The time of fed-batch fermentation is keeping in 72 ~ 78 h. The f_{gl} is collected real-timely at the speed of 1/min. The fermentation liquid is collected once in every 4 h and substrate concentration S is analyzed by off-line checking.

(2) Simulation results and analysis

There are 10 batch original data acquired from the experiment. And get 184 sets of input and output sample data. After being processed, there are 16 sets of data that have serious errors are rejected. The other 168 sets of data are divided into two groups, which 90 % of data is used as training sample sets $\{(\mathbf{x_i}, y_i)|i=1,2,\ldots,151\}$ and the rest 10 % is regarded as testing sample sets.

The GPC prediction model is built, and the system output is substrate concentration S. The parameters of LS-SVM are given by $[\gamma,\ \sigma^2]=[100,\ 3.6\,]$. The orders of the input and output prediction model are chosen as $n_u=3$, $n_y=2$. In order to show the high performance of GPC model based on LS-SVM, the original GPC model is also used to train and test the same samples. We denote the actual system output as Y, and the model output as Y_m. The prediction results of the two models are shown as the following Figs. 62.1 and 62.2. Compared with Fig. 62.2, it's clear that the LS-SVM generalized predictive control in the Fig. 62.1 has better predictive ability than another control model. The error between actual system and model output is shown in Fig. 62.3. We can see the error value is in the range [−0.03, +0.03], and the maximum error is 0.02995. That is, the above model has a higher precision and can be used well to fit nonlinear prediction system.

In order to test the approximation performance of the above model, let prediction horizon is $P = 4$, control horizon is $M = 2$, softness factor is $\alpha=0.3$, control factor is $\lambda = 0.7$, population $m = 40$. Then, the tracking curve of substrate concentration S that are described by two models of GPC are shown in Fig. 62.4. In Fig. 62.4, Y_r is reference trajectory, Y_1 is the output of the LS-SVM prediction model based on PSO rolling optimization, and Y_2 is the output of the LS-SVM prediction model. As known from Fig. 62.4, when reference trajectory is square-wave, the curve Y_1 shows faster corresponding speed, shorter adjusting time, and better stability. However, the curve Y_2 shows slower corresponding speed, and the response has been in a slight state of shock. The comparison indicates the above predictive algorithm based on LS-SVM could control the strongly nonlinear system by high response speed, small overshoot, tracking error, and good robustness.

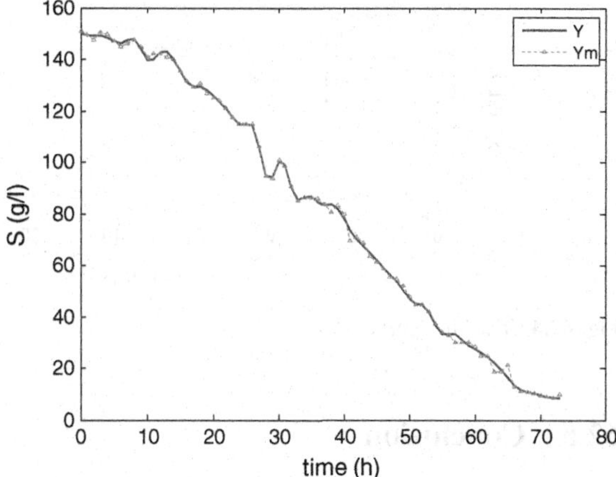

Fig. 62.1 GPC prediction model based on LS-SVM

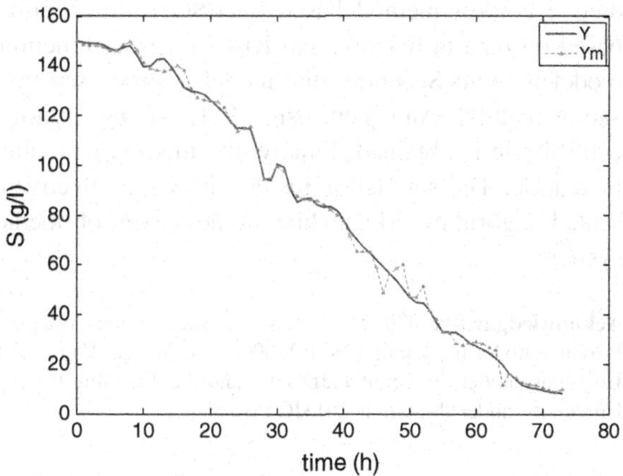

Fig. 62.2 Original GPC prediction model

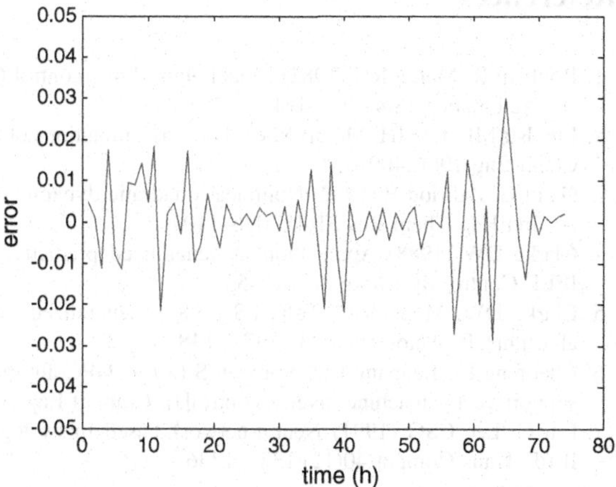

Fig. 62.3 Error curve

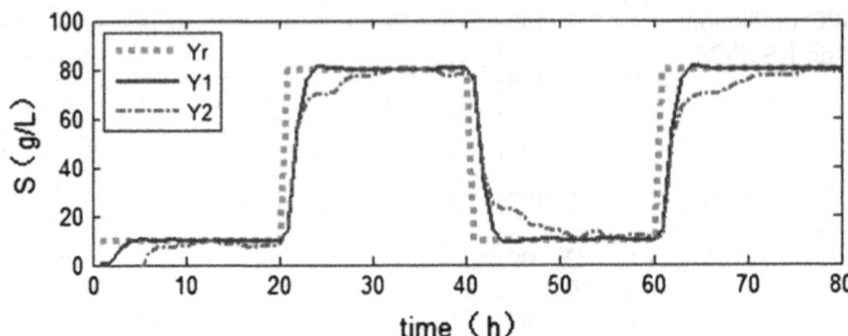

Fig. 62.4 Tracking curve

62.5 Conclusion

According to nonlinear fermentation process, a kind of LS-SVM generalized predictive control method based on PSO rolling optimization is presented. Firstly, based on data in reaction and RBF as kernel function, LS-SVM nonlinear control model is built. Secondly, the model is processed by means of subsection linearization method. And then, using PSO rolling optimization algorithm, the optimal control rate is obtained. Finally, the method is applied to lysine bio-chemical fermentation. The simulation results show the effectiveness and accuracy of the presented algorithm. Meanwhile, a new control method is supplied to nonlinear sys-tem.

Acknowledgments This research is supported by the Collegiate Natural Science Fund of Jiangsu Province under the Grant 12KJB210001, the Startup Fund for Distinguished Scholars of Jiangsu University under the Grant 12JDG108 and the Teaching Reform and Research Project of Jiangsu University under the Grant 2013JGYB004.

References

1. Rouhani R, Mehra RK (1982) Model algorithmic control (MAC); basic theoretical properties [J]. Automatica 18(4):401–414
2. Lundstr MP, Lee JH, Morari M et al (1995) Limitations of dynamic matrix control[J]. Comput Chem Eng 19(4):409–421
3. Gattu G, Zafiriou E (1992) Nonlinear quadratic dynamic matrix control with state estimation [J]. Ind Eng Chem Res 31(4):1096–1104
4. Clarke DW (1988) Application of generalized predictive control to industrial processes[J]. IEEE Control Syst Mag 8(2):49–55
5. Clarke DW, Mohtadi C, Tuffs PS (1987) Generalized predictive control—part I. the basic algorithm[J]. Automatica 23(2):137–148
6. Chaofeng LI, Jiangang LU, Youxian S (2011) GPC algorithm of nonlinear systems based on support vector machine inverse control[J]. Comput Eng Appl 47(2): 223–226.(in Chinese)
7. Lin C, Lee CSG (1991) Neural-network-based fuzzy logic control and decision system[J]. IEEE Trans Comput 40(12):1320–1336

8. Vapnik VN (1999) An overview of statistical learning theory[J]. IEEE Trans Neural Netw 10 (5):988–999
9. Suykens JA, Vandewalle J (1999) Least squares support vector machine classifiers[J]. Neural Process Lett 9(3):293–300
10. Arabloo M, Ziaee H, Lee M et al (2015) Prediction of the properties of brines using least squares support vector machine (LS-SVM) computational strategy[J]. J Taiwan Inst Chem Eng 50:123–130
11. Zhang C, Zhang HY (2015) Modelling and prediction of tool wear using LS-SVM in milling operation[J]. International J Comput Integr Manuf 1–16 (ahead-of-print)

Chapter 63
Delay Consensus of Second-Order Nonlinear Leader-Following Multi-Agent Systems

Jiezhi Wang, Qing Zhang and Hang Li

Abstract In this paper, the delay consensus of second-order nonlinear leader-following multi-agent systems is discussed. The considered multi-agent system has an active leader and the information exchange between two different agents possesses directional. A simple input control law is proposed. Based on the matrix theory and Lyapunov stability theory, the effectiveness of this control law is proved and a sufficient condition is obtained to realize delay consensus of the second-order multi-agent system.

Keywords Second-order multi-agent system · Delay · Consensus

63.1 Introduction

In recent years, multi-agent systems have an extensive application in many engineering fields, just as in formation control [1, 2], sampled-data [3], formation filtering [4], flocking [5] and so on. The most important and basic content is the consensus problem of the multi-agent systems. The study of consensus focused on the first-order [6–9] and second-order [10–13] leader-following or leaderless multi-agent systems. By now, consensus has been discussed under many different system conditions, just like with fixed topology or switching topologies [1, 7, 8, 11], with undirected graph or directed graph [11], with observer [7, 8] or with time-varying delays [9–11] and so on.

The above investigations are all under the complete consensus concept whose meaning is with time going on, the situations and velocities of each agent will all be

J. Wang (✉) · Q. Zhang
College of Science, Civil Aviation University of China, 300300 Tianjin, China
e-mail: wjzh197845@163.com

H. Li
Economics and Management College, Civil Aviation University of China, Tianjin 300300, China

© Springer-Verlag Berlin Heidelberg 2016

Y. Jia et al. (eds.), *Proceedings of the 2015 Chinese Intelligent Systems Conference*, Lecture Notes in Electrical Engineering 359,
DOI 10.1007/978-3-662-48386-2_63

the same at the same time. In fact, there must exist time delay in information exchange and information transfer. And in a complicated network, the consensus states of two different intelligent agents also have time delay. Based on these reasons, [14] proposed the concept of delay consensus in first-order leader-following multi-agent systems with directed graph for the first time and according to what we know, there is no another open published paper about delay consensus. Compare to the first-order multi-agent system, the research on consensus in second-order multi-agent systems is more complicated and difficult. In this paper, delay consensus in second-order nonlinear leader-following multi-agent systems with directed graph will be discussed.

The rest of this paper is organized as follows. Section 63.2 introduces the leader-following systems, the definition of delay consensus and the ordinary assumption. Section 63.3 designs an input control law to realize the delay consensus of the leader-following multi-agent system with directed graph. Section 63.4 draws the conclusions.

63.2 Problem Formulation and the Definition of Delay Consensus

The dynamics of agents are expressed by

$$\begin{cases} \dot{x}_i(t) = v_i(t), \\ \dot{v}_i(t) = f(x_i(t), v_i(t)) + \sum_{j=1}^{N} a_{ij} v_j(t) + u_i(t), \quad i = 1, 2, \ldots, N, \end{cases} \tag{63.1}$$

where $x_i(t) \in R^n$ and $v_i(t) \in R^n$ are the position and velocity of the i-th agent at time t. $f: R^n \times R^n \to R^n$ is a continuously differentiable vector-valued nonlinear function. $A = (a_{ij}) \in R^{N \times N}$ is the weighted adjacency matrix of the network, where $a_{ij} \geq 0$ ($a_{ij} > 0$ if the i-th agent can get information from the j-th agent) and $aii = -\sum_{j \neq i} a_{ij} \, (i = 1, 2, \cdots, N)$. Here, assume the information exchange between two agents has direction. So, the matrix A is not symmetric. $u_i(t) \in R^n \, (i = 1, 2, \cdots, N)$ is the control input.

The dynamics of the unique leader is expressed by

$$\begin{cases} \dot{x}_0(t) = v_0(t), \\ \dot{v}_0(t) = f(x_0(t), v_0(t)), \end{cases} \tag{63.2}$$

where $x_0(t) \in R^n$ and $v_0(t) \in R^n$ are the position and velocity of the leader at time t.

The adjacency matrix of the leader and agents is denoted by the diagonal matrix $B = \mathrm{diag}(b_1, b_2, \cdots, b_N)$, where

$$b_i \begin{cases} >0, & \text{if agent } i \text{ can get information from the leader,} \\ =0, & \text{otherwise.} \end{cases} \tag{63.3}$$

Lemma 1 *[7]: Suppose that a symmetric matrix is portioned as*

$$T = \begin{pmatrix} T_1 & T_2 \\ T_2^T & T_3 \end{pmatrix},$$

where T_1, T_3 *is square.* T *is positive definite if and only if both* T_1 *and* $T_3 - T_2^T T_1^{-1} T_2$ *are positive definite.*

Assumption 1 [15]: There exist constants $\gamma_1 \geq 0, \gamma_2 \geq 0, \forall x(t), y(t), v(t), z(t) \in R^n$, such that

$$(x-y)^T (f(x,v) - f(y,z)) \leq \gamma_1 ((x-y)^T (x-y) + (v-z)^T (v-z)), \tag{63.4}$$

$$(v-z)^T (f(x,v) - f(y,z)) \leq \gamma_2 ((x-y)^T (x-y) + (v-z)^T (v-z)). \tag{63.5}$$

Definition 1 Denote the error vector $\tilde{x}_i(t) = x_i(t) - x_0(t - \tau_i)$ and $\tilde{v}_i(t) = v_i(t) - v_0(t - \tau_i)$ $(i = 1, 2, \cdots, N)$, where $\tau_i \in R^+$. The **delay consensus** of system (1) and system (2) is said to be achieved if for any initial conditions,

$$\lim_{t\to\infty} \|\tilde{x}_i(t)\| = 0, \ \lim_{t\to\infty} \|\tilde{v}_i(t)\| = 0, \ i = 1, 2, \ldots, N. \tag{63.6}$$

Remark 1 By now, the researches on the consensus of second-order leader-following multi-agent systems with delays were still focus on achieving the complete consensus (i.e. $\lim_{t\to\infty} \|x_i(t) - x_0(t)\| = 0$, $\lim_{t\to\infty} \|v_i(t) - v_0(t)\| = 0$). The **delay consensus** means the consensus states of leader and followers also have time delay.

63.3 The Realization of the Delay Consensus

Here, only discuss the situation that all the delay interval $\tau_i = \tau \, (i = 1, 2, \cdots, N)$. Then,

$$\begin{aligned} \tilde{x}_i(t) &= x_i(t) - x_0(t - \tau), \\ \tilde{v}_i(t) &= v_i(t) - v_0(t - \tau), \quad i = 1, 2, \cdots, N' \end{aligned}$$

and

$$\dot{\tilde{x}}_i(t) = \dot{x}_i(t) - \dot{x}_0(t-\tau) = \tilde{v}_i(t), \quad i = 1, 2, \cdots, N.$$

Take

$$\tilde{x}(t) = \begin{pmatrix} \tilde{x}_1(t) \\ \tilde{x}_2(t) \\ \vdots \\ \tilde{x}_N(t) \end{pmatrix} = \begin{pmatrix} x_1(t) - x_0(t-\tau) \\ x_2(t) - x_0(t-\tau) \\ \vdots \\ x_N(t) - x_0(t-\tau) \end{pmatrix}, \tilde{v}(t) = \begin{pmatrix} \tilde{v}_1(t) \\ \tilde{v}_2(t) \\ \vdots \\ \tilde{v}_N(t) \end{pmatrix} = \begin{pmatrix} v_1(t) - v_0(t-\tau) \\ v_2(t) - v_0(t-\tau) \\ \vdots \\ v_N(t) - v_0(t-\tau) \end{pmatrix},$$

$$\tilde{y}(t) = \begin{pmatrix} \tilde{x}(t) \\ \tilde{v}(t) \end{pmatrix}.$$

So, $\dot{\tilde{x}}(t) = \tilde{v}(t)$ and

$$\begin{pmatrix} \dot{\tilde{x}}_i(t) \\ \dot{\tilde{v}}_i(t) \end{pmatrix} = \begin{pmatrix} \tilde{v}_i(t) \\ f(x_i(t), v_i(t)) + \sum_{j=1}^{N} a_{ij} v_j(t) + u_i(t) - f(x_0(t-\tau), v_0(t-\tau)) \end{pmatrix},$$
$$i = 1, 2, \cdots, N. \tag{63.7}$$

Let $H \overset{\Delta}{=} \frac{A+A^T}{2}$ and λ denote the largest eigenvalue of the symmetric matrix H. $b = \max\{b_1, b_2, \cdots, b_N\} > 0$ and b just is the largest eigenvalue of the diagonal matrix B.

Matrix $P > 0$ means P is positive definite. Matrix $P < 0$ means P is negative definite.

Theorem 1 *Suppose Assumption* 1 *holds. With*

$$\gamma_2 < \gamma\gamma_1, \tag{63.8}$$

$$-\frac{1}{2}(-I_N + (\gamma k - c)B + \gamma H)((\gamma\gamma_1 - \gamma_2)I_N + \gamma c B)^{-1}\frac{1}{2}(-I_N + (\gamma k - c)B + \gamma H)$$
$$+ (\gamma + \gamma\gamma_1 - \gamma_2 - \lambda - kb)I_N > 0, (0 < \gamma < 1) \tag{63.9}$$

the delay consensus of the second-order leader-following system (1) and system (2) can be achieved under the following input laws

$$u_i = c b_i \tilde{x}_i(t) + k b_i \tilde{v}_i(t), \quad i = 1, 2, \ldots, N, \tag{63.10}$$

where the nonnegative constants $c, k \in R^+$.

Proof Take the symmetric matrix

$$\tilde{P}=\begin{pmatrix} I_{nN} & -\gamma I_{nN} \\ -\gamma I_{nN} & I_{nN} \end{pmatrix},$$

where $0<\gamma<1$. According to Lemma 1, $\tilde{P}>0$. □

Now, chose a Lyapunov function $V(t)=\frac{1}{2}\tilde{y}^T(t)\tilde{P}\tilde{y}(t)$.
Calculating the derivative of $V(t)$, one gets

$$\begin{aligned}
\dot{V}(t) &= \frac{1}{2}\left(\dot{\tilde{y}}^T(t)\tilde{P}\tilde{y}(t)+\tilde{y}^T(t)\tilde{P}\dot{\tilde{y}}(t)\right) \\
&= \frac{1}{2}\left(\dot{\tilde{x}}^T(t)\tilde{x}(t)+\tilde{x}^T(t)\dot{\tilde{x}}(t)\right)-\frac{\gamma}{2}\left(\dot{\tilde{x}}^T(t)\tilde{v}(t)+\tilde{v}^T(t)\dot{\tilde{x}}(t)\right) \\
&\quad +\frac{1}{2}\left(\dot{\tilde{v}}^T(t)\tilde{v}(t)+\tilde{v}^T(t)\dot{\tilde{v}}^T(t)\right)-\frac{\gamma}{2}\left(\dot{\tilde{v}}^T(t)\tilde{x}(t)+\tilde{x}^T(t)\dot{\tilde{v}}(t)\right) \\
&\overset{\dot{\tilde{x}}_i(t)=\tilde{v}_i(t)}{=} \frac{1}{2}\left(\tilde{v}^T(t)\tilde{x}(t)+\tilde{x}^T(t)\tilde{v}(t)\right)-\gamma\tilde{v}^T(t)\tilde{v}(t) \\
&\quad +\frac{1}{2}\left(\dot{\tilde{v}}^T(t)\tilde{v}(t)+\tilde{v}^T(t)\dot{\tilde{v}}^T(t)\right)-\frac{\gamma}{2}\left(\dot{\tilde{v}}^T(t)\tilde{x}(t)+\tilde{x}^T(t)\dot{\tilde{v}}(t)\right)
\end{aligned}$$

where

$$\begin{aligned}
&\frac{1}{2}\left(\tilde{v}^T(t)\tilde{x}(t)+\tilde{x}^T(t)\tilde{v}(t)\right)-\gamma\tilde{v}^T(t)\tilde{v}(t) \\
&= \begin{pmatrix}\tilde{x}^T(t) & \tilde{v}^T(t)\end{pmatrix}\begin{pmatrix} 0 & \frac{1}{2}I_{nN} \\ \frac{1}{2}I_{nN} & -\gamma I_{nN}\end{pmatrix}\begin{pmatrix}\tilde{x}(t) \\ \tilde{v}(t)\end{pmatrix} \\
&= \begin{pmatrix}\tilde{x}^T(t) & \tilde{v}^T(t)\end{pmatrix}\begin{pmatrix} 0 & \frac{1}{2}I_{nN} \\ \frac{1}{2}I_{nN} & -\gamma I_{nN}\end{pmatrix}\begin{pmatrix}\tilde{x}(t) \\ \tilde{v}(t)\end{pmatrix} \\
&= \tilde{y}^T(t)\begin{pmatrix} 0 & \frac{1}{2}I_{nN} \\ \frac{1}{2}I_{nN} & -\gamma I_{nN}\end{pmatrix}\tilde{y}(t).
\end{aligned}$$

$$\begin{aligned}
&-\frac{\gamma}{2}\left(\dot{\tilde{v}}^T(t)\tilde{x}(t)+\tilde{x}^T(t)\dot{\tilde{v}}(t)\right) \\
&= -\frac{\gamma}{2}\left(\sum_{i=1}^{N}\dot{\tilde{v}}_i^T(t)\tilde{x}_i(t)+\frac{1}{2}\sum_{i=1}^{N}\tilde{x}_i^T(t)\dot{\tilde{v}}_i(t)\right) \\
&= -\gamma\left\{\sum_{i=1}^{N}\tilde{x}_i^T(t)\left[f(x_i(t),v_i(t))-f(x_0(t-\tau),v_0(t-\tau))\right.\right. \\
&\quad \left.\left.+\frac{1}{2}\sum_{j=1}^{N}(a_{ij}+a_{ji})\tilde{v}_j(t)+cb_i\tilde{x}_i(t)+kb_i\tilde{v}_i(t)\right]\right\} \\
&\overset{Assumption\ 1}{\leq} -\gamma\sum_{i=1}^{N}\gamma_1\left[(x_i(t)-x_0(t-\tau))^T(x_i(t)-x_0(t-\tau)+(v_i(t)-v_0(t-\tau))^T\right. \\
&\quad \left.(v_i(t)-v_0(t-\tau)\right]
\end{aligned}$$

$$
\begin{aligned}
&-\gamma \sum_{i=1}^{N}\sum_{j=1}^{N} \tilde{x}_i^T(t)\frac{a_{ij}+a_{ji}}{2}\tilde{v}_j(t) -\gamma \sum_{i=1}^{N} \tilde{x}_i^T(t)kb_i\tilde{v}_i(t) -\gamma \sum_{i=1}^{N} \tilde{x}_i^T(t)cb_i\tilde{x}_i(t)\\
&= -\gamma\gamma_1 \sum_{i=1}^{N} \left(\tilde{x}_i^T(t)x_i(t)+\tilde{v}_i^T(t)v_i(t)\right) -\gamma \sum_{i=1}^{N} \tilde{x}_i^T(t)cb_i\tilde{x}_i(t)\\
&\quad -\frac{\gamma}{2}\sum_{i=1}^{N}\sum_{j=1}^{N} \tilde{x}_i^T(t)\frac{a_{ij}+a_{ji}}{2}\tilde{v}_j(t) -\frac{\gamma}{2}\sum_{i=1}^{N}\sum_{j=1}^{N} \tilde{v}_i^T(t)\frac{a_{ij}+a_{ji}}{2}\tilde{x}_j(t)\\
&\quad -\frac{\gamma}{2}\sum_{i=1}^{N} \tilde{x}_i^T(t)kb_i\tilde{v}_i(t) -\frac{\gamma}{2}\sum_{i=1}^{N} \tilde{v}_i^T(t)kb_i\tilde{x}_i(t)\\
&= \tilde{y}^T(t)\begin{pmatrix} -\gamma(\gamma_1 I_N + cB)\otimes I_n & -\frac{\gamma}{2}(H+kB)\otimes I_n \\ -\frac{\gamma}{2}(H+kB)\otimes I_n & -\gamma\gamma_1 I_{nN}\end{pmatrix}\tilde{y}(t).
\end{aligned}
$$

$$
\begin{aligned}
&\frac{1}{2}\left(\dot{\tilde{v}}^T(t)\tilde{v}(t)+\tilde{v}^T(t)\dot{\tilde{v}}^T(t)\right)\\
&=\frac{1}{2}\sum_{i=1}^{N} \dot{\tilde{v}}_i^T(t)\tilde{v}_i(t)+\frac{1}{2}\sum_{i=1}^{N} \tilde{v}_i^T(t)\dot{\tilde{v}}_i(t)\\
&= \sum_{i=1}^{N} \tilde{v}_i^T(t)\left[f(x_i(t),v_i(t)) - f(x_0(t-\tau),v_0(t-\tau)) + \frac{1}{2}\sum_{j=1}^{N}(a_{ij}+a_{ji})\tilde{v}_j(t) + cb_i\tilde{x}_i(t) + kb_i\tilde{v}_i(t)\right]\\
&\leq^{\text{Assumption 1}} \sum_{i=1}^{N}\gamma_2\left[(x_i(t)-x_0(t-\tau))^T(x_i(t)-x_0(t-\tau)+(v_i(t)-v_0(t-\tau))^T(v_i(t)-v_0(t-\tau)\right]\\
&\quad +\frac{1}{2}\sum_{i=1}^{N}\sum_{j=1}^{N}\tilde{v}_i^T(t)(a_{ij}+a_{ji})\tilde{v}_j(t) + \sum_{i=1}^{N}\tilde{v}_i^T(t)kb_i\tilde{v}_i(t) + \sum_{i=1}^{N}\tilde{v}_i^T(t)cb_i\tilde{x}_i(t)\\
&=\gamma_2\sum_{i=1}^{N}\left(\tilde{x}_i^T(t)x_i(t)+\tilde{v}_i^T(t)v_i(t)\right) + \sum_{i=1}^{N}\sum_{j=1}^{N}\tilde{v}_i^T(t)\frac{a_{ij}+a_{ji}}{2}\tilde{v}_j(t) + \sum_{i=1}^{N}\tilde{v}_i^T(t)kb_i\tilde{v}_i(t)\\
&\quad +\frac{1}{2}\sum_{i=1}^{N}\tilde{v}_i^T(t)cb_i\tilde{x}_i(t)+\frac{1}{2}\sum_{i=1}^{N}\tilde{x}_i^T(t)cb_i\tilde{v}_i(t)
\end{aligned}
$$

Since

$$
\begin{aligned}
&\sum_{i=1}^{N}\sum_{j=1}^{N}\tilde{v}_i^T(t)\frac{a_{ij}+a_{ji}}{2}\tilde{v}_j(t) + \sum_{i=1}^{N}\tilde{v}_i^T(t)kb_i\tilde{v}_i(t)\\
&=\tilde{v}^T(t)H\tilde{v}(t)+k\tilde{v}^T(t)B\tilde{v}(t)\\
&\leq \lambda\tilde{v}^T(t)\tilde{v}(t)+kb\tilde{v}^T(t)\tilde{v}(t),
\end{aligned}
$$

then

$$
\begin{aligned}
&\frac{1}{2}\left(\dot{\tilde{v}}^T(t)\tilde{v}(t)+\tilde{v}^T(t)\dot{\tilde{v}}^T(t)\right)\\
&=\tilde{y}^T(t)\begin{pmatrix}\gamma_2 I_{nN} & \frac{1}{2}cB\otimes I_n\\ \frac{1}{2}cB\otimes I_n & \gamma_2 I_{nN}+(H+kB)\otimes I_n\end{pmatrix}\tilde{y}(t)\\
&\leq \tilde{y}^T(t)\begin{pmatrix}\gamma_2 I_{nN} & \frac{1}{2}cB\otimes I_n\\ \frac{1}{2}cB\otimes I_n & (\gamma_2+\lambda+kb)I_{nN}\end{pmatrix}\tilde{y}(t)
\end{aligned}
$$

One obtains

$$\begin{aligned}\dot{V}(t) &= \tilde{y}^T(t)\begin{pmatrix} (\gamma_2-\gamma\gamma_1)I_{nN}-\gamma cB\otimes I_n & \frac{1}{2}(I_N+cB-\gamma(H+kB))\otimes I_n \\ \frac{1}{2}(I_N+cB-\gamma(H+kB))\otimes I_n & (\gamma_2-\gamma-\gamma\gamma_1+\lambda+kb)I_{nN}\end{pmatrix}\tilde{y}(t). \\ &= -\tilde{y}^T(t)(Q\otimes I_n)\tilde{y}(t).\end{aligned}$$

Where

$$\begin{aligned} Q &= \begin{pmatrix} M_1 & M_2 \\ M_2^T & M_3 \end{pmatrix}, \\ M_1 &= (\gamma\gamma_1-\gamma_2)I_N+\gamma cB, \\ M_2 &= \frac{1}{2}(-I_N+(\gamma k-c)B+\gamma H), \\ M_3 &= (\gamma+\gamma\gamma_1-\gamma_2-\lambda-kb)I_N. \end{aligned}$$

Notice that M_1 is positive definite if and only if $\gamma\gamma_1-\gamma_2+\gamma cb_i>0$. One can take

$$\gamma_2<\gamma\gamma_1,$$

so $\gamma\gamma_1-\gamma_2+\gamma cb_i>0$.

Next, consider the symmetric matrix

$$\begin{aligned} & M_3-M_2^TM_1^{-1}M_2 \\ &= M_3-M_2M_1^{-1}M_2 \\ &= (\gamma+\gamma\gamma_1-\gamma_2-\lambda-kb)I_N \\ &\quad -\frac{1}{2}(-I_N+(\gamma k-c)B+\gamma H)((\gamma\gamma_1-\gamma_2)I_N+\gamma cB)^{-1}\frac{1}{2}(-I_N+(\gamma k-c)B+\gamma H) \end{aligned}$$

From Lemma 63.1, Q > 0 if and only if $\gamma_2<\gamma\gamma_1$ and

$$\begin{aligned} & (\gamma+\gamma\gamma_1-\gamma_2-\lambda-kb)I_N \\ & -\frac{1}{2}(-I_N+(\gamma k-c)B+\gamma H)((\gamma\gamma_1-\gamma_2)I_N+\gamma cB)^{-1}\frac{1}{2}(-I_N+(\gamma k-c)B+\gamma H)>0. \end{aligned}$$

Furthermore, $(Q\otimes I_n)>0$ if Q > 0. Then, $(-Q\otimes I_n)<0$ which means that $\dot{V}(t)<0$ for any $\tilde{y}(t)\neq 0$. According to the Lyapunov stability theory, it shows that the zero solution of the error system (7) is asymptotical stable. The delay consensus of the second-order leader-following system (1) and system (2) is realized.

For sufficient large γ_1 and the known matrices A and B, one can adjust the constants γ, γ_2, c and k to satisfy the conditions (8) and (9).

63.4 Conclusions

Because there still exist time delays in the consensus states of leader and followers, this paper proposed the definition of delay consensus of second-order nonlinear leader-following multi-agent systems and investigated the delay consensus of second-order nonlinear leader-following multi-agent system with directed graph. In theory, the proposed input control law could be used to make system (1) and system (2) reach the delay consensus. Since the adjacency matrix $\boldsymbol{A}$ is not symmetric, it needed some techniques during proving.

Next, what we will do is to give the corresponding simulation results. There are still some interesting and challenging works need to do, just like to discuss the delay interval τ_i is not equal and to design appropriate control law to realize the delay consensus of leader-following (or leaderless) multi-agent systems with switching topologies or more higher order leader-following multi-agent systems.

Acknowledgements This work was partially supported by the National Natural Science Foundation of China through Grant No. 11472298, the Fundamental Research Funds for the Central Universities through Grant Nos. ZXH2010D011 and ZXH2012K002 and the Scientific Research Foundation of Civil Aviation University of China through Grant No. 07QD05X.

References

1. Guodong Shi, Yiguang Hong (2012) K.H. Johansson. Connectivity and set tracking of multi-agent systems guided by multiple moving leaders [J]. IEEE Trans on Autom Control 57 (3):663–676
2. Jingang Lai, Shihua Chen, Xiaoqing. Lu (2012) Formation control for second-order multi-agent systems with time-varying delays [J]. CAAI Trans Intell Syst 7(2):135–141 (in Chinese)
3. Jingyuan Zhan, Xiang Li (2013) Consensus of sampled-data multi-agent networking systems via model predictive control [J]. Automatica 49:2502–2507
4. Casbeer DW, Randy B (2009) Distributed information filtering using consensus filters [J]. Am Control Louis 16(5):1882–1887
5. Housheng Su, Xiaofan Wang, Zongli Lin (2009) Flocking of multi-agents with a virtual leader [J]. IEEE Trans Autom Contol 54(2):293–307
6. Reza Olfati-Saber, Riched Murray (2004) Consensus problems in networks of agents with switching topology and time-delays [J]. IEEE Trans Autom Control 49(9):1520–1533
7. Hong Yiguang Hu, Jiangping Gao Linxin (2006) Tracking control for multi-agent consensus with an active leader and variable topology [J]. Automatica 42:1177–1182
8. Ke Peng, Yupu Yang (2009) Leader-following consensus problem with a varying-velocity leader and time-varying [J]. Phys A 388:193–208
9. Jianxiang Xi, Zongying Shi, Yisheng Zhong (2011) Consensus analysis and design for high-order linear swarm systems with time-varying delays [J]. Phys A 390:4114–4123
10. Wei Zhu, Daizhan Cheng (2010) Leader-following consensus of second-order agents with multiple time-varying delays [J]. Automatica 46:1994–1999
11. Song Li, Qinghe. Wu (2013) Average consensus of second-order multi-agent systems with time-delays and uncertain topologies [J]. Control Theory Appl 30(8):1047–1052

12. Wenwu Yu, Wei Ren, Weixing Zheng et al (2013) Distributed control gains design for consensus in multi-agent systems with second-order nonlinear dynamics [J]. Automatica 49:2107–2115
13. Kaien Liu, Guangming Xie, Long Wang (2014) Containment control for second-order multi-agent systems with time-varying delays [J]. Syst Control Lett 67:24–31
14. Yuanyan X, Yi W, Zhongjun M (2014) Delay consensus of leader-following multi-agent systems [J]. Acta Phys. Sin. 63(4):040202 (5 pages)
15. Hu Cheng Yu, Juan Jiang Haijun et al (2011) Synchronization of complex community network nonidentical nodes and adaptive coupling strength [J]. Phys Lett A 375(5):873–879

Chapter 64
An Investigation of Model-Based Design Framework for Aero-Engine Control Systems

Jinzhi Lu, DeJiu Chen, Jiqiang Wang and Wentao Li

Abstract Throughout the design of aero-engine control systems, modeling and simulation technologies have been widely used for supporting the conceptualization and evaluation. Due to the increasing complexity of such systems, the overall quality management and process optimization are becoming more important. This in particular brings the necessity of integrating various domain physical models that are traditionally based on different formalisms and isolated tools. In this paper, we present the initial concepts towards a model-based design framework for automated management of simulation services in the development of aero-engine control systems. We exploit EAST-ADL and some other existing state-of-the-art modeling technologies as the reference frameworks for a formal system description, with the content ranging from requirements, to design solutions and extra-functional constraints, and to verification and validation cases, etc. Given such a formal specification of system V&V (Verification and Validation) cases, dedicated co-simulation services will be developed to provide the support for automated configuration and execution of simulation tools. For quality management, the co-simulation services themselves will be specified and managed by models in SysML.

Keywords Aero-engine · Control system · System design · Model-based framework

J. Lu (✉) · D. Chen
Mechatronics, Department of Machine Design, KTH Royal Institute of Technology, 100 44 Stockholm, Sweden
e-mail: jinzhl@kth.se

J. Lu · W. Li
Aviation Industry Corporation of China Shenyang Engineer Design and Research Institute, Shenyang 110034, China

J. Wang
Jiangsu Province Key Laboratory of Aerospace Power Systems, Nanjing University of Aeronautics and Astronautics, Nanjing 210016, China

© Springer-Verlag Berlin Heidelberg 2016
Y. Jia et al. (eds.), *Proceedings of the 2015 Chinese Intelligent Systems Conference*, Lecture Notes in Electrical Engineering 359,
DOI 10.1007/978-3-662-48386-2_64

64.1 Introduction

Aero-engine control systems have been evolved continuously over the past decades from conventional mechanical systems to FADEC (Full Authority Digital Engine Control) systems. Meanwhile, due to the increasing functional and technical complexity, the design activities normally involve more designers and stakeholders than the past [1]. For the effectiveness, one prevailing scheme is based on the principle of divide-and-conquer [2]. Normally, the whole complex system will be separated into different parts with the associated engineering tasks allocated to different work groups or companies [3]. These groups often use the specific domain simulator tools to assist the specification, analysis and synthesis. Nevertheless, the performance of an aero-engine control system strongly depends on the characteristics of its subsystems and components being composed. So it's important for system integrators to conduct early system integration and thereby to understand the system behaviors and analyze system performance in the initial design phase. In current industrial practices, different CAD and CAE tools have been used for the development of subsystems and components, such as for requirement analysis, high level design, detailed level design, implementation, testing, etc. Since the subsystem designers often use their own tools to build models, it is very different for the system integrators to integrate the detailed models and to predict the whole system performance. This calls for a complete integration framework for different kinds of models and tools while taking the process management perspective into consideration.

This paper describes an investigation of model-based design framework for aero-engine control systems in regard to the methodology and technical roadmap. In such a framework, the graphical models are used to provide a formal system description that is shared among various stakeholders, including project managers, system engineers, modeler, system designer and simulation testers. The content of system information being described and integrated includes requirements, parameter setting, interface contracts, and tool specific information for co-simulation. Also, parameters related to optimized simulation behaviors can also be added. The paper has three main sections. Section 64.2 provides a brief introduction of simulation and co-simulation technologies, as well as the Modelica approach. Section 64.3 compares several well-known system modeling approaches, including EAST-ADL, AADL, and SysML. It also discusses the support for a co-simulation procedure of fuel distributor design. Section 64.4 analyzes the transformation between system models and physical models. The Sect. 64.5 presents the conclusion and future work.

64.2 Simulation Technologies in System Development

Time consumption and R&D cost have become more important, since more new technologies are used for the aero-engine design. Nowadays, computation design and simulation technology can help to improve the former key factors [4]. In the aero-engine design phase, model-based design is very important to predict the system performance in advance. Designers have used many types of modeling technologies. Software program was firstly used to design the computational aero-engine models [5]. After that with the help of modularly structured code, the code of this specified software program for aero-engine became reused and objected-oriented [6]. Then in [7], the object-oriented and visual modeling method advanced to predict the system performance. Lately in [4], Modelica language is used to build the aero-engine model. Many component libraries can be built in the simulator tools which strongly improve the efficiency of building models and model reuse. In [8], AMESim is used for fuel system modeling and in [9], hydro-mechanical controller component model is built in AMESim and Simulink is used for aero-engine controller model. Then co-simulation is used for predicting the performance of hydro-mechanical controller behaviors. This chapter discusses Modelica, co-simulation technology and approach of model scaling to separate physical models of aero-engine.

64.2.1 Modelica

Modelica language is a non-proprietary, object-oriented, equation based language to conveniently model complex physical systems [10]. In the Modelica platform, the aero-engine control system model can be separated into models for different areas. These models can be mathematically described by differential algebraic equations and discrete equations. And these models are object-oriented and schematics. From the point of view in [4], Modelica has been used for aero-engine system design.

64.2.2 Co-simulation

Co-simulation is a technology which can solve the multi-physics simulation integration [11]. It means a particular case of simulation scenario which there are at least two simulators solve coupled-algebraic equations and exchange the data with each other during simulation [12]. In [13], co-simulation can be defined based on the different behaviors of solvers. Figure 64.1 demonstrates 3 different types of co-simulation.

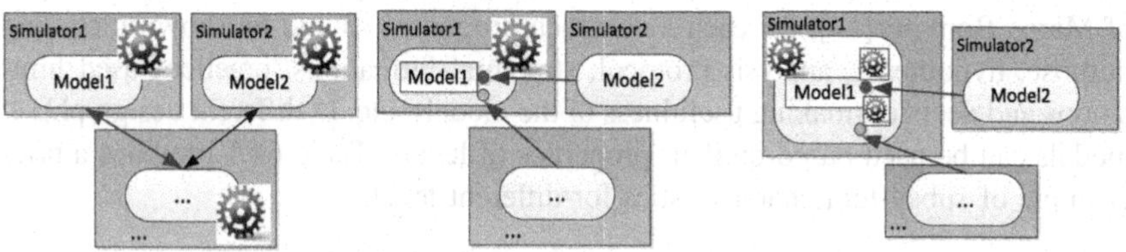

Fig. 64.1 Co-simulation, coupled-simulation and solver-coupled simulation

- Co-simulation

Co-simulation means simulator tools use their own solvers to calculate their own models and communicate with each other at the fixed time step point.

- Coupled-simulation

Coupled-simulation represents subordinated simulators transform their models into a mathematical form which the main simulator can read and it will be integrated into the main simulator's model. Then the main simulator uses its own solvers to calculate the coupling models.

- Solver-coupled simulation

Solver-coupled simulation is a co-simulation method which subordinated simulators can transform their models and solvers into a form which the main simulator can 'understand'. Then the main simulator solver can call their own other simulator solvers to calculate corresponding models. In Fig. 64.1, the icon of gear represents simulator solver.

From [9], we know, co-simulation has also been one technical solution for aero-engine control system design. However, no matter Modelica or co-simulation, simulation technology is used to help system designers to take decisions based on different design goals. When system designers run a 'project', there are different models used in the whole life cycle, even though the modeling object is fixed [1]. So it's very different to manage the models, because models present their builders' option on the matter. So a new standards or technologies should be provided to solve these problems.

64.2.3 Model Scaling and Management

Traditionally, the model-based design procedure of aero-engine control system is separately disintegrated process of different subsystems. Currently, there are no industrial standards to define the aero-engine control system model types or interfaces. But, in fact, we can distinguish (a) "control logic", (b) fuel system or hydraulic system, (c) mechanical system, (d) CFD for aerodynamic, (e) FEM for structure design, (f) thermodynamics of combustor, (g) electronic system and many more. So from model management view, we need to define the model libraries based on the design phase and detail level of models [14, 15]. From the perspective of Marco Bonvini [14], for each level, such properties should be shown including purpose, hypotheses, analysis protocol, structural limitations, practice-based limitations and decision-making usefulness of the models, etc. In different design phase, models can be used based on their properties of levels. Table 64.1 provides a brief example of subsystem characteristics for different levels.

Table 64.1 Overview of subsystem model properties for detailed level setting

Subsystem	Level 1	Level 2	Level 3	Level …
Control logic	Basic control logic	Basic control logic	Control scheme	…
Fuel system	Sample model (no fluid leakage, incompressible flow, Laminar orifice)	Sample model (no fluid leakage, incompressible flow, Laminar orifice)	More realistic model (fluid leakage, compressible flow, Laminar orifice)	…
Hydraulic system	Sample model (no fluid leakage, incompressible flow, Laminar orifice)	Sample model (no fluid leakage, incompressible flow, Laminar orifice)	More realistic model (fluid leakage, compressible flow, Laminar orifice)	…
Mechanical system	Kinematic	Kinematic	Kinematic and multi-body dynamic	…
Electronic system	Principle models	Basic element models	Complex system model	…
Aerodynamics	Fixed operating point (parameter)	Map based		…
FEM for structure	Fixed operating point (parameter)	Map based		…
Thermodynamics	Fixed operating point (parameter)	Map based, input filtering		…

64.3 System Modeling Approach

From the perspective of system engineering, physical system model is not the unique concern. Though there are a lot of complex and completed model libraries or model management platforms existing, it's also different for system designers in each layer to understand the whole physical system model and share the model information with each other except for documents or reports. In this chapter, several system modeling tools, (e.g., SysML, AADL, EAST-ADL, etc.) are investigated. Based on different system modeling language, one graphical model describing simulation for aero-engine control system is designed.

- SysML

SysML is a general-purpose modeling language for systems engineering with a subset of UML2 and additional extensions to satisfy the demands of the language description. SysML can provide powerful and simple constructs for modeling systems engineering problems of requirements, structure, behavior, allocations, and constraints [16]. In [17], Henson Graves provided an approach of SysML using in the aerospace design. So SysML has the capabilities to describe the simulation behaviors.

- AADL

AADL is formal modeling concepts for description and analysis of systems architecture in terms of individual components and their interactions including software, hardware and system component abstractions. It can specify and analyze real-time system, embedded system, complex SOS (system of system) and map software onto computational elements in hardware. So AADL has the potential to describe the architecture and behaviors of the aero-engine control system [18].

- EAST-ADL

EAST-ADL is a language to describe automotive electrical and electronic systems. It can grasp all the detailed information for documentation, design, analysis and synthesis from the top level characteristics to tasks and interface and communication framework [19]. As we know, there are many similar features between aero-engine control system and automotive electrical and electronic system, so EAST-ADL also has the potential to describe the system characteristic of aero-engine.

In Sect. 64.2, we have introduced that physical system model should be scaled and managed in different phase and simulation setting with interface or solve also need to be shared with other system designers. So ontology needs to describe such characteristics and features. Next part, we will discuss about the capabilities of different language and build a system model as a case for an integrated model of co-simulation method.

- Comparison between SysML, AADL and EAST-ADL

This graphical model should be able to describe information for behaviors of physical system modeling procedure, constructs of aero-engine models, simulation requirements, parameter setting, simulation setting, optimization behaviors and validation and verification, etc. Table 64.2 shows the different capabilities three kinds of language which can describe such characteristics. System integrators can choose the models in different levels based on simulation targets and requirements.

From the Table 64.2, we can see the characteristics which a simulation of physical system models have can be described by EAST-ADL. Next part, system model extended from EAST-ADL and SysML will be shown to describe an integrated model of co-simulation for aero-engine control system.

In Fig. 64.2, we can see a modeling structure. Requirement diagram is used for describing the requirements of stakeholders. Simulation diagram can contains simulation setting and solver setting. Structure diagram can describe the physical system models in different tools. Data diagram can be used for managing data from simulation. Feature diagram can be used for describe the behaviors of simulation testers and stakeholders.

Table 64.2 Behaviors of simulations and capabilities of three languages

Language Model type	SysML	AADL	EAST-ADL
Simulation requirements	Requirements	Annex libraries (AL)	Requirements
Constructs for models	Blocks	System Structure and Instantiation	No
Modeling procedure	Activity (no specified)	No	SystemModeling
Simulation setting	No	No	FeatureModeling
Interface setting	Point	Point	FunctionConnector
Validation and Verification	Activity (no specified)	No	VerificationValidation
Optimization behaviors	Activity (no specified)	No	ComputationConstraint

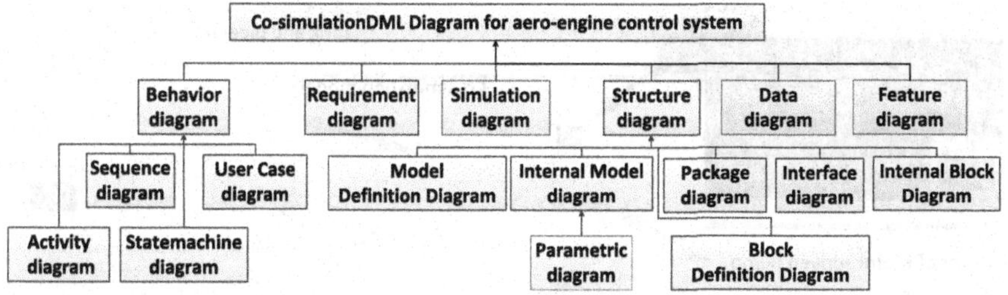

Fig. 64.2 Extensive structure of system model

- User case for system model describing physical system model

From [12], the author uses co-simulation between Saber, Simplorer, Flowmaster, AMESim and Simulink to predict the performance of fuel distributor in the aero-engine. Based on this case, system model for co-simulation procedure is built. In Fig. 64.3, a co-simulation strategy has been provided. It shows models in different tools and the data flows among tools. In Fig. 64.4, models in different tools are shown.

Depending on the extensive structure of SysML and EAST-ADL, system model is built to describe the procedure of verifying if co-simulation integrated model can product the same simulation result from the same model only built in AMESim [12]. From Fig. 64.5, we can see, five design phases are separated, including system requirement analysis, model design, subsystem modeling, co-simulation and verificationValidation.

In the phase of system requirement analysis, requirements can be described as Fig. 64.6. The simulation behaviors is the operation of building models in different tools and simulation requirements can capture the characteristics of simulation method and simulation setting demand.

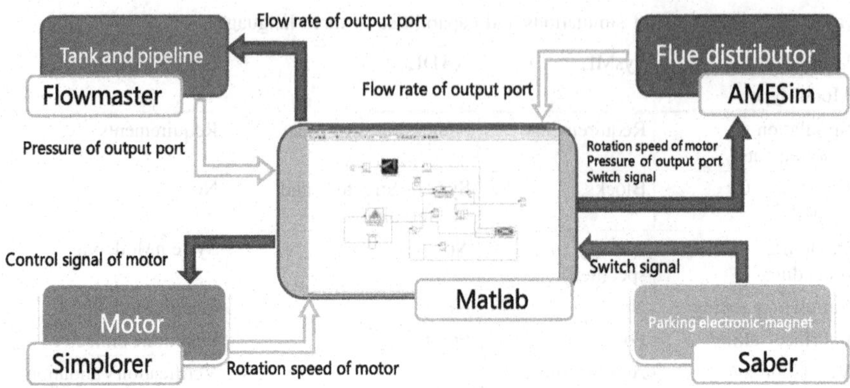

Fig. 64.3 Co-simulation strategy for flue distributor

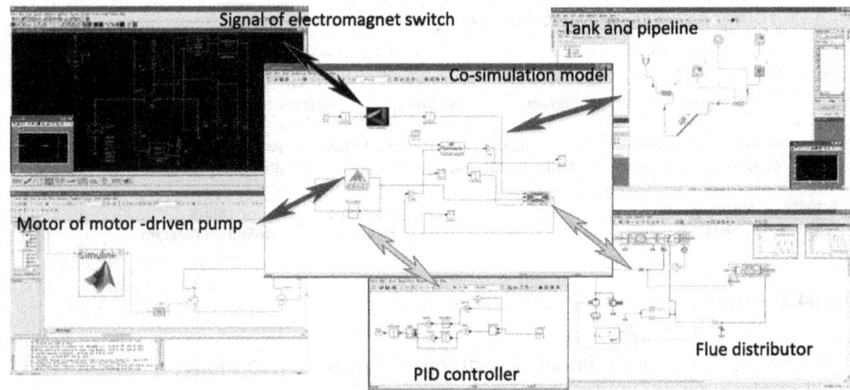

Fig. 64.4 Model of co-simulation for flue distributor

Fig. 64.5 System Model describe different design phase

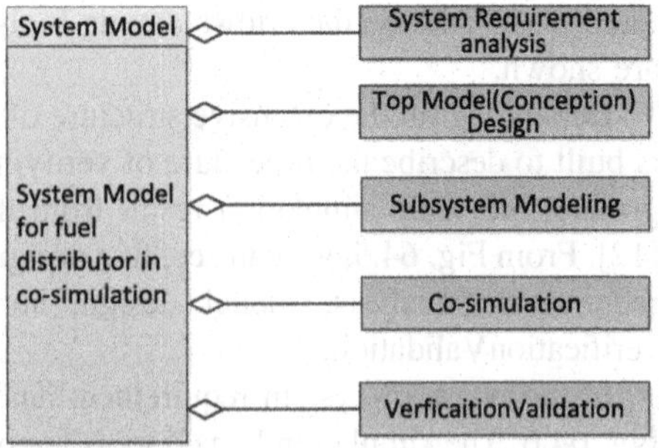

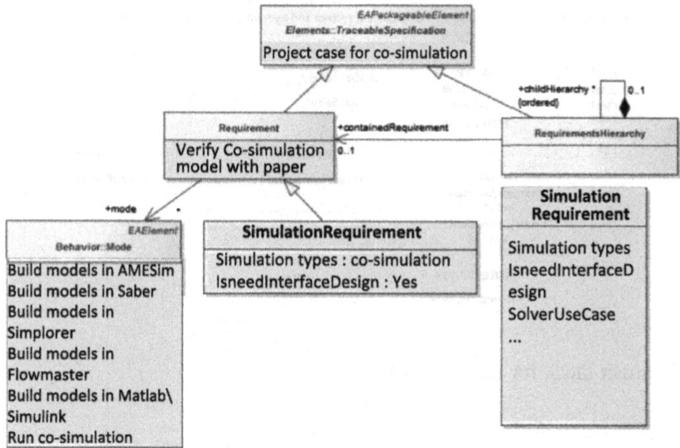

Fig. 64.6 Requirement description

In the model design phase, feature model can depict the properties of each component in the top system model and setting selection of the co-simulation in Fig. 64.4. In this phase, interface design strategy can be made and each subsystem model can be related to corresponding physical system model. In Fig. 64.7, the feature model can produce the construct model for the top system model (Fig. 64.8).

During subsystem model design, the construct feature of the physical system model can be shown. The components, lines and parameters in each subsystem model will be acted as the elementary block in the feature model, which is shown in Fig. 64.9. Based on the feature model, construct model for each tool can be built in Fig. 64.10. The model selection in different tools can be decided by modeling target

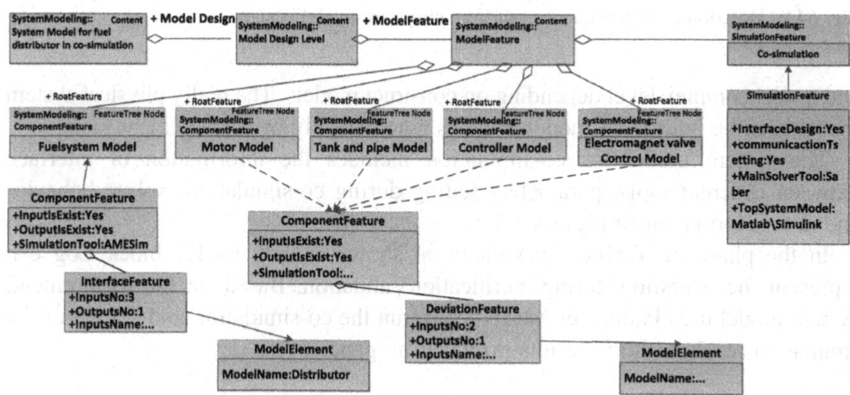

Fig. 64.7 Feature function describing model design

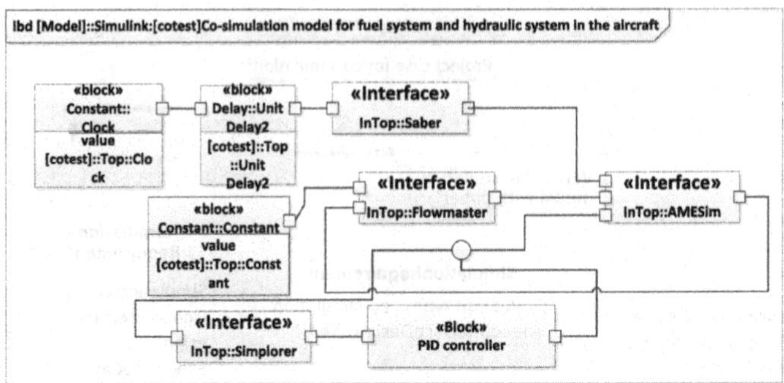

Fig. 64.8 Construct block for the top system model

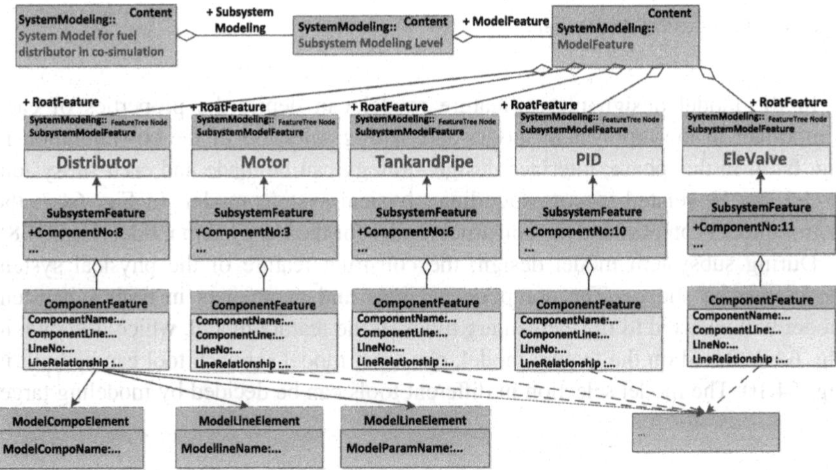

Fig. 64.9 Description of feature function for subsystem model design

and model complex level depending on construct models. The really physical system models can be built from model libraries automatically by management tools.

The system model for co-simulation includes the information of interface between different tools, parameters setting during co-simulation, solver behavior and data information in Fig. 64.11.

In the phase of VerificationValidation, shown in Fig. 64.12, block Log can represent the operation during VerificationValidation. Based on the requirement, system model uses Parameter Matrix (1) to run the co-simulation and compares the simulation result with the results provided by paper.

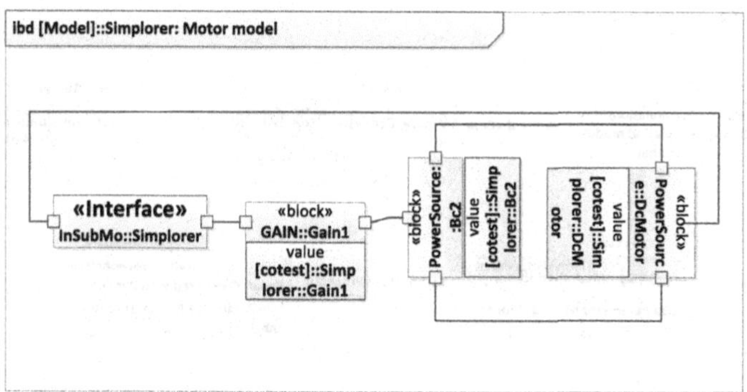

Fig. 64.10 Construct block for Simplorer models

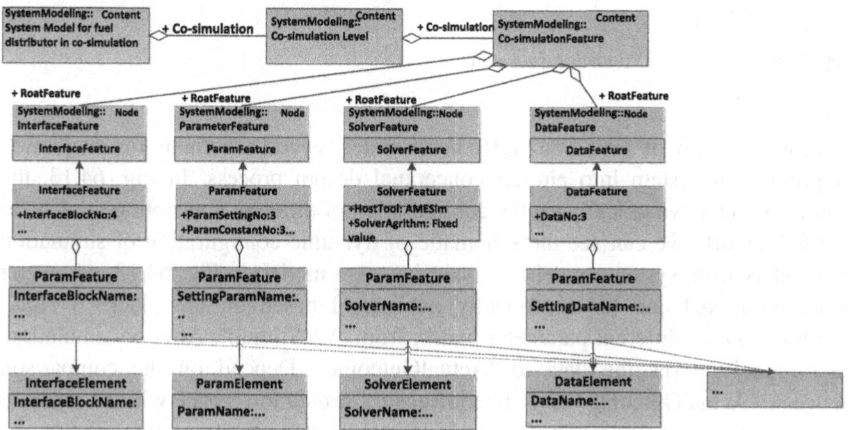

Fig. 64.11 Subsystem model in Simplorer

64.4 Transformation Between System Models and Physical System Models

During co-simulation, different parameter matrix will be used for different scene, scenarios and situation. Here, scene means a set of tasks of one simulation target. Scenarios signify different models in one simulation task. Situation represents different parameter matrix for one model in one simulation target. Specified parameter matrix for one situation is all the parameters for physical system models. When the parameter matrix for the physical system model is changed, this behavior can be described by VerificationValidation case in last section. And the capabilities of mapping from system model into physical system model are needed in order to run the simulation automatically.

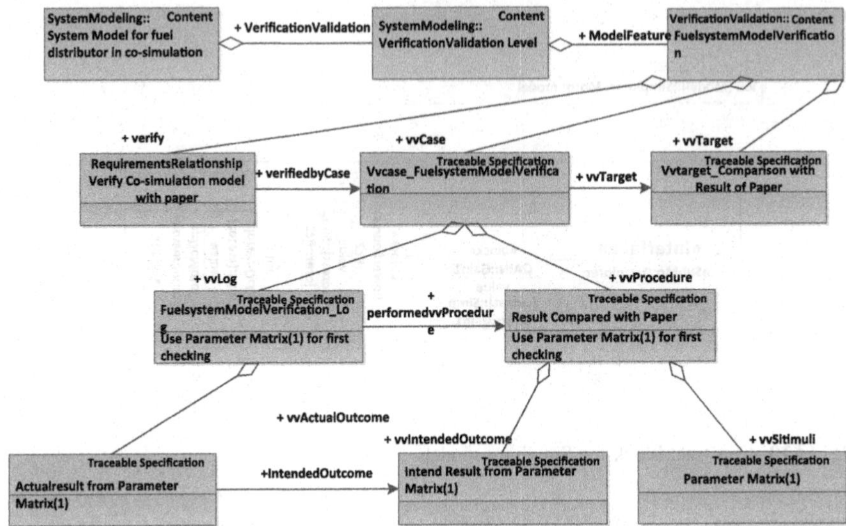

Fig. 64.12 VerificationValidation procedure

From the view of Arne Seitz [20], he provides several schematic for integration of propulsion system into aircraft conceptual design process. In Fig. 64.13, the framework can dynamically set the configuration of different conception models. In our framework, we can see the schematic of dynamic configuration of simulation parameters from system model to physical system model. In Fig. 64.13, parameter and data model can be dynamically configured based on the block 'VVlog'. 'VVlog' can set different parameter matrix from 'VVStimuli'. After co-simulation, data model returns the data to 'ActualOutcome'. Depend on the comparison between 'ActualOutcome' and 'IntendResult', simulation target will be checked and system will decide if the simulation model needs to continue the iteration.

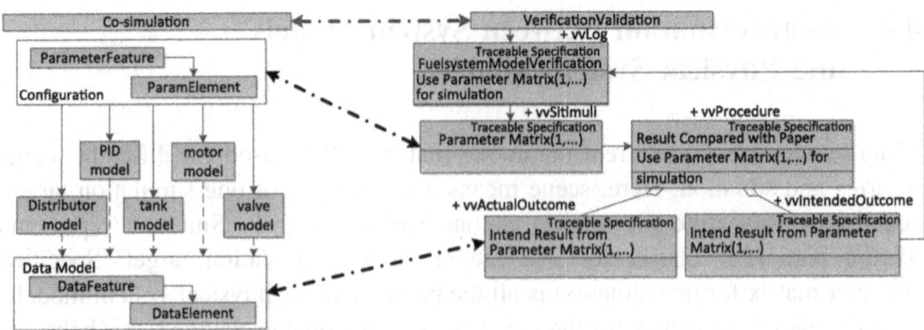

Fig. 64.13 Schematic for dynamic parameter setting

64.5 Conclusion and Future Work

From the test case, we can see state-of-the-art technologies like EAST-ADL and SysML can be used as the basis for describing simulation procedures. Dynamic parameter setting can capture all the system behaviors with parameter optimization. The initial concept models mentioned in this paper are the ones for physical system and control system analysis, which are tool- or language-specific. In the future, the model-based design framework will be further developed. As shown in Fig. 64.14, it is structured in four layers including the engineering management platform, business process model layer, system model layer and physical system model & data model layer. Stakeholders can manage the project on the engineering platform. Depending on the information of the engineering management platform, business process model can be built to manage the project, time plan and human resource. In different projects, system models capture the simulation behaviors and integrate them with the requirements, design, verification Validation and so on. Based on the specification provided by the system model, different data, component models or software modules will be selected from the data libraries and model libraries to build physical system model. In the future, the work would also provide a formal ontology for the integration of simulation technologies and system models. This paves the way for qualified and automated transformations among system models and related physical models.

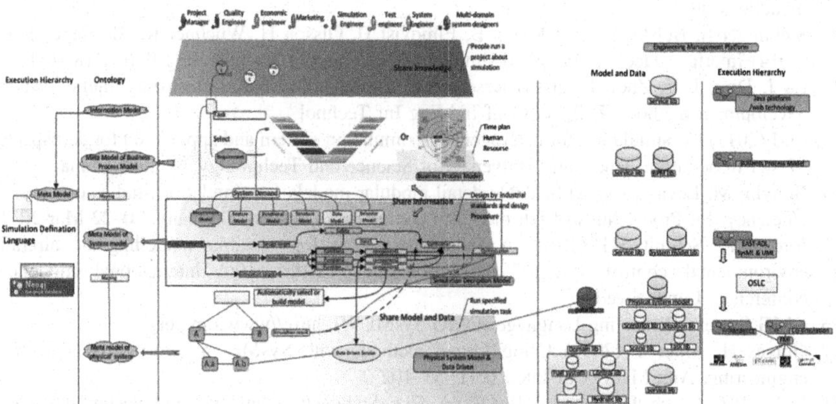

Fig. 64.14 Model-driven framework for aero-engine control system design

Acknowledgements We are grateful for the support of the Natural Science Foundation of Jiangsu Province (No. BK20140829); Jiangsu Postdoctoral Science Foundation (No. 1401017B).

References

1. Zhang D, Jin-zhi L, Wang L, Ii J (2014) Research of model-based aeroengine control system design structure and workflow. Asia Pac Int Symp Aerosp Technol 50(4):511–515
2. Ennsa D, Bugajskia D, Hendricka R, Steina G (1994) Dynamic inversion: an evolving methodology for flight control design. Int J Control 59(1):71–91
3. Prencipe A (1997) Technological competencies and product's evolutionary dynamics: a case study from the aero-engine industry. Res Policy 25:1261–1276
4. Cao Y, Jin XL (2006) Grid-based distributed simulation of an aero engine. Int J Adv Manuf Technol 27(7–8):631–637
5. Sellers JF, Daniele CJ (1975) DYNGEN: A program for calculating steady-state and transient performance of turbojet and turbofan engines. NASA
6. Schobeiri MT, Attia M, Lippke C (1994) GETRAN: A generic, modularly structured computer code for simulation of dynamic behavior of aero- and power generation gas turbine engines. J Eng Gas Turbines Power 116(3):483–494
7. Visser WPJ, Broomhead MJ (2000) A generic object-oriented gas turbine simulation environment. In: ASME Turbo Expo 2000, Munich, Germany, 2000
8. Kai P, Ding F, Bu Z, Li J, Yin F (2001) Failure analysis and parameter optimization for fuel distributor for aeroengine. J Propuls Technol 32(2) (Chinese)
9. Guo Y, Li K (2009) Modeling of aeroengine hydro-mechanical controller and analyzing of the control system performance. In: 2009 Fourth international conference on innovative computing, information and control, IEEE, pp 446–449
10. Modelica® (2012) A unified object-oriented language for systems modeling. https://www.modelica.org/
11. Sicklinger S, Belsky V, Engelmann B, Elmqvist H, Olsson H, Wüchner R, Bletzinger K-U (2014) Interface Jacobian-based co-simulation. Int J Numer Methods Eng 98(6):418–444
12. Lu J, Ding J-W, Zhou F (2014) Research of tool-coupling based electro-hydraulic system development method. IEEE Int Conf Ind Eng Inf Technol
13. Lu J (2011) Co-simulation for heterogeneous simulation system and application for aerospace. Master dissertation, Huazhong University of Science and Technology, Wuhan, China
14. Bonvini M, Leva A (2011) Scalable-detail modular models for simulation studies on energy efficiency. In: Proceedings of 8th modelica conference, Dresden, Germany, 20–22 Mar 2011
15. Jordan P, Schmitz G (2014) A modelica library for scalable modelling of aircraft environmental control systems. In: Proceedings of the 10th international modelica conference, Lund, Sweden
16. OMG Systems Modeling Language (OMG SysML™). http://www.omg.org
17. Graves H, Bijan Y (2011) Using formal methods with SysML in aerospace design and engineering. Ann Math Artif Intell 63(1):53–102
18. Feiler PH, Gluch DP, Hudak JJ (2006) The Architecture analysis and design language (AADL): an introduction. CMU/SEI-2006-TN-011
19. EAST-ADL-Specification_V2.1.12,EAST-ADL Association. http://www.east-adl.info/
20. Seitz Arne (2011) Advanced methods for propulsion system integration in aircraft conceptual design. Technischen Universität München, München

Chapter 65
Container Throughput Time Series Forecasting Using a Hybrid Approach

Xi Zha, Yi Chai, Frank Witlox and Le Ma

Abstract This paper proposed a novel two-stage hybrid container throughput forecasting model. Time series in reality exhibits both linear and nonlinear characteristics and individual models are not able to describe the two features simultaneously. Therefore, we combine linear model SARIMA (seasonal autoregressive integrated moving average) and nonlinear model ANN (artificial neural network). In order to break through the limitations of traditional hybrid models, based on the identified parameters of SARIMA in first stage, the structures of several ANN in second stage could be decided. Finally, we validate the proposed hybrid model 5 performs best with case study in Shanghai port.

Keywords Port throughput forecasting · SARIMA · ANN · Hybrid models

65.1 Introduction

In order to improve the competitiveness of a port, a forecasting model becomes a rather crucial solution to decision-making. Due to the constant monthly changes in volume of container throughput over time, these numbers could be considered as a time series process. Thus, a time series predictive model could be applied to forecast the future value. In the last several decades, many endeavours have been successfully made to the evolution and progress of the time series prediction methods. These time series forecasting methods, with difference of special

X. Zha · Y. Chai (✉) · L. Ma
College of Automation, Chongqing University, Chongqing 400044, China
e-mail: chaiyi@cqu.edu.cn

X. Zha
ITMMA, Institute of Transport and Maritime Management Antwerp, 2000 Antwerp, Belgium

F. Witlox
Department of Geography, Ghent University, Krijgslaan 281(S8), 9000 Ghent, Belgium

© Springer-Verlag Berlin Heidelberg 2016

Y. Jia et al. (eds.), *Proceedings of the 2015 Chinese Intelligent Systems Conference*, Lecture Notes in Electrical Engineering 359,
DOI 10.1007/978-3-662-48386-2_65

emphasis, can be divided into two models: (a) linear prediction and (b) nonlinear prediction model [1].

Linear prediction models are often referred to regression analysis method. Autoregressive (AR), moving average (MA), and autoregressive integrated moving average (ARIMA) belong to linear prediction. Compared with other linear models, ARIMA is the most popular and effective linear approach, which is widely used in short term load [2], electricity price [3], wind speed [4] and other domains.

The traditional linear methods above suppose that the time series comes from linear process. However, random time series with characteristic of nonlinearity [5], which means there is an underlying relationship between the previous data and the future data, is hard to be described with a linear regression precisely in real life. Aimed at solving this problem, artificial neural network (ANN) is applied into forecasting. ANN is a flexible nonparametric nonlinear computing frameworks in many practical types of time series forecasting these years [6].

Although both ANN and ARIMA models have demonstrated successful predictions in their respective fields. Nevertheless, neither ANN nor ARIMA is a general forecasting approach that is capable simultaneously to break through the shortcomings of each single models. Therefore, the combination of linear and nonlinear models is a valid solution to raise the forecasting performance. Here are some examples applied in literature, e.g. supply chain management [7], water quality [8], inspection [9].

The remainder of the thesis is organized as follows. Section 65.2, the principle of linear and nonlinear model will be introduced. In Sect. 65.3, different hybrid models of SAIRMA and ANN are present. Subsequently, the performance criteria will be shown. In Sect. 65.4, the independent and hybrid models are applied to container throughput time series in Shanghai port and their performances are compared. Lastly, Sect. 65.5 includes the concluding remarks.

65.2 Methodology

Time series in reality is hardly recognized as a pure linear or nonlinear structure. Thus, in this work, we regard the outstanding linear model SARIMA and nonlinear model ANN as the basic methods to construct the hybrid models through several different connections.

65.2.1 Scenario 1: SARIMA

The popular ARIMA model proposed by Box and Jenkins [10] is broadly applied to analyze stable and univariate time series. In order to consider seasonal factors, a SARIMA model, an extensive model of the ARIMA model, is often described as $SARIMA(p, d, q)(P, D, Q)s$. We formulate SARIMA model Eq. (65.1) as follows:

$$\phi_p(\mathrm{L})\Phi_P(\mathrm{L})\nabla^d\nabla_s^D y_t = \theta_q(\mathrm{L})\Theta_Q(\mathrm{L})\varepsilon_t \tag{65.1}$$

where y_t is the observed value at time; s refers seasonal term; L is a lag operator $L^m(y_t) = y_{t-m}$, $\nabla^d = (1-L)^d$ denotes the non-seasonal differencing operator, $\nabla_s^D = (1-L^s)^D$ is the seasonal counterpart; $\phi_p(\mathrm{L})$ denotes the non-seasonal AR term with order p, $\theta_q(\mathrm{L})$ is the MA term with order q, $\Phi_P(\mathrm{L})$ and $\Theta_Q(\mathrm{L})$ are seasonal counterparts, and ε_t is the white noise follows Gaussian distribution with zero mean and variance σ^2. The procedure of SARIMA contains the four major phrases: identification, estimation, diagnostic; and forecasting.

65.2.2 Scenario 2: ANN

ANN is a flexible computing model to learn from data and able to describe various different nonlinear processes. There is no requirement of pre-knowledge hypothesis of the process during the modelling procedure. On the contrary, the model primarily depends on the inherent characteristics of given time series. The mathematical expression Eq. (65.2) below shows the relations between outputs and inputs:

$$y_t = w_0 + \sum_{j=1}^{Q} w_j g(w_{0j} + \sum_{i=1}^{P} w_{i,j} x_i) + e_t \tag{65.2}$$

where x_i is the input data at time i, $w_{i,j}$ refers the weights liking weights between input neuron i and hidden neuron j, w_i denotes the weights liking weights between hidden neurons layer and output neurons, P and Q are the numbers of neurons in input and hidden layer respectively, g refers the activation function of the neuron. The selection of inputs plays a major role in deciding the structure. However, no general regulation exits in deciding the inputs. In following section, we proposed a novel idea on determining it. In supervised learning, the sample set could be divided randomly into three parts (i.e. training set, validation set and test set) to prevent in a situation of under-fitting or over-fitting [11]. Here, we apply the most successful tanning algorithm, back propagation, to reduce the error iteratively. After several trials, the network with optimal number of hidden neurons is chosen, which yields the minimum error.

65.3 Experimental Procedure

In the implementation procedure, we applied these data into seven scenarios shown in Fig. 65.1. Scenario 1 and Scenario 2 are individual SARIMA and ANN approaches respectively. Scenario 3 and Scenario 4 are traditional hybrid

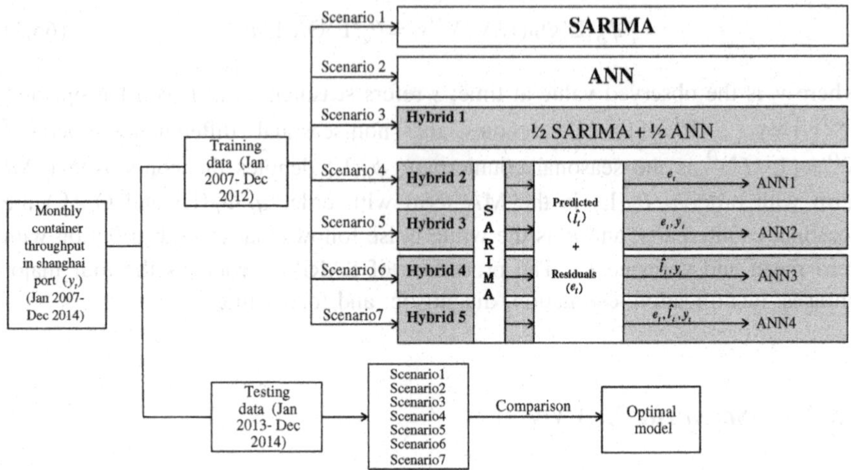

Fig. 65.1 The proposed forecasting methods based on SARIMA and ANN

approached. Scenario 5, Scenario 6 and Scenario 7 with different inputs for second stage are proposed novel hybrid models based on the parameters in SARIMA.

65.3.1 The Traditional Hybrid Models

Scenario 3: (a) Weighted combination of different models as Eq. (65.3)

$$\hat{y}_t = \sum_{k=1}^{n} w_k \hat{y}_t^k; \; \sum_{k=1}^{n} w_k = 1; \; 0 \le w_k \le 1. \tag{65.3}$$

where $\hat{y}_t^k$ denotes the predicted value using kth is model at time t; w_k denotes the weight of the different forecast methods; $\hat{y}_t$ is the final predicted value.

Scenario 4: (b) Additive relationship between linear and nonlinear part as Eq. (65.4)

$$\begin{aligned} y_t &= l_t + n_t; \\ e_t &= y_t - \hat{l}_t; \\ \hat{y}_t &= \hat{l}_t + \hat{n}_t + \varepsilon_t = \hat{l}_t + f(e_t) + \varepsilon_t \end{aligned} \tag{65.4}$$

where l_t is defined as the linear part, n_t is defined as the corresponding nonlinear part. Both of them should be formulated respectively. The forecasting value of linear part $\hat{l}_t$ is calculated by SARIMA model in the first step. e_t, denoting the residual at time t, can be obtained from the subtraction of the observed value y_t and the predicted value $\hat{l}_t$.

However, on one hand, the Scenario 3 would lead to reduce precision when a big error exist in the worse individual model, which is less accurate than the better individual one. On the other hand, Scenario 4 is under an assumption that the relationship between linear and nonlinear elements is additive. It will result in an underestimation the relationship if there is a different (e.g. multiplicative) association instead of additive connections [12]. In practice, Hibon and Evgeniou [13] proved that combined models are not more superior to individual models all the time. Therefore, the appropriate selection of hybrid model is a rather important process.

65.3.2 The Proposed Hybrid Models

The traditional hybrid approaches mentioned as above is not exact. In this work, we proposed three other hybrid models: Scenario 5, Scenario 6 and Scenario 7 as Eqs. (65.5)–(65.7) via a different connection inspired by the previously identified four parameters: p, d, P, D as well as the predicted value $\hat{l}_t$ from $SARIMA(p, d, q)(P, D, Q)s$.

Scenario 5:

$$y_t = f(h(y_t), e_{t-1}) + \varepsilon_t \tag{65.5}$$

Scenario 6:

$$y_t = f(\hat{l}_t, e_{t-1}) + \varepsilon_t \tag{65.6}$$

Scenario 7:

$$y_t = f(h(y_t), \hat{l}_t, e_{t-1}) + \varepsilon_t \tag{65.7}$$

Here $h(x_t) = \{x_{t-1}, x_{t-2}, \ldots x_{t-(p+d)}, x_{t-s}, x_{t-(s+1)} \ldots, x_{t-(s+P+D-1)}\}$, which depends on the parameters in $SARIMA(p, d, q)(P, D, Q)s$, f refers the nonlinear function, y_t denotes observed values at time t, $\hat{l}_t$ is the predicted values from SARIMA at time t, e_{t-1} refers the residual from the SARIMA at time $t-1$, ε_t is the random error.

These hybrid models break through the mentioned limitations of traditional hybrid models. Instead, they benefit from the respective capacities of SARIMA and ANN methods in linear and nonlinear elements without focus on the internal relationship between the two elements. Theoretically, taking use of the proposed hybrid forecasting methods without the uncertain assumption further decrease the residuals. Furthermore, the inputs for ANN also benefit from the identified parameters of SARIMA. In first stage of the framework, aiming to capture the linear characteristic with training data set, the optimal SARIMA model is identified and applied to forecast the future container throughput, which are prepared for second stage as the inputs. Afterwards, ANN is employed to deal with existing relationship of nonlinearity and existing linearity in preprocessed data and initial training data. By means of pre-process of SARIMA, we find the seasonal correlation between current value and past value. Hence, we depend on the extracted seasonal association rules and then decide the number and type of inputs to the networks. Then we predict the following values with the different chosen hybrid models. Furthermore, the outcomes of proposed hybrid models will be compared with those of the separated SARIMA, ANN and traditional hybrid models. Finally, we select three evaluation criteria (shown in Sect. 65.3.3) to evaluate the performance of the each model and then choose the optimal forecasting model.

65.3.3 Evaluation Criteria

For the purpose of evaluating each model's performance relatively, three different forecast consistency measures are taken into account (Table 65.1).

Where y_i and $\hat{y}_i$ are the monthly observed values and the monthly forecasted values in month *i* respectively, $\bar{y}_i$ and $\bar{\hat{y}}_i$ are the counterparts of average values. The values of R (standard correlation coefficient) range from −1 to 1 and the value close to 1 shows a better performance. It indicates how well a forecasting model fits observations by summarizing the differences between observed and the predicted container throughput. On the contrary, the lower SSE (sum square error) and RMSE (root mean square error) indicate a tighter fitting model with the actual time series.

Table 65.1 Three evaluation criteria

Index	Formula
R	$R = \sum_{i=1}^{n} (y_i - \bar{y}_i)*(\hat{y}_i - \bar{\hat{y}}_i) / \sqrt{\sum_{i=1}^{n} (y_i - \bar{y}_i)^2 * \sum_{i=1}^{n} (\hat{y}_i - \bar{\hat{y}}_i)^2}$
SSE	$SSE = \sum_{i=1}^{n} \lvert y_i - \hat{y}_i \rvert^2$
RMSE	$RMSE = \sqrt{\sum_{i=1}^{n} \lvert y_i - \hat{y}_i \rvert^2 / n}$

65.4 Data and Result

The monthly container throughput values in shanghai port from January 2007 to December 2012 is considered as the training data to estimate the model, while another period data from January 2013 to December 2014 is considered as the testing data to check the precision of predicted container throughput. These container throughput time series in Shanghai port reflects that it is a complex nonlinear system with three main characteristics: growing trend, seasonality, randomness shown in Fig. 65.2. Therefore, we conclude a single model that are not sufficient to describe such a complex system and thus a hybrid model of SARIMA and ANN is necessary to satisfy the requirement.

65.4.1 First Stage

(a) Identification
By inspecting the ACF (autocorrelation function) and PACF (partial autocorrelation function) figures of the time series shown in Fig. 65.3, several structures with different parameters are carried out. Based on the value of BIC (Bayesian information criterion), the best model $SARIMA(1,1,1)(0,1,1)_{12}$ is determined.

(b) Estimation
We estimate the parameters in Eq. (65.1) with maximum likelihood method as follows:

$$(1+0.663344L)(1-L)(1-L^{12})y_t=(1-0.0393171L)(1-0.771389L^{12})\varepsilon_t \tag{65.8}$$

(c) Diagnostic
Moreover, for the sake of checking if the residuals satisfy a white noise process, analysis of the residuals has been carried out in Fig. 65.4. As

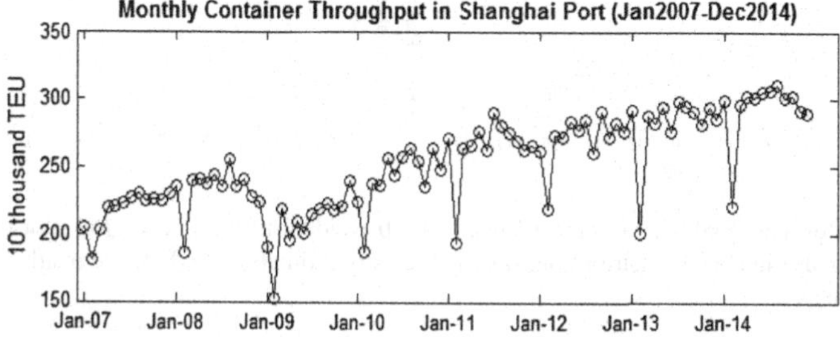

Fig. 65.2 Monthly container throughput in Shanghai Port (January 2007–December 2014)

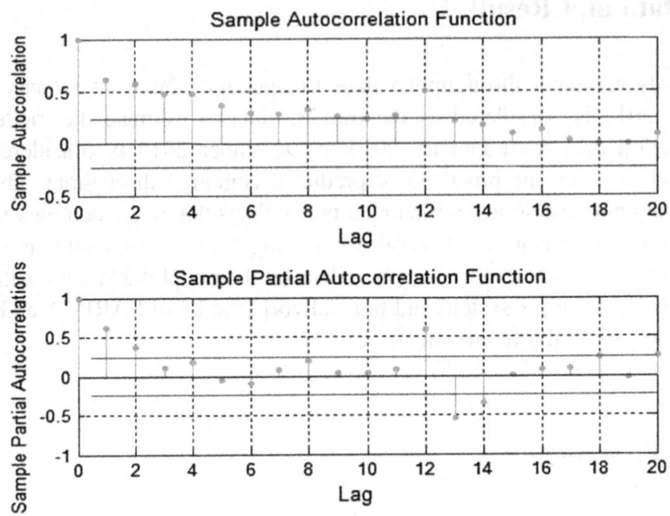

Fig. 65.3 ACF and PACF of monthly container throughput time series

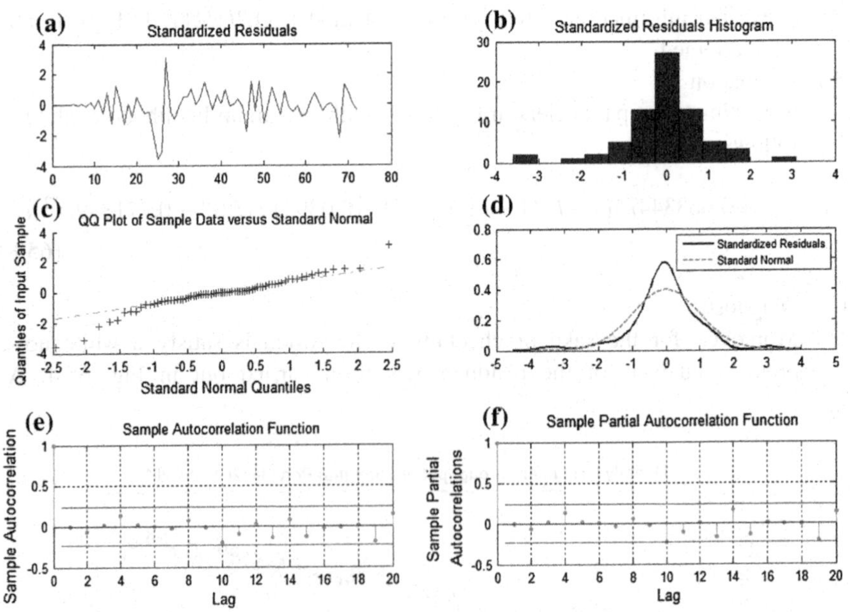

Fig. 65.4 Residual analysis: **a** standard residuals; **b** residual histogram; **c** Q-Q plot of standard residuals versus the normal distribution; **d** kernel density estimate; **e** ACF of the residuals; **f** PACF of the residuals

Fig. 65.4a, b displaying, the residuals of SARIMA appear a zero mean approximately. Furthermore, the quantile–quantile (Q-Q) plot in Fig. 65.4c as well as Kernel density estimation in Fig. 65.4d, indicates that the residual is under a normal distribution. In addition, the ACF and PACF of residuals shows no significant autocorrecting. As the analysis above, the defined is considered valid.

(d) Forecasting

Based on the decided SARIMA, predicted value $\hat{l}_t$ and error e_t are produced for the second stage.

65.4.2 Second Stage

Here we focus on the last scenario, Hybrid 5. y_{t-12}, y_{t-2}, y_{t-1} from observations, predicted values $\hat{L}_t$ and the residuals e_{t-1} from preprocess SARIMA are regards as the inputs for ANN shown in Fig. 65.5. By changing the number of hidden neurons from small to large, we find nine neurons is the best. When the number exceeds nine, the regression of the testing set in sample data becomes distorted. The actual data and predicted data are both shown in Fig. 65.6 and the analysis of error indicates that performance of Hybrid 5 satisfies error requirement.

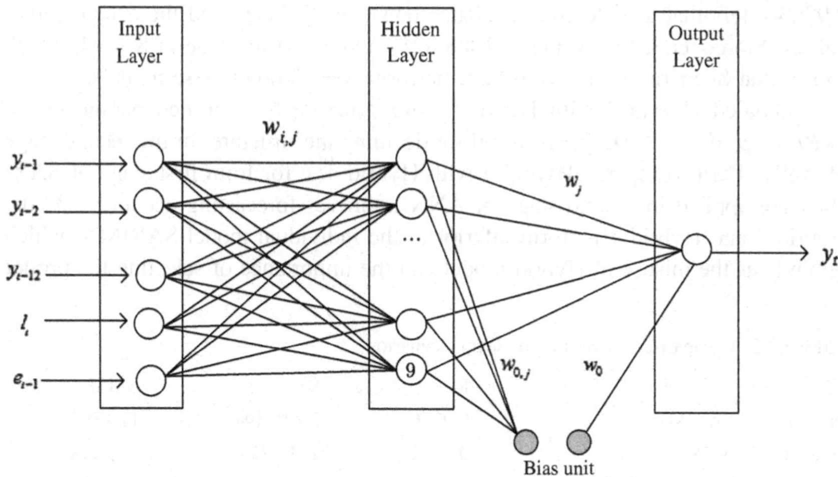

Fig. 65.5 The framework of the best-fitted ANN in hybrid 5

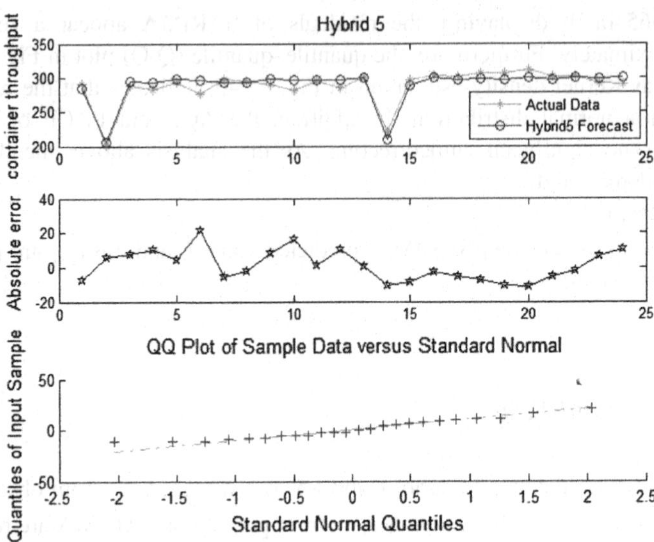

Fig. 65.6 Error analysis of hybrid 5

65.4.3 Comparison of Results for Seven Scenarios

Table 65.2 shows the comparison of the results obtained from the seven best models in the seven respectively distinct scenarios. Scenario 7 provides an R value of 0.9363 infers that a preferable relevance between predicted and the actual testing values. Moreover, a SSE value of 1888.8217 and a RMSE value of 8.8714, which shows that Scenario 7, also Hybrid 5, performs best among these models.

Compared Hybrid 3 with Hybrid 2, the structure of identified parameters in $SARIMA(p, d, q)(P, D, Q)s$ is useful for deciding the structure for the second stage of ANN. Then, compared Hybrid 5 with Hybrid 4, error from first stage of SARIMA are applied in second stage of ANN improves forecasting accuracy. While, Hybrid 2 and Hybrid 4 perform inferior to the individual model SARIMA, which reminds us the misuse of Hybrid model and the importance of selecting the proper

Table 65.2 Comparison of results for seven scenarios

	R	SSE	RMSE
Scenario 1: SARIMA	0.9276	2999.1086	11.1787
Scenario 2: ANN	0.9152	2813.8131	10.8278
Scenario 3: Hybrid 1 (Average)	0.9360	2043.7603	9.2280
Scenario 4: Hybrid 2 (Addition)	0.9315	3431.5348	11.9575
Scenario 5: Hybrid 3 (e_t, y_t)	0.9362	2261.8411	9.7079
Scenario 6: Hybrid 4 $(\hat{L}_t, y_t)$	0.9230	3194.4102	11.5369
Scenario 7: Hybrid 5 $(e_t, \hat{L}_t, y_t)$	**0.9363**	**1888.8217**	**8.8714**

inputs for the process of ANN framework. Not only should we the capture the nonlinearity left in residuals from SARIMA, but also the other inherent relationship except additivity between linear and nonlinear part is necessary to be taken into account to improve the accuracy of the entire model as well. To sum up, the proposed approach Hybrid 7 based on the previous structure of SARIMA are proved efficiency.

65.5 Conclusion

This paper provides a novel hybrid approach for time series forecasting. Based on the basic SARIMA and ANN model, the proposed hybrid model, Scenario 7, not only takes the advantages of the respective capacities of linear and nonlinear models, but also breakthroughs the limitation of traditional hybrid approaches. The identified parameters in SARIMA is useful to decide the structure or inputs for ANN in second stage and performs better than other six scenarios. As a result, the proposed hybrid model improves the forecasting accuracy and provides a meaningful predictor for decision-making or strategic management in seaport.

References

1. Deyang L (2014) Study on Combination forecasting model for time series based on SARIMA and BP [D]. University of Lanzhou, Lanzhou
2. Hagan MT, Behr SM (1987) The time series approach to short term load forecasting [J]. IEEE Trans Power Syst 2(3):785–791
3. Jakaša T, Andročec I, Sprčić P (2011) Electricity price forecasting—ARIMA model approach [C]. In: Energy market (EEM), 2011 8th international conference on the European, IEEE, pp 222–225
4. Li F, Wu X, Zhu Y (2015) Wind speed short-term forecast for wind farms based on ARIMA model [J]. Ind Electron Eng 93:97
5. Granger CW, Terasvirta T (1993) Modelling non-linear economic relationships. OUP Catalogue, Oxford
6. Granger CWJ, Terasvirta T (1993) Modelling non-linear economic relationships [J]. OUP Catalogue, Oxford
7. Aburto L, Weber R (2007) Improved supply chain management based on hybrid demand forecasts. Appl Soft Comput 7(1):136–144
8. Faruk DÖ (2010) A hybrid neural network and ARIMA model for water quality time series prediction. Eng Appl Artif Intell 23(4):586–594
9. Ruiz-Aguilar JJ, Turias IJ, Jiménez-Come MJ (2014) Hybrid approaches based on SARIMA and artificial neural networks for inspection time series forecasting. Transp Res Part E Logist Transp Rev 67:1–13
10. Box GEP, Jenkins GM (1976) Time series analysis: forecasting and control, revised ed [M]. Holden-Day, San Francisco

11. Ripley BD (1996) Pattern recognition and neural networks [M]. Cambridge University Press, Cambridge
12. Khashei M, Bijari M (2011) A novel hybridization of artificial neural networks and ARIMA models for time series forecasting [J]. Appl Soft Comput 11(2):2664–2675
13. Hibon M, Evgeniou T (2005) To combine or not to combine: selecting among forecasts and their combinations [J]. Int J Forecast 21(1):15–24

Chapter 66
A Dynamic Performance Analyzing Method of Intelligent Fire Control System Based on SCPN

Tianqing Chang, Junwei Chen, Bin Han, Kuifeng Su and Rui Wang

Abstract To describe tank future fire control system logic level generally and exactly, a modeling method is advanced which based on Stochastic Colored Petri Net (SCPN). Firstly, the constitutive structure, information structure and working flow of future fire control system are studied. Secondly, the tank future fire control system is formalized described. The rule which transform tank fire control system to Petri Net is established. Finally, Petri Net model of tank future fire control system is established. Its dynamic performance is analyzed. It proves the modeling method can describe tank future fire control system logic level effectively.

Keywords Intelligent fire control system · SCPN · System modeling · Dynamic performance analysis

66.1 Introduction

Tank fire control system should have the ability of rapid and accurate striking in high-mobile state of modern war. But because of the limit of physical condition of tank crew, the crew's ability of control, reaction and persistent working will decline in high-mobile state. It will lead to decline the combat efficiency of fire control system. So it is necessary to develop a kind of fire future control system to help tank crew accomplish combat mission in high-mobile state.

The first step of developing weapon system is modelling. Then the model should be demonstrated and improved repeatedly [1]. So the completeness and perfectibility of model is very important. In order to achieve the purpose, the system is usually divided into several levels and different models are built up for every level

T. Chang · J. Chen (✉) · B. Han · K. Su · R. Wang
Key Laboratory for All Electrification of Land Battle Platform, Department of Control Engineering, Academy of Armored Force Engineering, No. 21 Dujiakan Street, Fengtai District, Beijing 100072, People's Republic of China
e-mail: diegorevilo@163.com

© Springer-Verlag Berlin Heidelberg 2016
Y. Jia et al. (eds.), *Proceedings of the 2015 Chinese Intelligent Systems Conference*, Lecture Notes in Electrical Engineering 359,
DOI 10.1007/978-3-662-48386-2_66

[2]. The tank fire control system should be divided to three levels from top to bottom, including conceptual level, logical level and physical level. Conceptual level is defined to the set of definition, composition and function structure of the system. The diagrams of fire control system structure and system work flow in Ref. [3] are both the model of conceptual level. Physical Level is the set which can describe inner kinetic characteristic of the system, including gun dynamics model, control model and ballistic solving model in Ref. [4]. The logical level of fire control system was usually described indirectly by diagram of fire control system structure and flow of system work. But this kind of model is not only lack of necessary logical analysis elements, such as target and crew, but also cannot describe dynamic performance. Compare with existing tank fire control system, there is greater variety of information and more complex action logic among components in future tank fire control system. The traditional logical level description method will not satisfy the requests. It is need a full elements model, which can describe logical level of fire control system directly and exactly.

The logical level of tank future fire control system is composed with a numbers of sub-events, including target indication, threaten sequence, target identification, automatic tracking, ranging and resolving, firing and destruction judgment. State's change of fire control system is caused by those events which is discrete in time. So we can use discrete event modeling method to modeling the logical level. The modeling technique of Petri Net is applied in many dynamic and discrete-time systems [5–7]. It has the ability of describing many kinds of structure, including sequence, synchronize, concurrence and conflict. It is suitable to apply this modeling method in logical level of future tank fire control system.

66.2 The Analysis of Tank Future Fire Control System Structure

The composition and operate mode of exiting 'Hunter-Killer' fire control system are introduced in detail in Ref. [8], so we won't cover them in this paper. Fire future control system should be improved based on existing fire control system. It should improve intelligence level and automatic level and reasonably reduce number of crew in principle of not reducing function and performance. To achieve this goal, it should apply intelligence algorithm technique for improving intelligence level of system and reduce the burden and operation of tank commander. It should apply sensor technique, information technique and image processing technique to cancelling the gunner. Because the crew who is kept mainly completes the mission of tank commander, it also named tank commander. The tank commander, who is in future fire control system, takes the responsibility of searching, indicating and monitoring the work state of system, manual intervention if necessary. In brief, the work process is 'the tank commander in charge of capturing, the system in charge of annihilation'.

66.2.1 Structure of System

On the basis of original commander's vision block subsystem, the tank future fire control system add intelligent program computer to constitute commander's intelligent vision block subsystem. The intelligent program computer help tank commander to complete a series of mission, including target identification, choosing kind of shell, automatic tracking multi-targets, queuing up the level of threaten. It also gives the information of battlefield and system state to the tank commander. With the help of intelligent program computer, improved commander's vision block can acquainting image and automatic tracking target. On the basis of original gunner's sight, striking sight is improved with the function of acquainting image. On the basis of original fire control computer, intelligent striking computer is improved. The striking sight sends the image of battle to the intelligent striking computer, then the intelligent striking computer control aiming line to point to the area where intelligent program computer indicate. The intelligent striking computer automatically identify target, rang and resolve firing data based on the image information. After that, it would control gun system to strike the target and judge the degree of injury. Finally, it would decide whether strike it again or not. So the improved observing-sight subsystem, fire control computer subsystem and gun control subsystem make up intelligent striking subsystem. The tank future fire control system needs the information of position of itself. So positioning and navigation equipment is added based on the original sensor subsystem. The structure of system is shown in Fig. 66.1.

66.2.2 Structure of Information

Tank future fire control system's functions are achieved based on lots of information and correlative technique. The information transmission among components directly affects performance of fire control system. Connection of system and structure of information are built up based on functions which are mentioned before. They are shown in Fig. 66.2.

66.2.3 Work's Flow of System

In the work's mode of "the tank commander in charge of capturing, the system in charge of annihilation", the working flow is as follow in detail:

(1) Tank commander operates the control box to control commander's vision block for searching target. Tank commander indicates target to intelligent program computer when it is searched. At that time, tank commander obtain speed and range of target by using commander's vision block. And the

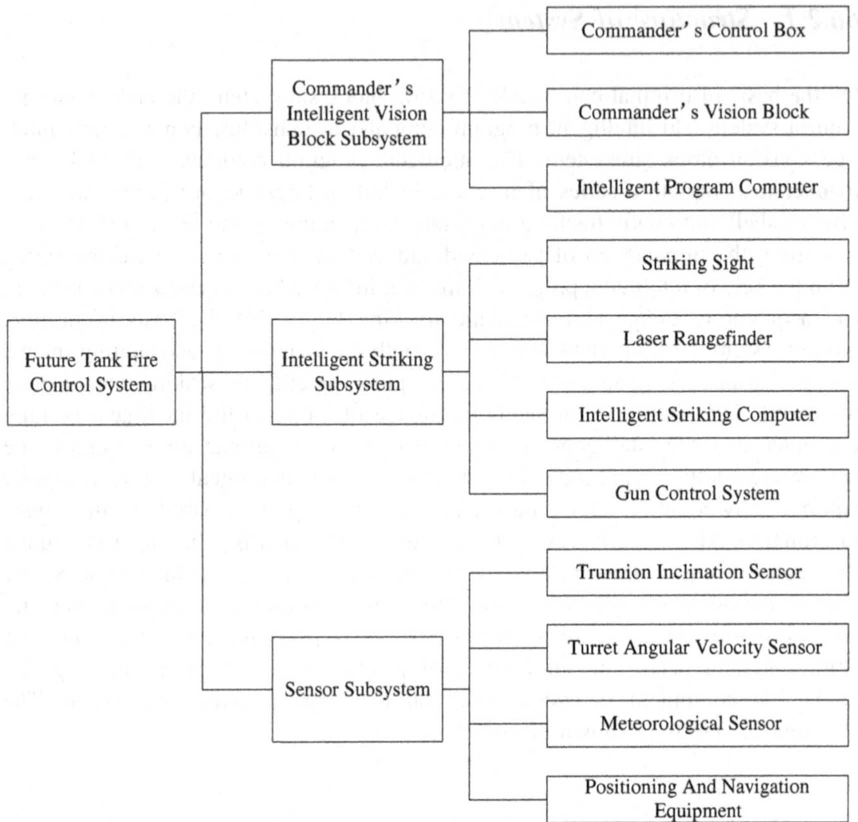

Fig. 66.1 Structure of tank future fire control system

commander help intelligent program computer to confirm the type of target and ammunition. After these, tank commander could search next target. At the same time, tank commander monitors the intelligent program computer and intelligent striking computer and gives manual intervention it, if necessary.

(2) Intelligent program computer tracks multi-targets and assesses threaten degree without artificial intervention. It also sends type, position of target and kind of shell to intelligent striking computer.

(3) Intelligent striking computer drives aiming line to target area. At the same time, it drives gun to follow aiming line. Intelligent striking computer independently identifies target, tracks, measures distance, resolves and fires. After firing, it would assess the degree of target destruction and decide whether strike it again. It also can ask for the commander's help, if necessary. When it decides to strike target again, it would identifies target, tracks, find range, resolves and fires again.

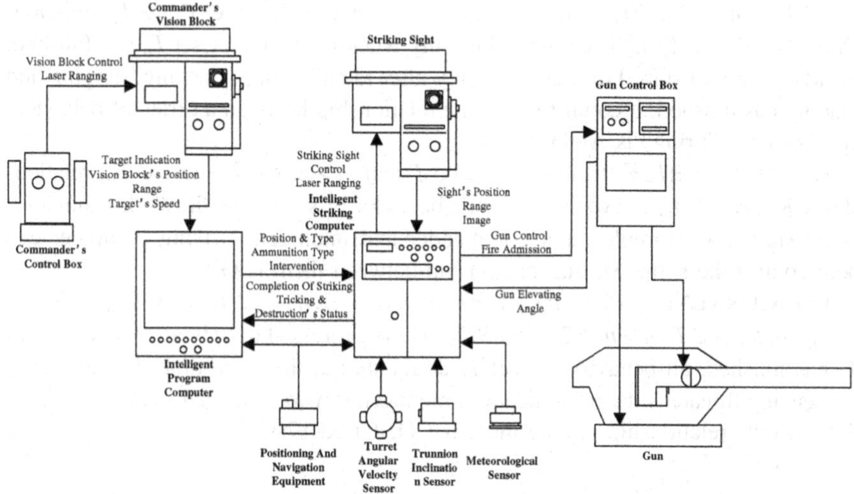

Fig. 66.2 Connection of system and structure of information

(4) Intelligent striking computer sends striking completion signal and destruction status to intelligent program computer in order to program the next target. It will delete the target from striking list, if target is destroyed. If not, it will strike target again, and continue to monitor and manage it.

The working flow is as follow (Fig. 66.3):

66.3 The Formalized Description Method of Tank Fire Control System Behavior

The research of formalized description method of fire control system can help us translate system structure to Petri Net model. Future fire control system can be formalized to a three-tuple $FC = \langle T, P, F \rangle . T = \{t_1, t_2, \ldots, t_n\}$. is expressed to set of fire control system behaviors. $P = \{p_1, p_2, \ldots, p_n\}$ is expressed to set of tank commander, targets and components. $F = \{f_1, f_2, \ldots, f_n\}$ is expressed to the set of logistic relationships among fire control system behaviors.

The logistic relationship among fire control system behaviors can be classified into four categories, they are SEQUENCE, AND, OR, CONCURRENCY. They can be described as follow:

SEQUENCE (*SeqR*): For $\forall t_i, t_j \in T\ (i,j = 1, 2, \cdots, n,$ 且 $i \neq j$, exist Sequence $SeqR(t_i, t_j)$. Only when behavior t_i has finished, behavior t_j would start. For example, only after fire control system has entered into firing gate, firing would carry out. So the relationship between behavior "entering into firing gate" and behavior "firing" is SEQUENCE.

AND (*AndR*): If $\exists T_p = (t_{i+1}, \ldots, t_{i+m}) \subset T$ and $m \geq 2$, $t_i \in T$, $t_i \notin T_p$, relationship AND $AndR(T_p, t_i)$ is expressed to only when the behavior set T_p has finished, behavior t_i could start. For example, only after identification, tracking, ranging and loading has finished, it could fire. So the relationship between a series of behaviors and behavior "firing" is AND.

OR (*OrR*): If $\exists T_p = (t_{i+1}, \ldots, t_{i+m}) \subset T$ and $m \geq 2$, $t_i \in T$ and $t_i \notin T_p$, relationship OR $OrR(T_p, t_i)$ is expressed to one behavior of set T_p has finished, behavior t_i could start. For example, tank commander and intelligent striking computer can both control the sight. So the relationship between them is OR.

CONCURRENCY (*ConcR*): If $\exists T_p = (t_{i+1}, \ldots, t_{i+m}) \subset T$, and $m \geq 2$, $\exists T_q = (t_{i+m}, \cdots, t_{i+n}) \subset T$, $n > m \geq 2$. $ConcR(T_p, T_q)$ is expressed to only after behavior set T_p has finished, all behaviors in set T_q could start at the same time. For example, sequencing threatens, target indication, monitoring work state can start at the same time. So the relationship among them is CONCURRENCY.

66.4 Translate Rules of Petri Net Model

66.4.1 Definition of Colored Stochastic Petri Net

Stochastic Colored Petri Net (SCPN) is eight-tuple [9]:

$$SCPN = (P, T; F, C, W, I, M, \lambda) \tag{1.1}$$

(*P*, *T*; *F*) is an original Petri Net. C is a finite colored set $C = \{c_1, c_2, \ldots, c_k\}$, $W: F \rightarrow L(C)_+$, $I: T \rightarrow L(C)_+$, $M: S \rightarrow L(C)$, $L(C)$ is non-negative integer-coefficient linear function which is defined in colored set C, $L(C)_+$ is expressed to coefficient which not all equal to zero, namely $L(C) = a_1c_1 + a_2c_2 + \ldots + a_kc_k$, $L(C)_+ = b_1c_1 + b_2c_2 + \ldots + b_kc_k$, a_i, $b_i(i = 1, 2, \ldots, k)$ are all non-negative integer, and $b_1 + b_2 + \ldots + b_k \neq 0$, $\lambda: T \rightarrow R_0$, it suppose $T = \{t_1, t_2, \ldots, t_n\}$. $\lambda(t) = a$ is expressed to transition t need time a to complete. a is a random variable. This net is named Stochastic Colored Petri Net.

66.4.2 Translation Rules of Fire Control System Resources and Behaviors

The resources (tank commander, target, component) in fire control system are translated into place, namely $P = \{p_1, p_2, \ldots, p_n\}$ is translated into the set of Petri Net place from set of tank commander, target and component. The fire control system behaviors are translated into transition, namely $T = \{t_1, t_2, \ldots, t_n\}$ is translated into the set of Petri Net transition from set of fire control system behaviors. The relationships among behaviors is translated into flow relation, namely $F = \{f_1, f_2, \ldots, f_n\}$ is translated into set of Petri Net flow relation from set of relationships. The information which transmits in fire control system is translated into token. The meaning of information is translated into color.

The creation method of Petri Net in detail is, let $\forall t_i \in T$, transition t_i is created firstly, namely the behaviors which happen in fire control system is confirmed. Let $\forall p_{ij} \in P$, p_{ij} which t_i is corresponding with is created, namely the behaviors which resources need. Let $f_{ij,i} \in F$, flow relation $f_{ij,i}$ and initial marking M_0 are created, and $N(f_{ij,i}) = (p_{ij}, t_j)$. N is the single mapping form set F to set $T \times P \cup P \times T$. It is expressed to the resource and aim of flow relation. Finally, tokens with different characteristics in the same place are endowed with different colors.

66.4.3 Translation Rules of Fire Control System Relationship Among Behaviors

The Petri Net translation rules of four behavior relationships in fire control system are as follow:

The translation rules of SEQUENCE (*SeqR*): if exist sequence relationship $SeqR(t_i, t_j)$, find place p_{ij}, flow relation $f_{i,ij}$ and $f_{ij,i}$ which fit to $N(a_{i,ij}) = (t_i, p_{ij})$, $N(f_{ij,i}) = (p_{ij}, t_j)$.

The translation rules of AND (*AndR*): if exist and relationship $AndR(T_p, t_i)$, transition t_{iu} is created. Then $\forall t_j \in T_p$, place p_{ij} is created. To find out transition t_{ij} and flow relation $f_{ij,ij}$ and $f_{ij,iu}$ which fit to $N(f_{ij,ij}) = (t_{ij}, P_{ij})$, $N(f_{ij,iu}) = (p_i, t_{iu})$.

The translation rules of OR (*OrR*): if exist or relationship (*OrR*), place P_i is created. To find out transition t_{iu} and t_{iv}, flow relation $f_{i,iu}$, $f_{iu,i}$ and $f_{i,iv}$ which fit to $N(f_{i,iu}) = (p_i, t_{iu})$, $N(f_{iu,i}) = (t_{iu}, p_i)$, $N(f_{i,iv}) = (p_i, t_{iv})$.

The translation rules of CONCURRENCY (*ConcR*): if exist concurrency relationship $ConcR(T_p, T_q)$, transition t_i is created. For $\forall t_j \in T_q$, place P_{ij} is created, to find out transition t_{ij} and flow relation $f_{i,ij}$ and $f_{ij,ij}$ which fit to $N(a_{i,ij}) = (t_i, p_{ij})$, $N(a_{ij,ij}) = (p_{ij}, t_{ij})$.

66.5 Creation and Analysis of Fire Control System

66.5.1 Creation of Petri Net Model

According to analysis of fire control system constitutive structure, information structure, working flow and translation rules of Petri Net, logic level of fire control

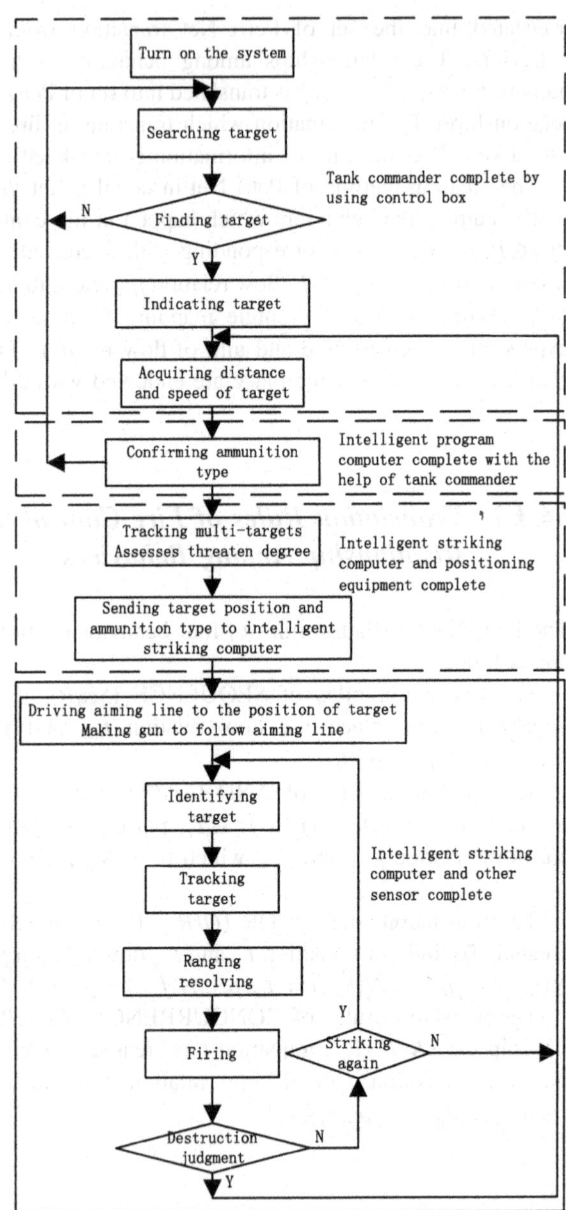

Fig. 66.3 Future fire control system work flow

system is translated into Petri Net by using CPN Tools [10]. The Petri Net is as Fig. 66.4.

Some places' capacity is limited. For example, striking sight only can track one target. But these operations are not supported by CPN Tools. So Anti-place is created to fix it by using translation rules of SEQUENCE (*SeqR*).

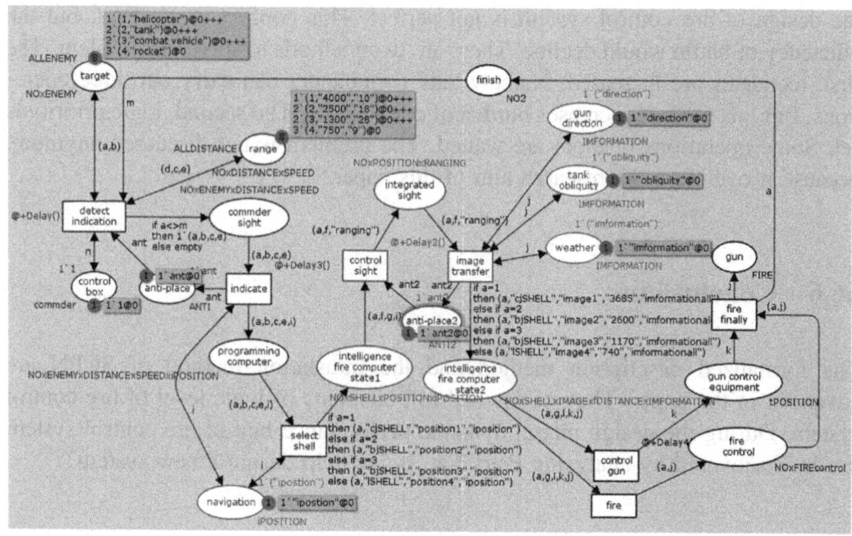

Fig. 66.4 Petri Net of fire control system logic level

It needs time to take place some transition. The time usually fit to normal distribution. So time function is created to fix it.

66.5.2 Dynamic Performance Analysis of Petri Net

Dynamic performance of fire control system can be analyzed by using analysis method of Petri Net.

Boundedness and security: each place in Petri Net of fire control system is boundedness. Bound of each place fits to $B(p) = MIN\{B | \forall M \in R(M_0): M(s) \le B\} = 1$. So Petri Net is bounded and security.

Liveness: M_0 is initial marking in Petri Net. For $\forall M \in R(M_0)$, exciting $M' \in R(M)$ fits to $M'[t>$. That means each transition is live. So the Petri Net is live.

Persistence: M_0 is initial marking in Petri Net. $(M[t_i > \wedge M[t_j > M') \to M'[t_i >$ fits to $\forall M \in R(M_0)$ and any $t_i, t_j \in T(t_i \neq t_j)$. So this Petri Net is continual system. There is new target, the system can work circularly.

Conflict: In Petri Net, for $\forall M \in R(M_0)$, $\forall M_i \in R(M_0)$, $\forall M_j \in R(M_0)$, $\forall t_x \in T$ $\forall t_y \in T$, exciting $M[t_x > M_i \to -M_i[t_y >$ or $M[t_y > M_j \to -M_j[t_x >$. So there is conflict in Petri Net.

There is a conflict in tank commander resource. Tank commander can carry out two operations at the same time (namely the capacity of place 'tank commander' is two). But if commander's vision block, control box and intelligent program computer ask tank commander to control at the same time, this conflict is coming. So the design of fire control system is not perfect. This conflict is not fatal, but the efficiency of battle would decline. There are two methods to solve the problem. The first, recourses are increased. Namely tank commander can carry out three operations. But this method increases burden of commander. The second, logic priority is set. Some operations of them are waited. The problem is not discussed anymore, because it oversteps the research aim of this paper.

66.6 Conclusions

The formalized description method and the translation method of SCPN are advanced in the paper. They can model full elements of logic level of fire control system. During the design phase, dynamic performance bug of fire control system can be discovered by analyzing Petri Net. It can help design a new system.

References

1. Song G, Shen R, Zhou W, Zou Q (2009) Weapon system engineering [M]. National Defense Industry Press, Beijing (In Chinese)
2. Booch G, Rumbaugh J, Jacobson J (2001). The unified modeling language user guide [M]. Addison-Wesley, Boston
3. Zhou Q, Liu C, Ge Y (2006) The development trail of architecture of modern tank fire control system [J]. Fire Control Command Control 10:4–7 (In Chinese)
4. Kang X, Ma C, Wei X (2003) Model theory of gun system [M]. National Defense Industry Press, Beijing (In Chinese)
5. Jiang C, Yu Y, Zhang L, Feng H (2006) Mission completion success probability simulation of materiel system based on Petri Net [J]. Comput Simul 1:29–32 (In Chinese)
6. nSong X, Ren W, Chen K, Wang Z, Wang H (2011) Modeling and optimizing on the bus of integrated electronic system based on Petri Net [J]. J Acad Armored Force Eng (10):38–43
7. Zhao X, Chen H, He M, Jiang Z (2009) Simulation validation method for capability requirement of weapon system of systems based on Petri Net [J]. J Syst Simul 2:1159–1163
8. Zhu J, Zhao B, Wang Q (2003) Modern tank fire control system [M]. National Defense Industry Press, Beijing (In Chinese)
9. Franceschinis G, Fumagalli A, Silinguelli A (1999) Stochastic colored petri Net models for rainbow optical networks [J]. Lect Notes Comput Sci 1:273–303
10. Jensen K, Kristensen LM (2007) Coloured petri Nets and cpn tools for modelling and validation of concurrent systems [M]. Springer, Berlin

Chapter 67
Forecast of Train Delay Propagation Based on Max-Plus Algebra Theory

Hui Ma, Yong Qin, Guoxing Han, Limin Jia and Tao zhu

Abstract The forecast of delay time is of great assistance to decision making in train operation adjustment when the schedule is disturbed either by infrastructure fault or natural hazard. This paper presents a railway delay propagation model to forecast the delay time, described by discrete event dynamic system (DEDS) and formulated by max-plus algebra theory. On the basis of the train operation regulations and headway constraints, a system matrix of max-plus algebra is acquired to illustrate the mechanism of delay propagation. And then a function to predict the delay time is proposed to solve the model, with two advantages: Firstly, the specific delay time is able to be calculated; secondly, the result of the prediction is comparatively precise due to the highly match of the model to the actual operation. Finally, by analysis of the prediction, this paper offers the decision support in train adjustment, from which the dispatcher can proactively conduct countermeasures to alleviate the propagation and even stop it.

Keywords Railway network modeling · Train delay propagation forecasting · Max-plus algebra theory

H. Ma · G. Han
School of Traffic and Tansportaion, Beijing Jiaotong University, Beijing 100044, China

H. Ma · Y. Qin (✉) · G. Han · L. Jia
State Key Lab of Rail Traffic Control & Safety, Beijing Jiaotong University, Beijing 100044, China
e-mail: yqin@bjtu.edu.cn

T. zhu
Information Technology Center of MOR, Beijing 100844, China

© Springer-Verlag Berlin Heidelberg 2016

Y. Jia et al. (eds.), *Proceedings of the 2015 Chinese Intelligent Systems Conference*, Lecture Notes in Electrical Engineering 359,
DOI 10.1007/978-3-662-48386-2_67

67.1 Introduction

As is prevalent in train operation, passenger train delay trends to bear a rather small tolerance of delayed time due to the elevated railway service. For better train operation management, it's crucial to alleviate the delay propagation by the forecast of the delay time. Currently, however, the forecast in China is estimated only by train dispatchers, depending on human experiences, which could be inaccurate.

In order to solve this problem, this paper presents a model to forecast the delay propagation in a systematic way, based on DEDS, in accordance to the essence of train operation—a series discrete train occupying the routes and stations in an explicit time sequence. As a preferable approach to calculate DEDS, max-plus algebra theory is considered to solve the prediction in the model.

Previous works on delay modeling are mainly stochastic method, which assumes initial delay as random variables with the purpose of estimating the distribution of delay in the networks [1–3]. However, this method can hardly provide the prediction of the train running time. Nevertheless, a deterministic method modeling the delay propagation as DEDS is able to solve the problem [4–6]. Braker [7] presents a recursive system of periodic timetable of Dutch railway network in max-plus algebra. Li [8] presents a modeling of urban rail transit focusing on the relationship between schedule and the non-block situation of the system.

In this paper, we first present the modeling of China railway network by DEDS in Sect. 67.2. And then the max-plus algebra theory is introduced to represent the train running procedure by acquiring the constraint matrix and formulating the system function. The forecast result is given in the last section to analyze the delay propagation with adjustment of train sequences discussed, for the purpose of decision support for train dispatching.

67.2 Modeling of the Real China Railway Networks

The speeding existing China railway is chosen to be the delay propagation target, in which the colored double-track railways in Fig. 67.1 are established for route modeling, with the mileage between stations marked above them. There are 4 train lines running in this network, all arranged to dwell at the junction station, Xuzhou, at the same platform. Consequently, these train lines are constrained with the adjacent trains for safety's concern. The timetable is shown in Table 67.1.

In order to build up the model based on Max-plus theory, we need to simplify the networks and transform it into a discrete event dynamic system (DEDS). Figure 67.2 is a simplification of the networks. The 5 stations, Zhengzhou, Xuzhou, Jinan, Bengbu, and Xinyi, are represented respectively by S1, S2, S3, S4, and S5, while the lines in different colors stands for different train lines.

Now, we use the DEDS [9] to describe the train running process as a series of transitions corresponding to events and places representing the process. In the

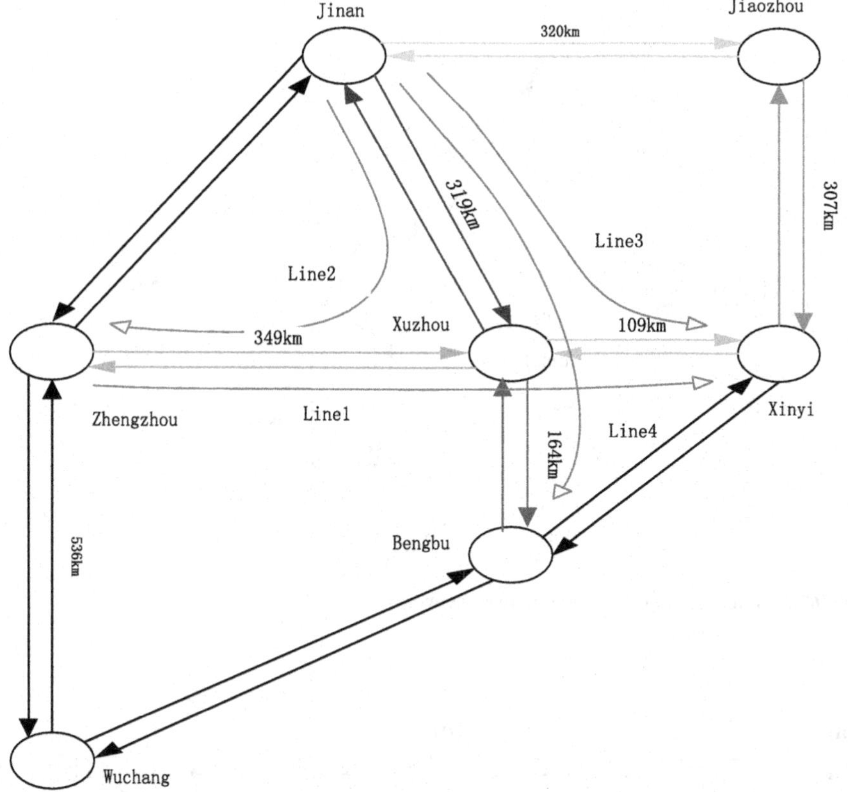

Fig. 67.1 A real China railway network

Table 67.1 The timetable of the chosen lines

Line1	1551	Line2	k15	Line3	k771	Line4	k491
Station	Time[1]	Station	Time	Station	Time	Station	Time
Zhengzhou	6:42	Jinan	7:08	Jinan	7:30	Jinan	11:16
Xuzhou	10:45	Xuzhou	11:20	Xuzhou	11:41	Xuzhou	15:05
Xuzhou	11:15	Xuzhou	11:26	Xuzhou	12:03	Xuzhou	15:19
Xinyi	12:42	Zhengzhou	15:39	Xinyi	13:33	Bengbu	17:12

[1]This timetable is been slightly adjusted for the sake of planning the trains at a same platform

representation of the network, the running at the links between stations and the stop at stations are modeled as places with a certain holding time(the running time and the dwell time), while the departure and the arrival at stations are regarded as transitions. A transition is enabled only when the incoming place has a token and

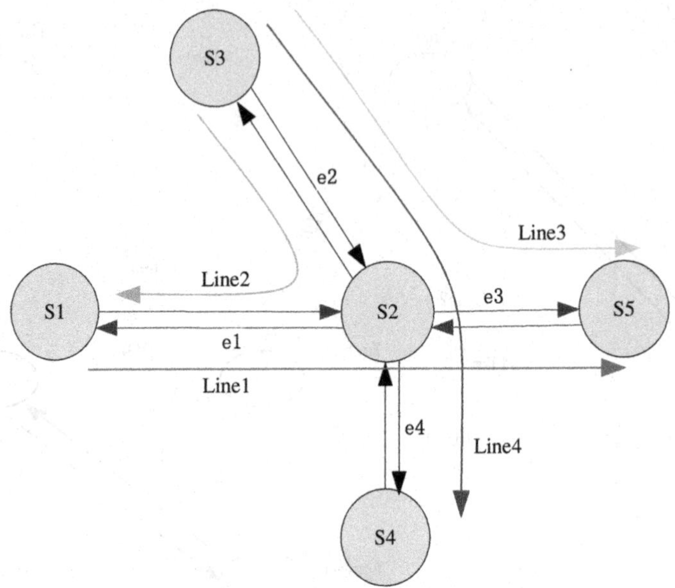

Fig. 67.2 Simplification of the real railway network

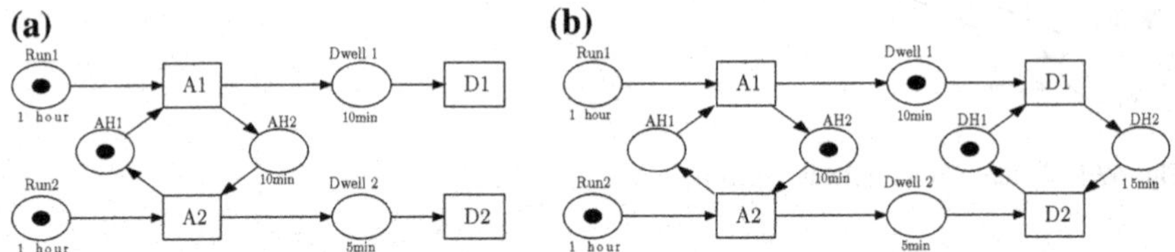

Fig. 67.3 DEDS representing two successive trains arriving at the same platform

the holding time has elapsed. Figure 67.3 interprets the representation by an example of two successive trains arriving at the same platform of a station.

Let $D_{i,j}$ to represent the departure time of the line i at the jth station alone the line, $P_{i,j}$ to represent the running time of line i between the jth station and the j+1th station, and $A_{i,j}$ to represent the arrival time of line i at the jth station. (i and j are integers). For example, $D_{1,1}$ means the departure time of line1 at the 1st station of its line. Then the train lines can be described as follows:

Line1: $D_{1,1}$, $P_{1,1}$, $A_{1,2}$, $D_{1,2}$, $P_{1,2}$, $A_{1,3}$; Line2: $D_{2,1}$, $P_{2,1}$, $A_{2,2}$, $D_{2,2}$, $P_{2,2}$, $A_{2,3}$
Line3: $D_{3,1}$, $P_{3,1}$, $A_{3,2}$, $D_{3,2}$, $P_{3,2}$, $A_{3,3}$; Line4: $D_{4,1}$, $P_{4,1}$, $A_{4,2}$, $D_{4,2}$, $P_{4,2}$, $A_{4,3}$

Table 67.2 Relative schedule of the real railway network

Line1-1551			Line2-K15			Line3-K771			Line4-K491		
State		Time	State		Time	State		Time	State		Time
$D_{1,1}(x_1)$	S1	0	$D_{2,1}(x_2)$	S3	44	$D_{3,1}(x_3)$	S3	66	$D_{4,1}(x_6)$	S3	292
$A_{1,2}(x_4)$	S2	261	$A_{2,2}(x_7)$	S2	296	$A_{3,2}(x_9)$	S2	317	$A_{4,2}(x_{13})$	S2	521
$D_{1,2}(x_5)$	S2	291	$D_{2,2}(x_8)$	S2	302	$D_{3,2}(x_{10})$	S2	339	$D_{4,2}(x_{14})$	S2	535
$A_{1,3}(x_{11})$	S5	378	$A_{2,3}(x_{15})$	S1	555	$A_{3,3}(x_{12})$	S5	429	$A_{4,3}(x_{16})$	S4	648

So far, the form of the timetable needs to be changed into a relative timetable in order to form the max-plus function in the next section. We pick up the earliest time 6:24 as the starting time. The relative schedule is obtained in Table 67.2.

Then, on the basis of the timetable and the representation of DEDS, we can have the constraints written as:

- Running time constraint:

$$\begin{aligned}&A_{1,2}\geq D_{1,1}+P_{1,1};A_{1,3}\geq D_{1,2}+P_{1,2};A_{2,2}\geq D_{2,1}+P_{2,1};A_{2,3}\geq D_{2,2}+P_{2,2};\\&A_{3,2}\geq D_{3,1}+P_{3,1};A_{3,3}\geq D_{3,2}+P_{3,2};A_{4,2}\geq D_{4,1}+P_{4,1};A_{4,3}\geq D_{4,2}+P_{4,2}\end{aligned}\tag{67.2.1}$$

- Departure headway constraint:

$$D_{3,2}\geq D_{1,2}+I\tag{67.2.2}$$

- Arrival headway constraint:

$$A_{2,2}\geq A_{1,2}+\tau_a,A_{3,2}\geq A_{2,2}+\tau_a,A_{4,2}\geq A_{3,2}+\tau_a;A_{3,3}\geq A_{1,3}+\tau_a\tag{67.2.3}$$

- Arrival/Departure headway constraint:

$$A_{2,2}\geq D_{1,2}+\tau_{ad},A_{3,2}\geq D_{2,2}+\tau_{ad},A_{4,2}\geq D_{3,2}+\tau_{ad}\tag{67.2.4}$$

- Dwell constraint:

$$D_{1,2}\geq A_{1,2}+t_s,D_{2,2}\geq A_{2,2}+t_s,D_{3,2}\geq A_{3,2}+t_s,D_{4,2}\geq A_{4,2}+t_s\tag{67.2.5}$$

I is the minimum inter-train tracking interval; τ_a is the arrival headway constraint; τ_{ad} is the arrival/departure headway constraint; t_s is dwell time.

In this network we have: I is 5 min; τ_a is 10 min; τ_{ad} is 5 min; the minimum dwell time is 2 min; the highest speed allowed is 100 km/h.

67.3 Max-Plus Algebra Theory to Represent the Delay Propagation

Here are the basic rules for max-plus algebra, in which $\varepsilon = -\infty$ and $e = 0$:

$$a \otimes b = a + b \tag{67.2.6}$$

$$a \oplus b = \max(a, b) \tag{67.2.7}$$

$$a \oplus \varepsilon = \varepsilon \oplus a = a; a \otimes \varepsilon = \varepsilon \otimes a = \varepsilon \tag{67.2.8}$$

$$a \otimes e = e \otimes a = a \tag{67.2.9}$$

A is a $m \times p$ matrix, B is a $p \times n$ matrix, the multiplication $\otimes$ of the matrix is defined as Eq. 67.2.10. The elements of the resulting matrix at (row *i*, column *j*) are determined by matrices A (row *i*) and B (column *j*).

$$[A \otimes B]_{ij} = \oplus_{k=1}^{p} (a_{ik} \otimes b_{kj}) = \max(a_{i1} + b_{1j}, \ldots, a_{ip} + b_{pj}) \tag{67.2.10}$$

According to the constraints in Eqs. 67.2.1–67.2.5 and the max-plus theory [10], we acquire a system of equations to predict the delay as follows (x_i is the initial state; d is the scheduled time; I_{e1}, I_{e2}, I_{e3}, and I_{e4} are the train running time at the links):

$$\begin{aligned}
&D_{1,1} = d; D_{2,1} = d; D_{3,1} = d \\
&A_{1,2} = \max(D_{1,1} + I_{e1}, d); D_{1,2} = \max(A_{1,2} + t_s, d) \\
&D_{4,1} = d; A_{2,2} = \max(D_{2,1} + I_{e2}, D_{1,2} + \tau_{ad}, d) \\
&D_{2,2} = \max(A_{2,2} + t_s, d); A_{3,2} = \max(D_{3,1} + I_{e2}, D_{2,2} + \tau_{ad}, d) \\
&D_{3,2} = \max(A_{3,2} + t_s, D_{1,2} + I, d); A_{1,3} = \max(D_{1,2} + I_{e3}, d) \\
&A_{3,3} = \max(D_{3,2} + I_{e3}, d); A_{4,2} = \max(D_{4,1} + I_{e2}, D_{3,2} + \tau_{ad}, d) \\
&D_{4,2} = \max(A_{4,2} + t_s, d); A_{2,3} = \max(D_{2,2} + I_{e1}, d) \\
&A_{4,3} = \max(D_{4,2} + I_{e4}, d)
\end{aligned} \tag{67.2.11}$$

Now we define matrix X to represent the train running state of the network, which elements are arranged in accordance with the sequence of the relative timetable in Table 67.2 in an increasing order:

$$\begin{aligned}
X = [&D_{1,1} \quad D_{2,1} \quad D_{3,1} \quad A_{1,2} \quad D_{1,2} \quad D_{4,1} \quad A_{2,2} \quad D_{2,2} \quad A_{3,2} \\
&D_{3,2} A_{1,3} A_{3,3} A_{4,2} \quad D_{4,2} \quad A_{2,3} A_{4,3}]^T \\
= &\{x_i | i = 1, 2, \ldots, 16\}^T
\end{aligned} \tag{67.2.12}$$

The matrix X' represents the actual running state of the train; matrix D represents the scheduled time; the sequences of the elements in X' and D are the same with X. By observing Eq. 67.2.11, we can find that each element can be written in as $x_{i1} = c \otimes x_{j1} \oplus d$. Here c is the constraint corresponding to x_{i1} and x_{j1}.

Therefore, combining the train state algebra and the scheduled time algebra, we can use max-plus algebra to illustrate the prediction as long as we obtain the constraints matrix C_0. Here's the steps to build up C_0:

1. Since the formula we set is $C_0 \otimes X$ and X is a 16×1 matrix, base on the rule of $\otimes$, C_0, is determined as a 16×16 matrix.
2. According to the multiplication rules, the resulting matrix is 16×1 dimension (j=1), as the outcome of train departure/arrival time corresponding to the constraints. Take $[C_0 \otimes X']_9$ for example:

$$[C_0 \otimes X']_9 = \oplus_{k=1}^{16}(c_{9k} \otimes x'_k) = \max(c_{9,1} + x'_1, \ldots, c_{9,16} + x'_{16})$$

In Eq. 67.2.11, $A_{3,2} = \max(D_{3,1} + I_{e2}, D_{2,2} + \tau_{ad}, d)$, where $A_{3,2}$ is X'_9, $D_{3,1}$ is X'_3 and $D_{2,2}$ is X'_8. So we have $C_{0(9,9)} = e$, $C_{0(9,3)} = I_{e2}$, and $C_{0(9,8)} = \tau_{ad}$. And then the row can be written as: $C_{0(9)} = [\varepsilon \ \ \varepsilon \ \ I_{e2} \ \ \varepsilon \ \ \varepsilon \ \ \varepsilon \ \ \varepsilon \ \ \tau_{ad} \ \ e \ \ \varepsilon \ \ \varepsilon \ \ \varepsilon \ \ \varepsilon \ \ \varepsilon \ \ \varepsilon \ \ \varepsilon]$.

Likewise, the elements in the each row of C_0 should be related to the elements in X' respectively, based on Eq. 67.2.11—if there's no constraint, mark ε; if it corresponds to itself, mark e; otherwise, put τ_a, τ_{ad}, I, or t_s based on the constraint.

Finally, we obtain the whole matrix C_0 as:

$$C_0 = \begin{bmatrix}
e & \varepsilon & \varepsilon & \varepsilon & \varepsilon & \varepsilon & \varepsilon & \varepsilon & \varepsilon & \varepsilon & \varepsilon & \varepsilon & \varepsilon & \varepsilon & \varepsilon & \varepsilon \\
\varepsilon & e & \varepsilon & \varepsilon & \varepsilon & \varepsilon & \varepsilon & \varepsilon & \varepsilon & \varepsilon & \varepsilon & \varepsilon & \varepsilon & \varepsilon & \varepsilon & \varepsilon \\
\varepsilon & \varepsilon & e & \varepsilon & \varepsilon & \varepsilon & \varepsilon & \varepsilon & \varepsilon & \varepsilon & \varepsilon & \varepsilon & \varepsilon & \varepsilon & \varepsilon & \varepsilon \\
I_{e1} & \varepsilon & \varepsilon & e & \varepsilon & \varepsilon & \varepsilon & \varepsilon & \varepsilon & \varepsilon & \varepsilon & \varepsilon & \varepsilon & \varepsilon & \varepsilon & \varepsilon \\
\varepsilon & \varepsilon & \varepsilon & t_s & e & \varepsilon & \varepsilon & \varepsilon & \varepsilon & \varepsilon & \varepsilon & \varepsilon & \varepsilon & \varepsilon & \varepsilon & \varepsilon \\
\varepsilon & \varepsilon & \varepsilon & \varepsilon & \varepsilon & e & \varepsilon & \varepsilon & \varepsilon & \varepsilon & \varepsilon & \varepsilon & \varepsilon & \varepsilon & \varepsilon & \varepsilon \\
\varepsilon & I_{e2} & \varepsilon & \varepsilon & \tau_{ad} & \varepsilon & e & \varepsilon & \varepsilon & \varepsilon & \varepsilon & \varepsilon & \varepsilon & \varepsilon & \varepsilon & \varepsilon \\
\varepsilon & \varepsilon & \varepsilon & \varepsilon & \varepsilon & \varepsilon & t_s & e & \varepsilon & \varepsilon & \varepsilon & \varepsilon & \varepsilon & \varepsilon & \varepsilon & \varepsilon \\
\varepsilon & \varepsilon & I_{e2} & \varepsilon & \varepsilon & \varepsilon & \varepsilon & \tau_{ad} & e & \varepsilon & \varepsilon & \varepsilon & \varepsilon & \varepsilon & \varepsilon & \varepsilon \\
\varepsilon & \varepsilon & \varepsilon & \varepsilon & I & \varepsilon & \varepsilon & \varepsilon & t_s & e & \varepsilon & \varepsilon & \varepsilon & \varepsilon & \varepsilon & \varepsilon \\
\varepsilon & \varepsilon & \varepsilon & \varepsilon & I_{e3} & \varepsilon & \varepsilon & \varepsilon & \varepsilon & \varepsilon & e & \varepsilon & \varepsilon & \varepsilon & \varepsilon & \varepsilon \\
\varepsilon & \varepsilon & \varepsilon & \varepsilon & \varepsilon & \varepsilon & \varepsilon & \varepsilon & \varepsilon & \varepsilon & I_{e3} & e & \varepsilon & \varepsilon & \varepsilon & \varepsilon \\
\varepsilon & \varepsilon & \varepsilon & \varepsilon & \varepsilon & \varepsilon & I_{e2} & \varepsilon & \varepsilon & \varepsilon & \tau_{ad} & \varepsilon & e & \varepsilon & \varepsilon & \varepsilon \\
\varepsilon & \varepsilon & \varepsilon & \varepsilon & \varepsilon & \varepsilon & \varepsilon & \varepsilon & \varepsilon & \varepsilon & \varepsilon & \varepsilon & t_s & e & \varepsilon & \varepsilon \\
\varepsilon & \varepsilon & \varepsilon & \varepsilon & \varepsilon & \varepsilon & \varepsilon & I_{e1} & \varepsilon & \varepsilon & \varepsilon & \varepsilon & \varepsilon & \varepsilon & e & \varepsilon \\
\varepsilon & \varepsilon & \varepsilon & \varepsilon & \varepsilon & \varepsilon & \varepsilon & \varepsilon & \varepsilon & \varepsilon & \varepsilon & \varepsilon & \varepsilon & \varepsilon & I_{e4} & e
\end{bmatrix} \tag{67.2.13}$$

And then the prediction can be described by the max-plus system as:

$$X = C_0 \otimes X' \oplus D \tag{67.2.14}$$

67.4 The Result of Delay Propagation Forecasting

67.4.1 Forecasting Delay Time Under Incident

Assuming that a local torrential rain struck Zhengzhou from 7:00 to 13:00 and then the running speed of the link between Zhengzhou and Xuzhou is limited for safety's concern, causing the train of line1 and line 2 directly delayed. By means of matlab simulation, we acquire the forecast delay time to predict whether the initial delay will propagate and the range of the propagation in Fig. 67.4.

Figure 67.4 shows that 1551 is delayed premarily, with the propagation to k15 and then to k771, while K491 stays out of interference. We can learn that when there is speed limit, the delay will occur in a train group and will propagate to the normal running train though previous delayed train. In this case, 1551 is the source, K15 is the intermediary of propagation, and K771 is the vulnerable train.

In this situation, by rearranging the platform of k15, we dismiss the constraints of those subsequent trains. Figure 67.5 illustrates that after changing the platform, the delay only effects in its own line without propagating to others.

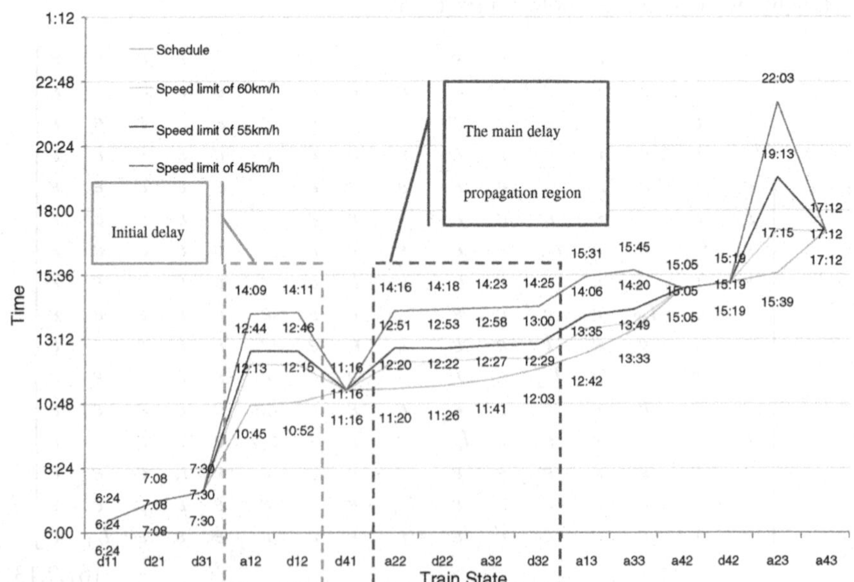

Fig. 67.4 The forecast of train delay under incident

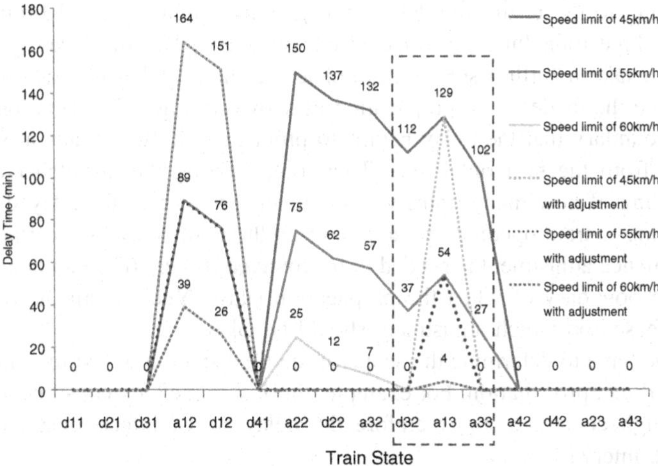

Fig. 67.5 The forecast of after-adjustment delay time compared to the no-adjustment delay time

67.4.2 *Forecasting Delay Time with Train Adjustment Analysis*

Now we look into to the delay propagation with train adjustment involved [11]. In this scenario, we assume 1551 has initial departure delay of different degrees. The result of the forecast delay time is shown in Fig. 67.6:

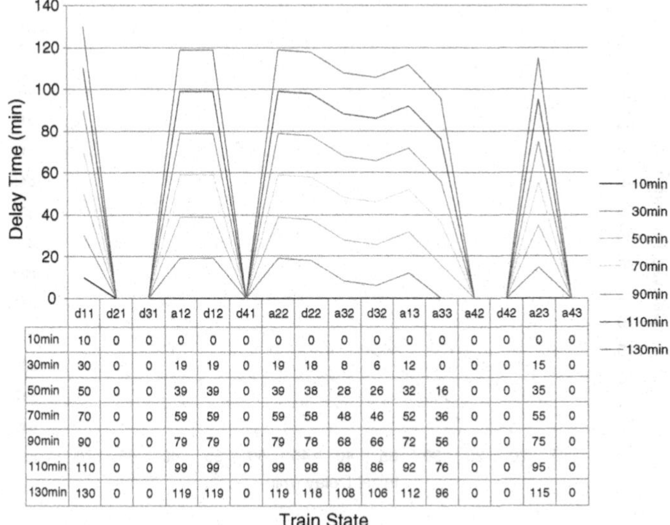

	d11	d21	d31	a12	d12	d41	a22	d22	a32	d32	a13	a33	a42	d42	a23	a43
10min	10	0	0	0	0	0	0	0	0	0	0	0	0	0	0	0
30min	30	0	0	19	19	0	19	18	8	6	12	0	0	0	15	0
50min	50	0	0	39	39	0	39	38	28	26	32	16	0	0	35	0
70min	70	0	0	59	59	0	59	58	48	46	52	36	0	0	55	0
90min	90	0	0	79	79	0	79	78	68	66	72	56	0	0	75	0
110min	110	0	0	99	99	0	99	98	88	86	92	76	0	0	95	0
130min	130	0	0	119	119	0	119	118	108	106	112	96	0	0	115	0

Fig. 67.6 The forecast of the delay time

According to the result, the delay propagation is minor under 10 min's initial delay and three train lines get infected except line4 when the delay goes above 30 min. To take a further step, we simulate the initial delay increasing in every 5 min to see the tendency of propagation as shown in Fig. 67.7. This figure gives the time boundary that the delay begins to propagate on other trains at where the lines start from the horizontal axis. Therefore, offering the dispatcher a general understanding of how many trains will be infected with specific delayed time.

According to train operation principle, when the train delays at a certain degree, a train sequence adjustment is needed to be involved. Figure 67.7 can help us find a lower time boundary of when the propagation starts. Yet, we still need a higher boundary based on which adjustment should be taken.

First, we have to determine the scope of the time span in which the initial delay will cause direct propagation. For example, line1 and line2 are constrained by each other through constraint τ_{da}; therefore the delay won't infect line2 unless it is beyond the interval:

$$\begin{aligned} &A_{2,2} - D'_{1,2,} \geq \tau_{da}, D'_{1,2,} - t_s = A'_{1,2,}, A'_{1,2,} - A_{1,2} = t_1; \\ &t_1 \leq A_{2,2} - \tau_{da} - t_s - D_{1,1} - I_{e1} \end{aligned} \tag{67.2.15}$$

t_1 in Eq. 67.2.15 is the minimum delay that begins to propagate, so the lower boundary of the direct propagation time span is calculate as:

$$t_{1\min} = A_{2,2} - \tau_{da} - t_s - D_{1,1} - I_{e1} \tag{67.2.16}$$

As long as line1 arrives earlier than line2, the train sequence is needless to be changed, so the upper boundary of the direct propagation time span is:

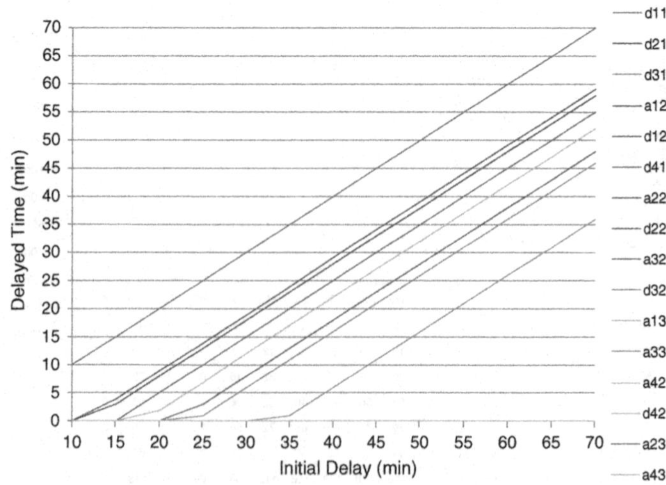

Fig. 67.7 The tendency of delay propagation

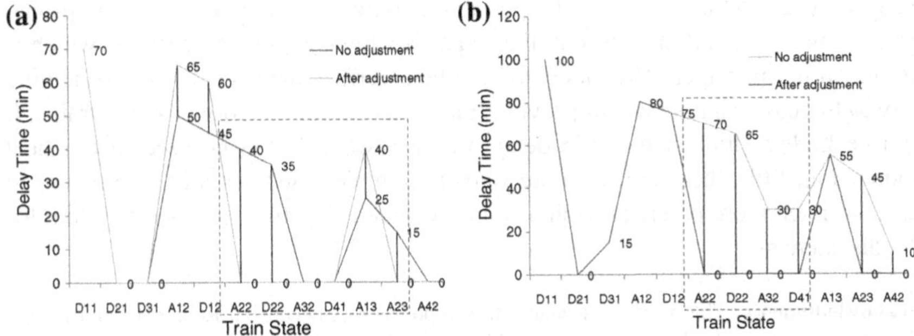

Fig. 67.8 The compare between after-adjustment and no-adjustment delay prediction under 70 min' initial delay (*left*) and 100 min' initial delay (*right*)

$$t_{1\max} = A_{2,2} - A_{1,2} + (A_{2,2} - I_{e2}) \tag{67.2.17}$$

In this network, we have $t_{1\min} = 10$ min and $t_{1\max} = 55$ min.

By the time span we just obtain, we rearrange the train sequences. When the train of line1 arrives at S2 later than 11:20, we let line2 arrive at S2 first. However, line1 will have additional delay. Therefore, it's helpful to let the dispatcher know whether making such adjustment is beneficial. The plot in Fig. 67.8 shows the differences. When the initial delay of line1 is 70 min, it adds up 10 min' delay, yet stops the propagation to line2. In contrast, the prediction without adjustment shows that the delay propagates to line2. It becomes more serious when the initial delay increases to 100 min that every train gets infected.

To summarize the delay propagation model, the achievements are as follows:

- The arrival and departure time of the train in the following stations is obtained.
- The trains that are infected by the initial delay propagation are detected with specific knocked-on delay time.
- The delay propagating time span determines the conditions for train adjustment.
- The forecast of train delay offers the dispatcher decision support according to the principle of train adjustment.

67.5 Conclusion

In daily railway operating, the route and the platform for each train is set in the station stage operating plan, which is delivered to the station 3–4 h earlier before the arrival of the train. When an initial delay happens in the system, the disturbed operating order will result in conflicts of the time and space in the station among different trains, finally leading to delay propagation. Hence, the purpose of studying delay propagation is to predict the possible conflicts through acquiring the propagating time span, and then prompt the dispatcher to take countermeasures, which is discussed in this paper. The model offered here verifies the necessity of a scientific way to forecast train delay and provides the possible way to reorganize the trains. It will be highly efficient for train delay recovery when it can be used in practical application. Still, there are far more steps to complete and optimize the model to make it much more practical, which is finally intended to form a support system for the dispatcher.

Acknowledgments This work was financially supported by Specialized Research Fund for the Train Fault Diagnosis and Potential Dangers Identification Research Based on Failure Cause-effect Chain (RCS2014ZT24).

References

1. Wendler E (2007) The scheduled waiting time on railway lines. Transp Res Part B 41:148–158
2. Yuan J (2006) Stochastic modeling of train delays and delay propagation in stations. Ph. D Dissertation, Delft University of Technology. Delft, The Netherlands
3. Huisman T, Boucherie RJ, Van Dijk NM (2002) A solvable queuing network model for railway network and its validation and applications for the Netherlands. Eur J Oper Res 142:30–51
4. Butkovic P (2010) Max-linear systems: theory and algorithms. Springer, Berlin, Germany
5. Subiono (2002) On classes of min-max-plus systems and their applications. Ph.D. Dissertation, Delft University of Technology, TRAIL Thesis Series, T2000/2. Delft, The Netherlands
6. Daamen W, Goverde RMP, Hansen IA (2008) Non-Discriminatory automatic registration of knock-on train delays. Springer, Berlin, Germany
7. Braker JG, (1991) Max-algebra modeling and analysis of time-table dependent transportation networks. In: Proceeding of 1st European Control Conference Grenoble, France, pp 1831–1836
8. Li A, Tang Z (2006) Study on urban rail transit model based on the max-plus algebraic theory. Railw Transp Econ 28(10):48–51 (in Chinese)
9. Ye Y, Jia L (2001) Petri Net with Objects and Its Application on Modeli ng Train Operation. China Railw Sci 22(3):15–20 (in Chinese)
10. Goverde RMP (2010) A delay propagation algorithm for large-scale railway traffic networks. Transp Res Part C 18:269–287
11. D'Ariano A, Pranzo M (2008) An advanced real-time train dispatching system for minimizing the propagation of delays in a dispatching area under severe disturbances. Springer, Berlin, Germany

Printed in the United States
By Bookmasters